Lecture Notes in Computer Science 16985

Founding Editors

Gerhard Goos
Juris Hartmanis

AF737077

The series Lecture Notes in Computer Science (LNCS), including its subseries Lecture Notes in Artificial Intelligence (LNAI) and Lecture Notes in Bioinformatics (LNBI), has established itself as a medium for the publication of new developments in computer science and information technology research, teaching, and education.

LNCS enjoys close cooperation with the computer science R & D community, the series counts many renowned academics among its volume editors and paper authors, and collaborates with prestigious societies. Its mission is to serve this international community by providing an invaluable service, mainly focused on the publication of conference and workshop proceedings and postproceedings. LNCS commenced publication in 1973.

Giovanni Iacca · Giorgia Nadizar · Anil Yaman ·
Doina Bucur · Antonio Della Cioppa · Ting Hu ·
Eric Medvet · Sarah L. Thomson
Editors

Parallel Problem Solving from Nature – PPSN XIX

19th International Conference, PPSN 2026
Trento, Italy, August 29 – September 2, 2026
Proceedings, Part I

Springer

Editors
Giovanni Iacca
University of Trento
Trento, Italy

Giorgia Nadizar
University of Toulouse Capitole
Toulouse, France

Anil Yaman
Vrije Universiteit Amsterdam
Amsterdam, The Netherlands

Doina Bucur
University of Twente
Enschede, The Netherlands

Antonio Della Cioppa
University of Salerno
Fisciano, Italy

Ting Hu
Queen's University
Kingston, ON, Canada

Eric Medvet
University of Trieste
Trieste, Italy

Sarah L. Thomson
Edinburgh Napier University
Edinburgh, UK

ISSN 0302-9743 ISSN 1611-3349 (electronic)
Lecture Notes in Computer Science
ISBN 978-3-032-36213-1 ISBN 978-3-032-36214-8 (eBook)
https://doi.org/10.1007/978-3-032-36214-8

This Springer imprint is published by the registered company Springer Nature Switzerland AG
The registered company address is: Gewerbestrasse 11, 6330 Cham, Switzerland

Preface

The International Conference on Parallel Problem Solving from Nature (PPSN) continued its long-standing tradition in 2026 by bringing together researchers from across the world in Trento, Italy. Hosted by the University of Trento, the conference provided an inspiring setting in the heart of the Alps, where the evolutionary computation community once again gathered to exchange ideas, present recent advances, and discuss future directions of nature-inspired computing.

PPSN XIX attracted a strong response from the community, with a total of 527 submissions received, out of which 443 were sent out for review. This number set a new record for PPSN. After a rigorous and extensive peer-review process, 165 papers were accepted for presentation in the proceedings, corresponding to an acceptance rate of approximately 31%, the lowest in the history of PPSN. Each submission received multiple expert reviews, resulting in a total of 1468 reviews (an average of 3.31 single-blind reviews per submission) provided by 517 members of the international Program Committee. This careful evaluation process ensured that the final program reflected high scientific quality and diversity across all areas of evolutionary computation and related fields.

The accepted papers were organized into 17 thematic clusters reflecting the broad scope of contemporary research in evolutionary computation and related areas: *Theory and Runtime Analysis*; *Evolution Strategies and Continuous Black-Box Optimization*; *Multi-Objective Optimization*; *Genetic Programming and Symbolic Regression*; *Bayesian and Surrogate-Assisted Optimization*; *Combinatorial Optimization*; *Reinforcement Learning, Hyper-Heuristics and Algorithm Selection*; *Swarm Intelligence and Ant Colony Optimization*; *Quantum Optimization and Computing*; *Large Language Models and Evolutionary Computation*; *Landscape Analysis and Benchmarking*; *Quality-Diversity and Novelty Search*; *Neural Architecture Search, AutoML and Neuroevolution; Robotics, Morphology and Design Encoding*; *Evolutionary Computation for Machine Learning, Data Science and Domain Applications*; *General Evolutionary Algorithm Methodology, Operators and Parallel Evolutionary Computation*; and *Coevolution and Evolutionary Dynamics.*

This edition of PPSN clearly reflected the growing impact of machine learning-driven approaches within evolutionary computation. The increasing integration of large language models with evolutionary algorithms, as well as advances in reinforcement learning, surrogate modeling, and automated algorithm design, highlight a notable shift toward hybrid, data-driven optimization systems. At the same time, traditional areas such as theory, continuous optimization, combinatorial problems, and multi-objective optimization remain central pillars of the field, demonstrating the continued breadth and balance of the PPSN community.

Following the tradition of PPSN, all accepted papers were presented as posters. The program was structured into multiple poster sessions, designed to maximize interaction across research areas and encourage discussion beyond disciplinary boundaries. This format continued to foster the open exchange of ideas that has long characterized the PPSN community.

The scientific program was complemented by five workshops and twenty-two tutorials, covering both foundational topics and emerging research directions. Workshops addressed Bayesian and surrogate-assisted optimization, explainable AI in evolutionary computation, high-performance computing approaches, quantum-classical optimization synergies, and the integration of domain knowledge into optimization methods. The tutorial program offered an exceptionally broad overview of the field, ranging from introductory lectures on evolutionary algorithms, CMA-ES, Bayesian optimization, landscape analysis, and theory, to advanced topics such as evolutionary multi-objective optimization, genetic improvement, constraint handling, GPU-based genetic programming, linkage learning, and the use of large language models in automated algorithm design. This strong tutorial lineup further reflects the increasing importance of AI-driven and data-driven approaches within evolutionary computation.

We were honored to welcome three distinguished keynote speakers: Lorenzo Bruzzone (University of Trento, Italy), Emma Hart (Edinburgh Napier University, UK), and Sebastian Risi (IT University of Copenhagen, Denmark), who each provided insightful perspectives on current and future challenges in applications of AI in space exploration, as well as optimization, machine learning, and adaptive systems.

The success of PPSN XIX would not have been possible without the contributions of many individuals. We sincerely thank all authors for submitting their work, the members of the Program Committee and external reviewers for their thorough and timely evaluations, and the workshop and tutorial organizers for their valuable contributions to the program. We are also grateful to the local staff at the University of Trento for their excellent support in preparing and hosting the event.

We gratefully acknowledge the support of Hexaly, our conference sponsor, as well as the continued backing of the PPSN steering committee and the broader research community, whose commitment ensures the continued success of this conference series. We would like to remember particularly Günter Rudolph, a pioneer of the field and member of the steering committee, who unfortunately passed away a few months before the conference.

Finally, we are pleased that more than 220 participants joined us in Trento, contributing to a vibrant and intellectually stimulating atmosphere throughout the conference. We hope that PPSN XIX will continue to inspire new ideas and collaborations in the years to come.

September 2026

Giovanni Iacca
Giorgia Nadizar
Anil Yaman
Doina Bucur
Antonio Della Cioppa
Ting Hu
Eric Medvet
Sarah L. Thomson

Organization

General Chair

Giovanni Iacca	Università degli Studi di Trento, Italy

Program Committee Chairs

Doina Bucur	University of Twente, The Netherlands
Antonio Della Cioppa	Università degli Studi di Salerno, Italy
Ting Hu	Queen's University, Canada
Eric Medvet	Università degli Studi di Trieste, Italy
Sarah L. Thomson	Edinburgh Napier University, UK

Proceedings Chairs

Giorgia Nadizar	University of Toulouse Capitole, France
Anil Yaman	Vrije Universiteit Amsterdam, The Netherlands

Tutorials Chair

Yuri Lavinas	University of Toulouse Capitole, France

Workshops Chair

Sarah L. Thomson	Edinburgh Napier University, UK

Publicity Chairs

Jessica Mégane	University of Coimbra, Portugal
Jéssica Parente	University of Coimbra, Portugal

Local Support Chairs

Elia Cunegatti	Università degli Studi di Trento, Italy
Stefano Genetti	Università degli Studi di Trento, Italy
Erik Nielsen	Università degli Studi di Trento, Italy
Chiara Camilla Rambaldi Migliore	Università degli Studi di Trento, Italy

Steering Committee

Michael Affenzeller	University of Applied Sciences Upper Austria, Austria
Thomas Bäck	Leiden University, Netherlands
Bogdan Filipič	Jožef Stefan Institute, Slovenia
Carlos Fonseca	University of Coimbra, Portugal
Emma Hart	Edinburgh Napier University, UK

Advisory Committee

David W. Corne	Heriot-Watt University, UK
Carlos Cotta	Universidad de Málaga, Spain
Kenneth De Jong	George Mason University, USA
Agoston E. Eiben	Vrije Universiteit Amsterdam, Netherlands
Juan Julián Merelo Guervós	University of Granada, Spain
Thomas P. Runarsson	University of Iceland, Iceland
Robert Schaefer	University of Krakow, Poland
Marc Schoenauer	Inria, France
Xin Yao	University of Birmingham, UK, Southern University of Science and Technology, China

Program Committee

Sumit Adak	Technical University of Denmark, Denmark
Michael Affenzeller	University of Applied Sciences Upper Austria, Austria
Alexandru Agapie	Bucharest University of Economic Studies, Romania
Hernan Aguirre	Shinshu University, Japan
Aishwaryaprajna	University of Exeter, UK
Imène Ait Abderrahim	Sorbonne University, France

Youhei Akimoto	University of Tsukuba, Japan
Mohammad Majid al-Rifaie	University of Greenwich, UK
Ines Alaya	Université Paris Nanterre, France
Wissam Albukhanajer	Roke Manor Research Limited, UK
Hanan Alsouly	Imam Mohammad Ibn Saud Islamic University, Saudi Arabia
Joan Alza	Robert Gordon University, UK
Marie Anastacio	Leiden University, Netherlands
Maciej Antczak	Poznań University of Technology, Poland
Denis Antipov	Sorbonne University, France
Kirill Antonov	Leiden University, Netherlands
Carlos Henggeler Antunes	University of Coimbra, Portugal
Daan Apeldoorn	TU Dortmund University, Germany
Claus Aranha	University of Tsukuba, Japan
Aluizio Araújo	Universidade Federal de Pernambuco, Brazil
Dirk V. Arnold	Dalhousie University, Canada
Etor Arza	Norwegian University of Science and Technology, Norway
Hirad Assimi	University of Adelaide, Australia
Nathanael Aubert-Kato	Ochanomizu University, Japan
Anne Auger	Inria, France
Kehinde Babaagba	Edinburgh Napier University, UK
Jaume Bacardit	Newcastle University, UK
Thomas Bäck	Leiden University, Netherlands
Marco Baioletti	University of Perugia, Italy
Wolfgang Banzhaf	Michigan State University, USA
Matthieu Basseur	Université du Littoral Côte d'Opale, France
Vitor Basto-Fernandes	ISCTE-University Institute of Lisbon, Portugal
João Batista	RIKEN-CCS, Japan
Lucas Batista	Universidade Federal de Minas Gerais, Brazil
Jakob Baumann	University of Passau, Germany
Slim Bechikh	University of Tunis, Tunisia
Andrej Belcer	Jožef Stefan Institute, Slovenia
Heder Bernardino	Universidade Federal de Juiz de Fora, Brazil
Ying Bi	Zhengzhou University, China
Rafał Biedrzycki	Warsaw University of Technology, Poland
Bernd Bischl	Ludwig Maximilian University of Munich, Germany
Maria J. Blesa	Universitat Politècnica de Catalunya, Spain
Aymeric Blot	Université de Rennes, France
Christian Blum	Artificial Intelligence Research Institute, Spain
Anna Sergeevna Bosman	University of Pretoria, South Africa

Peter Bosman	Centrum Wiskunde & Informatica, Netherlands
Amine Boumaza	Université de Lorraine, France
Jürgen Branke	University of Warwick, UK
Dimo Brockhoff	École Polytechnique, France
James Brookhouse	Manchester Metropolitan University, UK
Alexander Brownlee	University of Stirling, UK
Maxim Buzdalov	Aberystwyth University, UK
Stefano Cagnoni	Università degli Studi di Parma, Italy
Salvatore Calderaro	Università degli Studi di Palermo, Italy
Fabio Caraffini	Swansea University, UK
Giulia Caselli	Università degli Studi di Modena e Reggio Emilia, Italy
Pedro A. Castillo	University of Granada, Spain
Giancarlo Catalano	University of Stirling, UK
Diego Cattaruzza	Università degli Studi di Udine, Italy
Ying-ping Chen	National Yang Ming Chiao Tung University, Taiwan
Qi Chen	Victoria University of Wellington, New Zealand
Aaron Chen	Victoria University of Wellington, New Zealand
Kazuhisa Chiba	University of Electro-Communications, Japan
Francisco Chicano	Universidad de Málaga, Spain
Darren Chitty	University of Exeter, UK
Alexandre Chotard	Université du Littoral de la Côte d'Opale, France
Farhan Tanvir Chowdhury	Ohio State University, USA
Anders Lyhne Christensen	University of Southern Denmark, Denmark
Lee Christie	Robert Gordon University, UK
Tinkle Chugh	University of Exeter, UK
Carlos A. Coello Coello	CINVESTAV, Mexico
Mark Coletti	Oak Ridge National Laboratory, USA
Feijoo Colomine	Universidad Nacional Experimental del Táchira, Venezuela
Rebeka Čorić	University of Osijek, Croatia
João Correia	University of Coimbra, Portugal
Gabriel Cortês	University of Coimbra, Portugal
Victor Costa	University of Coimbra, Portugal
Ernesto Costa	University of Coimbra, Portugal
Carlos Cotta	Universidad de Málaga, Spain
Matthew J. Craven	University of Plymouth, UK
Jorge Mario Cruz-Duarte	Centre Inria de l'Université de Lille, France
Oliver Cuate	ESFM-IPN, Mexico
Elia Cunegatti	Università degli Studi di Trento, Italy

António Cunha	Universidade de Trás-os-Montes e Alto Douro, Portugal
Leonardo Lucio Custode	Independent Researcher, Italy
Francesca Da Ros	University of Udine, Italy
Zaineb Chelly Dagdia	University of Lille, France
Zakaria Abdelmoiz Dahi	Universidad de Málaga, Spain
Chedli Damak	University of Sfax, Tunisia
Duc-Cuong Dang	University of Passau, Germany
Grégoire Danoy	University of Luxembourg, Luxembourg
George De Ath	University of Exeter, UK
Kenneth Ege de Bruin	University of Oslo, Norway
Patrick De Causmaecker	KU Leuven, Belgium
Kenneth De Jong	George Mason University, USA
Luis Gerardo de la Fraga	CINVESTAV, Mexico
Camilo De La Torre	University of Toulouse Capitole, France
Andrea De Lorenzo	Università degli Studi di Trieste, Italy
Alex de Sa	University of Queensland, Australia
Roy de Winter	TU Delft, Netherlands
Antonio Della Cioppa	Università degli Studi di Salerno, Italy
Bobin Deng	Kennesaw State University, USA
Bilel Derbel	Centre Inria de l'Université de Lille, France
André Deutz	Leiden University, Netherlands
Clarisse Dhaenens	Université de Lille, France
Grant Dick	University of Otago, New Zealand
Federico Divina	Pablo de Olavide University, Spain
Anh Do	University of Adelaide, Australia
Carola Doerr	CNRS, Sorbonne University, France
Benjamin Doerr	École Polytechnique, France
Bernabe Dorronsoro	University of Cádiz, Spain
Daniela Scherer dos Santos	University of Coimbra, Portugal
John Drake	University of Leicester, UK
Rafał Dreżewski	AGH University of Krakow, Poland
Qiqi Duan	Jiangxi University of Finance and Economics, China
Mateja Đumić	University of Osijek, Croatia
Marko Đurasević	University of Zagreb, Croatia
Tome Eftimov	Jožef Stefan Institute, Slovenia
Agoston E. Eiben	Vrije Universiteit Amsterdam, Netherlands
Michel El Saliby	Università degli Studi di Trieste, Italy
AbdElRahman ElSaid	University of North Carolina Wilmington, USA
Achiya Elyasaf	Ben-Gurion University of the Negev, Israel
Michael Emmerich	Leiden University, Netherlands

Madalina-Andreea Ene	Universitatea de Vest din Timişoara, Romania
Andries P. Engelbrecht	Stellenbosch University, South Africa
Anton Eremeev	Sobolev Institute of Mathematics, Russia
Fatma Corut Ergin	Marmara University, Türkiye
Lutton Evelyne	INRAE, France
Yingjie Fan	Leiden University, Netherlands
Davide Farinati	IRCCS Ospedale San Raffaele, Italy
Lavinia Ferariu	Gheorghe Asachi Technical University of Iaşi, Romania
Francisco Fernández de Vega	Universidad de Extremadura, Spain
Eliseo Ferrante	New York University Abu Dhabi, United Arab Emirates
Pedro Ferreira	University of Lisbon, Portugal
Paola Festa	Università degli Studi di Napoli Federico II, Italy
João António Fialho Vilas Boas Duro	University of Sheffield, UK
Jonathan Fieldsend	University of Exeter, UK
Bogdan Filipič	Jožef Stefan Institute, Slovenia
Katarzyna Filus	Polish Academy of Sciences, Poland
Manon Flageat	University of Cambridge, UK
Gianluigi Folino	ICAR-CNR, Italy
Francesco Fontanella	Università degli Studi di Cassino e del Lazio Meridionale, Italy
Olivier Francon	Cognizant, USA
Martin Fyvie	Robert Gordon University, UK
Thomas Gabor	LMU Munich, Germany
Marcus Gallagher	University of Queensland, Australia
Edgar Galvan	Maynooth University, Ireland
Adán José García	Université de Lille, France
Carlos García-Martínez	University of Córdoba, Spain
Abel García-Nájera	Universidad Autónoma Metropolitana Cuajimalpa, México
José García-Nieto	Universidad de Málaga, Spain
Jochen Garcke	Universität Bonn, Germany
Andoni Irazusta Garmendia	University of the Basque Country, Spain
Noémi Gaskó	Babeş-Bolyai University, Romania
Andrea Gasparin	Università degli Studi di Trieste, Italy
Stefano Genetti	Università degli Studi di Trento, Italy
Mario Giacobini	Università degli Studi di Torino, Italy
Kyriakos Giannakoglou	National Technical University of Athens, Greece
Francisco Javier Gil-Gala	University of Oviedo, Spain
Tobias Glasmachers	Ruhr University Bochum, Germany

Kyrre Glette	University of Oslo, Norway
Kevin Godin-Dubois	Vrije Universiteit Amsterdam, Netherlands
Adrien Goëffon	University of Angers, France
Yue-Jiao Gong	South China University of Technology, China
Sebastian Rojas Gonzalez	Universiteit Gent, Belgium
Olivier Goudet	University of Angers, France
Daniele Gravina	Malta College of Arts, Science and Technology, Malta
David Greiner	Universidad de Las Palmas de Gran Canaria, Spain
Tong Guo	Beihang University, China
Boukthir Haddar	Université de Sfax, Tunisia
Alexander Hagg	Bonn-Rhein-Sieg University of Applied Sciences, Germany
Hisashi Handa	Kindai University, Japan
Julia Handl	University of Manchester, UK
Jin-Kao Hao	University of Angers, France
Tomohiro Harada	Saitama University, Japan
Kyle Harrison	University of Newcastle, Australia
Emma Hart	Edinburgh Napier University, UK
Mir Hassan	Università degli Studi di Trento, Italy
Nathan Haut	Michigan State University, USA
Michael Heider	Universität Augsburg, Germany
Ayah Helal	Manchester Metropolitan University, UK
Michael Hellwig	Vorarlberg University of Applied Sciences, Austria
Erik Hemberg	MIT, USA
Luis Miramontes Hercog	Eclectic Systems, Mexico
Daniel Hernández	Instituto Tecnológico de Tijuana, México
Carlos Ignacio Hernández Castellanos	Universidad Nacional Autónoma, México
Jesús-Antonio Hernández-Riveros	Universidad Nacional de Colombia, Colombia
Leticia Hernando	University of the Basque Country, Spain
Ishara Hewa Pathiranage	University of Adelaide, Australia
Malcolm Heywood	Dalhousie University, Canada
Martin Holeňa	Institute of Computer Science, Czech Republic
Wenjing Hong	Shenzhen University, China
Manar Hosny	King Saud University, Saudi Arabia
Guangyan Huang	Deakin University, Australia
Junhao Huang	Victoria University of Wellington, New Zealand
Martin Hurta	Brno University of Technology, Czech Republic
Ekhine Irurozki	Télécom Paris, France

Hisao Ishibuchi	Southern University of Science and Technology, China
Fuyuki Ishikawa	National Institute of Informatics, Japan
Christian Jacob	University of Calgary, Canada
Domagoj Jakobović	University of Zagreb, Croatia
Thomas Jansen	Aberystwyth University, UK
Andrzej Jaszkiewicz	Poznań University of Technology, Poland
Ya Jia	South China University of Technology, China
Taran Cyriac John	Victoria University of Wellington, New Zealand
Anju Johnson	University of Huddersfield, UK
Laetitia Jourdan	Université de Lille, France
Krzysztof Jurczuk	Bialystok University of Technology, Poland
Roman Kalkreuth	RWTH Aachen University, Germany
Masahiro Kanazaki	Tokyo Metropolitan University, Japan
George Karakostas	McMaster University, Canada
Kokila Kasuni	Adelaide University, Australia
Ed Keedwell	University of Exeter, UK
Pascal Kerschke	TU Dresden, Germany
Marie-Eléonore Kessaci	University of Lille, France
Ahmed Kheiri	University of Manchester, UK
Hadi Akbarzadeh Khorshidi	University of Melbourne, Australia
Joshua Knowles	University of Birmingham, UK
Anna Kononova	Leiden University, Netherlands
Lars Kotthoff	University of Wyoming, USA
Timo Kötzing	Hasso Plattner Institute, University of Potsdam, Germany
Oswin Krause	University of Copenhagen, Denmark
Martin Krejca	École Polytechnique, Institut Polytechnique de Paris, France
Marek Kretowski	Bialystok University of Technology, Poland
Gabriel Kronberger	University of Applied Sciences Upper Austria, Austria
Karthik Kuber	University of Toronto, Canada
Jakub Kůdela	Brno University of Technology, Czech Republic
William B. Langdon	University College London, UK
Adriana Lara	Instituto Politécnico Nacional, México
Giomara Larraga Maldonado	University of Jyväskylä, Finland
Antonio LaTorre	Universidad Politécnica de Madrid, Spain
Yuri Lavinas	University of Toulouse Capitole, France
Leni Le Goff	Edinburgh Napier University, UK
Clément Legrand	Université de Lille, France
Per Kristian Lehre	University of Birmingham, UK

Johannes Lengler	ETH Zurich, Switzerland
Miqing Li	University of Birmingham, UK
Nan Li	Northeastern University, China
Arnaud Liefooghe	Université du Littoral Côte d'Opale, France
Shishen Lin	University of Birmingham, UK
Fei Liu	City University of Hong Kong, China
Yiping Liu	Hunan University, China
Songbai Liu	Shenzhen University, China
Giosuè Lo Bosco	Università degli Studi di Palermo, Italy
Fernando Lobo	University of Algarve, Portugal
Daniele Loiacono	Politecnico di Milano, Italy
Gonçalo Lopes	University of Coimbra, Portugal
Manuel López-Ibáñez	University of Manchester, UK
Quintino Francesco Lotito	Central European University, Austria
Nuno Lourenço	University of Coimbra, Portugal
Simone Ludwig	North Dakota State University, USA
Herve Luga	IRIT-UT1C, France
Rodica Lung	Babeş-Bolyai University, Romania
Ngoc Hoang Luong	University of Information Technology, Vietnam
Gabriel Luque	Universidad de Málaga, Spain
Zimeng Lyu	Kean University, USA
Hui Ma	Victoria University of Wellington, New Zealand
João Macedo	University of Coimbra, Portugal
Penousal Machado	University of Coimbra, Portugal
Tokunbo Makanju	New York Institute of Technology, USA
Tina Malalanirainy	African Institute for Mathematical Sciences, Rwanda
Katherine Malan	University of South Africa, South Africa
Rammohan Mallipeddi	Kyungpook National University, South Korea
Jacek Mańdziuk	Warsaw University of Technology, Poland
Luca Manzoni	Università degli Studi di Trieste, Italy
Carolina Marcelino	Federal University of Rio de Janeiro, Brazil
Mariapia Marchi	ESTECO SpA, Italy
Valentin Margraf	LMU Munich, Germany
Luca Mariot	University of Twente, Netherlands
Michalis Mavrovouniotis	Cyprus University of Technology, Cyprus
Helmut Mayer	Universität Salzburg, Austria
Elisabeth Mayrhuber	University of Applied Sciences Upper Austria, Austria
Michael Mc Gettrick	University of Galway, Ireland
James McDermott	University of Galway, Ireland
Jessica Mégane	University of Coimbra, Portugal

Jörn Mehnen	University of Strathclyde, UK
Yi Mei	Victoria University of Wellington, New Zealand
Nouredine Melab	Inria, France
Ole Jakob Mengshoel	Norwegian University of Science and Technology, Norway
Juan J. Merelo Guervós	University of Granada, Spain
Marjan Mernik	University of Maribor, Slovenia
Olaf Mersmann	Hochschule des Bundes für öffentliche Verwaltung, Germany
Silja Meyer-Nieberg	Universität der Bundeswehr München, Germany
Efrén Mezura-Montes	University of Veracruz, Mexico
Krzysztof Michalak	Wrocław University of Economics, Poland
Gabriela Minetti	Universidad Nacional de La Pampa, Argentina
Edmondo Minisci	University of Strathclyde, UK
Mustafa Misir	Duke Kunshan University, China
Hyunho Mo	UMC Utrecht, Netherlands
Elizabeth Montero	Universidad Técnica Federico Santa María, Chile
Gladston Moreira	Universidade Federal de Ouro Preto, Brazil
Matthew Andres Moreno	Michigan State University, USA
Somnath Mukhopadhyay	Assam University, India
Mario Andrés Muñoz	University of Melbourne, Australia
Aidan Murphy	University College Dublin, Ireland
Jonathan Mwaura	Northeastern University, USA
Pawel Myszkowski	Wrocław University of Science and Technology, Poland
Jakub Nalepa	Silesian University of Technology, Poland
Emanuele Nardone	University of Eastern Finland, Finland
Asef Nazari	Deakin University, Australia
Ferrante Neri	University of Surrey, UK
Mehdi Neshat	University of Technology Sydney, Australia
Aneta Neumann	University of Adelaide, Australia
Frank Neumann	University of Adelaide, Australia
Duc Manh Nguyen	Hanoi National University of Education, Vietnam
Tai Nguyen	University of St Andrews, UK
Miguel Nicolau	University College Dublin, Ireland
Erik Nielsen	Università degli Studi di Trento, Italy
Ricardo Nieto	Centro de Investigación en Matemáticas, México
Adel Nikfarjam	University of Adelaide, Australia
Eleni Nisioti	IT University of Copenhagen, Denmark
Marco Salvatore Nobile	Università Ca' Foscari Venezia, Italy
Yusuke Nojima	Osaka Metropolitan University, Japan
Masahiro Nomura	Institute of Science Tokyo, Japan

Thambo Nyathi	University of Pretoria, South Africa
Michael O'Neill	University College Dublin, Ireland
Una-May O'Reilly	MIT, USA
Colm O'Riordan	University of Galway, Ireland
Gabriela Ochoa	University of Stirling, UK
Diego Oliva	Universidad de Guadalajara, México
Pietro S. Oliveto	Southern University of Science and Technology, China
Fabricio Olivetti de Franca	Universidade Federal do ABC, Brazil
Andre Opris	University of Passau, Germany
José Carlos Ortiz-Bayliss	Monterrey Institute of Technology and Higher Education, Mexico
Ender Özcan	University of Nottingham, UK
Ben Paechter	Edinburgh Napier University, UK
Denis Pallez	Université Côte d'Azur, France
Venkatesh Pandiri	Indian Institute of Information Technology, Design and Manufacturing, Kancheepuram, India
Lie Meng Pang	Southern University of Science and Technology, China
Gregor Papa	Jožef Stefan Institute, Slovenia
Gisele Pappa	Federal University of Minas Gerais, Brazil
Luís Paquete	University of Coimbra, Portugal
Andrew J. Parkes	University of Nottingham, UK
Konstantinos Parsopoulos	University of Ioannina, Greece
Antonio Parziale	Università degli Studi di Salerno, Italy
Alina Patelli	Aston University, UK
Thilina Pathirage Don	Adelaide University, Australia
Martín Pedemonte	Universidad de la República, Uruguay
David Pelta	University of Granada, Spain
Jordi Pereira	UPF — Barcelona School of Management, Spain
Francisco Pereira	University of Coimbra, Portugal
Miriam Pescador	Instituto Politécnico Nacional, México
Stjepan Picek	Radboud University, Netherlands
Gloria Pietropolli	Centrum Wiskunde & Informatica, Netherlands
Federico Pigozzi	Tufts University, USA
Martin Pilát	Charles University, Czech Republic
Zbyněk Pitra	Czech Technical University in Prague, Czech Republic
Valentina Poggioni	University of Perugia, Italy
Daniel Porumbel	CEDRIC, France
Petr Pošík	Czech Technical University in Prague, Czech Republic

Oliver Preuß	Paderborn University, Germany
Michal Przewozniczek	Wrocław University of Science and Technology, Poland
Robin Purshouse	University of Sheffield, UK
Chao Qian	Nanjing University, China
Xin Qiu	Cognizant, USA
Chiara Camilla Rambaldi Migliore	Università degli Studi di Trento, Italy
Elena Raponi	Leiden University, Netherlands
Khaled Rasheed	University of Georgia, USA
Kenneth Reid	University of Michigan, USA
Quentin Renau	Edinburgh Napier University, UK
Hendrik Richter	HTWK Leipzig University of Applied Sciences, Germany
María Cristina Riff	Universidad Técnica Federico Santa María, Chile
Denis Robilliard	Université du Littoral Côte d'Opale, France
Peter Rockett	University of Sheffield, UK
André Rodrigues da Cruz	Centro Federal de Educação Tecnológica de Minas Gerais, Brazil
Angel E. Rodriguez-Fernandez	Center for Research and Advanced Studies of the National Polytechnic Institute, Mexico
Eduardo Rodriguez-Tello	CINVESTAV, Mexico
Katya Rodriguez-Vazquez	IIMAS UNAM, Mexico
Andrea Roli	Università di Bologna, Italy
Jeroen Rook	University of Twente, Netherlands
Alejandro Rosales-Pérez	Centro de Investigación en Matemáticas, Mexico
Manou Rosenberg	Curtin University, Australia
Luigi Rovito	Università degli Studi di Trieste, Italy
Jonathan Rowe	University of Birmingham, UK
Francesco Rusin	Università degli Studi di Trieste, Italy
Conor Ryan	University of Limerick, Ireland
Cosmin Sabo	Technical University of Cluj-Napoca, Romania
Berfin Sakallioglu	Università degli Studi di Trieste, Italy
Carolina Salto	Universidad de La Pampa, Argentina
Roberto Santana	University of the Basque Country, Spain
Frederico J. J. B. Santos	Università degli Studi di Trieste, Italy
Valentino Santucci	Università per Stranieri di Perugia, Italy
Stefano Sarti	University of Stirling, UK
Hiroyuki Sato	University of Electro-Communications, Japan
Frédéric Saubion	Université d'Angers, France
Erdi Sayar	Paderborn University, Germany
Robert Schaefer	AGH University of Kraków, Poland
Lennart Schäpermeier	TU Dresden, Germany

Marc Schoenauer	Inria Saclay, CNRS, France
Jacob Schrum	Southwestern University, USA
Oliver Schuetze	CINVESTAV, Mexico
Dirk Schweim	University of Applied Sciences Mainz, Germany
Eduardo Segredo	Universidad de La Laguna, Spain
Carlos Segura	Centro de Investigación en Matemáticas, México
Lukas Sekanina	Brno University of Technology, Czech Republic
Yann Semet	Thales, France
Roman Senkerik	Tomas Bata University in Zlín, Czech Republic
Hadar Shavit	RWTH Aachen, Germany
Eliad Shem-Tov	Ben-Gurion University of the Negev, Israel
John Sheppard	Montana State University, USA
Jialong Shi	Xi'an Jiaotong University, China
Feng Shi	Central South University, China
Koji Shimoyama	Kyushu University, Japan
Ofer M. Shir	Tel-Hai College, Israel
Shinichi Shirakawa	Yokohama National University, Japan
María Sierra	University of Oviedo, Spain
Rodrigo Silva	Universidade Federal de Ouro Preto, Brazil
Sara Silva	University of Lisbon, Portugal
Anabela Simoes	University of Coimbra, Portugal
Lana Sinapayen	Sony Computer Science Laboratories, National Institute for Basic Biology, Japan
Alok Singh	University of Hyderabad, India
Hemant Singh	University of New South Wales, Australia
Moshe Sipper	Ben-Gurion University of the Negev, Israel
Adam Slowik	Koszalin University of Technology, Poland
Simon Smith	Edinburgh Napier University, UK
Zhenshou Song	Victoria University of Wellington, New Zealand
Victor-Adrian Sosa-Hernandez	ITESM-CEM, Mexico
Petr Sosík	Silesian University in Opava, Czech Republic
Giovanni Squillero	Politecnico di Torino, Italy
Adam Stanton	University of Sheffield, UK
Anthony Stein	University of Hohenheim, Germany
Sebastian Stich	CISPA, Germany
Catalin Stoean	University of Craiova, Romania
Thomas Stützle	Université Libre de Bruxelles, Belgium
Mihai Suciu	Babeş-Bolyai University, Romania
Dirk Sudholt	University of Passau, Germany
Jianyong Sun	Xi'an Jiaotong University, China
Leo Sünkel	LMU Munich, Germany
Andrew Sutton	University of Minnesota Duluth, USA

Reiji Suzuki	Nagoya University, Japan
Keiki Takadama	University of Tokyo, Japan
Ricardo Takahashi	Universidade Federal de Minas Gerais, Brazil
Ryoji Tanabe	Yokohama National University, Japan
Shoichiro Tanaka	University of Fukuchiyama, Japan
Ernesto Tarantino	ICAR-CNR, Italy
Sara Tari	Université du Littoral Côte d'Opale, France
Murat Taşkiran	Yıldız Teknik Üniversitesi, Turkey
Tomoaki Tatsukawa	Tokyo University of Science, Japan
Daniel R. Tauritz	Auburn University, USA
Roberto Tavares	Federal University of São Carlos, Brazil
Hugo Terashima-Marin	Tecnológico de Monterrey, México
Andrea Tettamanzi	Université Côte d'Azur, France
Dirk Thierens	Utrecht University, Netherlands
Dhananjay Thiruvady	Deakin University, Australia
Sarah L. Thomson	Edinburgh Napier University, UK
Renato Tinós	University of São Paulo, Brazil
Raca Todosijević	Polytechnic University of Hauts-de-France, France
Marco Tomassini	University of Lausanne, Switzerland
Alberto Tonda	INRAE, Université Paris-Saclay, France
Hao Tong	University of Birmingham, UK
Jamal Toutouh	Universidad de Málaga, Spain
Leonardo Trujillo	Instituto Tecnológico de Tijuana, México
Elio Tuci	University of Namur, Belgium
Tea Tušar	Jožef Stefan Institute, Slovenia
Alexandros Tzanetos	Jönköping University, Sweden
Kento Uchida	Yokohama National University, Japan
Neil Urquhart	Edinburgh Napier University, UK
Fumito Uwano	Okayama University, Japan
Gauri Vaidya	University of Limerick, Ireland
Inês Valentim	University of Coimbra, Portugal
Gara Miranda Valladares	Universidad de La Laguna, Spain
Niki van Stein	Leiden University, Netherlands
Leonardo Vanneschi	Universidade Nova de Lisboa, Portugal
Swetha Varadarajan	University of Cape Town, South Africa
Denis Vargas	Centro Federal de Educação Tecnológica de Minas Gerais, Brazil
Igor Vatolkin	RWTH Aachen University, Germany
Nadarajen Veerapen	Université de Lille, France
Sébastien Verel	Université du Littoral Côte d'Opale, France
Diederick Vermetten	Sorbonne University, France

Ana Viana	INESC TEC/Polytechnic of Porto, Portugal
Adriano Vinhas	University of Coimbra, Portugal
Alessandra Vitanza	National Research Council — Institute of Cognitive Sciences and Technologies, Italy
Vanessa Volz	Centrum Wiskunde & Informatica, Netherlands
Marcus Vukojevic	Università degli Studi di Trento, Italy
Markus Wagner	Monash University, Australia
David Walker	University of Exeter, UK
Hao Wang	Leiden University, Netherlands
Handing Wang	Xidian University, China
Shuai Wang	Sun Yat-sen University, China
Elizabeth Wanner	Aston University, UK
John Alasdair Warwicker	University of Sheffield, UK
Darrell Whitley	Colorado State University, USA
Simon Wietheger	TU Wien, Austria
Josh Wilkerson	Naval Air Systems Command, USA
Garnett Wilson	Dalhousie University, Canada
Dennis Wilson	ISAE-Supaero, France
Stephan Winkler	University of Applied Sciences Upper Austria, Austria
Carsten Witt	Technical University of Denmark, Denmark
Man Leung Wong	Lingnan University, China
Yue Xie	University of Cambridge, UK
Xingsi Xue	Fujian University of Technology, China
Bing Xue	Victoria University of Wellington, New Zealand
Ning Xue	University of Nottingham Ningbo, China
Takeshi Yamada	Kindai University, Japan
Anil Yaman	Vrije Universiteit Amsterdam, Netherlands
Xiankun Yan	University of Adelaide, Australia
Qiang Yang	Nanjing University of Information Science and Technology, China
Furong Ye	Leiden University, Netherlands
Iryna Yevseyeva	De Montfort University, UK
Tian-Li Yu	National Taiwan University, Taiwan
Martin Zaefferer	DHBW Ravensburg, Germany
Daniela Zaharie	West University of Timisoara, Romania
Aleš Zamuda	University of Maribor, Slovenia
Saúl Zapotecas-Martínez	Instituto Nacional de Astrofísica Óptica y Electrónica, México
Christine Zarges	Aberystwyth University, UK
Ciprian Zavoianu	Robert Gordon University, UK
Vašíček Zdeněk	Brno University of Technology, Czech Republic

Xingyi Zhang	Anhui University, China
Fangfang Zhang	Victoria University of Wellington, New Zealand
Hengzhe Zhang	Victoria University of Wellington, New Zealand
Mengjie Zhang	Victoria University of Wellington, New Zealand
Tian Zhang	Northeastern University, China
Huang Zhixing	Victoria University of Wellington, New Zealand
Jinghui Zhong	South China University of Technology, China
Yan Zhou	Shenyang Jianzhu University, China
Ryan Zhou	Queen's University, Canada
Adam Żychowski	Warsaw University of Technology, Poland

Contents

Theory and Runtime Analysis

An Analysis of the Focused Jump-and-Repair EA on Parameterized Max-2-Sat

Liam Gaeuman, Mahya Salimi Gamasaei, and Andrew M. Sutton(✉)

University of Minnesota Duluth, Duluth, MN 88512, USA
amsutton@d.umn.edu

Abstract. Evolutionary algorithms are often used in practice to tackle NP-hard optimization problems. However, comparatively few theoretical analyses exist for these types of problems since they are significantly more complex than simple benchmarks. Moreover, it is not always clear how best to design algorithm components for these domains. Recently, ideas from the field of parameterized algorithmics have been applied to the theoretical analysis of simple randomized search heuristics on NP-hard problems. By using a finer-grained analysis informed by problem parameters, these approaches can yield a sharper understanding of the influence of problem structure on algorithm behavior, and aid in the development of operators with provably good performance.

We consider the NP-hard MAX-2-SAT problem parameterized by the number of false clauses in an optimal solution. We design a probabilistic repair operator for 2SAT formulas as module for the (1+1) EA with focused jump and repair introduced by Branson and Sutton [FOGA, 2021]. We prove that the resulting algorithm solves MAX-2-SAT in expected time $O(12^k m \log n)$ where m is the formula size, n is the number of variables and k is the count of clauses that are false in an optimal solution. This establishes that the algorithm is an FPT randomized search heuristic for parameterized MAX-2-SAT. We complement our theoretical results with computational experiments.

1 Introduction

Evolutionary algorithms and related randomized search heuristics have been applied to solving maximum satisfiability (MAX-SAT) problems for decades [1, 4,6,9,13,23,25]. These techniques are a popular choice among practitioners [12, 26] for solving complex optimization problems like MAX-SAT because they are general-purpose programs that are relatively easy to deploy, and can quickly produce reasonable outcomes in the field.

Despite their prominent use in practice, there is a comparative lack of theoretical analyses for evolutionary algorithms on NP-hard problems like MAX-SAT. This is in part due to the fitness landscapes of such problems, which are considerably more complex and ill-conditioned than those of simple benchmark functions

G. Iacca et al. (Eds.): PPSN 2026, LNCS 16985, pp. 3–19, 2027.
https://doi.org/10.1007/978-3-032-36214-8_1

like OneMax and LeadingOnes. Nevertheless, there has been an effort in the research community to improve our theoretical understanding of the behavior of evolutionary algorithms on NP-hard problems [8,19,20,22,28].

For these kinds of problems, the performance of evolutionary algorithms can often be significantly improved by leveraging domain-specific problem structure. One strategy for incorporating domain knowledge into the search operators of an EA is the focused jump-and-repair framework introduced by Branson and Sutton [3] for solving parameterized graph problems. The jump-and-repair approach, the so-called (1+1) $\text{EA}^{\text{J+R}}$, utilizes a duplex genotype that encodes both a subgraph and a candidate solution to the subgraph. When an infeasible offspring is produced, a probabilistic jump followed by a domain-specific repair operation is attempted to fix the offspring.

In this paper, we consider the NP-hard maximimum 2-satisfiability (Max-2-Sat) problem parameterized by the number of false clauses in an optimal solution. We demonstrate how the (1+1) $\text{EA}^{\text{J+R}}$ can be adapted to this problem by designing a repair operator for infeasible offspring. We prove that this modification enables the (1+1) $\text{EA}^{\text{J+R}}$ to find optimal solutions on Max-2-Sat formulas with n variables and m clauses in expected time $O(12^k m \log n)$ where k is the number of clauses that are false in an optimal solution. From the perspective of parameterized complexity, this proves the (1+1) $\text{EA}^{\text{J+R}}$ is a randomized FPT algorithm for Max-2-Sat.

1.1 Related Work

In the context of satisfiable k-CNF formulas, Doerr et al. [8] proved that the (1+1) EA can find satisfying assignments to dense satisfiable k-SAT formulas in $O(n \log n)$ time, but fail when the clause-variable density is too low. Buzdalov and Doerr [5] followed up on this work and proved that the run time of the $(1+(\lambda,\lambda))$ GA with a careful setting for λ can overcome weaker fitness-distance correlation and solve in linear time formulas with polylogarithmic density. Neumann and Sutton [20] investigated the (1+1) EA on formulas generated by a community attachment model and proved that it solves asymptotically almost all formulas with constant constraint density for a wide range of modularities.

Maximum 2-satisfiability (Max-2-Sat) was shown to be NP-hard by reduction from 3-Sat [11]. The parameterized version, in which one is interested in finding an assignment with at most k false clauses, was shown to be fixed-parameter tractable (FPT) by Razgon and O'Sullivan [24], who presented a $O(15^k km^3)$ algorithm. The best FPT bound to date is $O^*(2.3146^k)$, attained by an algorithm due to Lokshtanov et al. [18]. The parameterized problem is also known as Almost-2-Sat, All-but-k 2-Sat, 2-CNF-Deletion.

The field of parameterized complexity can offer new perspectives to the run time analysis of evolutionary algorithms, as a more fine-grained analysis allows for more direct insight into how problem instance structure can influence run time on NP-hard problems [21]. The first parameterized run time analyses of evolutionary algorithms showed a $O(n^2 \log n + kn^2 + 4^k n)$ run time for Global SEMO to find vertex covers in graphs with optimal cover size k [15], and a $O(2^{15k^2 \log k})$

run time for a (1+1) EA solving maximum leaf spanning trees [14]. Most importantly, these works also offered clever insights into designing algorithmic components for these problems such as problem-tailored mutation operators and employing an LP-relaxation as an objective. Other parameterized analyses have been carried out for population-based GAs on the closest string problem [27], graph theoretic clustering problems [10] and vertex cover [17].

Our repair operator for MAX-2-SAT is based on identifying a directed graph associated with the 2-CNF formula and finding minimum cuts in this graph. This approach is rooted in the work of Kratsch and Wahlström [16] who derive a polynomial kernel for the problem.

2 Preliminaries

For any $n \in \mathbb{N}$, we denote $[n] := [1, n] \cap \mathbb{N}$. A propositional 2-CNF formula $\Phi = \{C_1, \ldots, C_m\}$ is a set of m disjunctive clauses over a set $\mathcal{V} = \{v_1, \ldots, v_n\}$ of n Boolean variables. Each clause $C_j = (\ell_{j1} \vee \ell_{j2})$ is a disjunction of two *literals*, where a literal $\ell \in \{v_i, \neg v_i : v_i \in \mathcal{V}\}$ is a variable or its negation. For any literal $\ell \in \{\neg v_i, v_i\}$, we define $\text{var}(\ell) := v_i$ to be the variable in ℓ. If $\ell = \text{var}(\ell)$, we say ℓ is a *positive literal*, otherwise if $\ell = \neg\,\text{var}(\ell)$, we say ℓ is a *negative literal*. For a clause $C = (\ell_1 \vee \ell_2)$, we denote $\text{vars}(C) := \text{var}(\ell_1) \cup \text{var}(\ell_2)$, and for a set of s clauses, $\text{vars}(\{C_1, \ldots, C_s\}) = \text{vars}(C_1) \cup \cdots \cup \text{vars}(C_s)$. Note that since we define a formula as a set of clauses we have $\text{vars}(\Phi) = \mathcal{V}$.

We associate the elements of $\{0,1\}^n$ with assignments to $\mathcal{V}$ in Φ. Given a length-n bit string $z \in \{0,1\}^n$, we say a clause $C \in \Phi$ is *true* under z if at least one of its literals evaluates to true, otherwise we say C is *false*. The objective of MAX-2-SAT is to find an assignment that maximizes the number of true clauses.

$$\max_{z \in \{0,1\}^n} \sum_{C \in \Phi} \mathbb{1}\{C \text{ is true under } z\}. \tag{1}$$

Given a bit string $z \in \{0,1\}^n$ and an index set $\mathcal{I} \subseteq [n]$ we denote as $z[\mathcal{I}]$ the subsequence of z selected by $\mathcal{I}$. This corresponds to a *partial assignment* to the variables of Φ. We will often refer to a subset of $[n]$ interchangeably as a subset of $\mathcal{V}$, i.e., an index $i \in [n]$ corresponds to the variable $v_i \in \mathcal{V}$, and the assignment bound to that variable under $z \in \{0,1\}^n$ is the Boolean value of z_i.

A parameterized problem is a language $L \subseteq \Sigma^* \times \mathbb{N}$ for a finite alphabet Σ. A problem L is *fixed-parameter tractable* (FPT) if $(x, k) \in L$ can be decided in time $g(k)\,|x|^{O(1)}$ for some function g that depends only on k. For a randomized search heuristic, the *optimization time* is characterized as a random variable T that measures the number of fitness function evaluations until an optimal solution is first visited. An algorithm is a *Monte Carlo FPT algorithm* for a parameterized problem L if it accepts $(x, k) \in L$ with probability at least $1/2$ in time $g(k)\,|x|^{O(1)}$ and accepts $(x, k) \notin L$ with probability zero. Any randomized search heuristic with a bound $\mathrm{E}[T] \leq g(k)\,|x|^{O(1)}$ on L can be transformed into a Monte Carlo FPT algorithm by stopping its execution after $2g(k)\,|x|^{O(1)}$ iterations. A randomized search heuristic runs in *randomized FPT time* on a parameterized problem of size n when $\mathrm{E}[T] \leq g(k) n^{O(1)}$.

The parameterized Max-2-Sat problem is, given a 2-CNF formula Φ and an integer k, determine if there is an assignment x to the variables of Φ such that at most k clauses are false under x.

2.1 Implication Flow Graph

Central to our repair operation is an *implication flow graph*, which is a weighted digraph corresponding to a satisfiable 2-CNF formula Φ with variables from $\mathcal{V}$ and a partial assignment $z[\mathcal{I}]$. The vertices of the implication flow graph are literals. Each variable of $\mathcal{V}$ appears either in a positive literal or a negative literal (but not both) depending on a satisfying assignment of Φ. The arcs of the graph correspond to implicative relationships between literals forced by certain clauses of Φ.

Given a satisfying assignment v^* to Φ, we construct a set $\mathcal{L} := \mathcal{L}(v^*)$ of literals, where a literal appears positively in $\mathcal{L}$ if its corresponding variable in $\mathcal{V}$ is set to false by v^*, otherwise it appears negatively. We also add to the vertex set $\mathcal{L}$ a *source* vertex s and a *sink* vertex t. The set of arcs is denoted $A \subseteq \mathcal{L} \times \mathcal{L}$. The procedure for constructing the implication flow graph $G = (\mathcal{L}, A)$ is listed in Algorithm 1.

Definition 1 (Flow Clause). *Let Φ be a satisfiable 2-CNF formula and let $G = (\mathcal{L}, A)$ be some implication flow graph constructed from Φ. A clause $C = (\ell_1 \vee \ell_2) \in \Phi$ is called a* flow clause *if $(\neg\ell_1, \ell_2) \in A$ or $(\neg\ell_2, \ell_1) \in A$. If C is not a flow clause, we refer to it as a* non-flow clause.

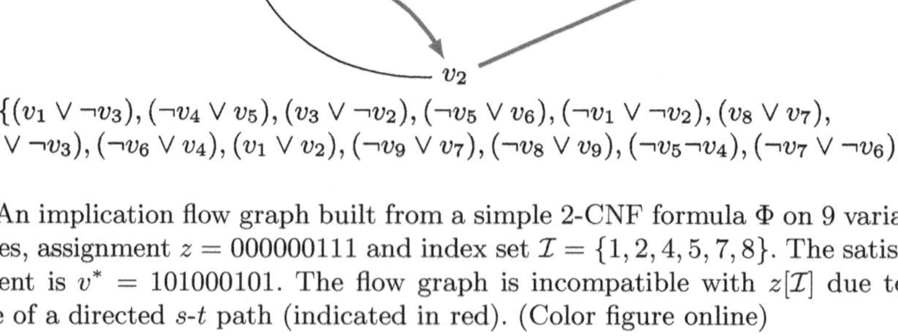

$$\Phi = \{(v_1 \vee \neg v_3), (\neg v_4 \vee v_5), (v_3 \vee \neg v_2), (\neg v_5 \vee v_6), (\neg v_1 \vee \neg v_2), (v_8 \vee v_7),$$
$$(\neg v_4 \vee \neg v_3), (\neg v_6 \vee v_4), (v_1 \vee v_2), (\neg v_9 \vee v_7), (\neg v_8 \vee v_9), (\neg v_5 \neg v_4), (\neg v_7 \vee \neg v_6)\}.$$

Fig. 1. An implication flow graph built from a simple 2-CNF formula Φ on 9 variables, 13 clauses, assignment $z = 000000111$ and index set $\mathcal{I} = \{1, 2, 4, 5, 7, 8\}$. The satisfying assignment is $v^* = 101000101$. The flow graph is incompatible with $z[\mathcal{I}]$ due to the presence of a directed s-t path (indicated in red). (Color figure online)

The implication flow graph (Fig. 1) is an important structure to our repair operation, because directed paths in the graph characterize chains of implications in Φ. This is captured by the following lemma.

Algorithm 1: BUILDIMPLICATIONFLOWGRAPH($\Phi, v^*, z, \mathcal{I}$)

input : A 2-CNF formula Φ with satisfying assignment v^*, a bit string $z \in \{0,1\}^n$ and an index set $\mathcal{I} \subseteq [n]$

1 $\mathcal{L} = \{i : i \in \mathcal{V}, v_i^* = 0\} \cup \{\neg i : i \in \mathcal{V}, v_i^* = 1\} \cup \{s, t\}$
2 $A \leftarrow \emptyset$
3 **foreach** *clause* $(\ell_1 \vee \ell_2) \in \Phi$ **do**
4   **if** $\neg\ell_1, \ell_2 \in \mathcal{L}$ **then** add arc $(\neg\ell_1, \ell_2)$ to A with unit weight
5   **if** $\ell_1, \neg\ell_2 \in \mathcal{L}$ **then** add arc $(\neg\ell_2, \ell_1)$ to A with unit weight
6 **foreach** $i \in \mathcal{I}$ **do**
7   **if** $z_i = 1$ **then**
8   **if** $i \in \mathcal{L}$ **then** add ∞-weight arc (s, i) to A
9   **else** add ∞-weight arc $(\neg i, t)$ to A
10   **if** $z_i = 0$ **then**
11   **if** $i \in \mathcal{L}$ **then** add ∞-weight arc (i, t) to A
12   **else** add ∞-weight arc $(s, \neg i)$ to A
13 **return** $G = (\mathcal{L}, A)$

Lemma 1. *Let Φ be a satisfiable 2-CNF formula, $z \in \{0,1\}^n$ a bit string and $\mathcal{I} \subseteq [n]$ an index set. If there is a directed s-t path in an implication flow graph built by Algorithm 1, then for every $x \in \{0,1\}^n$ with $x[\mathcal{I}] = z[\mathcal{I}]$, Φ is not satisfied by x.*

Proof. Let G be the implication flow graph built for Φ, z and $\mathcal{I}$ and assume there is a directed s-t path $(s, \ell_1), (\ell_1, \ell_2) \cdots (\ell_r, t)$ in G. By construction, every arc (ℓ_a, ℓ_b) corresponds to a clause $(\neg\ell_a \vee \ell_b) \in \Phi$ and thus encodes the implication $\ell_a \implies \ell_b$. Applying resolution to the clauses associated to the s-t path yields the implication $(\ell_1 \implies \ell_r)$. For every assignment that satisfies Φ, this implication must be true. Let $i := \text{var}(\ell_1)$ and $j := \text{var}(\ell_r)$. If both ℓ_1 and ℓ_r are positive literals, it follows that $v_i^* = v_j^* = 0$, as both i and j were added as positive literals in line 1 of Algorithm 1. Furthermore, since $\ell_1 = i$ is a positive literal, the arc $(s, \ell_1) = (s, i)$ c 8 and so $z_i = 1$. Similarly, since $\ell_r = j$ is a positive literal, the arc $(\ell_r, t) = (j, t)$ was added in line 8 so $z_j = 0$. But if both ℓ_1 and ℓ_r are positive, the implication $(\ell_1 \implies \ell_r)$ is equivalent to $i \implies j$ which is contradicted by any assignment x with $x_i = z_i = 1$ and $x_j = z_j = 0$.

If ℓ_1 and ℓ_r are both negative literals, it follows that $v_i^* = v_j^* = 1$ since $\neg i$ and $\neg j$ would be added as vertices of G in line 1. As $\ell_1 = \neg i$ is a negative literal, the arc $(s, \ell_1) = (s, \neg i)$ would have been added in line 12 and hence $z_i = 0$. Similarly, the arc $(\ell_r, t) = (\neg j, t)$ would have been added in line 12 and $z_j = 1$. In this case, the implication $(\ell_1 \implies \ell_r)$ is equivalent to $\neg i \implies \neg j \equiv j \implies i$ by contraposition. This is contradicted by any assignment x with $x_i = z_i = 0$ and $x_j = z_j = 1$.

Now suppose ℓ_1 is negative and ℓ_r is positive. By similar reasoning to above, $z_i = z_j = 0$ and the implication $(\ell_1 \implies \ell_r)$ is equivalent to $\neg i \implies j \equiv (i \vee j)$ which is false under all assignments x where $x_i = x_j = z_i = z_j = 0$. Similarly, if ℓ_1 is positive and ℓ_r is negative, $z_i = z_j = 1$ and $(\ell_1 \implies \ell_r)$ is equivalent to $(\neg i \vee \neg j)$ and is false under all assignments x where $x_i = x_j = z_i = z_j = 1$. □

We also reason in the other direction. If there is no s-t path in the implication flow graph for Φ, we can easily compute an assignment x consistent with $z[\mathcal{I}]$ so that all of the flow clauses of Φ are true under x.

Lemma 2. *Let Φ be a satisfiable 2-CNF formula, $z \in \{0,1\}^n$ a bit string and $\mathcal{I} \subseteq [n]$ an index set. If there is no directed s-t path in an implication flow graph $G = (\mathcal{L}, A)$ built by Algorithm 1, then there is an assignment $x \in \{0,1\}^n$ with $x[\mathcal{I}] = z[\mathcal{I}]$, such that every flow clause of Φ is true under x.*

Proof. We construct such an x as follows. For each vertex ℓ in $\mathcal{L} \setminus \{s,t\}$, if ℓ is reachable from s, set $x_{\text{var}(\ell)}$ so that ℓ is true under x. Otherwise, set $x_{\text{var}(\ell)}$ so that ℓ is false under x. Now let (ℓ_1, ℓ_2) be an arc in A such that ℓ_1 is reachable from s. By construction, this arc corresponds to a flow clause $(\neg\ell_1 \vee \ell_2)$ in Φ. Setting the two positions in x to ensure ℓ_1 and ℓ_2 are true ensures $(\neg\ell_1 \vee \ell_2)$ is true. Otherwise, if not reachable from s setting x so ℓ_1 and ℓ_2 are both false also ensures the clause is true. Finally, for each $i \in \mathcal{I}$, if $(s,i) \in A$, it was added because $z_i = 1$. To make i true, we set $x_i = 1 = z_i$. If $(s, \neg i) \in A$, then $z_i = 0$, and to make $\neg i$ true, we set $x_i = 0 = z_i$. Similarly, if (i,t) or $(\neg i, t)$ are arcs in G, to make the corresponding literal false, we would set $x_i = z_i$. □

Any path from s to t must contain at least one unit-weight arc that corresponds to a flow clause. The infinite weight arcs ensure that the minimum weight s-t cut consists only of arcs that correspond to flow clauses. Aside from s and t, the implication flow graph for a satisfiable formula Φ contains one vertex for each of its variables labeled either as a positive or negative literal. The following lemma establishes that no two vertices in $\mathcal{L}$ appear as literals in any clause of Φ.

Lemma 3. *Let Φ be a satisfiable 2-CNF formula and let G be an implication flow graph built by Algorithm 1. If $(\ell_1 \vee \ell_2) \in \Phi$, then $\neg\ell_1$ is a vertex of G or $\neg\ell_2$ is a vertex of G (or both).*

Proof. Let $G = (\mathcal{L}, A)$ be the implication flow graph. Suppose for contradiction that $(\ell_1 \vee \ell_2) \in \Phi$ but $\ell_1, \ell_2 \in \mathcal{L}$. Let $\text{var}(\ell_1) = i$ and $\text{var}(\ell_2) = j$. We argue that if ℓ_1 is a positive (respectively negative) literal, then $v_i^* = 0$ (respectively $v_i^* = 1$) because of how literals are added to the vertex set of G in line 1 of Algorithm 1. This implies that ℓ_1 is false under v^*. Symmetrically, ℓ_2 must be false under v^*. Therefore, $(\ell_1 \vee \ell_2)$ is false under v^*, contradicting the fact that v^* is a satisfying assignment to Φ. □

3 (1+1) EA with Jump-and-Repair

The (1+1) $\text{EA}^{\text{J+R}}$ is the simple (1+1) EA equipped with an extra operation to probabilistically repair infeasible offspring [3]. It can be regarded as a framework to which a problem-tailored repair operation can be added as a "plug-in" module when applied to different problem domains.

The design philosophy behind the (1+1) $\mathrm{EA}^{\mathrm{J+R}}$ is that, rather than searching directly the space of candidate solutions to maximize the objective function, the genotype is split into two components: a *subproblem* and a feasible *solution* to the subproblem. The search process then attempts to build increasingly larger subproblems while maintaining feasible solutions to those subproblems. In the context of problems from graph theory, a subproblem usually refers to an induced subgraph [3] of the original graph. For MAX-2-SAT, the most natural subproblem is a subset of propositional clauses from the original formula.

Given a formula $\Phi = \{C_1, C_2, \ldots, C_m\}$ over a set of n variables $\mathcal{V}$, we represent a candidate solution as a bit string $x \in \{0,1\}^{m+n}$. The length-m prefix $x_1 \cdots x_m$ of x selects a set of clauses that comprise a subformula of Φ and the length-n suffix $x_{m+1} \cdots x_{m+n}$ corresponds to an assignment to the n variables of $\mathcal{V}$. We write $\Phi[x] = \{C_i \in \Phi \colon x_i = 1\}$ to denote the subformula of Φ selected by the first m bits of x and $\alpha[x] = x_{m+1} \cdots x_{m+n}$ as the length-n suffix of x interpreted as an assignment to the n Boolean variables of Φ. We define the fitness function as

$$f(x) = \begin{cases} |\Phi[x]| & \text{if } |F| \leq k; \\ -|F| & \text{otherwise.} \end{cases} \tag{2}$$

$$\text{where } F := \{C \in \Phi[x] \colon C \text{ is false under} \alpha[x]\}$$

Assuming Φ has an assignment with at most k false clauses, then any x^* that maximizes $f(x^*) = m$ corresponds to such an assignment. This coincides with a solution to the parameterized MAX-SAT problem. The $-|F|$ component incurs a penalty for infeasible assignments that is commensurate to the number of clauses in the subformula that are violated.

Algorithm 2: (1+1) $\mathrm{EA}^{\mathrm{J+R}}$ for parameterized MAX-2-SAT.

input : A 2-CNF formula Φ on n variables and m clauses

choose x from $\{0,1\}^{m+n}$ uniformly at random
while $f(x) < m$ **do**
  Obtain y from x by flipping each bit with prob. $\frac{1}{m+n}$
  if $f(y) < f(x)$ **and** $f(y) < 0$ **then**
    $y \leftarrow$ JUMPANDREPAIR(y, Φ)
  if $f(y) \geq f(x)$ **then** $x \leftarrow y$

The focused jump-and-repair procedure, listed in Algorithm 3, takes an assignment that renders more than k clauses false, and attempts to *repair* it by translating it into one where at most k clauses are false. Note that Algorithm 3 is probabilistic in the sense that it may fail to produce a feasible offspring from a feasible parent. In such a case, the resulting offspring would not be accepted.

3.1 Focused Jump

A "focus set" $\mathcal{I}$ is computed as the set of variables that appear in false clauses. Each bit of $z[\mathcal{I}]$ is flipped with probability 1/2. The motivation for this jump is that it may be most beneficial to change the state of variables that are in false clauses. This is very similar to the so-called *alternative mutation operator* for vertex cover problems introduced by Kratsch and Neumann [15], which flips vertices incident to uncovered edges with probability 1/2.

In line 4 of Algorithm 3 we create a new formula φ by deleting the clauses that were false in the original assignment (note that the jump operation may have satisfied some originally false clauses as well as transformed some to false). However, φ must be a *satisfiable* 2-CNF formula because all of its clauses were true under the original assignment before the jump. The "jumped" assignment is subsequently passed to the repair operation, which we describe in Sect. 3.2.

Algorithm 3: JUMPANDREPAIR

input : A bit string $x \in \{0,1\}^{m+n}$ and a 2-CNF formula Φ

1 $z \leftarrow x_{m+1}x_{m+2}\cdots x_{m+n}$
2 $S \leftarrow \{C \in \Phi[x] \colon C \text{ is false under } z\}$
3 Flip z_i for each $i \in \text{vars}(S)$ with probability 1/2 `// Jump`
4 $\varphi \leftarrow \Phi[x] \setminus S$
5 $z \leftarrow$ REPAIR$(\varphi, z, \text{vars}(S))$ `// Repair`
6 **return** $x_1x_2\cdots x_m z_1 z_2 \cdots z_n$

3.2 Repairing Infeasible Assignments

We consider an assignment to a formula Φ to be *infeasible* when strictly more than k clauses of Φ are false under that assignment. The repair operator attempts to transform an infeasible assignment into a feasible one by making some educated guesses directed by the implication flow graph of the formula as defined in Sect. 2.1. The repair operation (Algorithm 4) takes a formula Φ on n variables, an assignment $z \in \{0,1\}^n$ and an index set $\mathcal{I} \subseteq [n]$. A critical precondition to the success of Algorithm 4 is that $\mathcal{I}$ provides the indexes of bits that are assumed to be set correctly with respect to an optimal solution. If this precondition is met, then the remaining bits of z simply need to be set to minimize the number of additional false clauses. To do this, the repair operation first removes clauses with both variables in $\mathcal{I}$, as we have assumed they are set correctly. It then uses the implication flow graph of the resulting formula φ to guess the remaining bits of z to produce a feasible assignment to Φ.

The repair operation is a Monte Carlo procedure in the sense that it is guaranteed to be efficient, but not always correct. A satisfying assignment to φ can be computed in linear time using the method of Aspvall, Plass and Tarjan [2] in

Algorithm 4: REPAIR($\varphi, z, \mathcal{I}$)

input : A satisfiable 2-CNF formula φ; string $z \in \{0,1\}^n$; index set $\mathcal{I} \subseteq [n]$

1 Compute v^* a satisfying assignment to φ
2 **repeat** *k times*
3     $G = (\mathcal{L}, A) \leftarrow$ BUILDIMPLICATIONFLOWGRAPH($\varphi, v^*, z, \mathcal{I}$)
4   Compute the minimum (s,t) cut $K \subseteq A$ of G
5   Let D be the flow clauses in φ corresponding to K
6   **foreach** *vertex* ℓ *in* $G \setminus K = (\mathcal{L}, A \setminus K)$ **do**
7   **if** ℓ *is reachable from* s *in* $G \setminus K$ **then**
8   Flip $z_{\text{var}(\ell)}$ so that ℓ is true under $z_{\text{var}(\ell)}$ // Satisfy the flow
9   **else**   // clauses of $\varphi \setminus D$
10   Flip $z_{\text{var}(\ell)}$ so that ℓ is false under $z_{\text{var}(\ell)}$
11   **if** $\varphi \setminus D$ *is satisfied by* z **then return** z
12   Choose a clause C in $\varphi \setminus D$ that is false under z
13   **if** RANDOM() $< 1/3$ **then**
14         $\varphi \leftarrow \varphi \setminus \{C\}$
15   **else**   // Guess the variables
16   choose $i \in \text{vars}(C)$ uniformly at random   // of a false clause
17         $z_i \leftarrow 1 - z_i$
18         $\mathcal{I} \leftarrow \mathcal{I} \cup \{i\}$
19 **return** z

line 1. The implication flow graph G is constructed in linear time by Algorithm 1, and a minimum s-t cut can be found in polynomial time by computing the maximum flow of G, e.g., with the push-relabel algorithm. After the flow clauses corresponding to the cut are removed, the algorithm guesses the state of variables in any remaining false clause. We later prove that only a short sequence of correct guesses suffices for a successful repair.

4 Fixed-Parameter Tractable Run Time

We now show that the (1+1) $\text{EA}^{\text{J+R}}$ is a randomized FPT algorithm for parameterized MAX-2-SAT. Our main result is stated in the following theorem.

Theorem 1. *Let Φ be a 2-CNF formula on n variables and m clauses such that*

$$\max_{z \in \{0,1\}^n} \sum_{C \in \Phi} \mathbb{1}\{C \textit{ is true under } z\} = m - k.$$

The (1+1) $\text{EA}^{\text{J+R}}$ *finds an optimal solution to Φ in expected time $O(12^k m \log n)$.*

Since there can be at most $4\binom{n}{2}$ unique disjunctive clauses on n variables, $m = O(n^2)$ and thus the expected run time of the (1+1) $\text{EA}^{\text{J+R}}$ on parameterized MAX-2-SAT is $O(12^k \operatorname{poly}(n))$, meaning it is a randomized FPT algorithm for the problem.

The proof of Theorem 1 relies on a judicious lower bound on the probability that the repair operation is successful, which we state in the following theorem.

Theorem 2. *Let Φ be a 2-CNF formula on at most n variables. Let $z^* \in \{0,1\}^n$ be an optimal assignment to the variables of Φ and the number of false clauses under z^* is at most k. Suppose there is a set $S \subseteq \Phi$ of clauses such that*

1. *$\Phi \setminus S$ is satisfiable, and*
2. *For all $i \in \text{vars}(S)$, $z_i = z_i^*$.*

Then Repair$(\Phi \setminus S, z, \text{vars}(S))$ *returns an optimal assignment to Φ with probability at least 3^{-k}.*

Proof. Denote as $X^* := \{C \in \Phi : C \text{ is false under } z^*\}$. By Condition 2 in the statement, the clauses of $S \setminus X^*$ are all true under z and the clauses of $S \cap X^*$ are all false under z. Thus it suffices to show that with the claimed probability, Repair flips bits of z in $[n] \setminus \text{vars}(S)$ so that at most $k - |S \cap X^*|$ clauses are false in $\Phi \setminus S$.

Let $\varphi := \Phi \setminus S$, $\mathcal{I} := \text{vars}(S)$, and $G = (\mathcal{L}, A)$ be the implication flow graph built and stored in line 3 of Algorithm 4. Denote as F the set of *flow clauses* of φ in the sense of Definition 1. Since every clause of φ outside of X^* is true under z^* and $z^*[\mathcal{I}] = z[\mathcal{I}]$, by the contrapositive of Lemma 1, $X^* \cap F$ must be an s-t cut of G. Therefore, the minimum s-t cut in G cannot be larger than k.

After determining the set D of flow clauses corresponding to a minimum cut, the loop from lines 6–10 sets the remaining variables of z as in the proof of Lemma 2; thus all the flow clauses in $\varphi \setminus D$ are guaranteed to be true under z. If all the clauses in $\varphi \setminus D$ are true, then z is optimal, as D is a minimum set of flow clauses that can be deleted.

We now guess the condition of any remaining clauses that are still false under z. Note that such clauses must be non-flow clauses. Let $C = (\ell_1 \vee \ell_2)$ be the false clause chosen in line 12 of Algorithm 4 and let $i = \text{var}(\ell_1)$ and $j = \text{var}(\ell_2)$. We argue that $\neg\ell_1, \neg\ell_2 \in \mathcal{L}$. By Lemma 3, both ℓ_1 and ℓ_2 cannot be in $\mathcal{L}$. Furthermore, if only one of the literals appears negated in $\mathcal{L}$ (but not the other), then by construction, C would have been a flow clause.

Since C is false, both $\neg\ell_1$ and $\neg\ell_2$ are true under z, and thus the vertices in G labeled $\neg\ell_1$ and $\neg\ell_2$ must be reachable from s. We make a case distinction on whether or not C belongs to X^*.

Case 1: $C \in X^*$. The correct action is to remove C from consideration. This occurs in line 14 with probability $1/3$.

Case 2: $C \notin X^*$. The correct action is to flip a variable in $\{\text{var}(\ell_1), \text{var}(\ell_2)\}$ to make C true. At least one of these variables does not match its value in the optimal solution z^*, as C is true under z^*. Without loss of generality, suppose this is $i = \text{var}(\ell_1)$. As argued above, the vertex labeled $\neg\ell_1$ must be reachable from s in G. Consider now what happens to the vertex labeled $\neg\ell_1$ when the implication flow graph is rebuilt after flipping z_i and adding i to $\mathcal{I}$.

Case 2a: z_i flipped from 1 to 0. For ℓ_1 to now be satisfied, it must be a negative literal, i.e., $\ell_1 = \neg i$ and so $\neg\ell_1 = i \in \mathcal{I}$. When G is rebuilt by Algorithm 1, an ∞-weight arc (i, t) is added in line 12, connecting the vertex labeled $\neg\ell_1$ to t.

Case 2b: z_i flipped from 0 to 1. For ℓ_1 to be satisfied, it must be a positive literal $\ell_1 = i$ and so $\neg\ell_1 = \neg i \notin \mathcal{I}$. When G is rebuilt, an ∞-weight arc $(\neg i, t)$ is added in line 8, connecting the vertex labeled $\neg\ell_1$ to t.
In both cases, the vertex labeled $\neg\ell_1$, which is reachable from s, is now also directly connected to t. This creates a new path from s to t disjoint from K. Thus the size of the minimum cut in G necessarily increases by at least one. A symmetric argument holds if the correct guess is to flip z_j where $j = \text{var}(\ell_2)$. In either case, the correct action is taken with probability at least 1/3, (2/3 to guess that $C \notin X^*$, and at least 1/2 to guess the correct literal).

We conclude the proof by showing a consecutive sequence of at most k correct actions suffices. Suppose that no incorrect actions occur before the repair is successful. Each correct action either (1) correctly identifies a non-flow clause in X^* in line 14, or (2) increases the size of the minimum s-t cut in G at line 16. Let a denote the number of times a non-flow clause is correctly identified in line 14, and let b denote the number of times the correct action is taken in line 16. Recall F is the set of flow clauses. Clearly $a \leq |X^* \setminus F|$. Moreover, the size of a minimum s-t cut in G is at least b. Again, since $X^* \cap F$ is an s-t cut, we have $b \leq |X^* \cap F|$. It follows that $a + b \leq |X^* \setminus F| + |X^* \cap F| = |X^*| \leq k$. Each action is correct with probability at least 1/3, which proves the claim. □

We point out that if the repair operation *fails*, then in the worst case, the resulting offspring is dominated by the parent and would simply be discarded. We are now ready to prove Theorem 1. Essentially, we show that when the current solution is feasible, it is possible for mutation to add a missing clause. If this happens to render the offspring infeasible, then the focused jump has a reasonable chance of flipping bits in the index set to good values. Furthermore, conditional on this occurrence, the repair operator is in a position to make the correct guesses to render the offspring feasible again. This allows for a bound on the expected increase in fitness in each step, which permits sufficient expected progress toward a global optimum.
Proof of Theorem. 1 We divide a run of the (1+1) EA$^{\text{J+R}}$ on Φ into two phases. The first phase consists of the steps during which the current point x has negative fitness. Let F denote the set of false clauses in $\Phi[x]$ under the assignment $\alpha[x]$. By (2), we have $|F| > k$ and $f(x) = -|F|$. During phase one, an improving mutation is always possible by flipping a bit in the length-m prefix of x corresponding to any clause in F (and no other bits), as this removes a false clause without introducing any other false clauses. This occurs with probability at least $\left(\frac{|F|}{m+n}\right) \cdot \left(1 - \frac{1}{m+n}\right)^{m+n-1} \geq \frac{|F|}{e(m+n)}$. As the fitness values cannot decrease, a simple multiplicative drift argument on the sequence of values for $|F|$ ensures that the expected length of phase one is $O((m+n)\log(m+n)) = O(m \log n)$, as $n = O(m)$ and $m = O(n^2)$.

The second phase consists of iterations where $f(x) \geq 0$, that is, all candidate solutions where x is feasible. Let x be the current solution and suppose $0 \leq f(x) < m$. By the definition (2) of f, we have $f(x) = |\Phi[x]|$.

We seek a lower bound on the probability a strictly improving offspring is created by mutation followed by jump and repair during phase two. Let y be the offspring created by mutation and denote as $\mathcal{E}_1$ the event that, in creating y, mutation flips exactly one of the $m - f(x)$ bits in the length-m prefix of x corresponding to a clause $C \notin \Phi[x]$ and no other bits. Clearly $\Pr(\mathcal{E}_1) = \left(1 - \frac{1}{m+n}\right)^{m+n-1} (m - f(x))/(m+n) \geq (m - f(x))/(e(m+n))$. If y is feasible, then $f(y) = f(x) + 1 > f(x)$, and y is already a strictly improving offspring. Otherwise, assume y is not feasible. In this case, $\Phi[y]$ must have exactly $k+1$ false clauses under y since it must have more than k false clauses, but only one new clause was added during the mutation from x to y. Since we now have $f(y) < 0$, the jump-and-repair operator is triggered in line 5. By the preceding arguments, $|S| = k+1$ in Algorithm 3 and since each clause contains two variables, we have $|\text{vars}(S)| = |\mathcal{I}| \leq 2(k+1)$.

Denote as $\{0,1\}^n \ni z = y_{m+1}y_{m+2}\cdots y_{m+n}$ the length-n suffix of offspring y and let $z^* \in \{0,1\}^n$ be an optimal assignment to Φ. Let $\mathcal{E}_2$ be the event that the focused jump operation in line 3 of Algorithm 3 sets, for each $i \in \mathcal{I}$, $z_i = z_i^*$. If this event occurs, both conditions of Theorem 2 will be satisfied. Let $\mathcal{E}_3$ be the event that repair operation is successful.

The joint event $\mathcal{E}_1 \cap \mathcal{E}_2 \cap \mathcal{E}_3$ is sufficient to ensure that $f(y) > f(x)$ after the repair operation, because the repaired offspring will again be feasible, but also contain an additional clause. We have $\Pr(\mathcal{E}_2 \mid \mathcal{E}_1) = (1/2)^{|\mathcal{I}|} \geq (1/4)^{k+1}$, and by Theorem 2, $\Pr(\mathcal{E}_3 \mid \mathcal{E}_2 \cap \mathcal{E}_1) \geq (1/3)^k$. It follows by the general product rule that

$$\Pr(\mathcal{E}_1 \cap \mathcal{E}_2 \cap \mathcal{E}_3) = \Pr(\mathcal{E}_1)\Pr(\mathcal{E}_2 \mid \mathcal{E}_1)\Pr(\mathcal{E}_3 \mid \mathcal{E}_1 \cap \mathcal{E}_2) \geq \frac{1}{12^k} \cdot \left(\frac{m - f(x)}{4e(m+n)}\right).$$

Let $(X_t)_{t \in \mathbb{N}}$ be the stochastic process that measures $m - f(x)$ where x is the candidate solution in step t. Since X_t decreases monotonically, we have $\mathrm{E}[X_t - X_{t+1} \mid X_t] \geq \frac{X_t}{12^k(4e)(m+n)}$. Denote as $T := \inf\{t \in \mathbb{N} \colon X_t = 0\}$. Applying the multiplicative drift Theorem [7], we bound the expected length of phase two as $\mathrm{E}[T] \leq 12^k(4e)(m+n)(\ln(m)+1)$. Summing the expected length of both phases, we conclude that the runtime of (1+1) EA$^{\text{J+R}}$ on Φ is at most $12^k(4e)(m+n)(\ln(m)+1) + O(m \log n) = O(12^k m \log n)$. □

5 Experiments

In this section we report the results of numerical simulations that compare the run time of the (1+1) EA$^{\text{J+R}}$ against a vanilla (1+1) EA on difficult MAX-2-SAT instances. We consider hand-designed formulas built up from collections of subformulas whose variables all must be synchronized to the same truth value. These formulas exhibit so-called *spin-flip symmetry* in which an incorrectly set subformula leads to a local optimum that is difficult for the (1+1) EA to escape.

We define a *ring* of size $s > 4$ as a set of s clauses on s variables defined as $R = (\neg x_1 \vee x_2), (\neg x_2 \vee x_3), \ldots, (\neg x_{s-1} \vee x_s), (\neg x_s \vee x_1)$. We define a *satisfiable ring* to be $R \cup \{(x_1 \vee x_2)\}$, and an *unsatisfiable ring* to be $R \cup \{(x_1 \vee x_2), (\neg x_1 \vee \neg x_2), (\neg x_3 \vee$

$\neg x_4)\}$. A satisfiable ring is satisfied by the assignment $x_1 = \cdots = x_s = 1$, whereas the assignment that maximizes the true clauses in an unsatisfiable ring is $x_1 = \cdots = x_s = 0$, where $(x_1 \vee x_2)$ is the only false clause.

We construct a *ring instance* as follows. For a particular $k > 1$, $s > 4$, $r > k$, we combined k unsatisfiable rings, each on s distinct variables, with $r - k$ satisfiable rings, also on s distinct variables. This results in a formula where an optimal assignment contains a 1 in every satisfiable ring, and a 0 in every unsatisfiable ring, yielding k false clauses.

A k-s-r ring instance has $n = s \times r$ variables and $m = r(s + 1) + 2k$ clauses. The spin-flip symmetry of the rings provides no gradient information to guide the (1+1) EA, and much of its behavior will degrade to a random walk.

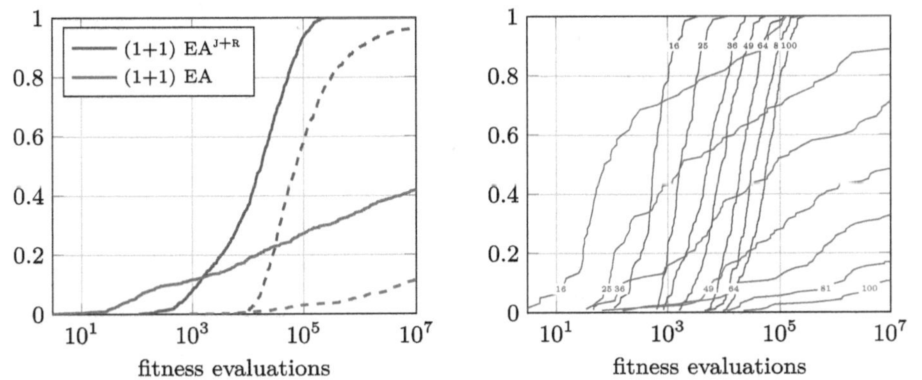

(a) Ring formulas (solid) and uniform random formulas (dashed).

(b) Ring formulas, separated by variable count $n = 16, 25, 36, 49, 64, 81, 100$.

Fig. 2. Empirical run time CDF for (1+1) EA and (1+1) EA^{J+R} with 10^7 cutoff.

For $s = \{4, 5, \ldots, 10\}$, $k = \{2, 3, \ldots, 10\}$ we constructed ring formulas with $s = r$ resulting in 42 distinct instances with n ranging from 16 to 100 and m ranging from 24 to 130. We ran both the (1+1) EA^{J+R} and the vanilla (1+1) EA on each formula 20 times with a budget of 10^7 fitness evaluations. Not all runs of the vanilla (1+1) EA were able to solve larger formulas within the allotted budget and we plot the empirical cumulative distribution function (CDF) of the run time by solid lines in Fig. 2a. All of the runs of the (1+1) EA^{J+R} on all formulas were successful within roughly 3.4×10^5 fitness evaluations, whereas only $352/840 \approx 42\%$ of the (1+1) EA were successful by the cutoff. To understand better how problem size affects this degradation, we separate the empirical run time distributions out for each value of n in Fig. 2b. In order to compare the run time behavior without the distortion effects of the cutoff, we present run time statistics (mean, median, standard deviation) with the failed runs of the (1+1) EA removed in Table 1. We performed a two-sample Kolmogorov-Smirnoff test and report in the rightmost column the p-values representing the significance levels for rejecting the null hypothesis that the run time of the (1+1) $\text{EA}^{\text{J+R}}$ is not stochastically smaller than that of the (1+1) EA.

Table 1. Run time statistics on ring instances. Reported *p*-values obtained from two-sample Kolmogorov-Smirnoff tests.

	(1 + 1) EA			(1 + 1) $\text{EA}^{1+\text{R}}$			
n	*mean**	*median**	*std. dev.**	*mean*	*median*	*std. dev.*	*p*-value
16	12334.33	83.00	29742.17	744.95	572.00	585.09	$3.8 \cdot 10^{-2}$
25	282604.82	1238.00	1014659.22	2355.09	1717.00	1769.62	$9.76 \cdot 10^{-6}$
36	864933.46	9676.00	2120522.16	7112.35	5361.50	5804.57	$5.62 \cdot 10^{-8}$
49	690278.53	62651.50	1697027.74	12732.80	9320.50	10367.65	$1.6 \cdot 10^{-11}$
64	1030500.33	356457.00	1999265.18	26667.70	20458.50	20235.85	$7 \cdot 10^{-15}$
81	1439786.33	751397.00	2065860.84	48079.35	34220.50	37564.94	$5.7 \cdot 10^{-10}$
100	2559628.47	1475801.00	2581427.42	72508.04	59124.00	51022.59	$3.04 \cdot 10^{-10}$

* Failed runs removed from calculation

We also generated uniform random 2-CNF formulas with $n = 100$ variables and $m \in \{170, 180, \ldots, 250\}$ clauses. For each m, we generated 100 instances and solved each for k using an integer linear program. From these formulas, we selected five for each $k = \{1, 2, \ldots, 9\}$. We ran both the (1+1) $\text{EA}^{\text{J+R}}$ and the vanilla (1+1) EA on each formula 20 times with a budget of 10^7 fitness evaluations and plot the empirical CDF of the run time as dashed lines in Fig. 2a. The (1+1) EA is only able to solve $101/880 \approx 11\%$ of the instances, while the (1+1) $\text{EA}^{\text{J+R}}$ solves $847/880 \approx 96\%$ of the instances within the allotted time.

6 Conclusion

We considered the (1+1) $\text{EA}^{\text{J+R}}$ applied to parameterized MAX-2-SAT. We introduced a probabilistic repair procedure as a "plug-in" to the existing algorithm. This repair module works by computing a so-called implication flow graph of the formula in order to guide the algorithm in making educated guesses about how to set variables to transform infeasible offspring into feasible offspring.

We proved that this approach guarantees the (1+1) $\text{EA}^{\text{J+R}}$ finds an optimal solution in $O(12^k m \log n)$ expected fitness evaluations where m is the number of clauses in the formula and n is the number of variables. This indicates that the (1+1) $\text{EA}^{\text{J+R}}$ is a randomized FPT algorithm for this problem. We also presented experiments comparing the (1+1) $\text{EA}^{\text{J+R}}$ to the vanilla (1+1) EA and observed that the jump-and-repair extension permits the optimal solution to be found significantly faster.

Natural directions for future work would be to explore techniques for reducing the base of the term exponential in k, and to extend the repair operation to handle other parameterized versions of maximum satisfiability.

Acknowledgments. This work was supported by NSF grant 2144080.

References

1. Ali, H.M., Mitchell, D., Lee, D.C.: MAX-SAT problem using evolutionary algorithms. In: 2014 IEEE Symposium on Swarm Intelligence, pp. 1–8 (2014). https://doi.org/10.1109/SIS.2014.7011783
2. Aspvall, B., Plass, M.F., Tarjan, R.E.: A linear-time algorithm for testing the truth of certain quantified Boolean formulas. Inf. Process. Lett. **8**(3), 121–123 (1979). https://doi.org/10.1016/0020-01907990002-4
3. Branson, L., Sutton, A.M.: Focused jump-and-repair constraint handling for fixed-parameter tractable graph problems closed under induced subgraphs. Theoret. Comput. Sci. **951**, 113719 (2023). https://doi.org/10.1016/j.tcs.2023.113719
4. Brownlee, A.E.I., McCall, J.A.W., Brown, D.F.: Solving the MAXSAT problem using a multivariate EDA based on Markov networks. In: Proceedings of the Ninth Annual Conference Companion on Genetic and Evolutionary Computation, pp. 2423–2428. GECCO '07, Association for Computing Machinery, New York, NY, USA (2007). https://doi.org/10.1145/1274000.1274005
5. Buzdalov, M., Doerr, B.: Runtime analysis of the $(1 + (\lambda, \lambda))$ genetic algorithm on random satisfiable 3-CNF formulas. In: Proceedings of the Genetic and Evolutionary Computation Conference, pp. 1343–1350. GECCO '17, Association for Computing Machinery, New York, NY, USA (2017). https://doi.org/10.1145/3071178.3071297
6. Das, R.P., Paul, A., Jena, J.J., Dash, B.B., De, U.C., Gourisaria, M.K.: Hybrid binary SGO-GA for solving MAX-SAT problem. Proc. Comput. Sci. **252**, 944–953 (2025). https://doi.org/10.1016/j.procs.2025.01.055, fourth International Conference on Evolutionary Computing and Mobile Sustainable Networks
7. Doerr, B., Johannsen, D., Winzen, C.: Multiplicative drift analysis. Algorithmica **64**, 673–697 (2012). https://doi.org/10.1007/s00453-012-9622-x
8. Doerr, B., Neumann, F., Sutton, A.M.: Time complexity analysis of evolutionary algorithms on random satisfiable k-CNF formulas. Algorithmica **78**(2), 561–586 (2016). https://doi.org/10.1007/s00453-016-0190-3
9. Framil, M., Cabalar, P., Santos, J.: A MaxSAT solver based on differential evolution (preliminary report). In: Marreiros, G., Martins, B., Paiva, A., Ribeiro, B., Sardinha, A. (eds.) Progress in Artificial Intelligence. Lecture Notes in Computer Science, vol. 13566, pp. 676–687. Springer International Publishing, Cham (2022). https://doi.org/10.1007/978-3-031-16474-3_55
10. Gaeuman, L., Sutton, A.M.: A fixed-parameter tractable GA for data clustering. In: Proceedings of the Eighteenth ACM/SIGEVO Conference on Foundations of Genetic Algorithms, pp. 49–60. FOGA '25, Association for Computing Machinery, New York, NY, USA (2025). https://doi.org/10.1145/3729878.3746632
11. Garey, M.R., Johnson, D.S., Stockmeyer, L.: Some simplified NP-complete graph problems. Theoret. Comput. Sci. **1**(3), 237–267 (1976). https://doi.org/10.1016/0304-3975(76)90059-1
12. Gogna, A., Tayal, A.: Metaheuristics: review and application. J. Exper. Theoret. Artif. Intell. **25**(4), 503–526 (2013). https://doi.org/10.1080/0952813X.2013.782347

13. Handa, H.: Robust solutions by using evolutionary computations on dynamic Max-Sat problems. In: Proceedings of the 2006 SICE-ICASE International Joint Conference, pp. 501–504 (2006). https://doi.org/10.1109/SICE.2006.315479
14. Kratsch, S., Lehre, P.K., Neumann, F., Oliveto, P.S.: Fixed parameter evolutionary algorithms and maximum leaf spanning trees: a matter of mutation. In: Schaefer, R., Cotta, C., Kołodziej, J., Rudolph, G. (eds.) Proceedings of the Eleventh International Conference on Parallel Problem Solving from Nature.,pp. 204–213. PPSN XI, Springer, Berlin, Heidelberg (2010). https://doi.org/10.1007/978-3-642-15844-5_21
15. Kratsch, S., Neumann, F.: Fixed-parameter evolutionary algorithms and the vertex cover problem. Algorithmica **65**(4), 754–771 (2012). https://doi.org/10.1007/s00453-012-9660-4
16. Kratsch, S., Wahlström, M.: Representative sets and irrelevant vertices: new tools for kernelization. J. ACM **67**(3), 16:1–16:50 (2020). https://doi.org/10.1145/3390887
17. Lee, J., Salimi Gamasaei, M., Sutton, A.M.: Evolving populations of solved subgraphs with crossover and constraint repair. In: Evolutionary Computation, pp. 1–28 (2026). https://doi.org/10.1162/EVCO.a.380
18. Lokshtanov, D., Narayanaswamy, N.S., Raman, V., Ramanujan, M.S., Saurabh, S.: Faster parameterized algorithms using linear programming. ACM Trans. Alg. **11**(2), 15:1–15:31 (2014). https://doi.org/10.1145/2566616
19. Neumann, F.: Expected runtimes of a simple evolutionary algorithm for the multi-objective minimum spanning tree problem. Eur. J. Oper. Res. **181**(3), 1620–1629 (2007). https://doi.org/10.1016/j.ejor.2006.08.005
20. Neumann, F., Sutton, A.M.: Evolving solutions to community-structured satisfiability formulas. In: Proceedings of the AAAI Conference on Artificial Intelligence **33**(01), 2346–2353 (2019). https://doi.org/10.1609/aaai.v33i01.33012346
21. Neumann, F., Sutton, A.M.: Parameterized complexity analysis of randomized search heuristics. In: Doerr, B., Neumann, F. (eds.) Theory of Evolutionary Computation: Recent Developments in Discrete Optimization, pp. 213–248. Natural Computing Series, Springer International Publishing (2020). https://doi.org/10.1007/978-3-030-29414-4_4
22. Oliveto, P.S., He, J., Yao, X.: Analysis of the (1+1) EA for finding approximate solutions to vertex cover problems. IEEE Trans. Evol. Comput. **13**(5), 1006–1029 (2009). https://doi.org/10.1109/TEVC.2009.2014362
23. Pinto, P.C., Runkler, T.A., Sousa, J.M.C.: An ant algorithm for static and dynamic Max-Sat problems. In: 2006 1st Bio-Inspired Models of Network, Information and Computing Systems. pp. 1–8 (2006). https://doi.org/10.1109/BIMNICS.2006.361793
24. Razgon, I., O'Sullivan, B.: Almost 2-SAT is fixed-parameter tractable. J. Comput. Syst. Sci. **75**(8), 435–450 (2009). https://doi.org/10.1016/j.jcss.2009.04.002
25. Sadowski, K.L., Bosman, P.A., Thierens, D.: On the usefulness of linkage processing for solving MAX-SAT. In: Proceedings of the Fifteenth Annual Conference on Genetic and Evolutionary Computation, pp. 853–860. GECCO '13, Association for Computing Machinery, New York, NY, USA (2013). https://doi.org/10.1145/2463372.2463474
26. Sarker, R., Mohammadian, M., Yao, X.: Evolutionary Optimization. In: International Series in Operations Research and Management Science, Springer (2002). https://doi.org/10.1007/b101816

27. Sutton, A.M.: Fixed-parameter tractability of crossover: steady-state gas on the closest string problem. Algorithmica **83**(4), 1138–1163 (2021). https://doi.org/10.1007/S00453-021-00809-8
28. Witt, C.: Worst-case and average-case approximations by simple randomized search heuristics. In: Diekert, V., Durand, B. (eds.) Proceedings of the Symposium on Theoretical Aspects of Computer Science, pp. 44–56. STACS 2005, Springer, Berlin, Heidelberg (2005). https://doi.org/10.1007/978-3-540-31856-9_4

Beyond IGO-Flow: Toward Convergence Analysis of IGO in Continuous Spaces

Ryosuke Kimura[1(✉)] and Youhei Akimoto[1,2]

[1] University of Tsukuba, Tsukuba, Japan
ryosuke.kimura@bbo.cs.tsukuba.ac.jp, akimoto@cs.tsukuba.ac.jp
[2] RIKEN Center for Advanced Intelligence Project, Tokyo, Japan

Abstract. Information-Geometric Optimization (IGO) provides a unified framework for black-box optimization by interpreting the adaptation of a search distribution as a natural gradient update. Despite its conceptual importance, the convergence theory of IGO remains limited: most existing results concern continuous-time idealizations such as the IGO flow, rather than discrete-time updates with non-infinitesimal learning rates. In this paper, we study discrete-time IGO in continuous spaces, formulated as natural gradient updates in the expectation-parameter coordinates of an exponential family. In particular, we analyze IGO over the multivariate Gaussian family on strongly convex quadratic objective functions. Our analysis covers a setting that simultaneously incorporates full covariance adaptation, a fixed positive learning rate, and quantile-based weights. In this setting, we prove that the covariance matrix converges to the zero matrix. We further show that the mean vector converges to the global optimum, provided that the condition number of the appropriately scaled covariance matrix is bounded at sufficiently frequent iterations. These results advance the convergence theory of IGO and help bridge the gap between the mathematical theory of IGO and practical covariance-adaptive search methods such as CMA-ES.

Keywords: Information-Geometric Optimization · Natural Gradient · Evolution Strategies · Convergence Analysis

1 Introduction

Many real-world optimization problems are black-box in the sense that only function evaluations are available and gradient information cannot be accessed [11]. Evolution strategies (ESs), especially CMA-ES [10], have demonstrated strong empirical performance across a wide range of black-box optimization (BBO) problems in continuous domains [7,12,14,16]. However, despite its practical success, the theoretical understanding of these methods remains limited.

A natural way to study ESs theoretically is through Information-Geometric Optimization (IGO), which provides a unified view of a broad class of evolutionary algorithms [15]. IGO formulates optimization as the evolution of a parametric search distribution updated along the natural gradient. This perspective

G. Iacca et al. (Eds.): PPSN 2026, LNCS 16985, pp. 20–35, 2027.
https://doi.org/10.1007/978-3-032-36214-8_2

is particularly appealing for rank-based stochastic search methods, including algorithms closely related to CMA-ES. Despite its conceptual importance, the convergence theory of IGO is still far from complete. Establishing theoretical guarantees for IGO as a unifying framework for evolutionary computation is of fundamental importance for improving its reliability, promoting its use in real-world optimization problems, and advancing research in evolutionary computation.

In this paper, we study discrete-time IGO in continuous domains, formulated in the expectation-parameter coordinates of an exponential family. We focus on IGO over the multivariate Gaussian family on strongly convex quadratic objective functions, while retaining three key features of practical covariance-adaptive search methods: full covariance adaptation, a fixed positive learning rate, and quantile-based weights. Although this setting captures essential aspects of practical covariance-adaptive search methods, it also makes the analysis substantially more difficult than in continuous-time idealizations or isotropic models.

Our main contributions are as follows. First, we prove that the covariance matrix generated by discrete-time IGO converges to the zero matrix under full covariance adaptation. Second, we show that the mean vector converges to the global optimum, provided that the condition number of the appropriately scaled covariance matrix is bounded at sufficiently frequent iterations. The present analysis does not yet give a fully unconditional convergence guarantee for the mean vector. Nevertheless, it isolates the remaining obstacle in terms of the conditioning of the scaled covariance matrix and thereby provides a concrete route toward a complete convergence theory. In this sense, our results help narrow the gap between the mathematical theory of IGO and practical covariance-adaptive search methods such as CMA-ES.

The rest of this paper is organized as follows. In Sect. 2, we outline related research and clarify the differences between prior studies and the present study. In Sect. 3, we explain IGO and present the formulation studied in this paper. In Sect. 4, we present the main convergence results. Specifically, we analyze the convergence of the covariance matrix and the mean vector in Sect. 4.2 and Sect. 4.3, respectively. Finally, we conclude the paper in Sect. 5.

2 Related Work

Information-Geometric Optimization (IGO) is a unified framework for black-box optimization based on natural gradient updates of a search distribution [15]. Existing analyses of IGO can be broadly divided into two lines: (1) analyses of discrete-time updates with a finite learning rate, and (2) analyses of idealized continuous-time trajectories obtained in the infinitesimal-step-size limit, such as the IGO flow.

For discrete-time IGO, existing analyses mainly establish monotone improvement guarantees, including recent divergence-based refinements [4,9]. There are also convergence analyses with finite learning rates for related algorithms [1,3], including an IGO-like natural gradient method with a simple objective transformation [1] and an ODE-based analysis proving geometric convergence for a simplified isotropic ES with step-size adaptation [3]. However, these results do not cover discrete-time IGO with both quantile-based weights and full covariance adaptation, which is the setting considered here.

Most convergence analyses of IGO have been developed for the continuous-time IGO flow [2,6,8]. While these studies are mathematically important, they analyze idealized trajectories rather than finite-step updates. In contrast, we study a less idealized and more practically relevant setting: discrete-time IGO in continuous spaces with multivariate Gaussian distributions, full covariance adaptation, a finite learning rate, and quantile-based weights.

3 Information-Geometric Optimization (IGO)

In this section, we present the IGO formulation studied in this paper. We first summarize the central idea of IGO and then introduce the specific formulation used in our analysis. In particular, we consider IGO on an exponential family of probability distributions, parameterized by expectation parameters, together with a quantile-based truncation weight.

3.1 Overview

Let $f : \mathcal{X} \to \mathbb{R}$ be an objective function on the search space $\mathcal{X}$, and consider the minimization of f. IGO is based on a parametric family of probability distributions $\{P_\theta\}_{\theta \in \Theta}$ on $\mathcal{X}$. Rather than optimizing f directly over $\mathcal{X}$, IGO iteratively updates the distribution parameter θ by natural gradient ascent on the surrogate objective

$$J_t(\theta) := \mathbb{E}_{x \sim P_\theta}\left[W_t^f(x)\right]. \tag{1}$$

Here, $W_t^f(x)$ is a rank-based weight defined with respect to the current distribution $P_t = P_{\theta_t}$. Typically, $W_t^f(x)$ is defined using a nonincreasing function $w : [0,1] \to \mathbb{R}$ as follows:

$$W_t^f(x) := w\Big(\Pr_{x' \sim P_t}\left(f(x') \leq f(x)\right)\Big). \tag{2}$$

Thus, $W_t^f(x)$ depends only on the rank (quantile level) of $f(x)$ under P_t.

The natural gradient with respect to the Fisher metric is defined by

$$\widetilde{\nabla}_\theta J_t(\theta) = F(\theta)^{-1} \nabla_\theta J_t(\theta), \tag{3}$$

where $F(\theta)$ denotes the Fisher information matrix [5]. The IGO update is then given by the natural gradient ascent

$$\theta_{t+1} = \theta_t + \tau \widetilde{\nabla}_\theta J_t(\theta_t), \tag{4}$$

with learning rate $\tau \in (0,1]$. Throughout this paper, we analyze the infinite-population model, i.e., the deterministic limit in which the expectations in the update are evaluated exactly rather than approximated from finite samples.

3.2 Quantile-Based Truncation Weight

Fix a quantile level $q \in (0, 1)$, and let κ_t^q denote the q-quantile of $f(x)$ under P_t, defined by

$$\kappa_t^q = \sup\Big\{u \in \mathbb{R} : \Pr_{x\sim P_t}\big(f(x) \leqslant u\big) \geq q \text{ and } \Pr_{x\sim P_t}\big(f(x) \geqslant u\big) \leq 1-q\Big\}. \quad (5)$$

In this paper, we use the quantile-based truncation weight

$$W_t^f(x) := \frac{1}{q}\,\mathbb{I}\{f(x) \leq \kappa_t^q\}. \quad (6)$$

For simplicity, we write $I_t^{\leq}(x) := \mathbb{I}\{f(x) \leqslant \kappa_t^q\}$. This weight corresponds to the case where $w(z) = q^{-1}\mathbb{I}\{z \leqslant q\}$ in (2).

When $f(x)$ has no atom at the threshold κ_t^q under P_t, we have

$$\Pr_{x\sim P_t}\big(f(x) = \kappa_t^q\big) = 0, \qquad \Pr_{x\sim P_t}\big(f(x) \leq \kappa_t^q\big) = q, \quad (7)$$

and therefore

$$\mathbb{E}_{x\sim P_t}\big[W_t^f(x)\big] = 1. \quad (8)$$

In Sect. 3.3, we formulate the IGO assuming normalized weights, as in (8). This assumption is satisfied in the analytical setting used in this study.

3.3 IGO in Expectation Parameterization

Assume that $\{P_\theta\}_{\theta\in\Theta}$ is an exponential family with respect to a reference measure ν on $\mathcal{X}$, with density of the form

$$p_\theta(x) = \exp\big(\beta^\top T(x) - \psi(\beta)\big)h(x), \quad (9)$$

with natural parameter β, sufficient statistic $T : \mathcal{X} \to \mathbb{R}^n$, and log-partition function

$$\psi(\beta) = \log \int \exp\big(\beta^\top T(x)\big)h(x)\nu(dx). \quad (10)$$

We use the expectation parameter

$$\eta_t = \mathbb{E}_{x\sim P_t}[T(x)] = \nabla_\beta\psi(\beta). \quad (11)$$

For exponential families, the natural gradient of J_t with respect to the expectation parameter η has the simple form [15]

$$\widetilde{\nabla}_\eta J_t(\theta_t) = \nabla_\beta J_t(\theta_t) = \mathbb{E}_{x\sim P_t}\big[W_t^f(x)(T(x) - \eta_t)\big]. \quad (12)$$

Hence, with learning rate $\tau \in (0, 1]$, the parameter update reads

$$\eta_{t+1} = \eta_t + \tau\widetilde{\nabla}_\eta J_t(\theta_t) \quad (13a)$$

$$= (1-\tau)\eta_t + \tau\mathbb{E}_{x\sim P_t}\big[W_t^f(x)T(x)\big]. \quad (13b)$$

In the Gaussian case $P_t = \mathcal{N}(m_t, C_t)$, define the weighted mean vector $m_t^\star$ and the weighted covariance matrix $C_t^\star$ by

$$m_t^\star := \mathbb{E}_{x\sim P_t}\big[W_t^f(x)x\big], \quad C_t^\star := \mathbb{E}_{x\sim P_t}\big[W_t^f(x)(x - m_t^\star)(x - m_t^\star)^\top\big]. \tag{14}$$

Then the parameter update (13b) reduces to

$$\begin{cases} m_{t+1} = (1-\tau)m_t + \tau m_t^\star, \\ C_{t+1} = (1-\tau)C_t + \tau C_t^\star + \tau(1-\tau)(m_t^\star - m_t)(m_t^\star - m_t)^\top. \end{cases} \tag{15}$$

Hereafter, we assume that $P_t = \mathcal{N}(m_t, C_t)$ and focus on the update rule (15).

4 Convergence Analysis of IGO in Continuous Spaces

We consider the convergence of the IGO update (15) on a d-dimensional strongly convex quadratic objective function $f : \mathbb{R}^d \to \mathbb{R}$ of the form

$$f(x) = x^\top Ax \qquad (A \succ 0). \tag{16}$$

Since $P_t = \mathcal{N}(m_t, C_t)$ on $\mathbb{R}^d$ and f is continuous, the random variable $f(x)$ under $x \sim P_t$ has a continuous distribution. In particular,

$$\Pr_{x\sim P_t}\left(f(x) = \kappa_t^q\right) = 0, \tag{17}$$

so (7) and (8) hold in this setting.

In Sect. 4.1, we establish a bound on the one-step progress of the expected objective value, which in particular implies its monotone improvement. This result is used in Sect. 4.2 to establish the convergence of the covariance matrix. In Sect. 4.3, we combine the fact $\lim_{t\to\infty} C_t = O$ with Assumption 1 and prove by contradiction that the mean vector converges to the optimal solution, i.e., $\lim_{t\to\infty} m_t = 0$.

4.1 Monotone Improvement

Define $\alpha := (0, A)$. The sufficient statistic for the Gaussian distribution is $T(x) = (x, xx^\top)$. Hence, the objective can be written as

$$f(x) = \langle \alpha, T(x)\rangle = \langle A, xx^\top\rangle = x^\top Ax. \tag{18}$$

Let $V(P_t) := \mathbb{E}_{x\sim P_t}[f(x)]$ denote the expected objective value under P_t. Since $V(P_t) = \mathbb{E}_{x\sim P_t}[\langle \alpha, T(x)\rangle] = \langle \alpha, \eta_t\rangle$, premultiplying both sides of (13b) by α yields

$$\langle \alpha, \eta_{t+1}\rangle = (1-\tau)\langle \alpha, \eta_t\rangle + \tau\big\langle \alpha, \mathbb{E}_{x\sim P_t}\big[W_t^f(x)T(x)\big]\big\rangle \tag{19}$$

$$\implies V(P_{t+1}) = (1-\tau)V(P_t) + \tau\mathbb{E}_{x\sim P_t}\big[W_t^f(x)f(x)\big]. \tag{20}$$

Therefore, since W_t^f is defined by (6), the evolution of $V(P_t)$ is determined by the average objective value over the selected q-quantile region.

Proposition 1. *Consider the sequence* $\{V(P_t)\}_{t\geqslant 0}$ *generated by the IGO update* (15) *applied to the strongly convex quadratic objective function* (16). *Define* $\mu_{t,(i)} := \mathbb{E}_{x\sim P_t}\left[(f(x) - V(P_t))^i\right]$ *for* $i = 2$ *and* 4. *Then, for all* $t \geqslant 0$, *the following inequality holds:*

$$V(P_{t+1}) \leqslant V(P_t) - \tau(1-q)\sqrt{\frac{1}{\mu_{t,(4)}/\mu_{t,(2)}^2+3}}\sqrt{\mu_{t,(2)}} \tag{21}$$

That is, $V(P_t)$ *is nonincreasing, i.e.,* $V(P_{t+1}) \leqslant V(P_t)$ *for all* $t \geqslant 0$.

Proof. We first show

$$V(P_{t+1}) \leqslant V(P_t) - \frac{\tau}{2q}\mathbb{E}_{x,y\overset{\text{i.i.d.}}{\sim}P_t}\left[|I_t^{\leqslant}(x) - I_t^{\leqslant}(y)|\right]\sqrt{\frac{\mu_{t,(2)}^3}{\mu_{t,(4)}+3\mu_{t,(2)}^2}}. \tag{22}$$

Let $x, y \overset{\text{i.i.d.}}{\sim} P_t$. Since $W_t^f(x) = q^{-1}I_t^{\leqslant}(x)$, (20) becomes

$$V(P_{t+1}) = (1-\tau)V(P_t) + \frac{\tau}{q}\mathbb{E}\left[I_t^{\leqslant}(x)f(x)\right]. \tag{23}$$

Since P_t is Gaussian and $A \succ 0$, the scalar random variable $f(x) = x^\top Ax$ has a continuous distribution under P_t. Therefore, we have $\Pr\left(f(x) = f(y)\right) = 0$,

and $I_t^{\leqslant}(x)$ is an almost surely nonincreasing function of $f(x)$. In particular, for $(P_t \otimes P_t)$-almost everywhere,

$$\left(I_t^{\leqslant}(x) - I_t^{\leqslant}(y)\right)\left(-f(x) + f(y)\right) \geqslant 0. \tag{24}$$

Thus, the assumptions of the improved Chebyshev sum inequality [3, Theorem 21] are satisfied by $I_t^{\leqslant}(x)$ and $-f(x)$. Then, applying [3, Theorem 21 with $K = 2$], we obtain

$$\begin{aligned} &-\mathbb{E}\left[I_t^{\leqslant}(x)f(x)\right] \\ &\quad\geqslant -\mathbb{E}\left[I_t^{\leqslant}(x)\right]\mathbb{E}\left[f(x)\right] + \frac{1}{4}\mathbb{E}\left[|I_t^{\leqslant}(x) - I_t^{\leqslant}(y)|\right]\mathbb{E}\left[|f(x) - f(y)|\right]. \end{aligned} \tag{25}$$

Since $\mathbb{E}[I_t^{\leqslant}(x)] = q$ and $\mathbb{E}[f(x)] = V(P_t)$, this yields

$$\mathbb{E}\left[I_t^{\leqslant}(x)f(x)\right] \leqslant qV(P_t) - \frac{1}{4}\mathbb{E}\left[|I_t^{\leqslant}(x) - I_t^{\leqslant}(y)|\right]\mathbb{E}\left[|f(x) - f(y)|\right]. \tag{26}$$

Moreover, since $f(x)$ is a quadratic form of a Gaussian random vector, it has a finite fourth moment. Hence [3, Proposition 22] applies and gives

$$\mathbb{E}\left[|f(x) - f(y)|\right] \geqslant 2\sqrt{\frac{\mu_{t,(2)}^3}{\mu_{t,(4)}+3\mu_{t,(2)}^2}}. \tag{27}$$

Substituting this into (26), we obtain

$$\mathbb{E}\left[I_t^{\leqslant}(x)f(x)\right] \leqslant qV(P_t) - \frac{1}{2}\mathbb{E}\left[|I_t^{\leqslant}(x) - I_t^{\leqslant}(y)|\right]\sqrt{\frac{\mu_{t,(2)}^3}{\mu_{t,(4)}+3\mu_{t,(2)}^2}}. \tag{28}$$

Finally, plugging this bound into (23) yields (22).

Now we prove (21) by explicitly evaluating the term

$$\mathbb{E}_{x,y \overset{\text{i.i.d.}}{\sim} P_t}\left[|I_t^{\leqslant}(x) - I_t^{\leqslant}(y)|\right] \tag{29}$$

appearing in (22). This term is evaluated as follows. Since $\Pr\left(f(x) = \kappa_t^q\right) = 0$, we have $I_t^{\leqslant}(x) \in \{0,1\}$ almost surely for $x \sim P_t$. Hence, $|I_t^{\leqslant}(x) - I_t^{\leqslant}(y)| = \mathbb{I}\{I_t^{\leqslant}(x) \neq I_t^{\leqslant}(y)\}$. Therefore, $\mathbb{E}\left[|I_t^{\leqslant}(x) - I_t^{\leqslant}(y)|\right] = \Pr\left(I_t^{\leqslant}(x) \neq I_t^{\leqslant}(y)\right)$.

Since $x, y \overset{\text{i.i.d.}}{\sim} P_t$,

$$\Pr\left(I_t^{\leqslant}(x) = 1,\ I_t^{\leqslant}(y) = 0\right) = \Pr\left(I_t^{\leqslant}(x) = 1\right)\Pr\left(I_t^{\leqslant}(y) = 0\right) = q(1-q). \tag{30}$$

Similarly, since $\Pr\left(I_t^{\leqslant}(x) = 0,\ I_t^{\leqslant}(y) = 1\right) = (1-q)q$, we obtain

$$\mathbb{E}\left[|I_t^{\leqslant}(x) - I_t^{\leqslant}(y)|\right] = q(1-q) + (1-q)q = 2q(1-q). \tag{31}$$

By substituting (31) into (22), we obtain (21). This completes the proof. □

4.2 Convergence of the Covariance Matrix

Now we establish the convergence of the covariance matrix C_t. Proposition 2 shows that, once $\mu_{t,(4)}$ is controlled by $\mu_{t,(2)}^2$, the variance of f converges, and the convergence of the covariance matrix C_t then follows immediately.

Proposition 2. *Consider IGO updates* (15) *applied to the strongly convex quadratic objective function* (16). *There exists a constant* $L \geqslant 1$ *such that*

$$\mu_{t,(4)} \leqslant L\mu_{t,(2)}^2 \qquad \text{for all } t \geqslant 0. \tag{32}$$

In fact, we may take $L = 15$. *Then, we have* $\lim_{t\to\infty} \mu_{t,(2)} = 0$, *implying*

$$\lim_{t\to\infty} AC_t = O. \tag{33}$$

Proof. The existence of a constant $L = 15$ can be derived in a straightforward manner from classical results on Gaussian quadratic forms; details are given in Appendix A of the supplementary material.[1]

First, we show that the second centered moment $\mu_{t,(2)}$ (i.e., the variance of f) converges to 0. By inequality (21), we obtain

$$V(P_{t+1}) \leqslant V(P_t) - \tau(1-q)\sqrt{\tfrac{1}{L+3}}\sqrt{\mu_{t,(2)}}. \tag{34}$$

Let $\delta := \tau(1-q)(L+3)^{-1/2} \in (0,\tau)$. Rearranging (34), we obtain

$$\delta\sqrt{\mu_{t,(2)}} \leq V(P_t) - V(P_{t+1}). \tag{35}$$

[1] Supplementary material is available at https://doi.org/10.5281/zenodo.19648468.

Summing from $t = 0$ to n yields

$$\delta \sum_{t=0}^{n} \sqrt{\mu_{t,(2)}} \leqslant \sum_{t=0}^{n} (V(P_t) - V(P_{t+1})) = V(P_0) - V(P_{n+1}) \leqslant V(P_0). \quad (36)$$

Since $V(P_0) < \infty$,

$$\sum_{t=0}^{\infty} \sqrt{\mu_{t,(2)}} \leqslant \frac{V(P_0)}{\delta} < \infty. \quad (37)$$

Since $\mu_{t,(2)} \geqslant 0$ for all $t \geqslant 0$, it follows that $\lim_{t\to\infty} \sqrt{\mu_{t,(2)}} = \lim_{t\to\infty} \mu_{t,(2)} = 0$.
Second, we show that $AC_t \to O$. Since

$$\mu_{t,(2)} = \mathrm{Var}_{x\sim P_t}[f(x)] = 2\,\mathrm{Tr}(AC_tAC_t) + 4m_t^\top AC_tAm_t, \quad (38)$$

we have $\mu_{t,(2)} \geqslant 2\,\mathrm{Tr}(AC_tAC_t)$. Hence $\mu_{t,(2)} \to 0$ implies $\mathrm{Tr}(AC_tAC_t) \to 0$. Noting that $\mathrm{Tr}(AC_tAC_t) \geqslant \lambda_{\max}^2(AC_t)$, we have $\lim_{t\to\infty} \lambda_{\max}^2(AC_t) = 0$, implying $\lim_{t\to\infty} AC_t = O$. This completes the proof. □

We note that the proof of $\mu_{t,(2)} \to 0$ does not use any Gaussian-specific property. Under (7), for IGO updates (15), this part of the argument only requires the moment bound (32).

4.3 Convergence of the Mean Vector

We prove the convergence of the mean vector to the optimal solution as follows. First, Lemma 1 shows that $m_t^\top Am_t$ has a limit $V' \geqslant 0$. Therefore, it suffices to exclude the case $V' > 0$. We prove it by contradiction. Lemma 3, together with Assumption 1, implies that $\mathrm{Tr}(AC_t)$ must grow asymptotically. This contradicts Proposition 2, and the resulting contradiction argument is formalized in Proposition 3.

To establish the convergence of the mean vector, we need an additional control on the shape of the covariance matrix, summarized in Assumption 1. While the previous subsection showed that $C_t \to O$, this does not prevent the covariance from becoming extremely anisotropic during the collapse. A trivial way to avoid this difficulty is to assume uniform boundedness of the condition number of the covariance matrix during the whole optimization process, which is, however, not valid. Instead, we construct a more realistic assumption on the shape of the covariance matrix.

For this purpose, consider the change in the trace $\mathrm{Tr}(AC_t)$ at each iteration. From the IGO update (15) for the covariance matrix, let $\theta = (m, C) \in \Theta$ and define

$$\Delta_C(\theta) := \tau C^\star + \tau(1-\tau)(m^\star - m)(m^\star - m)^\top - \tau C. \quad (39)$$

Then $\Delta_C(\theta) \succeq -\tau C$, and we obtain $\mathrm{Tr}(AC_{t+1}) = \mathrm{Tr}(AC_t) + \mathrm{Tr}(A\Delta_C(\theta_t))$, which is lower-bounded by $(1-\tau)\,\mathrm{Tr}(AC_t)$.

Based on this observation, we define

$$\gamma := \inf_{\theta \in \Theta} \left(1 + \frac{\mathrm{Tr}(A \Delta_C(\theta))}{\mathrm{Tr}(AC)} \right). \tag{40}$$

It is guaranteed that $\gamma \geqslant 1 - \tau$. With this quantity, we pose the following assumption on the condition number $\mathrm{Cond}(AC)$.

Assumption 1. *Let $\{C_t\}_{t \geqslant 0}$ be the sequence of covariance matrices generated by the IGO updates* (15) *applied to the strongly convex quadratic objective function* (16)*. Assume that there exists $R < \infty$ such that*

$$\liminf_{T \to \infty} \frac{1}{T} \sum_{t=0}^{T-1} \mathbb{I}\{\mathrm{Cond}(AC_t) \leqslant R\} > \frac{-\log(\gamma)}{\log(\rho) - \log(\gamma)}, \tag{41}$$

where $\rho > 1$ denotes the constant appearing in Lemma 3.

The assumption states that $\mathrm{Cond}(AC_t)$ need not be bounded during optimization, but is bounded by a constant R with higher frequency than the right-hand side of (41) would suggest. This is a more reasonable assumption than the boundedness of the condition number over the optimization process, as the bound can be violated infinitely many times.

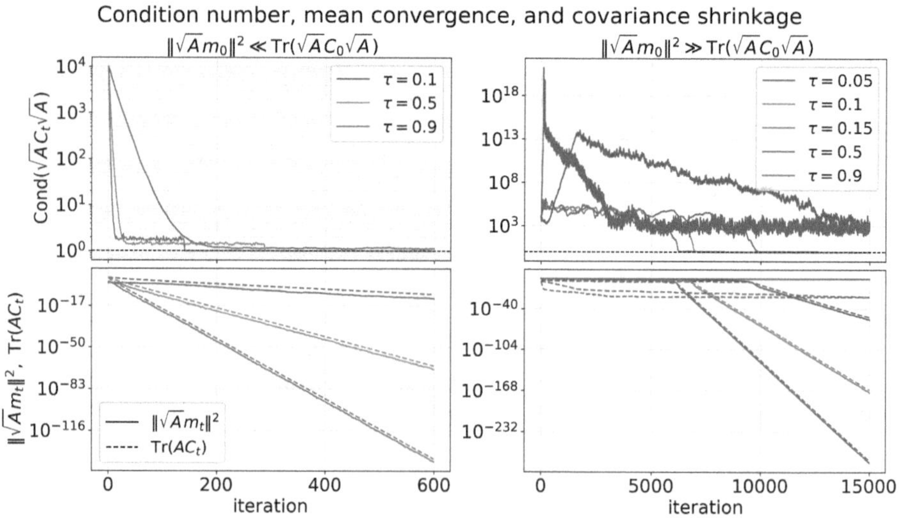

Fig. 1. Empirical behavior of the condition number, mean convergence, and covariance shrinkage. Experiments are performed on the quadratic objective $f(x) = x^\top Ax$ with $d = 10$, $A = \mathrm{diag}(10^0, 10^{4/9}, \ldots, 10^4)$, and the quantile level $q = 0.3$. We evaluate two initializations where the ratio of $\|\sqrt{A}m_0\|^2$ to $\mathrm{Tr}(AC_0)$ is approximately 10^{-4} and 10^4, respectively.

To clarify the scope of Assumption 1, we visualize the algorithmic behavior under two distinct initialization conditions in Fig. 1. Here, $\sqrt{A}$ is defined by $\sqrt{A} = Q\Lambda^{1/2}Q^\top$, where $A = Q\Lambda Q^\top$ is the eigendecomposition of A and $\Lambda^{1/2}$ is the diagonal matrix of the square roots of the eigenvalues. In the regime where $\|\sqrt{A}m_0\|^2 \ll \mathrm{Tr}(\sqrt{A}C_0\sqrt{A})$, $\mathrm{Cond}(\sqrt{A}C_t\sqrt{A})$ rapidly approaches 1, demonstrating that the covariance matrix converges to a scalar multiple of A^{-1}. Consequently, the assumption is empirically validated under this condition. In contrast, when $\|\sqrt{A}m_0\|^2 \gg \mathrm{Tr}(\sqrt{A}C_0\sqrt{A})$, serious anisotropy can emerge prior to mean convergence if a large learning rate is employed. However, given a sufficiently small learning rate, the evolution of the shape of the covariance is better regulated, and the condition number eventually approaches approximately 1. Therefore, this assumption serves to exclude overly aggressive initial conditions for a given learning rate, where the collapse of the covariance is much faster and more anisotropic than mean convergence.

Lemma 1. *Consider the sequences* $\{P_t\}_{t\geqslant 0}$ *generated by the IGO updates* (15) *applied to the strongly convex quadratic objective function* (16)*. There exists* $V' \geqslant 0$ *such that* $\lim_{t\to\infty} V(P_t) = \lim_{t\to\infty} m_t^\top A m_t = V'$.

Proof. By Proposition 1, the sequence $\{V(P_t)\}_{t\geqslant 0}$ is monotone nonincreasing. Moreover, since $f(x) \geqslant 0$, we have $V(P_t) \geqslant 0$ for all $t \geqslant 0$, and hence the sequence is bounded below. Therefore, $\{V(P_t)\}_{t\geqslant 0}$ converges to some $V' \geqslant 0$.

Recall that, for $P_t = \mathcal{N}(m_t, C_t)$ and $f(x) = x^\top A x$, $V(P_t) = m_t^\top A m_t + \mathrm{Tr}(AC_t)$. By Proposition 2, we have $AC_t \to O$, and hence $\lim_{t\to\infty} \mathrm{Tr}(AC_t) = 0$. Therefore, $\lim_{t\to\infty} m_t^\top A m_t = \lim_{t\to\infty} V(P_t) - \lim_{t\to\infty} \mathrm{Tr}(AC_t) = V'$. This completes the proof. □

Lemma 2. *Consider the IGO updates* (15) *applied to the strongly convex quadratic objective function* (16) *with the quantile level* $q \in (0, 1/2)$*. Define*

$$\alpha_q := \Phi^{-1}(q), \qquad \lambda_q := \frac{\phi(\alpha_q)}{q}, \tag{42}$$

where ϕ *and* Φ *are the density and cumulative distribution functions of the standard normal distribution, respectively. Let* V' *be the limit given by Lemma 1, and assume that* $V' > 0$*. For each* t*, let*

$$u_t := \sqrt{A}C_t A m_t, \qquad v_t := \sqrt{C_t} A m_t. \tag{43}$$

Then, for t *sufficiently large, if* $\mathrm{Cond}(AC_t) \leqslant R$ *for some constant* $R < \infty$*, the IGO update* (15) *for the covariance matrix in the* $\sqrt{A}$*-transformed coordinates satisfies*

$$\sqrt{A}C_{t+1}\sqrt{A} = \sqrt{A}C_t\sqrt{A} - \left(\tau\alpha_q\lambda_q + \tau^2\lambda_q^2\right)\frac{u_t u_t^\top}{\|v_t\|^2} + o\big(\lambda_{\max}(AC_t)\big). \tag{44}$$

Proof (Sketch). We give a proof sketch focusing on the key geometric idea, while deferring the rigorous details to Appendix B of the supplementary material.[2]

[2] Supplementary material is available at https://doi.org/10.5281/zenodo.19648468.

The proof has two conceptual steps. First, we show that in standardized coordinates the quadratic truncation boundary becomes asymptotically flat, so the selection event is asymptotically equivalent to a Gaussian half-space. Second, we evaluate the corresponding truncated Gaussian moments and plug them into the IGO updates (15).

Let $X_t = m_t + \sqrt{C_t} Z_t$, where $Z_t \sim \mathcal{N}(0, I)$, and define

$$B_t := \sqrt{C_t} A \sqrt{C_t}, \qquad e_t := \frac{v_t}{\|v_t\|}, \qquad \alpha_t := \frac{\kappa_t^q - m_t^\top A m_t}{2\|v_t\|}. \tag{45}$$

Then the truncation event can be rewritten as

$$f(X_t) \le \kappa_t^q \iff e_t^\top Z_t + \frac{Z_t^\top B_t Z_t}{2\|v_t\|} \le \alpha_t. \tag{46}$$

The key point is that the quadratic correction is asymptotically negligible. Indeed, writing $M_t := \sqrt{A} m_t$ and $\Sigma_t := \sqrt{A} C_t \sqrt{A}$, we have $\|M_t\|^2 \to V' > 0$, $\mathrm{Cond}(\Sigma_t) \le R$, and $AC_t \to O$, which implies $\|B_t\| / \|v_t\| \to 0$. Hence, on the Gaussian scale $Z_t = O_p(1)$, the curved boundary becomes asymptotically flat in standardized coordinates, so the selected region is asymptotically equivalent to a half-space orthogonal to e_t; see Fig. 2a.

More precisely, letting $Y_t := e_t^\top Z_t \sim \mathcal{N}(0, 1)$, we can write

$$S_t := \{Y_t + R_t \le \alpha_t\}, \qquad R_t := \frac{Z_t^\top B_t Z_t}{2\|v_t\|} = o_p(1). \tag{47}$$

Since κ_t^q is the q-quantile of $f(X_t)$, the selected mass is exactly q, which forces

$$\alpha_t \to \alpha_q := \Phi^{-1}(q) < 0. \tag{48}$$

Therefore, the true selection event is asymptotically equivalent to the Gaussian half-space $H_t := \{Y_t \leqslant \alpha_q\}$; see Fig. 2b.

To evaluate the truncated Gaussian moments, write

$$Z_t = Y_t e_t + U_t, \qquad U_t = (I - e_t e_t^\top) Z_t. \tag{49}$$

Since the events S_t and H_t are asymptotically equivalent, the corresponding selected moments differ only by $o(1)$. Moreover, under the half-space truncation $H_t = \{Y_t \le \alpha_q\}$, the above decomposition reduces the computation to the standard one-dimensional truncated Gaussian moments in the e_t-direction.

$$\psi_t := \mathbb{E}\left[\frac{1}{q} \mathbb{I}\{S_t\} Z_t\right] = -\lambda_q e_t + o(1), \tag{50}$$

and

$$\Psi_t := \mathbb{E}\left[\frac{1}{q} \mathbb{I}\{S_t\} Z_t Z_t^\top\right] = I - \alpha_q \lambda_q \, e_t e_t^\top + o(1), \tag{51}$$

where $\lambda_q := \phi(\alpha_q)/q$.

Finally, substituting these asymptotic moments into the weighted mean vector and the weighted covariance matrix (14) yields

$$m_t^\star = m_t + \sqrt{C_t}\psi_t, \qquad C_t^\star = \sqrt{C_t}(\Psi_t - \psi_t\psi_t^\top)\sqrt{C_t}. \tag{52}$$

Using

$$\sqrt{A}\sqrt{C_t}e_t = \frac{\sqrt{A}C_t A m_t}{\|v_t\|} = \frac{u_t}{\|v_t\|}, \tag{53}$$

together with (50) and (51), we obtain the claimed expansion (44). This completes the proof sketch. □

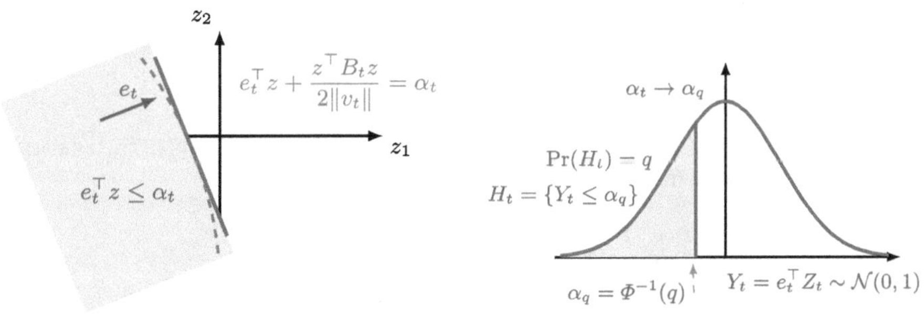

(a) The quadratic truncation boundary is locally approximated by the half-space $e_t^\top z \leqslant \alpha_t$ in standardized coordinates.

(b) Because the selected mass is q, the threshold converges to $\alpha_q = \Phi^{-1}(q)$, and the selection event is asymptotically equivalent to the half-space $H_t = \{Y_t \leq \alpha_q\}$.

Fig. 2. Geometric interpretation of the proof. Left: Local linearization of the truncation boundary. Right: Convergence of the threshold to the Gaussian q-quantile.

Lemma 3. *Consider the IGO updates* (15) *applied to the strongly convex quadratic objective function* (16) *with the quantile level* $q \in (0, 1/2)$. *Define*

$$\tau_{\text{crit}}(q) := -\frac{\alpha_q}{\lambda_q}, \qquad r_t := \operatorname{Tr}(AC_t). \tag{54}$$

Suppose $\tau \in (0, \tau_{\text{crit}}(q))$. *Let* V' *be the limit given by Lemma 1, and assume that* $V' > 0$. *Then, for sufficiently large* t, *if* $\operatorname{Cond}(AC_t) \leqslant R$ *for some constant* $R < \infty$, *there exists a constant* $\rho > 1$ *such that* $r_{t+1} \geqslant \rho r_t$.

Proof. Let $\beta_q(\tau) := -(\tau\alpha_q\lambda_q + \tau^2\lambda_q^2)$.

Since $\tau < \tau_{\text{crit}}(q) = -\alpha_q/\lambda_q$, we have $\beta_q(\tau) > 0$. By Lemma 2, we have

$$\sqrt{A}C_{t+1}\sqrt{A} = \sqrt{A}C_t\sqrt{A} + \beta_q(\tau)\frac{u_t u_t^\top}{\|v_t\|^2} + o(\lambda_{\max}(AC_t)). \tag{55}$$

Taking the trace yields

$$r_{t+1} = r_t + \beta_q(\tau)\frac{\|u_t\|^2}{\|v_t\|^2} + o\big(\lambda_{\max}(AC_t)\big). \tag{56}$$

By the definitions of u_t and v_t in (43), we have

$$\frac{\|u_t\|^2}{\|v_t\|^2} = \frac{m_t^\top AC_tAC_tAm_t}{m_t^\top AC_tAm_t} \geqslant \lambda_{\min}(AC_t). \tag{57}$$

Substituting this bound (57) into (56), we obtain

$$r_{t+1} \geqslant r_t + \beta_q(\tau)\lambda_{\min}(AC_t) + o\big(\lambda_{\max}(AC_t)\big). \tag{58}$$

Since $\mathrm{Cond}(AC_t) \leqslant R$ and $r_t \leqslant d\,\lambda_{\max}(AC_t)$, we have

$$\lambda_{\min}(AC_t) \geqslant \frac{1}{R}\lambda_{\max}(AC_t) \geqslant \frac{1}{Rd}r_t. \tag{59}$$

Moreover, since $\lambda_{\max}(AC_t) \leqslant r_t$, we have $o(\lambda_{\max}(AC_t)) = o(r_t)$. Substituting this into (58), we obtain

$$r_{t+1} \geqslant \left(1 + \frac{\beta_q(\tau)}{Rd}\right) r_t + o(r_t). \tag{60}$$

Dividing both sides by $r_t > 0$, we get

$$\frac{r_{t+1}}{r_t} \geqslant 1 + \frac{\beta_q(\tau)}{Rd} + o(1). \tag{61}$$

Hence, for all sufficiently large t, we can set $\rho := 1 + \frac{\beta_q(\tau)}{2Rd} > 1$. This completes the proof. □

Proposition 3. *Consider the IGO updates* (15) *applied to the strongly convex quadratic objective function* (16) *with the quantile level* $q \in (0, 1/2)$.

Suppose that $\tau \in (0, \tau_{\mathrm{crit}}(q))$ *and that Assumption 1 holds. Then*

$$\lim_{t\to\infty} m_t = 0. \tag{62}$$

Proof. We prove the proposition by contradiction. Assume, for contradiction, that $V' > 0$.

Let $\bar{T}$ be sufficiently large. Then, for $t \geqslant \bar{T}$, if $\mathrm{Cond}(AC_t) \leqslant R$, we have $\mathrm{Tr}(AC_{t+1}) \geqslant \rho\,\mathrm{Tr}(AC_t)$ in light of Lemma 3. If not, we know that $\mathrm{Tr}(AC_{t+1}) \geqslant \gamma\,\mathrm{Tr}(AC_t)$ from (40). Therefore, for $t \geqslant \bar{T}$, we have

$$\mathrm{Tr}(AC_{t+1}) \geqslant \rho^{\mathbb{I}\{\mathrm{Cond}(AC_t)\leqslant R\}}\gamma^{1-\mathbb{I}\{\mathrm{Cond}(AC_t)\leqslant R\}}\,\mathrm{Tr}(AC_t). \tag{63}$$

Taking logarithms in (63) and summing the resulting inequalities over $i = 0, \dots, T-1$, we obtain

$$\begin{aligned}
&\log \operatorname{Tr}(AC_{t+T}) - \log \operatorname{Tr}(AC_t) \\
&\quad = \textstyle\sum_{i=0}^{T-1} \big(\log \operatorname{Tr}(AC_{t+i+1}) - \log \operatorname{Tr}(AC_{t+i}) \big) && \text{(64a)} \\
&\quad \geqslant \textstyle\sum_{i=0}^{T-1} \big(\mathbb{I}\{\operatorname{Cond}(AC_{t+i}) \leqslant R\} \log(\rho) \\
&\qquad\qquad + (1 - \mathbb{I}\{\operatorname{Cond}(AC_{t+i}) \leqslant R\}) \log(\gamma) \big) && \text{(64b)} \\
&\quad = \textstyle\sum_{i=0}^{T-1} \mathbb{I}\{\operatorname{Cond}(AC_{t+i}) \leqslant R\} \big(\log(\rho) - \log(\gamma) \big) + T \log(\gamma). && \text{(64c)}
\end{aligned}$$

Then Assumption 1 leads to

$$\begin{aligned}
&\liminf_{T\to\infty} \tfrac{1}{T} \big(\log \operatorname{Tr}(AC_{t+T}) - \log \operatorname{Tr}(AC_t) \big) \\
&\geqslant \liminf_{T\to\infty} \tfrac{1}{T} \textstyle\sum_{i=0}^{T-1} \mathbb{I}\{\operatorname{Cond}(AC_{t+i}) \leqslant R\} \big(\log(\rho) - \log(\gamma) \big) + \log(\gamma) && \text{(65a)} \\
&> 0, && \text{(65b)}
\end{aligned}$$

which implies $\lim_{T\to\infty} \operatorname{Tr}(AC_{t+T}) = +\infty$, contradicting $\lim_{t\to\infty} \operatorname{Tr}(AC_t) = 0$. Therefore, $V' = 0$. Hence, by Lemma 1, $\lim_{t\to\infty} m_t = 0$. This completes the proof. □

Finally, Corollary 1 combines the convergence results established in Sect. 4.2 and Sect. 4.3.

Corollary 1. *Consider the IGO updates* (15) *applied to the strongly convex quadratic objective function* (16)*.*

1. *For any* $q \in (0, 1)$ *and* $\tau \in (0, 1]$, *we have* $\lim_{t\to\infty} C_t = O$.
2. *Suppose, in addition, that Assumption 1 holds. For any* $q \in (0, 1/2)$ *and* $\tau \in (0, \tau_{\text{crit}}(q))$, *with* $\tau_{\text{crit}}(q)$ *defined in Lemma 3, we have* $\lim_{t\to\infty} m_t = 0$. *Consequently,* $\lim_{t\to\infty}(m_t, C_t) = (0, O)$.

5 Conclusion

In this paper, we studied discrete-time Information-Geometric Optimization in continuous spaces. Focusing on IGO over the multivariate Gaussian family on strongly convex quadratic objective functions, we analyzed a practically relevant setting that combines full covariance adaptation, quantile-based weights, and natural-gradient updates in the expectation-parameter coordinates of an exponential family with a fixed positive learning rate.

In this setting, our analysis established the convergence of the covariance matrix to the zero matrix. In addition, under Assumption 1, we proved that the mean vector converges to the global optimum. These results constitute a mathematically meaningful first step toward a complete convergence theory of IGO in continuous domains.

Our results also clarify the main remaining obstacle: controlling the evolving shape of the covariance matrix, rather than only its overall collapse. In the current analysis, this issue is addressed through Assumption 1, thereby highlighting the gap to a fully unconditional convergence guarantee.

An important direction for future work is therefore to remove this assumption and establish a fully unconditional convergence guarantee for the mean vector. We hope that the present analysis provides a useful foundation for further progress in this direction.

Acknowledgment. This study was partly funded by JSPS KAKENHI 26K02993.

Disclosure of Interests. The authors have no competing interests to declare that are relevant to the content of this article.

References

1. Akimoto, Y.: Analysis of a natural gradient algorithm on monotonic convex-quadratic-composite functions. In: Proceedings of the 14th Annual Conference on Genetic and Evolutionary Computation, pp. 1293–1300 (2012)
2. Akimoto, Y., Auger, A., Hansen, N.: Convergence of the continuous time trajectories of isotropic evolution strategies on monotonic $\mathcal{C}^2$-composite functions. In: Parallel Problem Solving from Nature - PPSN XII, pp. 42–51 (2012)
3. Akimoto, Y., Auger, A., Hansen, N.: An ode method to prove the geometric convergence of adaptive stochastic algorithms. Stochastic Processes Appl. **145**, 269–307 (2022)
4. Akimoto, Y., Ollivier, Y.: Objective improvement in information-geometric optimization. In: Foundations of Genetic Algorithms - FOGA XII, pp. 1–10 (2013)
5. Amari, S.I.: Natural gradient works efficiently in learning. Neural Comput. **10**(2), 251–276 (1998)
6. Beyer, H.G.: Convergence analysis of evolutionary algorithms that are based on the paradigm of information geometry. Evol. Comput. **22**(4), 679–709 (2014)
7. Fujii, G., Takahashi, M., Akimoto, Y.: Cma-es-based structural topology optimization using a level set boundary expression-application to optical and carpet cloaks. Comput. Methods Appl. Mech. Eng. **332**, 624–643 (2018)
8. Glasmachers, T.: Convergence of the igo-flow of isotropic gaussian distributions on convex quadratic problems. In: Parallel Problem Solving from Nature - PPSN XII, pp. 1–10 (2012)
9. Guilmeau, T., Chouzenoux, E., Elvira, V.: A divergence-based condition to ensure quantile improvement in black-box global optimization. IEEE Trans. Evol. Comput. **29**(4), 1017–1028 (2025)
10. Hansen, N., Ostermeier, A.: Completely derandomized self-adaptation in evolution strategies. Evol. Comput. **9**(2), 159–195 (2001)
11. Larson, J., Menickelly, M., Wild, S.M.: Derivative-free optimization methods. Acta Numer **28**, 287–404 (2019)
12. Maki, A., Sakamoto, N., Akimoto, Y., Nishikawa, H., Umeda, N.: Application of optimal control theory based on the evolution strategy (cma-es) to automatic berthing. J. Mar. Sci. Technol. **25**(1), 221–233 (2019)
13. Mathai, A.M., Provost, S.B.: Quadratic Forms in Random Variables: Theory and Applications, Statistics: Textbooks and Monographs, vol. 126. Marcel Dekker, (1992)
14. Nomura, M., Watanabe, S., Akimoto, Y., Ozaki, Y., Onishi, M.: Warm starting cma-es for hyperparameter optimization. In: Proceedings of the AAAI Conference on Artificial Intelligence 35(10), pp. 9188–9196 (2021)

15. Ollivier, Y., Arnold, L., Auger, A., Hansen, N.: Information-geometric optimization algorithms: A unifying picture via invariance principles. J. Mach. Learn. Res. **18**(18), 1–65 (2017)
16. Varelas, K., et al.: A comparative study of large-scale variants of cma-es. In: Parallel Problem Solving from Nature – PPSN XV, pp. 3–15 (2018)

Equilibrium Analysis of a Variable Metric Evolution Strategy

Stephan Frank[1(✉)] and Youhei Akimoto[2]

[1] Fakultät für Informatik, Institut für Neuroinformatik, Ruhr-Universität Bochum, Bochum, Germany
stephan.frank@ini.rub.de
[2] Institute of Systems and Information Engineering/Center for Artificial Intelligence Research, University of Tsukuba, Tsukuba, Japan
akimoto@cs.tsukuba.ac.jp

Abstract. We study the expected one-step dynamics of a simplified rank-μ CMA-ES on the two-dimensional sphere in normal form, with emphasis on the highly anisotropic limit $\kappa \to \infty$. In this regime, selection collapses to a one-dimensional truncation geometry, making the dynamics analytically tractable. We derive explicit stationarity conditions for covariance adaptation and step-size adaptation and show that both are governed by the same reduced scaling variable $s = \sqrt{2}\cos(\alpha)/\sigma$, so that the corresponding stationary thresholds depend only on the selection rate ρ. We then analyze the equilibrium of the normalized step-size in normal form and interpret it as a self-similar stationary regime in which the canonical relation between distance to the optimum and effective step-size is restored in expectation after one update and renormalization. Our analysis isolates a tractable extreme of CMA-ES and clarifies the interaction of covariance adaptation, step-size adaptation, and normal-form renormalization near stationarity.

1 Introduction

The covariance matrix adaptation evolution strategy (CMA-ES) [11–13] is among the most successful and widely used black-box optimization methods in continuous domains. Although recent work has substantially advanced the theoretical analysis of CMA-ES, including convergence results based on normalized state representations [10], its theoretical understanding remains comparatively limited, especially when one aims at quantitative descriptions of the full adaptive dynamics and of the interaction between mean, step-size, and covariance matrix adaptation.

By contrast, the isotropic regime is much better understood. In particular, evolution strategies with weighted recombination and isotropic Gaussian sampling have been analyzed on the sphere and on general convex quadratic functions, yielding explicit characterizations of quality gain and optimal normalized step-sizes [2,4,7] and linear convergence guarantees [1,3,5,6,14,15]. In the language of the present paper, this provides a natural baseline corresponding to the

G. Iacca et al. (Eds.): PPSN 2026, LNCS 16985, pp. 36–51, 2027.
https://doi.org/10.1007/978-3-032-36214-8_3

regime without covariance anisotropy. What remains much less transparent is the opposite extreme, where the covariance becomes strongly anisotropic.

This motivates the present work. We study a simplified rank-μ update CMA-ES [9] on the two-dimensional sphere in normal form and focus on the highly anisotropic limit, corresponding to an infinitely ill-conditioned covariance matrix. In this regime, selection collapses to a one-dimensional truncation geometry, which makes it possible to derive explicit stationarity conditions for the covariance update and for the step-size update, and to study their interaction in the normal-form dynamics.

Our analysis shows that both covariance stationarity and algorithmic step-size stationarity are governed by the same reduced scaling variable $s = \frac{\sqrt{2}\cos(\alpha)}{\sigma}$, where α denotes the angle between the mean vector and the principal eigenvector of the covariance matrix. Consequently, the corresponding stationary thresholds depend only on the selection rate ρ. Building on this reduction, we then analyze the equilibrium of the normalized step-size in normal form. This equilibrium can be interpreted as a self-similar stationary regime in which the canonical relation between distance to the optimum and effective step-size is restored in expectation after one update and renormalization.

Overall, the present paper isolates a particularly anisotropic regime of CMA-ES in which the dynamics becomes tractable enough to expose the stationary geometry induced by covariance adaptation and to quantify how covariance adaptation, step-size adaptation, and normal-form renormalization interact near stationarity.

2 Setup

We analyze the infinite-population dynamics of a simplified version [9] of the rank-μ update CMA-ES [12] on the two-dimensional sphere

$$f(x) = \tfrac{1}{2}\|x\|^2. \tag{1}$$

Algorithm. The algorithm considered uses truncation selection with $\mu = \lfloor \rho\lambda \rfloor$ and equal recombination weights $w_i = 1/\mu$. Offspring are sampled as

$$x_i = m + \sigma y_i, \qquad y_i \sim \mathcal{N}(0, C). \tag{2}$$

Here and throughout, for $a_i \in \{x_i, y_i, z_i\}$, the notation $a_{i:\lambda}$ denotes the quantity associated with the i-th best offspring among the λ sampled offspring, ranked according to their objective values. In particular, $x_{i:\lambda}$ is the i-th best offspring, $y_{i:\lambda}$ its mutation step, and $z_{i:\lambda}$ the corresponding whitened step, with $y_{i:\lambda} = C^{1/2} z_{i:\lambda}$. The mean update is

$$m' = m + \sigma \sum_{i=1}^{\mu} w_i \, y_{i:\lambda}. \tag{3}$$

The step-size and covariance updates are given by

$$\sigma' = \sigma \cdot \exp\left(\frac{\sqrt{\mu}\, \|\sum_{i=1}^{\mu} w_i z_{i:\lambda}\|}{\mathbb{E}\|Z\|} - 1 \right), \tag{4}$$

$$C' = (1 - c_\mu) C + c_\mu \sum_{i=1}^{\mu} w_i\, y_{i:\lambda} y_{i:\lambda}^\top, \tag{5}$$

where $z_i \sim \mathcal{N}(0, I)$ and $y_i = C^{1/2} z_i$.

2.1 Normal Form

The normal form used in this paper builds on the normalized-state representation introduced in [8], but employs a trace-based normalization of the covariance matrix. In dimension two, this yields the convention $\mathrm{tr}(C) = 2$.

Let $\theta = (m, C, \sigma)$ be an arbitrary state with $m \in \mathbb{R}^2$, $C \succ 0$, and $\sigma > 0$. Exploiting rotational and scaling invariances of the sphere function, we construct a canonical representative as follows. First, we rotate the coordinate system such that C becomes diagonal with eigenvalues $\lambda_1 \geq \lambda_2 > 0$, and transform the mean accordingly. Second, we rescale the coordinate system such that $\|m\| = 1$. Third, we factor out a scalar from C such that $\mathrm{tr}(C) = 2$ and absorb this factor into the step-size. After optional axis flips and permutation of coordinates, the state can be written in the canonical form

$$m = \begin{pmatrix} \cos(\alpha) \\ \sin(\alpha) \end{pmatrix}, \qquad D = \begin{pmatrix} \frac{2\kappa}{1+\kappa} & 0 \\ 0 & \frac{2}{1+\kappa} \end{pmatrix}, \qquad \alpha \in [0, \tfrac{\pi}{2}], \qquad \kappa \geq 1.$$

Throughout the paper, we use C to denote a general covariance matrix, while D denotes its diagonal representative in normal form. Under this normalization, it is convenient to distinguish between the algorithmic step-size and the normalized step-size induced by the normal-form representation. We denote them by σ_{alg} and σ_{NF}, respectively, where

$$\sigma_{\mathrm{NF}} = \sigma_{\mathrm{alg}} \frac{\sqrt{\mathrm{tr}(C)/2}}{\|m\|}. \tag{6}$$

2.2 Derivation Under $\kappa \to \infty$

For simplicity, we consider the limit $\kappa \to \infty$. Since throughout this section the current state is represented in normal form, we have $\|m\| = 1$ and $\mathrm{tr}(C) = \mathrm{tr}(D) = 2$, and therefore $\sigma_{\mathrm{NF}} = \sigma_{\mathrm{alg}}$. We thus write simply σ in this section and consider only expected one-step dynamics in the infinite-population limit.

Sample Fitness. The objective function (1) can be written as

$$\begin{aligned} f(m+\sigma\sqrt{D}z) &= \frac{1}{2}\|m\|^2 + m^\top(\sigma\sqrt{D}z) + \frac{1}{2}\sigma^2 z^\top D z \\ &= \frac{1}{2} + \sigma\sqrt{\frac{2\kappa}{1+\kappa}}\cos(\alpha)\, z_1 + \sigma\sqrt{\frac{2}{1+\kappa}}\sin(\alpha)\, z_2 \\ &\quad + \sigma^2\frac{\kappa}{1+\kappa}z_1^2 + \sigma^2\frac{1}{1+\kappa}z_2^2. \end{aligned}$$

In the limit $\kappa \to \infty$, this expression simplifies to

$$f(m+\sigma\sqrt{D}z) \xrightarrow{\kappa\to\infty} \frac{1}{2} + \sqrt{2}\sigma\cos(\alpha)\, z_1 + \sigma^2 z_1^2, \tag{7}$$

and the fitness depends exclusively on the first coordinate z_1.

Ranking and Truncation. Since the fitness value is independent of z_2, the ranking induced by the fitness values—and thus the selection—is entirely determined by z_1. The second coordinate z_2, although normally distributed prior to selection ($z_2 \sim \mathcal{N}(0,1)$), does not affect the ranking. Let $\mathcal{A} \subset \mathbb{R}$ denote the region of z_1 corresponding to selected samples. Selection can then be interpreted as conditioning on the event $z_1 \in \mathcal{A}$, which yields

$$p(z_1, z_2 \mid z_1 \in \mathcal{A}) = p(z_1 \mid z_1 \in \mathcal{A})\, p(z_2).$$

Specifically, we define

$$\mathcal{A} := \left[-\frac{\sqrt{2}\cos(\alpha)}{\sigma} - \frac{\beta}{\sigma}, \frac{\beta}{\sigma}\right] \subset \mathbb{R},$$

where the truncation parameter $\beta > 0$ is chosen such that the interval $\mathcal{A}$ carries probability mass ρ under $z_1 \sim \mathcal{N}(0,1)$. For fixed ρ, this determines β implicitly as a function of α and σ.

Since $z_1 \sim \mathcal{N}(0,1)$ prior to selection, this condition can be written as

$$\boxed{\mathbb{P}(z_1 \in \mathcal{A}) = \Phi\left(\frac{\beta}{\sigma}\right) - \Phi\left(-\frac{\sqrt{2}\cos(\alpha)}{\sigma} - \frac{\beta}{\sigma}\right) = \rho}, \tag{8}$$

where Φ denotes the cumulative distribution function of the standard normal distribution. Thus, β/σ is determined by ρ and $-\sqrt{2}\cos(\alpha)/\sigma$.

We now define the random vector $\tilde{z}$ to denote a sample conditioned on selection,

$$\tilde{z} := (z_\rho, z_2), \qquad z_\rho := z_1 \mid z_1 \in \mathcal{A},$$

where $z \sim \mathcal{N}(0, I)$.

Substitutions. To express the selection geometry in a scale-invariant form, we introduce dimensionless ratios corresponding to the truncation boundaries. We define

$$s := \frac{\sqrt{2}\cos(\alpha)}{\sigma}, \qquad t := \frac{\beta}{\sigma}, \tag{9}$$

where t represents the *right truncation boundary*, while s denotes the offset between the two boundaries. We further denote the *left truncation boundary* by

$$u := -s - t.$$

With these definitions, the selection interval can be written in the scale-invariant form

$$\mathcal{A} = [u,\, t].$$

The resulting truncation geometry in whitened coordinates is illustrated in Fig. 1.

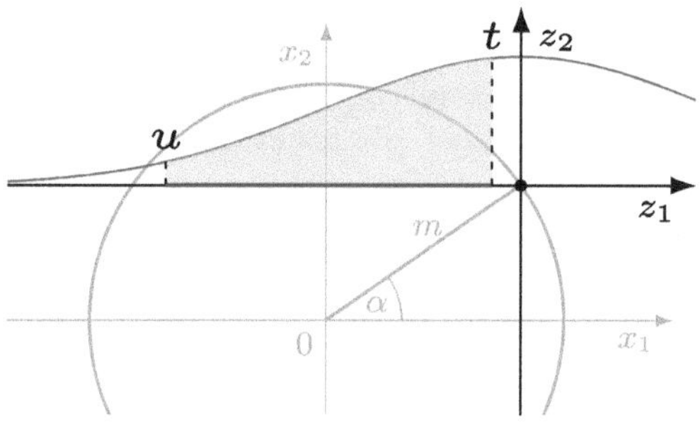

Fig. 1. Geometry of the selected region in whitened coordinates. The interval $\mathcal{A} = [u, t]$ on the z_1-axis induces the vertical strip $\mathcal{A} \times \mathbb{R}$ in the (z_1, z_2)-plane.

The selection rate constraint becomes

$$\boxed{\rho = \Phi(t) - \Phi(u)} \quad \text{(truncation condition)} \tag{10}$$

and the step-size can be recovered from the truncation geometry via

$$\sigma = \frac{\sqrt{2}\cos(\alpha)}{-(u+t)}. \tag{11}$$

Thus, all selection-induced expectations depend only on the truncation boundaries (u, t) and the selection rate ρ, while the original parameters (σ, α, β) appear only through these relations.

3 Covariance Update

We analyze the expected covariance drift in normal form. In the regime $\kappa \to \infty$, selection acts only on z_1. This reduces the covariance question to a one-dimensional truncation problem.

Proposition 1 (Stationary Covariance Threshold). *For fixed selection rate ρ, there exists a quantity $s_\kappa(\rho)$, depending only on ρ, such that the expected logarithmic condition-number drift vanishes in the regime $\kappa \to \infty$ precisely when*

$$\frac{\sqrt{2}\cos(\alpha)}{\sigma} = s_\kappa(\rho).$$

Equivalently,

$$\sigma = A_\kappa(\rho)\,\sqrt{2}\cos(\alpha),$$

where $A_\kappa(\rho)$ depends only on ρ.

Proof. We consider the one-step behavior in the diagonal normal form and rewrite the rank-μ update accordingly. The covariance update reads

$$C' = (1 - c_\mu)D + c_\mu \frac{1}{\lfloor\rho\lambda\rfloor} \sum_{i=1}^{\lfloor\rho\lambda\rfloor} \frac{(x^i - m)}{\sigma} \frac{(x^i - m)^\top}{\sigma}.$$

With the representation

$$x^i = m + \sigma\sqrt{D}\, z^i, \qquad z^i \sim \mathcal{N}(0, I),$$

this becomes

$$C' = \sqrt{D}\left((1 - c_\mu)I + c_\mu \frac{1}{\lfloor\rho\lambda\rfloor} \sum_{i=1}^{\lfloor\rho\lambda\rfloor} z^i (z^i)^\top \right) \sqrt{D}.$$

Although the update does not preserve diagonality, the resulting matrix can be mapped back to normal form in the subsequent step.

We now characterize the condition under which the covariance update decreases the condition number κ. The rank-μ update mixes the previous covariance matrix with the empirical covariance of the selected normalized steps, using the learning rate c_μ. In expectation, the covariance matrix will be

$$\mathbb{E}[C'] = \sqrt{D}\left((1 - c_\mu)I + c_\mu \mathbb{E}[\tilde{z}\tilde{z}^\top]\right)\sqrt{D}. \tag{12}$$

Since selection depends only on z_1, the components z_ρ and z_2 are independent, and $E[\tilde{z}\tilde{z}^\top]$ is diagonal with second diagonal entry equal to 1. In whitened coordinates, the covariance update therefore decreases the condition number if and only if $E[z_\rho^2] < 1$. The relevant second moment is

$$E[z_\rho^2] = \rho^{-1} \int_A z_1^2 \phi(z_1)\, dz_1 = \rho^{-1} \int_u^t z_1^2 \phi(z_1)\, dz_1.$$

Using

$$\int_{-\infty}^{a} z^2 \phi(z)\, dz = \Phi(a) - a\phi(a),$$

we obtain

$$E[z_\rho^2] = \rho^{-1}\big(\Phi(t) - t\phi(t) - \Phi(u) + u\phi(u)\big) = 1 + \rho^{-1}\big(u\phi(u) - t\phi(t)\big),$$

where we used the truncation condition (10).

At stationarity, the condition number remains unchanged, so that

$$E[z_\rho^2] = 1.$$

Using the expression above, this is equivalent to

$$u\phi(u) = t\phi(t), \qquad \text{(covariance stationarity condition)} \tag{13}$$

Since Φ is strictly monotone, the truncation condition (10) determines the right boundary as

$$t(u) = \Phi^{-1}\big(\rho + \Phi(u)\big). \tag{14}$$

Thus, the truncation condition reduces the truncation region to a single degree of freedom, which may be parametrized by the scalar variable u. Consequently, substituting $t = t(u)$ into the covariance stationarity condition (13) yields, for fixed ρ, a scalar equation in u. Any solution of this equation corresponds to a stationary truncation geometry. We denote the corresponding stationary left truncation boundary by $u^* = u^*(\rho)$, define the right truncation boundary by $t^* := t(u^*(\rho))$, and set

$$s = -u^* - t^* =: s(\rho),$$

so that the stationary truncation geometry, and hence s, is determined by the selection rate ρ alone. Recalling the definition $s = \sqrt{2}\cos(\alpha)/\sigma$, it follows that

$$\frac{\sqrt{2}\cos(\alpha)}{\sigma} = s(\rho), \qquad \text{or equivalently} \qquad \sigma = A_\kappa(\rho) \cdot \sqrt{2}\cos(\alpha).$$

This proves the claim. □

3.1 Numerical Analysis of the Threshold

The same numerical setup is used for all numerical results shown in this paper. In all cases, we use $\lambda = 6$ and $\mu = 3$, corresponding to the selection rate $\rho = \mu/\lambda = 1/2$, with equal recombination weights $w_i = 1/\mu$ for $i = 1, \dots, \mu$. We consider $\alpha \in [0, \pi/2]$ and $\sigma \in [0.01, 10]$, each sampled on a grid of resolution 1024. For each grid point, the corresponding expected one-step drift is estimated from 10^5 samples. The highly anisotropic reference case uses $\kappa = 10^6$; where indicated, we also show $\kappa = 20$ to illustrate finite-κ deviations (Fig. 2).

Approximation of the Proportionality Constant. The numerical zero-threshold contour of the expected logarithmic condition-number drift is well approximated by a relation of the form $\sigma = A_\kappa\sqrt{2}\cos(\alpha)$. Empirically, for the selection rate

Fig. 2. Common color convention: blue indicates expected decrease, red expected increase, and the white contour the stable threshold. (Color figure online)

$\rho = 1/2$ used in all numerical experiments and for $\kappa = 10^6$, this suggests the simple approximation

$$A_\kappa \approx \frac{1}{\pi}, \qquad \text{that is,} \qquad \sigma \approx \frac{\sqrt{2}}{\pi} \cos(\alpha). \tag{15}$$

This approximation matches the numerical threshold very well in the highly anisotropic regime. The threshold appears to depend on κ for finite values, whereas in the isotropic case $\kappa = 1$ it is independent of α. As shown in Fig. 3, visible deviations from the asymptotic prediction remain for $\kappa = 20$, while the agreement is much closer for $\kappa = 10^6$.

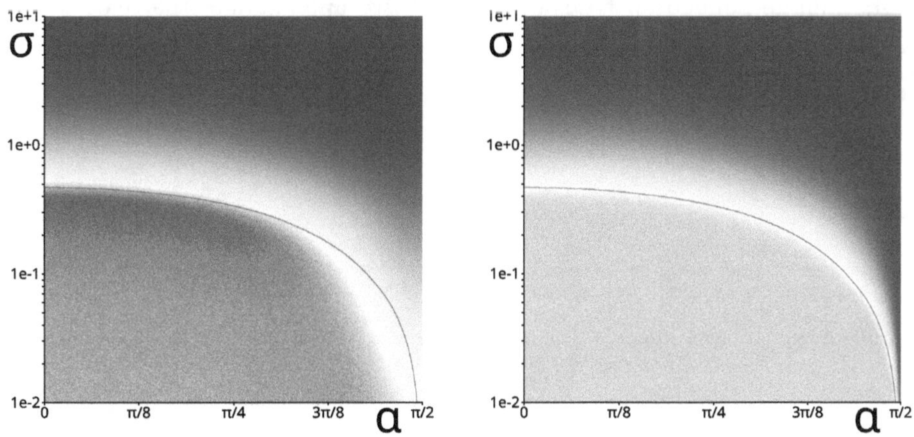

(a) $\kappa = 20$ (b) $\kappa = 10^6$

Fig. 3. Numerical illustration of the zero contour of the expected logarithmic condition-number drift for finite and asymptotic anisotropy. Blue indicates expected decrease, red expected increase, and the white contour marks the stable threshold. The black curve shows the asymptotic theoretical prediction $\sigma = A_\kappa \sqrt{2} \cos(\alpha)$. (Color figure online)

4 Step Size Update

We now analyze the expected logarithmic drift of the step-size. As in the covariance analysis, the regime $\kappa \to \infty$ reduces the stationarity condition to a one-dimensional truncation problem.

Proposition 2 (Stationary Logarithmic Step-Size Threshold). *For fixed selection rate ρ, there exists a quantity $s_\sigma(\rho)$, depending only on ρ, such that the expected logarithmic step-size drift vanishes in the regime $\kappa \to \infty$ precisely when*

$$\frac{\sqrt{2}\cos(\alpha)}{\sigma} = s_\sigma(\rho).$$

Equivalently,

$$\sigma = A_\sigma(\rho)\,\sqrt{2}\cos(\alpha),$$

where $A_\sigma(\rho)$ depends only on ρ.

Proof. For dimension 2, the random variable $\|Z\|$ follows a Rayleigh distribution, and its expectation admits the closed-form expression $\mathbb{E}\|Z\| = \sqrt{\pi/2}$.

From the step-size update (4), for equal recombination weights $w_i = 1/\mu$, we have

$$\log\frac{\sigma'}{\sigma} = \frac{\sqrt{\mu}}{\mathbb{E}\|Z\|}\left\|\frac{1}{\mu}\sum_{i=1}^{\mu} z_{i:\lambda}\right\| - 1.$$

In the infinite-population truncation model, we approximate the selected offspring steps by i.i.d. copies $\tilde{z}^{(1)}, \ldots, \tilde{z}^{(\mu)}$ of the selected-step distribution $\tilde{z}$. Accordingly,

$$\log\frac{\sigma'}{\sigma} = \frac{\sqrt{\mu}}{\mathbb{E}\|Z\|}\left\|\frac{1}{\mu}\sum_{j=1}^{\mu} \tilde{z}^{(j)}\right\| - 1.$$

Defining

$$S_\mu := \frac{1}{\sqrt{\mu}}\sum_{j=1}^{\mu} \tilde{z}^{(j)},$$

we obtain

$$\log\frac{\sigma'}{\sigma} = \frac{\|S_\mu\|}{\mathbb{E}\|Z\|} - 1.$$

Hence

$$\mathbb{E}\left[\log\frac{\sigma'}{\sigma}\right] = \frac{\mathbb{E}\|S_\mu\|}{\mathbb{E}\|Z\|} - 1,$$

and the stationary condition is

$$\mathbb{E}\|S_\mu\| = \mathbb{E}\|Z\| = \sqrt{\frac{\pi}{2}} \qquad \text{(logarithmic step-size stationarity condition).} \quad (16)$$

As in the covariance analysis, for fixed ρ the truncation condition (10) determines t as a function of u. Hence the logarithmic step-size stationarity condition becomes

$$\mathbb{E}\|S_\mu\| = \int_u^t \cdots \int_u^t \gamma\left(\frac{z_1 + \cdots + z_\mu}{\sqrt{\mu}}\right) \prod_{j=1}^{\mu} \frac{1}{\rho}\phi(z_j)\, dz_1 \cdots dz_\mu = \sqrt{\frac{\pi}{2}}. \quad (17)$$

where

$$\gamma(x) := \mathbb{E}_{G\sim\mathcal{N}(0,1)}\left[\sqrt{x^2+G^2}\right].$$

For fixed ρ, the truncation geometry is again described by a single parameter, which we take to be u. The logarithmic step-size stationarity condition (16) therefore becomes a scalar equation in u. Any solution defines a stationary truncation geometry, and hence a quantity $s = s(\rho)$. Recalling $s = \sqrt{2}\cos(\alpha)/\sigma$, we obtain

$$\frac{\sqrt{2}\cos(\alpha)}{\sigma} = s(\rho), \qquad \text{or equivalently} \qquad \sigma = A_\sigma(\rho)\cdot\sqrt{2}\cos(\alpha).$$

This proves the claim. □

4.1 Numerical Analysis of the Threshold

The numerical zero contour of the expected logarithmic step-size drift is well approximated by a relation of the form $\sigma = A_\sigma\sqrt{2}\cos(\alpha)$. Empirically, for the selection rate $\rho = 1/2$ used in all numerical experiments and for $\kappa = 10^6$, this is close to the simple approximation

$$A_\sigma \approx \frac{\pi}{4}, \qquad \text{that is,} \qquad \sigma \approx \frac{\pi}{2\sqrt{2}}\cos(\alpha). \tag{18}$$

This approximation matches the numerical threshold well in the highly anisotropic regime. This is illustrated in Fig. 4a, which compares the finite-κ case $\kappa = 20$ with the highly anisotropic reference case $\kappa = 10^6$.

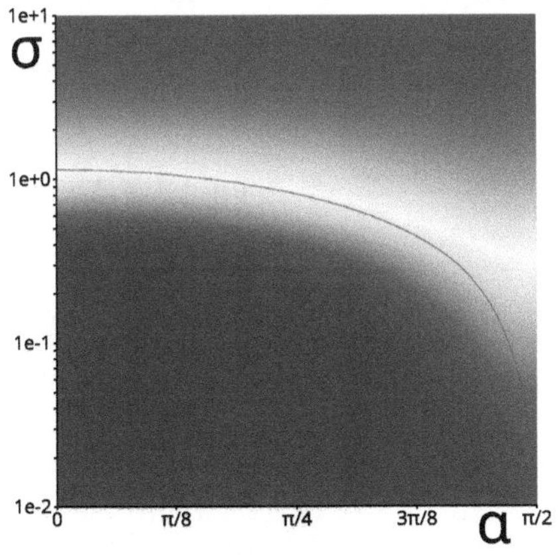

(a) $\kappa = 20$

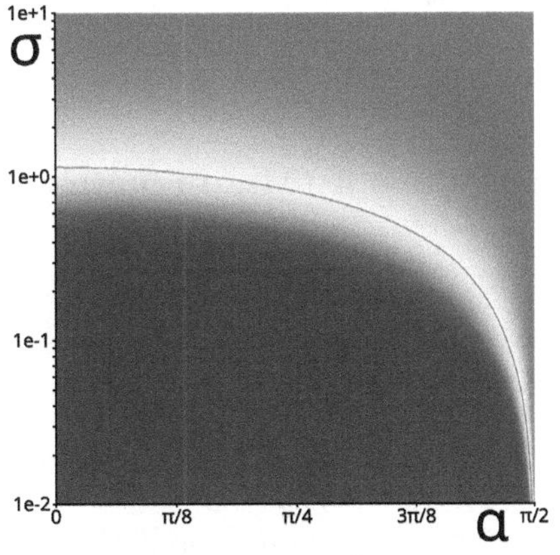

(b) $\kappa = 10^6$

Fig. 4. Numerical illustration of the zero contour of the expected logarithmic step-size drift for $\kappa = 20$ and $\kappa = 10^6$. The black curve shows the asymptotic theoretical prediction $\sigma = A_\sigma\sqrt{2}\cos(\alpha)$.

5 Equilibrium of the Step-Size in Normal Form

The previous sections derived zero-drift conditions for the covariance update and for the logarithm of the algorithmic step-size. We now turn to the equilibrium of the normalized step-size in normal form and derive the corresponding fixed-point equation, which will subsequently serve as the starting point for controlled approximations and their empirical validation.

In the infinite-population setting, we interpret σ'_{NF} as the normalized step-size after one expected update step. The fixed-point condition is therefore written as

$$\sigma'_{\mathrm{NF}} = \sigma_{\mathrm{NF}}. \tag{19}$$

Since the current state is represented in normal form, we have $\sigma_{\mathrm{NF}} = \sigma$ at the current point. After one update step, however, the normalized step-size is modified by three factors: the algorithmic step-size update (known from Sect. 4), the covariance renormalization (known from Sect. 3), and the renormalization induced by the new mean norm, which will be analyzed below. Accordingly, the fixed-point condition can be written as the product of the algorithmic step-size update, the covariance renormalization, and the mean renormalization:

$$\frac{\sigma'_{\mathrm{alg}}}{\sigma} \cdot \sqrt{\frac{\mathrm{tr}(C')}{2}} \cdot \frac{1}{\|m'\|} = 1. \tag{20}$$

We refer to such a fixed point as a self-similar stationary regime: although the raw state continues to evolve, the normalized relation between distance to the optimum and effective step-size is preserved in expectation after renormalization. Since only the product of the three factors must equal one, the first two are already determined once the truncation geometry is parametrized by u. The only new contribution is therefore the renormalization induced by the updated mean norm.

Proposition 3 (Mean-renormalization Term). *Let $\bar{y}$ denote the recombined selected step and define*

$$d := -\bar{y}_1 \geq 0.$$

Then, in the regime $\kappa \to \infty$,

$$\|m'\|^2 = 1 - 2d\,\sigma\cos(\alpha) + d^2\sigma^2.$$

Proof. In the infinite-population setting, the mean update is described by its expectation. Writing

$$m' = m + \sigma\bar{y},$$

where $\bar{y}$ denotes the expected recombined selected step, we obtain

$$\|m'\|^2 = \|m + \sigma\bar{y}\|^2 = \|m\|^2 + 2\sigma\, m^\top \bar{y} + \sigma^2\|\bar{y}\|^2.$$

Since the current state is represented in normal form, we have $\|m\| = 1$, and therefore

$$\|m'\|^2 = 1 + 2\sigma\, m^\top \bar{y} + \sigma^2 \|\bar{y}\|^2.$$

In the regime $\kappa \to \infty$, selection acts only on the first coordinate, so that the expected selected step is aligned with the first axis. Hence

$$\bar{y} = (\bar{y}_1, 0), \qquad m = (\cos(\alpha), \sin(\alpha)),$$

and therefore

$$m^\top \bar{y} = \cos(\alpha)\, \bar{y}_1, \qquad \|\bar{y}\|^2 = \bar{y}_1^2.$$

Substituting this into the previous expression gives

$$\|m'\|^2 = 1 + 2\sigma \cos(\alpha)\, \bar{y}_1 + \sigma^2 \bar{y}_1^2.$$

Since selection favours offspring closer to the optimum along the first coordinate, we have $\bar{y}_1 \leq 0$, with strict inequality for $\alpha < \pi/2$. We therefore write

$$d := -\bar{y}_1 \geq 0.$$

Then

$$\|m'\|^2 = 1 - 2d\,\sigma \cos(\alpha) + d^2\sigma^2. \tag{21}$$

This proves the claim. □

5.1 Approximating the Equilibrium

Rather than yielding a rigorous closed-form solution of the fixed-point equation, the following argument motivates an effective approximation of the normal-form equilibrium. For this purpose, we neglect the residual covariance renormalization and use as a baseline an effective proportional relation motivated by the zero contour of the expected logarithmic algorithmic step-size drift from Sect. 4.1. We write this relation as

$$\sigma \approx A \cos(\alpha), \tag{22}$$

where $A > 0$ denotes an effective proportionality constant. Substituting this into

$$\|m'\|^2 = 1 - 2d\,\sigma \cos(\alpha) + d^2\sigma^2$$

gives

$$\|m'\|^2 \approx 1 - 2dA \cos^2(\alpha) + d^2 A^2 \cos^2(\alpha).$$

With $B := dA$, this simplifies to

$$\|m'\|^2 \approx 1 - B(2 - B) \cos^2(\alpha).$$

We thus arrive at the approximation

$$\sigma^*_{\mathrm{NF}}(\alpha) \approx \frac{A \cos(\alpha)}{\sqrt{1 - B(2 - B)\cos^2(\alpha)}}. \tag{23}$$

The square-root expression above is based on the baseline approximation (22), namely

$$\sigma \approx A\cos(\alpha),$$

which identifies the equilibrium with raw logarithmic step-size stationarity. In general, however, normal-form equilibrium is not determined by raw logarithmic step-size stationarity alone. Even if covariance renormalization is neglected, the algorithmic step-size change and the renormalization induced by the updated mean norm enter multiplicatively and may compensate each other. As a consequence, the normal-form equilibrium need not lie on the raw logarithmic stationary step-size curve, and the resulting correction is not, in general, captured by the reduced square-root closure. For this reason, we also consider the simpler first-order approximation

$$\sigma^*_{\mathrm{NF}}(\alpha) \approx \frac{A\cos(\alpha)}{1 - b\cos^2(\alpha)}. \tag{24}$$

Here, b denotes an empirical correction parameter. Empirically, this first-order approximation provides a better fit than the square-root expression, and we therefore use it as the preferred effective description of the normal-form equilibrium.

5.2 Numerical Validation

For the empirical approximation of the normal-form equilibrium, we use $b \approx \frac{1}{\sqrt{2}}$, and $A \approx \frac{\pi}{2\sqrt{2}}$. Note that the constant A used here differs from the raw step-size constant A_σ in Sect. 4 by a factor of $\sqrt{2}$, since the present parametrization

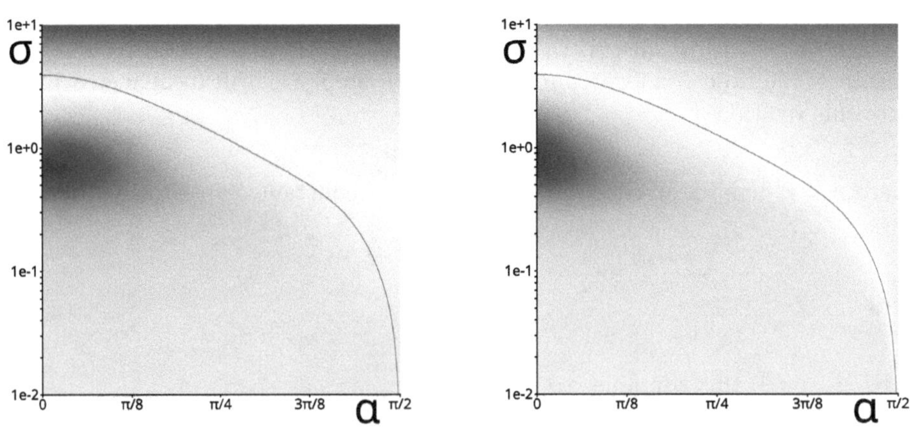

(a) $\kappa = 20$ (b) $\kappa = 10^6$

Fig. 5. Numerical validation of the normal-form equilibrium for $\kappa = 20$ and $\kappa = 10^6$. The white contour marks the stable equilibrium curve, and the black curve the approximation $\sigma^*_{\mathrm{NF}}(\alpha) \approx \frac{A\cos(\alpha)}{1-b\cos^2(\alpha)}$, with $b \approx 1/\sqrt{2}$ and $A \approx \pi/(2\sqrt{2})$.

absorbs the factor $\sqrt{2}$ into A. As shown in Fig. 5, the first-order approximation captures the stable equilibrium curve well in the highly anisotropic reference case $\kappa = 10^6$, while visible deviations remain for the finite-κ case $\kappa = 20$.

6 Conclusion

In this paper, we analyzed the expected one-step dynamics of a simplified rank-μ CMA-ES on the two-dimensional sphere in normal form, with particular focus on the limit $\kappa \to \infty$. In this regime, both covariance stationarity and logarithmic algorithmic step-size stationarity reduce to the same one-dimensional truncation geometry. Equivalently, both are governed by the normal-form scaling variable $s = \frac{\sqrt{2}\cos(\alpha)}{\sigma}$, which at stationarity depends only on the selection rate ρ.

For the covariance dynamics, our analysis shows that even in the regime of very large, or formally infinite, eccentricity, there always exists a step-size threshold below which the covariance becomes even more eccentric. At the same time, the algorithmic step-size update exhibits its own threshold at which the expected logarithmic step-size drift vanishes. Empirically, the corresponding constants satisfy $A_\kappa < A_\sigma$, which indicates that step-size adaptation tends to drive the algorithm into a region in which covariance adaptation acts in the stabilizing direction and reduces anisotropy rather than amplifying it.

We then turned to the equilibrium of the normalized step-size in normal form. The resulting stationary point can be interpreted as an invariant relation between distance to the optimum and effective step-size. Although the raw state of the algorithm continues to evolve, this normalized relation is preserved in expectation at equilibrium, and the dynamics appears to drive the system toward it. In this sense, the normal-form equilibrium describes a self-similar stationary regime of the algorithm.

Taken together, these results provide a tractable description of a highly anisotropic regime of CMA-ES and clarify how covariance adaptation, step-size adaptation, and normal-form renormalization interact near stationarity. Through the standard transformation of a convex-quadratic function to sphere coordinates, the regime considered here also suggests an interpretation of early-stage behavior on highly ill-conditioned problems, before the covariance matrix has adapted to the inverse Hessian structure.

Future Work. A natural direction for future work is to study more systematically how the thresholds and equilibrium relations identified here depend on the remaining algorithmic parameters, in particular the recombination weights and the population size. Another important question is to what extent the normal-form picture developed in this work persists beyond the simplified infinite-population setting and beyond the two-dimensional sphere. Finally, it would be of clear interest to extend the present analysis from the limit $\kappa \to \infty$ to finite values of κ, thereby interpolating between the maximally anisotropic regime studied here and the isotropic case $\kappa = 1$.

Acknowledgments. This publication is based upon work from COST Action ROAR-NET, CA22137, supported by COST (European Cooperation in Science and Technology).

Disclosure of Interests. The authors have no competing interests to declare that are relevant to the content of this article.

AI Use. A large language model assisted with drafting and presentation. The authors verified all results and conclusions and take full responsibility for the manuscript.

References

1. Akimoto, Y., Auger, A., Glasmachers, T., Morinaga, D.: Global linear convergence of evolution strategies on more than smooth strongly convex functions. SIAM J. Optim. **32**(2), 1402–1429 (2022)
2. Akimoto, Y., Auger, A., Hansen, N.: Quality gain analysis of the weighted recombination evolution strategy on general convex quadratic functions. In: Proceedings of the 14th ACM/SIGEVO Conference on Foundations of Genetic Algorithms, pp. 111–126 (2017)
3. Akimoto, Y., Auger, A., Hansen, N.: An ODE method to prove the geometric convergence of adaptive stochastic algorithms. Stochastic Process. Appl. **145**, 269–307 (2022)
4. Arnold, D.V.: Optimal weighted recombination. In: Proceedings of the 8th International Conference on Foundations of Genetic Algorithms, pp. 215–237 (2005)
5. Auger, A.: Convergence results for the $(1, \lambda)$-SA-ES using the theory of φ-irreducible markov chains. Theoret. Comput. Sci. **334**(1–3), 35–69 (2005)
6. Auger, A., Hansen, N.: Linear convergence of comparison-based step-size adaptive randomized search via stability of Markov chains. SIAM J. Optim. **26**(3), 1589–1624 (2016)
7. Beyer, H.: The Theory of Evolution Strategies. Springer Science & Business Media (2001)
8. Frank, S., Glasmachers, T.: A potential function for a variable-metric evolution strategy. In: International Conference on Parallel Problem Solving from Nature, pp. 221–235. Springer (2024)
9. Gissler, A.: Evaluation of the impact of various modifications to CMA-ES that facilitate its theoretical analysis. In Proceedings of the Companion Conference on Genetic and Evolutionary Computation, pp. 1603–1610 (2023)
10. Gissler, A.: Linear Convergence of Evolution Strategies with Covariance Matrix Adaptation. PhD thesis. Institut Polytechnique de Paris (2024)
11. Hansen, N., Kern, S.: Evaluating the CMA evolution strategy on multimodal test functions. In: Parallel Problem Solving from Nature - PPSN VIII, pp. 282–291 (2004)
12. Hansen, N., Müller, S.D., Koumoutsakos, P.: Reducing the time complexity of the derandomized evolution strategy with covariance matrix adaptation (CMA-ES). Evol. Comput. **11**(1), 1–18 (2003)

13. Hansen, N., Ostermeier, A.: Completely derandomized self-adaptation in evolution strategies. Evol. Comput. **9**(2), 159–195 (2001)
14. Jägersküpper, J.: Algorithmic analysis of a basic evolutionary algorithm for continuous optimization. Theoret. Comput. Sci. **379**(3), 329–347 (2007)
15. Morinaga, D., Fukuchi, K., Sakuma, J., Akimoto, Y.: Convergence rate of the (1+1)-ES on locally strongly convex and Lipschitz smooth functions. IEEE Trans. Evol. Comput. **28**(2), 501–515 (2024)

First Hitting Times for Multiple Processes and Targets

Timo Kötzing[1(✉)] and Jonathan E. Rowe[2]

[1] Hasso Plattner Institute, Potsdam, Germany
timo.koetzing@hpi.de
[2] University of Birmingham, Birmingham, UK

Abstract. Given n processes and upper bounds on their respective first hitting times, we give an upper bound on the time until all processes have visited their target states. The first main result requires time-homogeneous Markov chains and a runtime bound independent of starting points for each individual process. Our second main result assumes a bound based on the fitness level method. We illustrate our results with various examples.

1 Introduction

Suppose we have a randomised process, or collection of processes, and a number of events. We are interested in the time it takes for all the events to occur, as a function of the expected time for each event individually to happen. For example, in the classic *coupon collector's* problem, the expected time to collect any one coupon is n (where n is the total number of distinct coupons), and the expected time to collect all of the coupons is nH_n (where H_n is the nth harmonic number). In a previous paper [1], it is shown that if the expected time for each individual event is bounded above by τ, then the time for them all to occur is, with high probability, $O(\tau \log n)$ (see Theorem 1 for the formal statement). Here, we generalise the result so that if events $i \in \{1, 2, \dots, n\}$ have expected times upper bounded by τ_i, and we index the events such that $\tau_1 \geq \tau_2 \geq \dots \geq \tau_n$, then the expected time for them all to happen is no more than $\sum_{i=1}^{n} \tau_i / i$ (see Theorem 5 for the formal statement).

It can be seen that this general result immediately gives us the correct runtime for the coupon collector's problem. We also give some further applications, including the analysis of an evolutionary algorithm on the integer-valued OneMax problem [4] (see Theorem 9).

If we know more about the individual processes than just their expected first hitting times, then we can give better bounds. Suppose we have n identical parallel processes that are defined on a number of levels. The processes move forward towards the target level, and we have bounds on the probability of leaving each level and progressing. Using the standard *fitness level* argument [12], we can derive an upper bound τ on the expected runtime of each process. We then derive an upper bound for all n processes to reach their targets. This bound

G. Iacca et al. (Eds.): PPSN 2026, LNCS 16985, pp. 52–66, 2027.
https://doi.org/10.1007/978-3-032-36214-8_4

is significantly better than the general one, due to the fact that the runtime of each process is strongly concentrated. We use this bound to show that the time taken to collect m copies of each of n coupons is bounded above by $n\ln(n) + (m-1)n\ln(\ln(n)) + O(n)$, a result first derived in [9] (see Corollary 13). We further illustrate this result by deriving a bound on the runtime of a population of $(1+1)$ EAs running in parallel (see Corollary 14).

Note that some proofs are ommitted due to space constraints.

1.1 Related Work

In the following, for any $n \in \mathbb{N}$, we let $[n] = \{1, \ldots, n\}$.

Wormald [13] gave a method for tracking the progress of n random processes, based on systems of differential equations. This method is notoriously hard to apply, since differential equations are hard to solve and a number of boundary constraints need to be met.

A much simpler to apply version was given in [10], for the case of multiplicative drift. [5] gives a two-dimensional follow-up.

A first more general result was given in [1], which we state as follows.

Theorem 1. *Let $(X^1(t))_{t\in\mathbb{N}}, \ldots, (X^n(t))_{t\in\mathbb{N}}$, be n random processes (not necessarily independent) that run synchronously in parallel. For each $i \in [n]$, let there be an event of interest U_i and let T_i denote the first hitting time of event U_i by process $(X^i(t))_{t\in\mathbb{N}}$. That is,*

$$T_i = \inf\left\{t \in \mathbb{N} \mid X^i(t) \in U_i\right\}.$$

Let τ be an upper bound for the expected time each process takes to reach its respective event of interest, regardless of the starting point. That is, for any time t and any state s, if the process $X^i(t)$ is in state s at time t, the expected time for event U_i to occur is at most $t+\tau$. Let $T^* = \max_{i\in[n]} T_i$. Then, for any $\kappa \in \mathbb{R}_+$,

$$\Pr\left[T^* \geq e(\kappa+1)\tau\ln n\right] \leq \frac{1}{n^\kappa}.$$

Note that [7] considered parallel processes and used similar ideas, with results based on moment generating functions.

It is easy to see that this implies an expected run time of $O(\tau \log n)$. We see that this is asymptotically tight in both n and τ with the following example.

Proposition 2. *Let $\tau \geq 4$ and $n \in \mathbb{N}$ be given. Let $(Y^i)_{i\in[n]}$ be n independent geometrically distributed random variables with parameter $1/\tau$. For all $i \in [n]$, let X^i be the process such that, for all $t \in \mathbb{N}$, $X^i(t) = 1$ if $t < Y^i$ and $X^i(t) = 0$ otherwise. For all $i \in [n]$ let*

$$T_i = \inf\left\{t \in \mathbb{N} \mid X^i(t) = 0\right\}$$

and $T^* = \max_{i\in[n]} T_i$. Then

$$E\left[T^*\right] \geq \tau(\ln(n) - 2)/6.$$

2 General Results for Hitting Times for Multiple Processes

This section is dedicated to showing our general result for multiple targets and multiple processes (Theorem 5). But first we will give two trivial bounds.

Theorem 3. *Let $n \in \mathbb{N}_+$ and $(X^i(t))_{t\in\mathbb{N},i\in[n]}$ be n time-homogeneous Markov-chains on a set $\mathcal{X}$. For each $i \in [n]$, fix a target set $S_i \subseteq \mathcal{X}$ and define $T_i = \inf\{t \in \mathbb{N} \mid X^i(t) \in S_i\}$. Let T be the first time such that all n processes have visited their respective goal sets at least once. Then we have*

$$E[T] \leq \sum_{i=1}^{n} E[T_i].$$

We can also give a trivial lower bound as follows.

Theorem 4. *Let $n \in \mathbb{N}_+$ and $(X^i(t))_{t\in\mathbb{N},i\in[n]}$ be n stochastic processes on a set $\mathcal{X}$. For each $i \in [n]$ fix a target set S_i and define $T_i = \inf\{t \in \mathbb{N} \mid X^i(t) \in S_i\}$. Let T be the first time such that all n processes have visited their respective goal sets at least once. Then we have*

$$E[T] \geq \max_{i\in[n]} E[T_i].$$

Now we show our improved upper bound using similar assumptions on the processes as used in [1] (see Theorem 1).

Theorem 5. *Let $\mathcal{X}$ be a set and $(X(t))_{t\in\mathbb{N}}$ be a time-homogeneous Markov chain on $\mathcal{X}$. Let $S_1, \ldots, S_n \subseteq \mathcal{X}$ be target sets. For each i, let $\tau_i > 0$ be an upper bound on the expected time to reach S_i regardless of the starting state, and suppose these are indexed such that $\tau_1 \geq \tau_2 \geq \cdots \geq \tau_n$. Let T^* be the minimum time such that all the target sets have been visited. Then*

$$E[T^*] \leq \sum_{i=1}^{n} \frac{\tau_i}{i}.$$

Proof. For each $i \in [n]$, let $T_i(x)$ be the random variable which is the time to reach subset S_i from state $x \in \mathcal{X}$ (using that we have a time-homegeneous Markov chain). We have an upper bound τ_i for the expected time to reach the target S_i, regardless of the initial state, so that $E[T_i(x)] \leq \tau_i$ for all $x \in \mathcal{X}$.

It is possible that, once the process reaches one of its target sets, it might then leave it again, before it has reached all the other targets. We therefore need a mechanism to keep track of when the different targets have been hit. We augment the process to one defined on the state space $\mathcal{X} \times \{0,1\}^n$. We assume, as a worst case, that $X(0) \notin S_i$, for all i, so the initial state is $(X(0), 1^n)$. We define the process on this new state space as follows. If the original process moves from state x to state y then the augmented process moves from a state (x, u) to state (y, v) where the string v is defined by:

$$v_i = \begin{cases} 0, & \text{if } y \in S_i; \\ u_i, & \text{otherwise.} \end{cases}$$

We will denote the random sequence of bitstrings so generated by $U(t)$, for $t \geq 0$, so that our augmented sequence is $(X(t), U(t))$.

It can be seen that the bitstring keeps track of which of the target states have been visited, and we are interested in the expected time until the bitstring is 0^n. A state $(x, u) \in \mathcal{X} \times \{0,1\}^n$ is *valid* if $x \in S_i \implies u_i = 0$. We denote the set of all valid states by $\mathcal{Y}$.

The following lemma establishes a potential and drift for individual components; we will later use this to construct our potential function.

Lemma 6. *For each* $i \in [n]$, *define* $\beta_i \colon \mathcal{Y} \to \mathbb{R}, (x, u) \mapsto u_i E[T_i(x)]$. *Then*

$$E[\beta_i(X(t), U(t)) - \beta_i(X(t+1), U(t+1)) \mid X(t) = x, U(t) = u] = u_i.$$

Proof. If $u_i = 0$ then it follows that $U(t+1)_i = 0$ and both terms are zero and the claim holds. From now on, assume that $u_i = 1$ and, in particular, $X_t \notin S_i$. Then

$$\begin{aligned} &E[\beta_i(X(t), U(t)) - \beta_i(X(t+1), U(t+1)) \mid X(t) = x, U(t) = u] \\ &\quad = E[u_i E[T_i(x)] - U(t+1)_i E[T_i(X(t+1))] \mid X(t) = x, U(t) = u] \\ &\quad = E[T_i(x)] - E[U(t+1)_i E[T_i(X(t+1))] \mid X(t) = x, U(t) = u] \end{aligned}$$

We have that

$$T_i(X(t+1)) = \min_{k \in \mathbb{N}} X(t+1+k) \in S_i.$$

Since $X(t) \notin S_i$, we get

$$\min_{k \in \mathbb{N}} X(t+1+k) \in S_i = \left(\min_{k \in \mathbb{N}} X(t+k) \in S_i\right) - 1 = T_i(X(t)) - 1.$$

Therefore

$$E[T_i(X(t+1)) \mid X(t) = x, U(t) = u] = E[T_i(x) - 1].$$

If $X(t+1) \notin S_i$, then $U(t+1)_i = 1$, and so

$$\begin{aligned} &U(t+1)_i E[T_i(X(t+1)) \mid X(t) = x, U(t) = u] \\ &\quad = E[T_i(X(t+1)) \mid X(t) = x, U(t) = u] \\ &\quad = E[T_i(x) - 1]. \end{aligned}$$

If $X(t+1) \in S_i$, then $U(t+1)_i = 0 = E[T_i(X(t+1)) \mid X(t) = x, U(t) = u]$, and so

$$\begin{aligned} &U(t+1)_i E[T_i(X(t+1)) \mid X(t) = x, U(t) = u] \\ &\quad = E[T_i(X(t+1)) \mid X(t) = x, U(t) = u] \\ &\quad = E[T_i(x) - 1]. \end{aligned}$$

So, overall, assuming $u_i = 1$,

$$\begin{aligned}
&E[\beta_i(X(t), U(t)) - \beta_i(X(t+1), U(t+1)) \mid X(t) = x, U(t) = u] \\
&\quad = E[T_i(x)] - E[U(t+1)_i E[T_i(X(t+1))] \mid X(t) = x, U(t) = u] \\
&\quad = E[T_i(x)] - E[T_i(x) - 1] \\
&\quad = 1 = u_i.
\end{aligned}$$

This concludes the proof of the lemma. □

Now we are ready to define our potential function for the process on $\mathcal{Y}$. Define $d\colon \mathcal{Y} \to \mathbb{R}$ by

$$d(x, u) = \sum_{i=1}^{n} \beta_i(x, u) = \sum_{i=1}^{n} u_i E[T_i(x)]$$

and notice $d(x, u) = 0$ if and only if $u = 0^n$, and that $0 \leq d(x, u) \leq \sum_{i=1}^{n} \tau_i$.

We will show a lower bound for the drift of this process employing the following function g, which counts the smallest number of non-visited targets that can correspond to an expected hitting time. Define

$$g\colon d(\mathcal{Y}) \to \mathbb{N}, z \mapsto \min_{(x,u)\in\mathcal{Y}} \left\{ \sum_{i=1}^{n} u_i \;\middle|\; d(x, u) = z \right\}.$$

The following lemma shows that g indeed gives a lower bound on the drift.

Lemma 7. *For all $t \in \mathbb{N}$ and all $(x, u) \in \mathcal{Y}$ it holds that*

$$E[d(X(t), U(t)) - d(X(t+1), U(t+1)) \mid X(t) = x, U(t) = u] \geq g(d(x, u)).$$

Proof. Let $\zeta\colon \mathcal{Y} \to \mathbb{R}$ be

$$\zeta(y, v) = \sum_{i=1}^{n} v_i$$

so that, for any $(y, v) \in \mathcal{Y}$,

$$\zeta(y, v) \in \{\zeta(x, u) \mid d(x, u) = d(y, v)\}$$

which means

$$\zeta(y, v) \geq \min_{(x,u)\in\mathcal{Y}} \{\zeta(x, u) \mid d(x, u) = d(y, v)\} = g(d(y, v)).$$

Then, using Lemma 6 in the third equality,

$$\begin{aligned}
&E[d(X(t), U(t)) - d(X(t+1), U(t+1)) \mid X(t) = x, U(t) = u] \\
&\quad = E\left[\sum_{i=1}^{n} \beta_i(x, u) - \sum_{i=1}^{n} \beta_i(X(t+1), U(t+1)) \mid X(t) = x, U(t) = u\right] \\
&\quad = \sum_{i=1}^{n} E[\beta_i(x, u) - \beta_i(X(t+1), U(t+1)) \mid X(t) = x, U(t) = u] \\
&\quad \text{(by linearity of expectation)} \\
&\quad = \sum_{i=1}^{n} u_i = \zeta(x, u) \geq g(d(x, u)).
\end{aligned}$$

□

For each $i \in [n]$, let $\chi_i(z)$ be the indicator function

$$\chi_i(z) = \begin{cases} 1, & \text{if } z > \sum_{j=1}^{i-1} \tau_j, \\ 0, & \text{otherwise.} \end{cases}$$

Define

$$h(z) = \sum_{i=1}^{n} \chi_i(z).$$

Lemma 8. *For all $z \in d(\mathcal{Y})$ it holds that*

$$g(z) \geq h(z)$$

Proof. First note that $0 \leq d(\mathcal{Y}) \leq \sum_{i=1}^{n} \tau_i$. Then the function

$$h(z) = \sum_{i=1}^{n} \chi_i(z)$$

has $h(0) = 0$ and, for $0 \leq k \leq n-1$, we have $h(z) = k+1$ for $\sum_{i=1}^{k} \tau_i < z \leq \sum_{i=1}^{k+1} \tau_i$.

Clearly $g(0) = 0 = h(0)$. Now consider the case where $0 < z \leq \tau_1$. Then, for all states (x, u) such that $d(x, u) = z$,

$$0 < \sum_{i=1}^{n} u_i E[T_i(x)] \leq \tau_1$$

and hence $g(z) \geq 1 = h(z)$.

Now consider $\tau_1 < z \leq \tau_1 + \tau_2$. For all states (x, u) with $d(x, u) = z$ we have

$$\tau_1 < \sum_{i=1}^{n} u_i E[T_i(x)] \leq \tau_1 + \tau_2$$

and so $\sum_{i=1}^{n} u_i \geq 2$, exploiting that no single process can contribute more than τ_1 to the potential, hence $g(z) \geq 2 = h(z)$.

In general, for all z with $\sum_{i=1}^{k} \tau_i < z \leq \sum_{i=1}^{k+1} \tau_i$, if $d(x, u) = z$ then

$$\sum_{i=1}^{k} \tau_i < \sum_{i=1}^{n} u_i E[T_i(x)] \leq \sum_{i=1}^{k+1} \tau_i$$

and so $\sum_{i=1}^{n} u_i \geq k+1$, hence $g(z) \geq k+1 = h(z)$. □

Note that $h(z)$ is non-negative for $z > 0$ and is a monotonic increasing function. Also note that

$$d(x, u) \neq 0 \Longrightarrow d(x, u) \geq 1$$

for all $(x, u) \in \mathcal{Y}$. We therefore apply the variable drift theorem [6,8] to get

$$E[T^*] \leq 1 + \int_1^{\sum_{i=1}^n \tau_i} \frac{1}{h(s)} ds.$$

Recall that, for $0 \leq k \leq n-1$, we have $h(z) = k+1$ for $\sum_{i=1}^k \tau_i < z \leq \sum_{i=1}^{k+1} \tau_i$. It follows that

$$\begin{aligned} E[T^*] &\leq 1 + \int_1^{\tau_1} \frac{1}{h(s)} ds + \int_{\tau_1}^{\tau_1+\tau_2} \frac{1}{h(s)} ds + \cdots + \int_{\tau_1+\cdots+\tau_{n-1}}^{\tau_1+\cdots+\tau_n} \frac{1}{h(s)} ds \\ &= 1 + (\tau_1 - 1) + \frac{\tau_2}{2} + \cdots + \frac{\tau_n}{n} \\ &= \sum_{i=1}^n \frac{\tau_i}{i} \end{aligned}$$

as required. □

3 Example Applications

In this section we discuss several applications of our Theorem 5, for which we turn to variants on the coupon collector problem. For an overview of known results, see [3].

3.1 Coupon Collector's Problem

In the coupon collector's problem, there are n different types of coupon to be collected, and we receive one random coupon at each time step. There are therefore n events of interest, corresponding to collecting each type of coupon.

We model this with n different processes, where process $i \in [n]$ is 1 while the coupon was never drawn and 0 otherwise. The probability for process i to change from 1 to 0 in any given iteration is $1/n$, so the expected time to collect any one particular type i of coupon is $\tau_i = n$. Our theorem then tells us that the expected time to collect all the coupons is at most nH_n. It is well-known that this bound is precise.

3.2 Coupon Collecting with Varying Probability of Success

Suppose the probability of collecting a coupon of type i is given by p_i, such that $p_1 \leq p_2 \leq \cdots \leq p_n$, and $\sum_{i=1}^n p_i \leq 1$ (so there may be a chance of not collecting any coupons in a turn). Then the expected time to collect a coupon of type i is $1/p_i$, and our theorem tells us that the time to collect all the coupons is at most

$$\sum_{i=1}^n \frac{1}{ip_i}.$$

Note that this gives an example where the bound from Theorem 5 can be tighter than the bound derived from Theorem 1: Consider $p_1 = 1/n^2$ while for all $i \in [n] \setminus \{1\}$ we set $p_i = 1/n$. Then Theorem 5 gives a (tight) bound of $O(n^2)$, while Theorem 1 gives $O(n^2 \log n)$.

Collecting Multiple Copies of Each Coupon

Suppose that we want to collect multiple coupons of each type. Specifically, for type $i \in [n]$ we want to collect $m_i \geq 1$ copies, with $m_1 \geq m_2 \geq \cdots \geq m_n$. The expected time to collect m_i coupons of type i is $m_i n$. So our theorem gives an upper bound for the time to collect the right number of each coupon as

$$\sum_{i=1}^{n} \frac{m_i n}{i}.$$

In the case where $m_i = m$ for all i, this gives us mnH_n. The true runtime is $n \ln n + (m-1)n \ln \ln n + O(n)$ (see [9]). Our bound is therefore somewhat disappointing, the more so as it could be simply derived just by waiting nH_n timesteps until we have collected at least 1 of each coupon, then a further nH_n to collected the second and so on. However, it does give us a very easy upper bound for the case where the m_i are unequal. We can also generalise further and consider that the probability of drawing coupon type i is p_i, to get an upper bound of

$$\sum_{i=1}^{n} \frac{m_i}{ip_i}$$

where we now index such that m_i/p_i is a decreasing sequence.

We will return to this example later, where we will use a method based on fitness levels to give a tight upper bound (see Corollary 13).

Collecting Coupons Without Replacement

Suppose now that there are c_i copies of coupon i, such that $c_1 \leq c_2 \leq \cdots \leq c_n$, and $\sum_{i=1}^{n} c_i = N$. At each time step we draw a coupon at random, and do not replace it. We seek the expected time until we have at least one coupon of each type. Now the expected time to draw coupon i is $(N+1)/(c_i+1)$. So our upper bound becomes

$$\sum_{i=1}^{n} \frac{N+1}{i(c_i+1)}.$$

Multiple Collectors

Suppose we have μ people collecting coupons at the same time. We might be interested in:

1. How long until all coupons have been collected by all people?
2. How long until each coupon has been collected by at least one person?
3. How long until each coupon has been collected by a majority of people?

It is now simple to get an upper bound for the first question. If there are n coupons, it takes one person nH_n attempts to collect them all. So the time for all μ people to collect all n coupons is bounded above by nH_nH_μ using

Theorem 5 on μ parallel process with an expected time of at most nH_n each from any starting point. However, we can get a better bound by considering events $E_{i,j}$, where person i collects coupon j. The expected time for this to happen is n, and there are μn such events. Therefore, we get an upper bound of $nH_{\mu n}$ using Theorem 5 on μn parallel process with an expected time of at most n each from any starting point.

For the second question, notice that the probability of a particular coupon being collected at least once at any given time step is

$$1 - \left(1 - \frac{1}{n}\right)^{\mu} \geq \frac{\mu}{n + \mu}$$

(using the inequality of Rowe & Sudholt [11]). So the expected time for at least one person to collect it is at most $n/\mu + 1$. Since there are n coupons, the time until they have all been collected by at least one person is at most $nH_n/\mu + H_n$ again using Theorem 5. In particular, we note that a population of size $\log n$ can collect all the coupons between them in linear time, and a population of size n can collect them in logarithmic time.

For the third question, consider the time it takes for one person to collect, say, 3/4 of the coupons (or any constant fraction greater than a half). This is, in expectation, $\Theta(n)$. Thus the time it takes for all the collectors to achieve this is $O(n \log \mu)$ using Theorem 5 on μ parallel process with an expected time of at most n. Further, we can use Chernoff's inequality to show the probability that a particular coupon has not, by this time, been collected by more than half the collectors, to be at most $\exp(-\mu/8)$. Setting $\mu = c \log n$ for some constant $c > 1$, we can the use the union bound to show that all the coupons have been collected by the majority of collectors with high probability in time $O(n \log \log n)$.

As a corollary to this last example, we have shown that applying a bitwise vote to a population of $c \log n$ independent copies of RLS on OneMax or Jump (where the gap can be any fraction of n greater than a half) finds the optimum in $O(n \log \log n)$ generations.

Generalised Collecting

Suppose we have a set S of objects, and a collection of subsets $S_1, \ldots, S_n \subseteq S$, ordered such that $|S_1| \leq |S_2| \leq \cdots \leq |S_n|$. We draw objects randomly from S (with or without replacement). The expected time until we have seen an element of each subset is no more than

$$\sum_{i=1}^{n} \frac{|S|}{i|S_i|}$$

if we are replacing objects, and no more than

$$\sum_{i=1}^{n} \frac{|S| + 1}{i(|S_i| + 1)}$$

if we are not. N.B. If the size of the subsets is small, compared to the total number of them, then this bound might not be helpful, as it could be bigger than $|S|$.

For example, suppose we have an undirected graph on n vertices, and we select random vertices, without replacement, until we find a dominating set. If the degree of vertex k is $d(k)$, and we order the vertices by degree (smallest first), then the time taken to find a dominating set is at most

$$\sum_{k=1}^{n} \frac{n+1}{k(d(k)+2)}$$

since the size of the set that dominates vertex k is $d(k)+1$. In the case of a regular graph where $d(k) = d$ for all k, this gives us $(n+1)H_n/(d+2)$.

Whether these bounds are helpful depends on the size of the vertex degrees, compared to the number of vertices.

4 Integer-Valued OneMax Using RLS

In this section we give sample applications from the analysis of randomized search heuristics. We consider the setting of finding an n-dimensional integer vector starting at the all-0 vector, where fitness is the 1-distance (sum of all distances of the components). This is known as the integer-valued OneMax problem [4]. We consider Random Local Search (RLS) as optimization algorithm which, in each iteration, changes one position chosen uniformly at random and steps a random *step size* in a random direction. It accepts that change if it improves fitness and otherwise reverts it.

Theorem 9. *Consider an integer-valued OneMax problem with target $a \in \mathbb{N}^n$, starting at 0^n. Using RLS, let the step size be S, where*

$$\Pr(S = s) = \frac{1}{\zeta(3/2)s^{3/2}}.$$

Note that $\zeta(3/2) = \sum_{z=1}^{\infty} z^{-3/2}$. The time to hit the target is $O(n\sqrt{|a|_\infty} \log n)$.

In [4], some other step sizes were analyzed, for example constantly usgin $s = 1$ or very heavy-tailed distributions. Before proving this theorem, we show the following lemma.

Lemma 10. *For any $n \in \mathbb{N}$ it holds that*

$$\sqrt{n} \leq \sum_{k=1}^{n} \frac{1}{\sqrt{k}} \leq 2\sqrt{n}.$$

Proof. This follows from Riemann sum approximations. □

Now we show Theorem 9.

Proof. Suppose we are trying to solve an integer-valued OneMax problem on n variables, with target $(a_1, \ldots, a_n)$, and starting at 0^n. We use a random step size S chosen according to the distribution

$$\Pr(S = s) = \frac{1}{cs^{3/2}}$$

where $c = \zeta(3/2) \approx 2.612$. Consider a particular variable k, and suppose it is currently at distance d from the target a_k. The expected jump size is at least

$$\sum_{s=1}^{d} \frac{s}{2cns^{3/2}} = \frac{1}{2cn} \sum_{s=1}^{d} \frac{1}{s^{1/2}} \geq \frac{\sqrt{d}}{2cn}$$

using the lemma. By the variable drift theorem [6,8], the expected time to reach the target a_k is bounded above by

$$\sum_{d=1}^{|a_k|} \frac{2cn}{\sqrt{d}} \leq 4cn\sqrt{|a_k|},$$

where we again use the lemma. If, without loss of generality, we assume the variables are indexed such that $|a_1| \geq |a_2| \geq \cdots \geq |a_n|$, then we can use Theorem 5 to bound the overall runtime to discover the target vector by

$$4cn \sum_{k=1}^{n} \frac{\sqrt{|a_k|}}{k} = O(n\sqrt{|a|_\infty} \log n).$$

□

Corollary 11. *The runtime of RLS on the integer-valued OneMax problem is* $O(n\sqrt{|a|_1})$.

Proof. Applying the Cauchy-Schwarz inequality as follows:

$$\sum_{k=1}^{n} \frac{\sqrt{|a_k|}}{k} \leq \sqrt{\sum_{k=1}^{n} |a_k|} \sqrt{\sum_{k=1}^{n} \frac{1}{k^2}} \leq \frac{\pi}{\sqrt{6}} \sqrt{|a|_1}$$

gives a runtime bound of $O(n\sqrt{|a|_1})$. □

5 Fitness Levels

The so-called fitness level method was introduced in [12] as one of the earliest tools to perform a run time analysis for evolutionary algorithms. Abstractly speaking, the method considers a monotone non-increasing process on $\{0, \ldots, n\}$ and bounds the time to reach 0 by the sum of the inverses of the probability to leave a state $s \in \{0, \ldots, n\}$. We extend this idea to multiple such processes in the following theorem.

Theorem 12. *Consider n processes, not necessarily independent, executing in parallel. For each process, the set of states is $\{0, 1, \ldots, m\}$. The initial state is m. The final state is 0. For each $k > 0$, let $p(k)$ be a lower bound on the probability that a process leaves state k and moves to a smaller state, otherwise it stays in state k (N.B. the lower bound must hold regardless of the states of the other processes). Let $p^* = \min_{k>0} p(k)$. For all $w > 1$, the expected time until all n processes have reached state 0 is bounded above by*

$$E[T] \leq \left(1 + \frac{1}{w-1}\right)\left(\frac{1+\ln n}{p^*} + \sum_{j=2}^{m} \frac{1}{p(j)} \ln w\right).$$

Proof. Let $x \in \{0, 1, \ldots, m\}^n$ denote the states of the whole system, and let $X(t)$ be the (random) state at time t. We will define a potential function of the form $g\colon \{0, 1, \ldots, m\}^n \to \mathbb{R}, x \mapsto \sum_{i=1}^{n} [x_i > 0] f(x_i)$ where $f\colon \mathbb{R} \to \mathbb{R}$ is a monotonic increasing function to be determined. For $x \in \{0, 1, \ldots, m\}^n$ and all $i \in [n]$ we define A to be the event that $X(t+1)_i < x_i$ and compute

$$\begin{aligned}
&E[[X(t+1)_i > 0] f(X(t+1)_i) \mid X(t)_i = x_i, A] \\
&= \sum_{k=0}^{x_i - 1} E[[k > 0] f(k) \mid X(t) = x, X(t+1)_i = k] \Pr(X(t+1)_i = k \mid X(t) = x_i, A) \\
&= \sum_{k=0}^{x_i - 1} [k > 0] f(k) \Pr(X(t+1)_i = k \mid X(t) = x_i, A) \\
&\leq \sum_{k=0}^{x_i - 1} [x_i - 1 > 0] f(x_i - 1) \Pr(X(t+1)_i = k \mid X(t) = x_i, A) \\
&= [x_i - 1 > 0] f(x_i - 1) \sum_{k=0}^{x_i - 1} \Pr(X(t+1)_i = k \mid X(t) = x_i, A) \\
&= [x_i - 1 > 0]\; f(x_i - 1) \leq [x_i > 0]\; f(x_i - 1).
\end{aligned}$$

Note that $\Pr[A] \geq p(x_i)$. We now compute the drift for a single component by employing the law of total expectation on the event A (where the term in the expectation for the opposite event has a value of 0):

$$\begin{aligned}
&E[[x_i > 0] f(x_i) - [X(t+1)_i > 0] f(X(t+1)_i) \mid X(t) = x] \\
&\quad = \Pr[A]\, E\left[[x_i > 0] f(x_i) - [X(t+1)_i > 0] f(X(t+1)_i) \mid X(t) = x, A\right] \\
&\quad \geq p(x_i) \left([x_i > 0] f(x_i) - [x_i > 0] f(x_i - 1)\right) \\
&\quad = p(x_i) \left(1 - \frac{f(x_i - 1)}{f(x_i)}\right) [x_i > 0]\; f(x_i).
\end{aligned}$$

Using this to compute drift, we have

$$\begin{aligned}
&E[g(X(t)) - g(X(t+1)) \mid X(t) = x] \\
&\quad = E\left[\sum_{i=1}^{n} [x_i > 0] f(x_i) - [X(t+1)_i > 0] f(X(t+1)_i) \mid X(t) = x\right] \\
&\quad \geq \sum_{i=1}^{n} p(x_i) \left(1 - \frac{f(x_i - 1)}{f(x_i)}\right) [x_i > 0]\, f(x_i).
\end{aligned}$$

Let $0 < \alpha < 1$, and let $p^* = \min_{k>0} p(k)$. We define

$$f(k) = \prod_{j=1}^{k} \frac{1}{1 - \alpha p^*/p(j)}$$

so that

$$\frac{f(k-1)}{f(k)} = 1 - \frac{\alpha p^*}{p(k)}$$

giving us

$$E[g(X(t)) - g(X(t+1)) \mid X(t) = x] \geq \alpha p^* \sum_{i=1}^{n} [x_i > 0] f(x_i) = \alpha p^* g(x).$$

The maximum value of $g(x)$ is

$$g(m, \ldots, m) = \sum_{i=1}^{n} \prod_{j=1}^{m} \frac{1}{1 - \alpha p^*/p(j)} = \frac{n}{\prod_{j=1}^{m} 1 - \alpha p^*/p(j)}$$

and the minimum value for $x \neq 0$ is

$$\frac{1}{1 - \alpha p^*/p(1)}.$$

So by the multiplicative drift theorem [2], the time, T, taken to reach state $x = 0$ has expectation bounded above by

$$\begin{aligned}
E[T] &\leq \frac{1}{\alpha p^*} \left(1 + \ln\left(\frac{n}{\prod_{j=2}^{m} 1 - \alpha p^*/p(j)}\right)\right) \\
&= \frac{1}{\alpha p^*} \left(1 + \ln n + \sum_{j=2}^{m} \ln\left(\frac{1}{1 - \alpha p^*/p(j)}\right)\right).
\end{aligned}$$

Now write $\alpha = 1 - 1/w$ for some $w > 1$. Then

$$1 - \frac{\alpha p^*}{p(j)} = \frac{1}{w}\left(\left(1 - \frac{p^*}{p(j)}\right) w + \frac{p^*}{p(j)}\right)$$

so that

$$\ln\left(\frac{1}{1-\alpha p^*/p(j)}\right) = \ln w - \ln\left(\left(1-\frac{p^*}{p(j)}\right)w + \frac{p^*}{p(j)}\right).$$

The term inside the second logarithm is a convex combination of w and 1, and so, since ln is a concave function, we have, by Jensen's inequality:

$$\ln\left(\left(1-\frac{p^*}{p(j)}\right)w + \frac{p^*}{p(j)}\right) \geq \left(1-\frac{p^*}{p(j)}\right)\ln w$$

which means

$$\begin{aligned} E[T] &\leq \frac{1}{(1-1/w)p^*}\left(1+\ln n + \sum_{j=2}^{m}\left(\ln w - \left(1-\frac{p^*}{p(j)}\right)\ln w\right)\right) \\ &= \frac{1}{(1-1/w)p^*}\left(1+\ln n + \sum_{j=2}^{m}\frac{p^*}{p(j)}\ln w\right) \\ &= \left(1+\frac{1}{w-1}\right)\left(\frac{1+\ln n}{p^*} + \sum_{j=2}^{m}\frac{1}{p(j)}\ln w\right). \end{aligned}$$

□

The best choice for w depends on the exact relationship between the number of processes n, the number of levels m, and the transition probabilities.

As a corollary, we can find the asymptotically tight bound for collecting m copies of each of n coupons [9].

Corollary 13. *Suppose we want to collect m sets each of n different coupons, where m is a constant. Let T be the first time such that all coupons are collected. Then we have*

$$E[T] \leq n\ln(n) + (m-1)n\ln(\ln(n)) + O(n).$$

Proof. Take $p(j) = 1/n$ for all $j \in [m]$ and $w = \ln n$. □

We finish with an application to parallel evolutionary algorithms.

Corollary 14. *Let there be a population of μ parallel copies of the $(1+1)EA$, each of which is solving OneMax, with bitstrings of length n. The expected time for them all to find the optimum is bounded above by $2en(1+\ln(n\mu))$.*

Proof. There are μ processes, and n levels. For each $j \in [n]$ the escape probability is bounded below by j/en. We take $w = 2$. □

References

1. Aishwaryaprajna, Rowe, J.E.: Evolutionary anytime algorithms. In: Krejca, M.S., Wagner, M. (eds.) Proceedings of the EvoCOP 2025. LNCS, vol. 15610, pp. 18–32. Springer (2025). https://doi.org/10.1007/978-3-031-86849-8_2
2. Doerr, B., Johannsen, D., Winzen, C.: Multiplicative drift analysis. Algorithmica **64**(4), 673–697 (2012). https://doi.org/10.1007/S00453-012-9622-X
3. Doumas, Aristides V., Papanicolaou, Vassilis G.: The coupon collector's problem revisited: generalizing the double dixie cup problem of Newman and Shepp. ESAIM: Probab. Stat. **20**, 367–399 (2016). https://doi.org/10.1051/ps/2016016
4. Harder, J.G., Kötzing, T., Li, X., Radhakrishnan, A., Ruff, J.: Run time bounds for integer-valued OneMax functions. In: Li, X., Handl, J. (eds.) Proceedings of the GECCO 2024. ACM (2024). https://doi.org/10.1145/3638529.3654091
5. Janett, D., Lengler, J.: Two-dimensional drift analysis: optimizing two functions simultaneously can be hard. Theoret. Comput. Sci. **971**, 114072 (2023). https://doi.org/10.1016/J.TCS.2023.114072
6. Johannsen, D.: Random Combinatorial Structures and Randomized Search Heuristics. Ph.D. thesis, Universität des Saarlandes (2010). https://doi.org/10.22028/D291-26016
7. Lehre, P.K., Sudholt, D.: Parallel black-box complexity with tail bounds. IEEE Trans. Evol. Comput. **24**(6), 1010–1024 (2020). https://doi.org/10.1109/TEVC.2019.2954234
8. Mitavskiy, B., Rowe, J.E., Cannings, C.: Theoretical analysis of local search strategies to optimize network communication subject to preserving the total number of links. Int. J. Intell. Comput. Cybern. **2**(2), 243–284 (2009). https://doi.org/10.1108/17563780910959893
9. Newman, D.J., Shepp, L.: The double dixie cup problem. Am. Math. Mon. **67**(1), 58–61 (1960). https://doi.org/10.2307/2308930
10. Rowe, J.E.: Linear multi-objective drift analysis. Theoret. Comput. Sci. **736**, 25–40 (2018). https://doi.org/10.1016/J.TCS.2018.02.011
11. Rowe, J.E., Sudholt, D.: The choice of the offspring population size in the (1, λ) evolutionary algorithm. Theoret. Comput. Sci. **545**, 20–38 (2014). https://doi.org/10.1016/J.TCS.2013.09.036
12. Wegener, I.: Theoretical aspects of evolutionary algorithms. In: Orejas, F., Spirakis, P.G., van Leeuwen, J. (eds.) Proceedings of the ICALP 2001. LNCS, vol. 2076, pp. 64–78. Springer (2001). https://doi.org/10.1007/3-540-48224-5_6
13. Wormald, N.C.: The differential equation method for random graph processes and greedy algorithms. Lectures on Approximation and Randomized Algorithms, pp. 73–155 (1999). https://doi.org/10.1214/aoap/1177004612

Improved Runtime Bound for the $(\mu + 1)$-EA on BinVal

Joris Belder, Johannes Lengler(✉), and Raghu Raman Ravi

ETH Zürich, Zürich, Switzerland
jbelder@student.ethz.ch, {johannes.lengler,raghu.ravi}@inf.ethz.ch

Abstract. We study the $(\mu+1)$ EA on the Binary Value function BinVal. We show that it needs at most $O(\mu \log \mu \cdot n \log n)$ function evaluations to find the optimum when $\mu = o(n/\log n)$. This substantially improves upon the recent upper bound of $O(\mu^5 n \log(n/\mu^4))$ by Krejca, Neumann and Witt. Our results hold for several mutation operators including standard bit mutation. In particular, our bound implies that the $(\mu+1)$ EA is at most a factor $O(\log \mu \cdot \log n)$ slower on BinVal than on OneMax.

Keywords: Evolutionary Algorithms · Population · Binary Value Function · Drift Analysis · Runtime Analysis

1 Introduction

The runtime analysis of evolutionary algorithms (EAs) studies the expected number of fitness evaluations until an optimal solution is first produced. Over the past two decades, this approach has yielded precise bounds for many algorithm–benchmark combinations and has clarified how design parameters such as the mutation rate or the population size affect performance; see the surveys [13,21] for an overview.

A classical result is that the $(1 + 1)$ EA with mutation rate $1/n$ optimizes any linear function $f(x) = \sum_{i\in[n]} w_i x_i$ with positive weights w_i in expected time $O(n \log n)$ [9]. Witt [24] later tightened this to $en \ln n + O(n)$. Subsequent work provided alternative proofs and extensions via multiplicative drift, adaptive drift, and related techniques [5,7,18]. While single-individual algorithms on linear functions are thus well understood, practical EAs typically maintain *populations* of candidate solutions.

The $(\mu + 1)$ EA, which keeps a population of μ individuals and in each generation creates one offspring by mutating a uniformly chosen parent, is a prototypical population-based elitist algorithm. Understanding the impact of the population size μ on the optimization time—the study of *population dynamics*—remains one of the most challenging aspects of EA theory [12,21]. The first rigorous runtime analysis of the $(\mu + 1)$ EA was carried out by Witt [22], who showed that the expected runtime on OneMax is $O(\mu n + n \log n)$, a bound later

G. Iacca et al. (Eds.): PPSN 2026, LNCS 16985, pp. 67–82, 2027.
https://doi.org/10.1007/978-3-032-36214-8_5

confirmed to be tight [1]. These results show that for OneMax, the population adds an overhead that is at most linear in μ. However, the effect of population size can be more dramatic in general: Witt [23] demonstrated that there exist functions where increasing μ by a constant factor reduces the runtime from exponential to polynomial, while conversely, Lengler and Zou [19] proved that larger populations can cause superpolynomial slowdowns on certain monotone functions.

The BinVal function, defined as $\text{BinVal}(x) = \sum_{i\in[n]} 2^{n-i}x_i$, assigns exponentially decreasing weights to the bit positions. Because an offspring is accepted if and only if it improves the most significant bit at which it differs from the current worst individual, improvements propagate from more significant to less significant positions. Especially with high mutation rates, fixing a significant bit can cause many detrimental mutations of less significant bits. In a population setting, even if all individuals have set a certain bit correctly, this fixation can be destroyed when a more significant bit is improved. Since this effect is most extreme for BinVal among all linear functions, it is a particularly interesting benchmark for studying population effects. While all linear functions have the same asymptotic runtime $O(n \log n)$ for the $(1+1)$ EA, this is no longer the case for population-based algorithms: Doerr and Künnemann [8] showed that for the $(1+\lambda)$ EA, different linear functions can have different asymptotic runtimes, with BinVal being an asymptotic worst case. The BinVal function has also received attention in the dynamic setting [15,16] and in anytime and fixed-target analyses [11], further establishing it as a key benchmark for population effects on weighted linear functions.

For the $(\mu+1)$ EA on BinVal, generic approaches such as fitness levels and drift-based analyses lead to bounds with an additive $\Theta(n^2)$ term, failing to recover the $O(n \log n)$ scaling even for constant μ; see the discussion in [12]. This was resolved recently by Krejca, Neumann, and Witt [12], who introduced a block-based analysis of population diversity within short contiguous blocks of bits. Their approach yields a runtime bound of $O(\mu^5 n \log(n/\mu^4))$ for standard bit mutation, matching $O(n \log n)$ for constant μ and eliminating the quadratic additive term. However, the fifth-degree polynomial dependence on μ appears to be an artifact of the proof technique—the experimental results in [12] suggest that the theoretical μ^5 factor is too pessimistic. For a more comprehensive account of the related literature, we refer the reader to [12] and the references therein, as well as the surveys [13,21].

1.1 Our Contribution

In this work, we substantially improve the upper bound on the expected runtime of the $(\mu+1)$ EA on BinVal. Recall that the $(\mu+1)$ EA maintains a population of μ individuals; in each iteration it selects a parent uniformly at random, creates an offspring by mutation, and retains the best μ individuals (see Sects. 2 and 3.3 for the precise definitions). We show runtime results for a generic class of mutation operators, which we call *conservative operators*. Those include in particular *0/1-bit mutation*, which flips exactly one uniformly chosen bit with probability $1 - p$ and leaves the string unchanged with probability p for a constant $p \in (0, 1)$; and *standard bit mutation*, which flips each bit independently with probability χ/n for any $\chi > 0$. Then our main result is the following.

Theorem 1. *Consider the $(\mu+1)$ EA on the* BinVal *function of dimension n, with population size $\mu \geq 2$. Under any conservative mutation operator or specifically either 0/1-bit mutation with constant cloning probability $p \in (0, 1)$, or standard bit mutation with rate χ/n for constant $\chi > 0$, the expected runtime is*

$$O(\mu \log \mu \cdot n \log n) .$$

More precisely, the expected runtime is $O(\mu \log \mu \cdot n \max\{1, \log(n/(\mu \log \mu))\})$, which is a stronger bound when μ grows with n. This reduces the dependence on the population size from μ^5 in the bound of [12] to $\mu \log \mu$, while also substantially relaxing the restriction on the range of admissible μ: the result of [12] requires $\mu = o((n/\log n)^{1/5})$, while our result improves the previous best bounds for any $\mu = o(n/\log n)$. Our upper bound also closes the gap to the known $\Omega(\mu n + n \log n)$ lower bound [22] up to logarithmic factors. For constant μ, the bound recovers the $O(n \log n)$ runtime of the $(1+1)$ EA on linear functions [24]. Finally, while [12] only considered mutation rate $1/n$ for standard bit mutation, our result extends to any mutation rate χ/n with $\chi = O(\sqrt{n}) \cap \Omega(1)$ at the cost of a factor e^{χ}.

Our improvement stems from a more refined analysis of the *interference probability* during the spreading of a beneficial mutation through the population. In the framework of [12], the bit string is partitioned into contiguous blocks, and within each block, mutations spread from a single copy to all μ individuals. During this spreading process, an interfering mutation at another position in the same block can disrupt this fixation. While [12] bounds this interference probability uniformly by the block size, we observe that it depends on the number k of remaining unfixed zeros in the current block. When few zeros remain, interference is less likely and mutations spread more easily. This k-dependent analysis, combined with Markov chain arguments for modeling population fixation and additive drift analysis for bounding macro-step progress, yields the improved bound for mutation rate at most $1/n$.

In order to extend the result to non-constant mutation rates, we show that every run of the $(\mu+1)$ EA with mutation rate χ/n on BinVal with $\chi > 1$ can be decomposed into a sequence of $\lceil \chi \rceil$ optimization processes of BinVal with smaller dimension and smaller mutation parameter $\chi' \leq 1$. Although a rather basic observation, we believe that this decomposition is of general interest, as it generalizes far beyond the $(\mu + 1)$-EA.

Finally, we present experimental results for the $(\mu + 1)$-EA on BinVal for both standard bit mutation and 0/1 mutation to provide empirical validation and to investigate the tightness of the analysis.

2 Preliminaries

Notation and Problem Setting. The (μ+1)-EA maintains a population P_t of size μ at every time step t for some $\mu \in \mathbb{N}$. The objective is to optimise a fitness function $f\colon \{0,1\}^n \to \mathbb{R}$. In each generation, a uniformly random individual x is chosen from the population, and an offspring y is produced from x through mutation. The offspring y is then added to the population, and the individual with the worst fitness (possibly y itself) is discarded. In this paper, we study the $(\mu+1)$-EA for $\mu \geq 2$ on the BinVal fitness function, defined for $x \in \{0,1\}^n$ as

$$\textsc{BinVal}(x) = \sum\nolimits_{i=1}^{n} 2^{n-i} \cdot x_i.$$

Because BinVal is injective, every distinct bit string yields a unique fitness value, ensuring that the removal of the worst individual is always unambiguous.

We analyse a class of *conservative mutation operators* that is defined in Sect. 3.1. It includes in particular 0/1 *mutation*, in which the offspring y is an exact clone of x with probability $p \in (0,1)$, and otherwise, a single bit of x is chosen uniformly at random and flipped to produce y. And it includes *standard bit mutation*, in which every bit of x is flipped independently with probability χ/n to produce y, for some *mutation parameter* $0 < \chi \leq 1$, which we later extend to $\chi > 1$ as well.

In our analysis, we partition each bit string into $\nu := n/n'$ blocks B_i of length n', for $i \in [\nu]$. We denote by $B_i(x)$ the substring corresponding to the i-th block of bits in x, specifically $B_i(x) := x_{(i-1)n'+1} \ldots x_{i \cdot n'}$. We call a position j *fixated* at time t if $\forall x \in P_t\colon x_j = 1$. Note that once a leading segment is fixated for BinVal, it remains fixated forever. We are interested in the expected time it takes for all positions in the current block to become fixated, conditional on finishing the previous block. Formally, we define $S_0 = 0$ and let S_i be the first time step where all bits in block B_i are fixated across the entire population:

$$S_i := \inf\{t \in \mathbb{N} \mid t \geq S_{i-1} \wedge \forall x \in P_t\colon B_i(x) = 1^{n'}\}.$$

We are interested in $\mathbb{E}[T_i]$, where $T_i := S_i - S_{i-1}$. For convenience, at time step t, we denote by B the *current block* B_i, where $i \in [\nu]$ such that $S_{i-1} \leq t < S_i$. The total time T for the process to produce an optimal individual is given by $T = \sum_{i=1}^{\nu} T_i$, and by linearity of expectation, $\mathbb{E}[T] = \sum_{i=1}^{\nu} \mathbb{E}[T_i]$.

At time t, we say the process is in a *flip phase* if $\forall x, y \in P_t\colon B(x) = B(y)$; otherwise, we say it is in a *spread phase.* Note that the process may only switch from a flip phase to a spread phase when there is at least one $0 \rightsquigarrow 1$ mutation in the current block B. Let q denote the position of the most significant non-fixated 1 in the population. We refer to position q as the *current mutation* with respect to a spread phase. During a spread phase, two failure events may occur:

- *Interference:* A $0 \rightsquigarrow 1$ mutation is accepted in an unfixated position $b \in B$ which is more significant than q, that is, $b < q$.
- *Destruction:* A previously fixated position becomes unfixated (flip $1 \rightsquigarrow 0$).

Our convention is that upon both interference and destruction a new spread phase starts. If, during a spread phase, position q becomes fixated, the process either switches back to a flip phase or, if there remains another non-fixated 1 in the population, a new spread phase starts. We refer to a *macro step* as the period from the end of a spread phase until the end of the subsequent spread phase. Hence, a macro step may consist of a flip phase and a spread phase, or possibly only of a spread phase. Intuitively, a macro step represents one continuous attempt to fixate a bit in the current block, where the attempt may fail due to interference or destruction.

Note 2. In the proof, we will later choose a block length of $n' = cn/(\mu \ln \mu)$, where c is a constant depending on the mutation operator. We will see that we may restrict to the case $\mu = o(n/\log n)$ in our proofs, which implies $n' \geq 1$ for sufficiently large n. For the sake of exposition, we will assume that n' is an integer. Otherwise, we may round down to the next integer. Similarly, it may be the case that n' does not divide n, forcing the final block to be shorter, and for the sake of exposition we assume that n' divides n.

Tools. In our analysis we will employ the following additive drift theorem: [10]

Theorem 3. *Let $(X_t)_{t \geq 0}$ be a sequence of non-negative random variables with a finite state space $\mathcal{S} \subseteq \mathbb{R}_{\geq 0}$ such that $0 \in \mathcal{S}$. Let $T := \inf\{t \geq 0 \mid X_t = 0\}$. If there exists $\delta > 0$ such that for all $s \in \mathcal{S} \setminus \{0\}$ and for all $t \geq 0$*

$$\Delta_t(s) := \mathbb{E}[X_t - X_{t+1} \mid X_t = s] \geq \delta,$$

then $\mathbb{E}[T] \leq \mathbb{E}[X_0]/\delta$.

We will also use the following basic observation.

Lemma 4. *Assume $\mu = o(n/\log n)$. Then $\mu \log \mu = o(n)$.*

Proof. For sufficiently large n we have $\mu \leq n$ and thus $\log \mu \leq \log n$. Hence, $\mu \log \mu = o(n/\log n \cdot \log n) = o(n)$. □

3 Proofs

3.1 Conservative Mutation Operators

In this section we will introduce the class of conservative mutation operators. Then we will show the main theorem, the runtime bound for the $(\mu + 1)$-EA with arbitrary conservative mutation operator.

Definition 5 (Conservative Mutation Operator). *A mutation operator is said to be* conservative *if there exist constants $c_1, c_2 > 0$ such that for all disjoint sets of bits $S, S' \subseteq \{1, \ldots, n\}$, the following conditions are satisfied:*

- *The probability of producing an exact clone, p_{clone}, is $\Omega(1)$.*

- *The conditional probability* p_{safe} *that in a single iteration only bits in* S *are flipped given that some bit in* S *is flipped, is* $\Omega(1)$*.*
- *The probability* $p_{flip,safe}$ *of flipping at least one bit in* S *and no bit in* $[n] \setminus S$ *in a single iteration satisfies* $p_{flip,safe} \geq c_1|S|/n$*.*
- *The expected number of bits in* S *that are flipped in a single iteration is at most* $\mathbb{E}[destroy] \leq c_2|S|/n$*. The bound remains true if we condition on flipping at least one bit in* S' *in that iteration.*

Given these conditions, we can systematically bound the expected optimization time. We first analyze the probability of a mutation successfully taking over the population.

Lemma 6 (Spread Probability). *During a spread phase, the probability that the current mutation successfully fixates across the population before a failure event occurs is at least* $(e\mu)^{-C}$*, where* $C = c_2 n'\mu/(np_{clone})$*.*

Proof We model the spread of the current mutation as a Markov Chain with transient states $i \in \{1, 2, \dots, \mu - 1\}$, a target absorbing state μ (successful fixation), and a failure absorbing state F. State i represents i individuals currently holding the mutation. The transition probabilities are bounded by:

- **Progress:** $p_{i,i+1} \geq p_{\text{clone}} \cdot i/\mu$ (the probability of cloning an individual with the current mutation).
- **Failure:** $p_{i,F} \leq c_2 n'/n$ (since the probability of flipping a bit in the current block is upper bounded by $\mathbb{E}[\text{destroy}]$ by Markov's inequality).
- **Staying Put:** $p_{i,i} = 1 - p_{i,i+1} - p_{i,F}$.

Let s_i be the probability of reaching state μ from state i, with $s_\mu = 1$ and $s_F = 0$. By the law of total probability,

$$s_i = p_{i,i+1}s_{i+1} + (1 - p_{i,i+1} - p_{i,F})s_i.$$

Rearranging yields $s_i = \frac{p_{i,i+1}}{p_{i,i+1}+p_{i,F}} s_{i+1}$. Clearly, this relation implies that s_1 is smallest of the probabilities. Thus, it suffices to bound s_1. We first expand s_1 via

$$s_1 = \prod_{i=1}^{\mu-1} \frac{p_{i,i+1}}{p_{i,i+1} + p_{i,F}} = \prod_{i=1}^{\mu-1} \frac{1}{1 + \frac{p_{i,F}}{p_{i,i+1}}}.$$

Using the bound $\frac{p_{i,F}}{p_{i,i+1}} \leq \frac{c_2 n'\mu}{np_{\text{clone}}i} = \frac{C}{i}$ and the standard inequality $\frac{1}{1+x} \geq e^{-x}$,

$$s_1 \geq \prod_{i=1}^{\mu-1} e^{-C/i} = e^{-C\sum_{i=1}^{\mu-1} 1/i} = e^{-CH_{\mu-1}}.$$

As $H_{\mu-1} \leq \ln\mu + 1$, the probability of fixating the current mutation is at least

$$\mathbb{P}[\text{fixate}] \geq e^{-C(\ln\mu+1)} = e^{-C} \cdot \mu^{-C} = (e\mu)^{-C}.$$

□

Lemma 7 (Phase Durations). *Let k be the number of non-fixated bits in the current block. The expected duration of a flip phase is $\mathbb{E}[T^k_{flip}] \le n/(c_1 k)$, and the expected duration of a spread phase is $\mathbb{E}[T_{spread}] \le \mu(\ln\mu + 1)/p_{clone}$.*

Proof Taking S to be the set of k non-fixated bits in the current block, $\mathbb{E}[T^k_{\text{flip}}]$ is stochastically dominated by a geometric distribution with success probability at least $c_1 k/n$, yielding $\mathbb{E}[T^k_{\text{flip}}] \le n/(c_1 k)$.

To bound the expected time of a spread phase, we assume the worst-case scenario where no failure events truncate the phase, relying purely on the progress probability $p_{i,i+1}$:

$$\mathbb{E}[T_{\text{spread}}] \le \sum_{i=1}^{\mu-1} \frac{1}{p_{i,i+1}} \le \sum_{i=1}^{\mu-1} \frac{\mu}{p_{\text{clone}} \cdot i} = \frac{\mu}{p_{\text{clone}}} H_{\mu-1} \le \frac{\mu(\ln\mu + 1)}{p_{\text{clone}}}.$$

□

We now evaluate the process in terms of macro steps. Let X_m denote the number of non-fixated bits in the current block at macro step m. To analyze the time it takes to reduce the non-fixated bits from k to $k-1$, we shift X_m to define a zero-state: $X^{k-1}_m = X_m - (k-1)$.

Lemma 8 (Macro Step Drift). *Assume $\mu = o(n/\log n)$. For a suitable constant $c > 0$ and block size $n' := cnp_{clone}/(c_2\mu\ln\mu)$, the expected drift $\Delta^k_m(s) := \mathbb{E}[X^{k-1}_m - X^{k-1}_{m+1} \mid X^{k-1}_m = s - (k-1)]$ is $\Delta^k_m(s) = \Omega(1)$.*

Proof The drift depends on the probability $P_{s,s-1}$ of successfully fixating a bit, and $P_{s,s+i}$ of unfixating i previously fixated bits:

$$\Delta^k_m(s) = P_{s,s-1} - \sum_{i=1}^{n'-s} i \cdot P_{s,s+i}.$$

We establish bounds for these terms:

- **Progress:** If the macro step starts in a flip phase, then consider the set S of unfixated bits in B. When one of them is flipped, the probability of not flipping any bit outside of S is at least p_{safe}. In this case a spread phase is entered, and the probability of successfully fixating the current mutation is at least $(e\mu)^{-C}$ by Lemma 6. Hence, $P_{s,s-1} \ge (e\mu)^{-C} \cdot p_{\text{safe}}$.
- **Destruction:** A macro step consists of at most one flip phase and one spread phase, and destruction can only occur in the spread phase or the final iteration of a flip phase. In the latter, if we define S' to be the set of all non-fixated positions in B then we condition on at least one bit flip in S'. The same bound on the expected number of bit flips applies to both cases:

$$\sum_{i=1}^{n'-s} i \cdot P_{s,s+i} \le \mathbb{E}[\text{destroy}] \cdot (\mathbb{E}[T_{\text{spread}}] + 1) \le c_2 \frac{n'}{n} \cdot \left(\frac{\mu(\ln\mu + 1)}{p_{\text{clone}}} + 1 \right).$$

Combining these bounds yields

$$\Delta_m^k(s) \geq (e\mu)^{-C} \cdot p_{\text{safe}} - c_2 \frac{n'}{n} \cdot \left(\frac{\mu(\ln \mu + 1)}{p_{\text{clone}}} + 1 \right).$$

Note that $n' \geq 1$ by Lemma 4, and recall from Lemma 6 that $C = \frac{c_2 n' \mu}{n p_{\text{clone}}} = \frac{c}{\ln \mu}$. Substituting this and n' yields

$$\begin{aligned}
\Delta_m^k(s) &\geq e^{-c(1+1/\ln \mu)} \cdot p_{\text{safe}} - c \cdot \frac{p_{\text{clone}}}{\mu \ln \mu} \cdot \left(\frac{\mu(\ln \mu + 1)}{p_{\text{clone}}} + 1 \right) \\
&\geq p_{\text{safe}} \left(1 - c \left(1 + \frac{1}{\ln \mu} \right) \right) - c \cdot \frac{(\ln \mu + 1)}{\ln \mu} - c \cdot \frac{p_{\text{clone}}}{\mu \ln \mu}.
\end{aligned}$$

The right hand side is of the form $p_{\text{safe}} - c \cdot O(1)$. Thus if the constant $c > 0$ is sufficiently small then the drift is $\Omega(1)$. □

Theorem 9. *The expected optimization time of the* $(\mu+1)$*-EA on* BinVal *with a conservative mutation operator is* $O(\mu \log \mu \cdot n \log n)$ *for* $\mu = o(n/\log n)$*, and* $O(\mu \log \mu \cdot n)$ *for* $\mu = \Omega(n/\log n)$*.*

Proof. First consider $\mu = o(n/\log n)$. Let $\mathbb{E}[T_i]$ be the expected time to fixate block i. Let $T_{i,k}$ be the first point in time when there are k fixated positions in block i. By Lemma 8 and the additive drift theorem, Theorem 3, we need in expectation $O(1)$ macro steps to go from time $T_{i,k}$ to time $T_{i,k-1}$. Note that the number of non-fixated positions is not necessarily decreasing, but the additive drift theorem also applies to non-monotone processes. While $T_{i,k-1}$ is not reached, each flip phase has k' non-fixated bits for some $k' \geq k$. Hence, by Lemma 7 its expected duration is $\mathbb{E}[T_{\text{flip}}^{k'}] \leq n/(c_1 k') \leq n/(c_1 k)$. Together with the bound $\mathbb{E}[T_{\text{spread}}] \leq \mu(\ln \mu + 1)/p_{\text{clone}}$ for a spread phase, and since each macro step consists of at most one flip phase and one spread phase, we obtain

$$\begin{aligned}
\mathbb{E}[T_i] &= \sum_{k=1}^{n'} \mathbb{E}[T_{i,k-1} - T_{i,k}] \leq O(1) \cdot \sum_{k=1}^{n'} \left(\frac{n}{c_1 k} + \frac{\mu(\ln \mu + 1)}{p_{\text{clone}}} \right) \\
&\leq O(1) \cdot \left(\frac{n}{c_1} (\ln n' + 1) + n' \cdot \frac{\mu(\ln \mu + 1)}{p_{\text{clone}}} \right).
\end{aligned}$$

The total time is bounded by $\mathbb{E}[T] = \sum_{i=1}^{\nu} \mathbb{E}[T_i] = \frac{n}{n'} \mathbb{E}[T_i]$. This gives

$$\mathbb{E}[T] \leq O(1) \cdot \left(\frac{n^2}{c_1 n'} (\ln n' + 1) + \frac{n\mu(\ln \mu + 1)}{p_{\text{clone}}} \right).$$

Now we substitute our chosen block size $n' = \Theta(\frac{n}{\mu \log \mu})$ into the first term:

$$\mathbb{E}[T] = O(1) \cdot \left(\frac{c_2 n \mu \ln \mu}{c_1 c p_{\text{clone}}} (\ln n' + 1) + \frac{n\mu(\ln \mu + 1)}{p_{\text{clone}}} \right) = O(\mu \log \mu \cdot n \log n').$$

Since $n' \leq n$, we obtain the final bound $\mathbb{E}[T] = O(\mu \log \mu \cdot n \log n)$.

For the case $\mu = \Omega(n/\log n)$, as already noted in [12], we can use blocks of length $n' = 1$, in which case no interference or destruction can occur. We then easily obtain a bound of $O(n\mu \log \mu + n^2)$, since each flip phase takes time n and each spread phase takes time $O(\mu \log \mu)$. The term n^2 can be omitted because $\mu \log \mu = \Omega(n)$. □

Note 10. The proof implies that the bound in Theorem 9 also holds for the time until *all* individuals in the population are at the optimum.

3.2 0/1 Mutation

The 0/1 mutation operator is defined as the stochastic choice between identity and a single-bit flip. Specifically, it produces a clone of the parent with a constant probability $p \in (0, 1)$, and performs an RLS mutation, flipping exactly one uniformly chosen bit, with probability $1 - p$.

Corollary 11. *The* 0/1 *mutation operator is a conservative mutation operator. Consequently, the expected optimization time of the* $(\mu+1)$*-EA on* BinVal *using* 0/1 *mutation is* $O(\mu \log \mu \cdot n \log n)$.

Proof. We verify the conditions of Definition 5 to apply Theorem 9.

By definition, $p_{\text{clone}} = p = \Omega(1)$. At most one bit is flipped per invocation, hence we trivially get $p_{\text{safe}} = 1$. With probability $1 - p$, a single bit is chosen uniformly at random and flipped. Therefore, also using the previous insight,

$$p_{\text{flip,safe}} = \mathbb{E}[\text{destroy}] = (1 - p)|S|/n,$$

resulting in the desired bounds for $p_{\text{flip,safe}}$ and $\mathbb{E}[\text{destroy}]$ with constants $c_1 = c_2 = 1 - p$. The bound on $\mathbb{E}[\text{destroy}]$ also holds when conditioning on flipping a bit in S' as at most 1 bit is flipped per iteration.

As all conditions for a conservative mutation operator are met, the complexity bound follows from Theorem 9. □

3.3 Standard Bit Mutation with $\chi \leq 1$

The standard bit mutation operator modifies an individual by flipping each bit independently with probability $p = \chi/n$, for some mutation rate $\chi > 0$.

Corollary 12. *For any constant mutation parameter* $\chi \leq 1$*, the standard bit mutation operator with rate* χ/n *is a conservative mutation operator. Consequently, the expected optimization time of the* $(\mu + 1)$*-EA on* BinVal *under standard bit mutation is* $O(\mu \log \mu \cdot n \log n)$.

Proof. We verify the conditions of Definition 5 to invoke Theorem 9.

- p_{clone}: A clone is produced if no bits are flipped, yielding $p_{\text{clone}} = (1 - \chi/n)^n = \Theta(1)$.

- p_{safe}: For a subset of bits $S \subseteq [n]$, a mutation is considered safe if no bits in $[n] \setminus S$ are modified. Due to the independence of bit flips, we have $p_{\text{safe}} = (1 - \chi/n)^{n-|S|} = \Theta(1)$.
- $p_{\text{flip,safe}}$: The probability $p_{\text{flip,safe}}$ of flipping at least one bit in S while flipping no bits in $[n] \setminus S$ is given by

$$p_{\text{flip,safe}} = \left(1 - (1 - \chi/n)^{|S|}\right) \cdot (1 - \chi/n)^{n-|S|}.$$

We use the inequality

$$1 - (1 - x)^k = x \cdot \sum_{i=0}^{k-1} (1 - x)^i \geq kx(1 - x)^{k-1},$$

for $x \in (0, 1)$, derived from the geometric sum formula, to lower bound the first term by $\frac{\chi|S|}{n}(1 - \frac{\chi}{n})^{|S|-1}$. Multiplying by the second term results in the lower bound $\frac{\chi|S|}{n}(1-\frac{\chi}{n})^{n-1} \geq \chi e^{-\chi}\frac{|S|}{n}$ for $\chi \leq 1$, hence there exists a constant c_1 such that $p_{\text{flip,safe}} \geq c_1|S|/n$.
- $\mathbb{E}[\text{destroy}]$: The number of bits flipped within S follows a binomial distribution $B(|S|, \chi/n)$. The expected number of such flips is $\mathbb{E}[\text{destroy}] = \chi|S|/n$. This satisfies the condition $\mathbb{E}[\text{destroy}] \leq c_2\frac{|S|}{n}$ for a constant $c_2 \geq \chi$. When conditioning on flipping at least one bit in S' the same bound holds as bits are flipped independently.

Since the operator is conservative, the result follows from Theorem 9. □

3.4 Standard Bit Mutation with $\chi \geq 1$

To bound the runtime for larger mutation parameters $1 < \chi < n$, we couple the $(\mu + 1)$-EA on BinVal with mutation rate χ/n to $\lceil \chi \rceil$ sequential optimization processes with mutation rate at most $1/n$.

Formally, we use the following proposition. For a (possibly randomized) algorithm A let the *lazy* version of A with idle probability p_{idle} be the algorithm which flips a coin in each step: with probability p_{idle} it does nothing, otherwise it performs the next step of A. So the trajectory of the idle version of A is the same as for A, but the idle version follows the trajectory at lower speed.

Proposition 13. *Let $0 < \chi < n$. Consider a run of the $(\mu + 1)$-EA with mutation rate χ/n on n-dimensional* BinVal. *Let $B = \{n_b, n_b + 1, \ldots, n_b + n' - 1\}$ be a consecutive block of length n', and assume that for any position $b < n_b$, $\forall x \in P_0 : x_b = 1$. Let $B(P_t)$ be restriction of the population P_t at time t to B.*

Now consider the lazy $(\mu + 1)$-EA on n'-dimensional BinVal *with $p_{idle} = 1 - (1 - \chi/n)^{n_b - 1}$ and with mutation rate χ'/n', where $\chi' = \chi n'/n$. Let P'_t be the population of this lazy version after t steps, and assume $P'_0 = B(P_0)$.*

Then $B(P_t)$ and P'_t follow the same distribution for all $t \geq 0$, i.e., for all populations $P \in \{0, 1\}^{n'}$ we have $\Pr[B(P_t) = P] = \Pr[P'_t = P]$.

In particular, let $T := \inf_t\{\forall x \in B(P_t) : x = 1^{n'}\}$ and $T' := \inf_t\{\forall x \in P'_t : x = 1^{n'}\}$. Then T and T' follow the same distribution.

Proof. We prove the statement by induction over t. It is true for $t = 0$ since $P'_0 = B(P_0)$, so assume that as induction hypothesis that it is true for some $t \geq 0$. We need to show that both algorithm perform equivalent update steps, i.e., for all $P_1, P_2 \in \{0,1\}^{n'}$ we have

$$\Pr[B(P_{t+1}) = P_2 \mid B(P_t) = P_1] = \Pr[P'_{t+1} = P_2 \mid P'_t = P_1], \tag{1}$$

because then for all P_2,

$$\begin{aligned}\Pr[B(P_{t+1}) = P_2] &= \sum_{P_1 \in \{0,1\}^{n'}} \Pr[B(P_{t+1}) = P_2 \mid B(P_t) = P_1] \cdot \Pr[B(P_t) = P_1] \\ &= \sum_{P_1 \in \{0,1\}^{n'}} \Pr[P'_{t+1} = P_2 \mid P'_t = P_1] \cdot \Pr[P'_t = P_1] = \Pr[P'_{t+1} = P_2].\end{aligned}$$

To show (1), consider one update step of the $(\mu+1)$-EA. With probability $1-(1-\chi/n)^{n_b-1}$ at least one of the first $n_b - 1$ bits is flipped. In this case the offspring is rejected and the population stays the same. This corresponds to the probability p_{idle} of the lazy $(\mu+1)$-EA. If none of the first $n_b - 1$ bits is flipped, then each bit in B is flipped with probability $\chi/n = \chi'/n'$, as for the lazy $(\mu+1)$-EA. The selection is based on the bits in B, since the whole population agrees on the more significant bits. Note that in case of ties within B, selection for the $(\mu+1)$-EA may be based on the less significant bits. However, this only breaks ties between individuals that are identical on B, hence $B(P_{t+1})$ is independent of which copy is chosen. Overall, mutation and selection are identical for the $(\mu+1)$-EA and the lazy $(\mu+1)$-EA, which proves (1).

Finally, the statement on T and T' follows, because T and T' are defined in terms of $B(P_t)$ and P'_t, which follow the same distribution. □

Although we have only formulated Proposition 13 for the $(\mu+1)$-EA, the principle is applicable to other algorithms as well. This includes the $(\mu+\lambda)$ EA, but also genetic algorithms with uniform crossover (in which case idle steps only occur for mutation, not for crossover). As we will see below, it gives a very generic argument for ruling out thresholds where the runtime jumps to a different asymptotic level when χ is increased by an additive constant. In other situations, such jumps do occur frequently [2,6,14,17,20].

Proposition 13 is a rather basic observation, but it has a surprising consequences. Consider the $(\mu+1)$-EA with large mutation parameter χ on BinVal. Then the time needed to optimize a block is coupled to a lazy optimization process of BinVal for a smaller value of n', and crucially, for a smaller mutation parameter χ'. As an immediate consequence, we get that the runtime of the $(\mu+1)$-EA with mutation rate χ/n for $\chi > 1$ can be dominated by the sum of $\lceil\chi\rceil$ runtimes of a lazy $(\mu+1)$-EA with mutation parameter $\chi \in [1/2, 1]$.

Theorem 14. *Let $T_{\chi,n,\max}$ be the worst-case expected time, taken over all initial populations, until the $(\mu+1)$-EA with mutation rate χ/n on n-dimensional BinVal creates a population that consists of copies of the optimum. Let $T_{[1/2,1],n,\max} := \sup_{\chi\in[1/2,1]} T_{\chi,n,\max}$. Then for all $1 \leq \chi < n$ and all $n \in \mathbb{N}$,*

$$T_{\chi,n,\max} \leq \left(1-\chi/n\right)^{-n} \cdot \lceil\chi\rceil \cdot T_{[1/2,1],\lceil n/\lceil\chi\rceil\rceil,\max}.$$

Proof. We split the string of length n into $\lceil\chi\rceil$ blocks of length $n' = \lceil n/\lceil\chi\rceil\rceil$, where we allow the last block to be shifted so that it overlaps with the penultimate block if n' does not divide n. By Proposition 13, after the i-th block is optimized, the time until the next block is optimized is bounded by the time until a lazy $(\mu+1)$-EA has optimized an n'-dimensional BinVal instance. The mutation parameter of the lazy $(\mu+1)$-EA is $\chi n'/n \in [1/2, 1]$. The idle steps increase the expected time by a factor $1/(1-p_{\text{idle}})$, where $1 - p_{\text{idle}} \geq (1-\chi/n)^n$. □

Since Corollary 12 gives us a bound on $T_{[1/2,1],\lceil n/\lceil\chi\rceil\rceil,\max} = O(\mu\log\mu\cdot(n/\chi)\cdot\log n)$, we immediately get a runtime bound for large values of χ.

Corollary 15. *For any mutation rate $1 \leq \chi < n$, the expected optimization time of the $(\mu+1)$-EA with standard bit mutation and mutation rate χ/n on BinVal is $O((1-\chi/n)^{-n}\cdot\mu\log\mu\cdot n\log n)$.*

If $\chi = O(\sqrt{n})$ then this runtime is $O(e^{\chi}\cdot\mu\log\mu\cdot n\log n)$.

Proof. The first formula is immediately implied by Theorem 14 and the bound $T_{[1/2,1],\lceil n/\lceil\chi\rceil\rceil,\max} = O(\mu\log\mu\cdot(n/\chi)\cdot\log(n/\chi))$ from Corollary 12, where we upper bound $\log(n/\chi) \leq \log(n)$.

For the second formula, the Taylor expansion of the function $\ln(1-x) = -x - O(x^2)$ for $x \to 0$ implies by exponentiating that $1 - x = e^{-x-O(x^2)}$. We plug in $x = \chi/n$ and raise both sides to power $-n$, which gives for $\chi = O(\sqrt{n})$,

$$(1-\chi/n)^{-n} = (e^{-\chi/n-O((\chi/n)^2)})^{-n} = e^{\chi+O(\chi^2/n)} = e^{\chi+O(1)} = O(e^{\chi}).$$

□

4 Experimental Results

We implemented the $(\mu{+}1)$-EA using lexicographical comparisons, which allowed us to avoid the large coefficients of the BinVal fitness function. This allows us to measure the runtime across for large problem sizes (specifically $n = 1000$). For each parameter configuration, the simulation is initialised uniformly at random and run until the optimum (the all 1 s string) is sampled. The reported runtime is averaged over 50 independent trials.

Standard Bit Mutation Dynamics. To evaluate the predictive accuracy of our theoretical models, we first simulate the expected runtime using the standard bit mutation operator, where each bit is flipped independently with probability χ/n. We tested three different mutation parameters: $\chi \in \{0.5, 1.0, 1.5\}$.

To clearly observe the penalty induced by population size μ, we normalise the empirical expected runtime by $\mu\cdot n$. The results are shown in Fig. 1 (a).

We observe that the expected runtime scaled by μn starts at a large value, especially for $\mu = 1$. This case is well understood, the runtime is $(1+o(1))\frac{e^{\chi}}{\chi}n\ln n$. Indeed, an *additive* term $(1+o(1))\frac{e^{\chi}}{\chi}n\ln n$ seems inevitable.

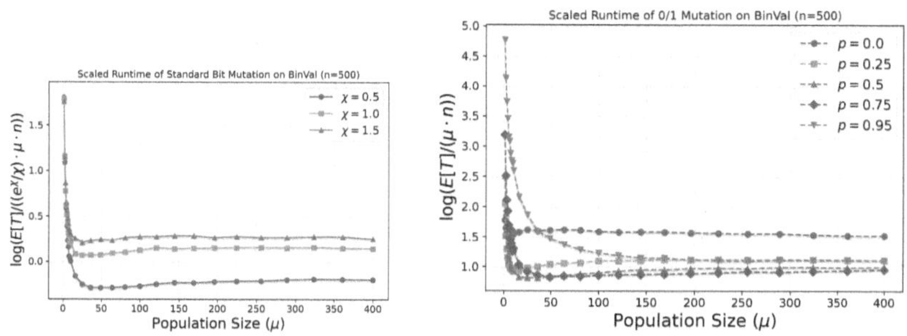

(a) (b)

Fig. 1. Logarithm of the expected runtime of the $(\mu + 1)$-EA on BinVal ($n = 500$) plotted against the population size. Also plotted is the 95% confidence interval as a shaded region. *(a):* Standard bit mutation across varying mutation parameters χ. The expected runtime is scaled by $\frac{e^\chi}{\chi}\mu n$ and then enclosed in a logarithm. *(b):* 0/1 Mutation for varying zero-flip probabilities p. The expected runtime is scaled by μn and then enclosed in a logarithm.

Consider the smallest Hamming distance from the optimum in the population. Reducing this Hamming distance from i to $i - 1$ plausibly takes expected time at least $\approx \frac{e^\chi}{\chi} \cdot \frac{n}{i}$ for small values of i (the time to reduce it by a single-bit flip), even in the optimistic case that *all* individuals have Hamming distance i from the optimum. Summing this over all i yields the term $(1 + o(1))\frac{e^\chi}{\chi} n \ln n$. However, the fact that the rescaled runtime in Fig. 1 (a) decays rapidly for $\mu > 1$ suggests that this is only an additive term which is not multiplied by μ in the real runtime.

As μ increases further beyond $\mu \approx 10$, the plot stabilizes to a small constant. This seems to suggest that the factor $\mu \log \mu$ in our theoretical bound is tight up to the $\log \mu$ term, which may be superfluous. In general, the experiments suggest a runtime of $\Theta(\mu n + n \ln n)$. We leave it to future work to investigate those hypotheses further.

0/1 Mutation. Figure 1 (b) shows our empirical evaluation for the 0/1 mutation operator. Under this regime, exactly one bit is flipped uniformly at random with probability $1 - p$, and zero bits are flipped with probability p. This operator explicitly parametrises the generation of exact clones, allowing us to isolate the effect of duplicate creation on the algorithmic slowdown. We simulate the process for varying no-flip probabilities $p \in \{0.0, 0.25, 0.5, 0.75, 0.95\}$.

We observe a similar scaling of the runtime as in the standard-bit mutation, which is line with the similar theoretical results that we got for this setting. For $\mu = 1$ we expect a runtime of $(1 + o(1))n \ln n/(1 - p)$, since the process is a variant of Random Local Search (RLS) that is slowed down by a factor of $(1-p)$. We observe that the runtime is minimized for an intermediate value of $p \approx 0.75$, while both smaller and larger values increase the runtime. Note that increasing p all the way to 1 would lead to infinite runtimes, since the algorithm could then only produce clones. As the value of μ increases, we observe that the plots seem to converge to a constant value. Similar to the standard bit mutation, the plot suggests a runtime of $\Theta(\mu n + n \log n)$.

5 Conclusion

We have improved the best known upper bound on the expected runtime of the $(\mu + 1)$ EA on BinVal from $O(\mu^5 n \log(n/\mu^4))$ [12] to $O(\mu \log \mu \cdot n \log n)$ for $\mu = o(n/\log n)$, a substantial reduction in the polynomial dependence on the population size μ. Our analysis builds on the block-based partitioning framework of [12] but refines the treatment of interference during the spreading of mutations through the population, by tracking the number of remaining unfixed bits in the current block rather than bounding the interference probability uniformly by the block size. This more precise analysis removes the high-degree polynomial factors that appeared in the earlier bound.

Our improved bound recovers the $O(n \log n)$ runtime of the $(1 + 1)$ EA on linear functions [24] for constant μ and grows only mildly with the population size. Since a lower bound of $O(\mu n + n \log n)$ is known for this problem [22], our upper bound is consequently tight up to some log factors. Whether the factor $\mu \log \mu$ is tight or can be further reduced—ideally closer to the additive μn overhead observed for OneMax [1,22]—remains an interesting open question. It would also be interesting to re-investigate to related questions that were studied in [12]: the runtime of the RLS algorithm, i.e. 0/1 mutation with $p = 0$; and for all algorithms the time to approximate the optimum so that only the least significant n/μ bits are non-optimal.

On the methodological side, it would be natural to investigate whether the techniques developed here extend to all linear functions, or whether the specific exponential weight structure of BinVal is essential for the analysis. Finally, it would be interesting to see whether the class of conservative mutation operators can be extended to also include crossover operators, so that the runtime bound generalizes to the $(\mu + 1)$ GA as well. Moreover, it would be interesting to study whether crossover can provably accelerate optimization of BinVal, as crossover has produced provable speedups on other benchmark families such as Jump [3,4], and related positive evidence exists in dynamic BinVal settings [15].

Acknowledgments. This project was supported by the Swiss National Science Foundation [grant number 0003390].

References

1. Antipov, D., Doerr, B.: A tight runtime analysis for the $(\mu + \lambda)$ EA. Algorithmica **83**(4), 1054–1095 (2021). https://doi.org/10.1007/S00453-020-00731-5
2. Antipov, D., Doerr, B., Yang, Q.: The efficiency threshold for the offspring population size of the (μ, λ) EA. In: Proceedings of the Genetic and Evolutionary Computation Conference, pp. 1268–1276. ACM (2019). https://doi.org/10.1145/3321707.3321851

3. Dang, D.C., et al.: Escaping local optima using crossover with emergent diversity. IEEE Trans. Evol. Comput. **22**(3), 484–497 (2018). https://doi.org/10.1109/tevc.2017.2724201
4. Doerr, B., Echarghaoui, A., Jamal, M., Krejca, M.S.: Runtime analysis of the $(\mu + 1)$ GA: provable speed-ups from strong drift towards diverse populations. Proc. AAAI Conf. Artif. Intell. **38**(18), 20683–20691 (2024). https://doi.org/10.1609/aaai.v38i18.30055
5. Doerr, B., Goldberg, L.A.: Adaptive drift analysis. Algorithmica **65**(1), 224–250 (2013). https://doi.org/10.1007/S00453-011-9585-3
6. Doerr, B., Jansen, T., Sudholt, D., Winzen, C., Zarges, C.: Mutation rate matters even when optimizing monotonic functions. Evol. Comput. **21**(1), 1–27 (2013). https://doi.org/10.1162/EVCO_A_00055
7. Doerr, B., Johannsen, D., Winzen, C.: Multiplicative drift analysis. Algorithmica **64**(4), 673–697 (2012). https://doi.org/10.1007/s00453-012-9622-x
8. Doerr, B., Künnemann, M.: Optimizing linear functions with the $(1 + \lambda)$ evolutionary algorithm–different asymptotic runtimes for different instances. Theoret. Comput. Sci. **561**, 3–23 (2015). https://doi.org/10.1016/j.tcs.2014.03.015
9. Droste, S., Jansen, T., Wegener, I.: On the analysis of the $(1+1)$ evolutionary algorithm. Theoret. Comput. Sci. **276**(1–2), 51–81 (2002). https://doi.org/10.1016/s0304-3975(01)00182-7
10. He, J., Yao, X.: A study of drift analysis for estimating computation time of evolutionary algorithms. Nat. Comput. **3**(1), 21–35 (2004). https://doi.org/10.1023/B:NACO.0000023417.31393.c7
11. Kötzing, T., Sander, J.: Anytime analysis on BinVal: adaptive parameters help (2026). https://doi.org/10.48550/ARXIV.2604.06976
12. Krejca, M.S., Neumann, F., Witt, C.: Population dynamics and improved runtime guarantees for the $(\mu + 1)$ EA on BinVal. In: Proceedings of the Foundations of Genetic Algorithms, pp. 142–153. FOGA 2025, ACM (2025). https://doi.org/10.1145/3729878.3746614
13. Lengler, J.: Drift analysis, pp. 89–131. Springer (2019). https://doi.org/10.1007/978-3-030-29414-4_2
14. Lengler, J.: A general dichotomy of evolutionary algorithms on monotone functions. IEEE Trans. Evol. Comput. **24**(6) (2019). https://doi.org/10.1109/TEVC.2019.2917014
15. Lengler, J., Meier, J.: Large population sizes and crossover help in dynamic environments. In: Bäck, T., et al. (eds.) Proceedings of the Parallel Problem Solving from Nature, pp. 610–622. LNCS, Springer (2020). https://doi.org/10.1007/978-3-030-58112-1_42
16. Lengler, J., Riedi, S.: Runtime analysis of the $(\mu + 1)$-EA on the dynamic BinVal function. SN Comput. Sci. **3**(4) (2022). https://doi.org/10.1007/s42979-022-01203-z
17. Lengler, J., Schaller, U.: The $(1 + 1)$-EA on noisy linear functions with random positive weights. In: Proceedings of the IEEE Symposium Series on Computational Intelligence, pp. 712–719. IEEE (2018). https://doi.org/10.1109/SSCI.2018.8628785
18. Lengler, J., Steger, A.: Drift analysis and evolutionary algorithms revisited. Comb. Probab. Comput. **27**(4), 643–666 (2018). https://doi.org/10.1017/S0963548318000275
19. Lengler, J., Zou, X.: Exponential slowdown for larger populations: The $(\mu+1)$-EA on monotone functions. Theoret. Comput. Sci. **875**, 28–51 (2021). https://doi.org/10.1016/j.tcs.2021.03.025

20. Lissovoi, A., Witt, C.: A runtime analysis of parallel evolutionary algorithms in dynamic optimization. Algorithmica **78**(2), 641–659 (2016). https://doi.org/10.1007/s00453-016-0262-4
21. Sudholt, D.: The benefits of population diversity in evolutionary algorithms: a survey of rigorous runtime analyses, pp. 359–404. Springer (2019). https://doi.org/10.1007/978-3-030-29414-4_8
22. Witt, C.: Runtime analysis of the $(\mu + 1)$ EA on simple pseudo-Boolean functions. Evol. Comput. **14**(1), 65–86 (2006). https://doi.org/10.1162/evco.2006.14.1.65
23. Witt, C.: Population size versus runtime of a simple evolutionary algorithm. Theoret. Comput. Sci. **403**(1), 104–120 (2008). https://doi.org/10.1016/j.tcs.2008.05.011
24. Witt, C.: Tight bounds on the optimization time of a randomized search heuristic on linear functions. Comb. Probab. Comput. **22**(2), 294–318 (2013). https://doi.org/10.1017/s0963548312000600

Never-Forgetting Genetic Algorithms on OneMax: First Insights

Maxim Buzdalov(✉) and Christine Zarges

Aberystwyth University, Aberystwyth SY23 3DB, UK
{mab168,chz8}@aber.ac.uk

Abstract. One possible, but rarely used way of managing populations in evolutionary algorithms is to never discard solutions and let the population grow, which puts much more emphasis on selection operators. A recent work proposed a particular framework for that, called *never-forgetting genetic algorithms*, and showed experimentally that it has noticeable benefits in terms of the expected number of evaluations until the optimum is hit on a number of pseudo-Boolean optimisation problems of theoretical importance. The working principles, however, were not understood well.

In this work, we make first important steps towards such an understanding by combined means of runtime analysis and experimental insights. These include proven recommendations for the mutation operator parameters following from some lower bounds on the runtime, experimental confirmation of importance of selection mechanisms, a strong indication that requiring a stochastically sampled minimum distance between crossover parents is crucial, and few insights into the structure of the population.

Keywords: Crossover · Population diversity · Selection operators

1 Introduction

Populations are a key aspect of evolutionary algorithms and many other randomised search heuristics. Most such algorithms only work with populations whose size is relatively small compared to the number of solutions (termed *individuals*) created and evaluated, and finding good ways to maintain such populations is one of the hottest research topics in the whole domain. Questions, such as whether an algorithm should be elitist or not [10,17,18], which population size is appropriate and when to change it [8,16,23,29], whether one needs to discard individuals that are there for too long [6,22], are far from receiving definite answers.

One of the possible strategies, which seems counter-intuitive at first, is to never discard any individual. In some areas, such as expensive optimisation, this happens naturally, and in particular in Bayesian optimisation one often keeps all individuals to build probabilistic models of the search space [9,26]. In theoretical

G. Iacca et al. (Eds.): PPSN 2026, LNCS 16985, pp. 83–98, 2027.
https://doi.org/10.1007/978-3-032-36214-8_6

research, some provably effective strategies require keeping most if not all individuals too: an example of this is solving the OneMax toy problem by making random queries until there remains only one potential optimum point that is consistent with all the queries and their function values, which is asymptotically only two times worse than the lower bound [1,15]. Generally, however, keeping all individuals is considered impractical at least for the reasons of algorithmic efficiency and memory consumption.

With the help of data structures that significantly reduce the computational requirements to maintain big populations [4,24], this strategy can be reconsidered. Though time and memory requirements are still polynomial in the number of individuals sampled, much faster evaluation can be seen in contexts of experiments with problems of theoretical importance solvable in polynomial time and with fixed-budget settings (cf. 25 000 function evaluations often seen in multiobjective benchmarking). Eventually, some complicated operations, such as finding the best individual at a given distance from a certain individual, may be accelerated with data structures supporting multidimensional range queries.

An extended abstract [5] proposed an idea of *never-forgetting genetic algorithms* (NFGA) and presented first promising experimental results, including what appears to be linear runtime on the famous toy problem OneMax. Here, we start where [5] ends and investigate, in detail, what happens when different variants of NFGA solve OneMax. So far, our theoretical understanding covers a limited set of aspects (Sect. 3). We complement this with an experimental study (Sect. 4), which includes an extensive ablation study and insights into the population structure.

2 Preliminaries

Basic Definitions. Throughout the paper, we denote as $[a..b]$ the range of integer numbers $\{a, a+1, \ldots, b\}$, the range of real numbers $a \leq x \leq b$ is denoted as $[a; b]$, and "u.a.r." means uniformly at random. We consider optimisation problems and algorithms working with bit strings of size n, the set of which is denoted as $\{0,1\}^n$. For a bit string $x \in \{0,1\}^n$, the value of the i-th bit in it $(1 \leq i \leq n)$ is written as x_i, and is either 0 or 1. The OneMax problem is defined as follows:

$$\textsc{OneMax}_n(x) = \sum\nolimits_{i=1}^{n} x_i.$$

Since all algorithms discussed in this paper are unbiased in the sense of [20], this simplified version of the definition is enough for our purposes.

Power-Law Distributions. For a finite n and a parameter $\beta > 0$, we say that we sample an integer $I \in [1..n]$ according to a power-law distribution if

$$\Pr[I = i] = \begin{cases} i^{-\beta} \cdot C_{n,\beta} & 1 \leq i \leq n \\ 0 & \text{otherwise,} \end{cases}$$

where $C_{n,\beta} = (\sum_{j=1}^{n} j^{-\beta})^{-1}$. Power-law distributions have been studied in the context of evolutionary algorithms, for example in [2,3,11,14]. Contrary to, say, binomial distributions, power-law distributions are much less concentrated around their mean (the probability of sampling a large value decreases polynomially and not exponentially), and so they are classified as heavy-tailed distributions, which have been used in real-valued algorithms too [30].

It is well known that $C_{n,\beta} = \omega(1)$ for $0 < \beta \le 1$ and $\Theta(1)$ for $\beta > 1$, which means that, although smaller values are always preferred by this distribution, the value of 1 is sampled with a constant probability only when $\beta > 1$, and with a less than constant probability otherwise. In the context of mutation operators, using standard bit mutation operator with $p = \xi/n$ where ξ is sampled from a power-law distribution, as done in [14]—or, which has very similar characteristics, sampling instead directly the number of bits to flip—performs local search with a constant probability when $\beta > 1$. Population sizes can be sampled from a power-law distribution too [3], and if one orders a collection of individuals by non-improving fitness, one can sample the rank of an individual from a power-law distribution, which would act as an effective selection operator [11].

Randomised Local Search. For bit strings, the simple optimisation algorithm called randomised local search (RLS) maintains the best known individual. On every iteration it chooses a bit index u.a.r., flips it, evaluates the resulting individual, and replaces the best one if the new one is at least as good. On OneMax, this is well-known to have an expected running time of $n \ln n + O(n)$ fitness evaluations. In terms of the performance it is a good, if a bit optimistic, representative of a wide class of randomised search heuristics on problems such as OneMax. For this purpose, we include it in all our experimental runtime comparisons.

The $(1 + (\lambda, \lambda))$ *Genetic Algorithm.* Proposed in [13], the $(1 + (\lambda, \lambda))$ GA is an optimiser of significant theoretical importance, which was the first black-box general purpose evolutionary algorithm to break the $\Omega(n \log n)$ barrier on OneMax: for the best static parameter choice λ its runtime is $o(n\sqrt{\log n})$, and either the self-adjusting strategy [12] or sampling λ using a power-law distribution [2] yields a linear runtime. This was shown to be possible due to a tight interaction between the mutation and the crossover operators as well as the population size. Specifically, on each iteration λ mutants are generated from the single parent, using the same Hamming distance sampled from the binomial distribution $\text{Bin}(n, \lambda/n)$, and then the parent and the best mutant give birth to λ more offspring using a "biased" crossover operator which only takes bits from the mutant with probability $1/\lambda$. Here, crossover effectively functions as mutation subsampling to borrow a few improving bits, which could appear in the mutant with an elevated probability compared to e.g. RLS, into the parent with a purpose to immediately increase its fitness.

The Diversity Exploiting Genetic Algorithm. Another example of a crossover-based algorithm of theoretical importance, DEGA [21] can solve another well-known problem, LeadingOnes, in expected time $O(n^{5/3}(\log n)^{2/3})$, which is faster than $\Omega(n^2)$ demonstrated by most other algorithms. This algorithm makes

Algorithm 1: The generic never-forgetting genetic algorithm

Input: problem size n, function to maximise $f : \{0,1\}^n \to \mathbb{R}$.
Setup: distributions $\mathcal{D}_{\text{pm}}$, $\mathcal{D}_{\text{dm}}$, $\mathcal{D}_{\text{p1}}$, $\mathcal{D}_{\text{p2}}$, $\mathcal{D}_{\delta}$, $\mathcal{D}_{\text{dc}}$, parameters p_m, $d_{\max}$.
1 Sample $x^{(0)}$ u.a.r. from $\{0,1\}^n$ and evaluate it;
2 Define population $X := \{x^{(0)}\}$;
3 **for** $t = 1, 2, \dots$ **do**
4   **if** *crossover is impossible* **or** *with probability* p_m **then**
5   Sample parent $x \in X$ using $\mathcal{D}_{\text{pm}}$;
6   Sample mutation distance d from $[1..n]$ using $\mathcal{D}_{\text{dm}}$;
7   Create y by copying x and flipping d bits in it chosen u.a.r.;
8   **else**
9   Sample first parent $x^{(1)} \in X$ using $\mathcal{D}_{\text{p1}}$;
10   Collect distances $\Delta := \{d = \mathcal{H}(x^{(1)}, x) \mid x \in X, d \geq 2, d \leq d_{\max}\}$;
11   **if** Δ is empty **then go to** line 9;
12   Sample minimum acceptable distance $\delta \in \Delta$ using $\mathcal{D}_{\delta}$;
13   Sample second parent $x^{(2)} \in \{x \in X \mid \mathcal{H}(x, x^{(1)}) \geq \delta\}$ using $\mathcal{D}_{\text{p2}}$;
14   Sample crossover distance d from $[1..\mathcal{H}(x^{(1)}, x^{(2)}) - 1]$ using $\mathcal{D}_{\text{dc}}$;
15   Determine differing bits $I := \{i \mid x_i^{(1)} \neq x_i^{(2)}\}$;
16   Create y by copying $x^{(1)}$ and flipping d bits chosen u.a.r. from I;
17   **if** $y \notin X$ **then**
18   Evaluate y and set $X \leftarrow X \cup \{y\}$;

use of the following observation. Assume one has two parents $x^{(1)}$ and $x^{(2)}$ and performs *uniform crossover* on them, which for every bit position takes a bit from $x^{(1)}$ with probability $1/2$, and from $x^{(2)}$ otherwise, and assume further that the resulting individual y replaces one of the parents in the population. Then the distance between the individuals that remain in the population shrinks by a factor of 2 in expectation, which is bad for diversity. The algorithms from the DEGA family address this by attempting to transfer only the minimum required number of bits to achieve a similar effect in terms of fitness, which is done by subsampling the change originally made by the uniform crossover. Here, however, the algorithm usually prefers to move a few bits from a better individual to a worse one, as opposed to the $(1 + (\lambda, \lambda))$ GA.

The Analysed Algorithm. We consider a generic version of the never-forgetting genetic algorithm (NFGA) originally proposed in [5], which is outlined as Algorithm 1. It keeps all individuals sampled so far as a set X. On every iteration it performs either mutation or crossover. The mutation is rather straightforward: first a parent is selected from X, then a mutant is generated at a Hamming distance which is sampled from $[1..n]$.

The crossover is more involved. First, the first parent, $x^{(1)}$, is sampled such that there exists another individual in X which is at a Hamming distance of at least 2 and at most $d_{\max}$ from it; if there is no such parent, the crossover is considered impossible and only mutation can be performed on this iteration. Then, all possible Hamming distances to $x^{(1)}$ to any other individual in X are collected, subject to belonging to $[2..d_{\max}]$. A distance δ is then sampled from this set, which will serve as the *minimum acceptable distance* between the crossover parents. The second parent, $x^{(2)}$, is then sampled from the portion of X that is at least δ bits away from $x^{(1)}$, and then the crossover is performed. This crossover, similarly to the previous two algorithms, is also difference subsampling, where the Hamming distance from the first parent is sampled from some distribution, subject to never sampling either of the parents again.

Several distributions are used in these operators, either to sample some distance or some index in a sorted set of entities. In [5] these distributions have been chosen in a particular way, but in this paper, we are going to consider a number of alternative choices. These distributions are as follows.

- The parent sampling distributions $\mathcal{D}_{\text{pm}}$ and $\mathcal{D}_{\text{p1}}$ sample the mutation parent and the first crossover parent, respectively. These take the population X, sort solutions in a non-increasing order of their fitness value, and choose the solution by sampling its index in the resulting list. In the case when there are multiple solutions with the same fitness as the chosen one, the solution is resampled u.a.r. from all solutions with this fitness. In [5], in both cases a power-law distribution with a parameter β_1 was used, with a default choice of $\beta_1 = 2.5$. With this, the best solution is sampled with a large constant probability.
- The mutation distance distribution $\mathcal{D}_{\text{dm}}$ samples the Hamming distance to the mutant, and in [5] this was again a power-law distribution with parameter β_m defaulting to 1.5. This also prefers shorter distances (and performs local search with a constant probability), but puts less emphasis on them. The value of 1.5 is also considered to be a default recommendation in the similar role in [14].
- The minimum acceptable crossover distance distribution $\mathcal{D}_\delta$ samples a distance δ from the set of the actual Hamming distances between $x^{(1)}$ and the rest of the population. It should never be less than 2, because one cannot perform a crossover on individuals at distance less than 2 such that the result is different from both of the parents. Here, we also introduce a parameter $d_{\max}$, which was absent in [5] and serves as a cap on the possible values of δ, but as we will see later, using a non-trivial cap is likely discouraged. Sampling a distance from the list of possible distances uses another power-law distribution in [5] with a parameter β_d defaulting to 1.5.
- The second parent sampling distribution $\mathcal{D}_{\text{p2}}$ samples the second crossover parent from the set of solutions at a distance not less than δ and otherwise works similarly to $\mathcal{D}_{\text{p1}}$. In [5], this is also a power-law distribution with parameter β_2 defaulting to 2.5.
- The crossover distance sampling distribution $\mathcal{D}_{\text{dc}}$ samples a distance from the first parent $x^{(1)}$ when performing crossover on $x^{(1)}$ and $x^{(2)}$ from the range between 1, inclusively, and $\mathcal{H}(x^{(1)}, x^{(2)})$, exclusively. In [5], two different distributions were used for this: first, a power-law distribution with parameter

β_c defaulting to 1.5, but applied in a symmetric way (implemented by swapping the parents with probability 1/2), and second, a uniform distribution. It was conjectured in [5], however, that uniform crossover (corresponding to a binomial distribution with probability 1/2 for $\mathcal{D}_{\text{dc}}$) may also work well.

In this paper, we consider by default a version of NFGA which follows Algorithm 1 and uses power-law distributions with parameters $\beta_1, \beta_m, \beta_d, \beta_2$ as described above, except for $\mathcal{D}_{\text{dc}}$ which corresponds to the uniform crossover. The probability of choosing mutation is set to $p_m = 0.5$, and the distance cap to $d_{\max} = \infty$. We denote this version as NFGA/UX (for "NFGA with uniform crossover"), and all other versions will be introduced when needed.

3 Theoretical Analysis

In this section, we provide theoretical analyses for some components of NFGA. Note that in this section, unlike in most of experiments, we work with the number of individuals sampled (that is, the number of iterations in Algorithm 1), and not with the number of unique queries made (the number of execution of lines 1 and 18). We leave rigorous analysis of the differences between these two quantities for future work, noting, however, that for OneMax they cannot be more than a constant factor apart.

Considering, for an arbitrary bit position i, the presence of individuals in the population that have this bit set to its optimum value (those x such that $x_i = 1$), we note that in Algorithm 1 only mutation can create one-bits, and crossover is only capable or recombining these from parent individuals, possibly creating better ones. In what follows, we analyse the one-bit creating by mutation and the final stages of recombining them that lead to the improvement of the globally best fitness value. These two parts, however, do not make a complete theoretical analysis of the algorithm.

One-Bit Creation. Let $X^{(t)}$ be the population at time t, where $X^{(0)} = \{x^{(0)}\}$ with $x^{(0)}$ sampled u.a.r. (line 1 of Algorithm 1). We denote as $B_j(t)$ the condition that there is at least one $x \in X^{(t)}$ such that $x_j = 1$, that is, the one-bit at index j is discovered. Note that once $B_j(t)$ holds, then it holds for all $t' \geq t$ too, as individuals are never discarded. Let $M^{(t)} = \{j \mid B_j(t) \text{ does not hold}\}$ be the index set of not yet created one-bits. With this, we have the following theorem[1].

Theorem 1. *Consider Algorithm 1 where the mutation distance distribution* $\mathcal{D}_{dm}$ *is a power-law distribution with parameter* $\beta_m > 0$ *and the probability of performing mutation is* $p_m > 0$*. Let* $T_{bits} = \min\{t \in \mathbb{N} \mid M^{(t)} = \emptyset\}$ *be the time*

[1] For the sake of saving space, all proofs are omitted from this version of the paper.

to create all one-bits. Then

$$\mathbb{E}\left[T_{bits} \cdot p_m\right] = \begin{cases} O(\log n) & \text{if } 0 < \beta_m \le 1; \\ O(n^{\beta_m - 1} \log n) & \text{if } 1 < \beta_m < 2; \\ O(n) & \text{if } \beta_m = 2; \\ \Theta(n \log n) & \text{if } \beta_m > 2. \end{cases}$$

For $\beta_m > 2$, Theorem 1 immediately provides a lower bound of $\Omega(n \log n)$ for Algorithm 1 on any function with a single global optimum, including OneMax. We believe that with some additional effort the same bound can be proven for a wider class of distributions $\mathcal{D}_{\mathrm{dm}}$, including the one emulating standard bit mutation. A recommendation hence follows to employ $\beta_m < 2$ as long as power-law distributions are used for sampling mutation distance. We remark here that low values of $\mathbb{E}\left[T_{\mathrm{bits}}\right]$ do not necessarily imply that the algorithm is efficient. For instance, one can trivially introduce all one-bits into the population by starting with two individuals: one sampled u.a.r., another a complement to the first one, similarly to what DEGA does [21], which of course does not guarantee a success.

Best Fitness Improvements. Now we present analysis of the process which, as we believe, very often constructs individuals that improve the best seen fitness. Note that [19, Lemma 1] is similar to Lemmas 1 and 3 applied to uniform crossover.

Lemma 1. *Let $x^{(1)}$ and $x^{(2)}$ be two different individuals with OneMax$(x^{(1)})$ = OneMax$(x^{(2)})$ with which the crossover operator is performed as in Algorithm 1. Then, regardless of the crossover distance distribution $\mathcal{D}_{dc}$, the probability of creating an offspring y with OneMax(y) > OneMax$(x^{(1)})$ is at least 1/6.*

Note that, if one follows the proof of the lemma, one can easily see that the probability of generating an offspring with OneMax$(y) \ge$ OneMax$(x^{(1)})$ while still being structurally different from both parents is at least $1/2$, and if the fitness values are equal, the offspring is closer to either of the parents than the parents themselves, which improves the probability of a successful global improvement in the next steps.

We complement this with the following lemma, which essentially says that if there are two best individuals in the population, it is easy to select them for crossover.

Lemma 2. *Assume Algorithm 1 optimises OneMax and the current population X has $f_{\max} = \max\{f(x) \mid x \in X\}$, $d_{\max} \ge 2$ and the following conditions are satisfied:*

(a) there are $x', x'' \in X$ such that $f(x') = f(x'') = f_{\max}$ and $x' \neq x''$;
(b) there are $y', y'' \in X$ such that $y' \neq y''$, $f(y') = f_{\max}$ and $\mathcal{H}(y', y'') \le d_{\max}$.
(c) both $\mathcal{D}_{p1}$ and $\mathcal{D}_{p2}$ select the best individual with probability $\Theta(1)$;
(d) $\mathcal{D}_{\delta}$ samples distance 2 with probability $\Theta(1)$.

Then the probability of an event that two individuals $x^{(1)} \neq x^{(2)}$ with $f(x^{(1)}) = f(x^{(2)}) = f_{\max}$ are selected for crossover is $\Theta(1 - p_m)$.

One can see that, with sane enough configuration of relevant components (such that using power-law distributions for $\mathcal{D}_{\text{p1}}$, $\mathcal{D}_{\text{p2}}$, $\mathcal{D}_\delta$ with $\beta > 1$) and with a constant nonzero probability of choosing crossover, Lemmas 1 and 2 essentially guarantee that, whenever there are several "best" individuals in the population, an improvement happens in $\Theta(1)$ iterations.

Seemingly, the probability of success at this step barely depends on the employed crossover distance distribution $\mathcal{D}_{\text{dc}}$. However, the scale of the possible improvement does depend on the choice of the distribution.

Lemma 3. *Let $x^{(1)}$ and $x^{(2)}$ be two different individuals with $\textsc{OneMax}(x^{(1)}) = \textsc{OneMax}(x^{(2)})$ with which the crossover operator is performed as in Algorithm 1. Let $2h = \mathcal{H}(x^{(1)}, x^{(2)})$ be the Hamming distance between these individuals. Assume an offspring y was created after sampling a crossover distance $d = 2\delta+1$, and $\Delta = \textsc{OneMax}(y) - \textsc{OneMax}(x^{(1)})$. Then*

$$\mathbb{E}[\Delta \mid \Delta > 0] = \Theta(\sqrt{\min\{\delta, h - \delta\}}) = \Theta(\sqrt{\min\{d, 2h - d\}}).$$

This lemma deals only with odd d values for simplicity, however, a similar behaviour is expected for even d too. Two observations can be made here. First, crossing over individuals separated by a larger Hamming distance can be beneficial for OneMax because of a potential for bigger improvements. Second, values of crossover distance d that are closer to half the distance between parents are capable of producing bigger improvements. One can estimate expected improvements possible with different crossover distance distributions $\mathcal{D}_{\text{dc}}$ as follows.

Lemma 4. *In conditions of Lemma 3, assuming H to be the Hamming distance between the parents, the following bounds hold for conditional improvements.*

$\mathbb{E}[\Delta \mid \Delta > 0]$	$\mathcal{D}_\delta$	*Comment*
$\Theta(1)$	$1 + \text{Bin}(H-2, 1/H)$	*Emulates the $(1 + (\lambda, \lambda))$ GA*
$\Theta(1)$	$H - 1 - \text{Bin}(H-2, 1/H)$	*Emulates DEGA*
$\Theta(1)$	$\beta_c > 1.5$	
$\Theta(\log H)$	$\beta_c = 1.5$	*Power-law distance sampling,*
$\Theta(H^{1.5-\beta_c})$	$1 < \beta_c < 1.5$	*including also symmetric*
$\Theta(\sqrt{H}/\log H)$	$\beta_c = 1$	*versions as in [5]*
$\Theta(\sqrt{H})$	$0 < \beta_c < 1$	
$\Theta(\sqrt{H})$	*u.a.r. from* $[1..H-1]$	*Uniform distance sampling*
$\Theta(\sqrt{H})$	$1 + \text{Bin}(H-2, 1/2)$	*Uniform crossover*

Strictly speaking, Lemmas 3 and 4 only apply to parents with the same fitness. We will see in the experiments, however, that some of these trends tend to carry over to the general case.

What Remains. In essence, we only considered the initial mandatory phase of acquiring all bits set to correct values and the last step preceding many global fitness improvements. However, the presence of different good individuals with equal fitness is not given, and cannot easily happen via e.g. mutation in later stages of optimisation. Similarly, analyses of the $(\mu + 1)$ GA [7,27] show that it is easy to recombine equally good individuals, but it is hard to construct them. In our case, we hypothesise that the structure of the entire population is maintained by crossover, mutation and appropriate individual selection in such a way that the probability of synthesising multiple similarly good individuals is elevated compared to pure mutation or to crossover on randomly chosen or randomly generated individuals.

To shed some light on how many insights from this section, and to which degree, carry over the entire optimisation process, and to get additional insights, we have to perform experiments, which are presented in the next section.

4 Experiments

For experiments, we used the implementation of patch-based genetic algorithms[2] available on GitHub. To make a fair comparison between NFGA, which puts a lot of effort into not re-sampling existing individuals regardless of the presence of external machinery that saves fitness evaluations, and other algorithms known to optimise OneMax reasonably well (here, RLS and $(1+(\lambda,\lambda))$ GA), we implemented the latter in the same framework and report the number of unique queries until the optimum is found. A subsection is dedicated to understanding whether this has a large impact on the numbers.

We optimise OneMax with problem size n equal to powers of two from $2^4 = 16$ to $2^{14} = 16384$. Powers of two were chosen to simplify plotting in logarithmic scale in problem sizes. The particular upper limit was chosen so that runs terminate within approximately 12 h on available hardware. For some variations of NFGA, we reduced the upper limit. For every algorithmic configuration and every problem size, we performed 51 independent runs.

The main algorithmic configurations are: the standard implementation of RLS, the heavy-tailed version of the $(1 + (\lambda, \lambda))$ GA with parameter $\beta = 2.5$, a choice most suited to solves OneMax to optimality [2]. For NFGA, the primary configuration is NFGA/UX, which uses uniform crossover, power-law $\mathcal{D}_{\text{pm}}, \mathcal{D}_{\text{p1}}, \mathcal{D}_{\text{p2}}$ with $\beta = 2.5$, power-law $\mathcal{D}_{\text{dm}}, \mathcal{D}_{\text{dc}}$ with $\beta = 1.5$, $p_m = 0.5$, $d_{\max} = \infty$. All other configurations will be introduced when needed.

4.1 Measuring Runtime: All Queries vs Unique Queries

How exactly does counting only unique fitness queries influence the experimental runtime measurements? Fig. 1 presents average normalised runtimes, measured either as the number of unique fitness queries until the optimum is found (this

[2] https://github.com/mbuzdalov/patch-based-ga/.

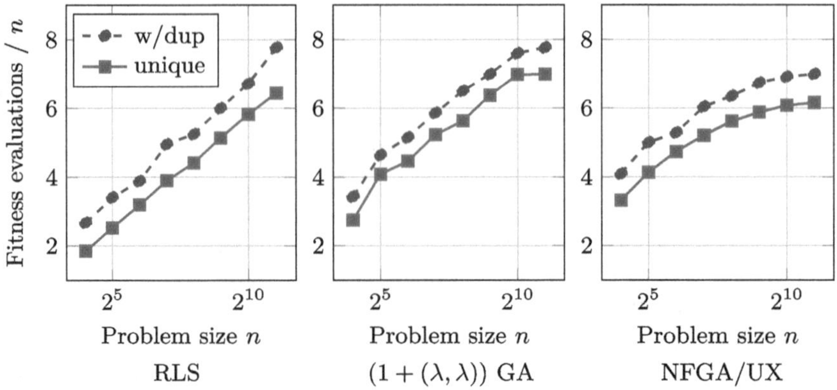

Fig. 1. Average runtimes until the optimum, counting either all queries (blue) or unique queries only (red), for the main algorithms. (Color figure online)

paper's default) or as the number of any fitness queries until the optimum is found (the standard way) for three different algorithms on OneMax with problem sizes up to 2^{11}.

From Fig. 1, the difference, while visually noticeable, does not seem significant, and in particular does not seem to change the perceived asymptotics. The ratio of the averages, for $n = 2^8 \dots 2^{11}$, is in the range of $[1.18; 1.21]$ for RLS, $[1.09; 1.15]$ for the $(1+(\lambda,\lambda))$ GA, and $[1.13; 1.15]$ for NFGA/UX, without noticeable dependency on n for this range. As these numbers are quite similar, we conjecture that all of our conclusions made in this section for unique fitness evaluations on OneMax will carry over to the standard way of evaluating the runtime. For non-unimodal problems, however, it is not difficult to come up with an example where these two approaches will produce vastly different results.

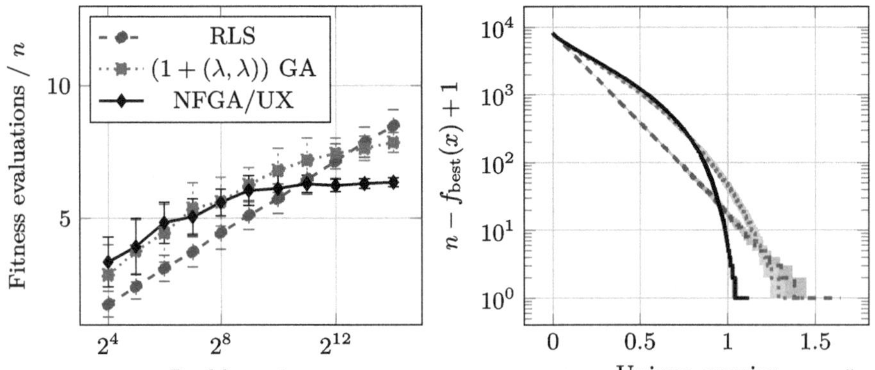

Fig. 2. Average runtimes and anytime performance of NFGA/UX compared to RLS and $(1+(\lambda,\lambda))$ GA (right: $n = 2^{14}$).

4.2 General Linear Runtime Trend for NFGA/UX

Figure 2 presents two plots for the main tested algorithms. On the left, the mean runtimes augmented by standard deviations are presented for the entire range of problem sizes. With logarithmic scaling of the x axis representing the problem size n, and with *linear* scaling of the y axis depicting the runtime normalised by n, algorithms that run in time $\Theta(n \log n)$ should show a linear plot with a non-zero slope coefficient, and algorithms that run in linear time should eventually produce horisontal plots. In Fig. 2 we see a straight sloped line for RLS, and both other algorithms indeed tend to level out. For the $(1+(\lambda, \lambda))$ GA, the $\Theta(n)$ runtime is proven in [2], and NFGA/UX appears even faster with an almost horisontal plot towards the upper end of the range. The Wilcoxon rank-sum test [28] as implemented in R [25] reports that NFGA/UX is statistically significantly faster than either of the other algorithms with $p < 3.3 \cdot 10^{-13}$ for each size in $\{2^{12}, 2^{13}, 2^{14}\}$.

On the right of Fig. 2, anytime performance plots of all three participating algorithms are shown. Here, on the x axis, the spent computational budget is shown, measured in unique queries made, and on the y axis, with logarithmic scaling, the closest seen distance to the optimum plus 1 (in other words, $n - \text{OneMax}(x) + 1$ for the best so far individual x) is presented. The problem size for this plot is $n = 2^{14}$. For every algorithm, for the given computational budget, the median fitness is plotted as a line, and the interquartile range is filled.

In these axes, RLS shows linear convergence, whereas two other algorithms show a different behaviour. Notably, they start as simply worse hill-climbers, but then the plots bend down, indicating acceleration of the search, and come to the bottom line in a much steeper way. As one can see, most of the benefit from the non-RLS algorithms is about optimising the last $\log n$ or so bits, where RLS spends a good portion of its budget and the other algorithms spend much less on it.

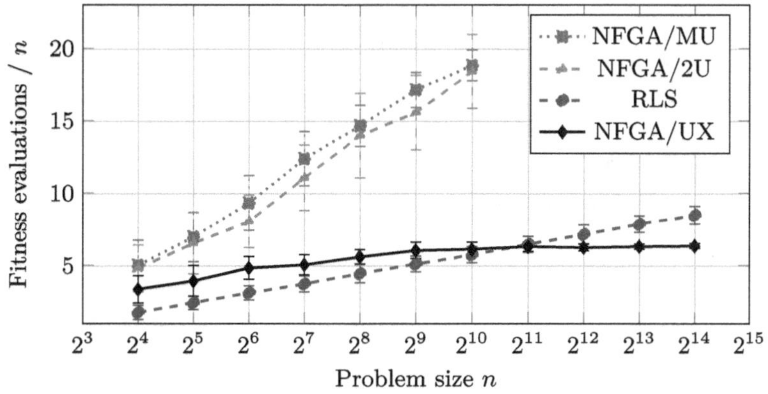

Fig. 3. Average runtimes of NFGA with uniform second crossover parent selection and uniform mutation parent selection, compared to other algorithms

4.3 Selection of the Parents Matters

In this subsection, we consider two variations of NFGA: NFGA/MU is configured in the same way as NFGA/UX, except that it samples the mutation parent uniformly (e.g. $\mathcal{D}_{\text{pm}}$ is uniform), and similarly, NFGA/2U samples the second crossover parent uniformly (via $\mathcal{D}_{\text{p2}}$). Figure 3 shows the mean runtimes for these algorithms in comparison to RLS and NFGA/UX. Both NFGA/MU and NFGA/2U are statistically significantly slower than NFGA/UX at $p < 1.1 \cdot 10^{-10}$ for all inspected problem sizes.

As one can see, disabling selection pressure for either of these components degrades the performance significantly. Both NFGA/MU and NFGA/2U still seem to stay within $\Theta(n \log n)$ runtime, which is not surprising for NFGA/2U because its mutation can still emulate actions of RLS with a constant probability. Why and how the uniform crossover can still somewhat make up for the lack of proper mutation in NFGA/MU, is not very clear at this moment.

Following our theoretical discussion on collecting one-bits (Theorem 1), we note that mutation in NFGA in all these cases uses a power-law distribution $\mathcal{D}_{\text{dm}}$ with $\beta = 1.5$ for sampling mutation distance, which means that all one-bits are collected in the population within some $\sqrt{n} \log n$ evaluations. The performance of NFGA/MU shows that just the presence of all one-bits is not enough for the full performance of the algorithm. What seems likely at this point is that it does not only matter whether the i-th one-bit is in the population, but also whether it occurs at high fitness levels, and also quite likely, whether it is present sufficiently often at *various* high enough fitness levels.

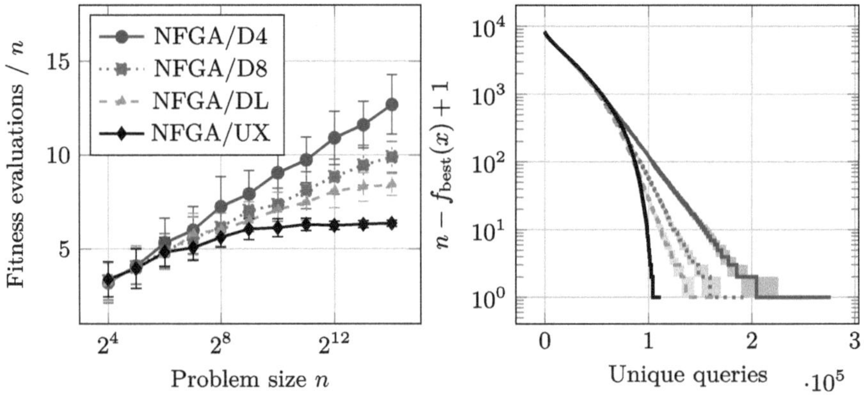

Fig. 4. Average runtimes and anytime performance of NFGA variations with capped minimum acceptable distance between crossover parents (right: $n = 2^{14}$).

4.4 Large Distances Between Crossover Parents Are Beneficial

Our theoretical results (Lemmas 1–4) only cover crossover application between equally fit parents, and internally only require short-distance crossovers to enable this kind of progress. A natural question is whether long-distance crossovers are needed, and whether the distribution $\mathcal{D}_\delta$ requires to be sufficiently heavy-tailed. To alleviate occasional large minimum acceptable distance samplings (not even crossovers of individuals at large distances), we introduced the parameter $d_{\max}$ in Algorithm 1 which caps the minimum acceptable distance from the above. We tested NFGA/D4 with $d_{\max} = 4$, NFGA/D8 with $d_{\max} = 8$, and NFGA/DL with $d_{\max} = \max\{2, \log_2(n+1)\}$, which are otherwise configured exactly like NFGA/UX.

From Fig. 4 left, it can be immediately seen that this kind of capping has a detrimental effect on the performance, although not as bad as uniform parent selection. Ordering of the algorithms is statistically significant pointwise with $p < 4.2 \cdot 10^{-5}$ for problem sizes $n \geq 2^{12}$. On the right of Fig. 4, the anytime performance plots show an interesting picture. Low values of $d_{\max}$ apparently prevent the algorithm from acceleration beyond some multiple of linear convergence speed. This somewhat resembles the behaviour of the $(1+(\lambda,\lambda))$ GA with fixed small λ, where the runtime scales as $O(\frac{n \log n}{\lambda})$. In the region of low fitness, however, there is some small but noticeable benefit from capping, which can be explained by saving up on query attempts that do not benefit the algorithm at this stage.

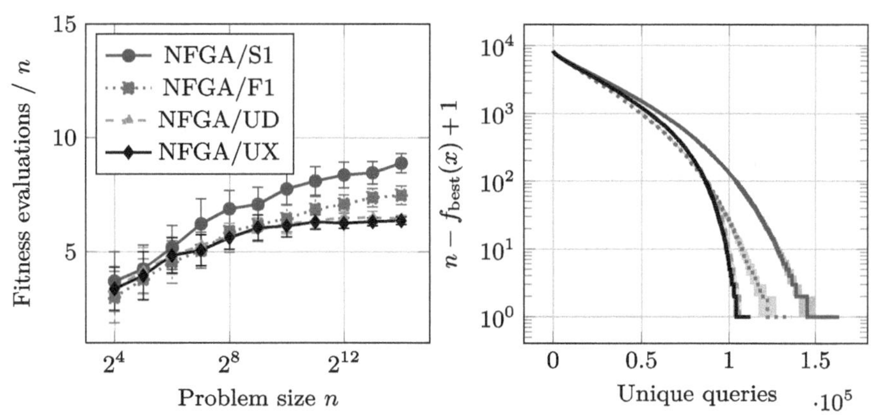

Fig. 5. Average runtimes and anytime performance of NFGA variations with different crossover distributions (right: $n = 2^{14}$).

4.5 Crossover Distribution Matters only Slightly

Following the discussion of Lemmas 3 and 4, we presented in Fig. 5 the results of NFGA configured identically to NFGA/UX apart from the crossover distance distribution. Namely, NFGA/UD has the uniform distribution as $\mathcal{D}_{dc}$, whereas NFGA/F1 always creates offspring at a distance of 1 to the first parent, and NFGA/S1 does it at a distance 1 to the second parent. These all are somewhat extreme cases for what is possible in the crossover. As we can see in Fig. 5, not all crossovers perform identically: while NFGA/UX and NFGA/UD are nearly indistinguishable, NFGA/F1 is noticeably slower (compared to NFGA/UX, with significance $p < 3.9 \cdot 10^{-8}$ at $n \geq 2^{11}$), and NFGA/S1 even more so (compared to NFGA/F1, with significance $p < 1.2 \cdot 10^{-7}$ at $n \geq 2^{7}$).

The anytime plots (Fig. 5, right) tell an interesting story nevertheless. In all cases, plots curve down, indicating acceleration of the search, compared to linear convergence typical to RLS, towards the end, which indicates that all of these crossovers are functioning well. When comparing NFGA/S1 to NFGA/D8 from the previous subsection, the former starts noticeably slower and crosses a distance of 10^3 later, but then catches up and has a better runtime overall (more precisely, NFGA/S1 is statistically faster than NFGA/D8 at $p < 1.3 \cdot 10^{-8}$ for $n \geq 2^{13}$). This all indicates that Lemma 1 somewhat carries over to the whole search process in the sense that any crossover works. Effects of Lemma 3 have some influence on the whole optimisation too, but nothing in it predicts the asymmetry between NFGA/F1 and NFGA/S1. Also, an interesting observation from Fig. 5 right is that NFGA/F1 is slightly faster at the start than NFGA/UX but NFGA/S1 is not.

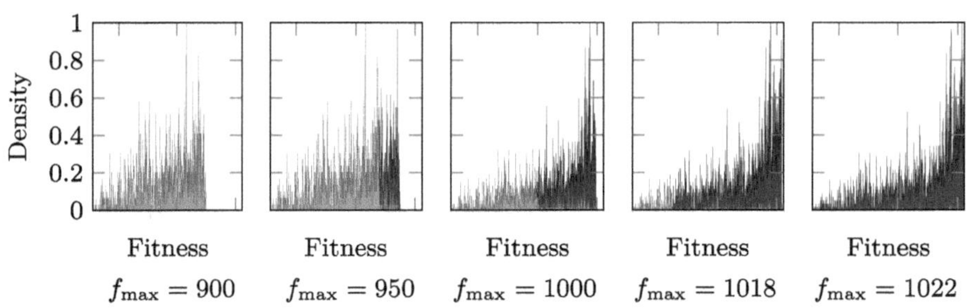

Fig. 6. Population structure for a run of NFGA/UX with $n = 1024$ at the end of phases with different maximum fitness. Colours are: blue: individuals generated by mutation with some "good" bits, violet: mutation and no "good" bits, green: crossover with some "good" bits, yellow: crossover with no "good" bits. (Color figure online)

4.6 Insights Into Population Structure

Finally, we present some basic insights into how the population is structured during the run. Figure 6 shows population densities at last iterations with various maximum fitness values, with four groups of individuals shown in different colours ranging by whether they were created by crossover or mutation, and by whether they contain any 1-bits which the best individual does not have (the "good" bits). One can see that there are usually a lot of "good" individuals with smaller fitness, but there are still a few "good" individuals with large fitness too, and our main conjecture is that the drift of "good" individuals towards the fitness level of the best individual is the key to understanding the behaviour of NFGA.

5 Conclusion

We presented the first reasonably in-depth analysis of *never-forgetting genetic algorithms*, a class of algorithms that never discard individuals and employ an intricate interplay of crossover, mutation and selection instead. We combined theoretical analysis of what can be seen as the first and the last stages of optimisation, tried to validate how the underlying ideas extend also to intermediate stages, and gained additional insights that may guide further research into this class of apparently promising algorithms.

Analysing NFGA appears to be challenging because of long-term interactions between many individuals in the ever-growing population that are difficult to quantify using a handful of numbers, as well as because the algorithm does not seem to progress in clearly defined steps. We hope that this algorithm will thus serve as a driving force towards developing more powerful analysis tools.

References

1. Anil, G., Wiegand, R.P.: Black-box search by elimination of fitness functions. In: Proceedings of Foundations of Genetic Algorithms, pp. 67–78 (2009)
2. Antipov, D., Buzdalov, M., Doerr, B.: Fast mutation in crossover-based algorithms. Algorithmica **84**(6), 1724–1761 (2022)
3. Antipov, D., Buzdalov, M., Doerr, B.: Lazy parameter tuning and control: choosing all parameters randomly from a power-law distribution. Algorithmica **86**(2), 442–484 (2024)
4. Buzdalov, M.: Improving time and memory efficiency of genetic algorithms by storing populations as minimum spanning trees of patches. In: Proceedings of Genetic and Evolutionary Computation Conference Companion, pp. 1873–1881 (2023)
5. Buzdalov, M.: Never-forgetting genetic algorithms: a promising architecture for theory and practice. In: Proceedings of Genetic and Evolutionary Computation Conference Companion (to appear) (2026)
6. Corus, D., Oliveto, P.S., Yazdani, D.: When hypermutations and ageing enable artificial immune systems to outperform evolutionary algorithms. Theoret. Comput. Sci. **832**, 166–185 (2020)
7. Corus, D., Oliveto, P.S.: Standard steady state genetic algorithms can hillclimb faster than mutation-only evolutionary algorithms. IEEE Trans. Evol. Comput. **22**(5), 720–732 (2018)
8. Corus, D., Oliveto, P.S.: On the benefits of populations for the exploitation speed of standard steady-state genetic algorithms. Algorithmica **82**, 3676–3706 (2020)
9. Couckuyt, I., Gonzales, S.R., Branke, J.: Tutorial: Bayesian optimization. In: Proceedings of Genetic and Evolutionary Computation Conference Companion, pp. 772–790 (2024)
10. Dang, D.C., Eremeev, A.V., Lehre, P.K.: Non-elitist evolutionary algorithms excel in fitness landscapes with sparse deceptive regions and dense valleys. In: Proceedings of Genetic and Evolutionary Computation Conference, pp. 1133–1141 (2021)
11. Dang, D.C., Eremeev, A.V., Lehre, P.K., Qin, X.: Fast non-elitist evolutionary algorithms with power-law ranking selection. In: Proceedings of Genetic and Evolutionary Computation Conference, pp. 1372–1380 (2022)

12. Doerr, B., Doerr, C.: Optimal static and self-adjusting parameter choices for the $(1 + (\lambda, \lambda))$ genetic algorithm. Algorithmica **80**(5), 1658–1709 (2018)
13. Doerr, B., Doerr, C., Ebel, F.: From black-box complexity to designing new genetic algorithms. Theoret. Comput. Sci. **567**, 87–104 (2015)
14. Doerr, B., Le, H.P., Makhmara, R., Nguyen, T.D.: Fast genetic algorithms. In: Proceedings of Genetic and Evolutionary Computation Conference, pp. 777–784 (2017)
15. Erdős, P., Rényi, A.: On two problems of information theory. Magyar Tudományos Akadémia Matematikai Kutató Intézet Közleményei **8**, 229–243 (1963)
16. Hansen, N.: Benchmarking a BI-population CMA-ES on the BBOB-2009 function testbed. In: Proceedings of Genetic and Evolutionary Computation Conference Companion, pp. 2389–2396. ACM (2009)
17. Jägersküpper, J., Storch, T.: When the plus strategy outperforms the comma strategy—and when not. In: Proceedings of IEEE Symposium on Foundations of Computational Intelligence, pp. 25–32 (2007)
18. Jorritsma, J., Lengler, J., Sudholt, D.: Comma selection outperforms plus selection on one max with randomly planted optima. Algorithmica **87**(12), 1804–1863 (2025)
19. Köetzing, T., Sudholt, D., Theile, M.: How crossover helps in pseudo-Boolean optimization. In: Proceedings of Genetic and Evolutionary Computation Conference, pp. 989–996 (2011)
20. Lehre, P.K., Witt, C.: Black-box search by unbiased variation. Algorithmica **64**, 623–642 (2012)
21. Lengler, J., Offermann, T.: Diversity-preserving exploitation of crossover. In: Foundations of Genetic Algorithms, pp. 154–165 (2025). https://arxiv.org/html/2507.01524v1
22. Oliveto, P.S., Sudholt, D.: On the runtime analysis of stochastic ageing mechanisms. In: Proceedings of Genetic and Evolutionary Computation Conference, pp. 113–120 (2014)
23. Oliveto, P.S., Sudholt, D., Witt, C.: Tight bounds on the expected runtime of a standard steady state genetic algorithm. Algorithmica **84**, 1603–1658 (2022)
24. Pitzer, E., Affenzeller, M.: Cheating like the neighbors: Logarithmic complexity for fitness evaluation in genetic algorithms. In: Proceedings of IEEE Congress on Evolutionary Computation, pp. 1431–1438 (2021)
25. R Core Team.: R: A language and environment for statistical computing (2025). https://www.R-project.org/
26. Shahriari, B., Swersky, K., Wang, Z., Adams, R.P., de Freitas, N.: Taking the human out of the loop: a review of Bayesian optimization. Proceedings of IEEE **104**(1), 148–175 (2015)
27. Sudholt, D.: How crossover speeds up building block assembly in genetic algorithms. Evol. Comput. **25**(2), 237–274 (2017)
28. Wilcoxon, F.: Individual comparisons by ranking methods. Biometrics Bull. **1**(6), 80–83 (1945)
29. Witt, C.: Runtime analysis of the $(\mu + 1)$ EA on simple pseudo-Boolean functions. Evol. Comput. **14**(1), 65–86 (2006)
30. Yao, X., Liu, Y.: Fast evolution strategies. In: Proceedings of International Conference on Evolutionary Programming, pp. 149–161. No. 1213 in Lecture Notes in Computer Science (1997)

On the Generalization Bounds of Symbolic Regression with Genetic Programming

Masahiro Nomura[1(✉)], Ryoki Hamano[2], and Isao Ono[1]

[1] Institute of Science Tokyo, Yokohama, Japan
{nomura,isao}@comp.isct.ac.jp
[2] CyberAgent, Shibuya, Japan
hamano_ryoki_xa@cyberagent.co.jp

Abstract. Symbolic regression (SR) with genetic programming (GP) aims to discover interpretable mathematical expressions directly from data. Despite its strong empirical success, the theoretical understanding of why GP-based SR generalizes beyond the training data remains limited. In this work, we provide a learning-theoretic analysis of SR models represented as expression trees. We derive a generalization bound for GP-style SR under constraints on tree size, depth, and learnable constants. Our result decomposes the generalization gap into two interpretable components: a structure-selection term, reflecting the combinatorial complexity of choosing an expression-tree structure, and a constant-fitting term, capturing the complexity of optimizing numerical constants within a fixed structure. This decomposition provides a theoretical perspective on several widely used practices in GP, including parsimony pressure, depth limits, numerically stable operators, and interval arithmetic. In particular, our analysis shows how structural restrictions reduce hypothesis-class growth while stability mechanisms control the sensitivity of predictions to parameter perturbations. By linking these practical design choices to explicit complexity terms in the generalization bound, our work offers a principled explanation for commonly observed empirical behaviors in GP-based SR and contributes towards a more rigorous understanding of its generalization properties.

Keywords: Symbolic regression · Genetic programming · Generalization bound

1 Introduction

Symbolic regression (SR) [19] aims to discover analytical expressions that describe relationships between input variables and target outputs directly from data. Unlike conventional regression methods, which assume a predefined model class and then optimize parameters within that class, SR simultaneously searches for both the functional form of the predictor and its numerical coefficients. Because the output of SR is an explicit mathematical expression, the resulting models are often interpretable and can sometimes reveal underlying mechanisms

G. Iacca et al. (Eds.): PPSN 2026, LNCS 16985, pp. 99–114, 2027.
https://doi.org/10.1007/978-3-032-36214-8_7

in the data. For this reason, SR has long been regarded as a promising approach for interpretable modeling and scientific discovery, where compact analytical expressions can provide insight in addition to predictive accuracy [8,33,35].

Genetic programming (GP) [18,29] is one of the most widely used approaches to SR. In GP-based SR, candidate models are typically represented as expression trees whose internal nodes correspond to operators and whose leaves correspond to variables or constants. Evolutionary search is then used to explore the space of symbolic structures through operations such as mutation, crossover, and selection. In addition, the numerical constants within a given structure are often refined using dedicated continuous optimization techniques, such as nonlinear least squares [17]. This hybrid nature—combining discrete structure search with continuous parameter optimization—makes GP particularly attractive for SR, as it enables open-ended exploration of functional forms without requiring prior commitment to a specific model class. However, the same flexibility that allows GP to discover concise symbolic models also raises a fundamental question: why should a model found by GP generalize beyond the training data? In learning-theoretic terms, this question concerns how well a model that fits the training data will perform on unseen data. This is typically formalized through the notion of a generalization gap, the difference between training error and expected prediction error, and the goal of deriving bounds that hold uniformly over a class of models. In practice, GP-based SR is known to be sensitive to design choices such as tree-size limits, maximum depth, constant optimization strategies, and the use of numerically stable operators, many of which are motivated by the need to control overfitting.

These observations naturally lead to the question of how such design choices influence generalization from a theoretical perspective. The GP literature has proposed various techniques intended to improve the generalization ability of SR models, including parsimony pressure, depth limits, and operator design aimed at improving numerical stability [15,23,30]. These approaches are widely used and often effective in practice. Nevertheless, their effects are typically justified empirically rather than through a unified theoretical framework. One reason is that SR induces a hypothesis class with a hybrid structure, combining discrete structural choices with continuous parameter optimization, which makes standard learning-theoretic tools difficult to apply directly. As a result, although many practical heuristics for controlling overfitting in GP-based SR have been developed, there remains limited learning-theoretic understanding of how these two aspects jointly influence generalization. Understanding this interplay is essential for developing principled methods for controlling model complexity in SR.

In this paper, we present a learning-theoretic analysis of SR models represented by expression trees. To the best of our knowledge, this is among the first works to derive an explicit generalization bound for GP-style SR models. Our main result provides a bound that decomposes the generalization gap into two interpretable components corresponding to structure selection and constant fitting. Specifically, the bound consists of a term that scales with the sensitivity of the model to its learnable constants and a term that depends on the logarithm of the number of admissible tree structures, reflecting the combinatorial complexity of structure selection. This decomposition offers a theoretical lens for understanding how common mechanisms used in GP-based SR influence generalization. In particular, structural control techniques such as parsimony pressure or depth limits restrict the space of admissible tree structures, while numerical-stability mechanisms such as interval-based restrictions help control the sensitivity of predictions to parameter perturbations. The resulting separation is reminiscent of the function- and parameter-complexity penalties used in MDL and Bayesian model-selection approaches to SR [2,3,10], but here it is obtained top-down from a generalization analysis rather than derived from coding-length or marginal-likelihood principles.

The remainder of this paper is organized as follows. Section 2 introduces the SR setting and the expression-tree representation considered in this paper. Section 3 presents our main theorem, the generalization bound for GP-based SR. Section 4 proves the main theorem using learning-theoretic analysis. Section 5 discusses the connection between our theoretical result and empirical successes in GP-based SR. Finally, Sect. 6 concludes the paper.

2 Setup

2.1 Data, Risks, and Loss

For SR, we consider a supervised regression setting. Let $(x, y) \sim \mathcal{D}$ with $x \in \mathcal{X} \subseteq \mathbb{R}^d$ and $y \in \mathcal{Y} \subseteq \mathbb{R}$. Given an i.i.d. sample $S = \{(x_i, y_i)\}_{i=1}^m$, define the population risk and the empirical risk:

$$L(f) = \mathbb{E}[\ell(f(x), y)] \ , \quad \widehat{L}_S(f) = \frac{1}{m}\sum_{i=1}^{m} \ell(f(x_i), y_i) \ , \tag{1}$$

where ℓ is a loss function measuring prediction error.

Definition 1 (Bounded Lipschitz loss). *We assume $\ell : \mathbb{R} \times \mathcal{Y} \to [0, 1]$ is 1-Lipschitz in its first argument:*

$$|\ell(a, y) - \ell(a', y)| \leqslant |a - a'| \ , \quad \forall a, a' \in \mathbb{R} \ , \ \forall y \in \mathcal{Y} \ . \tag{2}$$

This assumption is standard in statistical learning theory [25] and allows generalization bounds to be expressed in terms of the complexity of the function class. In practice, common losses such as mean squared error (MSE) can be accommodated through appropriate normalization or bounded-domain restrictions.

2.2 Expression-Tree Structures

Our predictors are represented as expression trees, as commonly used in GP-based SR. Internal nodes represent operators and leaves represent input variables or constants.

Let the operator vocabulary be partitioned by arity (number of arguments) up to $a_{\max}$:

$$\mathcal{O} = \bigcup_{k=1}^{a_{\max}} \mathcal{O}_k \ , \quad M_k = |\mathcal{O}_k| \ . \tag{3}$$

Typical examples include $\mathcal{O}_1 = \{\sin, \cos, \exp\}$ and $\mathcal{O}_2 = \{+, -, \times, \div\}$, where unary operators belong to $\mathcal{O}_1$ and binary operators belong to $\mathcal{O}_2$.

Leaves of the tree represent input variables and constants. Let

$$\mathcal{L}_0 = \{x_1, \ldots, x_d\} \cup \mathcal{C}_{\text{fix}} \ , \quad N_0 = |\mathcal{L}_0| \ , \tag{4}$$

where $\mathcal{C}_{\text{fix}}$ is an optional set of fixed constants (for example, $\{1, \pi\}$). In addition, we allow a special leaf symbol c, representing a learnable constant. During model fitting, this constant is optimized using the data.

Definition 2 (Expression-tree structure). *An expression-tree structure T is a rooted ordered tree with the following properties:*

- *Each internal node has arity k and is labeled by an operator in $\mathcal{O}_k$.*
- *Each leaf is labeled by a symbol in $\mathcal{L}_0 \cup \{c\}$.*

Let $p(T)$ be the number of leaves labeled by the learnable constant symbol c. We fix a deterministic ordering of these leaves (e.g., depth-first order). For a parameter vector $\theta \in \mathbb{R}^{p(T)}$, the function represented by the tree is $f_{T,\theta} : \mathcal{X} \to \mathbb{R}$ obtained by assigning the value θ_j to the j-th constant leaf.

Thus, each tree structure defines a family of functions parameterized by the constants θ.

2.3 Budgets and the Union Class

To control model complexity, we restrict both the size and depth of expression trees. Let size(T) be the number of nodes and depth(T) the maximum root-to-leaf path length measured in edges (with the root at depth 0).

Definition 3 (Budgets and induced classes). *Fix a size budget $s \in \mathbb{N}$ and a depth budget $D \in \mathbb{N}$ with $D \geqslant 1$. We define the set of admissible tree structures*

$$\mathcal{T}_{s,D} = \{T : \ \text{size}(T) \leqslant s, \ \text{depth}(T) \leqslant D\} \ . \tag{5}$$

For a parameter radius $R > 0$ and dimension $p \in \mathbb{N}$, define $\Theta_p(R) = \{\theta \in \mathbb{R}^p : \|\theta\|_2 \leqslant R\}$. For each tree $T \in \mathcal{T}_{s,D}$, we consider the class of functions

$$\mathcal{F}_T(R) = \{f_{T,\theta} : \ \theta \in \Theta_{p(T)}(R)\} \ . \tag{6}$$

Finally, the overall function class allowed under the budgets is the union

$$\mathcal{F}_{s,D}(R) = \bigcup_{T \in \mathcal{T}_{s,D}} \mathcal{F}_T(R) \ . \tag{7}$$

This class contains all expression-tree predictors whose structure satisfies the size and depth constraints and whose learnable constants lie within the radius-R ball.

2.4 Assumptions

To obtain generalization guarantees, we impose two standard conditions on the functions represented by expression trees. Intuitively, these assumptions ensure that predictions remain bounded and that small changes in the learnable constants lead to controlled changes in the output.

We assume that predictions produced by budgeted trees are uniformly bounded.

Definition 4 (Bounded predictions). *There exists $B > 0$ such that for all $T \in \mathcal{T}_{s,D}$, all $\theta \in \Theta_{p(T)}(R)$, and all $x \in \mathcal{X}$,*

$$|f_{T,\theta}(x)| \leqslant B \ . \tag{8}$$

This condition prevents expressions from producing arbitrarily large values. In practice, boundedness can be enforced by restricting the input domain, limiting parameter magnitudes, or using numerically stable operators.

We also assume that the output of the tree varies smoothly with respect to the learnable constants.

Definition 5 (Coefficient-Lipschitz sensitivity). *There exists a quantity $G > 0$ such that for all $T \in \mathcal{T}_{s,D}$, all $\theta, \theta' \in \Theta_{p(T)}(R)$, and all $x \in \mathcal{X}$,*

$$|f_{T,\theta}(x) - f_{T,\theta'}(x)| \leqslant G \ \|\theta - \theta'\|_2 \ . \tag{9}$$

The above condition is satisfied whenever the gradient with respect to the parameters is uniformly bounded: $\|\nabla_\theta f_{T,\theta}(x)\|_2 \leqslant G$. In SR practice, the constant G typically depends on the stability of operators, restrictions on the input domain, and scaling of the expressions.

3 Main Result

The following theorem provides a unified generalization bound for SR models represented by expression trees. The result explicitly decomposes the generalization error into two sources: structure selection and constant fitting.

Theorem 1 (Unified generalization bound for GP-SR). *Assume the bounded Lipschitz loss, bounded predictions, and coefficient-Lipschitz sensitivity assumptions. Fix a size budget s, a depth budget D, and a parameter radius $R > 0$. Then for any confidence level $\delta \in (0, 1)$, with probability at least $1 - \delta$ over the samples $S \sim \mathcal{D}^m$, the following inequality holds simultaneously for all $f \in \mathcal{F}_{s,D}(R)$:*

$$L(f) \ \leqslant \widehat{L}_S(f) + C_1 \, R \, G\sqrt{\frac{s}{m}} + C_2 \, B\sqrt{\frac{\log|\mathcal{T}_{s,D}|}{m}} + C_3\sqrt{\frac{\log(1/\delta)}{m}}, \tag{10}$$

where $C_1, C_2, C_3 > 0$ are absolute constants.

Moreover, in the unary/binary operator regime (arity at most 2), $\log|\mathcal{T}_{s,D}|$ *admits a depth-aware bound with exponential base*

$$\rho_D = 4\cos^2\left(\frac{\pi}{D+2}\right) < 4, \tag{11}$$

so that $\log|\mathcal{T}_{s,D}| = O(\log s + s\log(\rho_D(M_1+M_2)(N_0+1)))$ *up to lower-order terms.*

The first complexity term, $RG\sqrt{s/m}$, reflects the cost of fitting learnable constants within a fixed tree structure, while the second term, $B\sqrt{\log|\mathcal{T}_{s,D}|/m}$, captures the combinatorial cost of selecting a tree structure. The depth budget affects the latter term through ρ_D: smaller depth reduces the exponential growth rate of admissible structures.

4 Proof

This section proves the generalization bound stated in Theorem 1. The argument proceeds in three main steps. First, in Sect. 4.1, we control the Rademacher complexity of a finite union of function classes, which corresponds to selecting a tree structure among the candidates in $\mathcal{T}_{s,D}$. Second, in Sect. 4.2, we bound the Rademacher complexity of functions associated with a fixed tree structure, capturing the cost of fitting the learnable constants. Third, in Sect. 4.3, we bound the number of admissible tree structures, which determines the structure-selection term in the generalization bound. Finally, combining these results yields the main theorem (Sect. 4.4).

4.1 Finite Union Rademacher Complexity

Our function class $\mathcal{F}_{s,D}(R)$ is defined as a union over tree structures: $\mathcal{F}_{s,D}(R) = \bigcup_{T\in\mathcal{T}_{s,D}}\{f_{T,\theta} : \theta \in \Theta_{p(T)}(R)\}$. Thus, bounding its Rademacher complexity requires controlling the complexity of a finite union of function classes. This corresponds to the learning-theoretic cost of selecting a tree structure among the candidates in $\mathcal{T}_{s,D}$.

We use empirical Rademacher complexity [5,25]:

$$\mathfrak{R}_S(\mathcal{G}) = \mathbb{E}_\sigma\left[\sup_{g\in\mathcal{G}}\frac{1}{m}\sum_{i=1}^{m}\sigma_i g(x_i)\right], \tag{12}$$

where σ_i are i.i.d. Rademacher variables. The following standard lemma bounds the Rademacher complexity of a finite union of classes.

Lemma 1 (Finite union Rademacher bound). *Fix sample inputs* $x_1, \ldots, x_m$. *Let* $\{\mathcal{F}_j\}_{j=1}^{M}$ *be function classes such that* $|f(x_i)| \leqslant B$ *for all* $f \in \cup_j\mathcal{F}_j$ *and all* i. *Then*

$$\mathfrak{R}_S\left(\bigcup_{j=1}^{M}\mathcal{F}_j\right) \leqslant \max_{1\leqslant j\leqslant M}\mathfrak{R}_S(\mathcal{F}_j) + B\sqrt{\frac{2\log M}{m}}. \tag{13}$$

Proof. Define

$$Z_j(\sigma) := \sup_{f \in \mathcal{F}_j} \frac{1}{m} \sum_{i=1}^{m} \sigma_i f(x_i) \ , \quad Z(\sigma) := \max_{1 \leqslant j \leqslant M} Z_j(\sigma) \ . \tag{14}$$

Then $\mathfrak{R}_S(\mathcal{F}_j) = \mathbb{E}_\sigma Z_j(\sigma)$ and $\mathfrak{R}_S(\cup_j \mathcal{F}_j) = \mathbb{E}_\sigma Z(\sigma)$.

For each i, modifying only σ_i changes $Z_j(\sigma)$ by at most $2B/m$. Hence the centered variables $Z_j(\sigma) - \mathbb{E}Z_j(\sigma)$ are sub-Gaussian with proxy variance B^2/m. For sub-Gaussian variables with variance proxy σ^2, the standard maximal inequality [25] holds that

$$\mathbb{E}\left[\max_{1 \leqslant j \leqslant M}(Z_j - \mathbb{E}[Z_j])\right] \leqslant \sigma\sqrt{2\log M} \ . \tag{15}$$

With $\sigma = B/\sqrt{m}$, it yields

$$\mathbb{E}\left[\max_{1 \leqslant j \leqslant M}(Z_j - \mathbb{E}[Z_j])\right] \ \leqslant \ B\sqrt{\frac{2\log M}{m}} \ . \tag{16}$$

Therefore

$$\mathbb{E}[Z] \leqslant \max_j \mathbb{E}[Z_j] + B\sqrt{\frac{2\log M}{m}}, \tag{17}$$

which is the claim. □

Remark 1 (Structure selection as model choice). Lemma 1 captures the learning-theoretic cost of selecting a tree structure. Each class $\mathcal{F}_T(R)$ corresponds to a fixed symbolic structure with different numerical constants. The union over T therefore represents the discrete model-selection problem inherent in SR. The additional log term quantifies the combinatorial cost of searching over alternative symbolic forms.

4.2 Constant-Fitting Complexity

We now analyze the complexity of functions associated with a fixed tree structure T. In this setting, the tree structure is fixed, and the only source of variability comes from the learnable constants appearing in the tree.

Let $T \in \mathcal{T}_{s,D}$ be fixed and let $p = p(T)$ denote the number of constant leaves. To control the complexity of this parameterized class, we measure distances between parameter vectors through the induced prediction pseudometric

$$d_S(\theta, \theta') = \left(\frac{1}{m}\sum_{i=1}^{m}(f_{T,\theta}(x_i) - f_{T,\theta'}(x_i))^2\right)^{1/2} \ . \tag{18}$$

Lemma 2 (Pseudometric domination). *Under the coefficient-Lipschitz sensitivity assumption, for all $\theta, \theta' \in \Theta_p(R)$,*

$$d_S(\theta, \theta') \leqslant G\ \|\theta - \theta'\|_2 \ . \tag{19}$$

Proof. For each i, $|f_{T,\theta}(x_i) - f_{T,\theta'}(x_i)| \leqslant G\,\|\theta - \theta'\|_2$. Squaring, averaging over i, and taking square roots gives the claim. □

Remark 2 (Sensitivity to constants). The quantity G measures how strongly predictions change when the learnable constants are perturbed. Large values of G correspond to expressions whose outputs vary sharply with small parameter changes. Such expressions are often numerically unstable and prone to overfitting, a phenomenon frequently observed in GP-based SR when nested nonlinear operators are present.

We next bound the covering number of the parameter space under the pseudometric d_S. Intuitively, this measures how many parameter vectors are needed to approximate all possible constant settings up to prediction error ε on the sample.

Lemma 3 (Covering number under d_S). *Let $p = p(T)$. For any $\varepsilon > 0$,*

$$N(\varepsilon, \Theta_p(R), d_S) \;\leqslant\; \left(1 + \frac{2RG}{\varepsilon}\right)^p \;.$$

Proof. By Lemma 2, an (ε/G)-cover in Euclidean norm is an ε-cover under d_S. By the Euclidean ball covering bound [34] (scaled to radius R), $N(\eta, \Theta_p(R), \|\cdot\|_2) \leqslant (1 + 2R/\eta)^p$. Substitute $\eta = \varepsilon/G$. □

To convert this covering-number bound into a Rademacher complexity estimate, we use Dudley's entropy integral, which relates Rademacher complexity to the logarithm of covering numbers.

Theorem 2 (Dudley's entropy integral [11]). *There exists an absolute constant $C > 0$ such that for any class $\mathcal{G}$ of real-valued functions on $\{x_1, \ldots, x_m\}$, the empirical Rademacher complexity satisfies*

$$\mathfrak{R}_S(\mathcal{G}) \;\leqslant\; \frac{C}{\sqrt{m}} \int_0^\infty \sqrt{\log N(\varepsilon, \mathcal{G}, d_S)}\, d\varepsilon \;, \tag{20}$$

where d_S is the empirical L_2 pseudo-metric.

Applying this result to the class $\mathcal{F}_T(R)$ yields the following bound.

Theorem 3 (Fixed-structure constant-fitting bound). *For fixed T with $p = p(T)$, there exists an absolute constant $C_{\text{par}} > 0$ such that*

$$\mathfrak{R}_S(\mathcal{F}_T(R)) \;\leqslant\; C_{\text{par}}\, R\, G \sqrt{\frac{p}{m}} \;. \tag{21}$$

Consequently, since $p(T) \leqslant \text{size}(T) \leqslant s$,

$$\mathfrak{R}_S(\mathcal{F}_T(R)) \;\leqslant\; C_{\text{par}}\, R\, G \sqrt{\frac{s}{m}} \;. \tag{22}$$

Proof. By Theorem 2 and Lemma 3,

$$\mathfrak{R}_S(\mathcal{F}_T(R)) \leqslant \frac{C}{\sqrt{m}} \int_0^{\infty} \sqrt{p \log\left(1 + \frac{2RG}{\varepsilon}\right)}\, d\varepsilon \ . \tag{23}$$

Truncate the integral at $2RG$ using the diameter bound from Lemma 2. Then bound $\log(1+u) \leqslant u$ and integrate as in standard Dudley computations, yielding $\mathfrak{R}_S(\mathcal{F}_T(R)) \leqslant (4C)\, RG\sqrt{p/m}$. Set $C_{\text{par}} = 4C$. □

Remark 3 (Effect of learnable constants). The bound grows with the square root of the number of constant leaves. This reflects the intuition that introducing additional numerical parameters increases the flexibility of the expression even when its symbolic structure is fixed.

4.3 Structure-Selection Complexity

We now bound the number of admissible tree structures in $\mathcal{T}_{s,D}$. This quantity determines the structure-selection term in the generalization bound through $\log |\mathcal{T}_{s,D}|$.

The analysis proceeds in two steps. First, we bound the number of possible tree shapes with bounded size and depth. Second, we incorporate labelings by operators and terminals.

Definition 6 (Unary/binary operator regime). *Assume all operators have arity at most 2. Let $M_1 = |\mathcal{O}_1|$ and $M_2 = |\mathcal{O}_2|$ denote the number of unary and binary operators, respectively.*

We begin by bounding the number of rooted order tree shapes of bounded depth. Classical results on planted plane trees (see, e.g., [7]) imply that the number of trees with bounded depth admits the following asymptotic growth:

Theorem 4 (de Bruijn–Knuth–Rice asymptotic [7]). *Let $A_{s,D}$ denote the number of planted plane trees with s nodes and depth[1] at most D. Then for fixed D and $s \to \infty$,*

$$A_{s,D} \sim \frac{4^s}{D+2} \tan^2\left(\frac{\pi}{D+2}\right) \cos^{2\,s}\left(\frac{\pi}{D+2}\right) \ . \tag{24}$$

The asymptotic expression reveals the exponential growth rate of $A_{s,D}$. From this we obtain a simple exponential upper bound.

Corollary 1 (Exponential upper bound with explicit base). *For each fixed $D \geqslant 1$ there exists $c_D > 0$ such that for all $s \geqslant 0$,*

$$A_{s,D} \leqslant c_D\, \rho_D^s \ , \quad \rho_D := 4\cos^2\left(\frac{\pi}{D+2}\right) < 4 \ . \tag{25}$$

[1] While we use a depth measured in *edges* D (with the root at depth 0), [7] uses a height in *nodes* h (counting the root as 1). The two are related by $D = h - 1$.

Proof. From Theorem 4, the dominant exponential factor is $4^s \cos^{2s}(\pi/(D+2)) = (4\cos^2(\pi/(D+2)))^s$. The remaining factors depend only on D and can therefore be absorbed into the constant c_D, yielding the stated bound.

Combining the results yields the following estimate.

Theorem 5 (Depth-aware structure counting). *Assume the unary/binary operator regime. There exists $c_D > 0$ such that*

$$|\mathcal{T}_{s,D}| \leqslant c_D\,(s+1)\left(\rho_D(M_1+M_2)\,(N_0+1)\right)^s , \tag{26}$$

where

$$\rho_D = 4\cos^2\left(\frac{\pi}{D+2}\right) < 4 . \tag{27}$$

Consequently,

$$\log|\mathcal{T}_{s,D}| \leqslant O(\log s) + s\log\rho_D + s\log(M_1+M_2) + s\log(N_0+1) . \tag{28}$$

Proof. We derive an upper bound on $|\mathcal{T}_{s,D}|$ by overcounting admissible tree structures. For each $n \leqslant s$, let $A_{n,D}$ denote the number of rooted ordered tree shapes with n nodes and depth at most D. By Corollary 1, we have

$$A_{n,D} \leqslant c_D\rho_D^n . \tag{29}$$

For each such tree shape, each internal node can be labeled by one of at most M_1+M_2 operators, and each leaf by one of N_0+1 terminal symbols. Let i and ℓ denote the number of internal nodes and leaves, respectively, so that $i+\ell=n$. The number of labelings is at most

$$(M_1+M_2)^i(N_0+1)^\ell \leqslant (M_1+M_2)^n(N_0+1)^n . \tag{30}$$

Summing over $n \leqslant s$ gives

$$|\mathcal{T}_{s,D}| \leqslant \sum_{n=0}^{s} c_D\rho_D^n\,(M_1+M_2)^n\,(N_0+1)^n \tag{31}$$

$$= c_D\sum_{n=0}^{s}\left(\rho_D(M_1+M_2)(N_0+1)\right)^n \tag{32}$$

$$\leqslant c_D(s+1)\left(\rho_D(M_1+M_2)(N_0+1)\right)^s . \tag{33}$$

We used the elementary bound for a geometric series

$$\sum_{n=0}^{s} a^n \leqslant (s+1)a^s \quad (a \geqslant 1) \tag{34}$$

in the last step. Taking logarithms gives the stated bound. □

Remark 4 (Effect of depth constraints). The depth parameter D influences the growth rate of the hypothesis class through the exponential base

$$\rho_D = 4\cos^2\left(\frac{\pi}{D+2}\right) . \tag{35}$$

This quantity is strictly smaller than 4 and increases monotonically toward 4 as D grows. Consequently, smaller depth budgets reduce the exponential growth rate of admissible tree structures. Intuitively, depth constraints remove highly elongated trees with long chains of nested operators. Even when the size budget s is fixed, such trees constitute a large fraction of the combinatorial search space. Restricting depth therefore acts as a structural regularizer by limiting the variety of possible expression shapes.

4.4 Proof of the Main Theorem

We now combine the preceding ingredients. A standard generalization bound for bounded 1-Lipschitz losses [25] implies that, with probability at least $1-\delta$,

$$\sup_{f\in\mathcal{F}_{s,D}(R)} \left(L(f) - \widehat{L}_S(f)\right) \leqslant a\,\mathfrak{R}_S(\mathcal{F}_{s,D}(R)) + b\sqrt{\frac{\log(1/\delta)}{m}} , \tag{36}$$

for absolute constants $a, b > 0$. Since the hypothesis class decomposes as $\mathcal{F}_{s,D}(R) = \cup_{T\in\mathcal{T}_{s,D}}\mathcal{F}_T(R)$, Lemma 1 gives

$$\mathfrak{R}_S(\mathcal{F}_{s,D}(R)) \leqslant \max_{T\in\mathcal{T}_{s,D}} \mathfrak{R}_S(\mathcal{F}_T(R)) + B\sqrt{\frac{2\log|\mathcal{T}_{s,D}|}{m}} . \tag{37}$$

By Theorem 3, for every $T \in \mathcal{T}_{s,D}$,

$$\mathfrak{R}_S(\mathcal{F}_T(R)) \leqslant C_{\mathrm{par}}\, R\, G\sqrt{\frac{s}{m}} . \tag{38}$$

Hence the maximum over T satisfies the same bound.

In the unary/binary regime, Theorem 5 provides an explicit bound on $\log|\mathcal{T}_{s,D}|$, with exponential base

$$\rho_D = 4\cos^2\left(\frac{\pi}{D+2}\right) . \tag{39}$$

Substituting these bounds and absorbing absolute constants into C_1, C_2, C_3 yields the statement of Theorem 1. □

5 Connecting the Generalization Bound to GP Practice

We now relate the generalization bound in (10) to recurring design choices and empirical phenomena in GP-based SR. The key organizing principle is the two-term decomposition of the bound: a *structure-selection* term, governed by the size of the discrete hypothesis set, and a *constant-fitting* term, governed by the sensitivity of predictions to learnable constants. This perspective helps clarify how common GP heuristics act as forms of complexity control.

5.1 Structure-Selection Term

The structure-selection term reflects the combinatorial cost of choosing a tree structure from the hypothesis set $\mathcal{T}_{s,D}$. Accordingly, mechanisms that restrict structural complexity can be interpreted as reducing $\log |\mathcal{T}_{s,D}|$.

Bloat as Hypothesis-Class Growth. A well-known empirical phenomenon in GP is *bloat* [20,21,31,32], where program size increases even when improvements in training fitness have saturated. In our framework, increasing the tree-size budget enlarges the hypothesis set $\mathcal{T}_{s,D}$ and therefore increases the structure-selection term $\log |\mathcal{T}_{s,D}|$ in the bound. From this perspective, bloat can be interpreted as expanding the effective hypothesis class considered by the algorithm, which may increase the risk of overfitting.

Parsimony Pressure as Capacity Control. Parsimony pressure explicitly biases selection toward smaller trees, for example by adding a size penalty to fitness. The accuracy–parsimony tradeoff and methods for parsimony pressure are classical topics in GP [30,36], including lexicographic variants [23]. Under our decomposition, parsimony pressure acts as an algorithmic proxy for reducing structural complexity by shifting the search toward smaller-s structures. Consequently, it can be interpreted as reducing $\log |\mathcal{T}_{s,D}|$.

Depth Limits. Depth constraints are ubiquitous in GP practice and are widely observed to stabilize evolution and improve test performance. In our framework, depth constraints reduce the number of admissible structures and hence lower the growth rate of $|\mathcal{T}_{s,D}|$. They may also improve stability by limiting sensitivity amplification caused by deep compositions, although their most direct effect in the bound is through the structure-selection term.

5.2 Constant-Fitting Term

The constant-fitting term captures the complexity of optimizing continuous parameters within a fixed structure. In the bound, this contribution appears through the factor $RG\sqrt{s/m}$, where G measures the sensitivity of predictions to perturbations of learnable constants.

Constant Optimization and Sensitivity-Driven Overfitting. GP-based SR commonly introduces ephemeral random constants and may further optimize them via local search, nonlinear least squares [22,24], covariance matrix adaptation evolution strategy (CMA-ES) [13,14], or related procedures [17]. Constant optimization generally improves SR performance and can lead to smaller, better-generalizing trees. However, when the fitted expression is sensitive to its constants, especially due to unstable operators or highly nonlinear compositions, aggressive fitting may still increase overfitting risk. In the lens of (10), this behavior corresponds to a larger constant-fitting term when the induced function is highly sensitive to changes in constants, that is, when G is large. For example, expressions involving nested nonlinear operators such as $\exp(\exp(c_1 x))$ can exhibit high sensitivity to small changes in the learnable constants, leading to large values of G. In contrast, simpler expressions such as $\sin(c_2 x) + c_3 x$ tend to be more stable.

Interval Arithmetic and Interval-Aware Operators. Interval arithmetic can detect expressions that produce undefined or unbounded outputs on the input domain, enabling rejection or penalization of such trees. Keijzer showed that incorporating interval arithmetic can substantially improve the robustness of SR [15]. Under our decomposition, these techniques primarily improve generalization by ruling out expressions that produce undefined or excessively large outputs, thereby helping to control B, while also discouraging unstable expressions and helping to reduce G. Smooth operator definitions, such as analytic quotient operators, therefore fit naturally into our framework as mechanisms for limiting sensitivity amplification [27,28]. Interval-based pruning may also remove invalid structures from the search space, which can indirectly reduce the effective structural complexity.

Linear Scaling. Linear scaling, which fits an affine transformation of a candidate expression to the targets, is often effective in practice and is emphasized in interval-arithmetic-based SR [15,16]. In our framework, linear scaling can be interpreted as concentrating continuous fitting into a small and relatively well-conditioned set of parameters. This provides a structured way to reduce effective sensitivity without substantially enlarging the continuous search space.

5.3 Summary and Implications

Overall, the generalization bound highlights two complementary sources of complexity: the combinatorial growth of possible tree structures and the sensitivity of predictions to numerical constants. This perspective suggests that effective SR systems should balance structural expressiveness and numerical stability. Limiting tree size and depth reduces the combinatorial complexity of the hypothesis space, while stabilizing operators and controlling constant optimization reduce sensitivity to parameter perturbations.

6 Conclusion

In this paper, we presented a learning-theoretic analysis of SR models represented by expression trees. We derived a generalization bound for GP-based SR that explicitly accounts for both structural complexity and parameter fitting. Our result shows that the generalization gap can be decomposed into two interpretable components corresponding to the selection of the tree structure and the estimation of numerical constants. This perspective provides a principled way to understand how common mechanisms in GP-based SR influence generalization. In particular, our analysis offers a theoretical interpretation of widely used practices such as parsimony pressure, structural constraints, and the use of numerically stable operators. By relating these mechanisms to the complexity terms appearing in the bound, our work helps bridge the gap between empirical heuristics and learning-theoretic principles. We hope that this perspective will contribute to the development of more principled methods for controlling model complexity and improving generalization in SR.

Our analysis is based on worst-case complexity measures, including uniform sensitivity bounds and covering arguments over the full hypothesis class. As a result, the obtained generalization bound is data-independent and may be conservative in practice. Although the qualitative implications of our decomposition for model design remain largely unchanged, the resulting conservativeness may affect the quantitative tightness of the bound. An important direction for future work is to develop data-dependent bounds that better capture the effective complexity of SR models on observed data, as has been extensively studied in other areas of machine learning [1,6,12]. For example, replacing worst-case sensitivity parameters by empirical measures of parameter sensitivity, or characterizing the effective set of structures explored by the search process, could lead to substantially tighter guarantees. Such refinements may enable more nuanced trade-offs between expressiveness and stability, while preserving the core insights provided by our analysis.

Another promising direction is to leverage the proposed bound for the design of generalization-aware SR algorithms. In many areas of machine learning, generalization bounds have been used to guide model design and training, for instance through PAC-Bayes objectives [12], norm-based regularization in deep learning [4,26], and stability-driven analysis [6]. In SR, related ideas have been explored through structural risk minimization (SRM)-driven GP approaches [9], which incorporate complexity estimates, such as approximations of the VC dimension, into the evolutionary process. However, these methods typically rely on empirical or surrogate complexity measures and are not derived from explicit generalization bounds for GP-style models. In contrast, our analysis provides a unified learning-theoretic framework that explicitly decomposes generalization into structural and parametric components, which may serve as a principled basis for algorithm design. This perspective opens the possibility of directly optimizing surrogate objectives derived from the generalization bound.

Acknowledgement. We are grateful to the anonymous reviewers for their helpful comments and suggestions. We are also grateful to Peter Rockett for valuable feedback.

References

1. Arora, S., Ge, R., Neyshabur, B., Zhang, Y.: Stronger generalization bounds for deep nets via a compression approach. In: International Conference on Machine Learning, pp. 254–263. PMLR (2018)
2. Bartlett, D., Desmond, H., Ferreira, P.: Priors for Symbolic Regression. In: Proceedings of the Companion Conference on Genetic and Evolutionary Computation, pp. 2402–2411 (2023)
3. Bartlett, D.J., Desmond, H., Ferreira, P.G.: Exhaustive symbolic regression. IEEE Trans. Evol. Comput. **28**(4), 950–964 (2023)
4. Bartlett, P.L., Foster, D.J., Telgarsky, M.J.: Spectrally-normalized margin bounds for neural networks. Adv. Neural Inf. Process. Syst. **30** (2017)
5. Bartlett, P.L., Mendelson, S.: Rademacher and Gaussian Complexities: Risk Bounds and Structural Results. J. Mach. Learn. Res. **3**(Nov), 463–482 (2002)
6. Bousquet, O., Elisseeff, A.: Stability and Generalization. J. Mach. Learn. Res. **2**(Mar), 499–526 (2002)
7. de Bruijn, N.G., Knuth, D.E., Rice, S.: The average height of planted plane trees. In: Graph theory and computing, pp. 15–22. Elsevier (1972)
8. Brunton, S.L., Proctor, J.L., Kutz, J.N.: Discovering governing equations from data by sparse identification of nonlinear dynamical systems. Proc. Natl. Acad. Sci. **113**(15), 3932–3937 (2016)
9. Chen, Q., Zhang, M., Xue, B.: Structural risk minimization-driven genetic programming for enhancing generalization in symbolic regression. IEEE Trans. Evol. Comput. **23**(4), 703–717 (2018)
10. Desmond, H.: (Exhaustive) symbolic regression and model selection by minimum description length. Philosophical Trans. Roy. Soc. Math. Phy. Eng. Sci. **384**(2317), 20240584 (2026)
11. Dudley, R.M.: The sizes of compact subsets of Hilbert space and continuity of Gaussian processes. J. Funct. Anal. **1**(3), 290–330 (1967)
12. Dziugaite, G.K., Roy, D.M.: Computing Nonvacuous Generalization Bounds for Deep (Stochastic) Neural Networks with Many More Parameters than Training Data. arXiv preprint arXiv:1703.11008 (2017)
13. Hansen, N.: The CMA Evolution Strategy: A Tutorial. arXiv preprint arXiv:1604.00772 (2016)
14. Hansen, N., Ostermeier, A.: Completely derandomized self-adaptation in evolution strategies. Evol. Comput. **9**(2), 159–195 (2001)
15. Keijzer, M.: Improving Symbolic Regression with Interval Arithmetic and Linear Scaling. In: European Conference on Genetic Programming, pp. 70–82. Springer (2003)
16. Keijzer, M.: Scaled symbolic regression. Genet. Program Evolvable Mach. **5**(3), 259–269 (2004)
17. Kommenda, M., Burlacu, B., Kronberger, G., Affenzeller, M.: Parameter identification for symbolic regression using nonlinear least squares. Genet. Program Evolvable Mach. **21**(3), 471–501 (2020)
18. Koza, J.R.: Genetic Programming: On the Programming of Computers by Means of Natural Selection, vol. 1. MIT Press (1992)

19. Kronberger, G., Burlacu, B., Kommenda, M., Winkler, S.M., Affenzeller, M.: Symbolic Regression. Chapman and Hall/CRC (2024)
20. Langdon, W.B.: Evolving Data Structures with Genetic Programming. In: Proceedings of the 6th International Conference on Genetic Algorithms, pp. 295–302 (1995)
21. Langdon, W.B., Poli, R.: Fitness Causes Bloat. In: Soft Computing in Engineering Design and Manufacturing, pp. 13–22. Springer (1997)
22. Levenberg, K.: A method for the solution of certain non-linear problems in least squares. Q. Appl. Math. **2**(2), 164–168 (1944)
23. Luke, S., Panait, L.: Lexicographic Parsimony Pressure. In: Proceedings of the 4th Annual Conference on Genetic and Evolutionary Computation, pp. 829–836 (2002)
24. Marquardt, D.W.: An algorithm for least-squares estimation of nonlinear parameters. J. Soc. Ind. Appl. Math. **11**(2), 431–441 (1963)
25. Mohri, M., Rostamizadeh, A., Talwalkar, A.: Foundations of Machine Learning. MIT press (2018)
26. Neyshabur, B., Tomioka, R., Srebro, N.: Norm-Based Capacity Control in Neural Networks. In: Conference on Learning Theory, pp. 1376–1401. PMLR (2015)
27. Ni, J., Drieberg, R.H., Rockett, P.I.: The use of an analytic quotient operator in genetic programming. IEEE Trans. Evol. Comput. **17**(1), 146–152 (2012)
28. Nicolau, M., Agapitos, A.: Choosing function sets with better generalisation performance for symbolic regression models. Genet. Program Evolvable Mach. **22**(1), 73–100 (2021)
29. Poli, R., Langdon, W.B., McPhee, N.F., Koza, J.R.: A Field Guide to Genetic Programming. Lulu. com (2008)
30. Poli, R., McPhee, N.F.: Parsimony Pressure Made Easy. In: Proceedings of the 10th annual conference on Genetic and evolutionary computation, pp. 1267–1274 (2008)
31. Soule, T., Heckendorn, R.B.: An analysis of the causes of code growth in genetic programming. Genet. Program Evolvable Mach. **3**(3), 283–309 (2002)
32. Tackett, W.A.: Genetic Programming for Feature Discovery and Image Discrimination. In: Proceedings of the 5th International Conference on Genetic Algorithms, pp. 303–311 (1993)
33. Udrescu, S.M., Tegmark, M.: AI Feynman: A Physics-Inspired Method for Symbolic Regression. Sci. Adv. **6**(16), eaay2631 (2020)
34. Vershynin, R.: High-Dimensional Probability: An Introduction with Applications in Data Science, vol. 47. Cambridge university press (2018)
35. Wang, Y., Wagner, N., Rondinelli, J.M.: Symbolic regression in materials science. MRS communications **9**(3), 793–805 (2019)
36. Zhang, B.T., Mühlenbein, H.: Balancing accuracy and parsimony in genetic programming. Evol. Comput. **3**(1), 17–38 (1995)

On the Runtime Analysis of Reinforcement Learning Hyper-Heuristics

Pietro S. Oliveto(✉), Zhenyu Wang, Peizhou Wu, and Mengqing Xu

Southern University of Science and Technology, Shenzhen 518055, China
olivetop@sustech.edu.cn,
{wangzy20233,wupz2023,xumq2025}@mail.sustech.edu.cn

Abstract. Selection Hyper-heuristics (HHs) automate algorithmic design by selecting from a set of low-level heuristics which one to apply at each stage of the optimisation process. Several impressive results have been recently rigorously proven regarding the performance of selection hyper-heuristics (HHs) for standard benchmark functions. However, the learning mechanisms employed by these HHs are considerably simplified compared to the machine learning techniques typically used in real world applications. In this paper we analyse a Reinforcement Learning Hyper-heuristic (RLHH) from the literature. The only previous result available proved that for a wide range of parameter settings, RLHH does not learn to select heuristics appropriately for the standard LeadingOnes benchmark function. In this paper, we rigorously prove that with appropriate parameter values RLHH equipped with two random local search operators, RLS_1 and RLS_2 optimises the LeadingOnes benchmark function in the best possible expected runtime achievable with the two operators up to lower order terms. Experiments show that for realistic problem sizes it is faster than the Generalised Random Gradient HH which was previously proven to also have optimal expected runtime up to lower order terms.

Keywords: Hyper-heuristics · Reinforcement Learning · Runtime Analysis · Theory

1 Introduction

Hyper-heuristics (HHs) are automated algorithm design techniques for optimisation problems [9,23,47]. They work at a higher level of abstraction than randomised search heuristics by either generating offline a heuristic for the problem at hand using heuristic components (*generation* HH) or online by selecting which heuristic to apply directly at each stage of the optimisation process from a set of available low level heuristics (*selection* HH). The aim is to relieve the practitioner of the task of identifying which algorithm and related parameter settings to apply to the given optimisation problem. Furthermore, different heuristics, or even different parameter settings may perform better at different stages of the optimisation process.

G. Iacca et al. (Eds.): PPSN 2026, LNCS 16985, pp. 115–131, 2027.
https://doi.org/10.1007/978-3-032-36214-8_8

Numerous successful applications of HHs have been reported in several domains [2,8,29,30,33,38,50,51]. Although the field building a theoretical understanding of HHs is still in its infancy, several advances have been made recently regarding the theoretical foundations of selection HHs [5,16,19,36,37]. See [43] for a survey of earlier results.

Selection HHs typically consist of two components: 1) a heuristic methodology, also referred to as the *learning mechanism*, that selects which heuristic to be applied in the next step, and 2) a move-acceptance methodology to decide whether to accept the new solution. Concerning the move acceptance component it has been shown how simple move acceptance hyper-heuristics (MAHH) are very effective at escaping local optima. The MAHH switching with some probability p between an elitist strict improvement acceptance rule and a non-elitist rule that accepts any generated solution can optimise the standard CLIFF_d benchmark function in the best known expected runtime of $O(n^3/d^2+n\log n)$ [36]. A variant of the MAHH that switches between accepting strict improvements and only accepting worsening moves has been shown to optimise the JUMP_m function in an impressive expected runtime of $O((n^3/m)\log n)$ function evaluations [5].

Impressive results have also been shown when the performance of the HH depends on the learning mechanism used to determine the most appropriate low-level heuristic to be applied at each stage of the optimisation process. In particular, the Generalised Random Gradient (GRG) HH has been proven to optimise the standard LEADINGONES benchmark function in the optimal expected runtime (up to the leading constant) achievable by any unary unbiased search heuristic if equipped with a sufficient number of low-level Randomised Local Search (RLS) heuristics [35]. Since the result depends on setting its *learning period* τ parameter appropriately, an Adaptive Random Gradient (ARG) was introduced later that also runs in optimal expected runtime, dispensing the user from having to identify an appropriate value for τ [20]. These algorithms have also been shown to be effective for other classes of functions [34]. Indeed, GRG has the best known expected runtime for the multimodal TWORATES benchmark function [37].

All the mentioned HHs use very simple learning mechanisms, the most sophisticated being the random gradient one that selects an operator uniformly at random and continues applying it so long as the operator succeeds at identifying an improving solution within the learning period τ. These, however, are very different from the more sophisticated machine learning mechanisms typically employed by HHs in real-world applications [4,6,10,12,22,27,39,42,45,46,48]. One exception is the work by Alanazi and Lehre [3] which analysed a reinforcement learning HH (RLHH) from the HH literature [42]. The RLHH assigns weights to each low level heuristic which are increased by a rewarding additive term α if the applied heuristic identifies an improvement and are decreased by a penalty term β otherwise. Then the operator to be applied is decided at each step via roulette wheel selection.

The analysis of Alanazi and Lehre though only provided negative results by focusing on identifying parameter ranges for which the RLHH is inefficient. In particular, they analysed the RLHH for the LEADINGONES benchmark function equipped with m low-level heuristics among which only one has a positive probability of identifying improvements. In this scenario they derived parameter settings for which the RLHH provably cannot distinguish the effective operator from the rest, thus runs in the same expected runtime as an algorithm that at each step selects an operator uniformly at random (i.e., *simple random* HH [13]).

In this paper we aim to continue building the theoretical foundations of HH using reinforcement learning mechanisms. We prove that with appropriate parameter settings the RLHH equipped with two operators, RLS_1 which flips one bit and RLS_2 which flips two bits without replacement, runs in the best possible expected runtime achievable using the two operators (i.e. $\approx 0.42329n^2$) for LEADINGONES. An experimental analysis reveals that RLHH is faster than the GRG HH for realistic problem sizes.

For space limitations we cannot include the proofs in this manuscript.

2 Preliminaries

In this section, we formally define the Reinforcement Learning Hyper-Heuristic (RLHH) algorithm and the benchmark function used for the analysis. We also introduce the mathematical tools that will be used in our runtime analysis.

2.1 Algorithm

The Reinforcement Learning Hyper-heuristic (RLHH) framework analyzed in this work is a selection hyper-heuristic that dynamically manages a portfolio of m low-level heuristics $H = \{h_1, \ldots, h_m\}$. Unlike static selection mechanisms, RLHH aims to automatically learn the optimal heuristic to be applied at different stages of the optimization process by maintaining a weight w_i for each heuristic.

The pseudocode for the RLHH framework as considered in [3] is given in Algorithm 1. At each iteration, the algorithm selects a heuristic h_i with a probability proportional to its current weight $p_i = w_i / \sum_{j=1}^{m} w_j$. If the applied heuristic generates a strict fitness improvement (i.e., $f(s') > f(s)$), it is rewarded by increasing its weight by an additive term α, capped at a maximum value $w_{\max}$. Conversely, if the heuristic fails to strictly improve the fitness, it is penalized by subtracting a term β, bounded below by $w_{\min}$. Other choices for the update rules (e.g., multiplicativex updates rather than additive) and for the selection rules (e.g., selecting the heuristic with highest weight i.e., *max choice*) have been applied in the literature. Here we follow the scheme analysed by Alanazi and Lehre described in Algorithm 1 [3].

In our analysis, we consider a portfolio of two standard mutation operators (i.e., $m = 2$) commonly used in pseudo-Boolean optimization; $h_1 = RLS_1$ which flips exactly one bit chosen uniformly at random, and $h_2 = RLS_2$ which flips exactly two distinct bits chosen uniformly at random.

Algorithm 1. Reinforcement Learning Hyper-heuristic (RLHH) [3,42]

procedure RLHH($\alpha, \beta, w_{\min}, w_{\max}, w$)
  $s \leftarrow \text{Unif}(\mathcal{S}); w_{[m]}^{(0)} \leftarrow w$
  while stopping conditions not satisfied **do**
    Pick $i \in [m]$, with probability $p_i^{(t)} = \frac{w_i^{(t)}}{\sum_{j=1}^{m} w_j^{(t)}}$
    $s' \leftarrow h_i(s)$
    $w_i^{(t+1)} \leftarrow \begin{cases} \min(w_i^{(t)} + \alpha, w_{\max}), & \text{if } f(s') > f(s) \\ \max(w_i^{(t)} - \beta, w_{\min}), & \text{otherwise} \end{cases}$
    if $f(s') > f(s)$ **then**
      $s \leftarrow s'$
  return s

The only available runtime analysis of the RLHH has shown that if the parameters are set such that α is at most as large as β (i.e. $\alpha \leq \beta$), then the HH will not be able to distinguish between the low level heuristics, thus at each step chooses one uniformly at random [3]. For the hyper-heuristic to be effective, the reward term α should be able to counteract the weight decrease due to the expected number of failures before a success.

We configure the RLHH parameters as follows. Reward term $\alpha = 1+1/\log n$, penalty term $\beta = 1/n$, initial weight $w_1^{(0)} = w_2^{(0)} = \log n$, upper bound $w_{\max} = \frac{n}{\log^{1+\gamma} n}$, where $\gamma \in (0,1)$ is a strictly positive constant and lower bound $w_{\min} = 1$. Considering that a basic randomised local search heuristic (i.e., RLS_1) always has an improvement probability of at least $1/n$ unless it is trapped on a local optima, thus succeeds in n expected steps, we set the penalty term to $\beta = 1/n$ and the reward term to $\alpha = 1 + 1/\log n$ (slightly larger than n times β). Thus, as long as the HH is not trapped on a local optima the weight of RLS_1 always increases in expectation. Yet, if some other low-level heuristic has a higher success probability, its weight will increase faster. Since the operators are all initialised with the same weight value, we let the initial weight grow slightly with the problem size (i.e., $w_1^{(0)} = w_2^{(0)} = \log n$) so that the first operator to improve does not gain a large weight advantage which may strongly bias (maybe incorrectly) the process in its favour. An upper bound on the weights, $w_{\max}$, is required so that when the best operator for the current optimisation stage changes, the time for the respective weights to increase and decrease will not be too large. At the same time $w_{\max}$ should also not be too small, to avoid that the weight of RLS_1 eventually catches up with that of the current best operator by reaching the upper bound due to its positive drift (by design). In this case it would not be possible to distinguish the weights because, once at the boundary, the weight of the best operator cannot increase further. Our analyses will show that with a setting of $w_{\max} = \frac{n}{\log^{1+\gamma} n}$, $\gamma \in (0,1)$ a constant, the RLHH optimises LeadingOnes in optimal expected runtime up to lower order terms. The lower bound on the weight can be simply set to any constant and we choose $w_{\min} = 1$.

2.2 Benchmark Function

We analyze the performance of the RLHH on the same standard pseudo-Boolean benchmark function LEADINGONES (LO) used in the previous work in the literature [3]. The function is defined over the search space $\mathcal{S} = \{0,1\}^n$ as

$$\mathrm{LO}(s) = \sum_{i=1}^{n} \prod_{j=1}^{i} s_j.$$

where s_j is the value of the j-th bit in the bit string s. The LO value of a bit-string is the number of consecutive leading 1-bits in the bit-string, thus the global optimum is the all-ones string $s^* = (1, \ldots, 1)$ since we are maximizing the function.

To successfully improve a candidate solution s with current fitness $\mathrm{LO}(s) = i < n$, an algorithm must necessarily mutate the $(i+1)$-th bit (the first 0-bit in the string). Any mutations applied to the leading prefix of i ones will destroy the prefix and strictly decrease the fitness. Meanwhile, mutations applied to the trailing suffix (bits from position $i+2$ to n) are entirely invisible to the current fitness evaluation, acting as neutral moves. The unary unbiased black box complexity of LEADINGONES is $\Theta(n^2)$ [31] and evolutionary algorithms using standard bit mutation match this performance. The standard (1+1) EA (with mutation rate $\frac{1}{n}$) has an expected runtime of $\frac{e-1}{2}n^2 - o(n^2) \approx 0.85914n^2 - o(n^2)$ [7]. By implementing a fitness-dependent mutation rate, the expected runtime of the (1+1) EA decreases to $\approx (1 \pm o(1))0.68n^2$ [7]. RLS_1 has an expected runtime of $0.5n^2$ fitness function evaluations [11]. The best expected runtime achievable by unbiased (1+1) black box algorithms is $\approx (1 \pm o(1))0.388n^2$ [14,21]. The Generalised Random Gradient (GRG) and the Adaptive Random Gradient selection HHs with a sufficiently large constant sized low level heuristic set $|H| = k = O(1)$ have been shown to match this optimal performance [18,20,35]. If the heuristic set is limited to $H = \{\mathrm{RLS}_1, \mathrm{RLS}_2\}$, then the best possible expected runtime is $(1 + o(1)) \cdot 0.42329n^2$ [35]. Here we show that RLHH matches this optimal performance.

We point out that other classes of randomised search heuristics with higher arity operators, including Estimation of Distribution Algorithms and Convex Search Algorithms, can run in $o(n^2)$ expected runtime [1,17,26,41].

2.3 Mathematical Tools

To rigorously analyze the expected runtime and the stochastic dynamics of the weight adaptation mechanism of the RLHH algorithm, we rely on several fundamental tools from probability theory and stochastic processes.

Theorem 1 (Chernoff Bounds [40]). *Let $X = \sum_{i=1}^{n} X_i$ be the sum of independent Poisson trials such that $\mathbb{E}[X] = \mu$. Then for any $0 < \delta < 1$*

$$\Pr(X \leq (1-\delta)\mu) \leq \exp\left(-\frac{\delta^2 \mu}{2}\right),$$

and for any $\delta > 0$,

$$\Pr(X \geq (1+\delta)\mu) \leq \left(\frac{e^{\delta}}{(1+\delta)^{1+\delta}}\right)^{\mu}.$$

Theorem 2 (Doob's Maximal Inequality [24]). *Let $(Z_t)_{t\geq 0}$ be a stochastic process adapted to a filtration $(\mathcal{F}_t)_{t\geq 0}$. For any constant $\delta > 0$,*

(i) ***Submartingale bound:*** *If $(Z_t)_{t\geq 0}$ is a non-negative submartingale, then for any finite time horizon $T \geq 0$, the probability that the maximum of the process reaches or exceeds δ is bounded by*

$$\Pr\left(\max_{0\leq t\leq T} Z_t \geq \delta\right) \leq \frac{\mathbb{E}[Z_T]}{\delta}.$$

(ii) ***Supermartingale bound:*** *If $(Z_t)_{t\geq 0}$ is a non-negative supermartingale, then the probability that the supremum of the process reaches or exceeds δ over infinite time is bounded by*

$$\Pr\left(\sup_{t\geq 0} Z_t \geq \delta\right) \leq \frac{\mathbb{E}[Z_0]}{\delta}.$$

Theorem 3 (Freedman's Inequality [25], [49]). *Let $(Y_t)_{t\geq 1}$ be a martingale difference sequence with respect to a filtration $(\mathcal{F}_t)_{t\geq 0}$. Suppose there exists a constant R such that $Y_t \leq R$ almost surely for all $t \geq 1$. Let $M_k = \sum_{t=1}^{k} Y_t$ be the associated martingale, and let $V_k = \sum_{t=1}^{k} \mathrm{Var}(Y_t \mid \mathcal{F}_{t-1})$ be the predictable quadratic variation. For any $\Delta > 0$ and $v > 0$,*

$$\Pr\left(\exists k \geq 1 : M_k \geq \Delta \text{ and } V_k \leq v\right) \leq \exp\left(-\frac{\Delta^2}{2(v + R\Delta/3)}\right).$$

Theorem 4 (Additive Drift Theorem [28,32]). *Let $(X_t)_{t\geq 0}$ be a stochastic process adapted to a filtration $(\mathcal{F}_t)_{t\geq 0}$ over a state space $\mathcal{S} \subseteq \mathbb{R}$. Let T be the first hitting time of the region $(-\infty, a]$. If $X_0 \geq a$ and there exists a constant $\delta > 0$ such that for all $t < T$, $\mathbb{E}[X_t - X_{t+1} \mid \mathcal{F}_t] \geq \delta$, then*

$$\mathbb{E}[T \mid X_0] \leq \frac{X_0 - a}{\delta}.$$

Theorem 5 (Law of Total Expectation/Tower Property [40]). *Let $(\Omega, \mathcal{F}, \Pr)$ be a probability space, and let $\mathcal{F}_1$ and $\mathcal{F}_2$ be two sub-σ-algebras of $\mathcal{F}$ such that $\mathcal{F}_1 \subseteq \mathcal{F}_2$. For any random variable X with finite expectation ($\mathbb{E}[|X|] < \infty$), it holds almost surely that:*

$$\mathbb{E}[\mathbb{E}[X \mid \mathcal{F}_2] \mid \mathcal{F}_1] = \mathbb{E}[X \mid \mathcal{F}_1].$$

In the context of discrete-time stochastic processes with a natural filtration $(\mathcal{F}_t)_{t\geq 0}$, this implies that for any time steps $s \leq t$, conditioning on the historical state $\mathcal{F}_s$ absorbs the intermediate conditioning on $\mathcal{F}_t$,

$$\mathbb{E}[\mathbb{E}[X \mid \mathcal{F}_t] \mid \mathcal{F}_s] = \mathbb{E}[X \mid \mathcal{F}_s].$$

2.4 Notation

Throughout this paper, the optimization process is modeled as a discrete-time stochastic process. We denote the current time step, which exactly corresponds to the number of fitness evaluations, by $t \in \mathbb{N}_0$, and the candidate solution maintained by the algorithm at step t by $s^{(t)} \in \{0,1\}^n$. The fitness value of the current solution is typically denoted by $i = \mathrm{LO}(s^{(t)})$. To rigorously ground our probabilistic arguments, we let $\mathcal{F}_t$ be the natural filtration representing the complete historical trajectory of the algorithmic state (including all past solutions, weights, and operator selections) up to step t.

Regarding the operator selection mechanism, $w_j^{(t)}$ represents the current weight of heuristic h_j (where $j \in \{1,2\}$ respectively correspond to RLS_1 and RLS_2), and $p_j^{(t)} = w_j^{(t)}/(w_1^{(t)} + w_2^{(t)})$ defines the selection probability of that heuristic at step t. Furthermore, we define $P_j(i)$ as the *success* probability that applying heuristic h_j to a solution with fitness i yields a strict fitness improvement. Finally, Δ_j^{sel} denotes the conditional expected one-step change in the weight of heuristic h_j, given that it was selected for application at the current step. We say that an event occurs *asymptotically almost surely* (a.a.s.) if its probability is at least $1 - o(1)$.

3 RLHH Has Optimal Runtime for Leading Ones

In this section, we rigorously prove that the RLHH algorithm achieves the optimal expected runtime for the LeadingOnes function. To understand the algorithmic dynamics, we first evaluate the state-dependent success probabilities of the two mutation operators at any given fitness level $i = \mathrm{LO}(s)$.

For operator RLS_1, a strict fitness improvement occurs if and only if the first 0-bit is flipped. Thus, its success probability is uniformly bounded by $P_1(i) = 1/n$. For operator RLS_2, a success requires flipping the first 0-bit alongside any of the $n - i - 1$ trailing bits. The exact probability of this joint event evaluates to:

$$P_2(i) = \frac{(n-i-1)}{\binom{n}{2}} = \frac{2(n-i-1)}{n(n-1)}.$$

Comparing these transition probabilities reveals a strict phase transition in the optimal search strategy. Solving $P_2(i) > P_1(i)$ yields $i < n/2$. Consequently, RLS_2 is preferable during the first half of the optimization process, whereas RLS_1 becomes more effective once $i > n/2$. As established in the literature, the optimal strategy for an algorithm equipped with these two operators is to apply RLS_2 exclusively for optimizing the first $n/2$ bits, and then permanently switch to RLS_1 for the remainder of the optimization [35].

To demonstrate that the RLHH algorithm learns and executes this optimal operator sequence a.a.s., we divide our mathematical analysis into three subsequent subsections. First, we analyze Phase 1 ($i < n/2$), proving that the parameter adaptation mechanism rapidly establishes a strict dominance of RLS_2 a.a.s. during the initial stage of optimization. Next, we examine Phase 2 ($i \geq n/2$), demonstrating that as the fitness crosses the theoretical threshold, the algorithm successfully reverses its weight distribution to maintain the absolute dominance of RLS_1 a.a.s.. Finally, we synthesize the bounded times from these two distinct phases to compute the total expected optimization time, concluding that RLHH has the optimal expected runtime up to a lower-order $o(n^2)$ term.

3.1 Dominance of RLS_2 in Phase 1

In this subsection, we analyze the runtime behavior of the RLHH algorithm on the LeadingOnes problem during Phase 1, defined as the interval where the fitness satisfies $0 \leq \mathrm{LO}(s) < \frac{n}{2}$. In this region, the operator RLS_2 possesses a higher success probability than the operator RLS_1. The analysis demonstrates that the dynamic weight adaptation mechanism successfully identifies and exploits RLS_2 with a selection probability of $p_2 = 1 - o(1)$, incurring an $o(n^2)$ time overhead compared to the optimal expected runtime.

The stochastic dynamics of Phase 1 are established through a sequence of supporting lemmata. First, Lemma 1 derives the conditional drifts of the operator weights based on their state-dependent success probabilities, showing that when the respective operators are selected w_2 grows asymptotically faster than w_1 in expectation. Lemma 2 ensures that the selection probability $p_2^{(t)}$ remains bounded away from zero for Phase 1. Building on this, Lemma 3 proves that the weight w_2 rapidly reaches its upper limit $w_{\max} = \frac{n}{\log^{1+\gamma} n}$ within an $o(n^2)$ time overhead. Concurrently, Lemma 4 constructs a non-negative supermartingale to bound the multiplicative growth of w_1 during the climbing stage of w_2, ensuring w_1 remains sub-polynomial. Finally, Lemma 5 applies the law of total expectation, Doob's inequality and Freedman's inequality to show that the dominance of RLS_2 is robustly maintained a.a.s. until the fitness reaches $n/2$.

We put everything together in Theorem 6 to formally establish that RLHH optimises the first $n/2$ leading ones in expected time $T_{\mathrm{OPT}_{n/2}} + o(n^2)$, where $T_{\mathrm{OPT}_{n/2}}$ is the exact expected hitting time for the pure RLS_2 algorithm to traverse the same fitness distance.

We start by analysing the expected single-step changes of the operator weights conditional on the operators being selected for reproduction.

Lemma 1. *For the RLHH algorithm with reward parameter $\alpha = 1 + 1/\log n$ and penalty parameter $\beta = 1/n$ optimizing the LeadingOnes function, let Δ_1^{sel} and Δ_2^{sel} denote the expected one-step weight changes of the operators RLS_1 and RLS_2, conditioned on being selected. The weight drifts always satisfy*

$$\Delta_1^{sel} = \frac{1}{n \log n} + \frac{1}{n^2}, \quad \text{and} \quad \Delta_2^{sel} = \mathcal{O}\left(\frac{1}{n}\right).$$

Furthermore, for any fitness level $i \leq n/2 - \epsilon n$ with any small constant $\epsilon > 0$, there exists a constant $c > 0$ such that the weight drift of RLS_2 is lower-bounded by

$$\Delta_2^{sel} \geq \frac{c}{n}.$$

By defining a composite process $U_t = 2w_2^{(t)} - w_1^{(t)}$ and constructing an exponential potential function $Z_t = \exp(-\lambda U_t)$ that operates as a non-negative supermartingale, we rigorously bound the probability that w_1 ever exceeds twice the value of w_2 over a polynomial time horizon by applying Doob's maximal inequality (Theorem 2) to Z_t.

Lemma 2. *For the RLHH algorithm optimizing the* LeadingOnes *function with initial weights $w_1^{(0)} = w_2^{(0)} = \log n$, $\alpha = 1 + 1/\log n$, $\beta = 1/n$, consider a polynomial time horizon $T = \text{poly}(n)$. Assuming the fitness satisfies $LO(s^{(t)}) \leq n/2 - \epsilon n$ for a constant $\epsilon > 0$ for all $t \leq T$, the probability that $w_1^{(t)} \geq 2w_2^{(t)}$ occurs at any step $t \in [0, T]$ is bounded by*

$$\Pr\left(\exists t \in [0, T] : w_1^{(t)} \geq 2w_2^{(t)}\right) = n^{-\Omega(1)}.$$

To prove that the RLHH algorithm achieves the optimal Phase 1 runtime in Theorem 6, the selection probability of RLS_2 must rapidly approach 1. We first show that the weight w_2 climbs to its upper bound $w_{\max}$ in $o(n^2)$ steps with probability $1 - o(1)$. Within this time with overwhelming probability at most $\frac{n}{2} - \epsilon n$ leading ones are optimized for a constant $\epsilon > 0$. Lemma 1, thus ensures an $\Omega(1/n)$ drift for w_2 during the climbing stage.

Lemma 3. *For the RLHH algorithm optimizing the* LeadingOnes *function with parameters $w_{\max} = \frac{n}{\log^{1+\gamma} n}$ for a constant $0 < \gamma < 1$, $\alpha = 1 + 1/\log n$, $\beta = 1/n$, with probability at least $1 - \mathcal{O}\left(\frac{1}{\log^{\gamma/2} n}\right)$, the weight w_2 reaches $w_{\max}$ within $t_{\max} = \frac{n^2}{\log^{1+\gamma/2} n}$ steps, and the fitness satisfies $LO(s^{(t)}) \leq n/2 - \epsilon n$ during this entire climbing period.*

While Lemma 3 proves that the weight w_2 rapidly reaches $w_{\max}$, establishing a selection probability of $p_2 = 1 - o(1)$ also requires that w_1 remains comparatively small during the climbing stage of w_2. Relying on the $o(n^2)$ time bound established in Lemma 3, we analyze the multiplicative increase of w_1.

We divide the climbing process of the weight of RLS_2 from $w_2 = \log n$ to $w_{\max}$ into m phases of multiplicatively increasing length and for each phase we construct a non-negative supermartingale that allows us to prove using Freedman's inequality that with exponentially small probability in the length of the phase, the weight never drops below half the value it had at the beginning of the phase. Then we construct another martingale for each phase to show that the probability that w_2 reaches the end of the phase in more than twice its expected time is also exponentially small. Finally using the calculated bounds for w_2 we show that w_1 remains sub-polynomial i.e., much smaller than $w_{\max}$.

Lemma 4. *For the RLHH algorithm optimizing the* LeadingOnes *function with initial weights $w_1^{(0)} = w_2^{(0)} = \log n$, $\alpha = 1 + 1/\log n$, $\beta = 1/n$, let τ be the first time step where $w_2^{(\tau)}$ reaches $w_{\max} = \frac{n}{\log^{1+\gamma} n}$ for a constant $0 < \gamma < 1$. At time τ, the weight w_1 is bounded by $w_1^{(\tau)} = \mathcal{O}(\log^2 n)$ with probability at least $1 - \mathcal{O}\left(\frac{1}{\log n}\right)$.*

To complete the analysis of Phase 1, we prove that once the weight w_2 reaches $w_{\max}$, it stably maintains its dominant position until the algorithm successfully reaches the fitness level $n/2$. Because the dynamics of w_1 and w_2 are coupled, we analyze them sequentially: we first unconditionally bound the maximum possible growth of w_1 assuming w_2 does not drop by more than half its value. Afterwards we use this established bound on w_1 to prove via Freedman's inequality that w_2 still does not drop by half its value with high probability.

Lemma 5. *For the RLHH algorithm optimizing the* LeadingOnes *function with parameters* $w_{\max} = \frac{n}{\log^{1+\gamma} n}$ *for a constant* $0 < \gamma < 1$, $\alpha = 1 + 1/\log n$, $\beta = 1/n$, *suppose* w_2 *has reached* $w_{\max}$. *With probability at least* $1 - n^{-1/2+o(1)}$, *the weights maintain the strict bounds* $w_2^{(t)} \geq w_{\max}/2$ *and* $w_1^{(t)} \leq \sqrt{w_{\max}}$ *for all subsequent steps until Phase 1 concludes. Consequently, the selection probability robustly satisfies* $p_2^{(t)} = 1 - o(1)$ *throughout this entire period.*

We now prove the main result for Phase 1.

Theorem 6. *For the RLHH algorithm optimizing the* LeadingOnes *function with parameters* $w_1^{(0)} = w_2^{(0)} = \log n$, $w_{\max} = \frac{n}{\log^{1+\gamma} n}$ *for a constant* $0 < \gamma < 1$, $w_{\min} = 1$, $\alpha = 1 + 1/\log n$, $\beta = 1/n$, *let* T_{phase1} *be the total number of fitness evaluations required to reach a fitness of at least* $n/2$ *from a uniformly distributed random initial string. The expected runtime is strictly bounded by*

$$\mathbb{E}[T_{phase1}] = T_{OPT_{n/2}} + o(n^2),$$

where $T_{OPT_{n/2}}$ *is the expected runtime for the theoretically optimal* RLS_2 *algorithm to traverse the same fitness distance.*

3.2 Dominance of RLS_1 in Phase 2

In this section, we bound the expected runtime of RLHH during Phase 2. For the purpose of analysis, we reset the time counter to $t = 0$ at the transition point and pessimistically consider a worst-case initial weight distribution, where $w_1^{(0)} = w_{\min}$ and $w_2^{(0)} = w_{\max}$.

First, Lemma 6 uses the Artificial Fitness Level Method [44] to establish that the algorithm quickly reaches a sufficiently high fitness level where RLS_2 starts to lose its advantage quickly. Then, Lemma 7 proves that once the algorithm reaches this high-fitness region, at least one of the weights will rapidly hit a boundary. Once at least one boundary is hit, Lemma 5 guarantees that RLS_1 will rapidly become dominant and maintain its dominance until the end of the optimization process.

We put these lemmata together in Theorem 7 to establish that RLHH optimizes the remaining fitness levels from $n/2$ to n in expected time $T_{\mathrm{OPT}_{n/2\to n}} + o(n^2)$, where $T_{\mathrm{OPT}_{n/2\to n}}$ is the exact expected hitting time for the pure RLS_1 algorithm to traverse this fitness distance.

Lemma 6. *For the RLHH algorithm optimizing the* LeadingOnes *function, assume the algorithm has reached the beginning of Phase 2 with fitness* $LO(s^{(0)}) = n/2$. *The expected time to reach a fitness of at least* $n/2 + n/\log n$ *is bounded by* $\mathbb{E}[T_{buf}] \leq \frac{2n^2}{\log n}$.

We now bound the expected time required for either weight to reach its boundary. In particular, we define a potential function that captures the distance of the two weights from the respective boundaries. Then we apply the additive drift theorem to calculate the expected time for the potential function to reach zero which will imply that one of the two boundaries has been reached.

Lemma 7. *For the RLHH algorithm optimizing the* LeadingOnes *function with parameters* $w_{\max} = \frac{n}{\log^{1+\gamma} n}$ *for a constant* $0 < \gamma < 1$, $w_{\min} = 1$, $\alpha = 1 + 1/\log n$, $\beta = 1/n$, *assume the algorithm has reached the beginning of Phase 2 with fitness* $LO(s^{(0)}) = n/2$. *Let* τ *be the first time step where either the condition* $w_1^{(t)} = w_{\max}$ *or* $w_2^{(t)} = w_{\min}$ *is triggered. The expected hitting time is bounded by* $\mathbb{E}[\tau] = \mathcal{O}\left(\frac{n^2}{\log^{\gamma} n}\right)$.

In the following lemma, we prove that once either weight hits its respective boundary, the system rapidly stabilizes into a new regime that strictly favors RLS_1 (e.g., $p_1 = 1 - o(1)$). We split the proof into two cases.

For the case where w_1 has hit the upper boundary, we first use Freedman's inequality to show that w_1 never drops by more than half its value. Using this bound on w_1, we calculate the expected multiplicative decrease of w_2 assuming pessimistically that at the beginning of the phase its value is $w_{\max}$. Then the law of total expectation and a Markov inequality allow us to show that a.a.s. w_2 drops down to at least $o(w_{\max})$ in $o(n^2)$ steps. Finally, we apply Doob's inequality to show that w_2 remains $o(w_{\max})$ until the global optimum is found.

For the second case where w_2 has hit the lower boundary, we first use Doob's inequality to show that w_2 never increases above $\log^2 n$. Then we use this bound on the weight to calculate the additive drift of w_1 which allows us to derive the expected time for w_1 to reach a high enough value to guarantee a selection probability of $p_1 = 1 - o(1)$. A Markov inequality and a Freedman's inequality show that $p_1 = 1 - o(1)$ is sustained until the end of the phase.

Lemma 8. *For the RLHH algorithm optimizing the* LeadingOnes *function with parameters* $w_{\max} = \frac{n}{\log^{1+\gamma} n}$ *for a constant* $0 < \gamma < 1$, $w_{\min} = 1$, $\alpha = 1+1/\log n$, $\beta = 1/n$, *suppose at time* τ *the weights satisfy either* $w_1^{(\tau)} = w_{\max}$ *or* $w_2^{(\tau)} = w_{\min}$. *With probability at least* $1 - \mathcal{O}\left(\frac{1}{\log^2 n}\right)$, *within an additional* $t_{stab} = \frac{n^2}{\log^{\gamma/2} n}$ *steps, the algorithm establishes a dominance regime where the selection probability of* RLS_1 *satisfies* $p_1^{(t)} = 1 - o(1)$, *and this regime is maintained until Phase 2 concludes.*

We now put all the lemmata together to prove the main result for Phase 2.

Theorem 7. *For the RLHH algorithm optimizing the* LeadingOnes *function with parameters* $w_1^{(0)} = w_2^{(0)} = \log n$, $w_{\max} = \frac{n}{\log^{1+\gamma} n}$ *for a constant* $0 < \gamma < 1$, $w_{\min} = 1$, $\alpha = 1 + 1/\log n$, $\beta = 1/n$, *let* T_{phase2} *be the total number of fitness evaluations required to reach the global optimum* $\mathrm{LO}(s) = n$ *starting from a fitness of* $n/2$. *The expected runtime is strictly bounded by*

$$\mathbb{E}[T_{phase2}] = T_{OPT_{n/2 \to n}} + o(n^2),$$

where $T_{OPT_{n/2 \to n}}$ *is the exact expected hitting time for the theoretically optimal pure* RLS_1 *algorithm to traverse the second half of the search space.*

3.3 Total Expected Runtime

In this subsection, we summarize the runtime analyses of Phase 1 and Phase 2 to establish the overall expected runtime of the RLHH algorithm. We show that the weight adaptation successfully matches the performance of the theoretically optimal operator sequence, accumulating only a sub-quadratic time overhead across the entire optimization process.

Theorem 8. *For the RLHH algorithm optimizing the* LeadingOnes *function with parameters* $w_1^{(0)} = w_2^{(0)} = \log n$, $w_{\max} = \frac{n}{\log^{1+\gamma} n}$ *for a constant* $0 < \gamma < 1$, $w_{\min} = 1$, $\alpha = 1 + 1/\log n$, $\beta = 1/n$, *let* T *be the total number of fitness evaluations required to reach the global optimum* $LO(s) = n$ *from a uniformly distributed random initial string. The expected runtime is strictly bounded by*

$$\mathbb{E}[T] = T_{OPT_{n/2}} + T_{OPT_{n/2 \to n}} + o(n^2) = \frac{1 + \ln 2}{4} n^2 + o(n^2).$$

4 Experimental Evaluation

In this section, we empirically evaluate the performance of the RLHH algorithm on the LeadingOnes function for realistic problem sizes and compare it against the Greedy Random Gradient (GRG) HH which also has optimal expected runtime up to lower order terms [35]. We equip both HHs with the heuristic set $H = \{\mathrm{RLS}_1, \mathrm{RLS}_2\}$.

We simulate the algorithms across a wide range of problem sizes, from $n = 10^3$ to $n = 9 \cdot 10^9$ and for each problem size we take the average number of function evaluations over 700 runs. To allow to experiment with large problem sizes we apply the same inverse transform sampling method previously used for GRG in [35] and sample the expected waiting times to make an improvement of exactly d bits, for all possible values of d and for each operator instead of executing the operator itself.

For the RLHH algorithm, we utilize the theoretically derived parameter settings: $\alpha = 1 + 1/\log n$, $\beta = 1/n$, initial weights $w_1^{(0)} = w_2^{(0)} = \log n$, and the

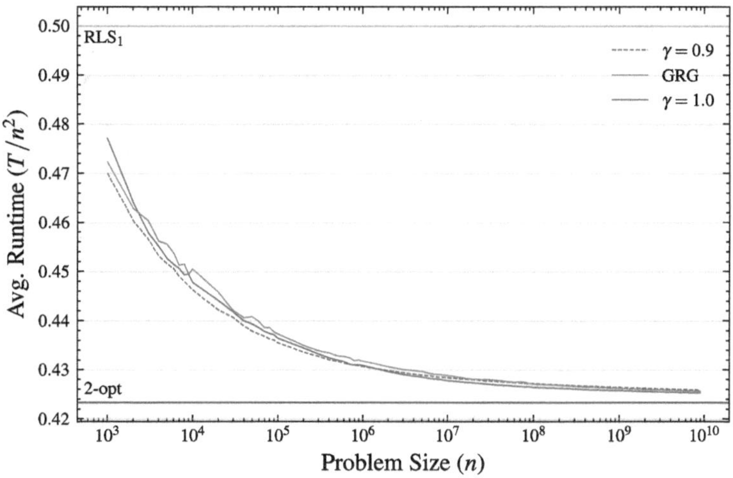

Fig. 1. Average runtime normalized by n^2 for the RLHH algorithm ($\gamma \in \{0.9, 1.0\}$) compared against GRG, RLS_1, and the theoretical 2-opt optimal strategy on the LeadingOnes function.

upper bound $w_{\max} = n/\log^{1+\gamma} n$. Our theoretical analysis indicates that $w_{\max}$ should be as small as possible, thus γ as close to 1 as possible. This respectively decreases the runtimes derived in Lemmata 3, 7 and 8 and the relevant failure probabilities. Indeed, preliminary tuning confirms this for these problem sizes. Thus, we test the algorithm with $\gamma = 0.9$ and $\gamma = 1.0$. For the GRG we set $\tau = 0.6n \ln n$ which provided the best performance in Lissovoi et al. [35].

Figure 1 displays the average runtime normalized by n^2 (i.e., the leading constants) versus the problem size. We observe that the runtimes of all three algorithms exhibit a clear, monotonic downward trend, steadily approaching the best possible expected runtime using the two operators of $\frac{1+\ln 2}{4}n^2 \approx 0.423n^2$ (i.e., 2-opt indicated in blue). The theoretical performance of the best randomised local search heuristic for the problem (i.e. RLS_1 with expected runtime $0.5n^2$) is depicted in yellow.

The configuration $\gamma = 1.0$ has the best average runtime, consistently outperforming GRG for all problem sizes greater than $n = 2 \cdot 10^3$. For the largest problem size, $n = 9 \cdot 10^9$, the two algorithms have average runtimes of $0.425n^2$ and $0.426n^2$ slightly larger than the best possible $0.423n^2$. While our formal proof conservatively requires $\gamma < 1$ to mathematically guarantee that $w_1 = o(w_{\max})$ at the conclusion of Phase 1, the practical growth rate of the weight (bounded by $e^{c \log^{\gamma} n}$) is sufficiently slow. Consequently, the experiments indicate that setting $\gamma = 1.0$ provides an excellent balance between rapid weight adaptation and the stability of the dominance regimes in practice. On the other hand, the configuration with $\gamma = 0.9$ consistently outperforms the other two algorithms for smaller problem sizes and eventually becomes only slightly worse than GRG.

The only other reinforcement learning heuristic for which a theoretical runtime analysis is available is the theory-driven designed one introduced by Doerr et al. [15]. The authors proved optimal performance for OneMax and provided experiments for LeadingOnes. The best average runtime they reported with their chosen parameters is $0.45n^2$ for $n = 10^4$ which is beaten by all three algorithms in Fig. 1.

5 Conclusion

In this paper, we presented a rigorous runtime analysis of the Reinforcement Learning Hyper-heuristic (RLHH) for the standard LeadingOnes benchmark function. This is the first time a rigorous positive result is presented regarding a hyper-heuristic using a reinforcement learning mechanism from the literature. The only previous theoretical analysis had considered the RLHH equipped with only one useful operator having positive improvement probability [3]. The authors proved that for a wide range of parameter values (i.e., $\alpha \leq \beta$) RLHH is ineffective for LeadingOnes by showing that its behaviour is similar to that of a Simple Random HH that at each step selects an operator uniformly at random (i.e. the algorithm does not learn to identify the only good operator).

Our analysis proves that RLHH equipped with a low-level heuristic set $H = \{\text{RLS}_1, \text{RLS}_2\}$ and appropriate parameter values adapts the weights optimally throughout the process, thus optimises LeadingOnes in an expected runtime of $\frac{1+\ln 2}{4}n^2 + o(n^2)$ perfectly matching, up to the leading constant, the best possible offline-designed sequence of these operators. The proof of the result relies on tracking the stochastic weight dynamics via various martingale processes and drift analyses. An empirical analysis shows that the algorithm is effective for realistic problem sizes even outperforming the Generalised Random Gradient (GRG) HH previously shown to also have optimal expected runtime.

Future research should extend the analysis to larger heuristic set sizes and to wider classes of functions. Furthermore the comparative performance using alternative update and selection rules from the literature should also be analysed such as multiplicative updates or the *max choice* selection rule.

References

1. Afshani, P., Agrawal, M., Doerr, B., Doerr, C., Larsen, K.G., Mehlhorn, K.: The query complexity of finding a hidden permutation. In: Brodnik, A., López-Ortiz, A., Raman, V., Viola, A. (eds.) Space-Efficient Data Structures, Streams, and Algorithms. LNCS, vol. 8066, pp. 1–11. Springer, Heidelberg (2013). https://doi.org/10.1007/978-3-642-40273-9_1
2. Ahmed, L., Mumford, C., Kheiri, A.: Solving urban transit route design problem using selection hyper-heuristics. Eur. J. Oper. Res. **274**(2), 545–559 (2019)
3. Alanazi, F., Lehre, P.K.: Limits to learning in reinforcement learning hyper-heuristics. In: Chicano, F., Hu, B., García-Sánchez, P. (eds.) EvoCOP 2016. LNCS, vol. 9595, pp. 170–185. Springer, Cham (2016). https://doi.org/10.1007/978-3-319-30698-8_12

4. Bai, R., Burke, E., Kendall, G.: Heuristic, meta-heuristic and hyper-heuristic approaches for fresh produce inventory control and shelf space allocation. J. Oper. Res. Soc. **59**(10), 1387–1397 (2008)
5. Bendahi, A., Doerr, B., Fradin, A., Lutzeyer, J.F.: Speeding up hyper-heuristics with markov-chain operator selection and the only-worsening acceptance operator. In: Proceedings of the of IJCAI, p. 25 (2025)
6. Blazewicz, J., Burke, E., Kendall, G., Mruczkiewicz, W., Oguz, C., Swiercz, S.: A hyper-heuristic approach to sequencing by hybridization of DNA sequences. Ann. Oper. Res. **207**, 27–41 (2011)
7. Böttcher, S., Doerr, B., Neumann, F.: Optimal fixed and adaptive mutation rates for the leadingones problem. In: Parallel Problem Solving from Nature (PPSN 2010), pp. 1–10 (2010)
8. Branke, J., Nguyen, S., Pickardt, C.W., Zhang, M.: Automated design of production scheduling heuristics: a review. IEEE Trans. Evol. Comput. **20**(1), 110–124 (2016)
9. Burke, E.K., et al.: Hyper-heuristics: a survey of the state of the art. J. Oper. Res. Soc. **64**(12), 1695–1724 (2013)
10. Burke, E., Kendall, G., Soubeiga, E.: A tabu-search hyperheuristic for timetabling and rostering. J. Heurist. **9**(6), 451–470 (2003)
11. Buzdalov, M., Buzdalova, A.: Can OneMax help optimizing LeadingOnes using the EA+RL method? In: IEEE Congress on Evolutionary Computation (CEC 2015), pp. 1762–1768. IEEE (2015)
12. Cowling, P., Chakhlevitch, K.: Hyperheuristics for managing a large collection of low level heuristics to schedule personnel. In: Proceedings of the IEEE Congress on Evolutionary Computation (CEC2003), pp. 1214–1221 (2003)
13. Cowling, P., Kendall, G., Soubeiga, E.: A hyperheuristic approach to scheduling a sales summit. In: Burke, E., Erben, W. (eds) PATAT 2000. LNCS, vol. 2079, pp. 176–190. Springer, Heidelberg (2001). https://doi.org/10.1007/3-540-44629-x_11
14. Doerr, B.: Better runtime guarantees via stochastic domination. In: Evolutionary Computation in Combinatorial Optimization (EvoCOP 2018), pp. 1–17 (2018)
15. Doerr, B., Doerr, C., Yang, J.: k-bit mutation with self-adjusting k outperforms standard bit mutation. In: Parallel Problem Solving from Nature (PPSN 2016), pp. 824–834 (2016)
16. Doerr, B., Dremaux, A., Lutzeyer, J., Stumpf, A.: How the move acceptance hyper-heuristic copes with local optima: drastic differences between jumps and cliffs. In: Proceedings of the Genetic and Evolutionary Computation Conference (GECCO 2023), pp. 990–999 (2023)
17. Doerr, B., Krejca, M.S.: Significance-based estimation-of-distribution algorithms. In: Proceedings of the Genetic and Evolutionary Computation Conference (GECCO 2018), pp. 1483–1490. ACM (2018)
18. Doerr, B., Lissovoi, A., Oliveto, P.S., Warwicker, J.A.: On the runtime analysis of selection hyper-heuristics with adaptive learning periods. In: Proc of GECCO 2018, pp. 1015–1022 (2018)
19. Doerr, B., Lutzeyer, J.F.: Hyper-heuristics can profit from global variation operators. arXiv preprint arXiv:2407.14237 (2024)
20. Doerr, B., Oliveto, P.S., Warwicker, J.A.: Selection hyper-heuristics can automatically adjust the learning period to optimally solve pseudo-Boolean problems. Artif. Intell. **357**, 104560 (2026)
21. Doerr, C., Wagner, M.: Simple on-the-fly parameter selection mechanisms for two classical discrete black-box optimization benchmark problems. In: Proceedings of

the Genetic and Evolutionary Computation Conference (GECCO 2018), pp. 943–950 (2018)
22. Dowsland, K., Soubeiga, E., Burke, E.: A simulated annealing hyperheuristic for determining shipper sizes for storage and transportation. Eur. J. Oper. Res. **179**(3), 759–774 (2007)
23. Drake, J.H., Kheiri, A., Özcan, E., Burke, E.K.: Recent advances in selection hyper-heuristics. Eur. J. Oper. Res. **285**(2), 405–428 (2020)
24. Durrett, R.: Probability: Theory and Examples, vol. 49. Cambridge University Press (2019)
25. Freedman, D.A.: On tail probabilities for martingales. Ann. Probab. 100–118 (1975)
26. Friedrich, T., Kötzing, T., Krejca, M.S.: EDAs cannot be balanced and stable. In: Proceedings of the Genetic and Evolutionary Computation Conference (GECCO 2016), pp. 1139–1146. ACM (2016)
27. Gibbs, J., Kendall, G., Özcan, E.: Scheduling English football fixtures over the holiday period using hyper-heuristics. In: Parallel Problem Solving from Nature (PPSN 2010), pp. 496–505 (2010)
28. He, J., Yao, X.: Drift analysis and average time complexity of evolutionary algorithms. Artif. Intell. **127**(1), 57–85 (2001)
29. Kendall, G., Li, J.: Competitive travelling salesmen problem: a hyper-heuristic approach. J. Oper. Res. Soc. **64**(2), 208–216 (2013)
30. Kheiri, A., Keedwell, E.: A hidden Markov model approach to the problem of heuristic selection in hyper-heuristics with a case study in high school timetabling problems. Evol. Comput. **25**(3), 473–501 (2017)
31. Lehre, P.K., Witt, C.: Black-box search by unbiased variation. Algorithmica **64**(4), 623–642 (2012)
32. Lengler, J.: Drift analysis. In: Doerr, B., Neumann, F. (eds.) Theory of Randomized Search Heuristics in Discrete Search Spaces, chap. 2, pp. 89–126. Springer, Heidelberg (2019)
33. Li, J., Kendall, G.: A hyperheuristic methodology to generate adaptive strategies for games. IEEE Trans. Comput. Intell. AI Games **9**(1), 1–10 (2017)
34. Lissovoi, A., Oliveto, P., Warwicker, J.A.: How the duration of the learning period affects the performance of random gradient selection hyper-heuristics. In: Proceedings of AAAI-20, pp. 2376–2383 (2020)
35. Lissovoi, A., Oliveto, P.S., Warwicker, J.A.: Simple hyper-heuristics control the neighbourhood size of randomised local search optimally for leadingones. Evol. Comput. **28**(3), 437–461 (2020)
36. Lissovoi, A., Oliveto, P.S., Warwicker, J.A.: When move acceptance selection hyper-heuristics outperform metropolis and elitist evolutionary algorithms and when not. Artif. Intell. **314**, 103804 (2023)
37. Ma, Y., Oliveto, P.S., Warwicker, J.: Random gradient hyper-heuristics can learn to escape local optima in multimodal optimisation. In: Proceedings of GECCO 2025, pp. 1622–1630 (2025)
38. Martins, M.S., El Yafrani, M., Delgado, M.R., Wagner, M., Ahiod, B., Lüders, R.: HSEDA: a heuristic selection approach based on estimation of distribution algorithm for the travelling thief problem. In: Genetic and Evolutionary Computation (GECCO 2017), pp. 361–368 (2017)
39. McClymont, K., Keedwell, E.: Markov chain hyper- heuristic (MCHH): an online selective hyper-heuristic for multi- objective continuous problems. In: Proceedings of the 13th Annual Conference on Genetic and Evolutionary Computation (GECCO 2011), pp. 2003–2010 (2011)

40. Mitzenmacher, M., Upfal, E.: Probability and Computing: Randomization and Probabilistic Techniques in Algorithms and Data Analysis. Cambridge University Press (2017)
41. Moraglio, A., Sudholt, D.: Principled design and runtime analysis of abstract convex evolutionary search. Evol. Comput. **25**(2), 205–236 (2017)
42. Nareyek, A.: Choosing search heuristics by non-stationary reinforcement learning. In: Resende, M.G.C., Sousa, J.P. (eds.) Metaheuristics: Computer decision-making, pp. 523–544. Springer, Bostond (2003). https://doi.org/10.1007/978-1-4757-4137-7_25
43. Oliveto, P.S.: Rigorous performance analysis of hyper-heuristics. In: Pillay, N., Qu, R. (eds.) Automated Design of Machine Learning and Search Algorithms. NCS, pp. 45–71. Springer, Cham (2021). https://doi.org/10.1007/978-3-030-72069-8_4
44. Oliveto, P.S.: Time complexity analysis of stochastic search algorithms. In: Marti, R., Pardalos, P.M., Resende, M.G.C. (eds.) Handbook of Heuristics, chap. 37, 2nd edn., pp. 1131–1172. Springer, Cham (2025)
45. Özcan, E., Misir, M., Ochoa, G., Burke, E.: A reinforcement learning – great-deluge hyper-heuristic for examination timetabling. Int. J. Appl. Metaheuristic Comput. **1**(1), 39–59 (2010)
46. Rattadilok, P., Gaw, A., Kwan, R.S.K.: Distributed choice function hyper-heuristics for timetabling and scheduling. In: Burke, E., Trick, M. (eds.) PATAT 2004. LNCS, vol. 3616, pp. 51–67. Springer, Heidelberg (2005). https://doi.org/10.1007/11593577_4
47. Pillay, N., Qu, R.: Hyper-Heuristics: Theory and Applications. Springer, Cham (2018)
48. Thabtah, F., Cowling, P.: Mining the data from a hyperheuristic approach using associative classification. Expert Syst. Appl.: Int. J. **34**(2), 1093–1101 (2008)
49. Tropp, J.A.: Freedman's inequality for matrix martingales. Electron. Commun. Probab. **16**, 262–270 (2011)
50. Walker, D.J., Keedwell, E.: Towards many-objective optimisation with hyper-heuristics: identifying good heuristics with indicators. In: Parallel Problem Solving from Nature (PPSN 2016), pp. 493–502 (2016)
51. Wu, X., Consoli, P., Minku, L., Ochoa, G., Yao, X.: An evolutionary hyper-heuristic for the software project scheduling problem. In: Parallel Problem Solving from Nature (PPSN 2016), pp. 37–47 (2016)

Runtime Analysis of the $(\mu + 1)$-ES in a Homogenous Progress Model

Johannes Lengler(✉) and Raghu Raman Ravi

Eidgenössische Technische Hochschule Zürich, Zürich, Switzerland
{johannes.lengler,raghu.ravi}@inf.ethz.ch

Abstract. We introduce a new simple model to study the fitness progress of Evolution Strategies (ES) in generic problems. In this model, we bypass the underlying fitness landscape and assume that the mutation of any individual produces an offspring whose fitness relative to the parent is given by an invariant distribution Z, such as a mean-shifted Gaussian. This serves as a prototypical model for the optimisation landscape when an evolution algorithm operates far from the global optimum. This simple model can be used to approximate the optimisation process for problems where it is intractable to model the exact fitness function, including tasks such as hyperparameter tuning in machine learning models.

We rigorously analyse the expected growth rate $\mathcal{R}_\mu$ of the continuous steady-state $(\mu+1)$-ES in this model. Unlike comma-selection strategies, the steady-state $(\mu+1)$-ES maintains overlapping generations, introducing complex mathematical dependencies among surviving parents that make it harder to analyse. We give a general technique to analyse the $(\mu+1)$-ES by constructing modified processes whose growth rates provably sandwich that of the original process. These modified processes are then easier to analyse but still close enough to the true process to give a tight bound on the expected growth rate. When $Z = \mathcal{N}(-\delta, 1)$ and $\mu \leq e^\delta$, we show that $\mathcal{R}_\mu = \frac{\log^{1+o(1)} \mu}{\mu} \mathcal{R}_1$.

Keywords: Evolution Strategies · Runtime Analysis · Population Size

1 Introduction

Evolutionary Algorithms are a prominent class of randomised optimisation heuristics in continuous search spaces intended to serve as a black-box optimisations tools. Inspired by the principles of biological evolution and natural selection, these algorithms iteratively maintain a population of candidate solutions, exploring the search space through stochastic mutation and exploiting favourable traits through a "survival of the fittest" selection mechanism.

In the discrete setting, algorithms such as the $(1+1)$-EA, the $(\mu+1)$-EA and the $(\mu+1)$-GA have been theoretically analysed for various benchmarks ranging from simple theoretical functions such as OneMax and LeadingOnes [8] to

G. Iacca et al. (Eds.): PPSN 2026, LNCS 16985, pp. 132–146, 2027.
https://doi.org/10.1007/978-3-032-36214-8_9

complex classes of functions such as monotone functions [18] or the SLO hierarchy [7]. Likewise, in the continuous setting, Evolution Strategies (ES) have been studied for various benchmarks, ranging from simple sphere functions to complex classes like locally strong convex and Lipschitz continuous functions [23]. Traditionally, the main focus has been on the convergence rate close to the optimum, culminating in the recent proof of linear convergence for the state-of-the-art CMA-ES [11,12] by Armand Gissler [9]. In recent years, attention has also turned to non-asymptotic analysis of ES [1] which yields insight into which improvements can be obtained far away from the optimum. Arguably, this is a domain that is very relevant for practice. For example, in tasks like hyper-parameter tuning for machine learning, each fitness evaluation corresponds to a costly training process, and thus optimisation is not pursued deeply into the basin of attraction of a local or global optimum. Unfortunately, non-asymptotic analysis remains challenging, and results could only be obtained for very simple benchmark functions.

Based on his experience with such machine learning applications, Esteban Real recently suggested a simple new model for the optimisation progress in a fitness region of steady, but slow progress [personal communication]. The model bypasses the need for an explicit fitness function. Instead, he assumes that the fitness of each individual (which we call its *phenotypical fitness* or simply *phenotype*) can be described as the sum of its inherent, but unknown *genotypical fitness* (or simply *genotype*) and some noise term, where the noise reflects the inherent uncertainty of the fitness evaluation process. Both phenotype and genotype are real numbers. The key assumption is that the distribution of the genotypical fitness improvement $Z := g_o - g_p$ is *translation-invariant* (i.e., independent of g_p), where g_p is the genotypical fitness of the parent and g_o is the genotypical fitness of the offspring. This captures the regime that slow, but steady progress can be achieved by mutation, which may be a reasonable assumption for progress while the optimisation process is not yet converging.

Concretely, Real suggests to use a Gaussian distribution $\mathcal{N}(-\delta, 1)$ for Z, where the negative mean reflects the fact that a random mutation typically makes the genotype worse. Moreover, he also suggests to model the noise as another Gaussian $\mathcal{N}(0, \sigma^2)$. Hence, starting from a parent of genotype g_p, each offspring will have genotype $g_o = g_p + \mathcal{Z}$ where $\mathcal{Z} \sim \mathcal{N}(-\delta, 1)$ and phenotype $p_o = g_o + \mathcal{M}$, where $\mathcal{M} \sim \mathcal{N}(0, \sigma^2)$. Note that only the phenotypical fitness is known to the algorithm, so selection is based on p_o, not on g_o. The quantity of interest for us then is the expected long term growth rate of the best genotype, which we denote by $\mathcal{R}_\mu$. We give a full formal description of the model in Sect. 2.

Interestingly, Real observed experimentally that for the $(\mu + 1)$-ES the progress rate is unimodal with respect to the population size μ, meaning that there is a non-trivial optimal choice for μ that depends on the parameters δ and σ of the two Gaussian distributions. He thus asks what the optimal population size is. We confirm his findings with our own experiments in Sect. 5. Unfortunately, in this work we cannot yet answer his question in the full model. Instead, we restrict ourselves to the noise-free case $\sigma = 0$. In this case, we determine the expected growth rate as a function of μ and δ, up to a factor of $O(\log\log\mu)$. The result suggests, confirmed by experiments, that the optimal population size in the noise-free case is $\mu = 1$, for which we give an exact formula.

Although the model seems simple, the analysis of the (μ+1)-ES is surprisingly non-trivial, even in the absence of noise. We develop novel techniques to analyse $\mathcal{R}_\mu$ for general distributions of Z (all in the absence of noise) and give tight bounds for it when the offspring distribution is given by a mean shifted Gaussian. We hope that those techniques can serve as a stepping stone for analyzing the general case with noise in future work.

1.1 Results and Proof Strategies

In this paper, we provide a mathematically rigorous analysis of the expected progress rate of the ($\mu+1$)-ES in the noise-free model when the offspring distribution Z is a mean-shifted standard Gaussian distribution $\mathcal{N}(-\delta, 1)$. For $\mu = 1$, the expected steady-state growth rate is exactly $\mathcal{R}_1(\delta) = \phi(\delta) - \delta\Phi^+(\delta)$, where ϕ and Φ^+ are respectively the probability density function (PDF) and the complementary cumulative density function (CDF) of the standard normal distribution.

For larger μ, the growth rate scales roughly as $(\log\mu)/\mu$ while $\mu \leq e^\delta$, and like $1/\mu$ afterwards. More precisely, let $m := \min\{\mu, e^\delta\}$. Then we prove that the expected growth rate satisfies $\mathcal{R}_\mu = O(\frac{\log m}{\mu}\mathcal{R}_1(\delta))$ and $\mathcal{R}_\mu = \Omega(\frac{\log m}{\mu\log\log m}\mathcal{R}_1(\delta))$. Note that upper and lower bound match up to a factor of $O(\log\log m)$. In the following paragraphs, we will discuss the more interesting case $\mu \leq e^\delta$, where $m = \mu$. The other case is simpler, but follows by similar arguments.

Intuition. Our proofs are rigorous and self-contained, but they were guided by an intuition for the distribution that the genotypes in a population of size μ should approximatively maintain: we expect that the ith fittest individual in the population has fitness roughly $(\log i)/\delta$ smaller than the maximal fitness g^* in the population. The reason is that from a typical parent, the probability of generating a fitness of at least $g^* - (\log i)/\delta$ is about i times larger than the probability of generating a fitness larger than g^*. Crucially, this ratio is (approximatively) *independent* of the exact fitness of the parent. Hence, in the range above $g^* - (\log i)/\delta$ we should expect i times as many individuals than in the range above g^*, which translates into the ith fittest individual having fitness roughly $g^* - (\log i)/\delta$. This has a remarkable consequence: the rate by which the ith fittest individual generates an offspring fitter than g^* scales like $1/(i\mu)$, where the factor $1/\mu$ stems from the rate $1/\mu$ at which each individual is picked for mutation. This translates into a growth rate of $\mathcal{R}_1(\delta)/(i\mu)$ from the ith fittest individual. Summing over all i yields the growth rate of $\approx \mathcal{R}_1(\delta)\log\mu/\mu$. Note that this rate is by a factor of $\log\mu$ higher than the rate that only mutations of the fittest individual would generate. So most offspring that improve over the fittest individual are not direct offspring of that individual, but rather the rate is composed of non-trivial contributions of all potential parents in the population.

Upper Bound. For the upper bound, we define a simpler process that always maintains a bounded gap of at most $(\log \mu)/\delta$ between the fittest and least fit individual, by artificially increasing the fitness to that level for weaker individuals. It is easy to show that the expected growth rate of this process strictly dominates that of the original process. Then, we introduce an exponential potential, $\Psi_t = \sum_{i=1}^{\mu} e^{\theta g_i}$, based on the fitness values g_i, and bound the expected increase in this potential over time. Consequently, we bound the growth rate of the fittest individual, which translates into an upper bound for the original process.

Lower Bound. For the lower bound, we again define a simpler process that always waits for the less fit individuals to catch-up before "accepting" an improvement to the best individual. More precisely, we wait until the ith fittest individual has fitness at least $g^* - (\log i)/\delta$, and in fact we artificially lower its fitness to that level. We show that this catch-up phase has a waiting time that scales like $\frac{\mu \log \log \mu}{\log \mu}$ with μ. After this phase, we keep the position frozen until some offspring is fitter than g^*, which starts the next catch-up phase. The waiting time for this last step scales like $\frac{\mu}{\log \mu}$, so it is up to a $\log \log \mu$ factor identical to the waiting time of the catch-up phase. The waiting times then translate into a growth rate of $\frac{\mathcal{R}_1(\delta) \log \mu}{\mu \log \log \mu}$. We show that this process gives a lower bound on the growth rate of the original process.

We note that both these techniques can be applied to more general distributions than just the Gaussian. In particular, in the lower bound we prove a general formula that holds for any distribution of Z. Moreover, the ideas could potentially be applied to discrete search spaces, too.

Note to Reviewers. As customary for theory submissions, we had to omit some proofs due to the page limit. Reviewers can find the full paper at [19] for their own discretion. They are not expected to consult or to review the full version.

1.2 Related Work

The study of expected improvement in fitness, termed progress rate, was highly popularised by Rechenberg [24] and further formalised by Beyer [4]. Classical models for Evolution Strategies include the sphere model [16] and quadratic functions [13]. For the $(1 + 1)$-ES [10,13,16] and $(1 + \lambda)$-ES [2,14], progress rates are well-documented because offspring generation steps are stochastically isolated. Over the past two decades, the theoretical computer science community has developed rigorous tools—most notably Drift Analysis [17]—to prove exact expected runtimes and progress rates. Jägersküpper [13] provided some of the first rigorous bounds for the $(1 + 1)$-ES on the sphere. More recently, Akimoto, Auger, and Glasmachers [1] successfully applied drift theory to continuous domains, analyzing the expected hitting time of the $(1 + 1)$ -ES with success-based step-size adaptation.

The $(\mu + 1)$-ES remains sparsely studied - Jägersküpper and Witt [15] provided the first rigourous analysis of the $(\mu + 1)$-ES on the sphere model with the

1/5-success rule. In contrast, the effect of Population on Evolutionary algorithms has been thoroughly investigated in discrete search spaces [20,21,27].

Our model is also closely related to the notion of N-particle branching random walks [3,5,6], which is a stochastic particle model for simulating wavefront propagation.

2 Notation and Preliminaries

We denote by $\mathbb{1}(E)$ the indicator random variable for the event E. We define $(x)_+ \equiv \max\{x, 0\}$. We denote by $\mathcal{N}(m, \sigma^2)$ a normal distribution with mean m and variance σ^2. The functions $\phi(\cdot)$ and $\Phi(\cdot)$ denote, respectively, the PDF and the CDF of the standard normal distribution. Furthermore, we denote by $\Phi^+(\cdot) \equiv 1 - \Phi(\cdot)$ the complementary CDF of the standard normal.

We now formally describe the general Progress Model for $(\mu+1)$ Evolutionary Algorithms. Consider a population of $\mu \in \mathbb{N}$ individuals. Each individual $i \in [\mu]$ is associated with a *genotype* $g_i \in \mathbb{R}$ and a *phenotype* $p_i \in \mathbb{R}$, which represent the inherent fitness and the observed fitness of the individual. At each time step $t \geq 1$, an individual i_t is selected uniformly at random from the population. This individual produces an offspring with genotype $g_o = g_{i_t} + Z$, where Z is random variable supported on the reals. Additionally, the phenotype of this offspring is given by $p_o = g_o + \mathcal{M}$, where $\mathcal{M}$ is random variable supported on the reals. Following reproduction, the population temporarily increases to $\mu + 1$ individuals. The individual with the lowest phenotype is then discarded, breaking ties arbitrarily, returning the population size to μ. Our objective is to analyse the evolution of the fittest individual's genotype over successive generations. In any generation t, we denote the genotype of the fittest individual as $g^*(t)$. We define the expected growth rate $\mathcal{R}_\mu$ of the process as $\liminf_{t\to\infty} \frac{\mathbb{E}g^*(t) - g^*(0)}{t}$.

Note 1. In this paper, we only study the noise-free version of the model, i.e., when $\mathcal{M} = 0$ and hence the genotypes and phenotypes of the individuals coincide. Moreover, we may identify an individual with its genotype, so we will treat g_i as an individual. We believe that this is only a first step towards studying the model in general, which exhibits a much richer and arguably more realistic behaviour.

We now introduce some general tools and lemmas that we use later in the analysis. We begin with a bound on the expecting waiting time for sampling sufficiently large values from a distribution. We use this lemma in the lower bound section.

Lemma 1. *Let N be a positive integer and for all $1 \leq i \leq N$, let x_i be real numbers. Let Z be a random variable and $F(\cdot)$ be a lower bound on its complementary CDF. Let $T = \max_i(i/F(x_i))$. Suppose that we draw independent samples of Z until the following event, E, occurs: for every $1 \leq i \leq N$, there are at least i samples of value at least x_i. Then, the expected waiting time for E is at most $8T$.*

Proof. First, fix an arbitrary $1 \leq i \leq N$, and let E_i be the event that there are at least i samples with a value of at least x_i. Suppose that we draw $4T$ samples. The expected number of values greater than x_i is at least $4TF(x_i) \geq 4 \cdot \frac{i}{F(x_i)} \cdot F(x_i) = 4i$. The event E_i fails to occur if and only if the number of samples is less than i. Thus, by applying Chernoff bounds, the failure probability of the event E_i is bounded by $\exp(-9i/8)$.

The failure probability of the global event $E = \cap_{i=1}^{N} E_i$ when drawing $4T$ samples can be bounded via a union bound:

$$\Pr[\neg E] \leq \sum_{i=1}^{N} e^{-9i/8} \leq \sum_{i=1}^{\infty} e^{-9i/8} = \frac{e^{-9/8}}{1 - e^{-9/8}} \leq \frac{1}{2}.$$

Since the probability of success in any block of $4T$ samples is at least $1/2$, the expected waiting time for the event E is at most $8T$. □

2.1 The Gaussian Distribution

In the following, we give many bounds pertaining to the Gaussian distribution which use extensively in our proofs.

We defer the proofs of these lemmas to the full version of the paper.

We begin with the following bound on the so called Mill's ratio, which is a direct consequence of the continued fraction representation given in [25].

Lemma 2. *For all $x > 0$,*

$$x < x + \frac{x}{x^2 + 2} < \frac{\phi(x)}{\Phi^+(x)} < x + \frac{1}{x}.$$

We next give some bounds on the ratios of $\Phi^+(.)$.

Lemma 3. *For any $x > y > 0$, the following bounds hold:*

1. $\frac{\Phi^+(x-y)}{\Phi^+(x)} \geq e^{xy - y^2/2}$
2. $\frac{\Phi^+(x+y)}{\Phi^+(x)} \geq e^{xy - y^2/2}\left(\frac{x}{x+y}\right)$

Next, we give the defining characterisation of the function $\xi(x) \equiv \phi(x) - x\Phi^+(x)$, which is central in our analysis.

Lemma 4.

$$\int_x^{\infty} \Phi^+(z)\, dz = \int_x^{\infty} (z - x)\, \phi(z)\, dz = \phi(x) - x\Phi^+(x) \equiv \xi(x).$$

Similar to Lemma 3, we now prove some bounds on the ratios of $\xi(.)$.

Lemma 5. *For any $x > y > 0$, we have:*

$$\frac{\xi(x+y)}{\xi(x)} \leq e^{-xy - y^2/2}\left(\frac{x}{x+y}\right)^2.$$

2.2 Single Individual

In the following, we analyse the case where $\mu = 1$. Under this regime, the expected growth rate of the genotype of the single individual (which is always the fittest, trivially) can be computed explicitly.

Lemma 6. *Suppose that* $Z \sim \mathcal{N}(-\delta, 1)$. *Then* $\mathcal{R}_1 = \xi(\delta)$.

Proof. Suppose that at any given time step, the individual has a genotype of g. We compute the expected increase in this value during this step. By definition, the offspring produced has a genotype of $g - \delta + \mathcal{N}$ (where we use the short hand $\mathcal{N} \equiv \mathcal{N}(0, 1)$). The improvement in a single step is, therefore, $(\mathcal{N} - \delta)_+$, and thus the expected improvement is

$$\mathbb{E}\left[(\mathcal{N} - \delta)\, \mathbb{1}_{\mathcal{N} \geq \delta}\right] = \int_{\delta}^{\infty} (x - \delta)\phi(x)dx = \xi(\delta),$$

where in the last step, we have used Lemma 4. □

3 Upper Bound

In the following, we establish an upper bound for the process. Our approach is to define an auxiliary process that is easier to analyse and whose growth rate provably dominates that of our original process. This modified process is nearly identical to the original, with one key exception: we continuously ensure that the genotype of each individual remains within a distance of $\Delta \geq 0$ from the fittest individual.

We now formally describe this modified process, denoted as $(\mu + 1)$-ES^{high}. We maintain a population of $\mu \in \mathbb{N}$ individuals, each with a genotype $g_i \in \mathbb{R}$. To keep our initial analysis as general as possible, let Z be a continuous random variable representing the mutation step.

In each step, an individual is chosen uniformly at random. If its genotype is g_p, it produces an offspring with genotype $g_o = g_p + Z$, where $Z \sim \mathcal{N}(-\delta, 1)$. We then select the μ individuals with the highest genotypes, breaking ties arbitrarily. Finally, if the new highest genotype is g^*, any individual with a genotype strictly less than $g^* - \Delta$ has its genotype artificially increased to $g^* - \Delta$.

We now establish a coupling between the $(\mu + 1)$-ES and $(\mu + 1)$-ES^{high} processes. Notice that at every time step $t \geq 1$, the transitions depend entirely on a uniform random choice $U_t \in [\mu]$ and a mutation realisation Z_t. Both processes are driven exclusively by these shared random variables.

Proposition 7. *Fix an arbitrary realisation of the random variables* $\{U_t, Z_t\}_{t \geq 1}$. *At any time step* $t \geq 1$, *let* $x_1(t) \leq x_2(t) \leq \cdots \leq x_\mu(t)$ *be the genotypes of the* $(\mu + 1)$*-ES and* $y_1(t) \leq y_2(t) \leq \cdots \leq y_\mu(t)$ *be the genotypes of the* $(\mu + 1)$-ES^{high}. *Then, we must have* $x_i(t) \leq y_i(t)$ *for all* $i \in [\mu]$.

Let $\mathcal{R}_\mu^{high}$ denote the expected growth rate of the $(\mu + 1)$-ES^{high}. Then, we immediately get the following corollary.

Corollary 8. $\mathcal{R}_\mu^{high} \geq \mathcal{R}_\mu$.

To facilitate the upper bound analysis, we define an exponential potential function that captures the growth of the process. Such a potential function is frequently used in the analysis of N-particle branching random walks [26]. Consider the $(\mu + 1)$-ES^{high} process. At any time step $t \geq 1$, let $g_1(t) \leq g_2(t) \leq \cdots \leq g_\mu(t)$ be the sorted genotypes. For a fixed $\theta > 0$, we define:

$$\Psi_t = \sum_{i=1}^{\mu} e^{\theta g_i}.$$

We first prove a bound on the expected growth of this potential for a general mutation variable Z.

Lemma 9. *Consider the* $(\mu + 1)$-ES^{high}. *At any time step* $t \geq 1$, *we have* $\mathbb{E}[\Psi_t] \leq \Psi_0 (1 + Q(\theta))^t$, *where*

$$Q(\theta) = \frac{\left(1 + \mu e^{-\theta \Delta}\right)}{\mu} \mathbb{E}\left[e^{\theta Z} \mathbb{1}(Z \geq -\Delta)\right].$$

Proof. Let us analyse the expected one-step increase in the potential, conditioned on its current state. Suppose that at the beginning of the time step $t \geq 1$, the individual i is chosen. The new offspring is produced with genotype $g_o = g_i(t-1) + Z$.

If $g_o < g_1(t-1)$, the offspring is rejected and the potential does not increase. If $g_o \geq g_1(t-1)$, the offspring is selected and the baseline increase in potential is at most $e^{\theta g_o}$. Additionally, if $g_o > g_\mu(t-1)$, this offspring becomes the new fittest individual. Consequently, any individual with a genotype strictly less than $g_o - \Delta$ is pulled up to $g_o - \Delta$. This artificial adjustment creates an additional potential increase of at most $\mu e^{\theta(g_o - \Delta)}$.

Thus, the total increase in potential when the offspring is selected is at most:

$$\left(1 + \mu e^{-\theta \Delta}\right) e^{\theta g_o} = \left(1 + \mu e^{-\theta \Delta}\right) e^{\theta g_i(t-1) + \theta Z}.$$

Taking the expectation over the uniform random choice of i yields the following:

$$\begin{aligned}\mathbb{E}\left[\Psi_t - \Psi_{t-1} \mid \Psi_{t-1}, Z\right] &\leq \frac{1}{\mu} \sum_{i=1}^{\mu} \left(1 + \mu e^{-\theta \Delta}\right) e^{\theta g_i(t-1) + \theta Z} \\ &= \frac{\Psi_{t-1}}{\mu} \left(1 + \mu e^{-\theta \Delta}\right) e^{\theta Z}.\end{aligned}$$

Next, we take the expectation with respect to the mutation variable Z, restricted to the domain where selection occurs (i.e., $g_o \geq g_1(t-1)$). This corresponds to the range $Z \geq g_1(t-1) - g_i(t-1)$. Because we are establishing an upper bound and the process strictly maintains a maximum gap of Δ between the highest and lowest genotypes, we can safely relax this integration range to $Z \geq -\Delta$.

Applying this relaxed range to the expectation over Z gives the following:

$$\mathbb{E}\left[\Psi_t \mid \Psi_{t-1}\right] \leq \Psi_{t-1}\left(1 + \frac{(1+\mu e^{-\theta\Delta})}{\mu}\mathbb{E}\left[e^{\theta Z}\mathbb{1}(Z \geq -\Delta)\right]\right).$$

Iterating this conditional expectation over all t time steps yields the desired claim. □

Next, we can use this to give a general upper bound for the process.

Lemma 10. *For any $\theta > 0$, $\mathcal{R}_\mu^{high} \leq \frac{Q(\theta)}{\theta}$.*

Proof. From Lemma 9, we have

$$\mathbb{E}\left[e^{\theta g_\mu(t)}\right] \leq \Psi_0\left(1 + Q(\theta)\right)^t \leq \exp\left(\log \Psi_0 + Q(\theta)t\right).$$

Now for every $A \geq 0$, applying Markov's inequality gives

$$\mathbb{P}\left[g_\mu(t) \geq A + \frac{\log \Psi_0 + Q(\theta)t}{\theta}\right] \leq \frac{\mathbb{E}\left[e^{\theta g_\mu(t)}\right]}{e^{\theta A + Q(\theta)t + \log \Psi_0}} \leq \exp\left(-\theta A\right).$$

In other words, the random variable $g_\mu(t) - \frac{Q(\theta)t + \log \Psi_0}{\theta}$ is sub-exponential with parameter θ. Consequently, we have

$$\mathbb{E}\left[g_\mu(t)\right] \leq \frac{Q(\theta)t + \log \Psi_0}{\theta} + \frac{1}{\theta},$$

which immediately implies the required lemma. □

We are now equipped to establish the upper bound on the growth rate by explicitly setting our mutation to be Gaussian.

Theorem 11. *Let $Z \sim \mathcal{N}(-\delta, 1)$ and $4 \leq \mu \leq \mu_{\max} = \lfloor e^\delta \rfloor$. Then, $\mathcal{R}_\mu^{high} \leq \frac{330\xi(\delta)\log\mu}{\mu}$.*

We use the same approach on the original process to obtain a bound for all μ, which in turn hold also for large μ.

Theorem 12. *Let $Z \sim \mathcal{N}(-\delta, 1)$ and $\mu \geq 4$. Then $\mathcal{R}_\mu^{high} \leq \frac{4\xi(\delta)\delta}{\mu}$.*

Proof. We can set $\Delta = \infty$ in Lemma 9 to obtain

$$Q(\theta) = \frac{\mathbb{E}\left[e^{\theta Z}\right]}{\mu} = \frac{e^{-\delta\theta}\mathbb{E}\left[e^{\theta\mathcal{N}}\right]}{\mu} = \frac{e^{-\delta\theta + \theta^2/2}}{\mu}.$$

Setting $\theta = \delta$, we further obtain

$$Q(\delta) = \frac{e^{-\delta^2/2}}{\mu} = \frac{\sqrt{2\pi}\phi(\delta)}{\mu} \leq \frac{\sqrt{2\pi}(\delta^2+3)\xi(\delta)}{\mu} \leq \frac{4\delta^2\xi(\delta)}{\mu}.$$

Finally, applying Lemma 10 gives us the required result. □

4 Lower Bound

Similarly to the upper bound, our approach is to define an auxiliary process that is easier to analyse and whose growth rate is provably dominated by our original process. We now formally describe the modified process, which we refer to as the $(\mu + 1)$-ES^{low}. We remark that this is specifically for the noiseless case, and hence we omit the phenotypes of the individuals. We still have $\mu \in \mathbb{N}$ individuals and each individual $i \in [\mu]$ has an associated *genotype* $g_i \in \mathbb{R}$. The process progresses in two alternating phases: the *catch-up* phase and the *improvement* phase. We also associate with the process parameters $0 = \Delta_1 \leq \Delta_2 \leq \cdots \leq \Delta_\mu$.

In the catch-up phase, the $(\mu + 1)$-ES^{low} process works exactly as the original process, except when there is an improvement to the fittest individual. If this happens, we artificially decrease the fitness of this new individual to that of the current fittest individual and simply continue until the end of the phase. The catch-up phase ends when the following condition is satisfied: for every $1 \leq i \leq \mu$, there are at least i individuals with a genotype of at least $g^* - \Delta_i$, where g^* is the genotype of the fittest individual. At the end of the catch-up phase, we artificially decrease the fitness of the individuals so that the ith fittest individual has a fitness of exactly $g^* - \Delta_i$. Without loss of generality, we assume that the initial configuration satisfies this condition. We will see later in the proof that the initial condition does not affect the asymptotic rate.

In the improvement phase, we reject all offspring that do not improve the fittest individual. Thus, the population remains the same throughout the phase and the phase ends when there is a new fittest individual.

We will formulate most of our analysis for arbitrary values of Δ_i, but set $\Delta_i = \log i/\delta$ in the end except for some truncation, see Theorems 16 and 17 below for details. As explained in Sect. 1.1, this choice is natural because the probability to generate a fitness at least $g^* - \Delta_i$ is proportional to i. Hence, in the real process we should expect approximatively i individuals of this fitness in the population, and this is what we aritficially enforce in the modified process.

In the next proposition, we establish a coupling between the two processes, demonstrating that the $(\mu + 1)$-ES^{low} process always maintains lower or equal genotypes compared to the coupled $(\mu + 1) - \text{ES}$. For each time step $t \geq 1$, we use a shared random choice of a uniform parent index, say U_t, and a shared sample from the mutation random variable, say Z_t.

Proposition 13. *Fix an arbitrary realisation of the random variables* $\{U_t, Z_t\}_{t \geq 1}$. *At any time step* $t \geq 1$, *let* $x_1(t) \leq x_2(t) \leq \cdots \leq x_\mu(t)$ *be the ordered genotypes of the* $(\mu + 1) - \text{ES}$ *and* $y_1(t) \leq y_2(t) \leq \cdots \leq y_\mu(t)$ *be the ordered genotypes of the* $(\mu + 1)$-ES^{low}. *Then, we must have* $x_i(t) \geq y_i(t)$ *for all* $1 \leq i \leq \mu$.

Let $\mathcal{R}_\mu^{low}$ denote the expected growth rate of the $(\mu + 1)$-ES^{low}. Then, we immediately get the following corollary.

Corollary 14. $\mathcal{R}_\mu^{low} \leq \mathcal{R}_\mu$.

We are now ready to investigate the expected growth rate of the $(\mu+1)$-ES^{low}.

Lemma 15. *Let $\mu \geq 2$, and let $\mathcal{D}$ be a probability distribution on $\mathbb{R}$ such that for $I \sim \mathcal{D}$ the tail probability at a point z is given by*

$$\Pr[I \geq z] = \frac{1}{\mu} \sum_{i=1}^{\mu} \Pr[Z > \Delta_i + z].$$

Then, $\mathcal{R}_\mu^{low} \geq \frac{\mathbb{E}[I \cdot \mathbb{1}(I \geq 0)]}{\mathbb{E}[W(I) \cdot \mathbb{1}(I \geq 0)]+1}$, where

$$W(I) \equiv 8\mu \cdot \max_{2 \leq i \leq \mu} \left(\frac{i}{\Pr[Z > -\Delta_i] + \sum_{j=1}^{\mu-1} \Pr[Z > I + \Delta_j - \Delta_i]} \right).$$

Proof. We consider a time frame that starts at the beginning of an improvement phase and ends at the end of the catch-up phase that immediately follows. Without loss of generality, we assume that the current fittest individual has a fitness of 0. First, notice that any given step in the improvement phase, the probability that a new fittest individual is sampled is exactly $\Pr[I \geq 0]$, where $I \sim \mathcal{D}$. This is because we know the exact positions of all the individuals throughout the phase. Thus, the expected waiting time for this event is exactly $\frac{1}{\Pr[I \geq 0]}$.

After the improvement phase ends, the new fittest individual has a fitness of I, where I is sampled from $\mathcal{D}$ conditioned on being non-negative. Additionally, the $(i+1)^{th}$ fittest individual now has a fitness of exactly $-\Delta_i$ for all $1 \leq i \leq \mu-1$. Now, the catch-up phase immediately following this continues until the $(i+1)^{th}$ fittest individual has a fitness of at least $I - \Delta_i$ for all $2 \leq i \leq \mu - 1$. Notice that at any given time step in this phase, the probability that an offspring with a genotype of at least z is produced is lower bounded by

$$\frac{1}{\mu}\left[\Pr[Z > z - I] + \sum_{j=1}^{\mu-1} \Pr[Z > \Delta_j + z]\right].$$

Thus, applying lemma 1, we conclude that the conditional expected waiting time for this phase to end is upper bounded by $W(I)$. Thus, the expected average progress made in this entire time period is lower bounded by

$$\frac{\mathbb{E}[I \mid I \geq 0]}{\mathbb{E}[W(I) \mid I \geq 0] + \frac{1}{\Pr[I \geq 0]}} = \frac{\mathbb{E}[I \cdot \mathbb{1}(I \geq 0)]}{\mathbb{E}[W(I) \cdot \mathbb{1}(I \geq 0)] + 1},$$

where in the last step, we have multiplied the numerator and the denominator with $\Pr[I \geq 0]$. Applying the Law of Large Numbers now concludes the Lemma. □

We now give an explicit lower bound when the mutations follow an offset Gaussian.

Theorem 16. *Let* $3 \leq \mu \leq e^{\delta}$ *and let* $Z \sim \mathcal{N}(-\delta, 1)$. *Then* $\mathcal{R}_{\mu}^{low} \geq \frac{\xi(\delta)\log(\mu)}{770\mu(1+\log\log\mu)}$.

Proof. We set $\Delta_i = \frac{\log i}{\delta} \leq 1$ for all $1 \leq i \leq \mu$ and apply Lemma 15. We then use Lemma 3 and Lemma 5 to appropriately bound the resulting expressions to obtain the result. □

Theorem 17. *Let* $\mu > e^{\delta}$ *and let* $Z \sim \mathcal{N}(-\delta, 1)$. *Then* $\mathcal{R}_{\mu}^{low} \geq \frac{\xi(\delta)\delta}{770\mu(1+\log\delta)}$.

Proof. We set $\Delta_i = \frac{\log i}{\delta} \leq 1$ for all $1 \leq i \leq e^{\delta}$, and $\Delta_i = \infty$ for all $i > e^{\delta}$. The rest of the proof is similar to Theorem 16. □

5 Experimental Results

Evaluating the exact steady-state growth rate of the $(\mu + 1)$-ES step-by-step is computationally intractable for environments with large threshold penalties δ. Because the acceptance probability decays exponentially as the penalty increases, simulating the algorithm generation-by-generation would require millions or billions of evaluations per successful offspring.

To bypass this without losing mathematical precision, we implemented a continuous-time Markov jump equivalent of the evolutionary process. Instead of simulating failures step-by-step, the algorithm mathematically "fast-forwards" directly to the next successful acceptance. Let $g_{min}(t)$ denote the fitness of the worst individual in the population at time t, and $g_j(t)$ denote the fitness of the j-th parent. For an offspring generated by the parent j with mutation $Z \sim \mathcal{N}(-\delta, 1)$ to be accepted, the following condition must be satisfied : $\mathcal{N}(0, 1) > g_{\min}(t) - g_j(t) + \delta$. We define this threshold as $T_j = g_{\min}(t) - g_j(t) + \delta$. The fast-forwarding relies on the following exact sampling techniques:

1. **Waiting Time Sampling:** The probability that the parent j generates a successful offspring is $p_j = \Phi^{+}(-T_j)$. The total probability of generating a successful offspring in a single generation is $P_{acc} = \frac{1}{\mu}\sum_{j=1}^{\mu} p_j$. The number of failed evaluations skipped before a success is sampled exactly using an exponential distribution with rate P_{acc}.
2. **Truncated Tail Sampling:** Once a success occurs, the responsible parent j is selected proportionally to its conditional probability of success: $\Pr(\text{Parent } j \mid \text{Success}) = p_j/(\sum_{k=1}^{\mu} p_k)$. The accepted mutation step is then sampled exactly from the right tail of the normal distribution, conditioned on $Z > T_j$, using Marsaglia's technique [22].

This methodology completely eliminates the computational bottleneck of low acceptance rates, allowing us to empirically measure the exact steady-state growth rate across extreme parameter regimes. For each parameter configuration, the simulation is run from an initial state of all zeros until the maximum true genotype in the population crosses a massive threshold ($g > 1000$), ensuring that the empirical average stabilises in the steady state.

5.1 Noiseless Translation-Invariant Regime

To evaluate the predictive accuracy of our theoretical model, we first simulated the steady-state growth rate in the noiseless environment over a wide range of population sizes, up to $\mu = 1024$. We tested four different penalty environments: $\delta \in \{2.5, 5.0, 10.0, 20.0\}$.

Our theorems establish that the growth rate is on the order of $\frac{\log^{1+o(1)} \mu}{\mu}$. In addition to the fact that $\mathcal{R}_1 = \xi(\delta)$, we compare the experimental results with the most plausible theoretical growth rate of H_μ/μ, where $H_\mu = \sum_{i=1}^{\mu} \frac{1}{i} \approx \log n + 0.57721$. Additionally, to demonstrate structural convergence, we normalise the empirical growth rates R_μ by $\xi(\delta)$ and overlay them against the idealised normalised curve H_μ/μ.

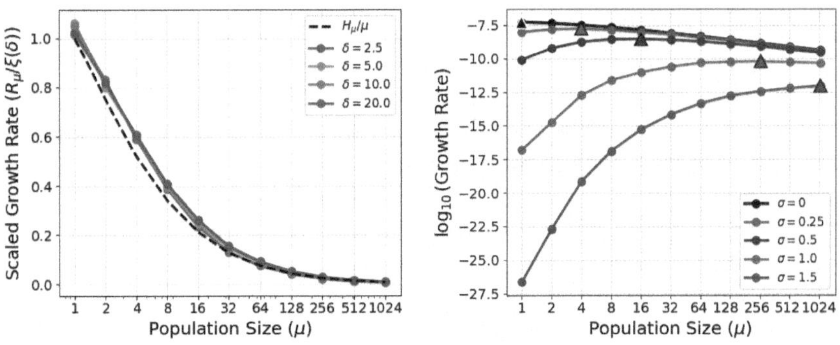

Fig. 1. Scaled steady-state growth rate $R_\mu/\xi(\delta)$ of the $(\mu+1)$-ES across varying population sizes. *Left:* The noiseless case $\sigma = 0$. The empirical dynamics collapse perfectly onto the theoretical idealised bound H_μ/μ, regardless of the penalty severity δ. *Right:* Varying noise levels σ for fixed $\delta = 5.0$. While severe noise heavily penalizes small populations, larger populations naturally recover parallel asymptotic scaling. Maximum for each curve indicated by triangle.

As shown in Fig. 1 (left), the empirical results validate the theoretical framework. By scaling the growth rate by $\xi(\delta)$, the curves for all δ values collapse into a single universal trajectory that closely traces the idealised H_μ/μ bound.

5.2 Impact of Evaluation Noise

We then extended our empirical evaluation to the noisy regime, where fitness evaluations are corrupted by Gaussian noise with variance σ^2. We fix the penalty at $\delta = 5.0$ and simulate the process for varying noise levels $\sigma \in \{0.5, 1.0, 2.0\}$, comparing them against the noiseless baseline ($\sigma = 0$).

The results, plotted on a logarithmic scale in Fig. 1 (right), illustrate the non-monotonic dependence on the population size μ. For small population sizes, large noise levels ($\sigma = 2.0$) trigger a catastrophic drop in the steady-state growth rate, spanning several orders of magnitude below the noiseless baseline. However, as the population size μ increases, the gap rapidly closes. This leads to a peak in the growth rate for a population size of $\mu > 1$. We also observe that this optimal μ is greater when the variance in the noise is larger.

6 Conclusion

In this paper, we have studied a new approach to model complex optimisation problems by representing fitness progress abstractly without explicit reference to a fitness function or a procedure for generating offspring. We investigate the noiseless case and develop novel approaches to upper and lower bound the expected growth rate in the fitness functions. When the offspring distribution is a mean shifted Gaussian $\mathcal{N}(-\delta, 1)$, we show that any non-trivial population $\mu \leq e^{\delta}$ incurs a penalty factor of $\frac{\log^{1+o(1)} \mu}{\mu}$. We also perform experiments to validate our results. Additionally, we also simulate the general noisy case to show that when there is non-zero noise, the optimal population size can be non-trivial.

The natural next step would be to investigate the noisy version of the model. Most intriguingly, does a non-trivial optimal population size exist, is it a true optimum in the long run or are we just observing a transient effect for small enough time periods? Moreover can we quantify the optimum analytically for a given offspring and noise distribution? One can also ask about the slow down in the growth rate due to noise, and to whether this slow down disappears for large enough population size.

Acknowledgments. This project was supported by the Swiss National Science Foundation [grant number 0003390]. The authors would also like to thank Esteban Real for the idea of the model.

References

1. Akimoto, Y., Auger, A., Glasmachers, T.: Drift theory in continuous search spaces: expected hitting time of the $(1 + 1)$-ES with $1/5$ success rule. In: Proceedings of the Genetic and Evolutionary Computation Conference, pp. 801–808 (2018)
2. Auger, A.: Convergence results for the $(1, \lambda)$-SA-ES using the theory of ϕ-irreducible Markov chains. Theoret. Comput. Sci. **334**(1–3), 35–69 (2005)
3. Bérard, J., Maillard, P.: The limiting process of N-particle branching random walk with polynomial tails. Electron. J. Probab. **19**(22), 1–17 (2014)
4. Beyer, H.G.: The Theory of Evolution Strategies. Springer, Heidelberg (2001)
5. Brunet, E., Derrida, B., Mueller, A.H., Munier, S.: Phenomenological theory giving the full statistics of the position of fluctuating pulled fronts. Phys. Rev. E **73**(5), 056126 (2006)
6. Brunet, E., Derrida, B., Mueller, A.H., Munier, S.: Effect of selection on ancestry: an exactly soluble case and its phenomenological generalization. Phys. Rev. E **76**(4), 041104 (2007)

7. Dang, D.C., Lehre, P.K.: The SLO hierarchy of pseudo-Boolean functions and runtime of evolutionary algorithms. In: Proceedings of the Genetic and Evolutionary Computation Conference, pp. 1551–1559 (2024)
8. Doerr, B., Neumann, F.: Theory of Evolutionary Computation: Recent Developments in Discrete Optimization. Springer (2019)
9. Gissler, A.: Linear convergence of evolution strategies with covariance matrix adaptation. Ph.D. thesis, Institut Polytechnique de Paris (2024)
10. Glasmachers, T.: Global convergence of the $(1+1)$ Evolution Strategy to a critical point. Evol. Comput. **28**(1), 27–53 (2020)
11. Hansen, N.: The CMA evolution strategy: a tutorial. arXiv preprint arXiv:1604.00772 (2016)
12. Hansen, N., Ostermeier, A.: Completely derandomized self-adaptation in evolution strategies. Evol. Comput. **9**(2), 159–195 (2001)
13. Jägersküpper, J.: How the $(1+1)$ ES using isotropic mutations minimizes positive definite quadratic forms. Theoret. Comput. Sci. **361**(1), 38–56 (2006)
14. Jägersküpper, J.: Probabilistic runtime analysis of $(1 +, \lambda)$ ES using isotropic mutations. In: Proceedings of the Genetic and Evolutionary Computation Conference, pp. 461–468 (2006)
15. Jägersküpper, J., Witt, C.: Rigorous runtime analysis of a $(\mu + 1)$ ES for the sphere function. In: Proceedings of the Genetic and Evolutionary Computation Conference, pp. 849–856 (2005)
16. Jiang, W., Qian, C., Tang, K.: Improved running time analysis of the $(1+1)$-ES on the sphere function. In: Proceedings of the International Conference on Intelligent Computing, pp. 729–739. Springer (2018)
17. Lengler, J.: Drift analysis. In: Theory of Evolutionary Computation: Recent Developments in Discrete Optimization, pp. 89–131. Springer (2019)
18. Lengler, J.: A general dichotomy of evolutionary algorithms on monotone functions. IEEE Trans. Evol. Comput. **24**(6), 995–1009 (2019)
19. Lengler, J., Ravi, R.R.: Runtime analysis of the $(\mu + 1)$-ES in an homogenous progress model (2026). https://doi.org/10.5281/zenodo.19259327
20. Lengler, J., Riedi, S.: Runtime analysis of the $(\mu + 1)$-EA on the dynamic BinVal function. In: Proceedings of the Evolutionary Computation in Combinatorial Optimization, pp. 84–99. Springer (2021)
21. Lengler, J., Zou, X.: Exponential slowdown for larger populations: the $(\mu+1)$-EA on monotone functions. In: Proceedings of the Foundations of Genetic Algorithm, pp. 87–101 (2019)
22. Marsaglia, G.: Generating a variable from the tail of the normal distribution. Technometrics **6**(1), 101–102 (1964)
23. Morinaga, D., Fukuchi, K., Sakuma, J., Akimoto, Y.: Convergence rate of the $(1+1)$-ES on locally strongly convex and Lipschitz smooth functions. IEEE Trans. Evol. Comput. **28**(2), 501–515 (2023)
24. Rechenberg, I., Evolutionsstrategien. In: Proceedings of the Workshop on Simulationsmethoden in der Medizin und Biologie, pp. 83–114. Springer (1978)
25. Shenton, L.: Inequalities for the normal integral including a new continued fraction. Biometrika **41**(1/2), 177–189 (1954)
26. Shi, Z., et al.: Branching Random Walks, vol. 2151. Springer (2015)
27. Witt, C.: Runtime analysis of the $(\mu+1)$ EA on simple pseudo-Boolean functions. Evol. Comput. **14**(1), 65–86 (2006)

Runtime Analysis of a Compact Genetic Algorithm on a Truly Multi-Valued OneMax Function

Martin S. Krejca[1(✉)] and Carsten Witt[2]

[1] Laboratoire d'Informatique (LIX), CNRS, École Polytechnique, IP Paris, Palaiseau, France
martin.krejca@polytechnique.edu
[2] DTU Compute, Technical University of Denmark, Kgs. Lyngby, Denmark
cawi@dtu.dtk

Abstract. Recently, the runtime analysis of multi-valued estimation-of-distribution algorithms in the framework of Ben Jedidia et al. (TCS 2024) has made significant advancements. However, almost all existing analyses are limited to multi-valued objective functions that in each dimension only distinguish between two types, also called categories, of values and hence can be treated with similar methods as pseudo-Boolean problems. Only recently, Adak and Witt (GECCO 2025) have presented a first runtime analysis of a multi-valued compact genetic algorithm (cGA) on the multi-valued OneMax function G–OneMax: $\{0, \ldots, r-1\}^n \rightarrow \mathbb{N}$ defined by G–OneMax$(\boldsymbol{x}_1, \ldots, \boldsymbol{x}_n) = \sum_{i=1}^{n} \boldsymbol{x}_i$ and truly depending on all r categories. We improve their runtime result from $O(nr^3 \log^2(n) \log(r))$ to $O(nr \log^3(n) \log^3(r))$, both for an optimal choice of the update strength K. Our result matches, up to polylogarithmic factors, the existing bound for the simpler r-valued OneMax function depending essentially only on two values and analyzed in several previous works. To show the new bound, we use improved drift theorems for processes with high self-loop probabilities and specifically derived concentration inequalities to analyze how probability mass in the multi-valued cGA moves into successively smaller and smaller intervals of the r-valued frequency matrix.

Keywords: Estimation-of-distribution algorithms · multi-valued search spaces · runtime analysis

1 Introduction

Recently, there has been growing interest in the runtime analysis of estimation-of-distribution algorithms (EDAs) on multi-valued search spaces. Classic EDAs work on the search space $\{0, 1\}^n$ of bit strings of length n, which is a simple and flexible representation. However, several combinatorial problems, such as routing and permutation problems, would not naturally be modeled as bit strings.

G. Iacca et al. (Eds.): PPSN 2026, LNCS 16985, pp. 147–162, 2027.
https://doi.org/10.1007/978-3-032-36214-8_10

To allow a more natural representation, recent studies have focused on the search space $R^n := \{0, 1, \dots, r-1\}^n$, i. e., where each component of the n-dimensional vector takes one of r different values. The domain size r can grow with the problem dimension n.

The first runtime analysis of an evolutionary algorithm (EA) for the search space R^n appeared 15 years ago [7] and since then, several studies of EAs have followed [5,8–10,14,17]. However, only recently have runtime analyzes of EDAs for the multi-valued search space R^n appeared. These analyses started with Jedidia et al. [13] laying a framework for multi-valued variants of classic univariate EDAs like the compact genetic algorithm (cGA [12]), univariate marginal distribution algorithm (UMDA [19]), and others. Their work was also the first to conduct a runtime analysis of a multi-valued EDA; more precisely, they proved that the multi-valued UMDA called r–UMDA optimizes an r-valued LEADINGONES function in time $O(n^2 r \log(n) \log^2(r))$ with high probability for appropriate choices of the algorithm parameters. Moreover, they gave a lower bound of similar order. Subsequently, several studies have considered a multi-valued variant of the cGA; in particular, Hamano et al. [11], independently of Jedidia et al. [13], proved runtime bounds for the multi-valued cGA (called *categorical compact GA* in their work) on variants of the well-known ONEMAX and BINVAL problems, including a bound of $O(nr \log^2(nr))$ on generalized ONEMAX holding with high probability for an optimized setting of the algorithm parameter. Independently of Hamano et al. [11], Adak and Witt [1] proved a runtime bound for technically the same multi-valued cGA, called r-cGA as in the framework of Jedidia et al. [13], on the same generalized ONEMAX function, albeit resulting in a weaker bound of $O(nr^2 \log^3(n) \log^2(r))$. In a follow-up work, Adak and Witt [2] improved this bound to $O(nr \log(n) \log(r))$, matching the bound $O(n \log n)$ for the binary case $r = 2$ [20], everywhere for a setting of the algorithm parameter that minimizes the more general runtime bound obtained in the work.

The generalizations of ONEMAX, LEADINGONES, and BINVAL studied in the above works essentially only distinguish between two different values per component. For example, the generalized ONEMAX is defined as r-ONEMAX$(\boldsymbol{x}) := \sum_{i=1}^{n} \mathbb{1}\{x_i = r-1\}$, i. e., the function only counts the number of components having the optimal value $r-1$, while all other values have the same contribution of 0. Similarly, for the multi-valued LEADINGONES function investigated with respect to the r–UMDA by Jedidia et al. [13] and with respect to the r-cGA by Adak and Witt [3], the function counts only the leading number of components having a certain value like $r-1$ and all other values are treated equivalently. Hence, essentially, when combining the values not contributing to fitness, the problem reduces to a binary one. This becomes especially evident in the considered multi-valued EDAs, which maintain a *frequency matrix* $p_{i,j}$, where $i \in \{1, \dots, n\}$ and $j \in \{0, \dots, r-1\}$, such that component i takes value j with probability $p_{i,j}$ in a sampling step of the algorithm. All entries of the matrix are initialized at $1/r$. Crucially, in all above-mentioned analyses, only the setting of the frequencies $p_{i,r-1}$ of sampling the optimal value $r-1$ at the components needs to be considered, and it does not matter how the remaining probability mass

$1 - p_{i,r-1} = \sum_{j=0}^{r-2} p_{i,j}$ is distributed on the other $r-1$ values. Hence, as already observed in Hamano et al. [11], the analysis on the multi-valued r-ONEMAX and the multi-valued r-LEADINGONES essentially proceeds in the same way as for the classic binary EDAs, with the only difference that the probability of sampling the optimal value per position starts at $1/r$ and not at $1/2$. Thus, it is debatable whether these functions are truly multi-valued and provide enough insight into the stochastic behavior of the probabilistic model of a multi-valued EDA.

Based on these insights, Adak and Witt [1] considered the multi-valued function G–ONEMAX: $R^n \to \mathbb{R}$ defined by G–ONEMAX$(\boldsymbol{x}) = \sum_{i=1}^{n} \boldsymbol{x}_i$ and already analyzed in the context of EAs [e. g., 5]. In contrast to r-ONEMAX, this function truly depends on all r possible values per component while still having the all $r-1$-string $(r-1)_{i=1}^{n}$ as the optimal solution. They only gave empirical evidence of that the r-cGA has a similar performance on G–ONEMAX as on the quasi-binary r-ONEMAX. Adak and Witt [2] then in a follow-up work provided the first runtime bound for the r-cGA on G–ONEMAX of $O(nr^3 \log^2(n) \log(r))$ for optimized settings of the algorithm parameter, which constitutes a significantly higher bound than the $O(nr \log(n) \log(r))$ obtained for the r-ONEMAX function.

Our Contribution. We revisit the runtime analysis of the r-cGA on the truly multi-valued G–ONEMAX function and prove with high probability a runtime bound of $O(nr \log^3(n) \log^3(r))$ if $r = O(n^{1/6-\varepsilon})$ for an arbitrarily small constant $\varepsilon > 0$ (Theorem 4). This improves the state of the art by a factor of $\Theta(r^2/(\log(n) \log^2(r)))$ for the considered range of r. To achieve these results, we provide new analytical arguments in two respects:

(1) A closer inspection of genetic drift compared to Adak and Witt [2]. More precisely, they pessimistically ignored that a frequency only changes with probability at most $1/r$ right after initialization, i. e., the frequency has a high probability of staying put. By using a genetic-drift theorem allowing for such self-loops of the stochastic process (Theorem 5), adapted from Jedidia et al. [13], who used a similar technique, we save a factor of $\Theta(r)$ in the runtime bound.
(2) A fine-grained analysis of the growth of frequencies, considering all possible $r-1$ values of the spectrum. More precisely, we show that for each position $i \in \{1, \dots, n\}$, the probability distribution $p_{i,\cdot}$ on the $r-1$ different values quickly concentrates on smaller and smaller intervals of $\{0, \dots, r-1\}$. While it is uniform to begin with, after time $O(nr \log^3(n) \log^2(r))$ with high probability, roughly speaking, all mass has moved to an upper interval $\{r/2, \dots, r-1\}$ of half the original size, and the probability mass $p_{i,j}$ is essentially monotonically increasing in j, as our intuition suggests, since sampling a higher value gives a higher fitness. Iterating this halving argument $O(\log r)$ times gives the final runtime bound.

To make this very rough idea formal, several obstacles have to be overcome. First of all, genetic drift may lead to minor distortions of the monotonicity hoped for. Here we again use the strong genetic drift theorems mentioned above to limit the error. More severely, even if a position is currently not exposed to genetic drift, the algorithm may be unlucky and more often than expected sample relatively small values within the target interval (like $\{r/2, \dots, r-1\}$) that the mass is moved into. This has an undesired self-reinforcing effect since if a value is sampled and its frequency is updated, the following steps give more preference to that value, and the monotonicity of the frequencies $p_{i,j}$ on the values does not hold any longer. However, for a sufficiently large choice of the update parameter K of the r-cGA, averaging effects limit this undesired reinforcement. We make this formal by deriving a new Chernoff bound for processes with time-dependent and thereby no longer independent success probabilities (Theorem 6).

Further technical challenges are solved by decomposing the process into genetic-drift steps and steps where a frequency experiences a fitness signal (so-called biased steps), and a refined analysis of the probability of a biased step. Unlike previous analyses, the refined analysis allows us to classify more steps as biased, including situations where the search points of the r-cGA do not sample exactly the same fitness in $n-1$ positions. Here a useful and novel auxiliary lemma (Lemma 1) dealing with the convolution of probability functions comes into play. Taken together, these tools avoid the pessimistic assumption of a sampling variance of $\Theta(nr)$ that Adak and Witt [2] made in their initial analysis. Here we save the second $\Theta(r)$ factor in the runtime bound (only incurring new factors being polylogarithmic in r).

Limitation. Our current proof relies heavily on the property that the relative difference of probability mass among well-chosen blocks of values remains almost the same throughout the run (Lemma 6). We prove this property via our concentration result for random processes with changing success probabilities (Theorem 6). However, since some of the blocks are small and thus initially only in the order of $\frac{1}{r}$, we only obtain a desired concentration for values of $r = O(n^{1/6-\varepsilon})$ for an arbitrarily small constant $\varepsilon > 0$.

Outline. We introduce our notation, setting, and mathematical tools in Sect. 2. In Sect. 3, we present our main result (Theorem 4) and discuss the outline of its proof in different levels of details, introducing various notation. In Sect. 4, we present the ideas discussed in Sect. 3 in rigorous detail and conclude with a proof sketch of our main result. We conclude our paper in Sect. 5.

Due to space restrictions, we only sketch the proofs of our results in this paper. A technical report containing the detailed proofs is available in [15].

2 Preliminaries

We first define the setting we consider, including the G–OneMax fitness function that we study. Afterward, we present the r-cGA, which is the algorithm that we analyze. Last, we present several mathematical tools that we use in our analysis.

Throughout this article, we let $\mathbb{N}$ denote the set of all natural numbers, including 0, and we let $\mathbb{R}$ denote the set of all reals. For all $a, b \in \mathbb{R}$, we let $[a..b] := [a, b] \cap \mathbb{N}$ as well as $[b] := [1..b]$. Moreover, given a random process $(X_t)_{t \in \mathbb{N}}$, we refer to the process simply as X for short.

Algorithm 1: r-valued Compact Genetic Algorithm (r-cGA) for the maximization of $f : [0..r-1]^n \to \mathbb{R}$

$t \leftarrow 0$;
for $(i, j) \in [n] \times [0..r-1]$ **do** $p_{i,j}^{(t)} \leftarrow \frac{1}{r}$;
while termination criterion not met **do**
    $\boldsymbol{x}_i^{(t,1)}, \boldsymbol{x}^{(t,2)} \leftarrow$ two individuals independently sampled from $\boldsymbol{p}^{(t)}$;
    $\boldsymbol{y}^{(t,1)} \leftarrow$ individual with the higher fitness from $\{\boldsymbol{x}^{(t,1)}, \boldsymbol{x}^{(t,2)}\}$ (breaking ties uniformly at random);
    $\boldsymbol{y}^{(t,2)} \leftarrow$ individual from $\{\boldsymbol{x}^{(t,1)}, \boldsymbol{x}^{(t,2)}\} \setminus \{\boldsymbol{y}^{(t,1)}\}$;
    for $i \in [n]$ **do**
        for $j \in [0..r-1]$ **do**
            $\boldsymbol{p}_{i,j}^{(t+1)} \leftarrow \boldsymbol{p}_{i,j}^{(t)} + \frac{1}{K}(\mathbb{1}\{\boldsymbol{y}_i^{(t,1)} = j\} - \mathbb{1}\{\boldsymbol{y}_i^{(t,2)} = j\})$;
    $t \leftarrow t + 1$;

2.1 The r-Valued Optimization Setting

Let $n, r \in \mathbb{N}_{\geq 2}$. We consider the maximization of r-valued *fitness* functions, i. e., of functions $f\colon [0..r-1]^n \to \mathbb{R}$. We call each $\boldsymbol{x} \in [0..r-1]^n$ an *individual*, and we call $f(\boldsymbol{x})$ the *fitness of* $\boldsymbol{x}$. For each $i \in [n]$, we denote component i of $\boldsymbol{x}$ by $\boldsymbol{x}_i$.

We analyze the G–OneMax (GOM) fitness function [1], which is the sum of all the values in the individuals. Formally, we have GOM: $\boldsymbol{x} \mapsto \sum_{i\in[n]} \boldsymbol{x}_i$.

2.2 The r-Valued Compact Genetic Algorithm

The *r-valued compact genetic algorithm* (r-cGA) was introduced by Jedidia et al. [13] as an extension of the *compact genetic algorithm* (cGA) by Harik et al. [12], for the binary domain. It aims at maximizing an r-valued fitness function of dimension $n \in \mathbb{N}_{\geq 2}$.

The r-cGA (Algorithm 1) maintains a probability matrix $(\boldsymbol{p}_{i,j})_{i\in[n],j\in[0..r-1]}$ (the *frequency matrix*), where each row sums to 1 and represents a probability distribution over $[0..r-1]$. Initially, each row represents a uniform distribution. In each iteration, each row is used independently in order to create two samples such that for each $i \in [n]$ and each $j \in [0..r-1]$, *frequency* $\boldsymbol{p}_{i,j}$ denotes the probability to create value j at position i. Formally, for all random samples $\boldsymbol{y} \in [0..r-1]^n$ and all (deterministic) individuals $\boldsymbol{x} \in [0..r-1]^n$, defining $0^0 := 1$, we have

$$\Pr[\boldsymbol{y} = \boldsymbol{x} \mid \boldsymbol{p}] = \prod_{i\in[n]} \prod_{j\in[0..r-1]} (\boldsymbol{p}_{i,j})^{\mathbb{1}\{\boldsymbol{y}_i = j\}}.$$

Given two samples, the r-cGA updates each row $i \in [n]$ of the frequency matrix by increasing the frequency of the value at position i of the better sample, and reducing the frequency of the value of the worse sample by the same amount. This amount is determined by the algorithm parameter $K \in \mathbb{N}_{\geq 1}$, known as the *hypothetical population size*. In order to not introduce mathematical border cases, we augment the terminology introduced by Doerr [4] and say that K is *well-behaved* if and only if K is divisible by r. This ensures that updates of $\frac{1}{K}$ to frequencies may exactly reach the extreme values 0 and 1, and are not cut off.

We note that the r-cGA by Jedidia et al. [13] is more general and allows to restrict the frequency values to a closed subset of $[0, 1]$ in order to prevent the frequencies from reaching the extreme values 0 and 1. We do not consider such non-trivial borders, as they make the already challenging analysis even more challenging. However, we believe that our arguments carry over to the setting of non-trivial borders, although with sufficiently additional effort.

2.3 Mathematical Tools

We make use of the following theorems, each dealing with a different aspect.

The first theorem provides us with strong concentration bounds for martingales, based on the variance of the process. Since the frequencies of the r-cGA all start at $\frac{1}{r}$, their initial variance is small. Hence, such a theorem provides us with stronger bounds than ones that do not incorporate the variance.

Theorem 1 ([18, Theorem 3.15]). *Let $(X_t)_{t\in\mathbb{N}}$ be a bounded martingale with respect to a filtration $(\mathcal{F}_t)_{t\in\mathbb{N}}$. Further, for all $t \in \mathbb{N}_{\geq 1}$, denote the deviation by $\mathrm{dev}_t := |X_t - X_{t-1}|$. In addition, let $b = \sup_{t\in\mathbb{N}} \mathrm{dev}_t$, and assume that b is finite. Last, for all $t \in \mathbb{N}$, let $\hat{v}_t = \sup \sum_{s\in[t]} \mathrm{Var}[X_s - X_{s-1} \mid \mathcal{F}_{s-1}]$. Then for all $t \in \mathbb{N}$ and all $\varepsilon \in \mathbb{R}_{\geq 0}$, we have*

$$\Pr[\exists s \in [0..t] \colon |X_t - \mathrm{E}[X_0]| \geq \varepsilon] \leq 2\exp\left(-\frac{\varepsilon^2}{2\hat{v}_t + 2b\varepsilon/3}\right).$$

The next theorem is the runtime result by Adak and Witt [2], which we use in order to cover some easy cases in our analysis.

Theorem 2 ([2, Theorem 4.6]). *Let $r \in \mathbb{N}_{\geq 2}$ with $r = \mathrm{poly}(n)$, and let $c^* \in \mathbb{R}_{>0}$ be a sufficiently large constant. Consider the r-cGA optimizing* GOM, *with a well-behaved $K \geq c^* r^2 \sqrt{n} \ln n$ with $K = \mathrm{poly}(n)$. Then with probability at least $1 - \mathrm{O}(\frac{1}{n})$, the r-cGA finds the global optimum after $O(Kr\sqrt{n}(\log r + \log K))$ iterations and function evaluations.*

The last theorem follows well known concentration results for the multiplicative drift theorem [e. g., 6,16]. Since we only have bounds on the steps of our process for a finite time horizon, we adapt these results to the setting that incorporates such stochasticity. We note that Eq. (1) in the following theorem essentially is a conditional drift condition, very similar to the drift theorems developed by Hamano et al. [11].

Theorem 3. *Let $(X_t)_{t\in\mathbb{N}}$ be an integrable random process over $\mathbb{R}_{\geq 0}$, adapted to a filtration $(\mathcal{F}_t)_{t\in\mathbb{N}}$. Let $x_{\min} \in \mathbb{R}_{>0}$ and $T = \inf\{t \in \mathbb{N} \mid X_t < x_{\min}\}$. Furthermore, let $t' \in \mathbb{N} \cup \{\infty\}$ and $q \in (0, 1]$, and let $(A_t)_{t\in[0..t']}$ be a collection of* *events*[1]*. Assume that there is a* $\delta \in (0,1)$ *such that for all* $t \in [0..t']$*, we have*

$$\mathrm{E}[(X_t - X_{t+1}) \cdot \mathbb{1}\{t < T \wedge \cap_{s\in[0..t']}A_s\} \mid \mathcal{F}_t] \geq \delta X_t \cdot \mathbb{1}\{t < T\} \Pr[\cap_{s\in[0..t']}A_s \mid \mathcal{F}_t]. \tag{1}$$

Then we have for all $t \in [0..t']$ *that*

$$\begin{aligned}\Pr[T > t] &\leq \Pr[X_t \cdot \mathbb{1}\{t < T\} \geq x_{\min} \wedge \cup_{s\in[0..t']}\overline{A_s}] \\ &\quad + (1-\delta)^t \cdot \frac{\mathrm{E}[X_0 \cdot \mathbb{1}\{T > 0 \wedge \cap_{s\in[0..t']}A_s\}]}{x_{\min}} \\ &\leq \Pr[\cup_{s\in[0..t']}\overline{A_s}] + (1-\delta)^t \cdot \frac{\mathrm{E}[X_0 \cdot \mathbb{1}\{T > 0\}]}{x_{\min}}.\end{aligned}$$

In particular, if $T > 0$*, and there is a* $q \in [0,1]$ *such that* $\Pr[\cup_{s\in[0..t']}\overline{A_s}] \leq q$*, then for all* $\gamma \in \mathbb{R}_{>0}$*, if* $t' \geq (\ln(\mathrm{E}[X_0]/x_{\min}) + \gamma)/\delta =: \beta$*, then* $\Pr[T > \beta] \leq q + \mathrm{e}^{-\gamma}$*.*

3 Main Result and Proof Overview

Our main result bounds the runtime of the r-cGA on the GOM fitness function with high probability while only depending linearly on r, up to polylogarithmic factors. As mentioned above, it requires, for any constant $\varepsilon \in \mathbb{R}_{>0}$, that $r = \mathrm{O}(n^{1/6-\varepsilon})$, as well as a sufficiently large K. For the best choice of K, it improves the previously best bound by Adak and Witt [2] (Theorem 2) by a factor of order $r^2/(\log(n)\log^2(r))$, which is asymptotically larger than 1 for $r = \Omega(\log n)$.

Theorem 4. *Let* $r \in \mathbb{N}_{\geq 2}$ *with* $r = \mathrm{O}(n^{1/6-\varepsilon})$ *for a constant* $\varepsilon \in \mathbb{R}_{>0}$*, and let* $c^* \in \mathbb{R}_{>0}$ *be a sufficiently large constant. Consider the r-cGA optimizing* GOM*, with a well-behaved* $K \geq c^* r\sqrt{n}\ln^2(n)\ln^2(r)$ *with* $K = \mathrm{poly}(n)$*. Then with probability at least* $1 - \mathrm{O}(\frac{1}{n})$*, the r-cGA finds the global optimum after* $O(Kr\sqrt{n}(\log r + \log K))$ *iterations and function evaluations.*

For the lowest admissible value of K*, this results in a runtime bound of* $O(nr\log^3(n)\log^3(r))$ *iterations and function evaluations.*

As explained in the introduction, we save a factor of r^2 in comparison to the result by Adak and Witt [2] by saving a factor of r in two occasions: (1) We conduct a tighter analysis for how frequencies behave in case of only receiving random noise as input for their update (known as *genetic drift*). (2) We conduct a tighter analysis for how likely it is that frequencies do receive a non-noisy input (known as *biased steps*).

High-level Proof Idea. The main idea of the proof of Theorem 4 applies the ideas of the proof technique for the binary case (as done by Sudholt and Witt [20]) in an iterative fashion. Roughly speaking, for each position $i \in [n]$, we create blocks of consecutive frequencies of initially approximately the same probability

[1] In case that $t' = \infty$, we interpret the set $[0..t']$ as $\mathbb{N}$.

mass that take on the roles of the two values 0 and 1 in the binary case. Once all probability mass from the block resembling 0 is removed, we apply our argument recursively to the remaining frequencies. We call the time that it takes for this shift of probability mass a *phase*. Since we show that the block resembling the value 0 initially represents a constant fraction of the remaining frequencies with positive mass, we require overall roughly $\ln r$ phases.

In order to combine our recursive argument for all n positions, we utilize that the process during each phase (per position) exhibits multiplicative drift, which is strongly concentrated around its expectation. Hence, by relying on Theorem 3, we obtain that each phase for each position lasts with high probability not longer than order $K\sqrt{n}\ln n$. A union bound over all n positions yields that a single phase for all n positions also only lasts order $K\sqrt{n}\ln n$ with high probability. Iterating this argument for all phases, as discussed above, yields the final result.

More Detailed Proof Idea. In order to make our ideas more formal, we first formalize the concept of *random-walk steps* and *biased steps*, which dates back to the analysis by Sudholt and Witt [20] for the binary case. In a nutshell, when considering a single position $i \in [n]$, the update performed receives a meaningful input if the difference in fitness of both samples, excluding position i, is smaller than the difference between the higher value and the lower value sampled at position i. In this case, the value at position i determines the order of the two samples, and the algorithm learns that higher values of GOM are better than lower ones. In the other cases, the value at position i does not determine the ranking of the two samples, and any value at position i may receive an update (in particular, a higher value may be ranked worse than a lower value).

An update, and consequently an iteration, where position i determines the ranking of the two samples is called a *biased step*. All other updates are called *random-walk steps*.

More formally, let $t \in \mathbb{N}$ be an iteration of the r-cGA optimizing GOM, and let $i \in [n]$ be a row index of the frequency matrix. Using the notation from Algorithm 1, we define

$$\begin{aligned} X_{\smallsetminus i}^{(t,)1} &= \textstyle\sum_{k \in [n] \smallsetminus \{i\}} \boldsymbol{x}_k^{(t,1)} \quad \text{and} \\ X_{\smallsetminus i}^{(t,)2} &= \textstyle\sum_{k \in [n] \smallsetminus \{i\}} \boldsymbol{x}_k^{(t,2)} \quad \text{as well as} \\ D_i^{(t)} &= X_{\smallsetminus i}^{(t,)1} - X_{\smallsetminus i}^{(t,)2}. \end{aligned} \tag{2}$$

We have a biased step in iteration t if $\max\{\boldsymbol{x}_i^{(t,1)}, \boldsymbol{x}_i^{(t,2)}\} - \min\{\boldsymbol{x}_i^{(t,1)}, \boldsymbol{x}_i^{(t,2)}\} > D_i^{(t)}$. In all other cases, we have random-walk steps.

In order to carefully follow the progress of the r-cGA, we group the frequencies at each position $i \in [n]$ into different blocks so that it is likely to observe a frequency increase in the block of the highest frequencies. Roughly speaking, our grouping splits the frequencies into four blocks: (1) The lowest-value frequencies that are already at 0 and do not contribute anything to the algorithm anymore since sampling from the corresponding values has become impossible. (2) The next-higher frequencies whose maximum value is chosen as explained below in Eq. (3). (3) Some intermediate-value frequencies whose maximum value is also chosen as explained below. (4) And the highest-value frequencies that have at least an overall constant probability mass.

The idea behind this grouping is that the difference in values of blocks (4) and (2) are so large that it is likely to be higher than the difference D_i in fitness of the two samples in all other positions. Vaguely, we separate blocks (4) and (2) by so much that their difference is at least as large as the standard deviation of the sampling variance, call it σ, in the fitness difference D_i. We show that, similarly to the binary case, the probability to have a sufficiently small difference D_i is in the order of σ^{-1} (Lemma 2). Given our grouping, σ is in the order of $\sqrt{n}$, resulting in a multiplicative drift with a factor of order $(K\sqrt{n})^{-1}$ at position i (Lemma 4).

We note that we prove Lemma 2 via a lemma that carefully studies the convolution of rather arbitrary functions and may thus be of independent interest (Lemma 1).

Formally, for our grouping, let $\kappa^* = \lceil \log_{3/2}(r-1) \rceil$, noting for $r > 2$ that $\kappa^* > \log_{3/2}(r-1)$, because any positive-integer power of $\frac{3}{2}$ is not an integer, as 2 and 3 are both prime. We define for all $\kappa \in [0..\kappa^* - 1]$ the index

$$\ell_\kappa = \left\lceil \left(1 - \left(\frac{2}{3}\right)^\kappa\right)(r-1) \right\rceil \text{ as well as the intervals} \tag{3}$$
$$K_\kappa = [\ell_\kappa..\ell_{\kappa+1} - 1] \text{ with } K_{\kappa^*} = \{r-1\} \text{ and}$$
$$S_\kappa = \bigcup\nolimits_{\nu \in [\kappa..\kappa^*]} K_\nu.$$

Note that $(K_\kappa)_{\kappa \in [0..\kappa^*]}$ is a partition of $[0..r-1]$, where it is important to check that $(1 - (\frac{2}{3})^{\kappa^*})(r-1) > r-2$. Hence, K_{κ^*-1} stops at $r-2$, which is exactly by 1 smaller than the start of K_{κ^*}.

In order to refer to the probability mass to the intervals defined in Eq. (3), we define for all $I \subseteq [0..r-1]$ and all $i \in [n]$ the notation

$$\mu_i^{(t)}(I) := \sum\nolimits_{j \in I} \boldsymbol{p}_{i,j}^{(t)}. \tag{4}$$

Given this terminology and notation, we prove a concentration result for random-walk steps (Theorem 5), which shows that the overall probability mass that we lose is overall negligible (Lemma 5).

Afterward, we show that the overall fraction of neighboring upper intervals from Eq. (3) remains per phase roughly the same, except for the blocks whose probability mass is lowest and shifted to higher blocks (Lemma 6). This guarantees that we have roughly the same starting setup for each phase.

We prove Lemma 6 via an interesting theorem that is similar to classic Chernoff bounds but allows for a change in the success probability (Theorem 6), which reflects the update in probability mass that the intervals receive.

In the end, our main result is essentially a combination of our drift bound (Lemma 4) and our result that each phase behaves roughly the same, only with the probability mass shifting to higher blocks (Lemma 6).

4 The Building Blocks of Our Main Proof

We provide the mathematically rigorous details to the concepts discussed in Sect. 3. In the statements, we refer to the definitions that we gave above, but we also introduce specific notation in particular cases that is mostly only needed for local instances.

4.1 Probability of Biased Steps

The r-cGA makes progress to the optimum thanks to the above-mentioned biased steps, which implies a drift of (sums of) frequencies. We bound the probability of an event required for a biased step in Lemma 2 below. Its proof is crucially based on the following, more general auxiliary statement dealing with a convolution.

Lemma 1. *Let $\rho \in \mathbb{N}$, $m \in \mathbb{N}_{\geq 1}$, $\rho^+ = \max\{1, \rho\}$, and $q\colon [m\rho^+] \to \mathbb{R}_{\geq 0}$ be arbitrary. Moreover, let $M \in \mathbb{R}_{\geq 0}$ be such that $\sum_{i \in [m\rho^+]} q(i) = M$. Then*

$$\textstyle\sum_{d \in [0..\rho]} \sum_{i \in [m\rho^+ - d]} q(i)q(i + d) \geq \frac{M^2}{2m}. \tag{5}$$

The following lemma bounds the probability of certain outcomes of the difference in fitness for all positions but position $i \in [n]$ (Eq. (2)).

Lemma 2. *Let $r, n \in \mathbb{N}_{\geq 2}$. Consider the r-cGA optimizing* GOM, *with arbitrary well-behaved K. Let $i \in [n]$, let $\Delta \in [0..(n-1)(r-1)]$, and let $t' \in \mathbb{N}$ denote a random iteration. Last, using the notation from Eq. (2), let $\sigma \in \mathbb{R}_{\geq (\Delta+2)/4}$ such that for all $\ell \in \{1, 2\}$, we have $\mathrm{Var}[X_{\setminus i}^{(t',)\ell} \mid \boldsymbol{p}^{(t')}] \leq \sigma^2$ as well as $\lfloor \mathrm{E}[X_{\setminus i}^{(t',)\ell} \mid \boldsymbol{p}^{(t')}] - 2\sigma \rfloor + 1 \geq 0$ and $\lceil \mathrm{E}[X_{\setminus i}^{(t',)\ell} \mid \boldsymbol{p}^{(t')}] + 2\sigma \rceil - 1 \leq (n-1)(r-1)$. Then*

$$\Pr\left[D_i^{(t')} \in [0..\Delta]\right] \geq \frac{9 \max\{1, \Delta\}}{32(4\sigma - 1)}.$$

The following lemma is an auxiliary result that bounds the variance of the sampling distribution, which typically enters our bounds on probabilities of biased steps (e. g., Lemma 2 above) and drift bounds. The lemma allows that certain subintervals of frequency values are empty, which, as explained above, is at the core of our main proof strategy.

Lemma 3. *Recall the notation from Eq. (2). Let $r \in \mathbb{N}_{\geq 2}$ and $n \in \mathbb{N}_{\geq 2}$. Consider the r-cGA optimizing* GOM, *with arbitrary well-behaved K. Let $i \in [n]$ and let $j^\star \in [0..r-1]$. Furthermore, let $t' \in \mathbb{N}$ be a random iteration such that for each $k \in [n] \setminus \{i\}$ the sum of all frequencies at position k with values in $[j^\star..r-1]$ is 1. Then we have for $\ell \in \{1, 2\}$ that*

$$\mathrm{Var}[X_{\setminus i}^{(t',)\ell} \mid \boldsymbol{p}^{(t')}] \leq (n-1)(r-1-j^\star)^2.$$

Large Biased Steps and Drift Bound. We precisely define the intervals of frequency values between which probability mass is moved, and we introduce a more specific type of large biased steps (Definition 1). With these tools at hand, we prove bounds on the drift of frequencies between subintervals in tandem with probabilities of large biased steps (Lemma 4). This statement is applied in different phases and subintervals in Sect. 4.2.

Definition 1. *Recall the notation from Eq.* (3)*. Given* $i \in [n]$*,* $t \in \mathbb{N}$*, and* $\kappa \in [0..\kappa^* - 2]$ *the event* $B_{i,\kappa,t}$*, called a* large biased step*, occurs at time* t *in position* i*, if and only if one sample in iteration* t *has a value in* K_κ *and the other one a value in* $S_{\kappa+2}$ *and the step is biased, i. e., the frequency update is performed with respect to the sample with its value in* $S_{\kappa+2}$*.*

Lemma 4. *Recall the notation from Eqs.* (2) *and* (4) *and Definition 1. Let* $r \in \mathbb{N}_{\geq 10}$ *and* $n \in \mathbb{N}_{\geq 4}$*. Consider the* r*-cGA optimizing* GOM*, with arbitrary well-behaved* K*. Let* $i \in [n]$*, let* t' *be a random iteration, and assume* $\mu_i^{t'}(S_\kappa) = 1$ *for some* $\kappa \geq 0$ *satisfying* $\ell_\kappa \leq r - 10$*. Assume furthermore that there is a constant* $c_{\text{drift}} > 0$ *(not depending on* i *and* t'*) such that* $\mu_i^{(t')}(S_{\kappa+2}) \geq c_{\text{drift}} \cdot \mu_i^{(t')}(S_{\kappa+1})$*. Moreover, let*

$$D_i^{(t')}(\kappa) := \frac{9c_{\text{drift}} \cdot \mu_i^{(t')}(S_{\kappa+1})(1 - \mu_i^{(t')}(S_{\kappa+1})) \max\{1, (r - 1 - \ell_\kappa)/2\}}{32(4\sqrt{(n-1)(r-1-\ell_\kappa)^2} - 1)}$$
$$\geq \frac{c_{\text{drift}} \cdot \mu_i^{(t')}(S_{\kappa+1})(1 - \mu_i^{(t')}(S_{\kappa+1}))}{29\sqrt{n}},$$

where the last estimate holds for $\ell_\kappa \leq r - 2$ *and sufficiently large* n*. Then we have for all* $t \geq t'$ *that*

$$\Pr[B_{i,\kappa,t'}] \geq D_i^{(t')}(\kappa) \quad \text{and}$$
$$\mathrm{E}[\mu_i^{(t'+1)}(S_{\kappa+2}) - \mu_i^{(t')}(S_{\kappa+2}) \mid \boldsymbol{p}^{(t')}] \geq \tfrac{D_i^{(t')}(\kappa)}{K}.$$

Moreover, the same bound holds for $\mathrm{E}[\mu_i^{(t'+1)}(S_{\kappa+1}) - \mu_i^{(t')}(S_{\kappa+1}) \mid \boldsymbol{p}^{(t')}]$*.*

4.2 Bounds for the Change of Frequency Sums in Phases

Recall the notation from Eq. (3). During each phase $\kappa \in [0..\kappa^* - 1]$, we show that for each $\kappa' \in [\kappa + 1..\kappa^*]$, each sum of frequencies in $S_{\kappa'}$ grows to about the fraction of $S_{\kappa'}$ in comparison to $S_{\kappa+1}$, removing all probability mass from K_κ with high probability. To this end, we study the contribution due to random-walk steps and due to biased steps separately, defining complementary sequences of increasing stopping times. The foundation are the events, similar to those defined in Definition 1, that define for each point in time whether it is a biased step or not. Formally, for all $i \in [n]$ and all $t \in \mathbb{N}_{\geq 1}$, recalling the notation of Algorithm 1 as well as of Eq. (2), we define the event for a random-walk step as

$$\mathcal{E}_i^{(t)} = \mathbb{1}\{\boldsymbol{y}_i^{(t-1,1)} - \boldsymbol{y}_i^{(t-1,2)} \leq D_i^{(t-1)}\}, \tag{6}$$

noting that $\mathcal{E}_i^{(t)}$ is measurable with respect to $\boldsymbol{p}^{(t)}$.

Let $t' \in \mathbb{N}$ be a random iteration, and let $i \in [n]$. Based on Eq. (6), we define the sequence of increasing stopping times for the random-walk steps at position i, starting from t', for all $s \in \mathbb{N}_{\geq 1}$ inductively as

$$R_{t'}^i[0] = t' \quad \text{and} \quad R_{t'}^i[s] = \inf\{t \in \mathbb{N} \mid t > R_{t'}^i[s-1] \wedge \mathcal{E}_i^{(t)} \text{ occurs}\}. \tag{7}$$

Similarly, for the biased steps, for all $s \in \mathbb{N}_{\geq 1}$, we inductively define the sequence

$$\overline{R}_{t'}^i[0] = t' \quad \text{and} \quad \overline{R}_{t'}^i[s] = \inf\{t \in \mathbb{N} \mid t > \overline{R}_{t'}^i[s-1] \wedge \mathcal{E}_i^{(t)} \text{ does not occur}\}. \tag{8}$$

Note that both sequences are strictly increasing by definition and define stopping times each with respect to the natural filtration of $(\boldsymbol{p}^{(t'+t)})_{t \in \mathbb{N}}$.

For a position $i \in [n]$ and an index $\kappa \in [\kappa^*]$ for an upper frequency sum, we define the *contribution* of the probability mass in S_κ in terms of random-walk and biased steps, starting from a random iteration $t' \in \mathbb{N}$. That is, for an increasing sequence of stopping times $(U_t)_{t \in \mathbb{N}}$, such as Eq. (7) or (8), we define

$$\left(\Delta_i^{(U_t)}(S_\kappa)\right)_{t \in \mathbb{N}_{\geq 1}} = \left(\mu_i^{(U_t)}(S_\kappa) - \mu_i^{(U_t - 1)}(S_\kappa)\right)_{t \in \mathbb{N}_{\geq 1}}, \tag{9}$$

noting that the time point $U_t - 1$ does not need to be an element from U. Last, using Eq. (9), we define the *change* of the frequency mass of S_κ as

$$\left(\mu_{i|U}^{(t)}(S_\kappa)\right)_{t \in \mathbb{N}} = \left(\mu_i^{(U_0)}(S_\kappa) + \sum\nolimits_{s \in [t]} \Delta_i^{(U_s)}(S_\kappa)\right)_{t \in \mathbb{N}}, \tag{10}$$

where we choose for U again either $R_{t'}^i$ or $\overline{R}_{t'}^i$.

Next, we analyze the change in random-walk and biased steps more closely.

Bounds on Frequency Decrease due to Genetic Drift. We now formulate two statements that we use to bound unwanted effects of genetic drift. The first one (Theorem 5) applies Theorem 1 to intervals of frequencies of the r-cGA under a martingale assumption, i. e., for random-walk steps. The second statement (Lemma 5) is a concrete application of Theorem 5 in the setting of our main theorem, considering the contribution of random-walk steps.

Theorem 5. *Let $r \in \mathbb{N}_{\geq 2}$, let f be an r-valued fitness function of dimension $n \in \mathbb{N}_{\geq 1}$. Consider the r-cGA optimizing f, with an arbitrary well-behaved K. Let $t' \in \mathbb{N}$ be a random iteration. Let $i \in [n]$, let $J \subseteq [0..r-1]$ with $J \neq \emptyset$, and for all $t \in \mathbb{N}$, let $P_t = \sum_{j \in J} \boldsymbol{p}_{i,j}^{(t'+t)}$. Furthermore, let $(U_t)_{t \in \mathbb{N}}$ be a sequence of increasing stopping times, each with respect to the natural filtration of $(\boldsymbol{p}^{(t'+t)})_{t \in \mathbb{N}}$, and assume that $(P_{U_t})_{t \in \mathbb{N}}$ is a martingale. Last, let $\alpha \in \mathbb{R}_{\geq 0}$ and $\beta = \max\{\min\{(1-\alpha)P_{U_0}, 1-(1-\alpha)P_{U_0}\}, \min\{(1+\alpha)P_{U_0}, 1-(1+\alpha)P_{U_0}\}\}$. Then, for all $t \in \mathbb{N}$, we have*

$$\Pr\left[\exists s \in [0..t] \colon |P_{U_s} - P_{U_0}| \geq \alpha P_{U_0} \mid P_{U_0}\right] \leq 2 \exp\left(-\frac{3(\alpha P_{U_0} K)^2}{4 \cdot \max\{6t\beta, \alpha P_{U_0} K\}}\right).$$

Lemma 5. *Recall the notation from Eqs.* (3), (4), (7) *and* (10). *Let* $r \in \mathbb{N}_{\geq 3}$, *and let* $c^*, c_{\text{stop}} \in \mathbb{R}_{>0}$ *be sufficiently large constants. Consider the* r*-cGA optimizing* GOM, *with a well-behaved* $K \geq c^* r\sqrt{n}\ln^2(n)\ln^2(r)$. *Let* $i \in [n]$, *let* $t' \in \mathbb{N}$ *be a random iteration, let* $\nu \in [0..\kappa^*]$, *where* $\kappa^* \geq 2$, *and assume that* $\mu_i^{(t')}(S_\nu) \geq \frac{e^{-3}}{r}$. *Then, abbreviating* $t_{\text{stop}} = c_{\text{stop}} K\sqrt{n}\ln n$, *we have*

$$\Pr\left[\exists s \in [0 \ldots t_{stop}] : \left|\mu_{i|R^i_{t'}}^{(t'+s)}(S_\nu) - \mu_{i|R^i_{t'}}^{(t')}(S_\nu)\right| \geq \frac{1}{\kappa^*}\mu_{i|R^i_{t'}}^{(t')}(S_\nu) \;\middle|\; \mu_{i|R^i_{t'}}^{(t')}(S_\nu)\right]$$
$$\leq 2n^{-c^*/(3857c_{stop})}.$$

Bounds on Increases due to Biased Steps. Next, we state the central Lemma 6. Roughly speaking, using the notation from Eq. (3), it shows that frequency mass in intervals S_ν and $S_{\nu+1}$ stays roughly in the same ratio throughout a phase that moves probability mass into S_ν from even smaller values. To prove this, the following Theorem 6 is crucial. It deals with the self-reinforcing effect of unwanted steps; more precisely, if probability mass is moved into $S_\nu \setminus S_{\nu+1}$, then this increases the probability of biased steps sampling a value in that interval and increasing this mass even further. However, Chernoff-type arguments can be used to limit this effect. The theorem is formulated in a general way dealing with time-dependent success probabilities and may be of independent interest.

Theorem 6. *Let* $p \in [0,1]$ *and* $\eta, \rho \geq 0$. *Let* X_t, $t \geq 1$, *be a sequence of independent Bernoulli trials. Let* $Z_t := \sum_{i=1}^{t} X_i$. *Assume that*

$$p_t := \Pr[X_t = 1 \mid Z_t] \geq \frac{\rho p + \eta Z_t}{\rho + \eta t}. \tag{11}$$

Let $0 < \delta < 1$ *and assume* $\frac{\eta t}{\rho + \eta t} \leq b < 1$. *Then for* $\mu := tp$ *it holds that*

$$\Pr[Z_t \leq (1-\delta)\mu] \leq t\exp(-(1-b\delta)(1-b)^2\delta^2\mu/2).$$

We now present the crucial lemma dealing with the relative frequency growth.

Lemma 6. *Recall the notation from Eqs.* (4) *and* (3). *Let* $r \in \mathbb{N}_{\geq 3}$ *with* $r = O(n^{1/6-\varepsilon})$ *for a constant* $\varepsilon \in \mathbb{R}_{>0}$, *and let* $c^*, c_{\text{stop}} \in \mathbb{R}_{>0}$ *be sufficiently large constants. Consider the* r*-cGA optimizing* GOM, *with a well-behaved* $K \geq c^* r\sqrt{n}\ln^2(n)\ln^2(r)$. *Let* $i \in [n]$, *let* $t' \in \mathbb{N}$ *be a random iteration, and assume that there is a* $\psi \in [0..\kappa^* - 1]$ *such that* $\mu_i^{(t')}(S_\psi) = 1$ *and* $\mu_i^{(t')}(K_\psi) > 0$. *Furthermore, assume for all* $\nu \in [\psi{+}1..\kappa^*{-}1]$ *that* $\mu_i^{(t')}(S_\nu) \geq (1-1/\kappa^*)^{3\kappa^*-3}/r$. *Then, abbreviating* $t_{\text{stop}} \leq c_{\text{stop}} K\sqrt{n}\ln n$, *for each constant* $c_{\text{phase}} \in \mathbb{R}_{>0}$, *we have for all* $\nu \in [\psi + 1..\kappa^* - 1]$ *and all* $t \in [0..t_{\text{stop}}]$ *that*

$$\Pr\left[\frac{\mu_i^{(t'+t)}(S_{\nu+1})}{\mu_i^{(t'+t)}(S_\nu)} < \left(1 - \frac{1}{\kappa^*}\right)^3 \cdot \frac{\mu_i^{(t')}(S_{\nu+1})}{\mu_i^{(t')}(S_\nu)} \mid t'\right]$$
$$\leq t_{stop}\exp(-\Omega(pn^{\varepsilon/2})) + 2t_{stop}n^{-c^*/(3857c_{stop})},$$

where $p = \frac{(1-1/\kappa^*)\mu_i^{(t')}(S_{\nu+1})}{(1+1/\kappa^*)\mu_i^{(t')}(S_\nu)}$.

4.3 Combining Everything

The proof of Theorem 4 proceeds as explained in Sect. 4. For space reasons, we only provide a proof sketch, recalling that we heavily rely on the notation from Eqs. (3) and (4).

As we explained before, the proof operates by considering κ^* phases, where a phase is defined per position $i \in [n]$ and per phase index $\kappa \in [0..\kappa^* - 1]$. A phase starts in iteration $t' \in \mathbb{N}$ once K_κ is the lowest-index interval with non-zero probability, i.e., $\mu_i^{(t')}(S_\kappa) = 1$ and κ is the smallest index such that $\mu_i^{(t')}(K_\kappa) > 0$. The phase ends once K_κ has a probability mass of 0.

For each phase but constantly many phases at the end, we apply Lemma 4, assuming that the conditions for drift being present are satisfied for order $K\sqrt{n} \ln n$ iterations. By an application of Theorem 3, we then obtain that the phase ends with high probability. Via a union bound over all n values for i, the same is true for all positions. For the remaining phases, where we only consider constantly many different values, we apply Theorem 2, which is better than our bound in these cases.

Last, we show that the assumptions of Lemma 4 are met, which we do by applying Lemma 6 inductively. For the base case, the ratio of probability mass of neighboring S-intervals is just the ratio of the interval sizes, since all frequencies start at $\frac{1}{r}$. For the other cases, we apply Lemma 5, showing that the probability mass in each S-interval we consider per phase is with high probability sufficiently large in order to apply Lemma 6 inductively. Thus, in each phase, we lose (per position) with high probability at most a fraction of $(1 - \frac{1}{\kappa^*})^3$ of the previous probability mass ratio. Since we have at most $\kappa^* - 1$ phases before the final application of the lemma, we lose with high probability overall no more than fraction of $(1 - \frac{1}{\kappa^*})^{3(\kappa^*-1)} \geq \mathrm{e}^{-3}$, which is a constant. Hence, the condition of Lemma 4 is satisfied for all phases (and all positions). Adding up all error probabilities via a union bound then yields our main result.

5 Conclusion

We provided an improved runtime guarantee of the r-cGA on the G–OneMax benchmark function. In comparison to the previous result by Adak and Witt [2], our result is better by a factor of order $r^2/\big(\ln(n)\ln^2(r)\big)$. Our main contribution is that we prove that the runtime does not scale cubicly in r but only linearly (up to polylogarithmic factors). We achieve this by conducting a more careful analysis of how frequencies behave when subject to random updates and how likely it is for a frequency to receive a useful, biased update.

In comparison to the well-known result of the (binary) cGA on the (binary) OneMax benchmark, we lose a logarithmic factor in n due to relying on a union bound for all positions via the multiplicative drift theorem. In turn, this results in a larger lower bound for K, which also affects the runtime.

Our scaling in r can likely be improved by combining phases more meticulously. In the binary case, the potential function considers the distance of all highest-value frequencies to their maximum value, which results in variable drift. In contrast, we still consider all positions independently.

With Theorem 3 and 6 and Lemma 1, we introduce in our analysis three general statements that are likely to be of independent interest.

Future work could focus on improving how to combine phases. In particular, our crucial Lemma 6 currently requires an upper bound on r that is very likely just an artifact of our analysis and can be improved.

Acknowledgments. Martin S. Krejca was supported by the French National Agency for Research (ANR) via the JCJC grant titled *MultiEDA* (ANR-25-CE23-2418-01) and by the EuroTech Visiting Researcher program 2025. Carsten Witt was supported by the Independent Research Fund Denmark via grant 10.46540/2032-00101B. This paper greatly benefitted from discussions at the Dagstuhl Seminar 25092.

References

1. Adak, S., Witt, C.: Runtime analysis of a multi-valued compact genetic algorithm on generalized OneMax. In: Affenzeller, M., et al. (eds.) Parallel Problem Solving from Nature – PPSN XVIII. PPSN 2024. LNCS, vol. 15150, pp. 53–69. Springer, Cham (2024). https://doi.org/10.1007/978-3-031-70071-2_4
2. Adak, S., Witt, C.: Improved runtime analysis of a multi-valued compact genetic algorithm on two generalized OneMax problems. In: Proceedings of the Genetic and Evolutionary Computation Conference (GECCO) 2025, pp. 1585–1593. ACM (2025)
3. Adak, S., Witt, C.: A runtime analysis of the multi-valued compact genetic algorithm on generalized LeadingOnes. In: Krejca, M.S., Wagner, M. (eds.) Evolutionary Computation in Combinatorial Optimization. EvoCOP 2025. LNCS, vol. 15610, pp. 1–17. Springer, Cham (2025). https://doi.org/10.1007/978-3-031-86849-8_1
4. Doerr, B.: The runtime of the compact genetic algorithm on jump functions. Algorithmica **83**(10), 3059–3107 (2021)
5. Doerr, B., Doerr, C., Kötzing, T.: Static and self-adjusting mutation strengths for multi-valued decision variables. Algorithmica **80**, 1732–1768 (2018)
6. Doerr, B., Goldberg, L.A.: Drift analysis with tail bounds. In: Schaefer, R., Cotta, C., Kołodziej, J., Rudolph, G. (eds.) PPSN 2010. LNCS, vol. 6238, pp. 174–183. Springer, Heidelberg (2010). https://doi.org/10.1007/978-3-642-15844-5_18
7. Doerr, B., Johannsen, D., Schmidt, M.: Runtime analysis of the (1+1) evolutionary algorithm on strings over finite alphabets. In: Proceedings of the 11th Workshop Proceedings on Foundations of Genetic Algorithms, pp. 119–126, ACM Press (2011)
8. Doerr, B., Krejca, M.S., Rudolph, G.: Runtime analysis for multi-objective evolutionary algorithms in unbounded integer spaces. In: Proceedings of the AAAI Conference on Artificial Intelligence, vol. 39, no. 25, pp. 26955–26963 (2025)
9. Doerr, B., Pohl, S.: Run-time analysis of the (1+ 1) evolutionary algorithm optimizing linear functions over a finite alphabet. In: Proceedings of the Genetic and Evolutionary Computation Conference (GECCO 2012), pp. 1317–1324, ACM Press (2012)

10. Doerr, B., Sudholt, D., Witt, C.: When do evolutionary algorithms optimize separable functions in parallel? In: Proceedings of the Twelfth Workshop on Foundations of Genetic Algorithms (FOGA 2013), pp. 51–64 (2013)
11. Hamano, R., Uchida, K., Shirakawa, S., Morinaga, D., Akimoto, Y.: Tail bounds on the runtime of categorical compact genetic algorithm. Evol. Comput. **33**(2), 141–189 (2025)
12. Harik, G.R., Lobo, F.G., Goldberg, D.E.: The compact genetic algorithm. IEEE Trans. Evol. Comput. **3**(4), 287–297 (1999)
13. Jedidia, F.B., Doerr, B., Krejca, M.S.: Estimation-of-distribution algorithms for multi-valued decision variables. Theor. Comput. Sci. **1003**, 114622 (2024)
14. Kötzing, T., Lissovoi, A., Witt, C.: (1+1) EA on generalized dynamic OneMax. In: Proceedings of the ACM Conference on Foundations of Genetic Algorithms (FOGA 2015), pp. 40–51 (2015)
15. Krejca, M.S., Witt, C.: Runtime analysis of a compact genetic algorithm on a truly multi-valued OneMax function (2026). https://arxiv.org/abs/2605.29477, arXiv preprint 2605.29477
16. Lengler, J., Steger, A.: Drift analysis and evolutionary algorithms revisited. Comb. Probab. Comput. **27**(4), 643–666 (2018)
17. Lissovoi, A., Witt, C.: MMAS vs. population-based EA on a family of dynamic fitness functions. In: Proceedings of the Genetic and Evolutionary Computation Conference (GECCO 2014), pp. 1399–1406. ACM Press (2014)
18. McDiarmid, C.: Concentration. In: Habib, M., McDiarmid, C., Ramirez-Alfonsin, J., Reed, B. (eds.) Probabilistic Methods for Algorithmic Discrete Mathematics. Algorithms and Combinatorics, vol. 16, pp. 195–248. Springer, Berlin, Heidelberg (1998). https://doi.org/10.1007/978-3-662-12788-9_6
19. Mühlenbein, H., Paaß, G.: From recombination of genes to the estimation of distributions I. Binary parameters. In: Voigt, H.-M., Ebeling, W., Rechenberg, I., Schwefel, H.-P. (eds.) PPSN 1996. LNCS, vol. 1141, pp. 178–187. Springer, Heidelberg (1996). https://doi.org/10.1007/3-540-61723-X_982
20. Sudholt, D., Witt, C.: On the choice of the update strength in estimation-of-distribution algorithms and ant colony optimization. Algorithmica **81**, 1450–1489 (2019)

Runtime Analysis of Cartesian Genetic Programming in Evolving Boolean Functions

Duc-Cuong Dang[1](✉), Roman Kalkreuth[2], and Andre Opris[1]

[1] Chair of Algorithms for Intelligent Systems, University of Passau, Passau, Germany
duccuong.dang@uni-passau.de

[2] Chair for AI Methodology (AIM), Faculty of Computer Science, RWTH Aachen University, Aachen, Germany

Abstract. Cartesian Genetic Programming (CGP) is among the practical and popular forms of Genetic Programming as it uses a graph-based representation of programs. This paper presents a first runtime analysis of CGP in evolving Boolean functions using complete training sets. We prove an asymptotic bound $O(nD^5)$ for the expected number of fitness evaluations of CGP to construct a conjunction of n inputs using at most $D \geq n-1$ binary gates, a minimal function set, and even with a strict survival selection. When the non-strict selection is used, the bound is improved to $O(nD^4)$. Our analysis reveals interesting characteristics of CGP induced search, which have been only observed empirically. In particular, enabling the acceptance of equally good solutions, including those with connected gates non-contributing to fitness, can lead to a speedup, and consequently a better asymptotic time bound. In contrast to conjunctions, we also prove a negative result which shows that CGP requires exponential time to evolve an exclusive disjunction. Experiments evolving conjunctions complement our theoretical findings. The use of incomplete training sets is found to further reduce the average number of fitness evaluations while maintaining a good level of generalisation.

Keywords: Genetic Programming · Runtime Analysis · Graph Representation

1 Introduction

Genetic Programming (GP) is a search paradigm originally proposed for the automated discovery of computer programs. GP leverages evolutionary search using selection and variation to evolve programs given a specification, constraints, and input-output examples. Early work by Cramer [3] and Dickmanns et al. [5] applied Genetic Algorithms (GA) to evolve computer programs. However, Koza [17,18] later introduced a tree-based form of GP to evolve symbolic expressions, commonly implemented as LISP S-expressions at that time, which

G. Iacca et al. (Eds.): PPSN 2026, LNCS 16985, pp. 163–179, 2027.
https://doi.org/10.1007/978-3-032-36214-8_11

widely popularised the concept of GP. The traditional form of GP therefore uses tree data structures to model programs, and this is commonly referred to as Tree-based GP (TGP). The leaf nodes represent the inputs of the program and are assigned elements from a *terminal set* T, which consists of variables or constants. The root and internal nodes compute functions from a *function set* F, taking the arguments from the outputs of their respective child nodes. After evaluation, the value produced at the root node constitutes the final output of the modeled program. Over time, GP has been extended to incorporate alternative program representations, including linear sequences [31,32], graph-based structures [26,33], and rule-based models [34].

In this work, we are interested in Cartesian Genetic Programming (CGP) [25], which is, unlike TGP, a graph-based form of GP. CGP is a well established methodology in logic synthesis and evolvable hardware [40], with many successful applications such as in designing high order combinational circuits [39], complex approximate circuits [41], and cryptographic functions [11]. These successes are believed to be due to the high versatility of the graph-based representation. Learning Boolean functions was also among the motivating applications for the introduction of CGP [26]. However, it remains poorly understood when and why even simple CGP algorithms perform well in this problem domain, and a solid theoretical foundation for CGP in evolving Boolean functions is still missing.

Our Contribution: Inspired by the work of Mambrini and Oliveto [22] on TGP, we initiate a rigorous analysis of CGP for evolving Boolean functions by analysing the runtime of the (1+1) CGP and the strict variant (1+1) CGP*. We consider the popular single active-gene mutation (SAM) [9] and the minimal function sets. For constructing conjunctions of n variables with $D \geq n-1$ AND gates and the complete training set, we obtain expected runtimes of $\mathcal{O}(nD^4)$ and $\mathcal{O}(nD^5)$, respectively, using fitness-level arguments and Markov chain analysis. The faster performance of (1+1) CGPover (1+1) CGP* stems from its non strict selection, which accepts equally fit solutions with redundant active gates, a beneficial effect previously observed only empirically [35]. Prior counterpart TGP results [19,22] make no distinction between strict and non strict selection, and in comparison, CGP is slower but guarantees to only use $\mathcal{O}(D \log D)$ space. The original CGP [26] was not designed for speeding up TGP, but to offer a more general representation of programs with natural support for Boolean circuit applications. A CGP program in the Boolean domain is a logic circuit. The aim of our paper is not about showing a superior performance, but rather to understand how and when polynomial performance guarantees can be achieved.

In contrast to conjunctions, we prove that (1+1) CGP requires exponential time with exponentially high probability, to construct an exclusive disjunction of n inputs out of $(1+d)n \geq D \geq n-1$ XOR gates for some $d < 1$ and using the complete training set. Here we apply the negative drift theorem [29]. Experiments on conjunctions support the theory, showing clear gains by using non strict selection. We also conduct experiments using incomplete static training sets, and observe that this setting further reduces the number of fitness evaluations required by CGP while maintaining a good level of generalisation. Due to space restrictions, proofs are omitted or only sketched. A full version of the paper with all proofs included is available on arXiv [4].

Related Work: From the empirical side, several studies have examined search behaviour of CGP, primarly focusing on characteristics such as neutral search [44], redundancy [24], bloat [35] and evolvability in the fitness landscapes induced by logic synthesis problems [44]. Moreover, based on the results of empirical studies, it has been demonstrated and hypothesised that CGP-driven program search benefits from high levels of redundancy, enabled by the use of very large genotypes [24,35]. This also includes handling neutral fitness landscapes by neutral genetic drift (NGD), which has been shown to help escape local optima [36,44]. A step forward in understanding the complexity characteristics of CGP induced search in logic synthesis was made by the proposal of the General Boolean Function Suite (GBFS) [14]. This is a diverse benchmark of different types of Boolean functions commonly used to evaluate GP methods.

Mambrini and Oliveto [22] were the first to rigorously analyse GP in evolving Boolean functions by considering the same TGP systems as [16,27] using the minimal function sets. On the complete training set, they showed that the (1+1) TGP and (1+1) TGP* with the single-step (local) mutation constructs a conjunction of n inputs in $\mathcal{O}(n \log n)$ expected fitness evaluations. On a static incomplete training sampled uniformly at random of

rows, the time bound to fit the training set is reduced to $\mathcal{O}(\log n)$. Using the negative drift theorem, they also showed that the (1+1) TGP requires exponential time to construct an exclusive disjunction of n variables. Other setting were also considered in their paper, such as constructing exclusive disjunction with incomplete training set, and conjunction with negated input variables. These results were later extended for the algorithms with the multi-step (global) mutation in Lissovoi and Oliveto [19,20], which added the generalisation ability guarantee for conjunctions on incomplete training sets, along with many other results. Doerr et al. [7,8] pushed this line of research further by considering the settings of non-minimal function sets, unknown target function size, and introduced a new tool, called the super-multiplicative theorem, for their analysis.

For CGP, we are unaware of any theoretical research in learning Boolean functions. The only work on runtime analysis of CGP that we know of is from Kalkreuth and Droschinsky [12] on one function node. Finally, the book chapter [21] summarises the current state-of-the-art theoretical research on GP.

2 Preliminaries

We use the asymptotic notation with symbols Ω, $\mathcal{O}$, Θ [1]. The sets of integer and natural numbers are $\mathbb{Z}$ and $\mathbb{N}$ respectively. For $n \in \mathbb{N}$, define $[n] := \{1, \dots, n\}$ and $[n]_0 := [n] \cup \{0\}$. The natural logarithm is denoted $\ln(\cdot)$ and that of base-2 is $\log(\cdot)$. We use H_n to denote the n-th harmonic number, i. e. $\ln(n+1) < H_n := \sum_{i=1}^{n} \frac{1}{i} < 1 + \ln(n)$, thus $H_n = \mathcal{O}(\log n)$. Let $\mathrm{Li}_s(z) := \sum_{i=1}^{\infty} \frac{z^i}{i^s}$ be the polylogarithm (or Jonquière's) function for $s, z \in \mathbb{C}$, i. e. extended by analytic continuation. Some special cases are $\mathrm{Li}_{-2}(z) = \sum_{i=1}^{\infty} i^2 z^i = \frac{z(1+z)}{(1-z)^3}$ for $|z| < 1$, and $\mathrm{Li}_s(1) = \sum_{i=1}^{\infty} \frac{1}{i^s} =: \zeta(s)$ which is known as the Riemann's Zeta function. Note that $\zeta(2) = \pi^2/6$ and $\zeta(4) = \pi^4/90$. The logic AND, OR and exclusive OR are denoted $\wedge$, $\vee$ and $\oplus$ respectively. The uniform distribution over a set A is denoted $\mathrm{Unif}(A)$ and the geometric distribution with parameter $p \in [0, 1]$ is denoted $\mathrm{Geom}(p)$. For two random variables X, Y, $X \succeq Y$ means that X stochastically dominates Y, that is, for any $a \in \mathbb{R}$ it holds $\Pr(X \leq a) \leq \Pr(Y \leq a)$.

2.1 Cartesian Genetic Programming

CGP is capable of modelling programs with N_{in} inputs and N_{out} outputs, and uses N_{func} (internal) function nodes. The functions that can be used to compute the output of those nodes are defined in a lookup table of K entries indexed from 0 to $K - 1$. The function specified in each entry has an arity of at most k. A (valid) program for such a CGP is then represented by a directed acyclic graph $G = (V, E)$, referred to as the *phenotype*, which defines how the outputs are computed by composing functions of the lookup table and using the respective inputs. The inputs and the lookup table model the same concepts as the terminal set T and the function set F in Tree-based GP, thus any TGP program always has an equivalent CGP program.

The original form of CGP [26] was proposed to arrange the function nodes on a grid-like structure similar to an FPGA circuit. However, we consider the modern (simplified) approach in which the nodes are arranged in a single line, thus V can be indexed from 0 to $N - 1$ where $N := N_{\mathrm{in}} + N_{\mathrm{func}} + N_{\mathrm{out}}$ or simply $V = [N - 1]_0$, i. e. starting with the input nodes N_{in}, then the function nodes N_{func}, and finally the output ones N_{out}. Each input node has no incoming edges (or connections) while each output node has exactly one coming from a non-output node, i. e. with index in $[N - N_{\mathrm{out}} - 1]_0$. Each function node is labelled with a function of the lookup table and has exactly k incoming edges from nodes of strictly smaller indices, thus from the inputs or from the outputs of the preceding function nodes. These edges represent the sequence of arguments that the function at the node takes, thus there is an ordering among them. If the arity of the corresponding function in the lookup table is smaller k, then only some first edges matter.

It is possible for a CGP program to ignore some of the function nodes, which are referred to as *inactive* (function) nodes. More generally, a non-terminal node of V is called *active* if there is a path from this specific node to one of the output nodes, otherwise it is called inactive. By this definition, output nodes are always active. The *genotype* (or the encoding) of G is a vector of $N_{\mathrm{func}}(k+1) + N_{\mathrm{out}}$ integers. The sequence is divided into $N_{\mathrm{func}} + N_{\mathrm{out}}$ blocks indexed from N_{in} to $N - 1$, and these indices match those nodes in the phenotype that the blocks encode. A block with index $i < N_{\mathrm{in}} + N_{\mathrm{func}}$ (a so called function block) contains $k + 1$ integers (or genes), the first, referred to as the *function gene*, is a number in $[K - 1]_0$ (pointing to an index in the lookup table) while the rest, referred to as *function connection genes*, are integers in $[i - 1]_0$. Blocks with a larger index than i (output blocks) only contain one integer in $[N - N_{\text{out}} - 1]_0$, called the *output connection gene*. Thus the set of all genotypes (all valid CGP programs), denoted by $\mathcal{X}$, is an integer-valued space. Unlike TGP, CGP therefore has a fixed space complexity to store a program as stated in the lemma below. Figure 1 gives an example of a CGP program with its phenotype representation and the corresponding genotype encoding.

Lemma 1. *Only* $(N_{\text{func}}(k+1) + N_{\text{out}})\log(1 + \max\{K, N_{\text{in}} + N_{\text{func}}\})$ *storage space is required to store a CGP program.*

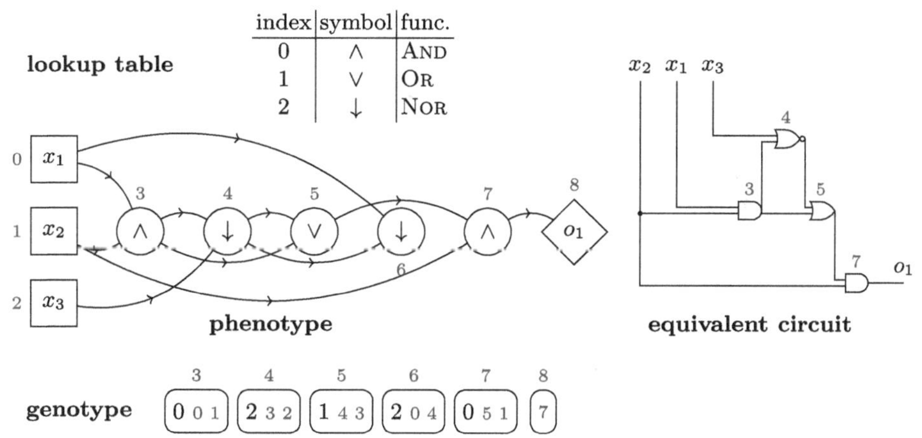

Fig. 1. Example of a CGP program that models a Boolean function. Node 6 is a so-called inactive node since no path touching this node leads to the output.

We consider the simplest CGP algorithm the so-called (1+1) CGP (Algorithm 1 for minimisation). Starting from a program initialised uniformly at random, in each iteration a new offspring program y is generated by mutating the current one. The offspring will replace the parent only if it has better or equally good fitness. Here $f(x)$ is referred to as the fitness of a program x which is problem-dependent and assumed given. By convention in runtime analysis, we let the algorithm run infinitely then estimate the expected time to reach some target state. Like TGP [22], we also refer to the variant of the (1+1) CGP where strict comparison is used instead in Line 4 as the (1+1) CGP*.

Algorithm 1: The (1+1)-CGP on the space $\mathcal{X}$ of programs.

1 $x \sim \text{Unif}(\mathcal{X})$;
2 **for** iter $:= 1 \to \infty$ **do**
3     $y := \text{mut}(x)$;
4   **if** $f(y) \leq f(x)$ **then** $x := y$;

Algorithm 2: NEWVALUE(x, i).

$b := \lfloor i/(k+1) \rfloor$ // block index
$j := i \mod (k+1)$ // index inside a block
if $i < N_{\text{func}}(k+1)$ **then**
    if $j = 0$ **then** $S := [K-1]_0$ // valid functions ;
    else $S := [N_{\text{in}} + b - 1]_0$ // valid input connections ;
else
    $S := [N_{\text{in}} + N_{\text{func}} - 1]_0$ // valid output connnections
if $|S \setminus \{x[i]\}| \geq 1$ **then**
    $S := S \setminus \{x[i]\}$ // avoid the old value if possible
$v \sim \text{Unif}(S)$;
return v

Algorithm 3: MUTATIONSAM(x).

Copy x to y;
active := false;
while ¬active **do**
    $i \sim \text{Unif}([N_{\text{func}}(k+1) + N_{\text{out}} - 1]_0)$;
    $y[i] :=$ NEWVALUE(y, i);
    active := ISACTIVE$(x, i) \wedge (y[i] \neq x[i])$;
return y

Mutations in CGP can alter any gene (component) of the genotype (vector) to a new value uniformly at random while respecting the constraint on their range, this is done with the help of the procedure NEWVALUE which is summarised in Algorithm 2. When enforcing the change of the gene value is not possible, for example if only one function is specified in the lookup table then the function genes are fixed to zero and cannot be mutated, the old value is returned instead.

The original paper [23] proposed *point mutation* operators in which a fixed or random number of genes (mutation points) are chosen uniformly at random for alternation. Nowadays, the so-called *single active-gene mutation* (SAM) [9] is more popular and is our focus. SAM is parameter-free and is described in Algorithm 3. It repeatedly modifies genes chosen uniformly at random until an active node is modified. The detection of active nodes can be done during the program evaluation, so we assume that there is a function ISACTIVE that can tell whether a given gene belongs to an active node or not in an evaluated program.

2.2 Learning Boolean Functions with GP

Learning Boolean functions or, more generally, logic circuits with heuristic methods is part of the wider field of logic synthesis [10]. In this section, we describe how the learning is evaluated for any GP.

The problem of learning a Boolean function $g\colon \textsc{Bool}^n \to \textsc{Bool}$ on n inputs where $\textsc{Bool} := \{\textsc{True}, \textsc{False}\}$ is to compose a program/circuit a set of elementary logic gates to exactly match or to approximates g using a set S of input-output examples of g referred to as the *training set*. To accommodate the construction of S by random sampling, we rather consider S as a multi-set, each of its elements is called a *row* which consists of a pair of example input assignment $\alpha \in \textsc{Bool}^n$ and the desired output $g(\alpha) \in \textsc{Bool}$. Formally, the set of input variables is $T := \{x_i \mid i \in [n]\}$ and coincides with the GP terminal set. Then each $\alpha\colon T \to \textsc{Bool}$ is a mapping of every input to a Boolean value. Like g, a given candidate program y is also a function $x_1, \ldots, x_n \mapsto y(x_1, \ldots, x_n)$ on the inputs. To calculated the fitness of y, the number of mismatched outputs of y on S is counted, that is,

$$f(y) := |\{(\alpha, g(\alpha)) \in S \mid y(\alpha(x_1), \ldots, \alpha(x_n)) \neq g(\alpha(x_1), \ldots, \alpha(x_n))\}|.$$

Thus the smaller $f(y)$ is the closer y is to fully fit the training set S.

In the so-called synthesis *with the complete training set*, the complete specification of g is available and is used to evaluate candidate programs, that is, $S = \mathcal{K}_g$ where $\mathcal{K}_g := \{(\alpha, g(\alpha(x_1), \ldots, \alpha(x_n)))\}_{\alpha \in \textsc{Bool}^n}$ is the complete truth table of g. Since the truth table grows exponentially in n, i. e. $|\mathcal{K}_g| = 2^n$, in practice to reduce the computational time of each fitness evaluation for large n, only a fraction of $\mathcal{K}_g$ is sampled and used as training set S. This is referred to as synthesis *with an incomplete training set*. We assume that S is sampled once and remains unchanged during evolution, i. e. S is called a *static* training set.

We only consider the complete training set in our theoretical analysis, the reason is discussed in the last section of the paper. However, in experiments, incomplete training sets are considered. Similarly to [7,22], we define the *generalisation error* $\varepsilon_g(y)$ of a program y in modelling a target function g as the probability that y returns a mismatched output of a row selected uniformly at random from $\mathcal{K}_g$, i. e. $\varepsilon_g(y) := \Pr(y(\alpha(x_1), \ldots, \alpha(x_n)) \neq g(\alpha(x_1), \ldots, \alpha(x_n)))$ where $\alpha \sim \mathrm{Unif}(\textsc{Bool}^n)$. The *generalisation ability* of y is the complement of its generalisation error, i. e. $G_g(y) := 1 - \varepsilon_g(y)$. Thus y fully meets the specification of g if and only if $G_g(y) = 1$. A value of $G_g(y)$ close 1 means that y generalises g well while a value close to $1/2$ means that it is only as good as a random guess. A program y with $G_g(y)$ close to 0 can be turned to $\bar{y}$ with $G_g(\bar{y})$ close to 1, thus y and $\bar{y}$ are equally informative of g.

The exact computation of $G_g(y)$ is of course expensive, however it can be approximated by considering a *validation set* (also called *test set*) S' which is sufficiently large and is created independently from S by sampling the rows of $\mathcal{K}_g$ uniformly at random with replacement. The fitness of y evaluated on S', for which y was not evolved to fit, gives an estimation of $\varepsilon_g(y)$ on S', and is denoted $\hat{\varepsilon}(y, S')$. The quantity $\hat{G}(y, S') := 1 - \hat{\varepsilon}(y, S')/|S'|$ is the maximum likelihood estimator, and hence an approximation, of $G_g(y)$.

3 Analysis of Evolving Conjunctions

The first target Boolean function that we analyse is the AND_n function of $n \geq 2$ inputs, i. e. $\text{AND}_n(x_1, \ldots, x_n) = x_1 \wedge x_2 \wedge \ldots \wedge x_n$, that must be built from at least $n-1$ binary AND gates. In GP terminology, we have the function set $F = \{\text{AND}\}$ and the terminal set $T := \{x_i \mid i \in [n]\}$. Similar to the naming convention for the MAX problem [16], but here without restricting the tree depth as it is not part of the problem definition, we adapt the name AND_n-$\{\text{AND}\}$-$\{x_i \mid i \in [n]\}$ for this logic synthesis problem. Regarding the complete training set of this function, only one row of $\mathcal{K}_{\text{AND}_n}$ has $\text{AND}_n(\alpha(x_1), \ldots, \alpha(x_n)) = \text{TRUE}$ when $\alpha(x_1) = \ldots = \alpha(x_n) = \text{TRUE}$, all the other rows have the desired output FALSE.

To implement this synthesis in CGP, it suffices to set $N_{\text{in}} = n$, $N_{\text{out}} = 1$, use a lookup table with the single binary function AND (i.e., $K = 1$ and $k = 2$), and choose $N_{\text{func}} \in \mathbb{N}$ such that $N_{\text{func}} \geq n - 1$. It follows from [22] that the fitness of programs for AND_n-$\{\text{AND}\}$-$\{x_i \mid i \in [n]\}$ is monotonically improved (here decreased for the minimisation problem) as a function of the number of input variables used/included (connected to active nodes in case of CGP).

Lemma 2 (Lemma 1 in[22]). *On AND_n-$\{\text{AND}\}$-$\{x_i \mid i \in [n]\}$, the truth table of a program currently using i distinct input variables differs from the target truth table of $x_1 \wedge x_2 \wedge \ldots \wedge x_n$ at exactly $f_i = 2^{n-i} - 1$ rows.*

We show that (1+1) CGP and (1+1) CGP* with $N_{\text{func}} := D \geq n - 1$, using SAM mutation and the complete training set, synthesise the target function of AND_n-$\{\text{AND}\}$-$\{x_i \mid i \in [n]\}$ in expected time polynomial in D. In particular, for $D \in [n-1, \infty) \cap O(n)$, the expected time is $\mathcal{O}(n^6)$ and the space required is $\mathcal{O}(n \log n)$.

Theorem 3. *The* (1+1) CGP *and* (1+1) CGP* *fit the complete training set of* AND_n-$\{\text{AND}\}$-$\{x_i \mid i \in [n]\}$ *using* $N_{\text{func}} := D \geq n - 1$, *and SAM mutation in* $\mathcal{O}(nD^5)$ *fitness evaluations in expectation and uses* $\mathcal{O}(D \log D)$ *space to store its programs. If* $D \geq 10$, *the probability that the number of fitness evaluations exceeds* $(1+\delta)20\pi^2(n-1)D^5$ *for any* $\delta \geq 1$ *is at most* $e^{-\delta/3}$.

Proof sketch. From Lemma 2 and the monotonicity of $2^{n-i} - 1$ in i, the number of input variables included in the current program cannot be decreased in future iterations. Hence, we apply the fitness-level method [42], using as levels the number of included input variables, to bound the expected time until all variables are included. When estimating the probability s_i to leave the current level to a higher one, we only consider strict improvements, thus the results hold for both (1+1) CGP and (1+1) CGP*. There are two cases, either there is a free (input) connection from an active node that can redirected to a new input variable, otherwise there are none. The former case gives a higher probability of $\frac{1}{3D+1} \cdot \frac{n-i}{D+n-1}$. In the latter case, the SAM mutation needs to re-arrange the connections of a non-active node so as to extend to a new input variable and to make sure that when the node is activated, the fitness is improved. The probability in this case is at least $\frac{2\cdot\alpha(n-i)\cdot(n-i)}{(3D+1)^3(2D)^3} = \frac{\alpha(n-i)^2}{(3D+1)^3(4D^3)}$ where $\alpha := D/(n-1)$. The expected running time is then bounded from above by

$$\sum_{i=1}^{n-1} \frac{(3D+1)^3(4D^3)}{\alpha(n-i)^2} = (3D+1)^3 4D^3\alpha^{-1}\sum_{i=1}^{n-1}\frac{1}{i^2} \leq (3D+1)^3 4D^3\alpha^{-1}\zeta(2)$$

and this is $(3D+1)^3D^3((n-1)/D)(2\pi^2/3) = \mathcal{O}\left(nD^5\right)$. The tail bound then follows by applying Theorem 1 of Witt [43]. $\square$

The bound $\mathcal{O}(n^6)$ of the minimal setting $D = n-1$ is tight and non-improvable for the (1+1) CGP* when the initial program almost already fits the training set.

Theorem 4. *There exist initial programs for which the* (1+1) CGP* *with* $N_{\text{func}} := n-1$, *and SAM mutation requires* $\Omega(n^6)$ *fitness evaluations in expectation to fit the complete training set of* AND_n-{AND}-$\{x_i \mid i \in [n]\}$.

In contrast, the upper bound for the expected running time of (1+1) CGP can be further improved by considering neutrality in the fitness landscape as follows. We need the following lemma on a three-state Markov chain. The same chain was used in [2,28] to analyse population diversity in Genetic Algorithms.

Lemma 5. *Consider the following Markov chain with three states with the absorbing state* 3 *and parameters* $p, q, r, s \in [0, 1)$ *where* $q, r > 0$.

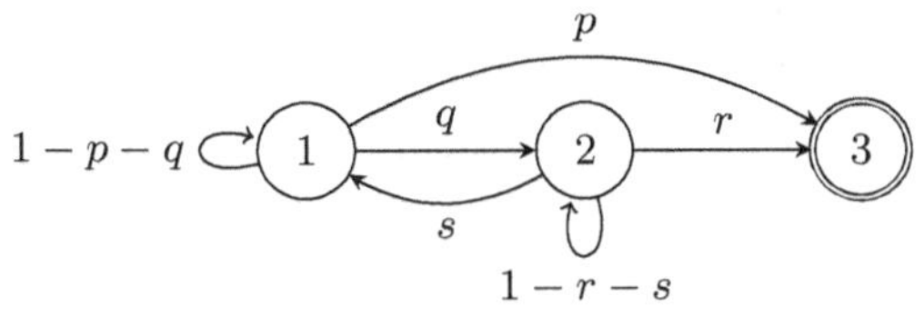

The expected time E_1 *to reach the absorbing state* 3 *starting from state* 1 *is*

$$E_1 = \frac{q+r+s}{rq+rp+sp} \leq \frac{1}{r} + \frac{1}{q} + \frac{s}{rq}.$$

Moreover, if we have $(p+q \geq q') \wedge (r \geq r') \wedge (s \leq s')$, *then the above upper bound also holds with the bounding estimates, that is,* $E_1 \leq \frac{1}{r'} + \frac{1}{q'} + \frac{s'}{r'q'}$.

Theorem 6. *The* (1+1) CGP *fits the complete training set of* AND_n-{AND}-$\{x_i \mid i \in [n]\}$ *using* $N_{\text{func}} := D \geq n-1$, *SAM mutation in* $\mathcal{O}(nD^4)$ *fitness evaluations in expectation.*

Proof sketch. Similarly to the proof of Theorem 3, we divide the search space into n levels according to the number of input variables included. But instead of estimating the probability to leave a level, we compute the expected waiting time τ_i for that same event to happen so that we can apply Lemma 5.

Therefore, we further divide each level i into two sublevels i^+ and i^- which match exactly the two cases in the proof of Theorem 3, i. e. there is at least a free connection versus there is none. The lemma is then applied for each level i with states $1, 2, 3$ being levels $i^-, i^+, i+1$ respectively, thus E_1 gives an upper bound on τ_i. Additionally, the probability of transiting from i^+ to $i+1$ is exactly the probability of leaving the level in easier case, thus at least $\frac{n-i}{(3D+1)(n+D-1)} \geq \frac{n-i}{(3D+1)(2D)} =: r'(i)$. The probability of leaving i^- is also higher compared that of the harder case of the previous proof as it suffices to include a new function node without the requirement on increasing the fitness, thus $p(i) + q(i) \geq \frac{2\alpha(n-i)}{(3D+1)^2(2D)^2} =: q'(i)$. Finally, we argue that there are at most two function nodes that can be deactivated to return to i^- from i^+, thus $s(i) \leq \frac{2}{3D+1} =: s'(i)$. Summing all τ_i gives the expected running time:

$$\begin{aligned}\sum_{i=1}^{n-1} \tau_i &\leq \sum_{i=1}^{n-1} \frac{1}{r'(i)} + \frac{1}{q'(i)} + \frac{s'(i)}{r'(i)q'(i)} \\ &\leq \left((3D+1)(2D) + (3D+1)^2(2D)^2(2\alpha)^{-1}\right) H_{n-1} \\ &\quad + 2(3D+1)^2(2D)^3(2\alpha)^{-1}\zeta(2).\end{aligned}$$

This is $\mathcal{O}(nD^4)$, and here we use $\alpha = D/(n-1)$ as in the previous proof. □

The analyses from Theorems 3 and 6 show that using more than the minimum $n-1$ function nodes can benefit CGP by increasing the bounds s_i in the harder case, and also $q'(i)$ for $p(i) + q(i)$ in the Markov chain.

4 Analysis of Evolving Exclusive Disjunctions

The second target function that we consider is the XOR_n function of $n \geq 2$ inputs where $\text{XOR}_n(x_1, \ldots, x_n) := x_1 \oplus x_2 \oplus \ldots \oplus x_n$ that must be built from at least $n-1$ binary XOR gates. This function is of interest for us because it is not evolvable [38] even though it is PAC-learnable [37] thus a candidate problem to show a negative result for CGP, especially for an advanced operator like SAM. We adopt the name XOR_n-{XOR}-$\{x_i \mid i \in [n]\}$ for this problem. A candidate CGP program is a function whose *raw form* may contain an input variable repeated multiple times, e. g. $x_1 \oplus x_1 \oplus x_2$ where x_1 is repeated twice. However, a variable that is repeated an even number of times is as if it is ignored, and it is repeated an odd number of times is as if it is repeated once. We refer to the function form after applying these simplifications as the *simplified form.*

For the XOR_n-{XOR}-$\{x_i \mid i \in [n]\}$ and using the complete training set $\mathcal{K}_{\text{XOR}_n}$, the fitness of a candidate program y can only take two possible values, that is, $f(y) = 0$ if the simplified form of y is exactly XOR_n, i. e. every input appears an odd number of times in the raw form, otherwise $f(y) = 2^{n-1}$, i. e. y is incorrect on a half of the rows (and correct on the other half) [22]. This property implies that if the use of an incomplete training set leads to a program that ignores some inputs in the simplified form, then the generalisation ability of such a program is always $2^{n-1}/2^n = 1/2$. This is as bad as a random guess, and justifies our focus on the setting of the complete training set $S = \mathcal{K}_{\text{XOR}_n}$.

We then expect CGP to perform a random walk in the space of programs until it realises an optimal program, i. e. having the simplified form XOR_n. To show an exponential time on reaching such a program, we first show that it is unlikely that such a program can emerge from a fixed subset L of function nodes. For a lower bound argument, it suffices to focus on two necessary properties of an optimal program built out of L: (i) every input is connected to L by an odd number of connections; (ii) the output of at least $n-2$ nodes of L is connected to other nodes in L. The second property is required because an optimal program has at least $n-1$ active function nodes, thus at least $n-2$ of those have their activeness maintained by L. Formally, an input variable x_i is called an *odd variable* with respect to L if it is connected to L by an odd number of connections, otherwise it is an *even variable* (with respect to L, we often omit this part when L is clear from the context). Thus we track the number of odd variables, denoted $V(L)$, and $U(L) := |\{\ell \in L \mid \text{the output of } \ell \text{ is connected to another node in } L\}|$ of the current program in CGP to reach their respective values n and $n-2$.

Lemma 7. *A randomly generated CGP program on* XOR_n-$\{\text{XOR}\}$-$\{x_i \mid i \in [n]\}$ *using* $N_{\text{func}} := D \geq n-1$ *has* $V(L) \leq 3n/4$ *with probability* $1 - e^{-\Omega(n)}$*, for any fixed set* L *of function nodes with* $m := |L| \in poly(n)$.

If an optimal program has not yet been found, then the (1+1) CGP always accepts offspring programs and all modifications by every iteration of the while-loop across SAM mutations are taken into account. We refer to each of these iterations as an *elementary step*. The following lemma, again using the negative drift theorem, shows that it is unlikely for a program with the two necessary properties, let alone for an optimal program, to emerge during the application of an exponential number of elementary steps to a randomly generated program.

Lemma 8. *Starting from a randomly generated CGP program on* XOR_n-$\{\text{XOR}\}$-$\{x_i \mid i \in [n]\}$ *using* $N_{\text{func}} := D \geq n-1$*, for a fixed set* L *of its function nodes with* $m := |L| \leq (1+d)n$ *for any* $d \leq 10^{-2}$*, the probability that a program with* $U(L) \geq n-2$ *and* $V(L) = n$ *emerges during the application of* $e^{\varepsilon n}$ *elementary steps of SAM for some constant* $\varepsilon > 0$ *is at most* $e^{-\Omega(n)}$.

Proof sketch. We apply the negative drift theorem, i. e. Theorem 2 in [30], using potential $Y_j = 2n-2-U_j-V_j$ where U_j is $\min(U(L), n-2)$ and V_j is $V(L)$ after applying j elementary steps to the initial solution. It follows from Lemma 7, that $V_0 \leq 3n/4$ with probability $1 - e^{-\Omega(n)}$, thus $Y_0 \geq 2n-2-(n-2)-3n/4 = n/4$ with that same probability. We estimate the drift $\mathrm{E}[Y_{j+1} - Y_j \mid Y_j]$ for $Y_j \in [n/256, n/128]$. Three events are of interest: (i) redirecting a connection from an odd input variable and to another odd input variable, i. e. with probability $\frac{V_j}{2m} \cdot \frac{V_j - 1}{n+m-1} > \frac{(127n/128)^2}{(2m)^2} > \frac{6n^2}{25D^2}$; (ii) choosing a connection linked to an even input variable for modification, i. e. with probability at most $(n/128 +$

$2dn)/(2m) \leq 1/128 + 2d$; (iii) redirect a connection not linked to an input variable to an even input variable or to a function node that is not counted in $U(L)$ yet, i.e. with probability at most $(n/128+dn)/n+n/(128n) \leq 1/64+d$. The first event contributes to the drift of 2, the two last contribute at most 3. The drift is $\mathrm{E}\left[Y_{j+1} - Y_j \mid Y_j \in [\frac{n}{256} - 2, \frac{n}{128} - 2]\right] > \frac{1}{1+d}\left(\frac{12}{25(1+d)} - \frac{3(1+d)}{42} - 9(1+d)d\right)$ which at least 3/10 after plugging $d = 10^{-2}$ in. The results then follow from the negative drift theorem. □

To extend the lemma to (1+1) CGP, we take a union bound over all subsets L, whose number can be estimated via the approximation of the standard binomial coefficient bounds using binary entropy. Finally, the SAM while-loop terminates once the output connection is selected for modification, which occurs with probability $1/(3D+1) = \Omega(1/n)$ for $D = \mathcal{O}(n)$. Hence, an exponential number of elementary SAM steps translates to the same order of SAM mutations (fitness evaluations) by a Chernoff bound.

Theorem 9. *Let $d < 10^{-2}$ be a sufficiently small constant. The probability that the* (1+1) CGP *fails to fit the complete training set of* XOR_n-$\{\text{XOR}\}$-$\{x_i \mid i \in [n]\}$ *within* $e^{\Omega(n)}$ *fitness evaluations using SAM mutation and* $N_{\text{func}} := D$ *for* $n-1 \leq D \leq n+dn$, *is* $1 - e^{-\Omega(n)}$.

Table 1. Configuration of GP and AND_n for the experiments.

AND_n param.	**Setting**	**GP param.**	**Setting**
Dimension n	$\{3,\ldots,15\}$ (compl.), $\{3,\ldots,50\}$ (incompl.)	Maximum iterations	1×10^6
Input i	$i \in \{0,1\}$	Search strategy	(1+1)
Function set	$F = \{\text{AND}\}$	Selection types	non-strict, strict
Terminal set	$T = \{x_1 \ldots x_n\}$	No. replications per n	30

5 Experiments

We evaluate CGP and TGP on AND_n through two sets of experiments. First, using the complete training set, we measure runtime to highlight the performance gap between the two approaches when evolving the exact solution. Second, with incomplete training and testing sets, we assess both runtime and generalisation ability to examine their generalisation performance.

Experimental Setup: We used TinyverseGP [13], a modular benchmarking Python library for GP that provides modules for various flavours of GP including CGP and TGP, to perform the experiments. Using the existing implementations `TinyTGP` and `TinyCGP` of the library, we derived the models `SimpleTGP` and

SimpleCGP which correspond to (1 + 1) TGP/CGP as described in Sect. 2.1 and [15,16]. The CGP model uses MutationSAM as described in Algorithm 3, while the TGP model uses the mutations from [15] which are based on the HVL-prime operator [27]. The implementations of the models and the AND_n as well as the raw data obtained in the experiments are provided in our repository[1].

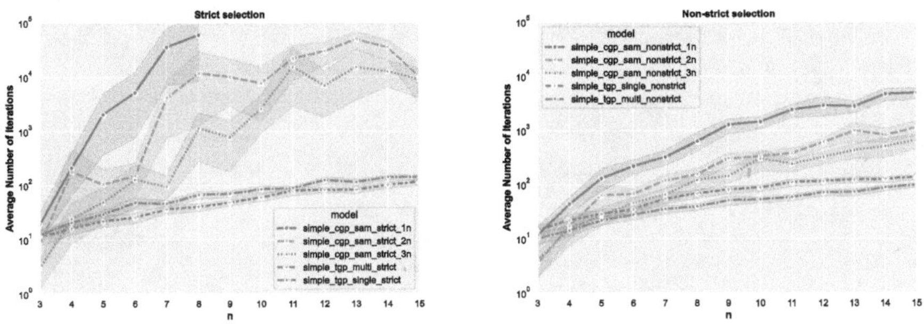

Fig. 2. Results of the comparison on AND_n using complete datasets.

We evaluated various settings of n in each scenario. For the scenario where the complete training set is used, we evaluated $n \in \{3, \ldots, 15\}$. However, since the evaluation of an incomplete training set is less expensive, we evaluated the performance for larger settings of $n \in \{3, \ldots, 50\}$ in this scenario. For each value of n, 30 runs were performed with different random seeds per model (CGP and TGP). The mean of each sample per n was then used for the illustration of our results. The GP configuration used in our experiments is summarised in Table 1. Running time was measured as the number of iterations for the algorithm to fit the training set. In our plots, we refer to the TGP models as simple_tgp_single and simple_tgp_multi and to the CGP model as simple_cgp_sam. For CGP, we defined the length of the genotype N_{func} with respect to the dimension n but also considered extra nodes to see if there is any gap in runtime. More precisely, we used $N_{\text{func}} \in \{n, 2n, 3n\}$, and refer to as 1n, 2n, 3n in our results.

We set the size of the incomplete training and testing sets S and S' as $|S| = |S'| = \lceil n^k \rceil$ with $k = 1.3$ for the settings of large n. Both sets were generated uniformly at random. Approximating the generalisation ability $G_g(y)$ by $\hat{G}(y, S')$ was done with the formula as explained in Sect. 2.2. Statistics are reported per setting of n excluding unsuccessful runs, i. e. exceeding the maximum number of iterations without fitting the training set.

Results: Figure 2 shows the results for the running time comparison on AND_n when the complete training set was used. It can be clearly seen that the considered TGP models perform generally better in this scenario. It can also be seen that increasing the number of function nodes generally improves the search performance of CGP while using strict selection deteriorates its performance.

[1] https://github.com/GPBench/TinyverseGP/tree/ppsn-2026.

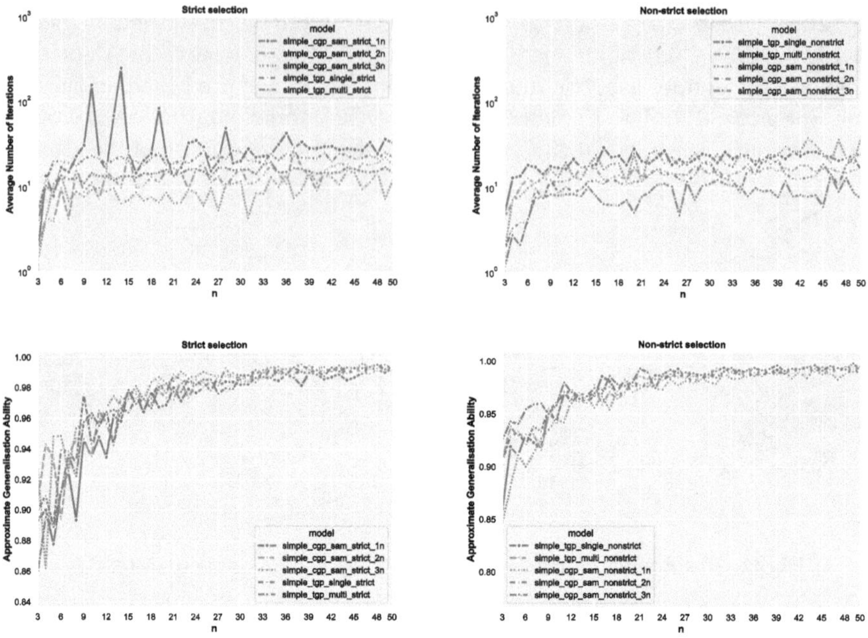

Fig. 3. Results of the comparison on AND_n using incomplete datasets.

Figure 3 show the results of our experiments with incomplete training and testing sets and it can also be seen that increasing the number of function nodes improves the search performance of CGP in terms of running time for both selection strategies. It is also evident that the generalisation error is relatively low for all models considered demonstrating good generalisation abilities on AND_n.

6 Conclusions and Discussions

We presented the first runtime analysis of CGP for evolving conjunctions and disjunctions, considering both complete and incomplete training sets. For AND_n with n variables and D nodes, we proved expected runtimes of $O(nD^5)$ for (1+1) CGP* and $O(nD^4)$ for (1+1) CGP with the complete training set. Our analysis revealed the characteristics of CGP induced search that enabling the acceptance of equally good solutions with active gates non-contributing to fitness is beneficial. We also showed that (1+1) CGP fails with overwhelming probability on the complete training set of the XOR_n. To complement our theoretical analysis, we conducted an empirical study examining the runtime performance and generalisation ability of CGP and TGP models for AND_n.

The theoretical analysis of incomplete training sets for XOR_n is left out for the reason that the generalisation ability one can hope for is only as good as a random guess. For AND_n, there is a technical obstacle that on an incomplete training set, our approach using the fitness-level method and Markov chain analysis no longer works with the current level partition. In particular, it is possible to swap a larger set of input variables for a smaller one, and such a change still improves the fitness. The same issue must occur for TGP under a global mutation operator, however we could not find any literature on this. We also believe drift analysis [6] is a better-suited tool for addressing such a setting, and leave the analysis for future work. Future work should also include extending the analysis to general Boolean formulas with AND/OR, identifying broader function classes where CGP is effective, and studying its behaviour on flat fitness landscapes. Achieving these, however, requires a deeper understanding of the underlying dynamics of CGP, which still is largely unexplored.

Acknowledgments. This work was supported by an Alexander von Humboldt Professorship in AI held by Holger Hoos. The authors are grateful to the anonymous reviewers of the conference for their thoughtful comments and pointers to the literature. One reviewer spotted a flaw in our attempt to show a speedup for incomplete training sets, leading to our retraction of that set of results, and the concluding discussion.

References

1. Cormen, T.H., Leiserson, C.E., Rivest, R.L., Stein, C.: Introduction to Algorithms. MIT Press (2009)
2. Corus, D., Oliveto, P.S.: Standard steady state genetic algorithms can Hillclimb faster than mutation-only evolutionary algorithms. IEEE Trans. Evol. Comput. **22**(5), 720–732 (2018)
3. Cramer, N.L.: A representation for the adaptive generation of simple sequential programs. In: Proceedings of the 1st International Conference on Genetic Algorithms, pp. 183–187. Lawrence Erlbaum Associates (1985)
4. Dang, D.C., Kalkreuth, R., Opris, A.: Runtime analysis of cartesian genetic programming in evolving Boolean functions. CoRR **abs/2606.15923** (2026). https://arxiv.org/abs/2606.15923
5. Dickmanns, D., Schmidhuber, J., Winklhofer, A.: Der genetische algorithmus: eine implementierung in prolog. Fortgeschrittenenpraktikum, Institut f ur Informatik, Lehrstuhl Prof. Radig, Technische Universit at Munchen (1987)
6. Doerr, B., Johannsen, D., Winzen, C.: Multiplicative drift analysis. In: GECCO '10: Proceedings of the 12th Annual Conference on Genetic and Evolutionary Computation, pp. 1449–1456. ACM (2010)
7. Doerr, B., Lissovoi, A., Oliveto, P.S.: Evolving Boolean functions with conjunctions and disjunctions via genetic programming. In: GECCO '19: Proceedings of the Genetic and Evolutionary Computation Conference, pp. 1003–1011. ACM (2019)
8. Doerr, B., Lissovoi, A., Oliveto, P.S.: (1+1) genetic programming with functionally complete instruction sets can evolve Boolean conjunctions and disjunctions with arbitrarily small error. Artif. Intell. **319**, 103906 (2023)
9. Goldman, B.W., Punch, W.F.: Reducing wasted evaluations in cartesian genetic programming. In: Krawiec, K., Moraglio, A., Hu, T., Etaner-Uyar, A.Ş, Hu, B. (eds.) EuroGP 2013. LNCS, vol. 7831, pp. 61–72. Springer, Heidelberg (2013)
10. Hachtel, G.D., Somenzi, F.: Logic Synthesis and Verification Algorithms. Kluwer (1996)

11. Husa, J., Sekanina, L.: Evolving cryptographic Boolean functions with minimal multiplicative complexity. In: IEEE Congress on Evolutionary Computation, CEC 2020. IEEE (2020)
12. Kalkreuth, R., Droschinsky, A.: On the time complexity of simple cartesian genetic programming. In: IJCCI 2019: Proceedings of the 11th International Joint Conference on Computational Intelligen, pp. 172–179. ScitePress (2019)
13. Kalkreuth, R., et al.: Tinyversegp: towards a modular cross-domain benchmarking framework for genetic programming. In: GECCO '25 Companion: Proceedings of the Genetic and Evolutionary Computation Conference Companion, pp. 127–130. ACM (2025)
14. Kalkreuth, R., Vašíček, Z., Husa, J., Vermetten, D., Ye, F., Bäck, T.: General Boolean function benchmark suite. In: Proceedings of the 17th ACM/SIGEVO Conference on Foundations of Genetic Algorithms, pp. 84–95 (2023)
15. Kötzing, T., Sutton, A.M., Neumann, F., O'Reilly, U.M.: The max problem revisited: the importance of mutation in genetic programming. In: GECCO '12: Proceedings of the 14th Annual Conference on Genetic and Evolutionary Computation, pp. 1333–1340. ACM (2012)
16. Kötzing, T., Sutton, A.M., Neumann, F., O'Reilly, U.M.: The max problem revisited: the importance of mutation in genetic programming. Theoret. Comput. Sci. **545**, 94–107 (2014)
17. Koza, J.R.: Hierarchical genetic algorithms operating on populations of computer programs. In: Proceedings of the 11th International Joint Conference on Artificial Intelligence - Volume 1, pp. 768–774. Morgan Kaufmann (1989)
18. Koza, J.R.: Genetic Programming: on the Programming of Computers by Means of Natural Selection. MIT Press (1993)
19. Lissovoi, A., Oliveto, P.S.: On the time and space complexity of genetic programming for evolving boolean conjunctions. In: Proceedings of the Thirty-Second AAAI Conference on Artificial Intelligence, AAAI Press (2018)
20. Lissovoi, A., Oliveto, P.S.: On the time and space complexity of genetic programming for evolving Boolean conjunctions. J. Artifi. Intell. Res. **66**, 655–689 (2019)
21. Lissovoi, A., Oliveto, P.S.: Computational complexity analysis of genetic programming. In: Theory of Evolutionary Computation - Recent Developments in Discrete Optimization, pp. 475–518. Natural Computing Series. Springer (2020)
22. Mambrini, A., Oliveto, P.S.: On the analysis of simple genetic programming for evolving boolean functions. In: Proceedings of the 19th European Conference on Genetic Programming, EuroGP 2016, pp. 99–114. Springer (2016)
23. Miller, J.F.: An empirical study of the efficiency of learning Boolean functions using a cartesian genetic programming approach. In: GECCO'99: Proceedings of the 1st Annual Conference on Genetic and Evolutionary Computation - Volume 2. vol. 2, pp. 1135–1142. Morgan Kaufmann (1999)
24. Miller, J.F., Smith, S.L.: Redundancy and computational efficiency in cartesian genetic programming. IEEE Trans. Evol. Comput. **10**(2), 167–174 (2006)
25. Miller, J.F., Thomson, P.: Cartesian genetic programming. In: Genetic Programming: European Conference, EuroGP 2000, pp. 121–132. Springer (2000)
26. Miller, J.F.: An empirical study of the efficiency of learning Boolean functions using a cartesian genetic programming approach. In: Proceedings of the 1st Annual Conference on Genetic and Evolutionary Computation - vol. 2, pp. 1135–1142. Morgan Kaufmann (1999)
27. Neumann, F., O'Reilly, U.M., Wagner, M.: Computational complexity analysis of genetic programming - initial results and future directions. In: Genetic Programming Theory and Practice IX, pp. 113–128. Springer (2011)

28. Oliveto, P.S.: Time complexity analysis of stochastic search algorithms. In: Handbook of Heuristics (Second Edition), pp. 849–884. Springer, Cham, Switzerland (2025)
29. Oliveto, P.S., Witt, C.: Simplified drift analysis for proving lower bounds in evolutionary computation. Algorithmica **59**(3), 369–386 (2011)
30. Oliveto, P.S., Witt, C.: Erratum: simplified drift analysis for proving lower bounds in evolutionary computation. CoRR **abs/1211.7184** (2012)
31. Openshaw, S., Turton, I.: Building new spatial interaction models using genetic programming. In: Evolutionary Computing, AISB workshop, pp. 11–13. Springer (1994)
32. Perkis, T.: Stack-based genetic programming. In: Proceedings of the First IEEE Conference on Evolutionary Computation, pp. 148–153. IEEE Press (1994)
33. Poli, R.: Parallel Distributed Genetic Programming. Technical Report CSRP-96-15, School of Computer Science, University of Birmingham (1996)
34. Ryan, C., Collins, J.J., O'Neill, M.: Grammatical evolution: evolving programs for an arbitrary language. In: Proceedings of the First European Workshop on Genetic Programming, pp. 83–96. Springer (1998)
35. Turner, A.J., Miller, J.F.: Cartesian genetic programming: why no bloat? In: Nicolau, M., et al. (eds.) EuroGP 2014. LNCS, vol. 8599, pp. 222–233. Springer, Heidelberg (2014)
36. Turner, A.J., Miller, J.F.: Neutral genetic drift: an investigation using cartesian genetic programming. Genet. Program Evolvable Mach. **16**(4), 531–558 (2015)
37. Valiant, L.G.: A theory of the learnable. In: STOC'84: Proceedings of the Sixteenth Annual ACM Symposium on Theory of Computing, pp. 436–445. ACM (1984)
38. Valiant, L.G.: Evolvability. J. ACM **56**(1), 3:1–3:21 (2009)
39. Vasícek, Z.: Cartesian GP in optimization of combinational circuits with hundreds of inputs and thousands of gates. In: Genetic Programming - 18th European Conference, EuroGP 2015, Copenhagen, Denmark, 8-10 April 2015, Proceedings, pp. 139–150. Springer (2015)
40. Vasícek, Z., Sekanina, L.: Evolutionary approach to approximate digital circuits design. IEEE Trans. Evol. Comput. **19**(3), 432–444 (2015)
41. Vasícek, Z., Sekanina, L.: Evolutionary design of complex approximate combinational circuits. Genet. Program Evolvable Mach. **17**(2), 169–192 (2016)
42. Wegener, I.: Methods for the analysis of evolutionary algorithms on pseudo-Boolean functions. In: Evolutionary Optimization, pp. 349–369. Kluwer (2002)
43. Witt, C.: Fitness levels with tail bounds for the analysis of randomized search heuristics. Inf. Process. Lett. **114**(1–2), 38–41 (2014)
44. Yu, T., Miller, J.: Neutrality and the evolvability of Boolean function landscape. In: Genetic Programming, pp. 204–217. Springer (2001)

Runtime Analysis of the (1+1) EA on Plateau Functions of Multi-valued Decision Variables

Liam Gaeuman and Andrew M. Sutton(✉)

University of Minnesota Duluth, Duluth, MN, USA
amsutton@d.umn.edu

Abstract. There has been a recent interest in the theory community to investigate the runtime of various randomized search heuristics optimizing functions of multi-valued decision variables. In this setting, it is not always clear how algorithm behavior changes when moving to higher cardinality alphabets. Moreover, the algebraic ring structure within each variable suggests new kinds of search operators. In this paper, we study the (1+1) EA optimizing functions over $\{0, 1, \ldots, q-1\}^n$ for any $q > 1$ in which large plateaus must be traversed. Recently, Doerr and Kelley [Algorithmica, 2024] connected discrete Fourier analysis to the plateau behavior of the (1+1) EA on functions over $\{0, 1\}^n$. We extend this work to q-ary strings for arbitrary q by expressing the mutation distribution in the orthogonal basis of complex Fourier functions which are characters of the abelian group $\mathbb{Z}_q^n$. We prove exact and tight asymptotic bounds for the (1+1) EA using standard uniform mutation on the multi-valued Needle and RoyalRoad functions. We adapt the analysis to prove coarser upper and lower bounds on the (1+1) EA using a less aggressive cycle mutation operator that recognizes the ring structure of each variable. We also conduct experiments to observe how the accuracy of the bounds on cycle mutation depend on q.

Keywords: Runtime analysis · Fourier analysis · Multi-valued strings

1 Introduction

Discrete optimization problems are characterized by a function $f \colon \mathcal{X} \to \mathbb{R}$ over a countable set $\mathcal{X}$, with the objective being to identify an optimal element of $\mathcal{X}$, i.e., one that minimizes or maximizes f. Historically, most runtime analyses of evolutionary algorithms and other randomized search heuristics fix the domain to be the set of length-n binary strings $\mathcal{X} = \{0, 1\}^n$. However, many practical optimization problems require the optimization of an objective function defined over strings of a larger finite alphabet. To address this, a number of theoretical works have focused on the analysis of randomized search heuristics on multi-valued strings [1,5,6,8,13,22], that is, the domain $\mathcal{X} = \{0, 1, \ldots, q-1\}^n$ for a constant $q \in \mathbb{N} \setminus \{1\}$. Often, this can be achieved by generalizing established

G. Iacca et al. (Eds.): PPSN 2026, LNCS 16985, pp. 180–195, 2027.
https://doi.org/10.1007/978-3-032-36214-8_12

proof techniques that work in the binary domain to extend the results to the q-ary domain.

In a recent paper, Doerr and Kelley [7] demonstrated how Fourier analysis over the elementary abelian 2-group $\mathbb{Z}_2^n$ can be used to get surprisingly tight bounds on the runtime of the (1+1) EA for searching plateaus in $\{0,1\}^n$. Their analysis expresses the mutation distribution as a linear combination of order-2 characters, which are also known as Walsh functions in other EC literature.

In this work, we extend this idea to multi-valued domain $\{0,\ldots,q-1\}^n$ for general $q \geq 2$. This extension is interesting because when $q > 2$, the characters become complex and extend beyond the real-valued Walsh functions. Moreover, we show how the approach can be adapted to different mutation operators that respect the ring structure of elements. We obtain precise asymptotic runtime bounds for the (1+1) EA using standard uniform mutation on the multi-valued NEEDLE function. We derive upper and lower bounds on the runtime of the EA using a slightly more restrictive *cycle mutation* operator that was first introduced in the context of dynamic multi-valued ONEMAX problems [13]. We also adapt the analysis to give upper bounds on the runtime of the (1+1) EA on multi-valued ROYALROAD functions.

Our results demonstrate how Fourier analysis over abelian groups can be further pushed to multi-valued discrete optimization, and provide tight asymptotic bounds for escaping plateaus in this setting. In addition to extending our understanding of the working principals of evolutionary algorithms to different domains, another motivation for studying the NEEDLE and ROYALROAD functions comes from applications in the field of post-quantum cryptography. For example, solving the NEEDLE problem over $\{0,\ldots,q-1\}^\ell$ can be characterized as a brute-force preimage attack on a hash function $h\colon \mathbb{Z}_q^\ell \to \mathbb{Z}_q^m$ such as those constructed by multivariate quadratic equations [3].

1.1 Related Work

Many applications in combinatorial domains require the optimization of functions over multi-valued strings, e.g., in scheduling [12] or representing rooted trees in graphs [17]. Several runtime analyses have been carried out explicitly for extensions of pseudo-Boolean benchmarks to corresponding functions[1] over $\{0,\ldots,q-1\}^n$. Using multiplicative drift, Doerr et al. [6] proved that the (1+1) EA solves linear functions over $\{0,1,2\}$ in $O(n\log n)$ times, whereas if q is too large, there is no single potential function that attains uniform positive drift for all linear functions, ruling out the possibility of using this proof technique for general q. Subsequently, Doerr and Pohl [8] used the adaptive potential function technique of Witt [21] to prove that the (1+1) EA has a runtime bounded above by $O(qn\log n) + O(q^3 n\log\log n)$ and below by $\Omega(qn\log n)$ on all linear functions over $\{0,\ldots,q-1\}^n$.

[1] Many of these works use the notation r (or even $r+1$) instead of q to denote the cardinality of the string alphabet. We restate their results here in our notation.

Kötzing et al. [13] extended the work of Droste [10] on dynamic OneMax to $\{0,\dots,q-1\}^n$. In this setting, the OneMax objective is defined as the distance to a target string, which changes in each iteration according to some probability. They also proved that the choice of q does not influence the ability of the (1+1) EA to track the optimal solution.

Doerr, Doerr and Kötzing [5] investigated multi-valued search spaces with a particular focus on how best to design mutation operators. The natural analogue of standard bit mutation is to simply change each element of the string with some probability to a distinct element. They pointed out that by respecting the algebraic structure of the elements, other natural mutation operators arise. This especially makes sense when the fitness function takes into account this structure. They proved that static mutation variants solve multi-valued OneMax functions over $\{0,\dots,q-1\}^n$ in time $\Theta(qn\log n)$ with standard mutation, in time $\Theta(n\log n + qn)$ with a cyclic mutation operator that changes each element by only one step, and in time $\Theta(n\log n\log q + n\log^2 q)$ for a harmonic mutation operator. They also proved that employing a self-adjusting mutation operator can improve the scaling with q by a logarithmic factor. The first precise runtime result for the (1+1) EA on the Needle function over $\{0,1\}^n$ is due to Garnier, Kallel and Schoenauer [11] who proved, using an analysis of a Markov chain over Hamming distance to the optimum, a runtime of $2^n\left(1+\frac{1}{e^c-1}\right)(1\pm o(1))$ when using standard bit mutation with rate $p=c/n$.

Plateau functions for binary strings were recently investigated by Lengler, Opris and Sudholt [14] to study the dynamics of population diversity under different crossover operators in fitness-neutral environments.

RoyalRoad functions were originally introduced by Mitchell, Forrest and Holland [15] to study the dynamics of crossover in genetic algorithms. A RoyalRoad function with block size ℓ is a collection of disjoint degree-ℓ monomials over $\{0,1\}^n$, and thus the (1+1) EA has a runtime of $\Theta\left(2^\ell (n/\ell)\log(n/\ell)\right)$ [9,20].

Discrete Fourier analysis has been utilized in the field of evolutionary computation for decades. The Walsh functions are a complete orthogonal basis in which any real function over $\{0,1\}^n$ can be represented, and were already used in 1981 in the Ph.D. work of Bethke [2] for constructing a class of deceptive fitness functions. Vose and Wright [19] employed Walsh transforms to analyze mutation and crossover in infinite population models of the simple genetic algorithm, with several other works extending to strings over arbitrary alphabets [16]. Fourier analysis has also been employed to characterize statistical features of fitness landscapes such as moments of fitness distribution over various search regions [4,18].

The first work connecting discrete Fourier analysis to runtime analysis was by Doerr and Kelley [7] who derived precise waiting times for escaping plateaus in $\{0,1\}^n$. Their work yields a succinct proof of the Needle result of Garnier, Kallel and Schoenauer [11], and gives bounds for the runtime of the (1+1) EA on the LeadingBlocks function. For the latter, they derive asymptotically optimal static mutation rates as well as optimal fitness-dependent mutation rates.

The remainder of this paper is structured as follows. In the next section, we present the algorithm, notational definitions, and tools from Fourier analysis over finite abelian groups. In Sect. 3, we use the analysis to derive runtime bounds for escaping plateaus in $\{0, \ldots, q-1\}^n$. In Sect. 4 we present some experiments for cyclic mutation to determine the tightness of our derived bound. We conclude the paper in Sect. 5.

2 Preliminaries

2.1 The (1+1) EA on Multi-Valued Strings

We consider the (1+1) EA maximizing a function over $\{0, 1, \ldots, q-1\}^n$. The (1+1) EA, defined in Algorithm 1, maintains a population of a single element, initially sampled uniformly at random, and proceeds with an iterative improvement search that employs a mutation operator to create an offspring which replaces the parent if it is no worse in fitness.

Algorithm 1: Multi-valued (1+1) EA

```
input : A function f: {0,1,...,q−1}^n → ℝ to maximize
Sample x ∈ {0,1,...,q−1}^n uniformly at random
while termination criteria is not met do
    y ← mutate(x,p)
    if f(y) ≥ f(x) then x ← y
```

We study two mutation operators for multi-valued decision variables. *Standard uniform mutation* is the typical mutation operator employed for multi-valued problems [6,8] and proceeds by replacing, with some probability, each element of the string with a distinct element. Doerr, Doerr and Kötzing [5] refer to this operator as the "uniform step" operator.

Definition 1. *Let* $x \in \{0, 1, \ldots, q-1\}^n$*. The* standard uniform mutation operator *with mutation rate* p *produces an offspring* $y \in \{0, 1, \ldots, q-1\}^n$ *by changing each element of* x *independently with probability* p*. In particular, for each index* $i \in \{1, \ldots, n\}$*, with probability* p*,* y_i *is set to a new element selected uniformly at random from* $\{0, 1, \ldots, q-1\} \setminus \{x_i\}$*. Otherwise* $y_i \leftarrow x_i$ *with probability* $(1-p)$*.*

Strictly speaking, standard uniform mutation does not require any algebraic structure on the values of the string, and is applicable to categorical decision variables. A less common mutation operation, which uses the cyclic group structure of $\mathbb{Z}_q$, is defined as follows.

Definition 2. *Let* $x \in \{0, 1, \ldots, q-1\}^n$*. The* cycle mutation operator *with mutation rate* p *produces an offspring* $y \in \{0, 1, \ldots, q-1\}^n$ *by changing each element of* x *independently with probability* p*. In particular, for each index* $i \in \{1, \ldots, n\}$*, with probability* p*,* $y_i \leftarrow x_i \pm 1 \pmod q$*, where the sign is chosen uniformly at random.*

This operator was first considered by Kötzing, Lissovoi and Witt [13] in the context of dynamic problems over multi-valued strings. Doerr, Doerr and Kötzing [5] also considered this operator, together with a related *harmonic* step operator modifies the jump-size rule: instead of always moving by one step, it allows a jump of size j with probability inversely proportional to j.

2.2 Fourier Analysis on Finite Abelian Groups

We assume the reader is familiar with the basic concepts of finite groups.

Definition 3 (Character). *Let G be a finite abelian group. A* character *of G is a function $\chi\colon G \to \mathbb{C} \setminus \{0\}$ that satisfies*

$$\chi(gh) = \chi(g)\chi(h) \tag{1}$$

for all $g, h \in G$. A character χ is called trivial *when $\chi(g) = 1$ for all $g \in G$.*

Definition 4 (Fourier transform). *Let G be a finite abelian group and let $f\colon G \to \mathbb{C}$ be a complex function of G. The Fourier transform of f is a function $\widehat{f}$ that maps the characters of G into $\mathbb{C}$.*

$$\widehat{f}(\chi) := \sum_{g \in G} f(g)\chi(g). \tag{2}$$

For $q > 1$, we treat $\{0, 1, \ldots, q-1\}^n$ as the direct product of cyclic groups $\mathbb{Z}_q^n = \mathbb{Z}_q \times \mathbb{Z}_q \times \cdots \times \mathbb{Z}_q$. The q^n characters of $\mathbb{Z}_q^n$ are of the form χ_α where $\alpha \in \mathbb{Z}_q^n$ defined as

$$\chi_\alpha(x) = \chi_{\alpha_1}(x_1) \cdots \chi_{\alpha_n}(x_n) = \omega_q^{\alpha_1 x_1} \cdots \omega_q^{\alpha_n x_n},$$

where $\omega_q = e^{2\pi i/q}$ is the complex q-th root of unity.

Definition 5 (Hamming norm). *Given an element $x \in \mathbb{Z}_q^n$ the* Hamming norm *of x is*

$$\|x\|_H := |\{i\colon x_i \neq 0\}|,$$

that is, the Hamming distance of x to the identity element $(0, 0, \ldots, 0)$ of $\mathbb{Z}_q^n$.

2.3 Mutation Distributions

We are interested in the probability distribution of the mutation process that creates an offspring. In a group-theoretic context, creating an offspring $y \in \{0, \ldots, q-1\}^n$ from a parent $x \in \{0, \ldots, q-1\}^n$ can be characterized as sampling a sequence of random elements of $\mathbb{Z}_q$. Using the additive notation for the binary operation on $\mathbb{Z}_q$, we can express the offspring y as $y_i = x_i + Z_i$, where Z_i is a $\mathbb{Z}_q$-random variable that depends on the mutation operator employed. For a particular mutation operator, we are interested in the mutation distribution $\mu\colon \mathbb{Z}_q^n \to [0,1]$ defined as a product distribution over $\mathbb{Z}_q^n$

$$\mu(z) = \prod_{i=1}^{n} \Pr(Z_i = z_i). \tag{3}$$

Mutation can thus be seen as a random step in $\mathbb{Z}_q^n$ distributed by μ and the probability of obtaining an offspring y from parent x is $\mu(yx^{-1})$.

The following lemma yields an exact expression for the Fourier transform of the mutation distribution. For notational convenience we write $\widehat{\mu}(\alpha)$ in place of $\widehat{\mu}(\chi_\alpha)$.

Lemma 1. *Let μ be a mutation distribution on $\mathbb{Z}_q^n$. The Fourier transform of μ can be written as*

$$\widehat{\mu}(\alpha) = \prod_{i=1}^{n} \left(\sum_{z \in \mathbb{Z}_q} \Pr(Z_i = z)\omega_q^{\alpha_i z} \right).$$

Proof. By Definition 4,

$$\begin{aligned}
\widehat{\mu}(\alpha) &= \sum_{x \in \mathbb{Z}_q^n} \mu(x)\chi_\alpha(x) = \sum_{x \in \mathbb{Z}_q^n} \mu(x)\, \omega_q^{\alpha_1 x_1} \cdots \omega_q^{\alpha_n x_n} \\
&= \sum_{x \in \mathbb{Z}_q^n} \left(\prod_{i=1}^{n} \Pr(Z_i = x_i) \right) \omega_q^{\left(\sum_{i=1}^{n} \alpha_i x_i\right)} \\
&= \prod_{i=1}^{n} \left(\sum_{z \in \mathbb{Z}_q} \Pr(Z_i = z)\omega_q^{\alpha_i z} \right).
\end{aligned}$$

□

We now consider particular mutation operators.

Theorem 1. *Let μ be the mutation distribution for standard uniform mutation with mutation rate p (Definition 1). The Fourier transform of μ is*

$$\widehat{\mu}(\alpha) = \left(1 - \frac{pq}{q-1}\right)^{\|\alpha\|_H}.$$

Proof. For standard uniform mutation, the i-th element of the string is not changed with probability $1-p$, otherwise, with probability p, the i-th element is replaced uniformly by one of the remaining elements. Thus we may characterize the distribution in Eq. (3) as

$$\Pr(Z_i = z) = \begin{cases} (1-p) & \text{if } z = 0, \\ p/(q-1) & \text{if } z \in \{1, \dots, q-1\}. \end{cases}$$

By Lemma 1,

$$\widehat{\mu}(\alpha) = \prod_{i=1}^{n} \left((1-p) + \frac{p}{q-1} \sum_{z \neq 0} \omega_q^{\alpha_i z} \right). \tag{4}$$

Note that when $\alpha_i = 0$,

$$(1-p) + \frac{p}{q-1} \sum_{z \neq 0} \omega_q^{\alpha_i z} = (1-p) + \frac{(q-1)p}{q-1} = 1,$$

and when $\alpha_i \neq 0$, by character orthogonality, $\sum_{z \in \mathbb{Z}_q} \omega_q^{\alpha_i z} = 0$, so $\sum_{z \neq 0} \omega_q^{\alpha_i z} = -\omega_q^{\alpha_i (0)} = -1$. Substituting this into (4) yields the claimed result. □

Theorem 2. *Let μ be the mutation distribution for cycle mutation with mutation rate $p < 1/2$ (Definition 2). The Fourier transform of μ is bounded by*

$$(1 - -2p)^{\|\alpha\|_H} \leq \widehat{\mu}(\alpha) \leq \left(1 - -2p \sin^2\left(\frac{\pi}{q}\right)\right)^{\|\alpha\|_H}.$$

Proof. For cycle mutation, the distribution in Eq. (3) can be written as

$$\Pr(Z_i = z) = \begin{cases} (1-p) & \text{if } z = 0, \\ p/2 & \text{if } z \in \{1, q-1\}, \\ 0 & \text{otherwise.} \end{cases}$$

By Lemma 1,

$$\widehat{\mu}(\alpha) = \prod_{i=1}^{n} \left((1-p) + \frac{p}{2} \left(\omega_q^{\alpha_i} + \omega_q^{\alpha_i (q-1)} \right) \right). \tag{5}$$

Note that $\frac{\omega_q^{\alpha_i} + \omega_q^{\alpha_i(q-1)}}{2} = \frac{e^{2\pi i \alpha_i/q} + e^{-2\pi i \alpha_i/q}}{2} = \cos(2\pi\alpha_i/q) = 1 - 2\sin^2(\alpha_i \pi/q)$. Substituting this into Eq. (5) yields

$$\widehat{\mu}(\alpha) = \prod_{i=1}^{n} \left(1 - 2p \sin^2(\alpha_i \pi/q)\right). \tag{6}$$

When $\alpha_i = 0$, $1 - 2p\sin^2(\alpha_i\pi/q) = 1$. Furthermore, in the interval $[1, q-1]$, the function $\sin(\pi x/q)$ attains a minimum at $x = 1$. Thus, as long as $p < 1/2$, we have for all $\alpha_i \neq 0$, $1 - 2p\sin^2(\alpha_i\pi/q) \leq 1 - 2p\sin^2(\pi/q)$, which proves the upper bound. For the lower bound, we use the trivial fact that $\sin^2(x) \leq 1$ yielding a factor of 1 in Eq. (6) for each $\alpha_i = 0$ and a lower bound of $(1 - 2p)$ for each $\alpha_i > 0$. □

3 Escaping Plateaus in $\{0, \ldots, q-1\}^n$

In this section, we use the above techniques to derive runtime analyses for the (1+1) EA on functions of $\{0, \ldots, q-1\}^n$ with large plateaus that must be escaped.

3.1 Runtime Analysis of the (1+1) EA on Multi-Valued Needle

The multi-valued Needle function is defined as follows. Given a target $z \in \{0,\ldots,q-1\}^n$,

$$\textsc{Needle}_z(x) = \begin{cases} 1 & \text{if } x = z, \\ 0 & \text{otherwise.} \end{cases} \tag{7}$$

The Needle function is a benchmark that poses the pathology of a large plateau of equal fitness with a single "escape". This allows us to analyze how the (1+1) EA equipped with different mutation operators handles this kind of pathology. A more application-oriented example of $\textsc{Needle}_z$ is illustrated by taking a hash function $h\colon \mathbb{Z}_q^n \to \mathcal{Y}$ over vectors of the finite field $\mathbb{Z}_q$ to some codomain $\mathcal{Y}$. A preimage attack on h is specified as follows. Given an element $y \in \mathcal{Y}$, find $z \in \mathbb{Z}_q^n$ such that $h(z) = y$. If we assume there is only one such element z, solving $\textsc{Needle}_z$ amounts to a brute-force preimage attack on h.

We characterize a run of the (1+1) EA on $\textsc{Needle}_z$ as a random walk on $\mathbb{Z}_q^n$ generated by the corresponding mutation distribution. Let $(X_t)_{t\geq 0}$ be a stochastic process on $\mathbb{Z}_q^n$ where X_t is the element in iteration t of Algorithm 1.

For $v, z \in \mathbb{Z}_q^n$, write $T_{v,z} = \min\{t \geq 0\colon X_t = z, X_0 = v\}$ for the first hitting time of z from v. The following theorem, due to Zhang [23], bounds $\mathrm{E}[T_{0,z}]$ for random walks on finite abelian groups. We restate it here in terms of $\mathbb{Z}_q^n$.

Theorem 3 (Theorem 3.1 in [23]). *Let μ be a probability distribution on $\mathbb{Z}_q^n$ and consider a random walk (X_t) on $\mathbb{Z}_q^n$ generated by μ starting at $X_0 = \mathbf{0}$. The expectation of $T_{0,z}$ is*

$$\mathrm{E}[T_{0,z}] = \sum_{\alpha \in \mathbb{Z}_q^n \setminus \{\mathbf{0}\}} \frac{1 - \chi_\alpha(z^{-1})}{1 - \widehat{\mu}(\alpha)},$$

where z^{-1} is the group inverse of z.

The initial string for the (1+1) EA is chosen uniformly at random from $\{0,\ldots,q-1\}^n$ (line 1 of Algorithm 1), so for our analysis we require the initial point X_0 of the random walk to be uniformly distributed over $\mathbb{Z}_q^n$. The expected hitting time to an element $z \in \mathbb{Z}_q^n$ under this walk can be expressed in terms of a sum over non-trivial points of the Fourier transform of μ.

Lemma 2. *Consider the stochastic process $(X_t)_{t\geq 0}$ where X_0 is chosen uniformly at random and X_t is produced from X_{t-1} by the mutation distribution μ. Let $T = \min\{t \geq 0\colon X_t = z\}$ be the hitting time of the process to z. The expected value of T is exactly*

$$\mathrm{E}[T] = \sum_{\alpha \in \mathbb{Z}_q^n \setminus \{0\}} \frac{1}{1 - \widehat{\mu}(\alpha)}.$$

Proof. The expected hitting time of the element z is

$$\mathrm{E}[T] = \sum_{v \in \mathbb{Z}_q^n} \mathrm{E}[T_{v,z}] \Pr(X_0 = v) = \frac{1}{q^n} \sum_{v \in \mathbb{Z}_q^n} \mathrm{E}[T_{v,z}].$$

By translation invariance of the random walk,

$$\mathrm{E}[T] = \frac{1}{q^n} \sum_{v \in \mathbb{Z}_q^n} \mathrm{E}[T_{v-v,z-v}] = \frac{1}{q^n} \sum_{w \in \mathbb{Z}_q^n} \mathrm{E}[T_{0,w}].$$

Applying Theorem 3, we obtain

$$\begin{aligned}\mathrm{E}[T] =& \frac{1}{q^n} \sum_{w \in \mathbb{Z}_q} \mathrm{E}[T_{0,z}] = \frac{1}{q^n} \sum_{w \in \mathbb{Z}_q^n} \sum_{\alpha \in \mathbb{Z}_q^n \setminus \{0\}} \frac{1 - \chi_\alpha(w^{-1})}{1 - \widehat{\mu}(\alpha)} \\ =& \frac{1}{q^n} \sum_{\alpha \in \mathbb{Z}_q^n \setminus \{0\}} \frac{1}{1 - \widehat{\mu}(\alpha)} \sum_{w \in \mathbb{Z}_q^n} (1 - \chi_\alpha(w^{-1})) = \sum_{\alpha \in \mathbb{Z}_q^n \setminus \{0\}} \frac{1}{1 - \widehat{\mu}(\alpha)}.\end{aligned}$$

The final step follows from the orthogonality of characters: for each $\alpha \neq 0$, $\sum_{w \in \mathbb{Z}_q^n} \chi_\alpha(w^{-1}) = 0$, and thus $\sum_{w \in \mathbb{Z}_q^n} (1 - \chi_\alpha(w^{-1})) = q^n$. □

The following lemma will be useful for obtaining tight asymptotic bounds.

Lemma 3. *Let $\gamma > 0$ be a constant and $q \in \mathbb{N} \setminus \{1\}$ be fixed. Then*

$$\sum_{j=1}^{n} \binom{n}{j} (q-1)^j \frac{1}{1 - (1 - \gamma/n)^j} = q^n \left(1 + \frac{1}{e^{(q-1)\gamma/q} - 1}\right) (1 \pm o(1))$$

where the asymptotic term is in n.

Proof. Taking the geometric expansion of $1/(1 - (1 - \frac{\gamma}{n})^j)$, we have

$$\sum_{j=1}^{n} \binom{n}{j} (q-1)^j \frac{1}{1 - \left(1 - \frac{\gamma}{n}\right)^j} = \sum_{j=1}^{n} \binom{n}{j} (q-1)^j \sum_{k=0}^{\infty} \left(1 - \frac{\gamma}{n}\right)^{jk}.$$

Rearranging and applying the Binomial Theorem, the sum is equal to

$$\sum_{k=0}^{\infty} \left(\sum_{j=1}^{n} \binom{n}{j} \left((q-1) \left(1 - \frac{\gamma}{n}\right)^k \right)^j \right) = \sum_{k=0}^{\infty} \left(\left(1 + (q-1) \left(1 - \frac{\gamma}{n}\right)^k \right)^n - 1 \right).$$

Factoring out q^n, we may write this infinite sum as

$$\sum_{k=0}^{\infty} \left(q^n \left(\frac{1}{q} + \frac{q-1}{q} \left(1 - \frac{\gamma}{n}\right)^k \right)^n - 1 \right) = q^n \sum_{k=0}^{\infty} a_k, \tag{8}$$

where

$$a_k = \left(1 - \frac{q-1}{q} \left(1 - \left(1 - \frac{\gamma}{n}\right)^k \right) \right)^n - q^{-n}.$$

Setting $\delta := (q-1)/q$, we can write the k-th term a_k of the series as

$$a_k = \left(1 - \delta \left(1 - \left(1 - \frac{\gamma}{n}\right)^k \right) \right)^n - q^{-n} = e^{-\delta\gamma k} + O\left(\frac{1}{n}\right) = e^{-\delta\gamma k}(1 \pm o(1)),$$

where we have applied the Taylor expansion of $(1-\delta(1-(1-\gamma x)^k))^{1/x}$ at $x=0$. The sum on the RHS of (8) can be written as $\sum_{k=0}^{\infty} a_k = (1 \pm o(1)) \sum_{k=0}^{\infty} e^{-\delta\gamma k}$, yielding a convergent geometric series $\sum_{k=0}^{\infty} e^{-\delta\gamma k}$ with closed form $1/(1-e^{-\delta\gamma}) = 1 + 1/(e^{(q-1)\gamma/q} - 1)$. □

Theorem 4. *The (1+1) EA using standard uniform mutation with rate $p = c/n$ for a constant $c > 0$ finds the optimum of the multi-valued* NEEDLE$_z$ *function in expected time*

$$q^n \left(1 + \frac{1}{e^c - 1}\right)(1 \pm o(1)).$$

Proof. Let T be the first time the (1+1) EA finds the optimum z on NEEDLE$_z$. Combining Lemma 2 and Theorem 1, we have

$$\mathrm{E}[T] = \sum_{\alpha \in \mathbb{Z}_q^n \setminus \{0\}} \frac{1}{1 - \widehat{\mu}(\alpha)} = \sum_{\alpha \in \mathbb{Z}_q^n \setminus \{0\}} \frac{1}{1 - \left(1 - \frac{pq}{q-1}\right)^{\|\alpha\|_H}}$$

$$= \sum_{j=1}^{n} \binom{n}{j} (q-1)^j \frac{1}{1 - \left(1 - \frac{pq}{q-1}\right)^j}. \tag{9}$$

Setting $p = c/n$ and applying Lemma 3 with $\gamma = cq/(q-1)$ yields the claim. □

We point out that Eq. (9) is the exact runtime of the (1+1) EA on the multi-valued NEEDLE as a function of n, p and q, when standard uniform mutation is employed. In the case of cycle mutation, we prove the following upper and lower bounds.

Theorem 5. *Let T be the runtime of the (1+1) EA using cycle mutation with rate $p = c/n$ for a constant $c > 0$ on* NEEDLE$_z$. *The expectation of T is bounded above and below as follows.*

$$q^n \left(1 + \frac{1}{e^{2c} - 1}\right)(1 \pm o(1)) \le \mathrm{E}[T] \le q^{n+2}\left(\frac{1}{4c} + \frac{1}{q^2}\right)(1 \pm o(1)).$$

Proof. Let T be the first time the (1+1) EA using cycle mutation finds the optimum z on NEEDLE$_z$. Combining Lemma 2 and Theorem 2, we have

$$\mathrm{E}[T] = \sum_{\alpha \in \mathbb{Z}_q^n \setminus \{0\}} \frac{1}{1 - \widehat{\mu}(\alpha)} \le \sum_{\alpha \in \mathbb{Z}_q^n \setminus \{0\}} \frac{1}{1 - \left(1 - 2p \sin^2\left(\frac{\pi}{q}\right)\right)^{\|\alpha\|_H}}$$

$$= \sum_{j=1}^{n} \binom{n}{j} (q-1)^j \frac{1}{1 - \left(1 - 2p\sin^2\left(\frac{\pi}{q}\right)\right)^j}.$$

For the upper bound, setting $p = c/n$ and applying Lemma 3 with $\gamma = 2c\sin^2(\pi/q)$ we have

$$\mathrm{E}[T] \le q^n \left(1 + \frac{1}{\exp\left(\frac{2c(q-1)\sin^2(\pi/q)}{q}\right) - 1}\right)(1 \pm o(1)).$$

Using the estimates $e^x \geq 1 + x$ and $\sin(\pi/x) \geq 2/x$ for $x \geq 2$,

$$\exp\left(\frac{2c(q-1)\sin^2(\pi/q)}{q}\right) - 1 \geq 8c(q-1)/q^3 \geq 4c/q^2,$$

for $q \geq 2$. It follows that

$$\left(1 + \frac{1}{\exp\left(\frac{2c(q-1)\sin^2(\pi/q)}{q}\right) - 1}\right) \leq 1 + \frac{q^2}{4c}.$$

For the lower bound, we estimate $\mathrm{E}[T]$ from below using Theorem 2 and applying Lemma 3 with $\gamma = 2c$ we obtain

$$\begin{aligned}\mathrm{E}[T] &= \sum_{\alpha \in \mathbb{Z}_q^n \setminus \{0\}} \frac{1}{1 - \widehat{\mu}(\alpha)} \geq \sum_{\alpha \in \mathbb{Z}_q^n \setminus \{0\}} \frac{1}{1 - (1-2p)^{\|\alpha\|_H}} \\ &\geq q^n \left(1 + \frac{1}{\exp(2c(q-1)/q) - 1}\right)(1 \pm o(1)).\end{aligned}$$

The claimed lower bound follows since $(q-1)/q \leq 1$. □

3.2 Multi-valued Royal Road Functions

The RoyalRoad class of functions was introduced by Mitchell, Forrest and Holland [15] to study the behavior of crossover in genetic algorithms. In the binary domain this function is usually defined [20] as a polynomial with n/ℓ non-overlapping degree-ℓ monomials $x \in \{0,1\}^n \mapsto \sum_{i=0}^{n/\ell-1} \prod_{j=1}^{\ell} x_{i \cdot \ell + j}$, where ℓ divides n. We generalize this to $\{0, \dots, q-1\}^n$ in the natural way

$$\textsc{RoyalRoad}(x) := \sum_{i=0}^{n/\ell-1} \prod_{j=1}^{\ell} \mathbb{1}_{\{q-1\}}(x_{i \cdot \ell + j}).$$

The choice of $q-1$ in the indicator is arbitrary. The behavior of the (1+1) EA (and our analysis) is identical with any other target component, or even expressed as a sum of n/ℓ length-ℓ Needle functions with different target strings.

The following theorem, due to Doerr, Sudholt and Witt [9], provides a bound for the (1+1) EA optimizing linearly separable Boolean subfunctions over $\{0,1\}^n$ in terms of the time to optimize a single subfunction. Their proof uses the fact that the expected point in time at which a particular subfunction is first solved is no more than a constant times the expected runtime of the (1+1) EA working on that subfunction alone. The result then follows by a multiplicative drift argument on the number of subfunctions solved. Both arguments rely only on conditioning on the event that mutation does not occur within the subfunction, which has a probability of at least $(1-1/n)^{n-1} \geq 1/e$. Thus their result can be immediately generalized to the multi-valued case, and we state it accordingly.

Theorem 6 (Theorem 3 in [9]). *Let $f = \sum_{i=1}^{k} f_i$ where $f_i\colon \{0,\ldots,q-1\}^n \to \{0,1\}$ and all f_i depend on pairwise disjoint sets of variables. Suppose for each f_i and every initial configuration x_i, the expected optimization time of the (1+1) EA working on f_i is bounded by T^*, then the expected optimization time of the (1+1) EA on f is at most $T^* \cdot 4e^2(1+\ln k)$.*

We may then leverage this result to bound the optimization time of the (1+1) EA on multi-valued RoyalRoad functions.

Theorem 7. *The (1+1) EA using standard uniform mutation with rate $p = c/n$ for a constant $c > 0$ finds the optimum of the multi-valued* RoyalRoad *function in expected time at most*

$$q^\ell \left(1 + \frac{n}{c\ell}\right) 4e^2(1 + \ln(n/\ell)).$$

Proof. We first consider the runtime of the (1+1) EA on a single block of length ℓ using mutation rate c/n. Let T_ℓ be the corresponding random variable. By Theorem 1 and Lemma 2,

$$\begin{aligned} \mathrm{E}[T_\ell] &= \sum_{\alpha \in Z_q^\ell \setminus \{0\}} \frac{1}{1 - \widehat{\mu}(\alpha)} = \sum_{\alpha \in Z_q^\ell \setminus \{0\}} \frac{1}{1 - \left(1 - \frac{pq}{q-1}\right)^{\|\alpha\|_H}} \\ &= \sum_{j=1}^{\ell} \binom{\ell}{j} (q-1)^j \frac{1}{1 - \left(1 - \frac{cq}{n(q-1)}\right)^j}. \end{aligned}$$

Applying the same steps as with the proof of Lemma 3 to obtain Eq. (8), we conclude that

$$\mathrm{E}[T_\ell] = q^\ell \sum_{k=0}^{\infty} \left(\left(1 - \frac{q-1}{q}\left(1 - \left(1 - \frac{cq}{n(q-1)}\right)^k\right)\right)^\ell - q^{-\ell} \right).$$

Setting $\delta := (q-1)/q$, we have

$$\mathrm{E}[T_\ell] = q^\ell \sum_{k=0}^{\infty} \left(\left(1 - \delta\left(1 - \left(1 - \frac{c}{\delta n}\right)^k\right)\right)^\ell - q^{-\ell} \right) \le q^\ell \sum_{k=0}^{\infty} \left(1 - \frac{c}{\delta n}\right)^{\delta k \ell},$$

where we have applied Bernoulli's inequality $1 - \delta x \le (1-x)^\delta$ for $x = 1 - (1 - c/(\delta n))^k$. Moreover, as $(1 - c/(\delta n))^\delta \le \exp(-c/n)$, we conclude

$$\mathrm{E}[T_\ell] \le q^\ell \sum_{k=0}^{\infty} \exp(-ck\ell/n) = q^\ell \left(1 + \frac{1}{e^{c\ell/n} - 1}\right) \le q^\ell \left(1 + \frac{n}{c\ell}\right).$$

This follows from the geometric series and the fact that $e^{c\ell/n} \ge 1 + c\ell/n$.

Applying Theorem 6, the runtime of the (1+1) EA is bounded by $\mathrm{E}[T_\ell] \cdot 4e^2(1 + \ln(n/\ell))$, which completes the proof. □

4 Numerical Simulations for Cycle Mutation

The bounds derived in Theorem 5 for the runtime of the (1+1) EA using cycle mutation on NEEDLE rely on somewhat coarse estimates of the Fourier coefficients from Theorem 2. These estimates bracket $\widehat{\mu}(\alpha) = \prod_{i=1}^{n}\left(1 - 2p\sin^2(\alpha_i\pi/q)\right)$ from above by $\left(1 - 2p\sin^2(\pi/q)\right)^{\|\alpha\|_H}$ and below by $(1-2p)^{\|\alpha\|_H}$. For certain vectors $\alpha \in \mathbb{Z}_q^n$, there may be significant slack. For example, suppose q is even and denote $\alpha^+ := (\frac{q}{2}, \frac{q}{2}, \ldots, \frac{q}{2})$ and $\alpha^- := (1, 1, \ldots, 1)$. The coefficient $\widehat{\mu}(\alpha^+) = (1-2p)^n$, which is sharp for the lower bound, but the $(1 - 2p\sin^2(\pi/q))^n$ estimate for the upper bound might be significantly larger, especially for larger q. Similarly, $\widehat{\mu}(\alpha^-)$ is sharp for the upper bound, but considerably underestimated by the lower bound. We expect these errors to accumulate and degrade both bounds as q increases.

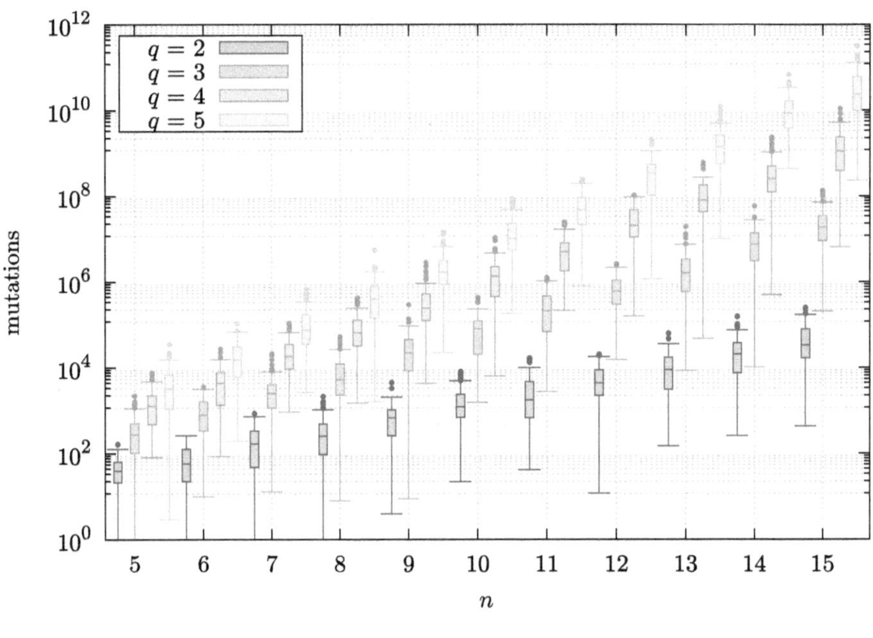

Fig. 1. Hitting time to target using cycle mutation on NEEDLE for various values of n and q.

We conducted numerical simulations to investigate the tightness of these bounds. For each $(n, q) \in \{5, \ldots, 15\} \times \{2, \ldots, 5\}$, we ran 100 random walks starting from a randomly selected point in $\{0, \ldots, q-1\}^n$, using cycle mutation with mutation rate $p = 1/n$, until an arbitrary target vector was hit. The median number of mutations until the target was hit over all n and q values was 55405.5, and the maximum was 3.13×10^{11}. We break down the statistics for each n and q in a box plot in Fig. 1. In Fig. 2 we plot the empirical mean number of mutations for each n and q together with the bounds provided by Theorem 5. As expected, the tightness of the bounds degenerate as q increases.

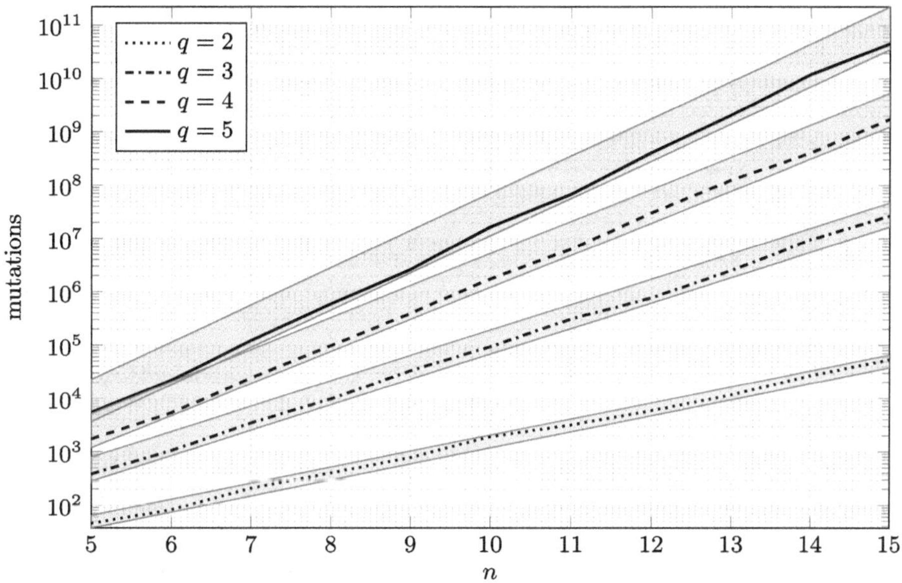

Fig. 2. EA runtime using cycle mutation on NEEDLE for various q. Shaded regions delineate the bounding interval $\left[q^n\left(1+\frac{1}{e^2-1}\right), q^{n+2}\left(\frac{1}{4}+\frac{1}{q^2}\right)\right]$ proved in Theorem 5.

5 Conclusions

We conducted a runtime analysis on the (1+1) EA escaping large plateaus in the multi-valued domain $\{0,\ldots,q-1\}^n$. We showed how the Walsh analysis for random walks on $\{0,1\}^n$ can be generalized to higher-order abelian groups for which the Fourier functions become complex. We used this to prove an exact runtime and an asymptotic bound for the (1+1) EA using standard mutation on the multi-valued NEEDLE function. For the less aggressive cycle mutation operator, we derived upper and lower bounds. We also calculated a straightforward upper bound on the standard (1+1) EA on multi-valued ROYALROAD functions. To investigate the tightness of the coarser bound for cycle mutation, we reported on empirical measurements of the hitting time for cycle mutation.

There are many opportunities for further work. The bounds on cycle mutation might be tightened by using more intricate estimates of individual Fourier coefficients, thereby improving the bounds for larger q. It would also be beneficial to analyze other mutation strategies such as the harmonic mutation operator of Doerr, Doerr and Kötzing [5], as well as investigating other algorithms and other types of functions over multi-valued decision variables.

Acknowledgments. This work was supported by NSF grant 2144080.

References

1. Ben Jedidia, F., Doerr, B., Krejca, M.S.: Estimation-of-distribution algorithms for multi-valued decision variables. In: Proceedings of the Genetic and Evolutionary Computation Conference, pp. 230–238. GECCO '2023, Association for Computing Machinery, New York, NY, USA (2023). https://doi.org/10.1145/3583131.3590523
2. Bethke, A.D.: Genetic Algorithms as Function Optimizers. Ph.D. thesis, University of Michigan (1981)
3. Billet, O., Robshaw, M.J.B., Peyrin, T.: On building hash functions from multivariate quadratic equations. In: Pieprzyk, J., Ghodosi, H., Dawson, E. (eds.) ACISP 2007. LNCS, vol. 4586, pp. 82–95. Springer, Heidelberg (2007). https://doi.org/10.1007/978-3-540-73458-1_7
4. Chicano, F., Sutton, A.M., Whitley, L.D., Alba, E.: Fitness probability distribution of bit-flip mutation. Evolutionary Comput. **23**(2), 217–248 (2015). https://doi.org/10.1162/EVCO_a_00130
5. Doerr, B., Doerr, C., Kötzing, T.: Static and self-adjusting mutation strengths for multi-valued decision variables. Algorithmica **80**(5), 1732–1768 (2018). https://doi.org/10.1007/s00453-017-0341-1
6. Doerr, B., Johannsen, D., Schmidt, M.: Runtime analysis of the (1+1) evolutionary algorithm on strings over finite alphabets. In: Beyer, H.G., Langdon, W.B. (eds.) Proceedings of the Eleventh Conference on Foundations of Genetic Algorithms, pp. 119–126. FOGA XI, Association for Computing Machinery, New York, NY, USA (2011). https://doi.org/10.1145/1967654.1967665
7. Doerr, B., Kelley, A.J.: Fourier analysis meets runtime analysis: precise runtimes on plateaus. Algorithmica **86**(8), 2479–2518 (2024). https://doi.org/10.1007/s00453-024-01232-5
8. Doerr, B., Pohl, S.: Run-time analysis of the (1+1) evolutionary algorithm optimizing linear functions over a finite alphabet. In: Proceedings of the Genetic and Evolutionary Computation Conference, pp. 1317–1324. GECCO '2012, Association for Computing Machinery, New York, NY, USA (2012). https://doi.org/10.1145/2330163.2330346
9. Doerr, B., Sudholt, D., Witt, C.: When do evolutionary algorithms optimize separable functions in parallel? In: Proceedings of the Twelfth Conference on Foundations of Genetic Algorithms, pp. 51–64. FOGA XII, Association for Computing Machinery, New York, NY, USA (2013). https://doi.org/10.1145/2460239.2460245
10. Droste, S.: Analysis of the (1+1) EA for a Dynamically Bitwise Changing OneMax. In: Cantú-Paz, E., et al., eds., GECCO 2003. LNCS, vol. 2723, pp. 909–921. Springer, Heidelberg (2003). https://doi.org/10.1007/3-540-45105-6_103
11. Garnier, J., Kallel, L., Schoenauer, M.: Rigorous hitting times for binary mutations. Evolutionary Comput. **7** (1999). https://doi.org/10.1162/evco.1999.7.2.173
12. Gunia, C.: On the analysis of the approximation capability of simple evolutionary algorithms for scheduling problems. In: Proceedings of the Genetic and Evolutionary Computation Conference, pp. 571–578. GECCO '2005, Association for Computing Machinery, New York, NY, USA (2005). https://doi.org/10.1145/1068009.1068106

13. Kötzing, T., Lissovoi, A., Witt, C.: (1+1) EA on generalized dynamic OneMax. In: Proceedings of the Thirteenth ACM Conference on Foundations of Genetic Algorithms, pp. 40–51. FOGA XIII, Association for Computing Machinery, New York, NY, USA (2015). https://doi.org/10.1145/2725494.2725502
14. Lengler, J., Opris, A., Sudholt, D.: Analysing equilibrium states for population diversity. Algorithmica **86**(7), 2317–2351 (2024). https://doi.org/10.1007/s00453-024-01226-3
15. Mitchell, M., Forrest, S., Holland, J.H.: The royal road function for genetic algorithms: fitness landscapes and GA performance. In: Varela, F.J., Bourgine, P. (eds.) Proceedings of the First European Conference on Artificial Life, pp. 245–254. MIT Press, Cambridge, MA (1992)
16. Rowe, J.E., Vose, M.D., Wright, A.H.: Structural search spaces and genetic operators. Evol. Comput. **12**(4), 461–493 (2004). https://doi.org/10.1162/1063656043138941
17. Scharnow, J., Tinnefeld, K., Wegener, I.: The analysis of evolutionary algorithms on sorting and shortest paths problems. J. Math. Modelling Algorithms **3**(4), 349–366 (2005). https://doi.org/10.1007/s10852-005-2584-0
18. Sutton, A.M., Chicano, F., Whitley, L.D.: Fitness function distributions over generalized search neighborhoods in the Q-ARY hypercube. Evol. Comput. **21**(4), 561–590 (2012). https://doi.org/10.1162/EVCO_a_00098
19. Vose, M.D., Wright, A.H.: The simple genetic algorithm and the walsh transform: Part i, theory. Evolutionary Comput. **6**(3), 253–273 (1998). https://doi.org/10.1162/evco.1998.6.3.253
20. Wegener, I., Witt, C.: On the optimization of monotone polynomials by simple randomized search heuristics. Combinatorics, Probability Comput. **14**, 225–247 (2005). https://doi.org/10.1017/S0963548304006650
21. Witt, C.: Tight bounds on the optimization time of a randomized search heuristic on linear functions. Comb. Probab. Comput. **22**(2), 294–318 (2013). https://doi.org/10.1017/S0963548312000600
22. Yu, Y., Qian, C., Zhou, Z.H.: Switch analysis for running time analysis of evolutionary algorithms. IEEE Trans. Evol. Comput. **19**(6), 777–792 (2015). https://doi.org/10.1109/TEVC.2014.2378891
23. Zhang, C.: Formulas for hitting times and cover times for random walks on groups (2017). https://arxiv.org/abs/2302.01963. [math.PR]

The $(1+1)$-EA in Dynamic Environments

Georg Hasebe, Johannes Lengler(✉), and Raghu Raman Ravi

Eidgenössische Technische Hochschule Zürich, Zürich, Switzerland
{georg.hasebe,johannes.lengler,raghu.ravi}@inf.ethz.ch

Abstract. We study the $(1+1)$-EA in dynamic linear environments, where in every generation selection is based on a freshly sampled linear function with positive weights. We consider the Dynamic Binary Value problem, where each generation uses a uniformly random permutation of $1, 2, 4, \ldots, 2^{n-1}$, and a Uniform weight variant, where the weights are drawn independently from Unif$(0, 1)$. Both of them have recently been integrated into the IOHprofiler platform and empirically studied.

For both models we prove a sharp threshold in the mutation parameter χ for mutation rate χ/n. Below the threshold, the expected optimisation time is $\mathcal{O}(n \log n)$, whereas above it the runtime becomes $2^{\Omega(n)}$.

For the Dynamic Binary Value problem in the exponential regime, we also quantify at what distance from the optimum the optimisation process stagnates. We show that there is a second threshold: a distance that is efficiently reached, but reaching any smaller distance takes exponential time. This quantifies and proves previous empirical findings.

Keywords: Evolutionary Algorithms · Mutation Rate · Drift Analysis · Dynamic Linear Functions

1 Introduction

Evolutionary algorithms (EAs) are widely used as general-purpose heuristics for black-box optimisation, where the objective function is accessible only through function evaluations. A central question in both theory and practice is how algorithmic design choices like the mutation rate translate into optimisation performance. While mutation rate effects on static benchmark functions such as OneMax or linear functions are well understood [32], the picture becomes more subtle in dynamic environments, i.e., settings in which the fitness function changes over time.

In order to study this question, Vermetten et al. [31] have recently integrated several dynamic benchmarks into the IOHprofiler framework [6] and performed large-scale benchmarking experiments on those environments. The benchmarks are a generalized version of dynamic linear functions as introduced in [24].[1] At any fixed point in time, selection is based on a pseudo-Boolean linear function with positive weights. Thus in particular, the global optimum of the search space

[1] They were called "noisy linear functions" in [24].

G. Iacca et al. (Eds.): PPSN 2026, LNCS 16985, pp. 196–211, 2027.
https://doi.org/10.1007/978-3-032-36214-8_13

$\{0, 1\}^n$ remains stable at the all-one string, there are no other local optima, and at each point in time the fitness landscape is monotone, meaning that flipping a zero-bit into a one-bit improves fitness. Selection is always comparison-based with respect to the current fitness function. Two central instances studied in [31] are the Dynamic Binary Value (DBV) benchmark introduced in [22] where the weights $2^0, 2^1, \ldots, 2^{n-1}$ are redistributed among the n bits in each round; and the dynamic linear function where weights are redrawn randomly in each generation from the uniform distribution Unif$(0, 1)$.

In parts, the simulation results in [31] confirmed previous theoretical predictions and experiments [22–24]. In particular, [24] had predicted for some dynamic linear functions that the performance of the $(1 + 1)$-EA should drop dramatically with the mutation rate for many dynamic linear functions, and this had been theoretically extended to DBV and experimentally confirmed in [22]. However, [31] also raised new questions. For mutation rates above the critical rate, it was observed that the fraction of correct bits quickly reaches a plateau at which progress comes to a complete halt. For DBV with mutation rate $3/n$, the $(1 + 1)$-EA remains below 80% correct bits even after the full budget of $1000n$ generations had been exhausted. It was noted in [31] that the distance from the optimum is surprisingly large, and the location of the plateau had not been quantified by theory before. The first novel contribution of this paper is an explicit formula for the location of this plateau. Moreover, we prove that indeed, while this plateau is reached in time $\mathcal{O}(n \log n)$ by the $(1 + 1)$-EA, any further progress towards the global optimum takes exponential time.

For the other benchmark, where weights are randomly redrawn from Unif$(0, 1)$, [31] was the first paper to empirically study this distribution. As for DBV, they observed a critical mutation rate above which the optimisation time increases dramatically. Our second contribution is to pinpoint the exact critical rate and prove rigorously that the optimisation time switches from time $\mathcal{O}(n \log n)$ to exponential at this rate.

Our Contribution in Detail

We will now discuss our results in more detail. As common in the theory community, due to space limitation this paper only present the results and some proof ideas. Reviewers can find a full version of the paper with complete proofs in [11], but they are not expected to read the full version or check omitted proofs. The following discussion is for the $(1 + 1)$-EA with standard bit mutation and mutation rate χ/n.

A mistake in Previous Proofs. Lengler and Schaller [24] have identified a sharp threshold $\chi_0 \approx 1.59362$ for the $(1+1)$-EA on the set of dynamic linear functions. More precisely, for every fixed $0 < \chi < \chi_0$ the $(1+1)$-EA optimises any dynamic linear function in $\mathcal{O}(n \log n)$ steps whereas for every fixed $\chi > \chi_0$ there are dynamic linear functions for which the optimisation time is superpolynomial. It was observed in [22] that the proof implies the same runtime results for DBV, which is why we will call the constant $\chi_0 = \chi_{\mathrm{dbv}}$ henceforth. However, as we will argue in Sect. 6, the proof in [24] contained a mistake, and thus the results in [24] and [22] only hold under the additional assumption that the optimisation is started sufficiently close to the optimum. Our proof fixes this issue and recovers the results from [24] and [22] as they were originally claimed. As we will see, the gap in the proof was substantial, and fixing it requires a non-trivial argument about convexity of the drift function. We also strengthen the original statements in several aspects: we show that the runtime in the negative cases is $2^{\Omega(n)}$ instead of just superpolynomial. Moreover, we give a unified proof for the efficient regime for a larger class of permutation-invariant dynamic environments that contains both dynamic linear functions and DBV. Those results are contained in Theorem 4 in Sect. 3.

DBV: Locating the Plateau. For DBV, other than previous work we can also pinpoint *where* the algorithm gets stuck for large mutation rates. Once $\chi > \chi_{\text{dbv}}$, we identify a constant $\alpha^*(\chi) \in (0, 1/2)$ such that the $(1+1)$-EA reaches distance αn from the optimum for any $\alpha > \alpha^*(\chi)$ in time $\mathcal{O}(n \log n)$, but reaching distance αn for any $\alpha < \alpha^*(\chi)$ takes exponential time. Thus we derive the exact location of the plateau at which the algorithm gets stuck. Our formula for $\alpha^*(\chi)$ is explicit, Eq. (7) in Sect. 4. For instance, for $\chi = 3$ the resulting bound corresponds to a stable plateau around a ratio of $1-\alpha^*(3) \approx 73\%$ correct bits. This confirms the empirical reports in [31, Figure 3] that progress of the $(1+1)$-EA stops at a surprisingly large distance from the optimum, even for rather moderate values of χ. Those results are given in Theorem 9 in Sect. 4. Finally, we also quantify the finite-size effects by computing how much the location of the plateau shifts for finite n.

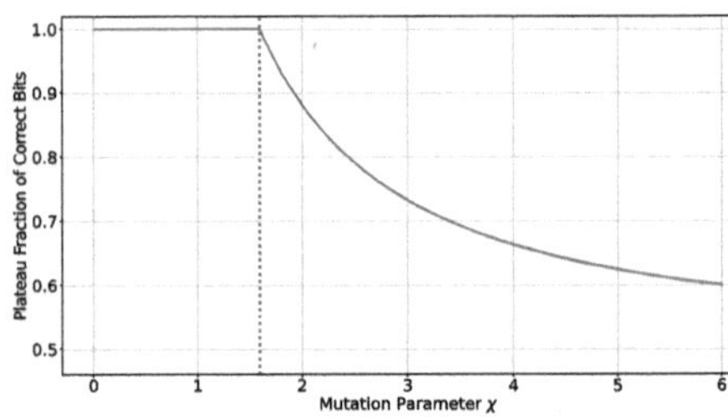

Fig. 1. Visualization of the DBV plateau location from Theorem 9 (3). For $0 < \chi < \chi_{\text{dbv}}$, the $(1+1)$-EA reaches the optimum efficiently, so the plotted plateau fraction of correct bits is 1. For $\chi > \chi_{\text{dbv}}$, the algorithm gets stuck at plateau fraction $1 - \alpha^*(\chi)$, where $\alpha^*(\chi)$ is given in Equation (7). The dotted line marks the critical threshold $\chi_{\text{dbv}} \approx 1.59362$.

Uniform Weights. For the Uniform weight model with weights in Unif$(0, 1)$ we obtain a substantially larger critical value $\chi_{\text{unif}} \approx 2.76531$ as the unique positive root arising in our drift analysis in Theorem 13. For every fixed $0 < \chi < \chi_{\text{unif}}$ the runtime is $\mathcal{O}(n \log n)$, whereas for every fixed $\chi > \chi_{\text{unif}}$ it is $2^{\Omega(n)}$. In particular, this identifies an intermediate regime $\chi_{\text{dbv}} < \chi < \chi_{\text{unif}}$ in which DBV already exhibits the pronounced slowdown, while the Uniform weight model is still efficiently optimisable.

Related Work

Mutation rate on Linear and Monotone Functions. In the static setting, the runtime of the $(1+1)$-EA on linear pseudo-Boolean functions is well understood. Droste, Jansen and Wegener [9] proved that every linear function is optimised in expected $\mathcal{O}(n \log n)$ time with mutation rate $1/n$. Using multiplicative drift analysis [5], Witt [32] sharpened this to the tight bound $(1+o(1))\, e\, n \ln n$ and extended the analysis to arbitrary mutation rates χ/n, showing that every constant χ still yields polynomial runtime. Thus on static linear functions the mutation parameter has no qualitative impact on the runtime order.

In other settings, the mutation rate is critical. This showed in particular in a line of work studying monotone functions, where flipping a zero-bit into a one-bit always increases fitness. Doerr, Jansen, Sudholt, Winzen and Zarges [4] proved that for $\chi < 1$ the $(1+1)$-EA optimises every monotone function in $\mathcal{O}(n \log n)$ time, while for sufficiently large constant χ some monotone functions require exponential time. Lengler [20] introduced the HotTopic construction as a specifically hard monotone function. He sharpened the upper threshold to $\chi > 2.13$ and extended the dichotomy to several algorithm variants including the $(\mu+1)$-EA and $(\mu+1)$-GA. For the latter, he found that a larger population size allows for arbitrarily large values of χ if the search is started sufficiently close to the optimum. Conversely, Lengler and Zou [25] showed that without crossover, sufficiently large populations cause exponential slowdown for the $(\mu+1)$-EA even for arbitrarily small χ, so the effect of populations is two-sided, and our picture is still very incomplete. In the other direction for general monotone functions, Lengler, Martinsson and Steger [21] showed via an entropy compression argument that the efficient regime extends slightly beyond $\chi = 1$. The critical value for the phase transition on monotone functions remains open.

Dynamic Environments. There is a long tradition of studying evolutionary algorithms in dynamic environments; see [2,27] for surveys. The first rigorous runtime analysis is due to Droste [7,8], who analysed the $(1+1)$-EA on dynamic OneMax, where the target string moves over time. The resulting runtime was polynomial if the target moved at most $\mathcal{O}(\log n/n)$ bits per generation and superpolynomial for $\omega(\log n/n)$ bits.

This started a systematic investigation of which evolutionary algorithms can track a moving optimum. Kötzing, Lissovoi and Witt [16] showed in an anytime analysis that even for larger mutation rates the $(1+1)$-EA stays close to the optimum of dynamic OneMax and generalisations. Various mechanisms have been shown to aid tracking: larger population sizes [3] and offspring population sizes [13]; diversity mechanisms [29]; and island models [18]. Alternative algorithmic paradigms such as Ant Colony Optimisation [17,26], Artificial Immune Systems [14], and Evolutionary Strategies [1] have also been studied in dynamic settings. For other base functions, tracking the optimum can be considerably more difficult [13,30].

Dynamic Linear Functions and Dynamic BinVal. The model of dynamic linear functions was introduced by Lengler and Schaller [24] and extended to DBV as a "limiting model" in [22]. In this model, in each generation t a set of weights $W_1^{(t)}, \dots, W_n^{(t)}$ is drawn independently and identically distributed (i.i.d.) from some distribution $\mathcal{D}$ with positive support, and the fitness function used for selection in generation t is then the linear function $f^{(t)}(x) = \sum_{i=1}^n W_i^{(t)} x_i$.

As mentioned before, [24] proved a sharp performance threshold in the mutation rate χ/n at $\chi_{\text{dbv}} \approx 1.59$, modulo the gap in the proof detailed in Sect. 6. Subsequent work has explored extensions beyond the (1 + 1)-EA on specific weight distributions. For $(\mu + 1)$-EA on DBV, it was shown in [23] that moderate population sizes increase the effective threshold near the optimum and, surprisingly, that the hardest region of optimisation lies far from the optimum rather than close to it. This was complemented by experiments in [22] indicating that crossover substantially extends the range of efficient mutation parameters. Kaufmann et al. [15] introduced Switching Dynamic BinVal (SDBV), proved it is drift-minimising among dynamic monotone functions for any mutation rate, and showed that the (1 + 1)-EA optimises it in $\Theta(n^{3/2})$ generations. Janett and Lengler [12] applied a two-dimensional drift framework to TwoLin, a minimal dynamic environment where only two linear functions alternate rather than a new permutation being drawn each generation, establishing that the threshold phenomenon persists beyond the DBV model. On the empirical side, Vermetten et al. [31] integrated dynamic linear functions and several DBV variants into the IOHprofiler framework and performed large-scale benchmarking, observing the mutation-rate sensitivity and benchmark divergence that motivate the present work.

2 Setup and Terminology

The (1 + 1)*-EA.* We study the (1 + 1)-EA with standard bit mutation and mutation rate χ/n where selection is based on the dynamic fitness environment $(f^{(t)})_{t\geq 0}$ as shown in Algorithm 1. Note that the (1 + 1)-EA bases its selection in generation t on the same fitness function $f^{(t)}$ for parent and offspring.

Algorithm 1: The (1 + 1)-EA with mutation rate χ/n maximising the dynamic environment $(f^{(t)})_{t\geq 0}$.

```
choose x^(0) ∈ {0,1}^n u.a.r.;
for t = 0, 1, ... do
    create y^(t) ∈ {0,1}^n by flipping each bit of x^(t) independently with
    probability χ/n;
    if f^(t)(y^(t)) ≥ f^(t)(x^(t)) then x^(t+1) ← y^(t); else x^(t+1) ← x^(t);
```

Permutation-Invariant Dynamic Functions. We will define a class of functions that generalize dynamic linear functions and DBV into a joint framework, the class of *permutation-invariant dynamic functions.*

Definition 1. *A permutation-invariant dynamic function is given as follows. In each generation t, an independent positive random weight vector $W^{(t)} = (W_1^{(t)}, \dots, W_n^{(t)}) \in \mathbb{R}^n_{>0}$ is sampled from a permutation-invariant distribution, i.e.*

$$(W_1^{(t)}, \dots, W_n^{(t)}) \overset{d}{=} (W_{\pi(1)}^{(t)}, \dots, W_{\pi(n)}^{(t)}) \quad \textit{for every permutation } \pi \in S_n.$$

The class of permutation-invariant dynamic functions includes the class of *dynamic linear functions* from [24] (with the name established later in [22]), where the weight $W_j^{(t)}$ are drawn independently and identically distributed from a distribution $\mathcal{D}$ with support in $\mathbb{R}_{>0}$. But our class also includes weight models such as DBV, where $W^{(t)}$ is a uniform random permutation of $(1, 2, 4, \dots, 2^{n-1})$. In this paper we restrict attention to two concrete instances from [31]:

- **DBV:** $W^{(t)}$ is a uniformly random permutation of $(2^0, 2^1, \dots, 2^{n-1})$.
- **Uniform:** $W_1^{(t)}, \dots, W_n^{(t)}$ are i.i.d. $\mathrm{Unif}(0, 1)$.

Runtime and Drift Analysis. Since all weights are non-negative, the global optimum remains fixed at $(1, \dots, 1)$. We define the runtime as $T := \min\{t \geq 0 : x^{(t)} = (1, \dots, 1)\}$. Let Y_t denote the number of zero-bits in $x^{(t)}$. Our main tool is drift analysis for the Markov process $(Y_t)_{t \geq 0}$. For $y \in \{0, 1, \dots, n\}$ we define the drift at state y as

$$\Delta(y) := \mathbb{E}[Y_t - Y_{t+1} \mid Y_t = y]. \tag{1}$$

Once the drift is computed, we use the multiplicative drift theorem [19, Theorem 2.4.5] for the $\mathcal{O}(n \log n)$ regime and the Simplified Drift Theorem [28, Theorem 2] to derive exponential lower bounds.

Notation and Preliminiaries. We write $\mathbb{N}_0 := \{0, 1, 2, \dots\}$ and $[n] := \{1, \dots, n\}$. We use the terminology of Algorithm 1, in particular the notation χ/n for the mutation rate and $x^{(t)}$ and $y^{(t)}$ for parent and offspring in generation t. We will assume that $\chi > 0$ is constant, so Landau notation $\mathcal{O}(\cdot)$, $\Omega(\cdot)$, $o(\cdot)$ may hide dependency on χ. A sequence of events $(A_n)_{n \geq 1}$ holds *asymptotically almost surely* (a.a.s.) if $\mathbb{P}[A_n] = 1 - o(1)$ as $n \to \infty$.

We denote by K the number of flipped bits in one mutation step and by I the number of flipped zero-bits (i.e. $0 \to 1$ flips). Note that $K \sim \mathrm{Bin}(n, \chi/n)$ and if we condition on $Y_t = y$ and $K = k$ then I follows the hypergeometric distribution $\mathrm{Hyp}(n, y, k)$ counting the number of successes when drawing k elements without replacement from a population of size n with y successes.

Let A be the event that the offspring is accepted in the selection step, and write $\mathbf{1}_A$ for its indicator. A key step will be to analyse the *acceptance probability*

$$p_A(k, i) := \mathbb{P}[A \mid (K = k,\, I = i)] \tag{2}$$

for $k > 0$ and $0 \leq i \leq k$. For convenience, we artificially set $p_A(0,0) := 0$ for the case that the parent and offspring are identical. This does not change the analysis since in this case the drift is trivially zero. The following two formulas for the drift can easily be shown, but we omit the proofs.

$$\Delta(y) = \mathbb{E}[(2I - K)\mathbb{1}_A \mid Y_t = y] \tag{3}$$

$$= \sum_{k=1}^{n} \mathbb{P}[K = k] \sum_{i=0}^{\min\{k,y\}} (2i - k)\, \mathbb{P}[I = i \mid (Y_t = y,\, K = k)]\, p_A(k,i). \tag{4}$$

Writing $\alpha := \alpha_t := Y_t/n \in [0,1]$ for the fraction of zero-bits, we can replace (4) by a smooth approximation in α that captures the limit for $n \to \infty$. This allows us to use standard calculus tools. For all $\chi > 0$ and $\alpha \in [0,1]$ we define

$$D(\chi,\alpha) := \sum_{k=1}^{\infty} e^{-\chi}\frac{\chi^k}{k!} \sum_{i=0}^{k} (2i - k)\binom{k}{i}\alpha^i(1-\alpha)^{k-i}\, p_A(k,i). \tag{5}$$

For fixed $\chi > 0$ we write $D_\chi(\alpha) := D(\chi,\alpha)$. This is the expression that (3) yields if the random variables were distributed as $K \sim \mathrm{Poi}(\chi)$ and $I \sim \mathrm{Bin}(k,\alpha)$, which holds in the limit for $n \to \infty$. We quantify the error in the following Proposition, again without proof as for most theorems and lemmas.

Proposition 2. *Fix $\chi \in (0,\infty)$ and let $y(\alpha) := \lfloor \alpha n \rfloor$. There exists a constant $c(\chi) > 0$, depending only on χ, such that uniformly for all $\alpha \in [0,1]$,*

$$|\Delta(y(\alpha)) - D_\chi(\alpha)| \leq c(\chi)/n.$$

In particular, for every fixed $\alpha \in [0,1]$ we have $\Delta(\lfloor \alpha n \rfloor) \to D_\chi(\alpha)$ as $n \to \infty$.

Close to the optimum we have $\alpha \approx 0$, i.e. $Y_t = \lfloor \alpha n \rfloor$ is small compared to n. In this regime, the sign of the drift is governed by the first-order term of $\alpha \mapsto D_\chi(\alpha)$ at $\alpha = 0$. Since $\alpha = 0$ is a boundary point, we work with the right derivative $\partial_+ D_\chi(0)$, which is given by the following proposition.

Proposition 3. *Let $\chi \in (0,\infty)$. For all permutation-invariant dynamic functions $\partial_+ D_\chi(\alpha)$ at $\alpha = 0$ exists and satisfies $\partial_+ D_\chi(0) = \sum_{k=1}^{\infty} e^{-\chi}\frac{\chi^k}{k!}\, k(2 - k)\, p_A(k,1)$.*

3 Permutation-Invariant Dynamic Environments

The main result of this section is the following.

Theorem 4. *Let $\chi_0 \approx 1.59362$ be the unique strictly positive root of $2-\chi-2e^{-\chi}$. For every constant $0 < \chi < \chi_0$ and any permutation-invariant dynamic function, the $(1+1)$-EA with mutation rate χ/n has runtime $\mathcal{O}(n \log n)$ in expectation and a.a.s.*

While this result was claimed for dynamic linear functions in [24], and by extension for DBV in [22], the proof there contained a severe gap, as we discuss in detail in Sect. 6 below: the drift was analysed in the limit $\alpha \to 0^+$, and a coupling argument was given to show that the drift is positive everywhere if its limit for $\alpha \to 0^+$ is positive. However, this coupling was wrong, and we fix that with Theorem 4, and at the same time generalize the result to all permutation-invariant dynamic functions. We do not see a way to rescue the coupling approach, so we employ a different strategy by finding an explicit lower bound for the drift for all $\alpha \in (0, 1]$, and then show convexity of this lower bound. This is actually the main part of our proof, so the extension compared to the incomplete proof in [24] is very substantial.

The remainder of the section will be devoted to proving Theorem 4. We start with a useful bound on the acceptance probability.

Lemma 5. *For all $k \geq 1$ and $0 \leq i \leq k$,*

$$\max\left\{0, \tfrac{2i}{k} - 1\right\} \leq p_A(k, i) \leq \min\left\{1, \tfrac{2i}{k}\right\}. \tag{6}$$

Moreover, for all $k \geq 3$ we have $p_A(k, 1) \leq 1/k$.

Proof. Let F be the set of flipped positions and $G \subseteq F$ the set of improving flips. Set $S := \sum_{j \in G} W_j^{(t)}$ and $T := \sum_{j \in F \setminus G} W_j^{(t)}$. Acceptance is the event $A = \{S \geq T\}$, or with $R := S/(S + T) \in [0, 1]$, we may equivalently write $A = \{R \geq 1/2\}$.

By permutation-invariance, conditioned on the values of K and I,

$$\mathbb{E}[R \mid K = k, I = i] = i/k.$$

For the upper bound in (6), apply Markov's inequality to the nonnegative random variable R:

$$p_A(k, i) = \mathbb{P}[R \geq 1/2] \leq \mathbb{E}[R]/(1/2) = 2i/k.$$

For the lower bound, note that $R \leq 1$ on A and $R \leq 1/2$ on $\neg A$, hence

$$\mathbb{E}[R] \leq 1 \cdot \mathbb{P}[A] + \tfrac{1}{2} \cdot \mathbb{P}[\neg A] \leq \tfrac{1}{2} + \tfrac{1}{2}\mathbb{P}[A],$$

which rearranges to $\mathbb{P}[A] \geq 2\mathbb{E}[R] - 1 = 2i/k - 1$. This yields (6).

Now let $i = 1$ and $k \geq 3$. If the offspring is accepted, then the single improving weight is at least the sum of the remaining $k - 1$ weights, and must hence be the strict maximum of all weights. By permutation-invariance, even if there is a strict maximum among the weights, the probability that it is the improving weight is $1/k$. □

A Lower Bound for $D_\chi(\alpha)$. We now come to the heart of the proof, a lower bound for the drift that holds for all α and that is convex. This is the point where we substantially deviate from the previous approach in [24].

For integers $k \geq 0$ and $0 \leq i \leq k$, define

$$\widetilde{p}_A(k,i) := \begin{cases} 0, & i = 0, \\ 1/k, & i = 1 \text{ and } k \geq 3, \\ 2i/k, & 1 < i < k/2, \\ 0, & i = k/2, \\ 2i/k - 1, & k/2 < i \leq k. \end{cases}$$

Moreover, for convenience we set $\widetilde{p}_A(k,i) = 0$ for $k < i$, which corresponds to the impossible case $K < I$. Replace $p_A(k,i)$ in Eq. (5) with $\widetilde{p}_A(k,i)$ to get

$$\widetilde{D}_\chi(\alpha) := \sum_{k=0}^{\infty} e^{-\chi} \frac{\chi^k}{k!} \sum_{i=0}^{k} (2i-k) \binom{k}{i} \alpha^i (1-\alpha)^{k-i} \widetilde{p}_A(k,i).$$

One can verify that the formula indeed provides a lower bound for the drift:

Lemma 6. *Let $\alpha \in [0,1]$. For all permutation-invariant dynamic functions, $D_\chi(\alpha) \geq \widetilde{D}_\chi(\alpha)$. The right derivative of $\widetilde{D}_\chi(\alpha)$ at $\alpha = 0$ is $\partial_+ \widetilde{D}_\chi(0) = 2 - \chi - 2e^{-\chi}$.*

The above lemma shows that the lower bound $\widetilde{D}_\chi(\alpha)$ is sharp in the limit $\alpha \to 0^+$, i.e., there its derivate converges to the exact value of $\partial_+ D_\chi(0) = 2 - \chi - 2e^{-\chi}$, which is known from [24]. There it has also been shown that there is a unique positive root $\chi_0 \approx 1.59362$ of $2-\chi-2e^{-\chi}$. Moreover, $2-\chi-2e^{-\chi} > 0$ for $\chi \in (0, \chi_0)$ and $2 - \chi - 2e^{-\chi} < 0$ for $\chi \in (\chi_0, \infty)$.

The key step is now to show that the function $\widetilde{D}_\chi(\alpha)$ is convex in α.

Lemma 7. *For each fixed $\chi \in (0, \chi_0)$, the function $\alpha \mapsto \widetilde{D}_\chi(\alpha)$ is convex on $[0,1]$.*

Proof. We only give some key steps. We decompose $\widetilde{D}_\chi(\alpha) = \sum_{k=0}^{\infty} p_k(\chi)\, \widetilde{H}_k(\alpha)$, where $p_k(\chi) = e^{-\chi}\chi^k/k!$, $\widetilde{H}_k(\alpha) := \sum_{i=0}^{k} \widetilde{h}_k(i)\, b_{k,i}(\alpha)$ (where $b_{k,i}$ is the PMF of a binomial random variable) and $\widetilde{h}_k(i) := (2i-k)\,\widetilde{p}_A(k,i)$. We then show that the tail $k \geq 5$ of the second derivative of $\widetilde{D}_\chi(\alpha)$ can be uniformly bounded by $\left|\sum_{k=5}^{\infty} p_k(\chi)\, \widetilde{H}_k''(\alpha)\right| \leq \frac{86760}{1331}\, p_5(\chi)$. This allows us to relate $\widetilde{D}_\chi''(\alpha)$ to a polynomial of bounded degree in χ via

$$\widetilde{D}_\chi''(\alpha) \geq e^{-\chi}\chi^2(2 + \chi R(x)), \quad \text{where} \quad R(\chi) := 1 + \tfrac{7}{16}\chi - \tfrac{723}{1331}\chi^2.$$

The function R is a concave quadratic and is easily seen to be positive on $(0, \chi_0)$. Hence, $\widetilde{D}_\chi''(\alpha)$ is also positive, and $\widetilde{D}_\chi(\alpha)$ is convex. □

We are now ready to put everything together. As announced earlier, convexity implies that the drift is positive for all $\alpha \in (0,1]$.

Corollary 8. *Let $\chi \in (0, \chi_0)$. Then $\widetilde{D}_\chi(\alpha) \geq \alpha\, \partial_+ \widetilde{D}_\chi(0) = \alpha\,(2 - \chi - 2e^{-\chi})$ for all $\alpha \in [0,1]$, and in particular $\widetilde{D}_\chi(\alpha) > 0$ for all $\alpha \in (0,1]$.*

Since we have shown multiplicative drift on the whole optimisation range, Theorem 4 follows by applying the multiplicative drift theorem [19, Theorem 2.4.5].

4 Dynamic Binary Value

In this chapter, we analyse the $(1+1)$-EA on the dynamic binary value (DBV) problem [22], where in each generation the weight vector is a uniformly random permutation of $(1, 2, 4, \ldots, 2^{n-1})$. Vermetten et al. [31] had reported that for larger mutation rates the $(1+1)$-EA fails to get close to the optimum on DBV even with a generous evaluation budget of $1000n$. A structural explanation was given in [22,24], where it was shown that when the mutation rate χ is above the critical mutation parameter χ_{dbv}, the drift is negative close to the optimum and thus the algorithm does not come close to the optimum. However, those papers did not make any statement on where exactly the algorithm would stall. We give a substantially more refined picture. For $\chi > \chi_{\text{dbv}}$, define

$$\alpha^*(\chi) := \tfrac{2-\chi-2e^{-\chi}}{2-2\chi-2e^{-\chi}}. \tag{7}$$

Then we show that the $(1+1)$-EA with mutation rate χ/n reaches distance $\alpha^*(\chi)$ quickly, but then progress comes to a halt for an exponential time. We plot the plateau at $1-\alpha^*(\chi)$ in Fig. 1.

Theorem 9. *Let $\chi_{\text{dbv}} \approx 1.59362$ be the unique strictly positive root of $2-\chi-2e^{-\chi}$. Consider the $(1+1)$-EA with mutation parameter χ on DBV.*

(1) For every fixed $0 < \chi < \chi_{\text{dbv}}$, the runtime is $\mathcal{O}(n \log n)$ in expectation and a.a.s.
(2) For every fixed $\chi > \chi_{\text{dbv}}$, the runtime is $2^{\Omega(n)}$ in expectation and a.a.s.
(3) For fixed $\chi > \chi_{\text{dbv}}$, consider $\alpha^*(\chi)$ from Eq. (7). For every fixed $\alpha \in [0, \alpha^*(\chi))$ the hitting time of $\{0, 1, \ldots, \lfloor \alpha n \rfloor\}$ is $2^{\Omega(n)}$ in expectation and a.a.s., and for every fixed $\alpha \in (\alpha^*(\chi), 1]$ the hitting time of $\{0, 1, \ldots, \lfloor \alpha n \rfloor\}$ is $\mathcal{O}(n)$ in expectation and a.a.s.

Part (1) follows from Theorem 4. Part (2) follows essentially from [22,24], since this part of the statement was not affected from the gap in that proof. We show a slightly stronger version with exponential instead of superpolynomial runtime. As mentioned, there is no analogue in the literature to part (3) of the statement.

The proof of Theorem 9 relies on a better control on the acceptance probability $p_A(k,i)$. While we could only give bounds in the general case, here we can compute $p_A(k,i)$ exactly. Since all weights are distinct powers of two, the total fitness change is dominated by the largest-weight flip: the offspring is accepted if and only if the largest-weight flipped bit changes from 0 to 1. Conditional on flipping a fixed set of k positions, the largest weight among these positions is equally likely to be attached to any of the k flipped positions. Therefore, we have $p_A(k,i) = i/k$. This allows us to compute the continuous drift approximation $D(\chi, \alpha)$ from Equation (5) in closed form for DBV.

Proposition 10. *For DBV, for all $(\chi, \alpha) \in (0, \infty) \times [0, 1]$,*

$$D(\chi, \alpha) = \alpha\Big((2-\chi-2e^{-\chi}) - \alpha(2-2\chi-2e^{-\chi})\Big). \tag{8}$$

The expression (8) is a quadratic polynomial in α and is already factorised. In particular, $\alpha = 0$ is always a root. Solving the bracket in (8) for α yields the second root α^*, which is given in (7). Since $2 - 2\chi - 2e^{-\chi} < 0$ for every $\chi > 0$, the bracket is strictly increasing in α. It is therefore enough to understand the sign of $A(\chi) := 2 - \chi - 2e^{-\chi}$, which is exactly the value of the bracket at $\alpha = 0$. Let $\chi_{\text{dbv}} \approx 1.59362$ denote the unique strictly positive root of $A(\chi)$.

Corollary 11. *For $\chi \neq \chi_{\text{dbv}}$, the drift has the following sign structure on $\alpha \in (0, 1]$.*

(1) If $0 < \chi < \chi_{\text{dbv}}$, then $D_\chi(\alpha) > 0$ for all $\alpha \in (0, 1]$.
(2) If $\chi > \chi_{\text{dbv}}$, then $\alpha^*(\chi) \in (0, 1)$ and

$$D_\chi(\alpha) < 0 \quad \text{for} \quad \alpha \in (0, \alpha^*(\chi)), \quad D_\chi(\alpha) > 0 \quad \text{for} \quad \alpha \in (\alpha^*(\chi), 1].$$

In addition, $\alpha^*(\chi) \to \frac{1}{2}$ as $\chi \to \infty$.

Theorem 9 now follows from a standard application of drift theorems for additive and negative drift [19].

Finite-n Drift and Finite-n Plateau. A convenient property of DBV is that the exact drift admits a simple closed form for every finite n. This allows us to give an exact formula for the drift for finite n and quantify how the finite-n plateau differs from $\alpha^*(\chi)$ in the limit setting. We only give the results.

Proposition 12. *Fix $\chi > 0$ and $n \geq 2$. For DBV, for all $y \in \{0, 1, \ldots, n\}$ with $\alpha := y/n$,*

$$\Delta(y) = \tfrac{n}{n-1}\, \alpha\big((2 - \chi - 2r_n(\chi) - \tfrac{\chi}{n}) - \alpha(2 - 2\chi - 2r_n(\chi))\big). \tag{9}$$

The two roots are $\alpha = 0$ and $\alpha_n^*(\chi) := \frac{2-\chi-2(1-\chi/n)^n-\chi/n}{2-2\chi-2(1-\chi/n)^n}$.

Due to the precise formulas in Proposition 12, we can also quantify the finite-size effect on the location $\alpha_n^*(\chi)$ of the plateau. A simple calculation shows that $\alpha_n^*(\chi) = \alpha^*(\chi) \pm \mathcal{O}(1/n)$.

5 Uniform Weight Model

In this chapter, we analyse the (1 + 1)-EA in the Uniform weight model, where in each generation t the weights are drawn independently from the uniform distribution Unif(0, 1).

Motivation and Results. Vermetten et al. [31] observe a stark difference between the Uniform weight model and the other variants: for small mutation rates all considered settings behave similarly. For larger mutation rates, however, the Uniform weight model becomes visibly easier than the rest. Our results identify a critical value $\chi_{\mathrm{unif}} \approx 2.76531$ such that, for every fixed $\chi < \chi_{\mathrm{unif}}$, the $(1+1)$-EA optimises the uniform setting in $\mathcal{O}(n \log n)$ in expectation and a.a.s. This is consistent with [31, Figure 3], which suggests an intermediate range between $\chi_{\mathrm{dbv}} \approx 1.59362$ and $\chi_{\mathrm{unif}} \approx 2.76531$ in which the optimum is reached efficiently for $\mathrm{Unif}(0,1)$ weights, while the other variants fail to be optimised.

Theorem 13. *Let $\chi_{\mathrm{unif}} \approx 2.76531$ be the unique strictly positive root of $g(\chi)$ from Lemma 15. Consider the $(1+1)$-EA with mutation parameter χ on dynamic linear functions in the Uniform weight model. For every fixed $0 < \chi < \chi_{\mathrm{unif}}$ the runtime is $\mathcal{O}(n \log n)$ in expectation and a.a.s. On the other hand, for every fixed $\chi > \chi_{\mathrm{unif}}$ the runtime is $2^{\Omega(n)}$ in expectation and a.a.s.*

The analysis is similar to the case of DBV, but here the formula for the acceptance probability $p_A(k,i)$ is much more complicated. However, it is possible to rewrite $p_A(k,i)$ in terms of the *Irwin-Hall distribution*, which is the distribution of the sum of k i.i.d. $\mathrm{Unif}(0,1)$ random variables. The result of this analysis is the following proposition.

Proposition 14. *In the Uniform weight model, for all $k > 0$ and $0 \le i \le k$, the acceptance probability defined in Eq. (2) satisfies*

$$p_A(k,i) = \frac{1}{k!} \sum_{m=0}^{i} (-1)^m \binom{k}{m} (i-m)^k.$$

This formula allows us to compute the local expansion of the drift $D_\chi(\alpha)$ at $\alpha = 0$ to identify the critical mutation parameter χ_{unif}.

Lemma 15. *In the Uniform weight model, the right derivative of the drift at $\alpha = 0$ is*

$$\partial_+ D_\chi(0) = g(\chi), \quad g(\chi) := \sum_{k=0}^{\infty} p_k(\chi)\, k(2-k)\, \frac{1}{k!},$$

where $p_k(\chi) := e^{-\chi}\chi^k/k!$ denotes the Poisson probability mass function. The function g has a unique strictly positive root $\chi_{\mathrm{unif}} \approx 2.76531$ and satisfies $g(\chi) > 0$ for $\chi \in (0, \chi_{\mathrm{unif}})$ and $g(\chi) < 0$ for $\chi \in (\chi_{\mathrm{unif}}, \infty)$.

Lemma 15 implies that for $\chi \neq \chi_{\mathrm{unif}}$ and sufficiently small $\alpha > 0$, the drift $D_\chi(\alpha)$ has the same sign as $g(\chi)$. In particular, for sufficiently small α we have $D_\chi(\alpha) > 0$ for $\chi < \chi_{\mathrm{unif}}$ and $D_\chi(\alpha) < 0$ for $\chi > \chi_{\mathrm{unif}}$. However, for $\chi < \chi_{\mathrm{unif}}$ we need positive drift in the whole range $\alpha \in [0,1]$. To obtain this, we show that the drift is convex in the whole range $[0,1]$, similarly as in Sect. 3. Note that it would not suffice to use the lower bound $\widetilde{D}(\chi, \alpha)$ from Sect. 3, since that is negative for $\chi \in (\chi_{\mathrm{dbv}}, \chi_{\mathrm{unif}})$ if $\alpha > 0$ is small. Thus it would not allow us to show positive drift for all $\chi < \chi_{\mathrm{unif}}$.

Our approach is still similar to the one in Sect. 3, but now we work with $D(\chi, \alpha)$ instead of $\widetilde{D}(\chi, \alpha)$. We again decompose $D(\chi, \alpha) = \sum_{k=0}^{\infty} p_k(\chi) H_k(\alpha)$, but now we need a tail bound $\left|\sum_{k=10}^{\infty} p_k(\chi) H_k''(\alpha)\right| \leq \frac{7}{5} 1504 \cdot p_{10}(\chi)$ for the $k \geq 10$ part of $D''(\chi, \alpha)$ and study the first 9 summands of the sum explicitly. We achieve this with computer-assisted verifications, the details of which can be found in the full manuscript [11] and in a reproducible script available in a Zenodo repository [10]. We omit the details. The result is the following lemma.

Lemma 16. *In the Uniform weight model there exists a polynomial $P(\chi)$ such that for all $(\alpha, \chi) \in [0,1] \times (0, \chi_{\text{unif}})$ we have $P(\chi) > 0$ and*

$$\frac{\partial^2}{\partial \alpha^2} D(\chi, \alpha) \geq e^{-\chi} P(\chi) > 0.$$

In particular, for each $\chi \in (0, \chi_{\text{unif}})$ the function $\alpha \mapsto D_\chi(\alpha)$ is convex on $[0,1]$, and

$$D_\chi(\alpha) \geq g(\chi)\alpha > 0 \quad \text{for all} \quad \alpha \in (0,1) \text{ and } \chi \in (0, \chi_{\text{unif}}).$$

Having established the multiplicative drift, Theorem 13 for $0 < \chi < \chi_{\text{unif}}$ follows again by a standard application of the Multiplicative Drift Theorem [5]. Likewise, since we have negative drift around the optimum for $\chi > \chi_{\text{unif}}$, the exponential runtime follows from the Simplified Drift Theorem [28].

6 A Mistake in the Previous Proof

A key step in [24] was a coupling argument to extend the sign of the drift for small α to the whole range $\alpha \in [0,1]$. The proof of [24, Theorem 4 a)] introduces an auxiliary search point $z^{(t)}$ whose set of zero-bits is a subset of the set of zero-bits of $x^{(t)}$, and then claims that at each step the difference between Y_t and Y_t', the number of zero bits in $x^{(t)}$ and $z^{(t)}$ respectively, can only decrease or remain the same. We argue here that this is false.

Let $\mathcal{Z}(x) \subseteq [n]$ denote the set of zero-bits of a bit string x. Assume that for some time t we have $\mathcal{Z}(z^{(t)}) \subsetneq \mathcal{Z}(x^{(t)})$. Then choose indices $a \in \mathcal{Z}(x^{(t)}) \setminus \mathcal{Z}(z^{(t)})$ and $b, c \notin \mathcal{Z}(x^{(t)})$, and apply the same mutation set, the set of bits which have been mutated in this time step, $F = \{a, b, c\}$ to both search points.

For $x^{(t)}$, exactly one zero-bit and two one-bits are flipped, so $K = 3$ and $I = 1$. Hence, if the offspring is accepted, the number of zero-bits increases by $K - 2I = 1$. For $z^{(t)}$, all flipped bits are one-bits, so $I' = 0$. Hence only $1 \to 0$ flips occur, and the offspring of $z^{(t)}$ is rejected.

Now consider a step in which the realised weights satisfy $W_a > W_b + W_c$. Then the offspring of $x^{(t)}$ is accepted, whereas the offspring of $z^{(t)}$ is rejected. Consequently, $Y_{t+1} = Y_t + 1$ and $Y_{t+1}' = Y_t'$, and therefore $Y_{t+1} - Y_{t+1}' = (Y_t - Y_t') + 1$. Thus the difference increases, contradicting the claim in [24].

It is important to note that this gap concerns only the argument away from the optimum. The analysis in the near-optimal regime, where $y = o[n]$, remains valid, since there the drift estimate is used directly and no coupling argument is needed. The lower runtime bounds in [24] are also unaffected.

7 Conclusion

We studied the (1 + 1)-EA in dynamic linear environments, where a weight vector is redrawn in every generation. A central empirical observation reported by Vermetten et al. [31] is that for large mutation rates, the algorithm fails to get close to the optimum even under generous evaluation budgets, and instead stalls at a surprisingly large distance from the optimum. Our results locate this distance for Dynamic Binary Value (DBV) and identify the critical mutation rate for the Uniform (0, 1) distribution of weights. Moreover, we show that the critical mutation parameter $\chi_{\text{dbv}} \approx 1.59362$ for DBV is universal in the sense that any mutation parameter $\chi < \chi_{\text{dbv}}$ guarantees optimisation time $\mathcal{O}(n \log n)$ on all permutation-invariant dynamic functions. This fixes a bug in a previous proof and extends and strengthens the results that were previously claimed.

Open Problems. We left some questions unanswered in this paper. First of all, we do not locate the plateau by an explicit formula for the Uniform (0, 1) distribution. Moreover, since it is difficult to realize DBV in a standard dynamic framework due to floating point limitations (the DBV weights get forbiddingly large), Vermetten et al. [31] also introduced further weight distributions such as PowersOfTwo and Pareto as approximations of DBV. The performance differences to DBV were empirically found to be small, and it would be interesting if they could be analytically quantified.

Widening the scope to other distributions for dynamic linear functions, it would be interesting to understand which weight distributions lead to a critical mutation parameter for the (1 + 1)-EA and which don't. For example, we show that the uniform distribution on (0, 1) leads to a threshold of $\chi_{\text{unif}} \approx 2.76531$. On the other hand, the constant distribution with value 1 just recovers the OneMax function which is known *not* to have a critical mutation rate, since the (1+1)-EA is efficient on OneMax for every constant mutation parameter $\chi > 0$ [32]. Particularly intriguing is whether there are distributions with compact domain which show a dichotomy. In other words: are there weight distributions with domain $[1, C]$, $C > 1$ for which the (1 + 1)-EA needs exponential time on the corresponding dynamic linear function for large χ? If yes, what is the smallest (infimal) C for which that can happen?

Finally, the situation for more complex algorithms is far from being understood. While the behaviour close to the optimum has been studied for many algorithms in [20], it is known that this is not always the hardest part for optimisation, especially for non-trivial populations [23, 25]. Our understanding even for simple algorithms like the $(\mu + 1)$ GA is still very limited.

Acknowledgments. This project was supported by the Swiss National Science Foundation [grant number 0003390].

References

1. Arnold, D., Beyer, H.G.: Optimum tracking with evolution strategies. Evol. Comput. **14**(3), 291–308 (2006)
2. Branke, J.: Evolutionary Optimization in Dynamic Environments, vol. 1. Springer (2002)
3. Dang, D.C., Jansen, T., Lehre, P.K.: Populations can be essential in tracking dynamic optima. Algorithmica **78**(2), 660–680 (2017)
4. Doerr, B., Jansen, T., Sudholt, D., Winzen, C., Zarges, C.: Mutation rate matters even when optimizing monotonic functions. Evol. Comput. **21**(1), 1–27 (2013). https://doi.org/10.1162/EVCO_a_00055
5. Doerr, B., Johannsen, D., Winzen, C.: Multiplicative drift analysis. Algorithmica **64**(4), 673–697 (2012). https://doi.org/10.1007/s00453-012-9622-x
6. Doerr, C., Ye, F., Horesh, N., Wang, H., Shir, O.M., Bäck, T.: Benchmarking discrete optimization heuristics with IOHprofiler. Appl. Soft Comput. **88**, 106027 (2020). https://doi.org/10.1016/j.asoc.2019.106027
7. Droste, S.: Analysis of the (1+1) EA for a dynamically changing OneMax-variant. In: Proceedings of the Congress on Evolutionary Computation, vol. 1, pp. 55–60. IEEE (2002)
8. Droste, S.: Analysis of the (1+1) EA for a dynamically bitwise changing OneMax. In: Proceedings of the Genetic and Evolutionary Computation Conference, pp. 909–921. Springer (2003)
9. Droste, S., Jansen, T., Wegener, I.: On the analysis of the (1+1) evolutionary algorithm. Theoret. Comput. Sci. **276**(1–2), 51–81 (2002). https://doi.org/10.1016/S0304-3975(01)00182-7
10. Hasebe, G.: Numerical verification for the (1+1) EA on noisy linear functions with Uniform(0,1) weights. Zenodo (2026). https://doi.org/10.5281/zenodo.18620302
11. Hasebe, G., Lengler, J., Ravi, R.R.: The (1 + 1)-EA in dynamic environments: extended version (2026). https://doi.org/10.5281/zenodo.19233285
12. Janett, D., Lengler, J.: Two-dimensional drift analysis: optimizing two functions simultaneously can be hard. Theoret. Comput. Sci. **971**, 114072 (2023). https://doi.org/10.1016/j.tcs.2023.114072
13. Jansen, T., Schellbach, U.: Theoretical analysis of a mutation-based evolutionary algorithm for a tracking problem in the lattice. In: Proceedings of the Genetic and Evolutionary Computation Conference, pp. 841–848 (2005)
14. Jansen, T., Zarges, C.: Evolutionary algorithms and artificial immune systems on a bi-stable dynamic optimisation problem. In: Proceedings of the Genetic and Evolutionary Computation Conference, pp. 975–982 (2014)
15. Kaufmann, M., Larcher, M., Lengler, J., Sieberling, O.: Hardest monotone functions for evolutionary algorithms. SN Comput. Sci. **6**(5), 512 (2025). https://doi.org/10.1007/s42979-025-03999-y
16. Kötzing, T., Lissovoi, A., Witt, C.: (1 + 1) EA on generalized dynamic OneMax. In: Proceedings of the Foundations of Genetic Algorithms, pp. 40–51 (2015)
17. Kötzing, T., Molter, H.: ACO beats EA on a dynamic pseudo-Boolean function. In: Proceedings of the Parallel Problem Solving from Nature, pp. 113–122. Springer (2012)
18. Lässig, J., Sudholt, D.: The benefit of populations and migration in dynamic optimization. In: Proceedings of the Genetic and Evolutionary Computation Conference, pp. 1267–1274 (2010)

19. Lengler, J.: Drift analysis. In: Doerr, B., Neumann, F. (eds.) Theory of Evolutionary Computation: Recent Developments in Discrete Optimization, pp. 89–131. Springer, Cham (2020). https://doi.org/10.1007/978-3-030-29414-4_2
20. Lengler, J.: A general dichotomy of evolutionary algorithms on monotone functions. IEEE Trans. Evol. Comput. **24**(6), 995–1009 (2020). https://doi.org/10.1109/TEVC.2019.2917014
21. Lengler, J., Martinsson, A., Steger, A.: When does hillclimbing fail on monotone functions: an entropy compression argument. In: Proceedings of the Analytic Algorithmics and Combinatorics, pp. 94–102. SIAM (2019)
22. Lengler, J., Meier, J.: Large population sizes and crossover help in dynamic environments. Nat. Comput. **23**(1), 115–129 (2024). https://doi.org/10.1007/s11047-022-09915-0
23. Lengler, J., Riedi, S.: Runtime analysis of the $(\mu + 1)$-EA on the dynamic BinVal function. SN Comput. Sci. **3**(4), 324 (2022). https://doi.org/10.1007/s42979-022-01203-z
24. Lengler, J., Schaller, U.: The (1 + 1)-EA on noisy linear functions with random positive weights. In: Proceedings of the IEEE Symposium Series on Computational Intelligence, pp. 712–719 (2018). https://doi.org/10.1109/SSCI.2018.8628785
25. Lengler, J., Zou, X.: Exponential slowdown for larger populations: the $(\mu + 1)$-EA on monotone functions. Theoret. Comput. Sci. **875**, 28–51 (2021). https://doi.org/10.1016/j.tcs.2021.04.011
26. Lissovoi, A., Witt, C.: Runtime analysis of ant colony optimization on dynamic shortest path problems. Theoret. Comput. Sci. **561**(Part A), 73–85 (2015)
27. Nguyen, T.T., Yang, S., Branke, J.: Evolutionary dynamic optimization: a survey of the state of the art. Swarm Evol. Comput. **6**, 1–24 (2012). https://doi.org/10.1016/j.swevo.2012.05.001
28. Oliveto, P.S., Witt, C.: Erratum: simplified drift analysis for proving lower bounds in evolutionary computation (2012). https://arxiv.org/abs/1211.7184
29. Oliveto, P.S., Zarges, C.: Analysis of diversity mechanisms for optimisation in dynamic environments with low frequencies of change. In: Proceedings of the Genetic and Evolutionary Computation Conference, pp. 837–844 (2013)
30. Rohlfshagen, P., Lehre, P.K., Yao, X.: Dynamic evolutionary optimisation: an analysis of frequency and magnitude of change. In: Proceedings of the Genetic and Evolutionary Computation Conference, pp. 1713–1720 (2009)
31. Vermetten, D., Lengler, J., Rusin, D., Bäck, T., Doerr, C.: Empirical analysis of the dynamic binary value problem with IOHprofiler. In: Proceedings of the Parallel Problem Solving from Nature, pp. 20–35. Springer (2024)
32. Witt, C.: Tight bounds on the optimization time of a randomized search heuristic on linear functions. Comb. Probab. Comput. **22**(2), 294–318 (2013). https://doi.org/10.1017/S0963548312000600

Theoretical Analysis of the (1+1)-EA-ES with Reduced Success Rule for Mixed Discrete–Continuous Optimization

Anne Auger[1], Dimo Brockhoff[1], Timo Kötzing[2], and Jurek Sander[2](✉)

[1] Inria and IP Paris, Palaiseau, France
{anne.auger,dimo.brockhoff}@inria.fr
[2] Hasso Plattner Institute, University of Potsdam, Potsdam, Germany
{timo.koetzing,jurek.sander}@hpi.de

Abstract. We introduce and analyze the (1+1)-EA-ES for mixed discrete–continuous optimization, combining a bitflip operator for discrete coordinates with a Gaussian mutation and a step size adapted via success rules for continuous coordinates. We show that the discrete component of the search—especially when close to optimal—can hinder progress in the continuous part by reducing the overall success probability. This disrupts the step size adaptation mechanism of the continuous component and may lead to premature convergence.

We formally prove that on SphereOneMax, for large enough dimensions, a reduced target success probability can bypass this problem. More precisely, we prove for the (1+1)-EA-ES with reduced success probability a bound on the expected time to reach an epsilon neighborhood of the optimum in $\mathcal{O}(\max\{n_c, n_d\} \cdot (\log(1/\varepsilon) + n_d + 1/n_c))$ where n_c represents the number of continuous variables and n_d the number of discrete ones.

Keywords: mixed discrete–continuous optimization · drift analysis

1 Introduction

Mixed discrete–continuous problems arise in many real-world applications. Even their continuous relaxation, where discrete variables are relaxed to continuous ones, is often difficult to handle with classical optimization techniques due to their reliance on strong assumptions such as convexity or differentiability [17]. Consequently, randomized approaches are attractive alternatives because of their robustness and weaker dependence on problem structure.

In the domain of continuous optimization for small to moderate dimension, a randomized approach, CMA-ES [10], inherited from Evolution Strategies [20], emerged as the state-of-the-art approach [9] exhibiting linear convergence on wide classes of problems [5]. Unsurprisingly, several extensions of CMA-ES for mixed discrete–continuous optimization have emerged in the past years. Several variants mainly use the continuous operator modified with lower bounds

G. Iacca et al. (Eds.): PPSN 2026, LNCS 16985, pp. 212–227, 2027.
https://doi.org/10.1007/978-3-032-36214-8_14

and possibly additional mechanisms to mutate the discrete part [6,16]. While this approach leverages the full power of the continuous algorithm for the continuous part of the search space, it leads to an inadequate treatment of the discrete space, as well as a disruption of the continuous search by the handling of the discrete space. In contrast, the CatCMA approach [7] takes the Estimation of Distribution Algorithm [4,15] and IGO [18] perspective and encodes a discrete probability distribution from which the discrete part of a solution is sampled and a multivariate normal distribution from which the continuous variables are sampled. An information geometric optimization inspired step updates both probability distributions.

In this context, assuming the mixed search space $\{0,1\}^{n_d} \times \mathbb{R}^{n_c}$ where n_d is the number of discrete variables and n_c the number of continuous variables respectively, we study a mixed discrete–continuous optimization algorithm derived from Evolution Strategies, which incorporates a dedicated mutation operator for the discrete variables.

Our objective is to analyze the algorithm theoretically, which motivates our focus on the (1+1) setting. The analysis of randomized algorithms with adaptive components is indeed well known to be challenging, already considering either pure discrete or pure continuous search spaces. Therefore, we begin with the simplest possible setting, given the inherent difficulty of the task further complexified by the mixed structure of the problem.

Importantly, the (1+1) framework is the only algorithmic structure that has been successfully analyzed using the same class of theoretical tools for continuous and discrete search spaces [1,2]. See [14] for an in-depth discussion on drift theory and its use for discrete randomized search heuristics. These approaches rely on establishing a drift condition on a suitably defined state variable of the algorithm, which in turn yields estimates of the hitting time to a neighborhood of the global minimizer. In the purely discrete setting, this corresponds to the hitting time of the global minimizer itself, whereas in the continuous case one typically considers the hitting time to an ε-neighborhood of the optimum. In this context, it is worth noting that while other approaches exist to analyze the linear convergence of adaptive Evolution Strategies [3,21], drift analysis provides a unified perspective for both discrete and continuous settings.

Our contributions are as follows: We define the (1+1)-EA-ES that combines a bit-flip mutation for varying the discrete part and a step size adaptive Gaussian mutation for varying the continuous component. The adaptation of the step size uses a reduced success rule, as we explain that the target probability of success around 1/5, known to be close to optimal in the continuous setting, is too large and leads to premature convergence here. We analyze the expected hitting time of an ε-neighborhood of the minimum on the SphereOneMax function as

$$(x,y) \in \{0,1\}^{n_d} \times \mathbb{R}^{n_c} \mapsto \sum_{i=1}^{n_d} x_i + \sum_{i=1}^{n_c} y_i^2 = ||x||_1 + ||y||^2$$

by designing an appropriate Lyapunov function using previous work [1,12]. The results rely on asymptotic estimates in the dimension to be able to balance different drift terms and thus hold for large enough dimensions. The theoretical convergence analysis in [8] is based on the same Lyapunov function, but, unlike to them, we analyze not only the convergence of the mixed algorithm after the discrete part is fully optimized, but also over the entire optimization process for both dimensions. Due to the complexity of the analysis caused by the analysis of the mixed discrete-continuous optimization, the core paper does not include all the details of the proof; we refer to the full version instead.

Notations. A multivariate normal distribution is denoted by $\mathcal{N}(0, I)$. With Φ_1 we denote the cumulative density function of the standard normal distribution $\mathcal{N}(0, 1)$ on $\mathbb{R}$, and φ_d is the pdf of the standard normal distribution on $\mathbb{R}^d$. The indicator function of a condition C is denoted by $\mathbf{1}_{\{C\}}$.

2 The (1+1)-EA-ES with Reduced Success Rule

As discussed above, it is natural to tackle mixed discrete-continuous problems with different sampling mechanisms for the discrete and the continuous parts. Here, we suggest building one of the simplest of such algorithms by combining the independent bitflip mutation of the discrete (1+1)-EA and the sampling from a multivariate normal distribution with multiplicative step size adaptation from the continuous (1+1)-ES. The algorithm works as follows. Starting from an initial search point $(x_0, y_0) \in \{0, 1\}^{n_d} \times \mathbb{R}^{n_c}$, in each iteration a new search point is generated by independently varying the discrete part x and the continuous part y via the mentioned (1+1)-EA and (1+1)-ES operators. If the resulting new search point (x'_t, y'_t) at iteration t is strictly better than the parent (x_t, y_t), the parent is replaced by the new search point and the step size σ_t of the continuous search distribution is increased by a factor of α. In case the new search point is not better than the parent, the step size σ_t is decreased to $\sigma_t \alpha^{-\frac{p_{\text{target}}}{1-p_{\text{target}}}}$. A common choice for the target success probability p_{target} is $1/5$ as a compromise between the optimal choices on the sphere and corridor model [19]—explaining the common term $1/5$ success rule.

Figure 1 shows results from running this algorithm with the original $1/5$ success rule (in blue) on the SphereOneMax function fifty times. Both the objective function value (fitness, thick lines) and the corresponding step size (σ, thinner lines, with $\sigma_0 = 1$) are shown over time (in number of evaluated search points).

While we observe that for an instance with 40 discrete and 5 continuous variables (left plot of Fig. 1), each of the fifty runs converge, none of the fifty runs reaches function values below 10^{-7} in the case of 20 discrete and 20 continuous variables (middle plot) and none of the runs even reaches a function value below 1 if the number of continuous variables is much larger than the number of discrete variables (right plot). The reason for this premature convergence (despite the absence of local optima in the objective function) is that the actual probability of success is reduced by the additional changes in the discrete part of a solution when compared to the only-continuous case, where $1/5$ is a reasonable choice.

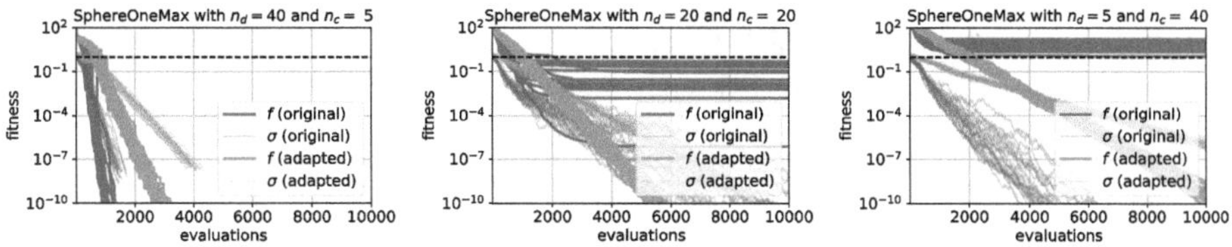

Fig. 1. Convergence plots, showing the function value (thicker) and the step size (thinner) for the original (1+1)-EA-ES with 1/5 success rule (in blue) and the (1+1)-EA-ES with reduced success probability $p_0/5$ (orange) on the SPHEREONEMAX function with varying number of discrete and continuous variables. (Color figure online)

In order to allow the algorithm to converge to the optimum of the SPHERE-ONEMAX function, we introduce p_0 as the probability that a discrete mutant x'_t at iteration t is equal to the current discrete solution x_t.

Definition 1. *Let $n_d \in \mathbb{N}_+$, $t \in \mathbb{N}$ and $x_t, x'_t \in \{0,1\}^{n_d}$. Consider the standard (1+1)-EA with mutation rate $1/n_d$. Then, in iteration t, the probability that the discrete mutant x'_t is equal to the current solution x_t is*

$$p_0 := \Pr[x_t = x'_t] = \left(1 - \frac{1}{n_d}\right)^{n_d} \leq e^{-1}.$$

We reduce the target success probability to $p_0/5$. The $p_0/5$ factor comes from our theoretical analysis. Algorithm 1 shows the pseudocode of the resulting algorithm that we will theoretically analyze in the remainder of the paper. Note that the only difference is the decrease of the step size in line 9, where the target success probability t_{target} is reduced to $p_0/5$.

3 (1+1)-EA-ES with Reduced Success-Rule: Asymptotic Estimate of the Hitting Time of an ε-Neighborhood of the Optimum

We have illustrated in Fig. 1 for different dimensions n_d and n_c of the discrete and continuous part of the objective that the (1+1)-EA-ES with reduced success probability converges linearly (see orange curve). Mathematically, it means that in a log-scale, the objective function decreases to minus infinity like $-t\text{CR}$ (with $\text{CR} > 0$), where $-\text{CR}$ represents the slope of the curve. For the (1+1)-ES with 1/5 success rule optimizing the sphere function (pure continuous setting), the scaling of CR with the dimension n_c is in $1/n_c$ [1].

In the continuous setting, there are typically two types of methods to analyze the linear convergence. The first one directly analyzes the stochastic process and shows that $\lim_{t\to\infty} \frac{1}{t} \ln \|y_t\|/\|y_0\|$ converges to $-\text{CR}$ (with CR a constant implicitly characterized) [3]. The second one investigates the hitting time of a ε-neighborhood of the minimizer. For a fixed dimension, linear convergence then

Algorithm 1: The (1+1)-EA-ES with reduced success probability integrating (1+1)-EA and (1+1)-ES with success-based step size adaptation to optimize $f\colon \{0,1\}^{n_d} \times \mathbb{R}^{n_c} \to \mathbb{R}$.

Input: fitness function $f\colon \{0,1\}^{n_d} \times \mathbb{R}^{n_c} \to \mathbb{R}$; success factor $1 < \alpha < 1.4$; independent bit-flip probability $1/n_d$

Initialize $x_0 \in \{0,1\}^{n_d}$, $y_0 \in \mathbb{R}^{n_c}$, $\sigma_0 > 0$, $p_0 = (1 - 1/n_d)^{n_d}$
$t \leftarrow 0$
while *stopping criterion not met* **do**
  $x'_t \leftarrow \text{discrete-mutation}_{1/n_d}(x_t)$ `// sample discrete`
  $y'_t \leftarrow y_t + \sigma_t \mathcal{N}(0, I_{n_c})$ `// sample continuous`
  if $f(x'_t, y'_t) < f(x_t, y_t)$ **then**
    $(x_{t+1}, y_{t+1}) \leftarrow (x'_t, y'_t)$ and $\sigma_{t+1} \leftarrow \sigma_t \cdot \alpha$
  else
    $(x_{t+1}, y_{t+1}) \leftarrow (x_t, y_t)$ and $\sigma_{t+1} \leftarrow \sigma_t \cdot \alpha^{-p_0/(5-p_0)}$
  end
  $t \leftarrow t + 1$
end
return (x_t, y_t, σ_t)

translates into a scaling in $\ln(1/\varepsilon)$ of this hitting time. The scaling in the dimension of CR in $1/n_c$ translates into a hitting time upper-bound in $\gamma n_c \ln(1/\varepsilon)$ for a constant γ independent of the dimension [1]. Prior to [1], analyses of an ε-neighborhood hitting time of the optimum were obtained assuming an asymptotic setting in the dimension. Those results allowed us to obtain the first hitting time bounds with an estimate of the dependency in the dimension [11,12], which were later on refined and proven for any finite dimension [1].

Building on previous works [1,12], we prove here an upper-bound on the expected hitting time of an ε-neighborhood of the optimum using a drift approach. Due to the complexity of the analysis, which must balance between the continuous and discrete parts, we succeed in estimating different parts of the drift by assuming an asymptotic setting in the dimension, similar to first expected hitting time results on Evolution Strategies [11].

Our main theorem gives an estimate (asymptotic in the dimension) of the hitting time of an epsilon neighborhood of the optimum. The result holds for the (1+1)-EA-ES with reduced success probability optimizing the SphereOneMax fitness function as defined in Algorithm 1 and reads as follows.

Theorem 2. *Suppose $n_c, n_d \in \mathbb{N}$ with $n_c, n_d \geq 2$ and assume the asymptotic setting, where $\min\{n_c, n_d\} \to \infty$. Consider the (1+1)-EA-ES with reduced success probability optimizing* SphereOneMax *as mixed-domain fitness function f with dimensions n_c and n_d. Let $\varepsilon < 1$ and $T_\varepsilon = \inf\{t \in \mathbb{N} \mid f(x_t, y_t) \leq \varepsilon\}$ be the first hitting time of $(-\infty, \varepsilon]$ by the fitness function value. With $p_0 := \Pr[x_t = x'_t]$ for all $t \leq T_\varepsilon$, the algorithm has the adjusted target probability of success of $p_0/5$ and the step size adaptation parameter fulfills $\alpha \in (1, 1.4)$ as defined in Algorithm 1. Then, the expected first hitting time of the (1+1)-EA-ES with reduced* *success probability starting from* $\theta_0 = (x_0, y_0, \sigma_0)$ *satisfies*

$$\mathrm{E}[T_\varepsilon] \in \mathcal{O}\left(\max\{n_c, n_d\} \cdot \left(\log\left(||y_0||^2 + ||x_0||_1\right) + \log(1/\varepsilon) + 1/n_c\right)\right).$$

4 Proof of Theorem 2

The proof of Theorem 2 relies on designing a Lyapunov (or potential) function $\theta \mapsto V(\theta)$ where θ denotes the state of the algorithm, namely at iteration t, $\theta_t = (x_t, y_t, \sigma_t)$ (see Algorithm 1) such that $Z_t = V(\theta_t)$ satisfies the drift condition expressed in the Additive Drift Theorem with Overshooting, originally introduced by Krejca in [13, Theorem 3.7], as stated in [14, Theorem 5.3].

Theorem 3. *Let* $(Z_t)_{t\in\mathbb{N}}$ *be random variables over* $\mathbb{R}$, *and denote the target by* $T = \inf\{t \in \mathbb{N} \mid Z_t \leq 0\}$. *Furthermore, suppose the following drift. There is a* $\delta > 0$ *such that, for all* $t < T$, *it holds that* $\mathrm{E}[Z_t - Z_{t+1} \mid Z_0, \ldots Z_t] \geq \delta$. *Then* $\mathrm{E}[T] \leq \frac{\mathrm{E}[Z_0]-\mathrm{E}[Z_T]}{\delta}$.

The structure of the proof of Theorem 2 is as follows. In Sect. 4.1, we define the success probability functions associated with the continuous part of the algorithm that are then needed to define parameters used in Sect. 4.2 to define the potential function in Sect. 4.3.

The potential function combines the joint fitness function value of the mixed-domain optimization objective with two terms that handle the step size adaptation when it is too small or too large, relative to the current value of the continuous fitness function, to make significant progress.

In Sect. 4.4 we present a sequence of lemmas: The first two, Lemmas 9 and 10, establish lower bounds of the positive drift in the case of a too small, or too large, step size. The drift in these cases stems from the parts of the potential function that control the step size adaptation. Next, Lemma 11 lower-bounds the negative drift introduced by the step size adaptation terms of the potential function when the step size is in a reasonable range.

Our main contribution is in the Lemmas 12 and 13, where we use the results of the previous lemmas and show additionally that we obtain a large enough positive drift if the step size is in a reasonable range. We distinguish in these lemmas two different cases. Lemma 12 handles the case that the continuous fitness function value is larger than the discrete, and establishes the lower bound of the expected drift through continuous progress. Lemma 13 covers the opposite case and obtains the required expected positive drift due to the discrete gain. Then, in Proposition 15, we prove that the two main lemmas cover all possible cases such that we obtain, in each iteration of the algorithm, the required positive drift in expectation. Lemma 16 lower-bounds this expected drift asymptotically.

In order to apply the additive drift theorem with overshooting, we lower-bound in Lemma 17 the expected amount our potential function overshoots its target value.

Then it is finally possible to combine all arguments and prove Theorem 2 by applying the drift theorem given in Theorem 3.

4.1 Probability of Continuous Gain or Loss with Rate r

In the following, we denote $d = \max\{n_c, n_d\}$ as the maximum of the continuous and discrete dimensions. Similar to [1, Section 3], we define the continuous success probability with rate $(1-r)$ for $r \in [0,1)$ given (y,σ) for the $||.||^2$-norm by squaring both sides of their inequality. We extend the definition such that it defines not only the success probability, but also, for $r \geq 1$, the probability that the fitness function value of the generated continuous offspring is at most r times the value of the current continuous solution. Thus, we define this probability for $r > 0$ and $y' \sim \mathcal{N}(y, \sigma^2 I)$ by $p_{r,d}^{\text{rate}}(y,\sigma) = \Pr_{y' \sim \mathcal{N}(y,\sigma^2 I)}\left(\|y'\|^2 < r^2 \|y\|^2\right)$. As a consequence of the isotropy of the multivariate normal distribution and of the sphere function, this success probability equals

$$p_{r,d}^{\text{rate}}(y,\sigma) = \Pr\left(\left\|e_1 + \frac{\sigma}{\|y\|^2}\mathcal{N}\right\|^2 < r^2\right),$$

where e_1 is the first unit vector of $\mathbb{R}^{n_c}$ and $\mathcal{N}$ a Gaussian vector in dimension n_c with mean vector 0 and identity covariance matrix. As in [1], we introduce the normalized step size $\overline{\sigma} = d\sigma/\|y\|$ and the function

$$p_{r,d}^{\text{rate}}(\overline{\sigma}) := \Pr\left(\left\|e_1 + \frac{\overline{\sigma}}{d}\mathcal{N}\right\|^2 < r^2\right). \tag{1}$$

Then $p_{r,d}^{\text{rate}}(\overline{\sigma}) = p_{r,d}^{\text{rate}}(y,\sigma)$. Note that for $r = 1$, $p_{1,d}^{\text{rate}}(\overline{\sigma})$ defines the continuous success probability.

Analogously to [1, Lemma 3.1], we obtain the following properties of the probability function.

Lemma 4. *Let $p_{r,d}^{\text{rate}}$ as defined in Eq. (1). Then the following holds.*

(1) For all $d \in \mathbb{N}$ and $r > 0$, $\overline{\sigma} \to p_{r,d}^{\text{rate}}(\overline{\sigma})$ is positive and continuous.
(2) For all $d \in \mathbb{N}$ and fixed $r \geq 1$, $p_{r,d}^{\text{rate}}(\overline{\sigma})$ is strictly monotonically decreasing and thus bijective.
(3) For all $d \in \mathbb{N}$, the image of $\overline{\sigma} \to p_{1,d}^{\text{rate}}(\overline{\sigma})$ is $(0, 1/2)$.

The next corollary can be directly retrieved from [1, Lemma 3.2] for $r \in (0,1]$.

Corollary 5. *Let $p_{r,d}^{\text{rate}}$ be defined as in Eq. (1) with $r \in (0,1]$. For $r = r(d)$ fulfilling $\lim_{d\to\infty} d\cdot(1-r(d)^2) = \rho$ the limit $p_\rho^{\text{rate}}(\overline{\sigma}) := \lim_{d\to\infty} p_{r,d}^{\text{rate}}(\overline{\sigma})$ exists, and it equals $p_\rho^{\text{rate}}(\overline{\sigma}) = \Phi_1\left(-\frac{\rho}{2\overline{\sigma}} - \frac{\overline{\sigma}}{2}\right)$ where Φ_1 is the cumulative density function of a standard normal distribution on $\mathbb{R}$. For $\rho = 0$, the function p_0^{rate} is continuous and strictly monotonically decreasing and the image of p_0^{rate} is $(0, 1/2)$.*

Corollary 5 shows that success is maximized when the step size is set to 0, while a non-trivial step size is needed for making significant progress.

We require the following property of $p_{r,d}^{\text{rate}}$ for the drift analysis.

Lemma 6. *For fixed $\overline{\sigma}$ and $d \in \mathbb{N}$, $p_{r,d}^{\text{rate}}(\overline{\sigma})$ is strictly monotonically increasing in r.*

4.2 Parameters

In the following, we introduce a sequence of parameters and constants that are required to define the potential function and to prove Theorem 2.

As given in the description of the (1+1)-EA-ES with reduced success probability, the parameter α which is used to adapt the step size of the continuous part of the algorithm fulfills $0 < \ln(\alpha) < \ln(1.4) = \frac{1}{2}\ln(1.96)$. Let $\varepsilon \ll 1$ be the target distance to the optimum as given in Theorem 2. We denote the dominant dimension of the optimization objective by $d = \max\{n_c, n_d\}$ and we define

$$A = \frac{1}{\max\{n_c, n_d\}} = \frac{1}{d}. \tag{2}$$

To handle the interaction between the discrete and continuous parts of the algorithm, let $p_0 = (1 - 1/n_d)^{n_d}$ be the probability that the discrete part does not change in an iteration, as already introduced in Definition 1. We denote, in every iteration $t \in \mathbb{N}$, the number of 1 bits in the bit string x_t by the random variable $X_t = ||x_t||_1$.

In order to lower- and upper-bound the probability of making continuous progress and determine the optimal range of making significant progress, we show with the next lemma that we can choose parameters such that certain properties are fulfilled.

Lemma 7. *Consider optimizing the* SphereOneMax *function f with dimensions $n_c, n_d \geq 2$ using the (1+1)-EA-ES with reduced success probability. Let $t \in \mathbb{N}$ and $\overline{\sigma}_t = d\sigma_t / ||y_t||$ be the normalized step size in iteration t. Then there are, referring to d, constants $0 < p_{cu} < p_u < p_l < 1$ and $u, l \in \Theta(1)$ such that the following holds.*

(i) $p^{\text{rate}}_{\sqrt{4},d}(u) = p_{cu}$ *and* $p^{\text{rate}}_{1,d}(l) = p_l$.
(ii) $0 < p_{cu} + \frac{4}{57}p_0 < p_u < \frac{p_0}{5} < \frac{1}{5} < p_l < \frac{1}{2}$.
(iii) $u/l \geq \alpha^{5/(5-p_0)}$.

Additionally, let $p_{su} := p^{\text{rate}}_{1,d}(u)$ and $p_{du} := p_u - p_{su}$ with $p_{su}, p_{du} \in \Theta(1)$. Using the parameters l and u, we define three ranges of the normalized step size $\overline{\sigma}$. For $\overline{\sigma} < l$, we consider the step size as "too small" to make significant progress, and for $\overline{\sigma} > u$ as "too large" to succeed with a large enough probability. In the range $\overline{\sigma} \in [l, u]$, we assume that the step size is reasonable to make significant progress with a sufficient probability.

In order to define the constant $v \in \Theta(1/d)$ for the potential function that we use to weight the adjustment of the step size adaptation in contrast to the gain of the fitness objective, we need first a few parameters that are later used in the calculation of the lower bounds of the continuous and discrete drift. For this purpose, let $r = \sqrt{2\exp(-A) - 1}$, and we lower-bound, for $\overline{\sigma}$ in the reasonable range $[l, u]$, the probability of making continuous progress with rate $1 - r$ by

$$p^* = \min_{\overline{\sigma} \in [l,u]} \left\{ p^{\text{rate}}_{r,d}(\overline{\sigma}) \right\}. \tag{3}$$

As we require it later to bound the expected drift asymptotically, we derive the following corollary that bounds p^* for $d \to \infty$ directly from the proof of [1, Proposition 4.3] using Corollary 5.

Lemma 8. *For p^* as defined in Eq. (3) and $d \to \infty$, it holds that*

$$\lim_{d\to\infty} p^* = \min\left\{\Phi_1\left(-\frac{1}{l}-\frac{l}{2}\right), \Phi_1\left(-\frac{1}{u}-\frac{u}{2}\right)\right\} \in \Theta(1).$$

In order to lower-bound the expected drift in the case that the discrete optimization dominates, we introduce the constants $0 < \gamma, \gamma', \gamma_1' < 1$ with $\gamma = \frac{(p_{su})^2}{2 \cdot e}$, $\gamma_1' = 0.04$, and $\gamma' = \frac{\gamma_1'^2}{(1+p_{su})\cdot 2 \cdot e}$.

As we are interested in the lower bound of different expected drift values that result from different cases, we set $v_{\min} = \min\{p^* \cdot p_0, \gamma, \gamma' \cdot p_{du}\} \in \Theta(1)$. Then it is possible to define the weight v for the potential function as

$$v = \frac{v_{\min}}{2 \cdot \ln(\alpha)} \cdot A = \frac{v_{\min}}{2 \cdot \ln(\alpha)} \cdot \frac{1}{d} \in \Theta(1/d). \tag{4}$$

4.3 Potential Function

Using the parameters in Sect. 4.2, it is now possible to define the potential function similar to [1]. We define $\ln^+(x) = \ln(x) \cdot \mathbf{1}_{\{x \geq 1\}}$ and introduce the potential function by combining additively different logarithmic terms. Since it is our goal to determine the expected number of iterations until the fitness function value is smaller than ε, the first term, $\ln(1/\varepsilon)$, shifts the potential function's target to zero. The next term, V_f, represents the joint fitness function value of SphereOneMax and therefore measures the optimization progress. The last term, V_s, handles the step size adaptation and combines V_l and V_u, which become active only when the step size is either "too small" or "too large". Using $d = \max\{n_c, n_d\}$, $v \in \Theta(1/d)$ as defined in Eq. (4) and $l, u \in \Theta(1)$ as given in Lemma 7, the potential function with input $\theta = (x, y, \sigma) \in \{0,1\}^{n_d} \times \mathbb{R}^{n_c} \times \mathbb{R}_{>0}$ is defined as follows.

$$V_f(\theta) = \ln\left(||x||_1 + ||y||^2\right) = \ln\left(f(\theta)\right); \tag{5}$$

$$V_l(\theta) = v \cdot \ln^+\left(\frac{\alpha \cdot l \cdot ||y||}{d \cdot \sigma}\right); \tag{6}$$

$$V_u(\theta) = v \cdot \ln^+\left(\frac{\alpha^{p_0/(5-p_0)} \cdot \sigma \cdot d}{u \cdot ||y||}\right); \tag{7}$$

$$V_s(\theta) = V_l(\theta) + V_u(\theta); \tag{8}$$

$$V(\theta) = \ln(1/\varepsilon) + V_f(\theta) + V_s(\theta). \tag{9}$$

The definition of the potential function ensures that $V(\theta) \leq 0$ implies $f(\theta) \leq \varepsilon$.

4.4 Proof

Now follows a sequence of lemmas that will be used to prove Theorem 2. The first lemma establishes a lower bound on the expected drift when the normalized step size is too small to make significant progress. In all what follows $\theta_t = (x_t, y_t, \sigma_t)$ denotes the state of the algorithm at iteration t.

Lemma 9. *Let* $T = \inf\{t \in \mathbb{N} \mid V(\theta_t) \le 0\}$. *Consider optimizing the* SPHERE-ONEMAX *function* f *with dimensions* $n_c, n_d \ge 2$ *using the (1+1)-EA-ES with reduced success probability. Let* $t \le T$, *and suppose* $\overline{\sigma}_t = d\sigma_t / ||y_t|| < l$. *Then we have for* V, *as defined in (9),*

$$\mathrm{E}\left[V(\theta_t) - V(\theta_{t+1}) \mid \theta_t\right] \ge v\frac{p_0(5p_l - 1)}{5 - p_0}\ln(\alpha) - o(1/d) = B_{\sigma,l}.$$

Analogously to the previous lemma, the next handles the case that the step size is too large.

Lemma 10. *Let* $T = \inf\{t \in \mathbb{N} \mid V(\theta_t) \le 0\}$. *Consider optimizing the* SPHERE-ONEMAX *function* f *with dimensions* $n_c, n_d \ge 2$ *using the (1+1)-EA-ES with reduced success probability. Let* $t \le T$, *and suppose* $\Pr\left[f(\theta_{t+1}) < f(\theta_t) \mid \theta_t\right] \le p_u$ *with* $p_u < \frac{p_0}{5}$ *and* $\overline{\sigma}_t = d\sigma_t / ||y_t|| > u$. *Then we have for* V, *as defined in (9),*

$$\mathrm{E}\left[V(\theta_t) - V(\theta_{t+1}) \mid \theta_t\right] \ge v\frac{p_0 - 5p_u}{5 - p_0}\ln(\alpha) - o(1/d) = B_{\sigma,u}.$$

For $t \in \mathbb{N}$ and $\overline{\sigma}_t$ in the reasonable range $[l, u]$, the third term of the potential function that handles the step size adaptation is zero and does not contribute to the expected difference. The term becomes only active when the adaptation of σ_t or $||y_t||$ results in $\overline{\sigma}_{t+1} \notin [l, u]$. With the next lemma, we lower-bound the introduced negative drift in this case in order to use it later in the analysis of the continuous and discrete drift in the reasonable range.

Lemma 11. *Let* $T = \inf\{t \in \mathbb{N} \mid V(\theta_t) \le 0\}$. *Consider optimizing the* SPHERE-ONEMAX *function* f *with dimensions* $n_c, n_d \ge 2$ *using the (1+1)-EA-ES with reduced success probability. Let* $t \le T$ *and* $\overline{\sigma}_t = d \cdot \sigma_t / ||y_t|| \in [l, u]$. *Then we have for* V_s, *as defined in Eq.* (8),

$$\mathrm{E}\left[V_s(\theta_t) - V_s(\theta_{t+1}) \mid \theta_t\right] \ge -v\frac{5}{5 - p_0}\ln(\alpha) - o(1/d) = -B_{\sigma,r}.$$

In the next two lemmas, reminding that $X_t = ||x_t||_1$, we distinguish between the two cases $||y_t||^2 > X_t$ and $||y_t||^2 \le X_t$, which describe whether the continuous or discrete part of the optimization objectives dominates and is therefore responsible for obtaining the required drift. We start with the case where the continuous fitness function value is larger.

Lemma 12. *Let $T = \inf\{t \in \mathbb{N} \mid V(\theta_t) \leq 0\}$. Consider optimizing the* SPHERE-ONEMAX *function with dimensions $n_c, n_d \geq 2$ using the (1+1)-EA-ES with reduced success probability. Let $X_t = ||x_t||_1$ be the random variable that denotes the number of* 1 *bits in the bit string x_t. Let $B_{\sigma,l}, B_{\sigma,u}, B_{\sigma,r}$ as defined in the Lemmas 9 to 11, and $B_{c,\sigma,r} = A \cdot p^* \cdot p_0 - B_{\sigma,r}$. Assume $||y_t||^2 > X_t$. For all $t \leq T$ that fulfill this constraint, we have for V, as defined in (9),*

$$\mathrm{E}\left[V(\theta_t) - V(\theta_{t+1}) \mid \theta_t\right] \geq \min\{B_{c,\sigma,r}, B_{\sigma,l}, B_{\sigma,u}\} = B_c.$$

Proof sketch. To obtain the lower bound for the expected drift in this case, we divide the possible values of θ_t depending on the normalized step size into three sets and show for each case that the expected drift can be lower-bounded by B_c.

Reasonable $\overline{\sigma}_t$ case: $\overline{\sigma}_t = \frac{d\sigma_t}{||y_t||} \in [l, u]$. Using the linearity of expectation, it is possible to divide the calculation of the expected difference of V into V_f and V_s. For V_f, we use the law of total expectation and the elitist property of the algorithm to obtain the drift conditional on $x_t = x'_t$, which occurs with probability p_0. Conditional on this event, we lower-bound the expected drift by calculating the probability that $V_f(\theta_t) - V_f(\theta_{t+1})$ is larger than $A = 1/d$ and multiplying it by A. Let $\delta = 2\exp(-A) - 1$ and $E_t := \left\{||y_{t+1}||^2 < \delta\,||y_t||^2\right\}$. Using $||y_t||^2 > X_t$ and conditional on $x_t = x'_t$, we get the sequence of events

$$\begin{aligned}
&\{V_f(\theta_t) - V_f(\theta_{t+1}) \geq A\} = \left\{\ln\left(\left(||y_t||^2 + X_t\right) / \left(||y_{t+1}||^2 + X_{t+1}\right)\right) \geq A\right\} \\
&= \left\{||y_t||^2 + (1 - \exp(A)) \cdot X_t \geq \exp(A) \cdot ||y_{t+1}||^2\right\} \\
&\supseteq \left\{(2\exp(-A) - 1) \cdot ||y_t||^2 > ||y_{t+1}||^2\right\} = E_t.
\end{aligned}$$

With $r = \sqrt{\delta}$ and p^*, as defined in Eq. (3), the probability of E_t can be lower-bounded by $\Pr\left[E_t \mid x'_t = x_t, \theta_t\right] \geq p^{\text{rate}}_{r,d}(\overline{\sigma}_t) \geq p^*$. Combining the lower-bound A of $V_f(\theta_t) - V_f(\theta_{t+1})$, its probability, and the error term for the step size adaptation term as lower-bounded in Lemma 11, results in

$$\mathrm{E}\left[V(\theta_t) - V(\theta_{t+1}) \mid \theta_t\right] \geq A \cdot p^* \cdot p_0 - B_{\sigma,r} \geq B_c.$$

Small $\overline{\sigma}_t$ case: $\overline{\sigma}_t = \frac{d\sigma_t}{||y_t||} < l$. Since $\overline{\sigma}_t < l$, it is possible to apply Lemma 9 to lower-bound the gain in the step size adaptation term while lower-bounding the expected difference in the fitness function value by 0, and we obtain the lower bound $\mathrm{E}\left[V(\theta_t) - V(\theta_{t+1}) \mid \theta_t\right] \geq B_{\sigma,l} \geq B_c$.

Large $\overline{\sigma}_t$ case: $\overline{\sigma}_t = \frac{d\sigma_t}{||y_t||} > u$. In order to use Lemma 10 to obtain the lower bound in this case, we have to show for the overall success probability that $\Pr\left[f(\theta_{t+1}) < f(\theta_t) \mid \theta_t\right] \leq p_u$ for $\overline{\sigma}_t > u$. Using $||x'_t||_1 \geq 0$ and $||y_t||^2 > X_t$, the probability that the generated mutant gets accepted can be upper-bounded by

$$\begin{aligned}
&\Pr\left[f(x'_t, y'_t) < f(x_t, y_t) \mid \theta_t\right] = \Pr\left[||x'_t||_1 + ||y'_t||^2 < X_t + ||y_t||^2 \mid \theta_t\right] \\
&\leq \Pr\left[||y'_t||^2 < X_t + ||y_t||^2 \mid \theta_t\right] \leq \Pr\left[||y'_t||^2 < 4\,||y_t||^2 \mid \theta_t\right] = p^{\text{rate}}_{\sqrt{4},d}(\overline{\sigma}_t).
\end{aligned}$$

Using Lemma 7 and $\overline{\sigma}_t > u$, the overall success probability is upper-bounded by $p^{\text{rate}}_{\sqrt{4},d}(\overline{\sigma}_t) \leq p_{cu} < p_u$. Then, by applying Lemma 10 to lower-bound the gain in the step size adaptation term while lower-bounding the expected difference of V_f by 0, we get $\mathrm{E}\left[V(\theta_t) - V(\theta_{t+1}) \mid \theta_t\right] \geq B_{\sigma,u} \geq B_c$.

In the next lemma, we handle the opposite case $||y_t||^2 \leq X_t$ where the discrete part is responsible for the progress.

Lemma 13. *Let* $T = \inf\{t \in \mathbb{N} \mid V(\theta_t) \leq 0\}$. *Consider optimizing the* SPHERE-ONEMAX *function with dimensions* $n_c, n_d \geq 2$ *using the (1+1)-EA-ES with reduced success probability. Let* $X_t = ||x_t||_1$ *be the random variable that denotes the number of* 1 *bits in the bit string* x_t. *Let* $B_{\sigma,l}, B_{\sigma,u}, B_{\sigma,r}$ *as defined in the Lemmas 9 to 11,* $B_{d,\sigma,r} = A \cdot \gamma - B_{\sigma,r}$, $B_{d,\sigma,u,\min} = \min\{\gamma' \cdot A, v/2 \cdot \ln(2 - \gamma_1')\}$, *and* $B_{d,\sigma,u} = p_{du} \cdot (B_{d,\sigma,u,\min} - v \cdot \ln(\alpha)) - o(1/d)$. *Assume* $||y_t||^2 \leq X_t$. *For all* $t \leq T$ *that fulfill this constraint, we have for* V, *as defined in Eq. (9),*

$$\mathrm{E}\left[V(\theta_t) - V(\theta_{t+1}) \mid \theta_t\right] \geq \min\{B_{d,\sigma,r}, B_{\sigma,l}, B_{\sigma,u}, B_{d,\sigma,u}\} = B_d.$$

Proof sketch. To obtain the lower bound for the expected drift in this case, we divide again the possible values of θ_t depending on the normalized step size into three sets and lower-bound for each case the expected drift by B_d.

Since we obtain the required drift in two of the sets due to the progress on the discrete side of the optimization objective, we show first in Claim 14 that it occurs under certain constraints. The intuition behind this claim is that the joint fitness function value decreases by $\gamma_1 \in (0,1)$ in expectation when, conditional on a discrete gain of 1, we lose on the continuous side at most $1 - \gamma_1$ in expectation. In the following, let $\Delta_c' = ||y_t||^2 - ||y_t'||^2$.

Claim 14. *Let* $\mathrm{E}\left[\Delta_c' \mid \Delta_c' > -1, \theta_t\right] \geq -(1 - \gamma_1)$, *and* $\Pr\left[\Delta_c' > -1 \mid \theta_t\right] \geq \gamma_2$ *with the constants* $\gamma_1, \gamma_2 \in (0,1)$. *Then it holds that*

$$\mathrm{E}\left[V_f(\theta_t) - V_f(\theta_{t+1}) \mid \theta_t\right] \geq \frac{\gamma_1 \cdot \gamma_2}{2 \cdot e} A.$$

Proof sketch. Let D_t denote the event that the discrete function value decreases by 1 and the generated mutant gets accepted. Using that the joint fitness function value decreases in expectation by γ_1 conditional on D_t, $-\ln(1-\alpha) \geq \alpha$ for $\alpha < 1$ and $||y_t||^2 \leq X_t$, we get

$$\mathrm{E}\left[V_f(\theta_t) - V_f(\theta_{t+1}) \mid D_t, \theta_t\right] = -\ln\left(\frac{f(\theta_t) - \gamma_1}{f(\theta_t)}\right)$$
$$\geq \frac{\gamma_1}{f(\theta_t)} = \frac{\gamma_1}{||y_t||^2 + X_t} \geq \frac{\gamma_1}{2 \cdot X_t}.$$

Since the probability of a discrete decrease by 1 is lower-bounded by $X_t/(en_d)$ and $\Pr\left[\Delta_c' > -1 \mid \theta_t\right] \geq \gamma_2$, we get $\Pr\left[D_t \mid \theta_t\right] \geq \gamma_2 X_t/(en_d) \geq \gamma_2 X_t/(ed)$.

Using the law of total expectation and the algorithm's elitist property results in

$$\begin{aligned}\mathrm{E}\left[V_f(\theta_t)-V_f(\theta_{t+1}) \mid \theta_t\right] &\geq \mathrm{E}\left[V_f(\theta_t)-V_f(\theta_{t+1}) \mid \theta_t, D_t\right]\cdot \Pr\left[D_t \mid \theta_t\right]\\ &\geq \frac{\gamma_1}{2\cdot X_t}\cdot\frac{\gamma_2\cdot X_t}{e\cdot d}=\frac{\gamma_1\cdot\gamma_2}{2\cdot e}A.\end{aligned}$$

□

Reasonable $\overline{\sigma}_t$ *case:* $\overline{\sigma}_t = \frac{d\sigma_t}{||y_t||} \in [l, u]$. With $\overline{\sigma}_t < u$ and that $p_{1,d}^{\text{rate}}$ is strictly monotonically decreasing, we can lower-bound the continuous success probability by $p_{1,d}^{\text{rate}}(u) = p_{su}$ in this case. It follows that $\mathrm{E}\left[\Delta_c' \mid \Delta_c' > -1, \theta_t\right] \geq -(1-p_{su})$ and $\Pr\left[\Delta_c' > -1 \mid \theta_t\right] \geq p_{su}$, which enables us to use Claim 14 in order to bound the expected difference of V_f. With the linearity of expectation and by applying Claim 14 and incorporating the error term for the step size adaptation term as bounded in Lemma 11, we obtain

$$\mathrm{E}\left[V(\theta_t)-V(\theta_{t+1}) \mid \theta_t\right] \geq \frac{(p_{su})^2}{2\cdot e}A - B_{\sigma,r} = \gamma\cdot A - B_{\sigma,r} \geq B_d.$$

Small $\overline{\sigma}_t$ *case:* $\overline{\sigma}_t = \frac{d\sigma_t}{||y_t||} < l$. Since $\overline{\sigma}_t < l$, it is possible to apply Lemma 9 to lower-bound the gain in the step size adaptation term while lower-bounding the expected difference in the fitness function value by 0, and we get the lower bound $\mathrm{E}\left[V(\theta_t)-V(\theta_{t+1}) \mid \theta_t\right] \geq B_{\sigma,l} \geq B_d$.

Large $\overline{\sigma}_t$ *case:* $\overline{\sigma}_t = \frac{d\sigma_t}{||y_t||} > u$. Since $\overline{\sigma}_t > u$ does not ensure for $||y_t||^2 \leq X_t$ that the joint success probability is smaller than p_u, we need a further case distinction. We distinguish between the following three cases.

Case I: $||y_t||^2 < 1$ and $\Pr\left[f(\theta_{t+1}) < f(\theta_t) \mid \theta_t\right] > p_u$

Case II: $||y_t||^2 < 1$ and $\Pr\left[f(\theta_{t+1}) < f(\theta_t) \mid \theta_t\right] \leq p_u$

Case III: $||y_t||^2 \geq 1$.

Since the calculation of the expected drift in Case I requires a detailed analysis, we provide here only its intuition. The complete proof can be found in the full version of the paper. In Case I, we obtain at least with the constant success probability $p_{du} = p_u - p_{su}$ progress in the fitness function value caused by a discrete gain. The intuition behind the proof of this case is that we then further distinguish between two possibilities. Either the continuous loss conditional on discrete success can be bounded such that it is possible to apply Claim 14 and we obtain the required drift due to the gain of the joint fitness function value. Or, the continuous fitness function value $||y_{t+1}||^2$ increases so much that the normalized step size $\overline{\sigma}_t$, which is defined by the ratio of the step size σ_t and $||y_t||$, decreases sufficiently such that the step size adaptation of V_u establishes the required drift. Using the linearity of expectation to combine both scenarios results in the lower bound $\mathrm{E}\left[V(\theta_t)-V(\theta_{t+1}) \mid \theta_t\right] \geq B_{d,\sigma,u} \geq B_d$.

For Case II is the condition of Lemma 10 already given by definition, and with $\Pr\left[||y_t'||^2 < 3 + ||y_t||^2 \mid \theta_t\right] \leq p_{\sqrt{4},d}^{\text{rate}}(\overline{\sigma}_t)$ and $\Pr\left[X_t - X_{t+1} \geq 4 \mid \theta_t\right] \leq \frac{4}{57}p_0$

we obtain for Case III that the joint success probability can be bounded by

$$\Pr\left[f(\theta_{t+1}) < f(\theta_t) \mid \theta_t\right] \leq p^{\text{rate}}_{\sqrt{4},d}(\overline{\sigma}_t) \cdot \Pr\left[X_t - X_{t+1} \leq 3 \mid \theta_t\right]$$
$$+ \Pr\left[X_t - X_{t+1} \geq 4 \mid \theta_t\right] \leq p_{cu} \cdot 1 + \frac{4}{57} p_0 \leq p_u.$$

Applying then Lemma 10 to lower-bound the gain in the step size adaptation term while lower-bounding the expected difference in the fitness function value by 0, results in $\mathrm{E}\left[V(\theta_t) - V(\theta_{t+1}) \mid \theta_t\right] \geq B_{\sigma,u} \geq B_d$.

Next, we combine the previous lemmas to obtain the overall expected drift.

Proposition 15. *Consider optimizing the* SphereOneMax *function with dimensions* $n_c, n_d \geq 2$ *using the (1+1)-EA-ES with reduced success probability. Let* B_c *and* B_d *as defined in the Lemmas 12 and 13. Let* $T = \inf\{t \in \mathbb{N} \mid V(\theta_t) \leq 0\}$, *and we have for all* $t \leq T$ *and* V, *as defined in (9),*

$$\mathrm{E}\left[V(\theta_t) - V(\theta_{t+1}) \mid \theta_t\right] \geq \min\{B_c, B_d\} = B.$$

Proof. Since Lemma 12 handles the case $||y_t||^2 > X_t$ and Lemma 13 the opposite case $||y_t||^2 \leq X_t$, they cover all possible cases. With B as the minimum of B_c and B_d and the law of total expectation, Proposition 15 follows.

After showing the overall expected drift in Proposition 15, we bound it asymptotically in the next lemma.

Lemma 16. *Let* $n_c, n_d \geq 2$, $d = \max\{n_c, n_d\}$), *and* B *as defined in Proposition 15. Then it holds that* $B \in \Theta(1/d)$.

Since the continuous part of the potential function V does not ensure that we hit its target directly, we lower-bound the expected value of V in the case that it becomes smaller than zero in the following lemma in order to apply the drift theorem with overshooting as defined in Theorem 3.

Lemma 17. *Let* $T = \inf\{t \in \mathbb{N} \mid V(\theta_t) \leq 0\}$. *Consider optimizing the* Sphere-OneMax *function with dimensions* $n_c, n_d \geq 2$ *using the (1+1)-EA-ES with reduced success probability. Then we have for* V, *as defined in Eq. (9),* $\mathrm{E}\left[V(\theta_T)\right] \in -\mathcal{O}\left(1/n_c\right)$.

Finally, it is now possible to prove Theorem 2.

Proof of Theorem 2. Using the potential function V, as defined in (9), Proposition 15, and the Lemmas 16 and 17, it is possible to apply the Additive Drift Theorem with Overshooting with $\mathrm{E}\left[V(\theta_t) - V(\theta_{t+1})\right] \geq B \in \Theta\left(1/\max\{n_c, n_d\}\right)$ and $\mathrm{E}\left[V(\theta_{T_\varepsilon})\right] \in -\mathcal{O}\left(1/n_c\right)$. It follows that

$$\mathrm{E}\left[T_\varepsilon\right] \leq \frac{\mathrm{E}\left[V(\theta_0)\right] - \mathrm{E}\left[V(\theta_{T_\varepsilon})\right]}{B}$$
$$\in \mathcal{O}\left(\max\{n_c, n_d\} \cdot \left(\log\left(||y_0||^2 + ||x_0||_1\right) + \log\left(1/\varepsilon\right) + 1/n_c\right)\right).$$

Acknowledgements. This research was (partially) funded by the HPI Research School on Foundations of AI (FAI).

References

1. Akimoto, Y., Auger, A., Glasmachers, T.: Drift theory in continuous search spaces: expected hitting time of the (1+1)-ES with 1/5 success rule. In: Proceedings of the Genetic and Evolutionary Computation Conference, pp. 801–808. GECCO '18, ACM (2018). https://doi.org/10.1145/3205455.3205606
2. Akimoto, Y., Auger, A., Glasmachers, T., Morinaga, D.: Global linear convergence of evolution strategies on more than smooth strongly convex functions. SIAM J. Optim. **32**(2), 1402–1429 (2022). https://doi.org/10.1137/20M1373815
3. Auger, A., Hansen, N.: Linear convergence of comparison-based step-size adaptive randomized search via stability of markov chains. SIAM J. Optim. **26**(3), 1589–1624 (2016). https://doi.org/10.1137/140984038
4. Baluja, S., Caruana, R.: Removing the genetics from the standard genetic algorithm. In: Machine Learning Proceedings, pp. 38–46 (1995). https://doi.org/10.1016/B978-1-55860-377-6.50014-1
5. Gissler, A.: Linear convergence of evolution strategies with covariance matrix adaptation. Ph.D. thesis, Institut Polytechnique de Paris (2024). https://hal.science/tel-04935192
6. Hamano, R., Saito, S., Nomura, M., Shirakawa, S.: CMA-ES with margin: lower-bounding marginal probability for mixed-integer black-box optimization. In: Proceedings of the Genetic and Evolutionary Computation Conference, pp. 639–647. GECCO '22, ACM (2022). https://doi.org/10.1145/3512290.3528827
7. Hamano, R., Saito, S., Nomura, M., Uchida, K., Shirakawa, S.: CatCMA: Stochastic optimization for mixed-category problems. In: Proceedings of the Genetic and Evolutionary Computation Conference, pp. 656–664. GECCO '24, ACM (2024). https://doi.org/10.1145/3638529.3654198
8. Hamano, R., Uchida, K., Shirakawa, S.: Convergence analysis of evolution strategies for mixed-integer optimization (2026). https://doi.org/10.48550/arXiv.2605.21000
9. Hansen, N., Auger, A., Ros, R., Finck, S., Pošík, P.: Comparing results of 31 algorithms from the black-box optimization benchmarking bbob-2009. In: Proceedings of the 12th Annual Conference Companion on Genetic and Evolutionary Computation, pp. 1689–1696. GECCO '10, ACM (2010). https://doi.org/10.1145/1830761.1830790
10. Hansen, N., Ostermeier, A.: Completely derandomized self-adaptation in evolution strategies. Evol. Comput. **9**(2), 159–195 (2001). https://doi.org/10.1162/106365601750190398
11. Jägersküpper, J.: Analysis of a Simple Evolutionary Algorithm for Minimization in Euclidean Spaces. In: Baeten, J.C.M., Lenstra, J.K., Parrow, J., Woeginger, G.J. (eds.) ICALP 2003. LNCS, vol. 2719, pp. 1068–1079. Springer, Heidelberg (2003). https://doi.org/10.1007/3-540-45061-0_82
12. Jägersküpper, J.: Algorithmic analysis of a basic evolutionary algorithm for continuous optimization. Theoret. Comput. Sci. **379**(3), 329–347 (2007). https://doi.org/10.1016/j.tcs.2007.02.042
13. Krejca, M.S.: Theoretical analyses of univariate estimation-of-distribution algorithms, Ph.D. thesis. Universität Potsdam (2019). https://doi.org/10.25932/publishup-43487
14. Kötzing, T.: Theory of Stochastic Drift. Habilitation thesis (2024). https://arxiv.org/abs/2406.14589

15. Lozano, J.A.: Towards a new evolutionary computation: advances on estimation of distribution algorithms, vol. 192. Springer Science & Business Media (2006). https://doi.org/10.1007/3-540-32494-1
16. Marty, T., Hansen, N., Auger, A., Semet, Y., Héron, S.: LB+IC-CMA-ES: Two simple modifications of CMA-ES to handle mixed-integer problems. In: Parallel Problem Solving from Nature, pp. 284–299. PPSN '24, Springer, Cham (2024). https://doi.org/10.1007/978-3-031-70068-2_18
17. Nocedal, J., Wright, S.J.: Numerical Optimization. Springer, 2nd edn. (2006). https://doi.org/10.1007/978-0-387-40065-5
18. Ollivier, Y., Arnold, L., Auger, A., Hansen, N.: Information-geometric optimization algorithms: A unifying picture via invariance principles. J. Mach. Learn. Res. **18**(18), 1–65 (2017). http://jmlr.org/papers/v18/14-467.html
19. Rechenberg, I.: Optimierung technischer Systeme nach Prinzipien der biologischen Evolution, Frommann-Holzboog Verlag (1973)
20. Schwefel, H.P.: Evolution strategies: A family of non-linear optimization techniques based on imitating some principles of organic evolution. Ann. Oper. Res. **1** (1984). https://doi.org/10.1007/BF01876146
21. Toure, C., Auger, A., Hansen, N.: Global linear convergence of evolution strategies with recombination on scaling-invariant functions. J. Global Optim. **86**(1), 163–203 (2023). https://doi.org/10.1007/s10898-022-01249-6

Combinatorial Optimization

A Bi-heuristic Initialized NSGA-II Algorithm for Multi-skilled Human-Robot Collaborative Disassembly Line Balancing Considering Robot Technology Tiers and Operator Experience Levels

Mohammad Ghasemi[1(✉)], Asef Nazari[1], Emadaldin Arabalibeik[2], Reza Shahabi-Shahmiri[2], and Dhananjay Thiruvady[1]

[1] School of Information Technology, Deakin University, Geelong, VIC 3216, Australia
{m.ghasemi,asef.nazari,dhananjay.thiruvady}@deakin.edu.au

[2] School of Industrial Engineering, College of Engineering, University of Tehran, Tehran, Iran
{emad.arab79,reza_shahabi011}@ut.ac.ir

Abstract. The rapid expansion of electric vehicles and renewable energy storage systems has made the efficient recovery of end-of-life batteries a critical global challenge. To support a sustainable circular economy, optimizing human-robot collaborative (HRC) disassembly lines is essential for maximizing resource recovery while controlling operational costs. However, the disassembly line balancing problem (DLBP) is inherently NP-hard, making the optimal assignment of tasks and resources an intractable computational task. In modern hybrid manufacturing environments, integrating multi-skilled (MS) resources with varying robot technology tiers (RTT) and operator experience levels (OEL) is vital, as these factors directly dictate execution speed, flexibility, and total costs. This study proposes a novel bi-objective mixed-integer linear programming model for the MS-HRCDLBP to simultaneously minimize cycle time and total disassembly cost. To solve this problem, the augmented epsilon-constraint (AUGMECON2) algorithm is used to validate the quality of solutions of smaller instances, while a customized hybrid non-dominated sorting genetic algorithm II (HNSGA-II) with a bi-heuristic initialization strategy is developed for large-scale instances. The proposed approach is validated using an industrial case study on EV battery disassembly, along with adapted benchmark datasets comprising up to 89 tasks. The computational results demonstrate that HNSGA-II effectively overcomes the computational limitations of the exact solver, successfully identifying high-quality Pareto-optimal trade-offs and providing decision-makers with diverse and cost-effective resource allocation strategies.

Keywords: Disassembly line balancing · Human-robot collaboration · Multi-skilled · Robot technology tier and operator experience level · Hybrid NSGA-II algorithm

G. Iacca et al. (Eds.): PPSN 2026, LNCS 16985, pp. 231–248, 2027.
https://doi.org/10.1007/978-3-032-36214-8_15

1 Introduction

Global electronic waste reached 53.6 million metric tonnes (Mt) in 2019 and is projected to hit 74.7 Mt by 2030 [5]. This escalation is largely driven by the energy transition; as governments mandate the phase-out of internal combustion engines to meet climate goals [15], electric vehicles (EVs) have become the primary, energy-efficient alternative [26]. With the global EV fleet exceeding 16.5 million units in 2021, lithium-ion batteries (LIBs) dominate as their power source, functioning for 8 to 10 years [30]. However, over 95% of spent LIBs currently face improper disposal, with only a fraction undergoing formal recovery or secondary repurposing [6]. Developing sustainable recycling methods is therefore urgent. Since batteries account for more than half the total cost of an electric car, manufacturers are actively embracing circular economy models to recover economic value and lower environmental damage. A central element of this strategy is the disassembly of end-of-life (EOL) products. This operation involves systematically taking apart EOL products to extract individual parts and raw materials for subsequent reuse [18].

Historically, disassembly operations have depended entirely on manual labor [20]. However, the toxic chemicals in EV batteries present severe safety risks, while highly repetitive tasks induce operator fatigue and drive up labor expenses [32]. To mitigate these challenges, modern facilities are adopting human-robot collaboration (HRC), pairing human adaptability with the speed and resilience of robots. Efficiently managing these collaborative environments requires dealing with the Disassembly Line Balancing Problem (DLBP), which allocates tasks across workstations to maximize resource utilization and reduce expenses [7]. Because EV batteries are exceptionally large and complex, traditional single-worker stations are inadequate, and multi-resource collaborative setups are essential. Despite this necessity, existing DLBP research largely overlooks comprehensive workforce planning. Current models frequently fail to account for the varying skill and experience levels of human operators, alongside the diverse technological capabilities of available robots. Furthermore, neglecting an operator's cognitive and physical workload directly degrades safety and productivity by increasing the likelihood of errors [12,17].

Therefore, addressing the multi-skilled Human-Robot Collaboration Disassembly Line Balancing problem (MS-HRCDLBP) is critical to maintaining safe work environments and maximizing efficiency. In this study, we develop a mathematical framework for the MS-HRCDLBP that systematically integrates operator experience level and specific robot technology tier. By explicitly incorporating these diverse resource parameters, the proposed model optimizes task allocation and scheduling to significantly reduce both the overall cycle time and total operational cost. To generate Pareto solutions in this work, the augmented epsilon-constraint method (AUGMECON2) is initially employed for solving small-scale instances to validate the correctness of the model. Then, a hybrid non-dominated sorting genetic algorithm (HNSGA-II) is utilized for comparative analysis and to find the solutions of large-scale benchmarks that AUGMECON2 (A2) is incapable of solving. Finally, an adapted case study of the Audi A3 Sportback e-tron

Hybrid LIB pack, alongside large-scale datasets, was used to verify the reliability and effectiveness of the proposed model and methodology.

The remainder of this paper is structured as follows. Section 2 reviews the relevant literature to highlight current research gaps and establish our primary contributions. Section 3 introduces the mathematical formulation of the proposed collaborative disassembly problem. The solution methodologies are detailed in Sect. 4. Section 5 describes the case study alongside the benchmark datasets, while Sect. 6 presents a comprehensive analysis of the experimental results. Finally, Sect. 7 concludes the study and outlines potential future research directions.

2 Literature Review

HRC has emerged as a transformative frontier in industrial manufacturing, particularly in disassembly operations [19]. By merging the cognitive adaptability, judgment, and perception of human operators [24] with the high speed, safety, and precision of industrial robots, HRC effectively overcomes the limitations of fully manual or fully automated systems [31]. Foundational research has explored various qualitative facets of this collaboration, including active compliance control [13], dynamic task classification based on respective skill sets [1], and structured evaluation frameworks encompassing vision, cognition, and execution stages [21]. The physical complexity of EOL products, such as the unpredictable degradation of EV batteries, necessitates a hybrid approach in which robots perform repetitive or high-risk tasks, while human workers address intricate and unpredictable variations in components. [11].

Building upon these operational concepts, recent literature has heavily focused on the mathematical modeling and optimization of the HRCDLBP [16]. Scholars have developed diverse mixed-integer programming (MIP) models to optimize multiple objectives, such as minimizing cycle time (CT), energy consumption, and operating costs, while maximizing profit and workload smoothness [10,14]. These models have been applied across various line configurations, including U-shaped [25], parallel [38], and two-sided layouts [33]. To solve these complex, NP-hard problems, studies frequently propose improved or hybrid metaheuristic algorithms, such as grey wolf [9], fruit fly [10], and moth-flame optimization [38], demonstrating superior performance in generating sequence-dependent disassembly plans and efficient task allocations [35,36].

A critical comparison of recent methodological advancements in HRCDLBP reveals a distinct evolution from rigid, static models toward highly adaptive frameworks capable of managing uncertainty and complex system dynamics. A significant gap in the literature remains the scarcity of comprehensive modeling for multi-skill workforces. While early models treated human and robotic resources homogeneously, recent studies differentiate between parallel and interactive collaboration [18], integrating dynamic workforce configurations that account for varying skill levels, workforce sizes, and salary grades [2]. Furthermore, the literature is increasingly concentrating on the inherent uncertainties of EOL products. To handle components suffering from severe aging or rust, researchers have introduced both physical interventions, such as partial destructive disassembly [8], and predictive mathematical frameworks, utilizing fuzzy set theory to absorb unpredicted task time variations without halting the disassembly line [37].

Despite significant advancements in modeling multi-skill capabilities, task scheduling, and workload balancing, a comprehensive summary of recent literature[1] highlights important research gaps. While existing studies successfully integrate multi-skill (MS) attributes, task assignment (TA), and task scheduling (TS) to optimize various economic and environmental objectives (e.g., [28,34,37]), they predominantly treat the workforce and robotic assets as static entities within their respective categories. Specifically, the current literature overlooks the micro-level heterogeneities that significantly impact execution speed and operational efficiency, in particular, robot technology tier (RTT) and operator experience level (OEL). None of the reviewed studies, including those optimizing CT and cost with workload balancing (e.g., [2,4,8,18,29,39]), simultaneously account for the varying technological tiers of robotic resources and the experience level of human operators.

To bridge these gaps, the primary contributions of this study are explicitly stated as follows:

- Proposing the first bi-objective mixed-integer linear programming (MILP) formulation for the MS-HRCDLBP that jointly incorporates RTT and OEL
- Deploying the A2 algorithm to establish a rigorous Pareto ground truth for small-to-medium problem instances, guaranteeing strongly efficient solutions.
- Integrating a detailed task-scheduling dimension alongside task assignment and workload balancing for the MS-HRCDLBP
- Developing HNSGA-II algorithm to overcome the computational intractability of large-scale instances. This algorithm features a problem-specific bi-heuristic initialization strategy and an explicit constraint-repair mechanism. Its methodological superiority is quantitatively validated against both the exact solver and a standard NSGA-II baseline using an industrial EV battery case study and comprehensive benchmark datasets.

3 Problem Definition and Formulation

This section outlines the problem definition and presents the mathematical formulation for the proposed disassembly system. Let N denote the set of all disassembly tasks, which are categorized into complex ($i_c \in N$), hazardous ($i_h \in N$), and collaborative ($i_l \in N$) subsets. Task characteristics dictate their allowable execution modes $m \in M$, where $m = 1$ represents human operators, $m = 2$ designates industrial robots, and $m = 3$ refers to collaborative robots (cobots).

[1] A detailed comparative table of the reviewed literature is available in the supplementary materials at: https://doi.org/10.5281/zenodo.19238017.

The processing time of each task, p_{im}, corresponding to the execution modes, is predetermined. Specifically, complex tasks are exclusively assigned to operators, hazardous tasks to robots, and collaborative tasks to cobots; however, any remaining unclassified tasks can be performed by any of the three modes. A set of resources $r \in R$ is available, each possessing specific skill sets $s \in S$. A critical feature of this framework is the incorporation of RTT and OEL $e \in E$.

Tasks are processed at utilized workstations $w \in W$, each bounded by a maximum capacity w^{cap}. Ultimately, the core decisions involve scheduling tasks subject to strict precedence relations ($\mathcal{P}_{ij}$), selecting the appropriate execution mode, assigning tasks and required resources to workstations, and minimizing absolute workload deviations (D_r) to ensure a balanced disassembly line.

To formulate the proposed problem, a novel bi-objective MILP model is presented in Eqs. (1) to (27). The nomenclature of the model is provided in Table 1. This model aims to minimize the CT and the total costs of the disassembly system, subject to constraints governing skill-dependent resource allocation, precedence relations, task assignment, and workload balancing. Simultaneously optimizing these two objectives is critical because they inherently conflict, as achieving a shorter cycle time typically requires investing in advanced robot technologies or experienced operators, highly skilled operators, or additional workstations, which increases total costs. Conversely, minimizing expenditures often leads to slower execution modes and reduced throughput, highlighting the need for a bi-objective approach to identify the optimal trade-off.

$$Z_1 = \min\ CT \tag{1}$$

$$Z_2 = \sum_{r\in\mathcal{R}}\sum_{e\in E}\sum_{w\in W} \beta_{rew}F_{re} + \sum_{r\in\mathcal{R}}\sum_{e\in E}\sum_{i\in N}\sum_{m\in M}\sum_{s\in S}\sum_{w\in W} x_{imsrew}V_{imsrew} + \sum_{r\in\mathcal{R}} D_r O_r + \sum_{w\in W} F_w \xi_w \tag{2}$$

$$y_{i1} = 1 \quad \forall\, i \in N_c \tag{3}$$

$$y_{i2} = 1 \quad \forall\, i \in N_h \tag{4}$$

$$y_{i3} = 1 \quad \forall\, i \in N_l \tag{5}$$

$$\sum_{m\in M} y_{im} = 1 \quad \forall\, i \in N - \{N_c \cup N_h \cup N_l\} \tag{6}$$

$$\sum_{r\in\mathcal{R}}\sum_{e\in E} x_{imsrew} = R_{ims}\delta_{imw} \quad \forall\, i, m, s, w \tag{7}$$

$$\sum_{m\in M}\sum_{s\in\mathcal{S}} x_{imsrew} \le \beta_{rew} \quad \forall\, i, r, e, w \tag{8}$$

$$z_{ijw} + z_{jiw} \le \lambda_{iw} \quad \forall\, i, j, w, i < j \tag{9}$$

$$\sum_{m\in M}\sum_{s\in\mathcal{S}}\sum_{e\in E} x_{imsrew} + \sum_{m\in M}\sum_{s\in\mathcal{S}}\sum_{e\in E} x_{jmsrew} \le 1 + z_{ijw} + z_{jiw} \quad \forall\, i, j, r, w, i < j \tag{10}$$

$$S_i + \sum_{m\in M} p_{im}y_{im} - \mathcal{M}\left(1 - z_{ijw}\right)\left(1 - \mathcal{P}_{ij}\right) \le S_j \quad \forall\, i, j, w, i < j \tag{11}$$

Table 1. Components of the model

Indices/Sets	Description
$i, j \in N$	Set of tasks,
$i \in N_c$	Set of complex tasks, $N_c \subset N$
$i \in N_h$	Set of hazardous tasks, $N_h \subset N$
$i \in N_l$	Set of collaborative tasks, $N_l \subset N$
$m \in M$	Set of task execution modes, ($m = 1, 2, 3$ in this work)
$s \in \mathcal{S}$	Set of skills
$r \in \mathcal{R}$	Set of resources
$w \in W$	Set of workstations
$e \in E$	Set of robot technology tier or operator experience level ($e = 1$ for traditional robots or junior operators; $e = 2$ for modern robots or senior operators)
Parameters	
R_{ims}	Number of resources r required with skill s in mode m to perform task i
$b_{res} \in \{0, 1\}$	1 if resource r masters skill s with technology tier or experience level e.
p_{im}	Standard processing time of i in mode m
$\mathcal{P}_{ij} \in \{0, 1\}$	1 if task i precedes task j.
w^{cap}	Maximum capacity of each workstation w
F_{re}	Fixed cost of operator or robot r with technology tier or experience level e
V_{imsrew}	Variable cost of using operator or robot r in mode m masters skill s with technology tier or experience level e to perform task i at workstation w
O_r	Overload penalty cost for resource r
F_w	Installation fixed expenditure of workstation w
AW	Average workload
$\mathcal{M}$	A big number
Variables	
Continuous	
S_i	The start time of task i
D_r	Absolute workload deviation of resource r from the average workload
CT	Cycle time
Binary	
y_{im}	If task i is assigned to mode m, 1, otherwise 0
β_{rew}	If resource r with technology tier or experience level e is hired at workstation w, 1, otherwise 0
ξ_w	If workstation w is utilized, 1, otherwise 0
λ_{iw}	If task i is assigned to workstation w, 1, otherwise 0
δ_{imw}	If task i in mode m performed at workstation w, 1, otherwise 0
z_{ijw}	If task i and j, where task i precedes j are assigned to the same workstation w, 1, otherwise 0
x_{imsrew}	If task i in mode m performed by resource r masters skill s with technology tier or experience level e at workstation w, 1, otherwise 0

$$x_{imsrew} \leq b_{res}\beta_{rew} \quad \forall\, i, m, s, r, e, w \tag{12}$$

$$y_{im} + \lambda_{iw} - 1 \leq \delta_{imw} \leq y_{im} \quad \forall i, m, w \tag{13}$$

$$\delta_{imw} \leq \lambda_{iw} \quad \forall i, m, w \tag{14}$$

$$\sum_{w \in W} \lambda_{iw} = 1 \quad \forall\, i \tag{15}$$

$$\sum_{w \in W} \beta_{rew} \leq 1 \quad \forall\, r, e \tag{16}$$

$$\sum_{i \in N} \lambda_{iw} \leq w^{cap} \quad \forall\, w \tag{17}$$

$$x_{imsrew} \leq \beta_{rew} \quad \forall\, i, m, s, r, e, w \tag{18}$$

$$x_{imsrew} \leq \lambda_{iw} \quad \forall\, i, m, s, r, e, w \tag{19}$$

$$\sum_{m \in M} \sum_{w \in W} \delta_{imw} = 1 \quad \forall\, i \tag{20}$$

$$\sum_{i\in N}\sum_{m\in M}\sum_{s\in\mathcal{S}}\sum_{r\in\mathcal{R}}\sum_{e\in E}\sum_{w\in W} x_{imsrew} - AW \le D_r \quad \forall\, r \tag{21}$$

$$AW - \sum_{i\in N}\sum_{m\in M}\sum_{s\in\mathcal{S}}\sum_{r\in\mathcal{R}}\sum_{e\in E}\sum_{w\in W} x_{imsrew} \le D_r \quad \forall\, r \tag{22}$$

$$\lambda_{iw} \le \xi_w \quad \forall i, w \tag{23}$$

$$\xi_{w-1} \ge \xi_w \quad \forall w \tag{24}$$

$$S_i \le CT \quad \forall\, i \tag{25}$$

$$S_i, D_r, CT \ge 0 \quad \forall\, i, r \tag{26}$$

$$y_{im}, \beta_{rew}, \xi_w, \lambda_{iw}, \delta_{imw}, z_{ijw}, x_{imsrew} \in \{0,1\} \quad \forall\, i, m, s, r, e, w \tag{27}$$

In this model, we aim to minimize the cycle time, Eq. (1), and the total cost Eq. (2). Constraints (3) to 5) assign predefined task categories to their mandatory execution modes, while constraint (6) ensures each remaining task is assigned to exactly one mode, and constraint (7) ensure that the skill sets of the resources are satisfied. Constraint (8) ensure that each resource of a given technology or experience level is allocated only once. The logical connections between the binary decision variables z_{ijw} at any given workstation w are enforced by constraint (9). Constraint (10) define the disjunctive resource allocation requirements. The finish-to-start precedence relations between disassembly tasks are formulated in constraint, where S_i represents a global continuous time across all workstations. (11). Constraint (12) enforce the logical dependency between task execution and resource allocation. Constraints (13) and (14) guarantee that task assignment to a specific mode and workstation occurs only when both the execution mode and the workstation are specifically chosen for that task. With stations, the allocation of each task to only one workstation, the association of the resource levels, and task-mode-workstation combinations are considered in constraints (15), (16), and (20) respectively. Constraint (17) ensure that the total number of tasks allocated to each workstation does not exceed its available capacity and constraints (18) and (19) enforce that the variable x_{imsrew} equals 1 only if both the resource allocation variable β_{rew} and the task assignment variable λ_{iw} are also equal to 1. The workload deviation is measured in constraints (21), and (23) ensure the availability of workstations for allocations. Constraint (24) order the consecutive utilization of workstations. In particular, workstation w can only be utilized if the preceding workstation $w-1$ is also utilized. Constraint (25) quantify the cycle time of the disassembly system. Finally, constraints (26) and (27) specify the domains for the variables.

4 Solution Strategies

4.1 AUGMECON2 Algorithm

Unlike weight-based methods that may fail to find solutions in the non-convex regions of a Pareto front, A2 systematically explores the objective space to generate the exact true Pareto optimal solutions. This is essential for accurately mapping the sensitive trade-offs between disassembly cycle times and operational expenditure [23]. The traditional epsilon-constraint method often yields weak Pareto optimal solutions, requiring additional computational tests to verify their efficiency. The A2 algorithm bypasses this flaw by introducing a slack variable penalty (augmented term) into the objective function, strictly guaranteeing that only strongly efficient solutions are produced at each grid point [22,23]. Since the proposed MS-HRCDLBP is highly constrained, large-scale instances cannot be solved to optimality within a reasonable computational timeframe using exact solvers. Therefore, A2 is deployed to solve small-to-medium scale instances, establishing a rigorous "ground truth" Pareto front. These exact solutions serve as a benchmark to evaluate the accuracy, convergence, and performance of the subsequently proposed hybrid metaheuristic algorithm.

In this study, for the proposed bi-objective MS-HRCDLBP, let $Z_1(x)$ represent the cycle time and Z_2 denote the total cost. The implementation of A2 proceeds as follows: First, a payoff table is constructed via lexicographic optimization to determine the ideal and nadir values, establishing the feasible range (r_2) of the secondary objective, Z_2. This range is divided into q_2 equal intervals to define a step size $(step_2 = r_2/q_2)$ and a set of right-hand side grid points (g_2). At each grid iteration, the model solves the modified single-objective problem, where the objective function is $Min \quad Z_1(x) - eps \times \left(\frac{s_2}{r_2}\right)$ subject to $\quad Z_2(x) + s_2 = g_2$. In this model, s_2 is a continuous surplus variable and *eps* is a sufficiently small constant (e.g., 10^{-3}). A major computational advantage of A2 is its dynamic bypass mechanism. After solving the problem at a specific grid point, a bypass coefficient $b = \lfloor s_2/step_2 \rfloor$ is calculated. If $b > 0$, it indicates that the current solution remains optimal for subsequent grid intervals; thus, the algorithm skips b redundant iterations, drastically reducing execution time while guaranteeing exact Pareto optimality.

4.2 Hybrid NSGA-II Algorithm

NSGA-II is selected for its established ability to approximate diverse Pareto-optimal trade-offs through fast non-dominated sorting and crowding-distance based diversity preservation [3]. In our approach, NSGA-II is enhanced with problem-specific heuristics, forming a hybrid metaheuristic (HNSGA-II) that accelerates convergence and improves solution quality while maintaining broad exploration of the search space. Each candidate solution represents a complete assignment of decisions for all tasks. The chromosome consists of two primary vectors defined over the task set, specifying the execution mode selected for each task and indicating the assigned workstation. Additional information, including resource allocations, resource usage indicators over time, and task start times, is maintained during decoding and repaired after crossover or mutation to preserve precedence, workstation-capacity, and skill-feasibility constraints.

The initial population is generated through a bi-heuristic initialization scheme designed to supply NSGA-II with strong starting solutions. This population is constructed by combining two distinct greedy strategies, each contributing

50% of the initial individuals. To prevent the generation of identical clones and preserve necessary population diversity, both strategies select assignments from a Restricted Candidate List (RCL) governed by a threshold parameter α. This parameter allows the algorithm to randomly select from a pool of near-optimal options rather than strictly choosing the single best available option. The first half of the population is produced using a cycle time-greedy logic, prioritizing execution modes with the shortest durations, and is later evolved through multi-point crossover and low-probability mutation, following the decision structure shown in Fig. 1a. The second subset is constructed using a cost-greedy approach, illustrated in Fig. 1b, where the RCL is built based on minimal cost evaluations and infeasible offspring are repaired before non-dominated sorting.

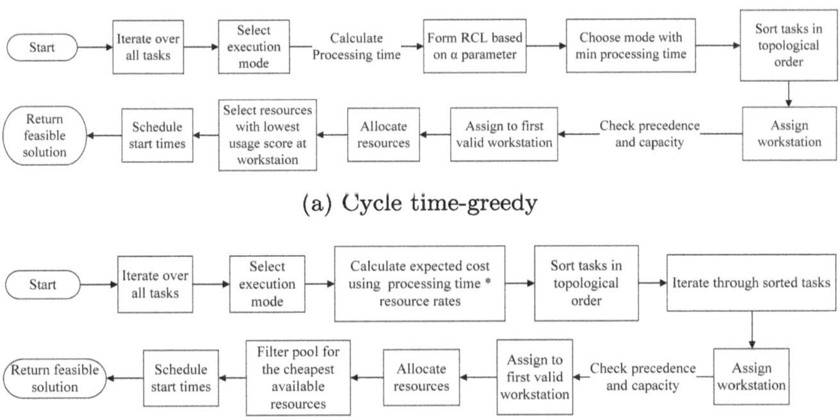

(a) Cycle time-greedy

(b) Cost-greedy

Fig. 1. Greedy heuristic initialization strategies.

5 Case Study and Dataset

To validate the reliability and effectiveness of the proposed MS-HRCDLBP model, computational experiments were conducted using both a real-world industrial case study and a set of widely recognized benchmark instances. The disassembly of an Audi A3 Sportback e-tron hybrid lithium-ion (Li-ion) high-voltage battery pack was selected as the primary case study. The main goal of disassembling EOL EV batteries (EVBs) is the safe and efficient extraction of intact battery modules (comprising 8 modules of 12 cells each), which hold significant residual value for secondary market applications, such as stationary household storage or forklift power systems. The EVB anatomy is highly complex, housing the Battery Management Controller (BMC), Battery Junction Box (BJB), high-voltage cabling, and a dedicated cooling system. The physical dismantling process, which includes removing the upper and lower housings, cutting high-voltage anchorages, disconnecting the BMC/BJB plugging cables, and unfastening the cooling plates and modules, is systematically mapped into a precedence-constrained network of disassembly tasks. The varied nature of these tasks (e.g., hazardous high-voltage disconnections versus delicate module removals) makes it an ideal candidate for human-robot collaborative optimization.

To further evaluate the performance of the proposed model across varying problem sizes, benchmark instances were adapted from the classical line balancing datasets compiled by [27]. Specifically, five widely used instances representing different product structures were considered: Bowman (8 tasks), Buxey (29 tasks, modified Sawyer), Gunther et al. (35 tasks, auto engine cradle), Hahn (53 tasks, hot water tank, excluding cycle-independent tasks), and Lutz (89 tasks, modified Lutz3 refrigerator). Since these benchmark datasets were originally developed for conventional single-mode assembly balancing, they do not contain the parameters required for the MS-HRCDLBP. Therefore, following standard practices in recent literature, HRC-specific attributes were systematically generated and integrated into the original precedence graphs, including task classifications (complex, hazardous, and collaborative), mode-dependent processing times associated with different RTT and OEL levels, skill requirements, and cost coefficients. The complete benchmark data, including precedence relations, generated task parameters, and the Audi EVB case study inputs, are publicly available on Zenodo[2] to support transparency and reproducibility.

To assess the contribution of the proposed hybridization strategy, HNSGA-II was additionally compared with a standard NSGA-II baseline using the same population size and stopping criterion. The baseline retained the same chromosome structure, crossover, and mutation operators, but replaced the proposed bi-heuristic initialization with fully random feasible initialization. Performance was evaluated on the representative large-scale benchmark instances (Hahn and Lutz) using hypervolume (HV) and inverted generational distance (IGD), together with computational runtime.

Table 2. Comparison of standard NSGA-II and HNSGA-II on representative large-scale instances

Instance	HV		Runtime (s)	
	NSGA-II	HNSGA-II	NSGA-II	HNSGA-II
Hahn (53)	15.94	17.83	2m18s	1m53s
Lutz (89)	0.82	0.90	4m06s	3m27s

As shown in Table 2, Compared with standard NSGA-II, HNSGA-II achieved higher hypervolume values on both representative large-scale instances, with improvements of approximately 12% for Hahn (53) and 10% for Lutz (89). The

[2] Dataset available at: https://doi.org/10.5281/zenodo.19235975.

hybrid initialization also reduced computational time, confirming that the bi-heuristic population seeding improves both convergence quality and search efficiency.

6 Experimental Results

This section presents the computational results for the industrial case study and the small- to large-scale benchmark instances using the A2 and HNSGA-II algorithms. The mathematical models and algorithms were implemented using the CPLEX solver within GAMS 24.1.2 and Python 3.10.0. All experiments were conducted on a laptop equipped with an 11th Gen Intel Core i7-11800H processor (2.30 GHz) and 16.0 GB of RAM under standard configurations. To ensure a fair and controlled computational environment, specific termination criteria were established for both methods: the exact CPLEX solver was constrained by a maximum time limit of 300 s per grid point with a relative optimality gap of zero, while the HNSGA-II metaheuristic was terminated upon reaching a maximum of 1,000 generations.

Table 3 provides the comparative results of the A2 algorithm applied to the case study under two distinct scenarios. Scenario (i) represents the proposed model utilizing the full spectrum of parameters, including advanced robot technologies and expert operators. Conversely, Scenario (ii) restricts the workforce entirely to traditional robots and novice operators. The results demonstrate that incorporating robot technology tiers and operator experience levels significantly impacts both the cycle time and the total disassembly cost. Furthermore, the number of Pareto solutions or grid points (GPs) generated in Scenario (i) is approximately twice that of Scenario (ii). On average, Scenario (i) yields an 84.25-second reduction in cycle time and a 29.65% decrease in total cost compared to Scenario (ii). However, as expected, the restricted solution space in Scenario (ii) dramatically reduces the computational burden, with the CPU time dropping from 2598.11 to 359.78 s. Furthermore, Fig. 2 illustrates the time-cost trade-offs (Figs. 2a and 2b) alongside the Gantt charts for the first grid point (Figs. 2c and 2d) under both scenarios. The trade-off curves, bounded by red payoff markers, confirm the conflicting dynamic where reducing cycle time inherently increases total costs. Comparing the Gantt charts visually demonstrates how advanced robotics and expert operators in Scenario (i) compress the overall cycle time. Specifically, higher RTT and advanced OEL significantly reduce individual task execution times, creating a denser and more efficient schedule than the restricted baseline.

To further analyze algorithm performance, A2 and HNSGA-II were evaluated across a variety of benchmark datasets, with the results detailed in Table 4. As indicated in the table, the exact solver exceeded the computational time limit for large-scale instances (i.e., 53 and 89 tasks) without guaranteeing optimality. However, for small and medium-sized problems, both algorithms produced highly similar cycle times and disassembly costs, successfully confirming the accuracy of the algorithms. As anticipated, the CPU time required by HNSGA-II is substantially lower than that of A2 across all evaluated instances, demonstrating its computational efficiency.

Table 3. Comparative results of the AUGMECON2 algorithm under two scenarios for the case study

GP	Scenario (i)			GP	Scenario (ii)		
	CT (s)	Cost (€)	CPU time (s)		CT (s)	Cost (€)	CPU time (s)
1	**1160.21**	**16576.59**		1	1229.53	23584.93	
2	**1171.96**	**16547.33**		2	1251.85	23423.59	
3	**1175.49**	**16513.91**		3	1284.75	23416.5	
4	**1179.01**	**16505.08**		4	1307.07	23346.78	
5	**1190.76**	**16477.82**		5	1307.07	23346.78	
6	**1194.29**	**16467.09**			-	-	
7	**1201.33**	**16441.43**			-	-	
8	**1213.08**	**16414.17**			-	-	
9	**1240.10**	**16360.91**			-	-	
Avg.	**1191.80**	**16478.26**	Total = 2598.11	Avg.	1276.05	23423.72	Total = **359.78**

Table 4. Comparative results of the AUGMECON2 and HNSGA-II algorithms for different problem instances

Algorithm	Dataset	NoP	CT (s)			Cost (€)			CPU
			Min	Avg.	Max	Min	Avg.	Max	time (s)
A2	Bowman	**4**	39	41.50	43	3908.92	**4212.36**	4212.36	**6.01**
	Buxey	7	123	**131.71**	s 147	11160.44	11223.17	**11291.15**	2109.76
	Gunther et al.	6	**148**	**165.50**	183	**20819.16**	20950.66	**21224.83**	1810.76
	Hahn				Time limit exceeded				
	Lutz 2				Time limit exceeded				
HNSGA-II	Bowman	3	39	**41**	43	3908.92	4313.51	4764.26	18.6
	Buxey	**10**	123	132.1	147	11160.4	11226.53	**11310.4**	**67.4**
	Gunther et al.	**13**	149	166.46	**182**	20820.2	**20935.28**	21280	**87.7**
	Hahn	**37**	**7271**	**8318.76**	**10135**	**3531.6**	**4851.27**	**10370.17**	**113**
	Lutz 2	**20**	**66**	**93.4**	**179**	**5830.57**	**8303.31**	**13313.24**	**207**

The experimental evaluation of the proposed bi-objective optimization approach focuses on the trade-offs between project cycle time and total cost. Furthermore, the evaluation analyzes the distribution of task execution modes and investigates the sensitivity of the algorithm to various hyper-parameters. The optimization is driven by the proposed HNSGA-II, which successfully identifies

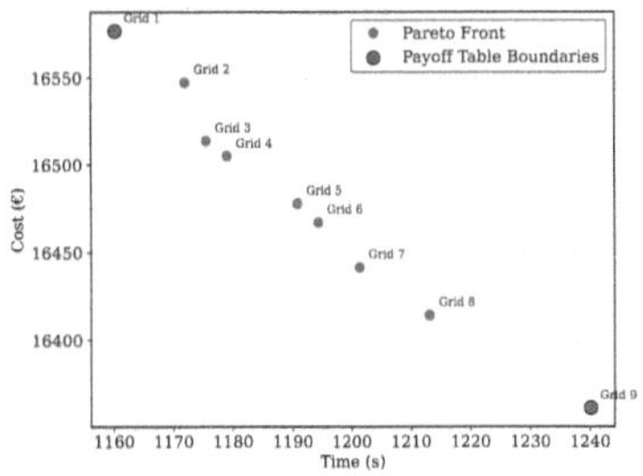

(a) Scenario (i)

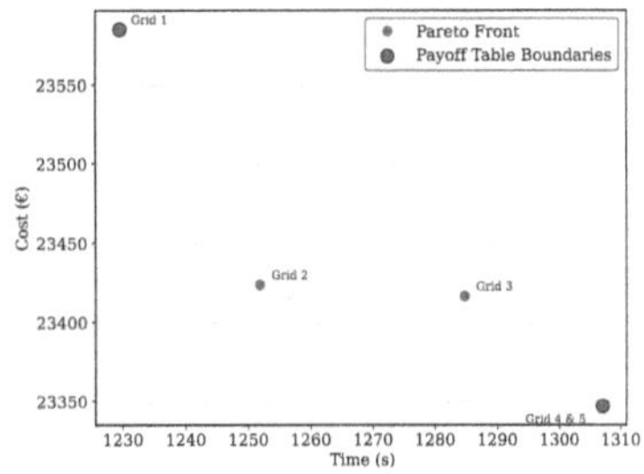

(b) Scenario (ii)

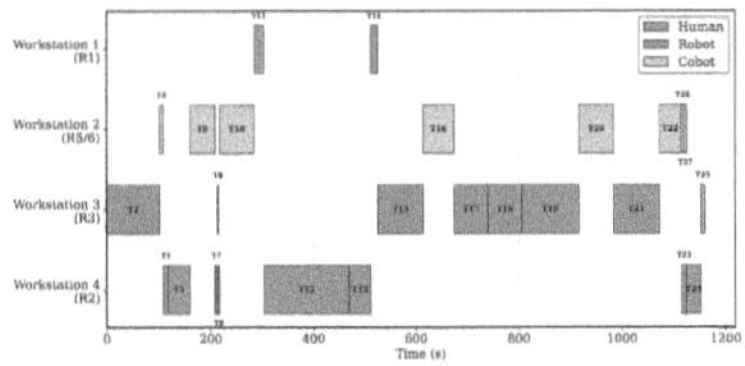

(c) Scenario (i)

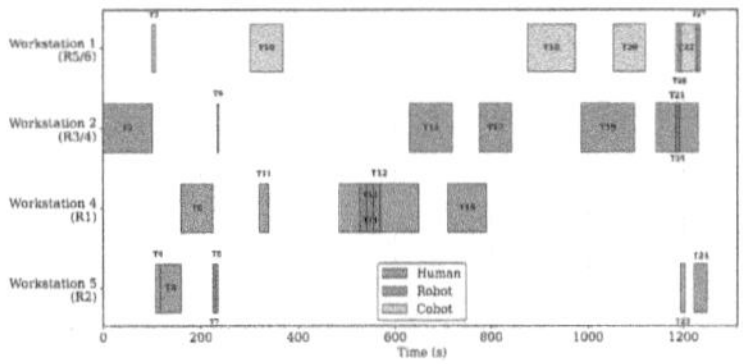

(d) Scenario (ii)

Fig. 2. Time-cost trade-off and First grid point ganttchart under two scenarios for the case study using the AUGMECON2 algorithm.

Pareto fronts outlining optimal operational strategies. Figures 3a and 3b illustrate these Pareto fronts for the 53- and 89-task instances, respectively, alongside the algorithmic convergence rates. The convergence plots demonstrate that the algorithm rapidly minimizes both the normalized cycle time and cost within the first few hundred generations before stabilizing, indicating efficient exploration of the search space. The resulting fronts reveal classic convex trade-off curves: achieving the absolute minimum cycle time requires a substantial financial investment, whereas significant cost reductions can be realized by accepting moderate extensions to the cycle time.

Beyond objective outcomes, a sensitivity analysis was conducted to determine the relative influence of key algorithmic and model parameters on the optimization metrics. Table 5 summarizes the most influential parameters across both objective functions. Algorithmic settings, particularly population size (*pop size*) and total generations (*total gens*), which act as the primary drivers of the Pareto front's overall quality and spread. For specific individual objectives, the parameter α has the highest impact on cycle time variance, producing an output swing (i.e., the absolute difference between the maximum and minimum recorded values) of 349.3 s. Meanwhile, population size dictates the most significant variation in overall cost, resulting in a maximum swing of 1,278.8 €. Problem-specific physical parameters, such as raw material cost (*Co raw*) and raw processing metrics (*P raw*), also introduce moderate variations. The findings indicate that while the algorithm is highly responsive to search density, it remains appropriately sensitive to underlying operational economics. For access

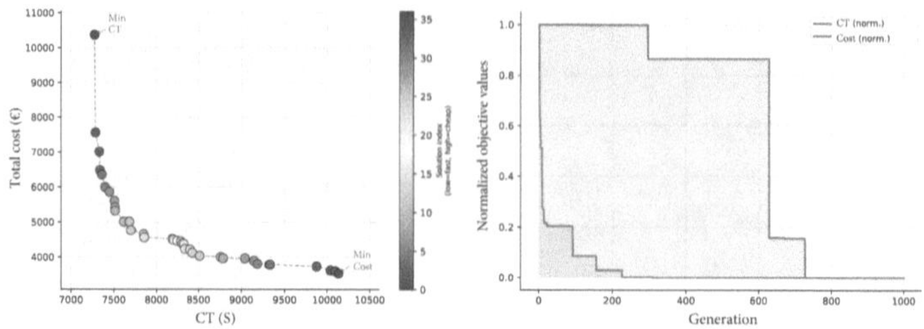

(a) Hahn - 53 tasks

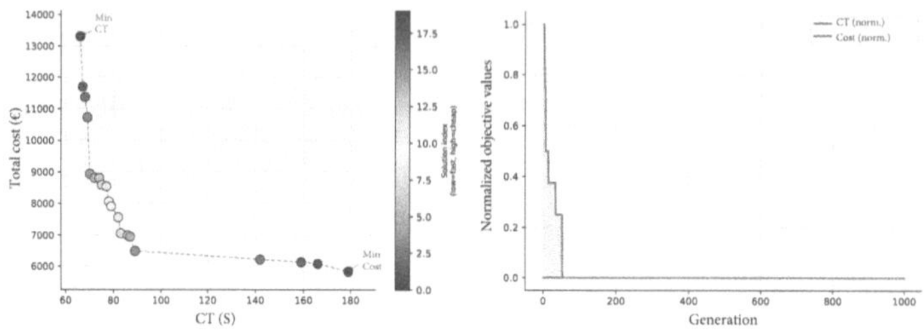

(b) Lutz2 - 89 tasks

Fig. 3. Trade-off between cycle time and total cost, and convergence of the HNSGA-II algorithm for the Hahn and Lutz2 benchmarks.

to the extended computational logs and further analysis outputs, readers are directed to the supplementary materials hosted on Zenodo[3].

Table 5. Maximum swing magnitude for highly sensitive parameters across optimization objectives

Parameter	CT Swing	Total Cost Swing	Hypervolume Swing
pop size	292.4	1278.8	7,865,741.4
mutation rate	233.4	579.8	2,267,613.9
P raw	65.2	446.1	741,101.3
total gens	264.4	1011.8	685,983.9
alpha	349.3	843.3	560,955.4
Co raw	166.2	825.1	170,989.5

7 Conclusion and Future Research Directions

This study addressed modern end-of-life product recovery by formulating a bi-objective mixed-integer linear programming model for the multi-skilled human-

[3] Dataset available at: https://doi.org/10.5281/zenodo.19236200.

robot collaborative disassembly line balancing problem (MS-HRCDLBP). The model minimizes cycle time and total costs by incorporating operational realities like robot technology tiers and operator experience levels. To solve this NP-hard problem, a hybrid metaheuristic (HNSGA-II) was developed, utilizing a 50/50 bi-heuristic (time- and cost-greedy) restricted candidate list initialization to generate a diverse, high-quality initial population. Computational experiments on an industrial case study and benchmark instances validated the efficiency of the approach. HNSGA-II consistently matched the exact Pareto fronts of the AUGMECON2 method for small-scale problems and rapidly generated comprehensive trade-offs for large-scale instances (e.g., 53 and 89 tasks) where the exact solver exceeded time limits. Results highlighted a clear convex trade-off: minimal cycle times require costly advanced robotics and expert operators, while traditional robots and novice operators offer substantial cost savings at the expense of longer durations. Furthermore, sensitivity analyses confirmed that population size and the greedy threshold (α) primarily dictate solution quality and spread. Despite its effectiveness, the model has limitations. First, it is restricted to a single-product, single-period environment, whereas realistic EV battery recovery facilities process multiple variants across dynamic planning horizons. Second, the model assumes deterministic processing times, static RTT and OEL levels, and error-free task execution, while real-world EV battery disassembly may involve dynamic operator learning, robot degradation, and operational uncertainty that affect task durations and resource utilization. To address these limitations, future research should extend the framework to multi-product and multi-period disassembly environments, incorporate dynamic adaptation under uncertainty, and further validate the proposed model through real industrial implementation.

References

1. Bruno, G., Antonelli, D.: Dynamic task classification and assignment for the management of human-robot collaborative teams in workcells. Int. J. Adv. Manuf. Technol. **98**(9), 2415–2427 (2018). https://doi.org/10.1007/s00170-018-2400-4
2. Chu, M., Chen, W.: A bi-objective discrete flower pollination algorithm for planning the collaborative disassembly of retired power batteries by humans and robots. Appl. Soft Comput. **177**, 113213 (2025). https://doi.org/10.1016/j.asoc.2025.113213
3. Deb, K., Pratap, A., Agarwal, S., Meyarivan, T.: A fast and elitist multiobjective genetic algorithm: Nsga-ii. IEEE Trans. Evol. Comput. **6**(2), 182–197 (2002). https://doi.org/10.1109/4235.996017
4. Durmaz, N., Budak, A.: An integrated bi-objective green vehicle routing and partial disassembly line problem for electronic waste: an industrial case study. Int. J. Comput. Integr. Manuf. **38**(3), 408–433 (2025). https://doi.org/10.1080/0951192X.2024.2335984
5. Forti, V., Baldé, C.P., Kuehr, R., Bel, G.: The global e-waste monitor 2020. United Nations University (UNU), International Telecommunication Union (ITU) & International Solid Waste Association (ISWA), Bonn/Geneva/Rotterdam **120**, 14 (2020). https://www.itu.int/en/itu-d/environment/documents/toolbox/gem_2020_def.pdf

6. Gerlitz, E., Greifenstein, M., Hofmann, J., Fleischer, J.: Analysis of the variety of lithium-ion battery modules and the challenges for an agile automated disassembly system. Procedia CIRP **96**, 175–180 (2021)
7. Gungor, A., Gupta, S.M.: A solution approach to the disassembly line balancing problem in the presence of task failures. Int. J. Prod. Res. **39**(7), 1427–1467 (2001). https://doi.org/10.1080/00207540110052157
8. Guo, L., Zhang, Z., Song, H., Li, Y.: Industrial application of a human-robot collaborative parallel two-sided destructive disassembly line balancing problem in multi-product, multi-line layouts. Robot. Comput.-Integr. Manuf. **99**, 103209 (2026). https://doi.org/10.1016/j.rcim.2025.103209
9. Guo, L., Zhang, Z., Wu, T., Zeng, Y., Zhang, Y., Xie, X.: A grey wolf optimization algorithm for solving partial destructive disassembly line balancing problem consider feasibility evaluation and noise pollution. Adv. Eng. Inform. **60**, 102418 (2024). https://doi.org/10.1016/j.aei.2024.102418
10. Guo, L., Zhang, Z., Wu, T., Zhang, Y., Zeng, Y., Xie, X.: Green and efficient-oriented human-robot hybrid partial destructive disassembly line balancing problem from non-disassemblability of components and noise pollution. Robot. Comput.-Integr. Manuf. **90**, 102816 (2024). https://doi.org/10.1016/j.rcim.2024.102816
11. Hellmuth, J.F., DiFilippo, N.M., Jouaneh, M.K.: Assessment of the automation potential of electric vehicle battery disassembly. J. Manuf. Syst. **59**, 398–412 (2021). https://doi.org/10.1016/j.jmsy.2021.03.009
12. Hopko, S., Wang, J., Mehta, R.: Human factors considerations and metrics in shared space human-robot collaboration: A systematic review. Front. Robot. AI **9**, 799522 (2022). https://doi.org/10.3389/frobt.2022.799522
13. Huang, J., et al.: A case study in human-robot collaboration in the disassembly of press-fitted components. Proc. Inst. Mech. Eng. Part B J. Eng. Manuf. **234**(3), 654–664 (2020). https://doi.org/10.1177/0954405419883060
14. Jiao, J., Feng, G., Yuan, G.: Research on the human-robot collaborative disassembly line balancing of spent lithium batteries with a human factor load. Batteries **10**(6), 196 (2024). https://doi.org/10.3390/batteries10060196
15. Kastanaki, E., Giannis, A.: Dynamic estimation of end-of-life electric vehicle batteries in the eu-27 considering reuse, remanufacturing and recycling options. J. Clean. Prod. **393**, 136349 (2023). https://doi.org/10.1016/j.jclepro.2023.136349
16. Kheirabadi, M., Keivanpour, S., Chinniah, Y.A., Frayret, J.M.: Human-robot collaboration in assembly line balancing problems: Review and research gaps. Comput. Ind. Eng. **186**, 109737 (2023). https://doi.org/10.1016/j.cie.2023.109737
17. Lagomarsino, M., Lorenzini, M., De Momi, E., Ajoudani, A.: An online framework for cognitive load assessment in industrial tasks. Robot. Comput.-Integr. Manuf. **78**, 102380 (2022). https://doi.org/10.1016/j.rcim.2022.102380
18. Laouini, O., Slama, I., Hnaien, F., Jemai, Z.: An integrated multi-objective approach to collaborative disassembly line balancing and planning with taog-based modelling. Int. J. Prod. Res. 1–26 (2025). https://doi.org/10.1080/00207543.2025.2584717
19. Lee, M.L., Behdad, S., Liang, X., Zheng, M.: Task allocation and planning for product disassembly with human-robot collaboration. Robot. Comput.-Integr. Manuf. **76**, 102306 (2022). https://doi.org/10.1016/j.rcim.2021.102306
20. Lee, M.L., Liang, X., Hu, B., Onel, G., Behdad, S., Zheng, M.: A review of prospects and opportunities in disassembly with human-robot collaboration. J. Manuf. Sci. Eng. **146**(2), 020902 (2024). https://doi.org/10.1115/1.4063992

21. Liu, B., Xu, W., Liu, J., Yao, B., Zhou, Z., Pham, D.T.: Human-robot collaboration for disassembly line balancing problem in remanufacturing. In: International Manufacturing Science and Engineering Conference, vol. 58745, p. V001T02A037. American Society of Mechanical Engineers (2019). https://doi.org/10.1115/MSEC2019-2919
22. Mavrotas, G.: Effective implementation of the ε-constraint method in multi-objective mathematical programming problems. Appl. Math. Comput. **213**(2), 455–465 (2009). https://doi.org/10.1016/j.amc.2009.03.037
23. Mavrotas, G., Florios, K.: An improved version of the augmented ε-constraint method (augmecon2) for finding the exact pareto set in multi-objective integer programming problems. Appl. Math. Comput. **219**(18), 9652–9669 (2013). https://doi.org/10.1016/j.amc.2009.03.037
24. Ogenyi, U.E., Liu, J., Yang, C., Ju, Z., Liu, H.: Physical human-robot collaboration: Robotic systems, learning methods, collaborative strategies, sensors, and actuators. IEEE Trans. Cybern. **51**(4), 1888–1901 (2019). https://doi.org/10.1109/TCYB.2019.2947532
25. Qin, S., Xiang, C., Wang, J., Liu, S., Guo, X., Qi, L.: Multiple product hybrid disassembly line balancing problem with human-robot collaboration. IEEE Trans. Autom. Sci. Eng. **22**, 712–723 (2024). https://doi.org/10.1109/TASE.2024.3418898
26. Rietmann, N., Hügler, B., Lieven, T.: Forecasting the trajectory of electric vehicle sales and the consequences for worldwide co2 emissions. J. Clean. Prod. **261**, 121038 (2020). https://doi.org/10.1016/j.jclepro.2020.121038
27. Scholl, A.: Data of assembly line balancing problems. schriften zur quantitativen betriebswirtschaftslehre. Technical report (1993)
28. Wang, J., Xi, G., Guo, X., Qin, S., Han, H.: Multi-objective advantage actor-critic algorithm for hybrid disassembly line balancing with multi-skilled workers. Information **15**(3), 168 (2024). https://doi.org/10.3390/info15030168
29. Wang, J., Li, M., Meng, F., Zhao, H., Sun, X.: Multi-objective reinforcement learning for dynamic balancing of two-sided human-robot collaborative disassembly lines. Comput. Ind. Eng. 111703 (2025). https://doi.org/10.1016/j.cie.2025.111703
30. Wu, S., Kaden, N., Dröder, K.: A systematic review on lithium-ion battery disassembly processes for efficient recycling. Batteries **9**(6), 297 (2023). https://doi.org/10.3390/batteries9060297
31. Wu, T., Zhang, Z., Yin, T., Zhang, Y.: Multi-objective optimisation for cell-level disassembly of waste power battery modules in human-machine hybrid mode. Waste Manag. **144**, 513–526 (2022). https://doi.org/10.1016/j.wasman.2022.04.015
32. Wu, T., Zhang, Z., Zeng, Y., Zhang, Y.: Mixed-integer programming model and hybrid local search genetic algorithm for human-robot collaborative disassembly line balancing problem. Int. J. Prod. Res. **62**(5), 1758–1782 (2024). https://doi.org/10.1080/00207543.2023.2201352
33. Wu, T., Zhang, Z., Zhang, Y., Zeng, Y.: Modelling and optimisation of two-sided disassembly line balancing problem with human-robot interaction constraints. Expert Syst. Appl. **230**, 120589 (2023). https://doi.org/10.1016/j.eswa.2023.120589
34. Xiao, L., Yang, X., Guo, X., Qin, S., Zhang, Q., Zhang, S.: An improved fireworks algorithm for the multi-skilled robot hybrid disassembly line balancing problem. In: 2024 International Conference on Networking, Sensing and Control (ICNSC), pp. 1–6. IEEE (2024). https://doi.org/10.1109/ICNSC62968.2024.10760023

35. Xu, W., Cui, J., Liu, B., Liu, J., Yao, B., Zhou, Z.: Human-robot collaborative disassembly line balancing considering the safe strategy in remanufacturing. J. Clean. Prod. **324**, 129158 (2021). https://doi.org/10.1016/j.jclepro.2021.129158
36. Yin, T., Zhang, Z., Zhang, Y., Wu, T., Liang, W.: Mixed-integer programming model and hybrid driving algorithm for multi-product partial disassembly line balancing problem with multi-robot workstations. Robot. Comput.-Integr. Manuf. **73**, 102251 (2022). https://doi.org/10.1016/j.rcim.2021.102251
37. Zhang, H., Huang, Z., Wang, D., Tian, G., Wang, W.: U-shaped disassembly line balancing problem under interval type-2 trapezoidal fuzzy set: Modeling and solution method. Eng. Appl. Artif. Intell. **145**, 110211 (2025). https://doi.org/10.1016/j.engappai.2025.110211
38. Zhang, Q., et al.: An improved moth-flame algorithm for human-robot collaborative parallel disassembly line balancing problem. Mathematics **12**(6), 816 (2024). https://doi.org/10.3390/math12060816
39. Zhang, W., Li, Y., Wang, K., Xu, W., Gao, L.: A green and efficient disassembly line balancing with human-robot collaboration and destructive disassembly. Robot. Comput.-Integr. Manuf. **97**, 103081 (2026). https://doi.org/10.1016/j.rcim.2025.103081

Accelerating Lagrangian Relaxation via Dual-Guided Genetic Search for UAV Trajectory and Camera Orientation Optimization

Ning Xue[1(✉)], Yifeng Sheng[1], Yiming Quan[2], Tianxiang Cui[1], and Ruibin Bai[1]

[1] School of Computer Science, University of Nottingham Ningbo China, 199 Taikang East Road, Ningbo 315100, Zhejiang, China
{ning.xue,yifeng.sheng,tianxiang.cui,ruibin.bai}@nottingham.edu.cn
[2] Ningbo Tianyi Design Research of Surveying and Mapping Co., Ltd., 100 Xinghai South Road, Ningbo 315040, China

Abstract. This paper addresses unmanned aerial vehicle (UAV) path optimization with integrated viewpoint selection and camera orientation planning. The proposed framework captures the intricate interdependencies among these decisions and optimizes them simultaneously for the first time, balancing surface coverage quality and flight energy consumption. To tackle this large-scale combinatorial problem, a Lagrangian relaxation approach decomposes the model into an orienteering subproblem and a coverage subproblem. A Genetic Algorithm (GA) with a dual-guided mutation operator efficiently solves the orienteering subproblem. Theoretical analysis and empirical validation are provided. Experimental results demonstrate that the dual-guided genetic search reduces runtime by 51.3% and improves the weighted objective by 9.3% over a standard genetic algorithm. Moreover, it matches or improves upon an exact mixed-integer linear programming solver while cutting runtime by over 90% on small instances and by 50–70% on larger ones.

Keywords: UAV path planning · Lagrangian relaxation · Dual-guided genetic algorithm · Viewpoint selection · Coverage optimization

1 Introduction

UAVs equipped with high-resolution cameras have become powerful platforms for large-scale aerial 3D scanning, exploiting agile low-altitude flights to capture imagery from diverse viewpoints. High-quality 3D reconstruction requires dense image coverage with sufficient angular diversity across all observed surfaces. Conventional flight planning methods typically extract building contours via semantic segmentation and then generate predefined grid-based or zigzag paths. However, these rigid trajectories often struggle to ensure adequate coverage and image quality for irregular structures or complex terrain, frequently

G. Iacca et al. (Eds.): PPSN 2026, LNCS 16985, pp. 249–264, 2027.
https://doi.org/10.1007/978-3-032-36214-8_16

necessitating manual refinement. This problem is further exacerbated by the limited flight time imposed by UAV battery constraints. The fundamental challenge is to jointly optimize three interdependent decisions: selecting which viewpoints to visit, determining the corresponding camera orientations, and computing the most efficient route between them. This forms a combinatorial optimization problem that is computationally intractable using exact methods under real-world scenarios. For instance, even a small instance with only 89 candidate viewpoints poses significant computational difficulty. Solved by Gurobi [2] on a standard PC, the problem requires roughly three days of runtime while still leaving an optimality gap of more than 20%, making real-time application impractical. To mitigate this complexity, many studies adopt a two-phase methodology [6]. In the first phase, candidate viewpoints are selected for reconstruction quality; in the second phase, path planning computes the shortest trajectory connecting the selected viewpoints. Although computationally manageable, this decomposition introduces suboptimality by decoupling interlinked objectives. A unified single-phase approach that simultaneously optimizes viewpoint selection and traversal order is therefore essential to achieve a globally balanced trade-off between reconstruction fidelity and travel efficiency.

Due to the computational complexity of the joint optimization problem, existing studies primarily rely on heuristic approaches, including population-based metaheuristics and trajectory-based search methods [3,6,7,9]. Nevertheless, such approaches often either decompose viewpoint selection and path planning into separate stages or rely heavily on stochastic search, which may lead to suboptimal solutions and inefficient exploration. Motivated by these limitations, we develop a unified optimization framework based on Lagrangian relaxation, which decomposes the original problem into coordinated orienteering and coverage optimization subproblems. This formulation jointly optimizes viewpoint selection, camera pose configuration, and tour construction, avoiding the suboptimality of conventional two-phase methods. To improve computational efficiency, the orienteering subproblem is solved using a GA. Furthermore, a dual-guided genetic search strategy is proposed, exploiting dual information from the Lagrangian relaxation to direct the search toward promising regions of the solution space and mitigate excessive stochasticity.

The main contributions of this study are summarized as follows:

- We present, to the best of our knowledge, the first framework that jointly optimizes the UAV trajectory and camera orientation in an integrated manner.
- We develop a Lagrangian relaxation formulation and propose a hybrid optimization framework that combines Lagrangian relaxation with genetic search.
- We design a dual-guided evolutionary operator that substantially accelerates the convergence of the GA, supported by both theoretical justification and empirical validation.

The remainder of this paper is organized as follows. Section 2 introduces the problem. Section 3 explores the separable structure of the problem and presents the Lagrangian relaxation formulation. Section 4 details the proposed hybrid

Lagrangian relaxation with genetic search framework. This section also includes the tailored standard GA and a dual-guided GA for solving the orienteering subproblem. Section 5 reports the experimental results, and Sect. 6 concludes the paper.

2 Problem Description

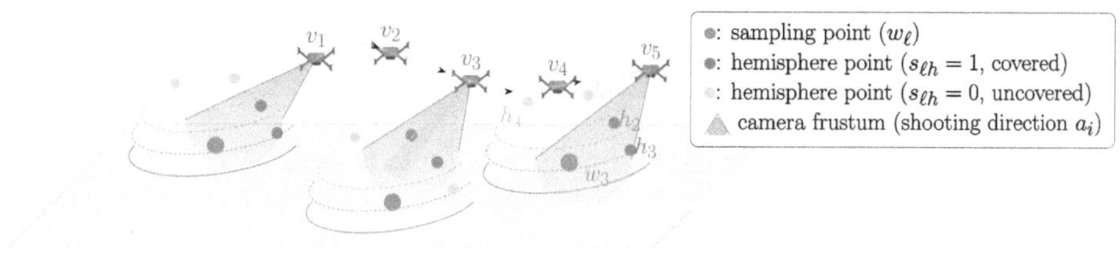

Fig. 1. UAV trajectory with camera frustums and viewpoints

We introduce the problem through the example illustrated in Fig. 1, with the relevant notations summarized in Table 1. In this example, a UAV must select a subset of viewpoints $\mathcal{V}$ to visit, choose shooting directions from the set $\mathcal{A}$ at those viewpoints, and plan a continuous tour, while ensuring that all sampling points $\mathcal{W}$ on the structure are imaged from multiple directions to enable reliable 3D reconstruction. Specifically, three viewpoints (v_1, v_3, v_5) are selected to capture sampling points via camera frustums. These viewpoints are connected by a flight path, whereas the intermediate viewpoints v_2 and v_4 are not selected.

To quantify observation quality, we employ a hemisphere representation [8] for each sampling point. For a given sampling point w_ℓ, the set $\mathcal{H}_\ell$ represents a discretized set of observation directions defined on the hemisphere aligned with the surface normal of the point. Figure 2 illustrates the coverage indicator matrix for sampling point w_1 and viewpoint v_2. In this example, the hemisphere $\mathcal{H}_1$ centered at w_1 is discretized into five sampling directions $h_1, \ldots, h_5$ (in our experiments we use 32 uniformly distributed points). From viewpoint v_2, five shooting directions $a_1, \ldots, a_5 \in \mathcal{A}$ are considered (the experiments use 14 unit direction vectors, including six axial and eight octant directions). The visibility of each hemisphere point is encoded by the coverage indicator $d_{\ell h}^{ka}$.

3 Problem Structure and Separability

The problem exhibits a natural separable structure, making it well suited to Lagrangian decomposition. The decision variables can be partitioned into two groups: (i) tour selection variables (x_{ij}, y_k), which determine the sequence of visited viewpoints, and (ii) camera orientation variables ($z_{ka}, s_{\ell h}$), which specify the photographs to be taken and the corresponding coverage of hemisphere

Table 1. Notation Summary

Symbol	Description
Sets	
$\mathcal{V}$	Set of candidate viewpoints $\{v_1, \ldots, v_n\}$
$\mathcal{T}$	Mandatory viewpoints (including start location)
$\mathcal{W}$	Sampling points $\{w_1, \ldots, w_m\}$ on the building surface
$\mathcal{A}$	Discrete camera shooting directions $\{a_1, \ldots, a_{14}\}$ available at each viewpoint
$\mathcal{S}_\ell$	Set of viewpoint–direction pairs that can cover sampling point w_ℓ
$\mathcal{H}_\ell$	Set of discrete hemisphere points for sampling point w_ℓ
Parameters	
e_{ij}	Energy cost for traveling from viewpoint v_i to v_j
$d_{\ell h}^{ka}$	Visibility coefficient: 1 if hemisphere point h of w_ℓ is visible from viewpoint v_k with direction a, otherwise 0
$\omega_{\ell h}$	Quality weight for hemisphere point h of w_ℓ, defined as $\cos\theta_h$ where θ_h is the angle between the hemisphere direction and the surface normal at w_ℓ
$N_{\ell h}$	Number of viewpoint–direction pairs covering hemisphere point h of w_ℓ
ρ	Minimum required coverage redundancy (each sampling point must be observed by at least ρ distinct camera poses)
Decision Variables	
x_{ij}	1 if the UAV traverses edge (i, j) in the tour, 0 otherwise
y_k	1 if viewpoint v_k is visited, 0 otherwise
z_{ka}	1 if a photo is taken at viewpoint v_k with direction a, 0 otherwise
$s_{\ell h}$	1 if hemisphere point h of w_ℓ is covered by at least one selected camera pose, 0 otherwise

points. These two variable sets interact only through linking constraints that enforce photographs to be captured exclusively at visited viewpoints:

$$\sum_{a \in \mathcal{A}} z_{ka} = y_k, \quad \forall v_k \in \mathcal{V} \tag{1}$$

These constraints ensure that exactly one camera direction is selected for each visited viewpoint and that no photographs are taken from unvisited viewpoints. For model simplicity, each visited viewpoint is associated with a single camera direction.

The routing variables satisfy the tour feasibility constraints, including connectivity, subtour elimination, and mandatory viewpoint visitation, defining the feasible region $(x, y) \in \mathcal{F}_{\text{tour}}$. The camera variables satisfy the multi-view coverage requirements, where each sampling point must be observed from at least ρ viewpoint–direction pairs and the coverage indicators $s_{\ell h}$ correctly represent

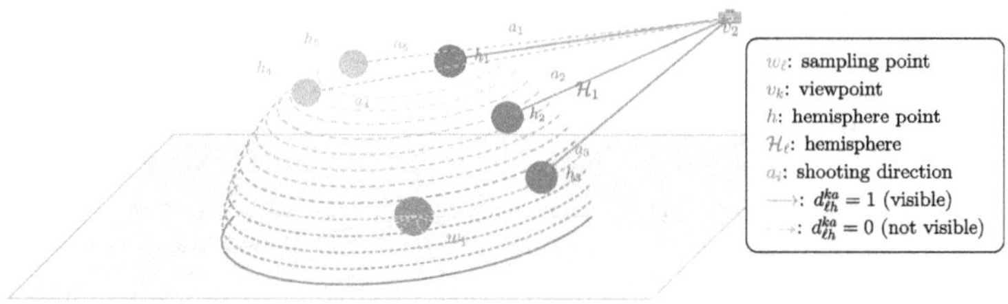

Fig. 2. Coverage indicator matrix for viewpoint v_2 and sampling point w_1

the resulting visibility relationships. These constraints define the feasible region $(z, s) \in \mathcal{F}_{\text{camera}}$.

3.1 Lagrangian Relaxation

The linking constraints in (1) constitute the only coupling between the two groups of decision variables. To manage this coupling, we employ a Lagrangian relaxation framework that transfers the linking constraints into the objective function by associating each constraint with a Lagrange multiplier. Let $\boldsymbol{\lambda} = (\lambda_1, \dots, \lambda_{|\mathcal{V}|}) \in \mathbb{R}^{|\mathcal{V}|}$ denote the vector of multipliers corresponding to the viewpoint set $\mathcal{V}$.

For a fixed vector of multipliers $\boldsymbol{\lambda}$, the Lagrangian relaxation of the original problem is defined as

$$L(\boldsymbol{\lambda}) = \min \quad \sum_{i \neq j} e_{ij} x_{ij} - \sum_{\ell,h} \omega_{\ell h} s_{\ell h} + \sum_{k} \lambda_k \left(\sum_{a} z_{ka} - y_k \right) \tag{2}$$

$$\text{s.t.} \quad (x, y) \in \mathcal{F}_{\text{tour}}, \quad (z, s) \in \mathcal{F}_{\text{camera}}. \tag{3}$$

By rearranging the objective function, we expose the problem's inherent separable structure, which allows the relaxed formulation to decompose into two independent subproblems, one governing tour selection and the other governing camera operation, coordinated through the price vector $\boldsymbol{\lambda}$:

$$L(\boldsymbol{\lambda}) = \underbrace{\min_{(x,y) \in \mathcal{F}_{\text{tour}}} \left\{ \sum_{i \neq j} e_{ij} x_{ij} - \sum_{k \in \mathcal{V}} \lambda_k y_k \right\}}_{\phi_1(\boldsymbol{\lambda})\text{:Tour selection subproblem}}$$
$$+ \underbrace{\min_{(z,s) \in \mathcal{F}_{\text{camera}}} \left\{ \sum_{k,a} \lambda_k z_{ka} - \sum_{\ell,h} \omega_{\ell h} s_{\ell h} \right\}}_{\phi_2(\boldsymbol{\lambda})\text{:Camera coverage subproblem}}. \tag{4}$$

We next present the formulation of the two subproblems. Although relaxing the linking constraints yields a separable Lagrangian, our implementation (Algorithm 1) solves the camera subproblem conditionally on the GA-selected tour y, thereby reintroducing coupling between the two components. Consequently, $L(\boldsymbol{\lambda})$ is not evaluated as an exact dual function but as an approximate Lagrangian value used as a surrogate objective to guide the search. Accordingly, the multipliers $\boldsymbol{\lambda}$ act as adaptive penalty weights influencing both the genetic search and camera selection. Standard projection and step-size updates are employed for stability, but no claim is made regarding exact dual ascent or valid dual bounds.

3.2 Subproblem 1: Orienteering Problem

This subproblem jointly determines viewpoint selection and tour construction, balancing travel energy costs e_{ij} against multiplier-adjusted viewpoint profits $-\lambda_k$. It corresponds to an Orienteering Problem and is solved over the feasible region $\mathcal{F}_{\text{tour}}$, whose complete formulation is given below:

$$\phi_1(\boldsymbol{\lambda}) = \min \quad \sum_{i \neq j} e_{ij} x_{ij} - \sum_{k \in \mathcal{V}} \lambda_k y_k \tag{5}$$

$$\text{s.t.} \quad \sum_{i<k} x_{ik} + \sum_{j>k} x_{kj} = 2y_k, \quad \forall k \in \mathcal{V} \tag{6}$$

$$\sum_{\substack{i \in S, j \notin S \\ \text{or } i \notin S, j \in S}} x_{ij} \geq 2y_t, \quad \forall S \subset \mathcal{V},\ 2 \leq |S| \leq |\mathcal{V}| - 2,\ \mathcal{T} \setminus S \neq \emptyset,\ \forall t \in S \tag{7}$$

$$y_k = 1, \quad \forall k \in \mathcal{T} \tag{8}$$

$$x_{ij} \in \{0,1\}, \quad y_k \in \{0,1\}, \quad \forall i,j,k. \tag{9}$$

Constraint (6) enforces that each visited viewpoint ($y_k = 1$) has exactly two incident edges, consistent with a closed tour structure. Subtour elimination constraints (7) guarantee a single connected tour, while (8) ensures that mandatory viewpoints are visited.

3.3 Subproblem 2: Coverage Optimization with Multiplier-Adjusted Costs

This subproblem optimizes camera direction selection and hemisphere coverage, with the cost of each photograph at viewpoint k adjusted by the corresponding multiplier λ_k. It corresponds to a weighted set covering problem, balancing coverage quality rewards $\omega_{\ell h} s_{\ell h}$ against multiplier-adjusted photograph costs. The full formulation expands $\mathcal{F}_{\text{camera}}$ as follows:

$$
\begin{aligned}
\phi_2(\boldsymbol{\lambda}) = \min \quad & \sum_{k\in\mathcal{V}}\sum_{a\in\mathcal{A}} \lambda_k z_{ka} - \sum_{\ell\in\mathcal{W}}\sum_{h\in\mathcal{H}_\ell} \omega_{\ell h} s_{\ell h} && (10)\\
\text{s.t.} \quad & \sum_{(v_k,a)\in\mathcal{S}_\ell} z_{ka} \geq \rho, \quad \forall \ell \in \mathcal{W} && (11)\\
& s_{\ell h} \leq \sum_{(v_k,a)\in\mathcal{S}_\ell} d_{\ell h}^{ka} z_{ka}, \quad \forall \ell \in \mathcal{W}, h \in \mathcal{H}_\ell && (12)\\
& \sum_{(v_k,a)\in\mathcal{S}_\ell} d_{\ell h}^{ka} z_{ka} \leq N_{\ell h} s_{\ell h}, \quad \forall \ell \in \mathcal{W}, h \in \mathcal{H}_\ell && (13)\\
& \sum_{a\in\mathcal{A}} z_{ka} \leq 1, \quad \forall k \in \mathcal{V} && (14)\\
& z_{ka} \in \{0,1\}, \quad s_{\ell h} \in \{0,1\}, \quad \forall k, a, \ell, h. && (15)
\end{aligned}
$$

The multiplier adjustment in (10) modulates photograph selection costs: a positive λ_k increases the cost of taking photographs at viewpoint k (discouraging selection), while a negative λ_k reduces the cost (encouraging selection). Constraint (11) enforces the minimum coverage redundancy requirement for each sampling point, where ρ is set to 1 in our experiments following standard photogrammetric practice. Constraints (12) and (13) link photograph selection z_{ka} to hemisphere coverage activation $s_{\ell h}$, ensuring that $s_{\ell h} = 1$ if and only if at least one covering viewpoint-direction pair is selected. Constraint (14) ensures at most one camera direction is selected per viewpoint.

4 Hybrid Lagrangian Relaxation Algorithm

Solving the Lagrangian relaxation model exactly is computationally intractable due to the complexity of its subproblems. In practice, high-quality approximate solutions that balance coverage performance and flight time are sufficient. Since the orienteering subproblem can be efficiently solved by genetic algorithms (GAs) [12], we employ a GA to approximately solve Subproblem 1, resulting in the Hybrid Lagrangian Relaxation with Genetic Search framework shown in Algorithm 1.

The algorithm starts by initializing the user-defined parameters, including the maximum number of generations $G_{\max}$, dual step-size bounds $(\alpha^0, \alpha^{\max})$, stagnation threshold $S_{\max}$, population size N, and a small constant $\varepsilon > 0$. The Lagrange multipliers $\boldsymbol{\lambda}$, generation counter t, stagnation counter cnt, and best primal objective value f^{best} are initialized accordingly.

The procedure iterates while $t < G_{\max}$ and $cnt < S_{\max}$. At each generation, the population is evaluated using the Lagrangian-based fitness function, and the best chromosome determines the viewpoint-selection vector y^t. Given y^t, Subproblem 2 is solved exactly via MILP to obtain the camera-orientation variables z^t and coverage variables s^t. The resulting linking-constraint violations define the subgradient vector $\mathbf{g}^t$, where $g_k^t = \sum_a z_{k,a}^t - y_k^t$. The corresponding

Algorithm 1. Hybrid Lagrangian Relaxation with Genetic Search

Require: $G_{\max}$, $\alpha^{\max}$, α^0, ε, $S_{\max}$, N
Ensure: (x^t, y^t, z^t, s^t)
Initialize $t \leftarrow 0$, $cnt \leftarrow 0$, $f^{best} \leftarrow +\infty$, $\boldsymbol{\lambda}^0 \leftarrow \mathbf{0}$, population $\mathcal{P}_0$
while $t < G_{\max}$ and $cnt < S_{\max}$ **do**
  Evaluate $\mathcal{P}_t$ using $\boldsymbol{\lambda}^t$ and select best individual i^*
  $y^t \leftarrow$ viewpoint selection of i^*
  $(z^t, s^t) \leftarrow$ MILPSolver(y^t)
  Compute primal objective $f^t \leftarrow f(x^t, y^t, z^t, s^t)$
  if $f^t < f^{best}$ **then**
    $cnt \leftarrow 0$
    $f^{best} \leftarrow f^t$
  else
    $cnt \leftarrow cnt + 1$
  end if
  Compute subgradient $g_k^t = \sum_a z_{k,a}^t - y_k^t$
  Compute Lagrangian value
  $\tilde{L}(\boldsymbol{\lambda}^t) \leftarrow f(x^t, y^t, z^t, s^t) + (\boldsymbol{\lambda}^t)^\top \mathbf{g}^t$
  Compute step size
  $$\alpha^t \leftarrow \frac{\max\left(0,\ \tilde{L}(\boldsymbol{\lambda}^t) - f^{best}\right)}{\|\mathbf{g}^t\|^2 + \varepsilon}$$
  $\alpha^t \leftarrow \min\{\alpha^{\max}, \max\{\alpha^0, \alpha^t\}\}$
  Update multipliers
  $\boldsymbol{\lambda}^{t+1} \leftarrow \boldsymbol{\lambda}^t + \alpha^t \mathbf{g}^t$
  Generate next population
  $\mathcal{P}_{t+1} \leftarrow$ Elitism$(\mathcal{P}_t,\ $ Mutation(Crossover(Selection$(\mathcal{P}_t)), \boldsymbol{\lambda}^t))$
  $t \leftarrow t + 1$
end while
$y^t \leftarrow$ best viewpoint selection found
$(z^t, s^t) \leftarrow$ MILPSolver(y^t)
return (x^t, y^t, z^t, s^t)

Lagrangian value is reconstructed as $\tilde{L}(\boldsymbol{\lambda}^t) = f(x^t, y^t, z^t, s^t) + (\boldsymbol{\lambda}^t)^\top \mathbf{g}^t$, and the duality gap is given by $\max(\tilde{L}(\boldsymbol{\lambda}^t) - f^{best}, 0)$.

The multipliers are then updated using the stabilized Polyak step size $\alpha^t = \max(\tilde{L}(\boldsymbol{\lambda}^t) - f^{best}, 0)/(|\mathbf{g}^t|^2 + \varepsilon)$, followed by $\boldsymbol{\lambda}^{t+1} = \boldsymbol{\lambda}^t + \alpha^t \mathbf{g}^t$, where α^t is clipped to $[\alpha^0, \alpha^{\max}]$. Genetic operators are subsequently applied to generate the next population, with mutation guided by the dual information to promote feasibility with respect to the relaxed constraints. The algorithm terminates when either the generation limit or stagnation threshold is reached, and the best viewpoint-selection vector is finally used to reconstruct the complete solution (x^t, y^t, z^t, s^t).

4.1 Standard GA for Subproblem 1

In this study, a customized GA is adopted as a baseline to solve Subproblem 1, which involves viewpoint selection and path sequencing. Each chromosome encodes a flight path, where each gene corresponds to a candidate viewpoint and stores its successor in the route. Unselected viewpoints are assigned a sentinel value of -1. As illustrated in Fig. 3, the selected viewpoints form an ordered path, and each viewpoint is associated with a predefined camera angle specifying the drone's orientation. The value stored at viewpoint v_i indicates its successor in the route (e.g., $v_1 \to v_5 \to v_7$), thereby defining the flight sequence over the selected viewpoints.

Based on this representation, the genetic operators are designed to preserve path feasibility and structural characteristics. During crossover, Edge Recombination Crossover (ERX) [13] is employed to retain adjacency information inherited from both parents, followed by a Lin–Kernighan local search procedure [5] to further improve route quality.

During mutation, a selective node replacement strategy is applied with probability p_m. A subset of viewpoints is removed and replaced with feasible alternatives while maintaining path connectivity and adjacency consistency, thereby introducing controlled perturbations without violating feasibility.

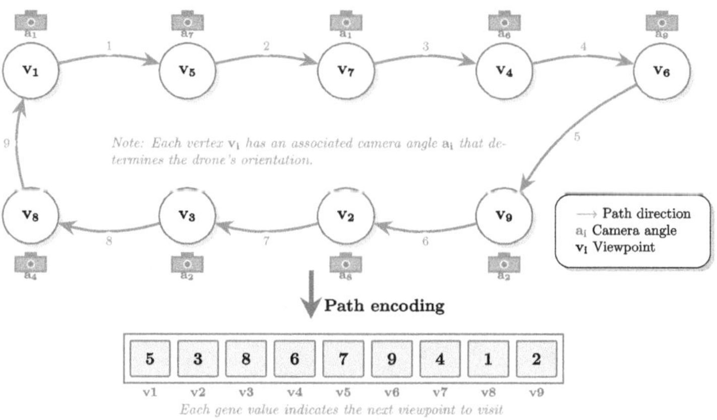

Fig. 3. Path encoding

4.2 Dual-Guided GA for Subproblem 1

Dual-Guided GA differs from the Standard GA (Sect. 4.1) only in its mutation operator (Algorithm 2), which exploits information from the Lagrangian relaxation through reduced costs. For each viewpoint v_k, the reduced cost is defined as $\tilde{c}_k = c_k^{base} - \lambda_k$, where λ_k is the associated dual variable and c_k^{base} represents the average transition cost to reachable neighbors:

$$c_k^{base} = \frac{1}{|N(v_k)|} \sum_{j \in N(v_k)} E(v_k, j),$$

where $N(v_k)$ is the set of reachable neighbors of v_k in the collision adjacency graph and $E(v_k, j)$ is the corresponding flight cost.

Given a chromosome at generation G, mutation probabilities are adjusted according to the reduced costs. For an unselected viewpoint ($y_k^G = 0$), the addition probability is $p_k^{\text{add}} = p_m \sigma(-\tilde{c}_k)$; for a selected viewpoint ($y_k^G = 1$), the removal probability is $p_k^{\text{drop}} = p_m \sigma(\tilde{c}_k)$, where p_m is the base mutation rate and $\sigma(\cdot)$ denotes the sigmoid function.

Algorithm 2. Dual-Guided Mutation Based on Reduced Cost

Require: Chromosome $y^{\mathcal{G}}$, reduced costs $\tilde{c}_k$, base rate p_m
Ensure: Mutated chromosome y^{new}

```
1: y^new ← y^G
2: for k ∈ V do
3:     p_k ← p_m · (y_k^new · σ(c̃_k) + (1 − y_k^new) · σ(−c̃_k))
4:     if rand() < p_k then
5:         y_k^new ← 1 − y_k^new
6:     end if
7: end for
8: return y^new
```

Theoretical Motivation for Dual-Guided Mutation. The dual-guided mutation is based on the reduced cost $\tilde{c}_k = c_k^{\text{base}} - \lambda_k^t$, where λ_k^t is the multiplier associated with viewpoint k at iteration t. Since the routing cost in the original problem is not separable by viewpoint, we approximate the contribution of each viewpoint by its average transition cost to neighbouring viewpoints, yielding the following linear surrogate objective:

$$F(y) = \sum_{k \in \mathcal{V}} \delta_k y_k + C, \qquad \delta_k = -\lambda_k^t + \gamma_k,$$

where

$$\gamma_k = c_k^{\text{base}} = \frac{1}{|N(k)|} \sum_{j \in N(k)} e_{kj},$$

and C is a constant independent of y.

Under this approximation, δ_k can be interpreted as the marginal contribution of viewpoint k to the surrogate objective. A negative δ_k indicates that selecting viewpoint k is beneficial, whereas a positive δ_k suggests that the viewpoint should be excluded. Consequently, the sign of δ_k provides a local search direction derived from the current dual information.

The proposed mutation operator increases the probability of flips that move y_k toward its preferred state (i.e., $y_k = 1$ when $\delta_k < 0$ and $y_k = 0$ when $\delta_k > 0$). Consider a simplified (1+1) elitist evolutionary process in which offspring replace the parent whenever the surrogate objective does not deteriorate. Let

$$Q^t = \sum_{k \in \mathcal{M}^t} |\delta_k|$$

denote a weighted error measure, where $\mathcal{M}^t$ contains the variables whose current values are inconsistent with the sign of δ_k.

The key observation is that each beneficial flip reduces both the surrogate objective and the error measure Q^t. Since the elitist acceptance criterion preserves non-deteriorating solutions, successful corrections accumulate over time. If beneficial flips occur with sufficiently high probability and mutation rates remain bounded, the expected decrease in Q^t is proportional to the number of remaining misaligned variables, leading to a contraction process in expectation. Under these assumptions, it can be shown that

$$\mathbb{E}[Q^{t+1}] \leq (1 - c^t)\,\mathbb{E}[Q^t],$$

for some contraction factor

$$c^t = p_b^t(1 - p_{\max}^t)^{n-1}.$$

Therefore, the expected error decreases geometrically over iterations. In contrast, standard uniform mutation yields a contraction factor proportional to $p_m(1 - p_m)^{n-1}$, which rapidly diminishes as the problem size increases. Consequently, the dual-guided mutation preserves stronger search pressure toward promising regions of the solution space and is expected to scale more effectively on large instances.

Although the complete UAV planning problem includes routing interactions that are not captured by the linear surrogate model, the reduced costs still provide informative estimates of variable desirability under the current dual prices. The analysis therefore establishes a theoretical link between dual information and mutation bias, while the computational results in Sect. 5 further demonstrate that the proposed mutation operator consistently accelerates convergence and improves solution quality across benchmark instances.

5 Experiments

5.1 Objective Function Normalization

The model in Sect. 3 combines two terms with different physical units and magnitudes: travel energy and reconstruction error. Direct optimization without scaling may allow one term to dominate, yielding mathematically optimal but practically undesirable solutions. To ensure balanced contributions from two objectives, we therefore employ the following normalized formulation:

$$\text{Minimize} \quad \underbrace{\frac{\sum e_{ij}x_{ij}}{E_{\max}}}_{\text{Normalized Energy}} + \underbrace{\frac{C_{\max} - \sum \omega_{\ell h}s_{\ell h}}{C_{\max}}}_{\text{Normalized Reconstruction Error}}$$

where the normalization constants are precomputed from problem data:

$$E_{\max} = \text{Energy of a full tour visiting all viewpoints}, \qquad C_{\max} = \sum_{\ell \in \mathcal{W}} \sum_{h \in \mathcal{H}_\ell} \omega_{\ell h}.$$

5.2 Experimental Setup

All experiments were conducted on a computer with an Intel Core i5 processor and 16 GB RAM. Algorithmic parameters were tuned using the Taguchi experimental design method [11]. The final settings were $P_0 = 100$, $G_{\max} = 300$, $S_{\max} = 40$, $p_c = 0.6$, $p_m = 0.2$, and $[\alpha_0, \alpha_{\max}] = [0.5, 5]$. The coverage redundancy parameter was set to $\rho = 1$, requiring each sampling point to be covered at least once.

To account for stochastic variability, each benchmark instance was independently run five times, and the mean and standard deviation were reported. All algorithms used identical initialization and termination conditions. The standard genetic algorithm (SGA) served as the baseline, while the dual-guided mutation genetic algorithm (DGM-GA) denotes the SGA enhanced with the proposed dual-guided mutation operator.

5.3 Dataset and Preprocessing

The building meshes used in this study are obtained from the Hong Kong Common Spatial Data Infrastructure [1], the UAV Pathplanning dataset [10], and the UrbanScene3D dataset [4]. Each instance contains sample points, candidate viewpoints with associated attributes, viewpoint adjacency information, and camera-pose coverage information. The viewpoint set is further augmented to ensure that each sample point is covered by multiple view frustums, yielding a complete instance that satisfies multi-view coverage requirements.

5.4 Results and Analysis

As shown in Table 2, DGM-GA consistently outperforms SGA across all tested instances, with greater performance gains observed on larger and more complex instances. A *Wilcoxon signed-rank test* [14] confirms the statistical significance of these improvements ($p < 0.001$). This demonstrates that the proposed mutation strategy is more effective in guiding the search process toward high-quality regions of the solution space.

In terms of computational efficiency, DGM-GA also exhibits a clear advantage, requiring substantially less runtime than SGA across all instances while achieving lower variability and greater stability. These findings indicate that the dual-guided mutation mechanism not only improves solution quality but also accelerates convergence. This aligns with the theoretical analysis (see Sect. 4.2), which shows that the dual-guided mutation maintains a contraction rate that does not diminish with problem size, unlike standard uniform mutation whose rate decays exponentially. Concretely, DGM-GA reduces average runtime by approximately 57% (small), 52% (medium), and 50% (large), demonstrating that the speedup persists even on large instances, as predicted by the theory.

To assess the absolute solution quality, we also solved the MILP formulation using Gurobi with a time limit of 6000 s per instance. This limit was chosen because the longest SGA run took approximately 5400 s, so we rounded up for

Table 2. Comparison of Standard Genetic Algorithm (SGA) and Dual-Guided Mutation GA (DGM-GA) with Gurobi Baselines

Instance	Gurobi		SGA		DGM-GA	
	MIP	Gap (%)	Fitness	Runtime (s)	Fitness	Runtime (s)
Small-scale ($\leq$ 170 viewpoints)						
building2	0.859027	20.72%	0.8616 ± 0.0342	440.22 ± 106.24	0.8531 ± 0.0114	239.42 ± 19.32
np_building3	0.845409	22.32%	0.8417 ± 0.0197	522.69 ± 92.57	0.8134 ± 0.0052	287.67 ± 24.53
Suzhou_293	0.994863	47.75%	0.9365 ± 0.0172	644.13 ± 306.37	0.8791 ± 0.0328	286.26 ± 90.24
Suzhou_353	0.94568	47.62%	0.8310 ± 0.0325	754.47 ± 264.16	0.8041 ± 0.0115	338.83 ± 114.59
building1	0.892978	21.44%	0.8551 ± 0.0170	772.32 ± 48.98	0.8363 ± 0.0122	333.56 ± 21.27
Shenzhen_163	1.072816	44.94%	1.0631 ± 0.0176	1153.84 ± 184.95	0.9204 ± 0.0149	361.34 ± 69.32
Mean (S)	*0.9351*	*34.13%*	$\mathbf{0.8982 \pm 0.0837}$	$\mathbf{714.61 \pm 282.63}$	$\mathbf{0.8511 \pm 0.0404}$	$\mathbf{307.85 \pm 67.96}$
Medium-scale (171–250 viewpoints)						
Shenzhen_44	0.910727	43.81%	0.9297 ± 0.0150	762.41 ± 267.46	0.8369 ± 0.0011	376.62 ± 39.00
Smith_NY	1.233463	38.14%	0.8641 ± 0.0105	1279.58 ± 381.03	0.8399 ± 0.0155	595.23 ± 22.97
Smith_GOTH	1.120369	31.36%	1.0421 ± 0.0024	2643.00 ± 3.56	0.9408 ± 0.0191	1531.41 ± 54.08
Suzhou_367	0.896232	37.38%	0.9305 ± 0.0183	1417.52 ± 1224.44	0.8805 ± 0.0124	524.08 ± 45.94
np_building1	0.995117	38.53%	1.0948 ± 0.0475	1394.03 ± 294.97	0.9150 ± 0.0020	684.04 ± 51.56
Suzhou_369	N/A	N/A	0.9854 ± 0.0175	1576.57 ± 329.71	0.9011 ± 0.0245	787.51 ± 210.18
np_building2	1.008322	37.83%	1.0415 ± 0.0079	2016.00 ± 633.55	0.9094 ± 0.0104	800.94 ± 119.44
Mean (M)	*1.0274*	*37.84%*	$\mathbf{0.9840 \pm 0.0815}$	$\mathbf{1584.25 \pm 642.78}$	$\mathbf{0.8891 \pm 0.0365}$	$\mathbf{757.12 \pm 358.44}$
Large-scale ($\geq$ 251 viewpoints)						
Suzhou_379	1.050578	49.59%	0.8612 ± 0.0117	1289.84 ± 373.05	0.8352 ± 0.0291	676.37 ± 237.10
Suzhou_363	1.52373	51.88%	0.8185 ± 0.0130	2920.88 ± 129.81	0.7911 ± 0.0448	2111.86 ± 584.30
Smith_UK	1.505698	43.67%	0.9223 ± 0.0290	5298.96 ± 84.75	0.9078 ± 0.0193	2950.77 ± 314.85
Suzhou_366	1.52072	61.97%	1.1178 ± 0.0590	1383.57 ± 52.81	0.9331 ± 0.0325	796.41 ± 70.49
Suzhou_378	1.489172	63.76%	1.0338 ± 0.0173	1321.21 ± 130.73	0.9173 ± 0.0181	780.75 ± 99.30
Suzhou_308	2.030927	74.59%	1.1870 ± 0.0938	1767.47 ± 602.49	0.9448 ± 0.0142	771.50 ± 91.95
Suzhou_377	2.251436	75.03%	1.0625 ± 0.0196	2405.26 ± 450.96	0.9596 ± 0.0498	957.07 ± 78.07
Suzhou_384	N/A	N/A	1.2099 ± 0.0177	3561.01 ± 1169.53	0.9406 ± 0.0029	1014.06 ± 133.43
Mean (L)	*1.6246*	*60.07%*	$\mathbf{1.0266 \pm 0.1478}$	$\mathbf{2493.53 \pm 1312.47}$	$\mathbf{0.9037 \pm 0.0621}$	$\mathbf{1257.35 \pm 796.21}$

a fair runtime comparison. Table 2 reports the best incumbent solution (MIP Obj) and the relative optimality gap (Gap), defined as 100 × (best bound − best incumbent)/best incumbent. When Gurobi could not find a feasible solution within the time limit or was terminated by the operating system, "N/A" is reported. In the small-scale group, Gurobi achieved an average MIP objective of 0.9351 with a mean gap of 34.13%, while SGA and DGM-GA obtained average fitness values of 0.8982 and 0.8511, respectively. Notably, DGM-GA outperforms the MIP incumbent in five of six small instances and matches or improves upon it in most medium and large cases. It achieves better or comparable fitness values while reducing runtime by more than 90% compared to the MIP time limit on small instances, and by 50–70% on larger ones. The mean gap rises to 37.84% for medium instances and 60.07% for large instances, indicating that Gurobi struggles to close the optimality gap as problem size grows. In contrast, DGM-GA maintains stable fitness values and low runtime variability, confirming that the proposed algorithm is both effective and scalable for large-scale 3D reconstruction planning.

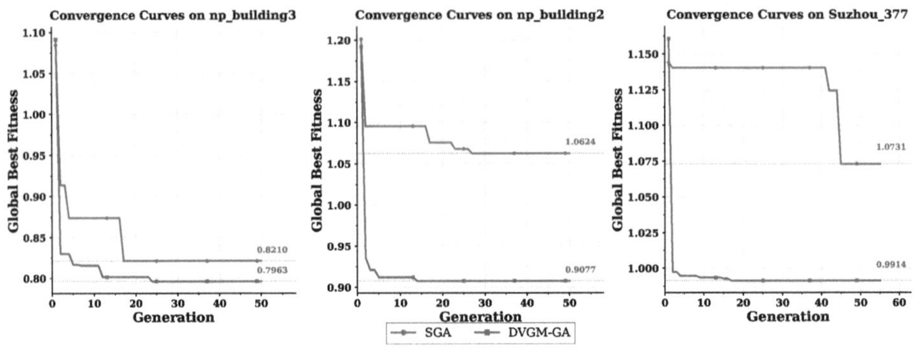

Fig. 4. The convergence graph of the genetic algorithm

Further insights can be obtained from the convergence behavior illustrated in Fig. 4, which shows the convergence status of instances np_building3, np_building2, and Suzhou_377. DGM-GA exhibits a faster convergence rate in early generations, while SGA shows slower improvement and tends to stagnate early. Moreover, DGM-GA continues to refine the solution and converges to a better final fitness value, whereas SGA remains at a higher plateau. This behavior indicates that the dual-guided mutation mechanism improves both exploration and exploitation. In addition, DGM-GA shows a smoother convergence trajectory with fewer fluctuations, reflecting better search stability. Overall, incorporating dual information into the mutation process enhances convergence efficiency and solution quality.

6 Conclusion

This paper presents a Lagrangian-relaxation-inspired framework for joint UAV trajectory and camera orientation optimization in 3D reconstruction, balancing coverage quality and energy efficiency. The hybrid framework, enhanced by a dual-guided evolutionary operator, substantially accelerates optimization. Compared to an exact MILP solver run to a time limit exceeding the longest GA runtime, our method consistently finds better or comparable solutions in small to medium instances, while Gurobi's optimality gap exceeds 60% on large problems. Specifically, DGM-GA achieves fitness values that are often lower (better) than the MIP incumbent, yet it finishes with a runtime reduction of over 90% compared to the MIP time limit. On average, the proposed DGM-GA reduces runtime by 51.3% and improves the weighted objective by 9.3% over a standard GA. The framework's principal strength lies in handling the complex interdependencies of path planning, viewpoint selection, and camera orientation. To our knowledge, this is the first work to simultaneously optimize these subproblems, overcoming the NP-hard intractability and enabling feasible large-scale applications such as aerial mapping and structural inspection.

Future work should aim to decouple the subproblems by solving Subproblem 2 independently, thereby enabling a direct comparison with a rigorous Lagrangian dual ascent and potentially yielding certified lower bounds.

Acknowledgments. This research was partially supported by Ningbo Tianyi Design Research of Surveying and Mapping Co. Ltd. under Project TYCH-2025-001 (Research on Trajectory Optimization Technology for UAV Nap-of-the-Object Photogrammetry).

References

1. Census and Statistics Department, The Government of the Hong Kong Special Administrative Region: Comprehensive social security assistance (2022). https://portal.csdi.gov.hk/csdi-webpage/, composite Social Data Infrastructure Portal
2. Gurobi Optimization, LLC: Gurobi Optimizer Reference Manual (2026). https://www.gurobi.com
3. Huang, R., Zou, D., Vaughan, R., Tan, P.: Active image-based modeling with a toy drone. In: 2018 IEEE International Conference on Robotics and Automation (ICRA), pp. 6124–6131. IEEE (2018)
4. Lin, L., Liu, Y., Hu, Y., Yan, X., Xie, K., Huang, H.: Capturing, reconstructing, and simulating: the urbanscene3d dataset. In: European Conference on Computer Vision, pp. 93–109. Springer, Cham (2022). https://doi.org/10.1007/978-3-031-20074-8_6
5. Lin, S., Kernighan, B.W.: An effective heuristic algorithm for the traveling-salesman problem. Oper. Res. **21**(2), 498–516 (1973)
6. Maboudi, M., Homaei, M., Song, S., Malihi, S., Saadatseresht, M., Gerke, M.: A review on viewpoints and path planning for uav-based 3-d reconstruction. IEEE J. Selected Top. Appl. Earth Observations Rem. Sens. **16**, 5026–5048 (2023)
7. Martin, R.A., Rojas, I., Franke, K., Hedengren, J.D.: Evolutionary view planning for optimized uav terrain modeling in a simulated environment. Rem. Sens. **8**(1), 26 (2015)
8. Roberts, M., et al.: Submodular trajectory optimization for aerial 3d scanning. In: 2017 IEEE International Conference on Computer Vision (ICCV), pp. 5334–5343 (2017). https://doi.org/10.1109/ICCV.2017.569
9. Sharma, O., Arora, N., Sagar, H.: Image acquisition for high quality architectural reconstruction. In: Graphics Interface, pp. 18–1 (2019)
10. Smith, N., Moehrle, N., Goesele, M., Heidrich, W.: Aerial path planning for urban scene reconstruction: a continuous optimization method and benchmark. ACM Trans. Graph. **37**(6) (2018). https://doi.org/10.1145/3272127.3275010
11. Taguchi, G.: Introduction to Quality Engineering: Designing Quality into Products and Processes, Asian Productivity Organization (1986)
12. Vansteenwegen, P., Souffriau, W., Van Oudheusden, D.: The orienteering problem: A survey. Eur. J. Oper. Res. **209**(1), 1–10 (2011)

13. Whitley, L.D., Starkweather, T., Fuquay, D.: Scheduling problems and traveling salesmen: The genetic edge recombination operator. In: International Conference on Genetic Algorithms (1989). https://api.semanticscholar.org/CorpusID:12884640
14. Wilcoxon, F.: Individual comparisons by ranking methods. Biometrics Bull. **1**(6), 80–83 (1945). https://doi.org/10.2307/3001968

Determining Destroy Sets in Large Neighborhood Search by Generative Flow Networks

Maria Bresich[1(✉)], Jingyi Peng[1], Günther R. Raidl[1], and Steffen Limmer[2]

[1] Institute of Logic and Computation, TU Wien, Vienna, Austria
{mbresich,raidl}@ac.tuwien.ac.at, e12402257@student.tuwien.ac.at
[2] Honda Research Institute Europe GmbH, Offenbach/Main, Germany
steffen.limmer@honda-ri.de

Abstract. Large Neighborhood Search (LNS) is an often applied metaheuristic for solving challenging combinatorial optimization problems. Typically, it relies on the scheme of iteratively removing parts of a solution by one or more *destroy* operators and cleverly augmenting the solution again by *repair* operators. Destroy operators are usually handcrafted and often a mixture of uniform random selection and problem-specific heuristics, and their choice can dramatically impact the performance. Inspired by the recent success of Generative Flow Networks (GFlowNets), we propose their usage to compile promising destroy sets. These generative deep learning models stand out in that their goal is not to learn a single high-reward structure but to produce diverse structures according to a complex probability density function. As considering many different but promising destroy sets is fundamental in LNS, we believe GFlowNets to be well suited for this purpose. We study an intentionally generic and simple LNS framework and apply it to three hard graph problems. The only problem-specific part is the repair operator that is realized by basic iterated random construction heuristics. When applying the best GFlowNet model, the effectiveness of the LNS is on average increased by 6.52% in terms of online performance compared to a corresponding baseline LNS. Moreover, a comparison to integer linear programming models solved by Gurobi shows that they are clearly dominated by our GFlowNet-guided LNS on the considered hard benchmark instances with a mean relative improvement of 20.73% in terms of final solution quality.

Keywords: Generative Flow Networks · Large Neighborhood Search

1 Introduction

Large neighborhood search (LNS) [2,27] is a highly popular metaheuristic for addressing challenging combinatorial optimization problems (COPs), with applications in vehicle routing, scheduling, timetabling, and location planning, see

G. Iacca et al. (Eds.): PPSN 2026, LNCS 16985, pp. 265–281, 2027.
https://doi.org/10.1007/978-3-032-36214-8_17

e.g., [5,14,20,21,26]. The key idea is to explore large neighborhoods of an incumbent solution instead of just small local changes as in local search. Most variants rely on the iterative *destroy and repair* scheme: a current solution is partially destroyed by removing a subset of its elements (or un-assigning corresponding decision variables), and then repaired either by heuristic or exact methods. Acceptance can be based on non-deterioration or more sophisticated criteria, and the process is repeated until a termination criterion is met. The performance of such an LNS depends heavily on the choice of its destroy and repair operators. A substantial variability of destroy sets across iterations is crucial to enable diversification and a good coverage of the search space. Destroy operators often use simple uniform random selection or are problem-tailored by domain experts.

The manual design and tuning of destroy operators, however, is time-consuming, and an "ideal" destroy operation may also depend more generally on the current state of the whole search. This limitation has fueled learning-based approaches for selecting the elements to destroy within an LNS [1,6,8,9,23,28,29]. Such neural LNS approaches typically use graph neural networks (GNNs) trained either in supervised fashion on labeled data obtained by time-expensive calculations or by reinforcement learning (RL) [30]. All these approaches suffer from what is known as *mode collapse* in generative machine learning [11]: For a given current solution, the learned model tends to repeatedly produce the same (or a similar) destroy set—more generally called *mode*—despite the existence of usually many different highly promising modes. This can hinder escaping local optima if this destroy set fails to improve the solution. The proposed methods therefore add random noise or use problem-specific tricks to increase destroy set variability, but this does not really address the issue of mode collapse. Ideally, the model should be able to produce diverse high quality destroy sets covering all or at least many promising modes on its own.

A relatively new family of generative models that is explicitly designed to produce structures according to a complex probability density function and thus avoid mode collapse is that of Generative Flow Networks (GFlowNets) [3,4]. Recent applications to COPs [17,33–35] have demonstrated their ability to sample high-quality solutions with high diversity. We therefore propose and investigate the usage of GFlowNets for constructing promising destroy sets within an LNS framework that we intentionally keep simple and generic to support diverse COPs. The only problem-specific components are the repair operator, realized via a simple iterated random construction heuristic, and the GFlowNet's underlying neural network. GFlowNets operate in an auto-regressive manner following a Markov decision process (MDP), which we tailor for use within the LNS. We intentionally minimize feature usage and investigate different reward assignments to enable effective training. Our GFlowNets use graph isomorphism networks and are trained offline on representative unlabeled problem instances. We demonstrate the benefits and generality of the approach through extensive experiments on a hard variant of the maximum independent set (MIS), the maximum cut (MAXCUT), and the densest k-subgraph (DkS) problems across three graph types. These COPs are classical benchmarks and many real-world problems can be modeled similarly as graph problems. We compare variants of the GFlowNet-guided LNS (GFN-LNS) against a classical baseline LNS and integer linear programming (ILP) models solved by Gurobi. GFN-LNS is shown to mostly outperform both baselines in terms of final solution quality and convergence speed (online performance) despite the relative simplicity of the setup, with the GFlowNet overhead justified by the gains. Additional tests on smaller and larger instances than those used for training further highlight the generalization capabilities of GFN-LNS. In summary, our main contributions are:

- A generic GFlowNet-guided LNS framework with model-based destroy set generation applicable to diverse COPs.
- A problem-agnostic MDP with effective reward assignments for training.
- A comprehensive evaluation of the approach on three hard graph problems across different graph types, demonstrating advantages over both baseline LNS and ILP approaches as well as strong generalization capabilities.

The next section provides necessary preliminaries and related work and specifies the COPs for benchmarking. Section 3 describes our GFlowNet-guided LNS framework, including the baseline LNS approach and detailing the MDP formulation and training procedure. Section 4 presents and discusses experimental results, and Sect. 5 concludes the paper, also outlining promising future work.

2 Preliminaries

We now provide background information on GFlowNets and related LNS approaches and introduce the benchmark problems we consider.

2.1 Generative Flow Networks

This relatively new family of stochastic generative deep learning models has been introduced in 2021 by Bengio et al. [3]. While bearing similarities with RL, they are trained to sample composite objects x of some object family $\mathcal{X}$ with probabilities proportional to a given non-negative reward function $R(x) \geq 0$. In contrast, standard RL seeks to maximize expected cumulative reward and thus tends to concentrate probability mass on a single high-reward output.

Formal Definitions. A GFlowNet operates on a directed acyclic graph (DAG) $\mathcal{G} = (\mathcal{S}, \mathcal{A})$, where $\mathcal{S}$ is a finite set of *states* and $\mathcal{A} \subseteq \mathcal{S} \times \mathcal{S}$ is a set of *transitions* (actions). A unique *source state* $s_0 \in \mathcal{S}$ has no incoming edges, while *terminal states* $x \in \mathcal{X} \subseteq \mathcal{S}$ have no outgoing edges. A trajectory $\tau = (s_0, s_1, \ldots, s_{n_\tau} = x)$ is a path in $\mathcal{G}$ from the source to a terminal state and represents the sequential construction of a specific object x. The model maintains a *forward policy* $P_F(s' \mid s)$, which is a probability distribution over the successor states s' that can be reached from each non-terminal state s, as well as a *backward policy* $P_B(s \mid s')$ indicating the probability that state s' is reached from one of its predecessor states s. A non-negative *flow function* $F : \mathcal{S} \to \mathbb{R}_{\geq 0}$ assigns an amount of flow to each state, with terminal states satisfying $F(x) = R(x)$ and the source node having the total flow $F(s_0) = Z = \sum_{x \in \mathcal{X}} R(x)$. Flow conservation requires that for every interior state $s \in S \setminus \{s_0\} \setminus \mathcal{X}$

$$\sum_{s' :\, (s,s') \in \mathcal{A}} F(s)\, P_F(s' \mid s) \;=\; \sum_{s'' :\, (s'',s) \in \mathcal{A}} F(s'')\, P_F(s \mid s'') \tag{1}$$

holds. A GFlowNet satisfying this condition provably samples each terminal state x with probability $P(x) = R(x)/Z$ [3].

Training Objectives. In practice, P_F, P_B, and F are parameterized by neural networks that take a state s as input and output transition probabilities or log-flows. GFlowNets are then trained by considering many trajectories in conjunction with their rewards and trying to establish the above conditions across these trajectories by updating the models' parameters using classical stochastic gradient descent. This can be done in different ways, and similarly to [34], we consider the following established loss functions as alternatives.

The *detailed balance (DB) loss* [4] aims at achieving $F(s)\, P_F(s' \mid s) = F(s')\, P_B(s \mid s')$ at each transition (s, s') of a considered trajectory. For numerical stability, this is expressed in log-domain and deviations are squared, yielding

$$\mathcal{L}_{\mathrm{DB}}(s, s') = \left(\log F(s) + \log P_F(s'|s) - \log F(s') - \log P_B(s|s')\right)^2 . \tag{2}$$

The *forward looking (FL) loss* [24] is a variation of the DB loss in which *intermediate rewards* are also considered, with the hope that they lead to faster convergence. It is defined as

$$\mathcal{L}_{\mathrm{FL}}(s, s') = \left(\log F(s) + \log P_F(s'|s) - \log F(s') - \log P_B(s|s') + \mathcal{E}(s \to s')\right)^2 , \tag{3}$$

where $\mathcal{E}(s \to s')$ is a log-reward contribution given for each transition in a trajectory; it is assumed that these contributions are additive and add up to $\log R(x)$.

The *trajectory balance (TB) loss* [22] avoids a general model for state flows $F(s)$ but just learns the total flow Z as scalar in addition to the forward and backward policies and tries to achieve consistency across complete trajectories by aiming for $Z \cdot \prod_{t=1}^{n_\tau} P_F(s_t \mid s_{t-1}) = R(x) \cdot \prod_{t=1}^{n_\tau} P_B(s_{t-1} \mid s_t)$. Expressing this in log-domain and squaring deviations yields

$$\mathcal{L}_{\mathrm{TB}}(\tau) = \left(\log Z + \sum_{t=1}^{n_\tau} \log P_F(s_t|s_{t-1}) - \log R(x) - \sum_{t=1}^{n_\tau} \log P_B(s_{t-1}|s_t)\right)^2 . \tag{4}$$

Training with TB may converge faster than with DB because the credit-assignment signal is distributed over an entire trajectory rather than individual edges [22]. On the downside, TB in general exhibits higher variance and can be less stable than DB.

Training Regimes. GFlowNets support both *off-policy* and *on-policy* training. For off-policy, trajectories are collected under a behavior policy that differs from P_F and are stored in a replay buffer for repeated use. In the on-policy setting, trajectories are generated by rolling out the current P_F. This flexibility and the general robustness of training are notable advantages over many RL algorithms, where off-policy training is not always possible or requires significant algorithmic overhead. Section 3.3 will describe our specific training procedure in the context of the LNS framework, which essentially is on-policy with small adaptations.

GFlowNets for Combinatorial Optimization. Zhang et al. [34] demonstrated that GFlowNets equipped with GNNs can generate high-quality heuristic solutions for NP-hard COPs on graphs, including the MIS, MAXCUT, and minimum dominating set problems. By training a GFlowNet to construct solutions autoregressively, with rewards derived from the objective value, the authors showed that the learned sampler produces competitive solutions while maintaining a high level of diversity. Building on this line of work, GFlowNets have since been applied to further combinatorial domains, including scheduling [33] and vehicle routing [35]. Extensions of GFlowNets to stochastic environments [15,25] are also relevant in the context of combinatorial optimization.

2.2 Machine Learning Based Destroy Operators in LNS

Related to our work, deep learning techniques have recently already been used for selecting which parts of a current solution to destroy within an LNS framework. Song et al. [28] proposed a general RL framework for learning which variables to fix or branch on within an ILP-guided LNS, while Addanki et al. [1] and Chen et al. [6] train (recurrent) GNNs to learn problem-specific destroy heuristics for integer programs and vehicle routing, respectively. Sonnerat et al. [29] and Oberweger et al. [23] train generative models based on GNNs via imitation learning to approximate an expert policy that optimally selects variables for re-optimization in LNS. A more advanced contrastive learning based approach is proposed by [12], while Cherrered et al. [7] employ graph attention networks to integrate node- and route-level features for more impactful removals. All these approaches potentially suffer from the issue of mode collapse, although they apply different tricks to increase the variability of the generated destroy sets. Moreover, they all rely on the node-wise logits from a single model pass to select a destroy set. While this is computationally efficient, it also implies that different modes are naively mixed, and specific dependencies between elements are not well captured. These weaknesses are directly addressed by the GFlowNet-based approach we propose. Another promising direction is the use of denoising diffusion models to generate destroy sets, as proposed in [8]. Like GFlowNets, they can also capture many modes, but a large number of necessary inferencing steps for the iterative denoising may lead to high computational overhead.

2.3 Considered Benchmark Problems

To assess the generality of our approach, evaluation is done on three different NP-hard COPs on graphs that are classical benchmarks in combinatorial optimization and cover different types of constraints and objective structures: a

MIS-BNB

$$\begin{aligned} \max\ & \textstyle\sum_{v\in V} x_v & \\ \text{s.t. } & x_u + x_v \le 1 & (u,v)\in E \\ & b_u \ge x_v & u\in V,\ v\in N(u) \\ & \textstyle\sum_{u\in V} b_u \le B & \\ & x_v, b_v \in \{0,1\} & v\in V \end{aligned}$$

MAXCUT

$$\begin{aligned} \max\ & \textstyle\sum_{(u,v)\in E} y_{u,v} & \\ \text{s.t. } & y_{u,v} \le x_u + x_v & (u,v)\in E \\ & y_{u,v} \le 2 - (x_u + x_v) & (u,v)\in E \\ & x_v \in \{0,1\} & v\in V \\ & y_{u,v} \in \{0,1\} & (u,v)\in E \end{aligned}$$

DkS

$$\begin{aligned} \max\ & \textstyle\sum_{(u,v)\in E} y_{u,v} & \\ \text{s.t. } & y_{u,v} \le x_u,\ y_{u,v} \le x_v & (u,v)\in E \\ & \textstyle\sum_{v\in V} x_v = k & \\ & x_v \in \{0,1\} & v\in V \\ & y_{u,v} \in \{0,1\} & (u,v)\in E \end{aligned}$$

Fig. 1. ILP models for the three considered graph problems.

variant of the MIS, the MAXCUT, and the DkS problem [10,16,19]. All problems are defined on a simple undirected graph $G = (V, E)$ with $n = |V|$ nodes and $m = |E|$ edges. We briefly explain the problems in the following, and corresponding basic ILP models, which we also use as a baseline for comparison in the experiments, are given in Fig. 1. In all these models, binary decision variables x_v indicate with value one that node $v \in V$ is part of solution set S.

Maximum Independent Set with Budgeted Neighborhood Blocking (MIS-BNB). An independent set (IS) is a subset $S \subseteq V$ where no two vertices are adjacent. The classical MIS problem seeks to maximize $|S|$ and is well solvable by modern ILP solvers and specialized algorithms such as KaMIS [18] on many practically relevant instance classes. We therefore consider a more challenging variant with a *blocking budget* $B \ge 0$ similar to the large secluded IS problem [31]. A vertex $u \notin S$ is *blocked* if it has at least one neighbor in S, and the goal is to find a MIS such that the total number of blocked vertices does not exceed B. The global budget constraint yields a weak LP relaxation that makes this variant substantially harder for ILP solvers as well as heuristics. In the ILP model, additional binary variables b_v are used to indicate blocked vertices.

Maximum Cut (MAXCUT). A cut of a graph is a subset of edges whose removal disconnects the graph into two components V_1 and $V_2 = V \setminus V_1$. The MAXCUT problem seeks a set V_1 that maximizes the number of edges in the cut, i.e., between V_1 and V_2. In the ILP model, additional binary variables $y_{u,v}$ indicate whether edge (u, v) is in the cut.

Densest k-Subgraph (DkS). Given a parameter $k \le n$, the DkS problem seeks a vertex set $S \subseteq V$ with $|S| = k$ that maximizes the number of edges in the induced subgraph $G[S]$ [19]. In the ILP model, additional binary variables $y_{u,v}$ indicate whether both endpoints of edge (u, v) are selected in S.

3 GFlowNet-Guided LNS (GFN-LNS)

We first describe the generic LNS framework that serves as our baseline and which will subsequently be enhanced by replacing the destroy operator with a learned GFlowNet. The Markov decision process (MDP) formulation for training the GFlowNet is then detailed, including the specific reward design that we use. The training process is explained, and we also provide details on the GFlowNet architecture and implementation.

3.1 Large Neighborhood Search Framework

We employ a standard LNS framework as described before, and in our setting, the decision variables correspond to nodes in the problem graph $G = (V, E)$. The destroy operator selects a subset $D \subseteq S$ of nodes from the current solution set $S \subseteq V$ and removes them, yielding a partial solution. The repair operator augments the remaining solution with additional nodes to obtain a new complete feasible solution, which is accepted if it is not worse than the current solution. As a baseline, we use a destroy operator that selects a fixed number of nodes from S uniformly at random. For the repair operator, we employ a simple and also rather generic *iterated random repair* strategy. In each iteration, this method considers all nodes not currently in the solution set in random order and inserts each node provided that the problem-specific feasibility constraints are not violated and the objective value does not decrease. The iterated random repair executes this random fill procedure r times independently, each time starting from the same destroyed partial solution, and returns the best solution obtained across all r repetitions. This repetition reduces the variance inherent in the randomized construction and increases the probability of finding a high-quality completion. The initial solution for the LNS is constructed by applying random repair to the empty solution.

3.2 GFlowNet-Based Destroy Set Selection

The main goal of this work is to improve upon the basic random destroy operator by selecting the subsets of nodes for removal more meaningfully and targeted. We train and apply a GFlowNet to choose in each iteration the elements to be destroyed one-by-one in an auto-regressive manner until a predefined number of nodes is reached. In this study, this number is fixed to keep the setup simple and comparable to the baseline. Note that auto-regressive construction of the destroy set allows the GFlowNet to capture complex dependencies between nodes, which we also see as a key advantage over approaches that select destroy sets based on the node-wise logits of a single GNN inference. The downside, however, is the increased computational overhead, which we also analyze in the experiments.

MDP Formulation and Reward Design. A GFlowNet trajectory corresponds to one isolated LNS-iteration consisting of a single destroy and repair operation as well as the acceptance decision for the candidate solution. The GFlowNet's DAG for construction of a destroy set is modelled as an MDP with states, actions, transitions, and a reward function. In this context, a *state* $s_t = (G, S, D_t)$ consists of the problem instance graph G, the current solution $S \subseteq V$, and the so far compiled (partial) destroy set $D_t \subseteq S$. Optionally, we provide information on earlier applied destroy sets as additional destroy history feature $\mathrm{EMA}^{\mathrm{sel}}$: Specifically, we store for each node an exponential moving average (EMA) of whether a node was selected for removal (value one) or not (value zero) in previous LNS iterations. The state representation and feature set are intentionally minimalistic and generic as we want to investigate the GFlowNet's abilities under raw basic conditions. However, providing further features either on global aspects of the search, such as the iteration number and aspects on convergence, or problem-specific features, such as centrality measures or the number of selected neighbors in S, would likely further improve the GFlowNet's performance.

The initial (source) state $s_0 = (G, S, D_0 = \emptyset, \mathrm{EMA}^{\mathrm{sel}})$ corresponds to the current solution of the LNS iteration and an empty destroy set. A terminal state is reached when $|D_t| = d$ for a predefined number of nodes d and is associated with the solution obtained from the repair operation after performing the destroy and the subsequent acceptance decision. An *action* $a_t = v \in S \backslash D_t$ represents the selection of a node v from the current solution that has not been picked yet for removal, and *transitions* always correspond to the update of the state by adding the selected node to the destroy set, i.e., $D_{t+1} = D_t \cup \{v\}$. Non-terminal successor states are deterministic, whereas terminal successor states are stochastic due to the randomized repair operator. This stochasticity is mitigated by employing iterated random repair which reduces the variance.

The *reward function* is a mapping $R : \mathcal{X} \to \mathbb{R}_{>0}$ of terminal states to positive numbers, defined as $R(x) = e^{-\mathcal{E}(S)/T}$ with $\mathcal{E}(s)$ denoting a problem specific energy function on the solution S associated with terminal state s and T denoting a temperature parameter [34]. If the candidate solution after the repair step is not accepted, the reward is set to one, which corresponds to an energy of zero. Otherwise, we base the reward and energy calculation on the improvement achieved by the repair step, i.e., the difference in objective value $\Delta = \mathrm{obj}(S) - \mathrm{obj}(S')$ between the repaired solution S and the initial solution S' of the iteration, where obj denotes the problem-specific objective function. In LNS, it is important to consider that finding large improvements over low quality (initial) solutions early on may be trivial whereas a small improvement over an already high quality solution near the end may be much harder to find and more valuable. Thus, the value of an improvement depends on the quality of the initial solution but also on the instance size, and rewards should ideally be normalized to be comparable across different instances and sizes. Using the negated raw improvement $-\Delta$ as energy does not reflect this context awareness and could be an insufficient base for reward calculation. As we train on homogeneous instances of equal size, a fixed global constant can be determined in advance as a baseline b for the objective value of high quality solutions. We thus propose a normalization procedure based on the remaining gap of the trajectory's initial solution to this baseline, $gap = b - \mathrm{obj}(S')$, with a lower bound of one to prevent shrinking rewards when $\text{obj}(S') > b$. The normalized improvement is also capped at one to treat improvements exceeding the gap equally, i.e., $\Delta_{\text{norm}} = \min(\Delta / \max(gap, 1), 1)$. To further ensure that even small improvements receive a noticeable reward and to provide a learning signal that clearly distinguishes between improving and non-improving actions, we extend the energy calculation to a hybrid approach using a weighted sum of the gap-based normalized improvement and a binary indicator for whether the solution was improved, such that

$$-\mathcal{E}(S) = \begin{cases} 0 & \text{if } S \text{ is not accepted,} \\ \alpha \cdot \mathbb{I}[\Delta > 0] + (1-\alpha) \cdot \Delta_{\text{norm}} & \text{if } S \text{ is accepted} \end{cases} \tag{5}$$

with $\alpha \in [0, 1]$ denoting a weight parameter. The case of $\alpha = 1$ corresponds to relying on the binary improvement indicator alone, to which we will also compare the hybrid approach in the evaluation.

3.3 GFlowNet Training

The training of the GFlowNet is done in an on-policy manner where trajectories are iteratively generated by running our GFN-LNS on always different training instances. In each iteration, the destroy set elements are sampled according to probabilities provided by the current forward policy P_F. Depending on the employed training objective, training samples are derived either from complete trajectories (when using TB) or by taking all transitions from each trajectory (for DB and FL) of a batch. As an LNS run usually performs a large number of iterations, it is expected that only a relatively small fraction of iterations will yield improvements, thus resulting in an excess of unsuccessful trajectories. To mitigate this data imbalance, we reduce the number of unsuccessful trajectories from an LNS run by first removing all trajectories after the last successful LNS iteration as we cannot be sure that any further improvement is possible. If sequences of unsuccessful iterations exceed a given length threshold γ (in our experiments $\gamma = 10$), we further only keep γ randomly sampled ones from these.

We consider the loss functions TB, DB, and FL already explained in Sect. 2.1. In case of the FL loss, we linearly interpolate from zero to the terminal reward for determining the corresponding log-reward contributions $\mathcal{E}(s \to s')$ for each trajectory. Before training, we run our baseline LNS (cf. Sect. 3.1) on a set of representative problem instances for two purposes: The initial states of successful trajectories are collected to be used as independent validation set for regularly evaluating the model during training. We also compute the mean objective value ($\overline{\text{obj}}$) and standard deviation ($\sigma(\text{obj})$) of the best solutions found across all runs and derive the baseline b for the energy calculation as $b = \overline{\text{obj}} + \sigma(\text{obj})$.

The GFlowNet is trained for a specified number of iterations. In each iteration, a new batch of instances is created and training trajectories are generated via on-policy rollouts of GFN-LNS. This coupling between data generation and the model to be trained may enable better and more fitting training data over time as well as a better trained model in the end. In each training iteration, a predefined number of update steps is performed on mini-batches of the collected trajectories employing one of the loss functions. To improve learning stability and convergence speed, we track an EMA of all seen rewards and center the rewards of each batch by subtracting this EMA. At regular iteration intervals, the GFlowNet is evaluated on the validation set and metrics such as the averages of loss, objective value, and rewards are measured. The training always completes all iterations and the final model to be used is selected based on the best validation performance across iterations, similarly to an early stopping approach.

GFlowNet Configuration. For the neural networks within the GFlowNet, we use an architecture similar to the one proposed by Zhang et al. [34] and refer to this work for more details. For the forward policy P_F and state flow function F when using DB or FL loss, we employ independent graph isomorphism networks (GINs) [32] with five convolutional layers, a hidden dimension of 256, and sum aggregation. The backward policy is defined as a uniform distribution over all possible feasible predecessor states. As in [34], we use the Adam optimizer with a learning rate of 10^{-3} for all learnable functions P_F, F, and Z. After preliminary tests, the temperature parameter T for reward calculation was set to 0.1. Training is done for 3000 iterations with 20 update steps each, using a batch size of 32, and we evaluate the model every 10 iterations.

4 Experiments

All solution approaches were implemented in Julia 1.12.5 and we used Gurobi 13.0.0[1] in single-threaded mode as ILP solver with a memory limit of 10 GB. The experiments were performed on a server with an NVIDIA A40 GPU with 45 GB memory and single cores of 4.05 GHz AMD EPYC 9274F processors. For training and benchmark instances, we consider random graphs of type Erdős-Rényi (ER), Random k-Regular (RR), and Barabási-Albert (BA) graphs, which are commonly used in the literature and cover diverse structural properties, and apply Julia's `Graphs.jl` package to create corresponding instances. We train a separate GFlowNet model for each combination of problem and graph type which takes between 6 and 47 h wall-clock time on the given hardware depending on the problem-graph configuration and especially the employed loss function, with TB loss being the slowest because of worse sample efficiency. Training graphs have $n = 300$ nodes and the following type-specific parameters: an edge probability $p_e = 0.1$ for ER, a degree $k = 4$ for RR (increased to $k = 20$ for DkS to get harder instances), and an attachment parameter $m = 10$ for BA. The values of problem-specific parameters, i.e., the blocking budget B for MIS-BNB and the subgraph size k for DkS, depend on the graph type and are chosen to yield instances that are generally hard to solve. The considered values are as follows: for MIS-BNB, we use $B = \lfloor \frac{2}{3}n \rfloor$ for ER, $B = \lfloor \frac{1}{3}n \rfloor$ for RR, and $B = \lfloor \frac{1}{2}n \rfloor$ for BA; for DkS, we use $k = 0.2n$ for ER and RR and $k = 0.4n$ for BA. The same

[1] https://www.gurobi.com.

configurations are used for generating 100 independent benchmark instances of each problem and graph type combination for evaluation.

We compare multiple variants of our GFN-LNS employing the different training objectives TB, DB, and FL to two baselines: the ILP models solved by Gurobi and the classic LNS with a uniform random selection of destroy sets, which we refer to as ILP and LNS-rand, respectively. Since finding high quality solutions fast is challenging but also essential in practice, we set a time limit of three seconds for all methods and instances. This is also to keep training efforts reasonable, but in general, consideration of longer run times would be possible. For the LNS-based approaches, a fixed destroy set size $d = 5$ is used in training as well as testing and across all configurations, and the number r of repetitions for the iterated random repair is varied depending on the problem to achieve a good balance between solution quality and run time; these values were determined in preliminary experiments. For testing, we use $r = 2000$ for MIS-BNB, $r = 1000$ for MAXCUT, and $r = 8000$ for DkS, and 1/10 of these iterations are used for training to speed up the process. Parameters specific to the GFlowNet training include the weight parameter $\alpha = 0.1$ for the energy calculation, and parameters for the EMA calculations for the history feature and the reward centering, which are set to 0.9 and 0.99. Again, these values were determined in preliminary experiments.

4.1 Results

In the following evaluations, we employ as main metrics the final solution quality, indicated by the mean objective value ($\overline{\text{obj}}$) and standard deviation ($\sigma(\text{obj})$) of the best solutions found across all 100 runs on the test instances, and the online performance in terms of the area under the curve (AUC) of the achieved objective value over time. In tables, bold values indicate best values, and also the average number of executed iterations is given as I^{total} for the LNS approaches. Shaded areas in the plots always illustrate the standard deviation across runs.

Table 1 compares the performance of our three main GFN-LNS variants employing the different loss functions to each other and the baselines across all combinations of problems and graph types. We observe that the ILPs yield the highest variance in all cases and perform notably worse than all LNS approaches, e.g., TB-based GFN-LNS achieves on average 20.73% better objective values. Even when increasing the time limit for Gurobi tenfold in additional experiments–given as ILP-30 in italics in the table and running out of competition due to the extended time limit–, it is still outperformed by GFN-LNS within three seconds in 2/3 of the cases. This confirms the practical hardness of the considered instances and motivates the usage of heuristic methods. Next, we observe that in 7/9 cases, GFN-LNS beats LNS-rand regarding final solution quality, with an additional win when considering the AUC, in terms of which GFN-LNS with TB loss performs on average 6.52% better. The first result shows that the benefit of the learned destroy sets is large enough to justify the time overhead induced by inferencing the models. This overhead is significant and decreases the number of executed iterations on average by 38% compared to LNS-rand, which

Table 1. Results on benchmark instances for all problems (MIS-BNB, MAXCUT, DkS) across three different types of graphs (ER, RR, and BA) with 300 nodes.

Prob	Method	ER				RR				BA			
		$\overline{\text{obj}}$	σ(obj)	AUC[%]	I^{total}	$\overline{\text{obj}}$	σ(obj)	AUC[%]	I^{total}	$\overline{\text{obj}}$	σ(obj)	AUC[%]	I^{total}
MIS-BNB	ILP	12.00	1.18	–	–	52.66	1.66	–	–	34.72	3.55	–	–
	ILP-30	*18.50*	*1.30*	–	–	*53.76*	*1.32*	–	–	*42.14*	*2.57*	–	–
	LNS-rand	**20.65**	0.83	88.27	241	53.49	1.37	90.54	339	42.58	2.61	92.03	254
	GFN-LNS-TB	20.60	0.85	**90.45**	155	53.23	1.32	89.33	185	**42.73**	2.56	**94.98**	157
	GFN-LNS-DB	20.53	0.95	87.99	157	53.45	1.38	90.12	185	42.36	2.54	90.92	157
	GFN-LNS-FL	20.36	0.97	83.22	154	**53.70**	1.35	**91.19**	184	42.17	2.50	88.37	160
MAXCUT	ILP	2470.87	41.20	–	–	489.40	10.00	–	–	1659.57	23.08	–	–
	ILP-30	*2470.87*	*41.20*	–	–	*513.30*	*3.28*	–	–	*1665.01*	*26.34*	–	–
	LNS-rand	2787.07	36.43	80.89	239	**507.06**	3.89	**85.14**	508	1867.25	11.12	84.36	300
	GFN-LNS-TB	**2799.79**	36.97	89.95	157	505.14	4.55	84.51	205	**1868.48**	13.47	**86.53**	175
	GFN-LNS-DB	2797.27	36.63	**90.42**	157	505.34	4.85	79.08	204	1864.40	12.51	82.42	177
	GFN-LNS-FL	2785.30	35.53	81.17	158	505.70	3.94	81.88	205	1864.99	11.95	81.44	177
DkS	ILP	300.63	11.48	–	–	178.06	7.62	–	–	977.32	68.74	–	–
	ILP-30	*329.84*	*11.09*	–	–	*221.63*	*6.08*	–	–	*1103.59*	*3.31*	–	–
	LNS-rand	339.22	8.25	80.28	170	224.32	4.34	70.41	182	1067.39	6.49	81.62	179
	GFN-LNS-TB	348.21	10.26	86.05	121	236.85	4.00	83.21	124	**1102.45**	3.09	**94.63**	126
	GFN-LNS-DB	**348.26**	8.74	**86.24**	120	**245.38**	4.27	**87.36**	124	1100.86	3.30	92.40	125
	GFN-LNS-FL	339.87	8.71	80.95	120	226.72	4.78	71.32	124	1100.61	3.31	93.46	125

is usually detrimental to LNS performance but the improvements enabled by the better selection of destroy sets make up for it. That larger improvements are found quicker by GFN-LNS is also reflected in the second observation regarding the better online performance, which is exemplarily illustrated in Fig. 2. The only case where no GFN-LNS variant outperforms LNS-rand is for MAXCUT on RR graphs, which we explain by their regular structure and consequently a lack of what a GFlowNet can learn to exploit. We further find that for MIS-BNB in general, there are only small performance differences among the LNS approaches and the variance is low across all methods. Regarding the other problem-graph combinations, GFN-LNS with TB loss performs best, closely followed by the variant employing DB loss, which is especially successful for DkS. Using FL loss often results in worse performance, which indicates that our simple intermediate reward calculation by linear interpolation is not informative enough.

Considering these results, we restrict further analyses to the GFN-LNS variant with TB loss, and to MAXCUT and DkS on ER and BA graphs. We evaluate the destroy set diversity exemplarily for MAXCUT-ER by computing the mean pairwise Jaccard distance $d_J(D', D'')$ [13] for destroy sets D' and D'' from different iterations within stagnation periods. These are sequences of iterations without improvement where the destroy sets stem from the same solution. Averaging this distance across all stagnation periods and runs yields an overall diversity measure with values close to one (zero) indicating high (low) diversity. Both LNS-rand and GFN-LNS-TB exhibit high values of 0.9829 and 0.9755, respectively, confirming the diversity of the produced destroy sets. In ablation studies, we investigate different components of our GFN-LNS, i.e., the design of the energy

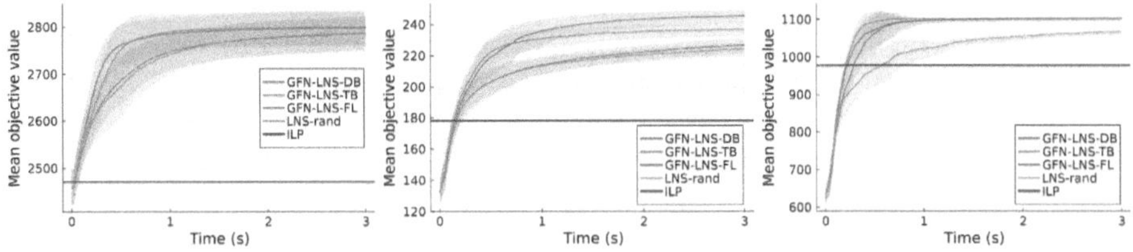

Fig. 2. Performance over time for MAXCUT on ER (left), DkS on RR (center), and DkS on BA (right). The line for ILP represents the final objective value.

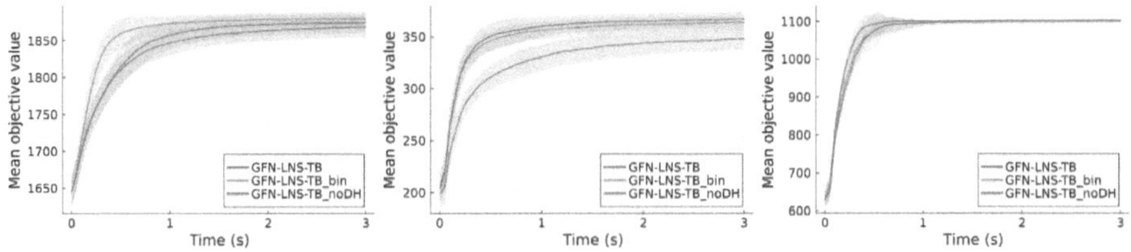

Fig. 3. Performance over time for different GFN-LNS variants for MAXCUT on BA (left), DkS on ER (center), and DkS on BA (right). Usage of the binary indicator for energy calculation is indicated by "bin" and "noDH" denotes the removal of the destroy history feature EMA^{sel}.

calculation and the usage of the destroy history feature EMA^{sel}, to get insights on their impact on the performance. The destroy set diversity is barely impacted by the binary approach for energy calculation with $d_J = 0.9754$ whereas removal of EMA^{sel} results in a slightly lower value of 0.9696. Complete results are too extensive to show here, but further interesting findings are illustrated in Fig. 3. According to pairwise one-sided Wilcoxon signed-rank tests (WSRT) with a threshold $p \leq 0.05$, removing EMA^{sel} significantly improves the performance for DkS-ER and MAXCUT-BA (but not in other cases), and for these as well as MAXCUT-ER, we observe that using the binary approach leads to significantly better performance than the hybrid normalization especially in the beginning. We believe this is due to the binary indicator providing a more robust and more distinct learning signal for improving actions and that the finer-grained consideration of objective gains can sometimes also be more distracting. Both results clearly underline that our GFlowNets are able to effectively learn without having to rely on more advanced reward design and features as often needed in other learning approaches. Nonetheless, for cases such as DkS-BA where results are close and variance is low, these enhancements seem to be slightly beneficial to capture important information about the search process.

Generalization. To test the generalization capabilities of GFN-LNS, we consider smaller and larger instances than those used for training, i.e., graphs with 200, 400, and 500 nodes. To maintain the structural properties of the graphs,

Table 2. Generalization results on instances of size 200 to 500 for different problem and graph type combinations. For each case, the GFlowNet model performing best on $n = 300$ is used which differ in whether the destroy history feature (Dest. Hist.) is used (y) or not (n) and which energy calculation approach is used ($\alpha = 0.1$ for the hybrid and $\alpha = 1$ for the binary approach). All models employ TB loss.

Problem/ Graph	Method	Dest. Hist.	α	200			300			400			500		
				$\overline{\text{obj}}$	σ(obj)	AUC[%]	$\overline{\text{obj}}$	σ(obj)	AUC[%]	$\overline{\text{obj}}$	σ(obj)	AUC[%]	$\overline{\text{obj}}$	σ(obj)	AUC[%]
MAXCUT/ ER	ILP	–	–	1640.25	27.15	–	2470.87	41.20	–	3306.22	47.85	–	4131.78	56.13	–
	LNS-rand	–	–	1853.28	28.26	88.42	2787.07	36.43	79.97	3705.95	44.07	72.27	4604.33	54.81	64.69
	GFN-LNS	y	1.0	**1858.10**	26.83	**94.74**	**2803.84**	35.63	**90.01**	**3741.30**	40.41	**84.53**	**4678.11**	50.68	**77.81**
MAXCUT/ BA	ILP	–	–	1084.07	17.25	–	1659.57	23.08	–	2240.53	23.07	–	2814.98	26.99	–
	LNS-rand	–	–	1227.07	9.78	89.43	1867.25	11.12	80.21	2503.83	13.57	75.06	3126.55	18.84	67.70
	GFN-LNS	y	1.0	**1231.05**	8.61	**94.59**	**1879.86**	9.67	**92.07**	**2528.99**	14.27	**87.71**	**3172.53**	15.92	**83.19**
DkS/ ER	ILP	–	–	201.57	7.92	–	300.63	11.48	–	376.86	18.67	–	434.98	17.89	–
	LNS-rand	–	–	226.71	5.94	78.82	339.22	8.25	71.48	445.35	11.12	64.83	546.83	12.79	58.55
	GFN-LNS	n	0.1	**238.80**	6.29	**94.54**	**366.64**	7.74	**92.08**	**493.12**	10.67	**88.66**	**618.73**	11.82	**84.82**
DkS/ BA	ILP	–	–	665.03	25.32	–	977.32	68.74	–	1316.66	92.73	–	1623.10	97.82	–
	LNS-rand	–	–	692.25	4.15	89.48	1067.39	6.49	81.62	1428.92	8.73	77.67	1809.73	8.83	71.96
	GFN-LNS	y	0.1	**704.13**	3.47	**96.54**	**1102.45**	3.09	**94.63**	**1501.65**	2.82	**90.77**	**1900.06**	2.88	**87.83**

we adjust the edge probability for ER graphs to $p_e = 0.15$, $p_e = 0.075$, and $p_e = 0.06$ for sizes 200, 400, and 500, respectively, but keep all other graph and LNS parameters the same as in training to provide a comparable setting. Results are shown in Table 2 and illustrate that GFN-LNS is able to generalize well to different graph sizes, consistently yielding significantly better solutions than the baselines again based on one-sided paired WSRT with $p \leq 0.05$. The performance improvements even increase for larger graphs where solution sets are usually larger and a better selection of destroy sets is thus more important and beneficial. This demonstrates that the learned policies are not only effective for the graph size they were trained on but are also applicable to other sizes, which is important in practice where training individual models for each size may not be feasible.

5 Conclusion

We proposed GFlowNet-guided Large Neighborhood Search (GFN-LNS), a generic framework for learning effective destroy operators in LNS that is applicable to diverse COPs on graphs without requiring problem-specific tuning. We introduced a problem-agnostic MDP formulation for auto-regressive construction of destroy sets via GFlowNets, enabling the models to capture complex dependencies between nodes. The effectiveness of GFN-LNS is demonstrated on three challenging graph problems (MIS-BNB, MAXCUT, and DkS) across three graph types (ER, RR, and BA). In most cases, GFN-LNS improves upon both, an LNS with random destroy set selection and Gurobi on respective ILP models, in terms of final solution quality and especially online performance. Ablation studies revealed that our approach is able to learn effective policies even with minimal feature engineering, indicating robustness and generality. Notably, the GFlowNets generalize well to different graph sizes not seen during training. On the downside, the computational overhead introduced by inferencing the GFlowNet models is significant and leads to fewer executed iterations, which can be detrimental to the performance especially when the repair operator is fast.

The lack of comparison to other learning-based approaches for selecting destroy sets within an LNS is a key limitation of our work to be addressed in the future. Several design choices such as the fixed destroy set size, the basic repair operator, and the minimalistic set of used features were intentionally kept simple to investigate the raw performance of our approach. While effective, these limitations clearly leave room for improvements. Future work could explore richer state representations and utilizing GFlowNets also for controlling other aspects of the LNS such as the destroy set size or which of a set of different repair operators should be applied.

Disclosure of Interests. The authors have no competing interests to declare that are relevant to the content of this article.

References

1. Addanki, R., Nair, V., Alizadeh, M.: Neural large neighborhood search. In: Learning Meets Combinatorial Algorithms at NeurIPS 2020 Workshop (2020). https://openreview.net/forum?id=xEQhKANoVW
2. Ahuja, R.K., Ergun, Ö., Orlin, J.B., Punnen, A.P.: A survey of very large-scale neighborhood search techniques. Discret. Appl. Math. **123**(1–3), 75–102 (2002)
3. Bengio, E., Jain, M., Korablyov, M., Precup, D., Bengio, Y.: Flow network based generative models for non-iterative diverse candidate generation. In: Advances in Neural Information Processing Systems, vol. 34, pp. 27381–27394. Curran Associates, Inc. (2021)
4. Bengio, Y., Lahlou, S., Deleu, T., Hu, E.J., Tiwari, M., Bengio, E.: GFlowNet foundations. J. Mach. Learn. Res. **24**(1), 210:10006–210:10060 (2023)
5. Bresich, M., Raidl, G.R., Limmer, S.: Revisiting large neighborhood search with on-the-fly charging station insertion for the electric autonomous dial-a-ride problem. ACM Trans. Evol. Learn. Optim. (2025)
6. Chen, M., Gao, L., Chen, Q., Liu, Z.: Dynamic partial removal: a neural network heuristic for large neighborhood search. arXiv preprint arXiv:2005.09330 (2020)
7. Cherrered, A., Oulamara, A., Cherif-Khettaf, W.R.: A deep reinforcement learning-based large neighborhood search for capacitated vehicle routing problems. Procedia Comput. Sci. **270**, 1568–1577 (2025)
8. Feng, S., Sun, Z., Yang, Y.: DIFUSCO-LNS: diffusion-guided large neighbourhood search for integer linear programming (2024). https://openreview.net/forum?id=9QV7Q9gKl9
9. Feng, S., Sun, Z., Yang, Y.: SPL-LNS: sampling-enhanced large neighborhood search for solving integer linear programs. arXiv preprint arXiv:2508.16171 (2025)
10. Garey, M.R., Johnson, D.S.: Computers and Intractability: A Guide to the Theory of NP-Completeness. W. H. Freeman (1979)
11. Goodfellow, I.J., et al.: Generative adversarial nets. In: Advances in Neural Information Processing Systems, pp. 2672–2680 (2014)

12. Huang, T., Ferber, A.M., Tian, Y., Dilkina, B., Steiner, B.: Searching large neighborhoods for integer linear programs with contrastive learning. In: Proceedings of the 40th International Conference on Machine Learning, vol. 202, pp. 13869–13890 PMLR (2023)
13. Jaccard, P.: Étude comparative de la distribution florale dans une portion des alpes et du jura. Bulletin de la Société Vaudoise des Sciences Naturelles **37**(142), 547–579 (1901). https://doi.org/10.5169/seals-266450
14. Jatschka, T., Rodemann, T., Raidl, G.R.: A large neighborhood search for a cooperative optimization approach to distribute service points in mobility applications. In: Metaheuristics and Nature Inspired Computing, vol. 1541, pp. 3–17. Springer, Cham (2022)
15. Jiralerspong, M., et al.: Expected flow networks in stochastic environments and two-player zero-sum games. arXiv preprint arXiv:2310.02779 (2024)
16. Karp, R.M.: Reducibility Among Combinatorial Problems, pp. 85–103. Plenum Press (1972)
17. Kim, M., Choi, S., Kim, H., Son, J., Park, J., Bengio, Y.: Ant colony sampling with GFlowNets for combinatorial optimization. In: Li, Y., Mandt, S., Agrawal, S., Khan, E. (eds.) Proceedings of The 28th International Conference on Artificial Intelligence and Statistics. Proceedings of Machine Learning Research, vol. 258, pp. 469–477. PMLR (2025)
18. Lamm, S., Sanders, P., Schulz, C., Strash, D., Werneck, R.F.: Finding near-optimal independent sets at scale. In: Proceedings of the 18th Workshop on Algorithm Engineering and Experiments, pp. 138–150. SIAM (2016)
19. Lanciano, T., Miyauchi, A., Fazzone, A., Bonchi, F.: A survey on the densest subgraph problem and its variants. ACM Comput. Surv. **56**(8), 1–40 (2024)
20. Limmer, S.: Bilevel large neighborhood search for the electric autonomous dial-a-ride problem. Transp. Res. Interdisc. Perspect. **21**, 100876 (2023)
21. Limmer, S., Varga, J., Raidl, G.R.: Large neighborhood search for electric vehicle fleet scheduling. Energies **16**(12) (2023)
22. Malkin, N., Jain, M., Bengio, E., Sun, C., Bengio, Y.: Trajectory balance: improved credit assignment in GFlowNets. In: Proceedings of the 36th International Conference on Neural Information Processing Systems, pp. 5955–5967. Curran Associates, Inc. (2022)
23. Oberweger, F.F., Raidl, G.R., Rönnberg, E., Huber, M.: A learning large neighborhood search for the staff rerostering problem. In: Integration of Constraint Programming, Artificial Intelligence, and Operations Research, pp. 300–317. Springer, Cham (2022)
24. Pan, L., Malkin, N., Zhang, D., Bengio, Y.: Better training of GFlowNets with local credit and incomplete trajectories. In: Proceedings of the 40th International Conference on Machine Learning, pp. 26878–26890. PMLR (2023)
25. Pan, L., Zhang, D., Jain, M., Huang, L., Bengio, Y.: Stochastic generative flow networks. In: Proceedings of the 39th Conference on Uncertainty in Artificial Intelligence, pp. 1628–1638. PMLR (2023)
26. Pisinger, D., Ropke, S.: Large neighborhood search. In: Gendreau, M., Potvin, J.Y. (eds.) Handbook of Metaheuristics, pp. 399–419. Springer, Cham (2010)
27. Shaw, P.: Using constraint programming and local search methods to solve vehicle routing problems. In: International Conference on Principles and Practice of Constraint Programming, pp. 417–431. Springer, Cham (1998)
28. Song, J., Lanka, R., Yue, Y., Dilkina, B.: A general large neighborhood search framework for solving integer linear programs. Adv. Neural. Inf. Process. Syst. **33**, 20012–20023 (2020)

29. Sonnerat, N., Wang, P., Ktena, I., Bartunov, S., Nair, V.: Learning a large neighborhood search algorithm for mixed integer programs. arXiv preprint arXiv:2107.10201 (2021)
30. Sutton, R.S., Barto, A.G.: Reinforcement Learning: An Introduction, 2nd edn. MIT Press (2018)
31. van Bevern, R., Fluschnik, T., Mertzios, G.B., Molter, H., Sorge, M., Suchý, O.: The parameterized complexity of finding secluded solutions to some classical optimization problems on graphs. Discret. Optim. **30**, 20–50 (2018)
32. Xu, K., Hu, W., Leskovec, J., Jegelka, S.: How powerful are graph neural networks? In: International Conference on Learning Representations (2019). https://openreview.net/forum?id=ryGs6iA5Km
33. Zhang, D.W., Rainone, C., Peschl, M., Bondesan, R.: Robust scheduling with GFlowNets. arXiv preprint arXiv:2302.05446 (2023)
34. Zhang, D., Dai, H., Malkin, N., Courville, A., Bengio, Y., Pan, L.: Let the flows tell: solving graph combinatorial optimization problems with GFlowNets. In: Proceedings of the 37th International Conference on Neural Information Processing Systems, pp. 11952–11969. Curran Associates, Inc. (2023)
35. Zhang, N., Yang, J., Cao, Z., Chi, X.: Adversarial generative flow network for solving vehicle routing problems. In: International Conference on Learning Representations (2025). https://openreview.net/forum?id=tBom4xOW1H

Effective Traveling for Metric Instances of the Traveling Thief Problem

Jan Eube[1], Kelin Luo[2], Aneta Neumann[3], Frank Neumann[3(✉)], and Heiko Röglin[1]

[1] University of Bonn, Bonn, Germany
[2] University at Buffalo, Buffalo, USA
[3] Optimisation and Logistics, School of Computer Science and Information Technology, Adelaide University, Adelaide, Australia
frank.neumann@adelaide.edu.au

Abstract. The Traveling Thief Problem (TTP) is a multi-component optimization problem that captures the interplay between routing and packing decisions by combining the classical Traveling Salesperson Problem (TSP) and the Knapsack Problem (KP). The TTP has gained significant attention in the evolutionary computation literature and a wide range of approaches have been developed over the last 10 years. Judging the performance of these algorithms in particular in terms of how close the get to optimal solutions is a very challenging task as effective exact methods are not available due to the highly challenging traveling component. In this paper, we study the tour-optimization component of TTP under a fixed packing plan. We formulate this task as a weighted variant of the TSP, where travel costs depend on the cumulative weight of collected items, and investigate how different distance metrics and cost functions affect computational complexity. We present an $O(n^2)$-time dynamic programming algorithm for the path metric with general cost functions, prove that the problem is NP-hard even on a star metric, and develop constant-factor approximation algorithms for star metrics. We complement our theoretical results with experimental evaluations on standard TTP instances adjusted to a path metric. Our experimental results demonstrate the practical effectiveness of our approaches by comparing it to solutions produced by popular iterative search algorithms. The results show that our methods are able to significantly improve the quality of solutions for some benchmark instances by optimizing the traveling part while pointing out the optimality of the travel component for other solutions obtained by iterative search methods.

1 Introduction

Many real-world optimization problems involve several NP-hard combinatorial optimization problems that need to be solved in combination to come up with high quality solutions for the complex problem at hand. Such problems being a combination of different hard combinatorial optimization problems are known

G. Iacca et al. (Eds.): PPSN 2026, LNCS 16985, pp. 282–297, 2027.
https://doi.org/10.1007/978-3-032-36214-8_18

as multi-component problems in the literature [3]. Iterative search algorithms such as different variants of local search, evolutionary algorithms and ant colony optimization have been applied to the well-known traveling thief problem (TTP) which is a multi-component problem that combines the traveling salesperson problem (TSP) and the knapsack problem (KP).

Understanding the interactions between these problems and how the combination of different subproblems impacts the complexity of solving them is a challenging task [17] and has been mainly studied from an empirical perspective [8,15]. With this paper, we contribute to this area of research from a theoretical perspective and investigate how to solve the class of weighted traveling salesman problems where travel costs increase monotonically with the weight of items collected while traveling. Our developed approaches allow to solve the traveling component of TTP in an optimal way in restricted settings. This allows to judge the quality of solutions obtained by popular iterative search algorithms for TTP and increase their theoretical understanding in terms of performance gaps.

1.1 Related Work

The traveling salesman problem (TSP) is one of the best studied combinatorial optimization problems. It has been investigated extensively both from a theoretical and practical perspective and a wide range of powerful TSP solvers are available. In real world settings, routing problems such as the TSP are often combined with other problems such as packing. In this paper, we consider the class of TSP problems with such packing components influencing the travel time when moving from a node i to node j in the TSP. Our setting is motivated by recently investigated problems such as the traveling thief problem (TTP) [2] and the weighted TSP [4] for which the travel cost when moving between cities increases monotonically with respect to the weight of items collected at previously visited cities. The TTP combines the TSP with the knapsack problem, and a combination of a tour and a packing plan needs to be found that optimizes an objective function which accounts for the profit of collected items and non-linear travel costs that depend on the weight of the collected items while visiting the cities. A wide range of approaches for the TTP and its variants have been proposed in the literature [8,12,14,17].

The case of finding an optimal packing plan for a fixed tour is already NP-hard. However, mixed integer programming [16] and dynamic programming [11] approaches are available and have been shown to solve the packing part when given a fixed tour exactly or approximately for up to 1000 items. Finding an optimal tour for a given packing plan on the other hand seems to be much more difficult and no effective exact approaches or approximation algorithms are currently available if the weighting of the travel distances is non-linear. If the packing plan is fixed, then the problem turns into a weighted TSP problem. Linear-weighted metric TSP problems have been examined in [4] and different constant factor approximation algorithms have been introduced. In the case that each node contributes a weight of 1, the relationship to the maximum latency problem [1] has been used to come up with constant-factor approximation algorithms and the results have also been generalized to polynomial weights. Improved approximation results for the case where the TSP only includes distance of 1 or 2 have been obtained by exploring the connection to approximation algorithms for the classical $\{1, 2\}$-TSP [13], which is already NP-hard as it is a generalization of the Hamiltonian cycle problem.

1.2 Our Contribution

With this paper, we contribute to solving the tour optimization part of such weighted TSP instances and provide algorithms with approximation guarantees and polynomial time dynamic programming approaches for important weighted TSP formulations. We first prove that the Weighted Traveling Salesman Problem (W-TSP) remains NP-hard, establishing that the problem is computationally challenging even in highly structured settings. Despite this hardness, we design an $(8 + \varepsilon)$-approximation algorithm running in polynomial time for this metric. This result demonstrates that strong approximation guarantees are achievable for nontrivial weighted variants of TSP and helps delineate the boundary between tractable and intractable cases. For W-TSP with a general cost function under a path metric, we develop polynomial-time dynamic programming algorithms for both fixed and arbitrary starting points, which form the core tour-optimization component of our experimental evaluation. By integrating this exact W-TSP solver into the TTP framework, we obtain substantially improved solution quality compared to existing approaches. Our experimental results show that the line-metric W-TSP algorithm consistently outperforms previously known heuristics on the considered TTP instances. To further enhance scalability, we introduce a grouping heuristic that accelerates the dynamic programming computation while preserving near-optimal solution quality, demonstrating the practical impact of our theoretical contributions and their effectiveness in advancing the state of the art for TTP.

The paper is structured as follows. In Sect. 2, we introduce the W-TSP which is the problems subject to our investigations and show in Sect. 3 that W-TSP is NP-hard even under restricted metric settings for the TSP part. Section 4 provides approximation and dynamic programming approaches for important settings. Our experimental investigations carried out in Sect. 5 show that the dynamic programming approach provides significant improvements when the underlying metric for the TSP is a path metric. Finally, we finish with some concluding remarks.

2 Problem Formulation

We introduce and study the *Weighted Traveling Salesman Problem* (W-TSP), which generalizes the classical TSP, and also provides a permutation-based solution to the *Traveling Thief Problem* (TTP) when the packing plan is fixed. The input is given as a TSP instance with cities $v_1, \ldots, v_n$ and distances $d(v_i, v_j)$

between them, and a set of n items with weights located at the different cities. We denote by w_{v_i} the weight of the item located at node v_i, and let $f(w)$ denote the cost per unit distance of traversing while carrying total weight w. To simplify writing we sometimes write w_i instead of w_{v_i}. The goal is to compute a TSP tour π that minimize the overall cost. We denote by $\pi = (\pi_1, \ldots, \pi_n)$ the permutation of the cities which determines the order in which the cities are visited. This presents a solution to the TSP part of the TTP problem.

In the classical TTP [2], the function f is a monotone function given by the inverse of the speed of the vehicle. The speed decreases linearly with respect to the weight of the items already collected. Furthermore, each tour starts at city v_1, and either no item is located at v_1, or the item located at v_1 is collected at the end of the tour.

We are interested in general monotone functions f that weight the distances which need to be traveled. Our goal is to find a permutation (also referred to as a tour) $\pi = (\pi_1, \pi_2, ..., \pi_n)$ which minimizes the travel cost given as

$$T(\pi) = f\left(\sum_{j=2}^{n} w_{\pi_j}\right) \cdot d(\pi_n, \pi_1) + \sum_{i=1}^{n-1} f\left(\sum_{j=2}^{i} w_{\pi_j}\right) \cdot d(\pi_i, \pi_{i+1}).$$

Related to this problem is the node weighted TSP problem introduced in [4]. Here, each node has a given weight and distances are weighted by the sum of the weights of the nodes that have already been visited except the first node. It is a special case of our formulation when the function of f returns the sum of the weight of the nodes visited until city π_j. In contrast, our work considers a more general cost function f, which does not necessarily satisfy the pure accumulated-weight property exploited in [4]. As a result, several algorithmic techniques and constant-factor approximation strategies developed for the node weighted TSP do not directly extend to our setting, and our approximation results apply to a strictly broader class of weighted TSP objectives.

3 NP-Hardness

Note that when $f(w) = 1$, the W-TSP reduces to the classical Traveling Salesman Problem (TSP), which is known to be NP-hard. For metric instances the TSP can not be approximation within a factor of less than 123/122 unless P=NP [10].

This section establishes that the W-TSP problem remains NP-hard even under restricted metric settings. Specifically, we show that W-TSP is NP-hard when the underlying distance metric is a special star metric.

We consider a special star metric, where a star represents a tree of height 1 with a root denoted as v_0 and leaves as $\{v_1, ..., v_n\}$. We define the metric on these nodes as follows: for all $i \in [n]$, the distance between v_0 and v_i is denoted as $d(v_0, v_i)$, and for all $i, j \in [n]$ where $v_i \neq v_j$, the distance between v_i and v_j is defined as $d(v_i, v_j) = d(v_0, v_i) + d(v_0, v_j)$.

Theorem 1. *Even when the metric is a star metric, the W-TSP problem is NP hard.*

Proof. We reduce the partition problem to the W-TSP with a star metric. For this, consider an instance of the partition problem, i.e., given is a set S of positive integers $\{s_1, s_2, ..., s_n\}$, and let $\lambda = \sum_{i\in[n]} s_i$, The question is if S can be partitioned into two subsets S_1 and S_2 such that the sum of the numbers in S_1 equals the sum of the numbers in S_2. This problem is NP-hard [7].

For the given instance of the partition problem, we create an instance of the W-TSP with a star metric as follows. The instance is defined on the star metric (V, E) with a root v_0 and leaves $\{v_1, ..., v_n, v_{n+1}\}$, $d(v_0, v_i) = s_i$ for all $i \in [n]$ and $d(v_0, v_{n+1}) = s_{\max} := \max_{i\in[n]} s_i$. Each node v_i has weight $w_i = s_i$ for all $i \in [n]$, node v_{n+1} has weight $w_{n+1} = s_{\max}$ and node v_0 has weight $w_0 = \lambda + s_{\max} + 1$. Finally, we specify the cost function $f(w)$ as a simple piecewise function as follows:

$$f(w) = \begin{cases} 0 & \text{if } w \leq \lambda/2 \\ 1 & \text{if } \lambda/2 < w \leq \lambda + s_{\max} \\ \infty & \text{if } w > \lambda + s_{\max} \end{cases} \tag{1}$$

Then the problem is to find a minimum W-TSP tour $\pi = (\pi_1, \ldots, \pi_{n+2})$ of V with minimum cost, i.e.,

$$d(\pi_{n+2}, \pi_1) \cdot f\Big(\sum_{j=2}^{n+2} w_{\pi_j}\Big) + \sum_{i=1}^{n+1} \Big(d(\pi_i, \pi_{i+1}) \cdot f\Big(\sum_{j=2}^{i} w_{\pi_j}\Big)\Big).$$

We show that there is a valid partition S_1, S_2 iff the W-TSP instance has a solution of cost at most $s_{\max} + \lambda$.

Let S_1, S_2 be a valid partition such that the sums of the elements in S_1 and S_2, respectively, are both $\lambda/2$. Then the following tour provides a solution for the W-TSP with travel cost $s_{\max} + \lambda$: starting from node v_0 (and pick up items in node v_0 at the end), first visit all nodes corresponding to the integer set S_1, which costs 0 as $f(w) = 0$ for $w \leq \lambda/2$. Then visit v_{n+1}, which increases the collected weight to $\lambda/2 + s_{\max}$. Then visit all nodes corresponding to the set S_2 and return to v_0 afterwards. The total length of the subtour starting at v_{n+1} and visiting all nodes corresponding to S_2 and returning to v_0 is $s_{\max} + 2\sum_{s_i\in S_2} s_i = s_{\max} + \lambda$. Since $f(w) = 1$ for $\lambda/2 < w \leq \lambda + s_{\max}$, the total contribution of this subtour to the costs is $s_{\max} + \lambda$. Finally, pick up node v_0, which incurs no additional cost.

For the other direction, assume we have a tour π which provides a solution with travel cost no more than $s_{\max} + \lambda$ for the W-TSP. Observe that the root v_0 must be the start node of the tour and is also the last node whose weight is collected. Otherwise, the tour would require returning to the start node after visiting v_0, during which the accumulated weight would include the additional weight $w_0 = \lambda + s_{\max} + 1$, resulting in $f(w) = \infty$ and hence an infinite travel cost. Therefore, without loss of generality, we assume that the tour starts at v_0. For the remaining nodes, without loss of generality, suppose node v_h with $h \in [n+1]$ is the latest visited node in π such that the total weight of all previous visited nodes is no more than $\lambda/2$, and the total weight of the previous visited nodes is X. Notice that $X \leq \lambda/2$ and $X + w_h > \lambda/2$. There are two cases: (1) $w_h = s_{\max}$. Then the total cost of the tour π is $s_{\max} + 2(\lambda - X)$, which is at most $s_{\max} + \lambda$ if and only if the partition instance admits a valid solution. Otherwise, $X < \lambda/2$, and thus $\lambda - X > \lambda/2$ implies $s_{\max} + 2(\lambda - X) > s_{\max} + \lambda$. (2) $w_h \neq s_{\max}$. Then the total cost of tour π is $s_h + 2(\lambda + s_{\max} - X - s_h) = \lambda + s_{\max} + (\lambda - 2X) + (s_{\max} - s_h) > \lambda + s_{\max}$ since $(\lambda - 2X) \geq 0$ and $s_{\max} - s_h > 0$. Thus, the constructed W-TSP instance admits a solution with total cost at most $s_{\max} + \lambda$ if and only if the original partition instance has a valid partition. This establishes the correctness of the reduction and completes the NP-hardness proof. □

4 Algorithms

In this section, we present three algorithms for constructing a visiting tour (also referred to as a permutation of all cities), each tailored to a different setting: two designed for special metrics—the star metric and the path metric—and one for a setting with a specialized linear cost function.

4.1 Star Metric: A Constant Approximation Algorithm

First, we simplify the problem by restricting attention to tours that start and end at the center node, with the center node's weight collected at the end. This restriction is without loss of generality. Indeed, consider any feasible tour π of the original problem that starts at a leaf node v and ends at v, collecting the weight of v at the end. We construct a corresponding tour π' that starts and ends at the center node by prepending the edge (v, center) to the beginning of π and appending the edge (center, v) to its end.

The additional cost incurred by this transformation is a fixed quantity determined solely by the distance between v and the center and the total accumulated weight at the end of the tour; in particular, it is independent of the order in which the remaining nodes are visited. Conversely, any tour starting and ending at the center can be transformed into a tour starting and ending at an arbitrary leaf node v by removing these two fixed segments, adjusting the total cost by the same constant. Therefore, optimizing over tours that start and end at the center is equivalent to optimizing over tours with arbitrary starting nodes, up to an additive constant. As a result, we may, without loss of generality, focus on the simplified setting in which the tour starts and ends at the center node. In this setting, visiting any other node consists of a trip from the center to that node followed by a return trip back to the center with the accumulated weight.

Observation 2. *Any optimal tour will have the same total distance, equal to twice the sum of all edge lengths in the star graph; Any tour will start with the same weight* 0 *and have the same total weight at the end, equal to the sum of the weights of all nodes.*

Given a set of leaf nodes J, let $D(J)$ denote the total distance required to traverse all nodes in J, starting and ending at the center. That is,

$$D(J) = \sum_{v \in J} (d_{\to v} + d_{\leftarrow v}),$$

where $d_{\to v}$ is the distance from the center to node v, and $d_{\leftarrow v}$ is the distance from node v back to the center.

Our algorithm follows the same structure as a scheduling algorithm developed by Epstein et al. which aims at obtaining a sequence of jobs that is robust against unforeseen changes in the speed of a given machine [5]. We end up with the following algorithm:

1. Scale the distances so that the minimum distance $d_v \geq 1$ for every $j \in J$.
2. For each $i \in \{1, \ldots, \lceil \log D(J) \rceil\}$, find a subset J_i^* of leaf nodes such that the total distance $D(J_i^*) \leq 2^i$ and the total weight $w(J_i^*)$ is maximized.
   Note: $J^*_{\lceil \log D(J) \rceil} = J$.
3. Construct a tour π as follows:
   - Initialize an empty tour.
   - For $i = \lceil \log D(J) \rceil$ down to 1:
     - Append the nodes in $J_i^* \setminus \bigcup_{k=0}^{i-1} J_k^*$ in any order to the end of the tour.

For a tour π, we decompose the tour into *round-trip segments*, where each segment consists of traveling from the center node to a leaf node and then returning from that leaf node back to the center. During each such round trip, the vehicle carries an accumulated weight corresponding to the total weight of all nodes collected before completing the return to the center.

For any weight threshold w, let $D^\pi(w)$ denote the total length of all round-trip segments in tour π whose return trips are completed while carrying an accumulated weight of at least w. Equivalently, $D^\pi(w)$ is the sum of the lengths of all center-to-leaf-to-center segments for which the accumulated weight on the return leg is at least w. Figure 1 illustrates this definition.

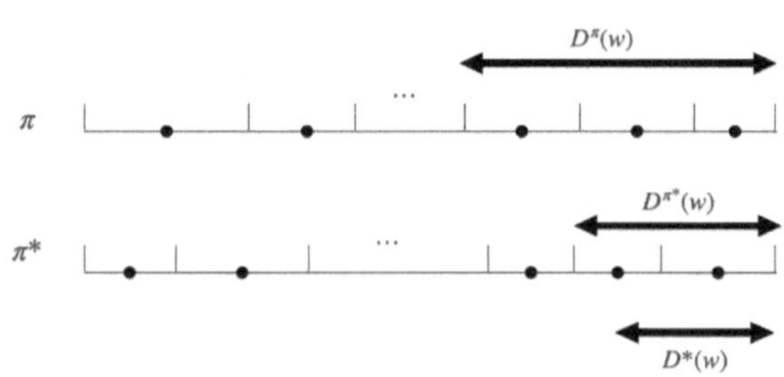

Fig. 1. Round-trip segment length with bounded weight.

Similar to the scheduling problem, we can prove that:

Lemma 1. *Consider an optimal tour π^* and a given tour π. If $D^\pi(w) \leq c \cdot D^{\pi^*}(w)$ holds for any w, then the total cost of tour π is at most $2 \cdot c$ times the total cost of the optimal tour.*

Proof. Let $J' \subset J$ be a subset of leaf nodes visited in the first part of tour π, and let w denote the total weight of J' together with the next leaf node visited in π. Define $D^*(w)$ to be the minimum total length traversed by the optimal tour π^* with accumulated total weight of at least w. Let v be the first leaf node in π^* such that the accumulated weight reaches or exceeds w. By definition, we have:

$$D^*(w) = D^{\pi^*}(w) - d_v \geq \frac{1}{2} D^{\pi^*}(w),$$

since $d_v \leq D^{\pi^*}(w) - D^*(w)$ and the minimum distance to a node is at least half the total length.

Given that $D^\pi(w) \leq c \cdot D^{\pi^*}(w)$, it follows that:

$$D^\pi(w) \leq 2c \cdot D^*(w).$$

By continuing with the the same steps as in the analysis by Epstein et al. [5] one then obtains that the total cost of π is at most $2c$ times that of π^*. □

Lemma 2. *If the metric is a star metric, our algorithm produces a tour π such that*

$$D^\pi(w) < 4D^{\pi^*}(w)$$

for all $w \geq 0$.

Proof. Let $0 < w < w(J)$ and let i be minimal such that $w(J_i^*) \geq w(J) - w$. By construction of π, leaf nodes $j \in \bigcup_{k=1}^{i} J_k^*$ have an accumulated weight greater than w when they are visited. Thus,

$$D^\pi(w) < \sum_{k=1}^{i} D(J_k^*) \leq \sum_{k=1}^{i} 2^k = 2^{i+1} - 2. \tag{2}$$

When $0 < w < w(J)$, it holds that $D^*(w) \geq 2$, because we scale the distances so that the minimum distance $d_{\rightarrow v} \geq 1$ for every $j \in J$.

In case $i = 1$, the claim is trivially true, and thus $D^*(w) \geq D^\pi(w) = 2$.

Suppose $i \geq 1$, then by our choice of i, it holds that $w(J_{i-1}^*) < w(J) - w$. Therefore, in any sequence π', the total distance (start and end with center) during which the accumulated weight on the return trip to the center is at least w is larger than 2^{i-1}, because otherwise we get a contradiction to the maximality of $w(J_{i-1}^*)$.

That is, $D^*(w) > 2^{i-1}$. Together with (1), this concludes the proof. □

By combining the previous two lemma one obtains:

Theorem 3. *Our algorithm computes an 8-approximation for the weighted TSP problem on star graphs.*

However, in Lemma 2, we implicitly ignored the computational complexity of the step asserting that the total distance (starting and ending at the center) for which the accumulated weight on the return trip to the center is at least w exceeds 2^{i-1}. In fact, Step (2) of the algorithm relies on a subroutine that solves a knapsack-type problem, namely maximizing the total collected weight of a subtour subject to a length constraint. This problem is NP-hard and admits only a $(1+\varepsilon)$-approximation in polynomial time [9].

Therefore, when restricting attention to polynomial-time algorithms, the above result should be stated as follows.

Theorem 4. *There exists an algorithm that computes an* $(8+\varepsilon)$*-approximation in polynomial time for the weighted TSP problem on star graphs.*

4.2 Path Metric: DP Algorithm

In this section, we consider a path metric with set of nodes $\{v_1, v_2, ...\}$ ordered from left to right. We define the metric on these nodes by $d(v_i, v_{i+1})$ for all $i \in [n-1]$, and $d(v_i, v_j) = d(v_j, v_i) = \sum_{h=i}^{j-1} d(v_h, v_{h+1})$ for all $i < j$.

We begin by solving the problem under the assumption that the starting node is fixed, and its weight is collected at the end of the traversal. Let t denote the starting node, where $t \in \{v_1, v_2, \ldots, v_n\}$. Our problem involves $n+1$ nodes, $\{v_1, v_2, \ldots, v_n\} \cup \{t\}$, where $\{v_1, v_2, \ldots, v_n\}$ are ordered from left to right. We will present a dynamic programming algorithm that computes an optimal solution in polynomial time. First we may observe that it does not make any sense to visit a node (except the starting node) when it is surrounded by two unvisited nodes because the agent will also pass by this node at a later point in time:

Lemma 3. *Given a set of uncollected nodes arranged on a path, there exists an optimal tour (i.e., a node permutation) such that for all* $i < j < k$*, the tour visits either* v_i *or* v_k *before visiting* v_j*.*

Proof. For every $i < j < k$ we say that v_j is visited prematurely if it is visited before v_i and v_k. Let π^* be an optimal tour minimizing the number of premature visits. Let us assume that there still exists a node v_j that is visited prematurely by π^*. Let us assume w.l.o.g. that the last node visited by π^* lies to the left of v_j. Let v_k be the last node to the right of v_j that is visited by π^* and let v_i be the node visited after v_k. Then we know that v_i lies to the left of v_j. Additionally, v_k is visited after v_j because v_j was visited prematurely.

Let us consider the tour π' where we remove the initial visit of v_j from π^* and visit v_j after v_k instead. One might note that π' is not longer than π^* because $d(v_k, v_j) + d(v_j, v_i) = d(v_k, v_i)$. Additionally, the segments of π^* and π' before π^* visits v_j and the segments after π' visits v_j are exactly the same. During the segment in between the visits of v_j by the two tours the only difference is that on tour π^* the agent has already picked up the weight of v_j while on π' this is not the case which means that the agent moves at least as fast on π' as on π^*. As a result we have that $T(\pi') \leq T(\pi^*)$. Since π' does not visit v_j prematurely anymore this contradicts that π^* minimizes the number of premature visits among all optimal tours. □

One might note that any tour that visits no node that is surrounded by two unvisited nodes can at any point in time only visit either the leftmost or the rightmost node. Thus:

Corollary 1. *On a path metric there exists an optimal tour, where at any point in time, the next node selected from the remaining unvisited (i.e., uncollected) nodes on the path is either the leftmost or the rightmost among the remaining unvisited nodes.*

This structural property forms the foundation of our dynamic programming algorithm. The core idea is to solve subproblems on subpaths and then combine their solutions. For each subpath, we consider a set of uncollected nodes within the interval $[a, b] \subseteq [v_1, v_n]$, where a and b denote the leftmost and rightmost nodes, respectively. Let s be the starting node of the current traversal, and let t be the ending node, whose weight will be collected at the end of the tour. For this subproblem, the total weight collected prior to visiting the nodes in the interval $[a, b]$ is equal to the sum of the weights of all nodes located outside this interval, excluding the end node t (whose weight is collected at the end). Based on the lemma, we know that the first node collected in any subpath must be either a or b, which allows us to recursively build the optimal tour by evaluating both options and choosing the one with minimum cost.

In the following, we formally define the subproblems and apply dynamic programming to solve the problem over the entire path.

Let $g(v_i, v_j, s, t)$ represent the optimal value of the subproblem defined on the subpath $[v_i, v_j]$, under the condition that all nodes except those on the subpath $[v_i, v_j]$ have already been visited, which means that the respective weights have been picked up. Here, $s \in \{v_i, v_j\}$ denotes the currently collected node, and t is the ending node, which will be collected last. The function is computed for every pair of nodes v_i and v_j where $v_i, v_j \in \{v_1, v_2, \ldots, v_n\}$ are the nodes on the path from v_1 to v_n.

There are $\sum_{i=1}^{n} \sum_{j=i}^{n} 2 = n^2 + n$ such relevant values, in which for each interval of $[x_i, x_j]$ with $i \leq j$, there are 2 choices for the starting nodes.

When $v_i = v_j$, the subpath consists of a single node. In this case, the tour starts at v_i, collects its weight, and then returns to t. In this case, $g(v_i, v_j, s, t) = d(v_i, t) \cdot f(w)$, where $w = \sum_{i=1}^{n} w_i$ is the total collected weight.

We are now ready to introduce the recurrence relation. Consider the subproblem $g(v_i, v_j, s, t)$, defined on the interval $[v_i, v_j]$ with $i < j$, where the remaining uncollected nodes lie within the interval, the traversal starts at node $s \in \{v_i, v_j\}$, and ends at node t. The collected weights before picking nodes in interval $[v_i, v_j]$ is $w = \sum_{\ell \in [1,i)} w_\ell + \sum_{\ell \in (j,n]} w_\ell$. To compute $g(v_i, v_j, s, t)$, we consider two possible cases:

- If $s = v_i$ and the next visited node is v_{i+1}, then the cost consists of the cost to visit v_{i+1} from s and the optimal cost of collecting all remaining nodes in the interval $[v_{i+1}, v_j]$. Thus, we have:

$$g(v_i, v_j, s, t) = f(w + w_i) \cdot d(v_i, v_{i+1}) + g(v_{i+1}, v_j, v_{i+1}, t).$$

Similarly, if $s = v_j$ and the next visited node is v_{j-1},

$$g(v_i, v_j, s, t) = f(w + w_j) \cdot d(v_j, v_{j-1}) + g(v_i, v_{j-1}, v_{j-1}, t).$$

- If $s = v_i$ and the next visited node is v_j, then the cost consists of the cost to visit v_j from s and the optimal cost of collecting all remaining nodes in the interval $[v_{i+1}, v_j]$. Thus, we have:

$$g(v_i, v_j, s, t) = f(w + w_i) \cdot d(v_i, v_j) + g(v_{i+1}, v_j, v_j, t).$$

Similarly, if $s = v_j$ and the next visited node is v_i,

$$g(v_i, v_j, s, t) = f(w + w_j) \cdot d(v_j, v_i) + g(v_i, v_{j-1}, v_i, t).$$

According to Corollary 1, the optimal tour for the subproblem $g(v_i, v_j, s, t)$ begins by visiting either the leftmost node v_i or the rightmost node v_j of the interval $[v_i, v_j]$. If $s = v_i$, then

$$g(v_i, v_j, s, t) = \min\{f(w + w_i) \cdot d(v_i, v_{i+1}) + g(v_{i+1}, v_j, v_{i+1}, t), f(w + w_i) \cdot d(v_i, v_j) + g(v_{i+1}, v_j, v_j, t)\}.$$

If $s = v_j$, then

$$g(v_i, v_j, s, t) = \min\{f(w + w_j) \cdot d(v_j, v_{j-1}) + g(v_i, v_{j-1}, v_{j-1}, t), f(w + w_j) \cdot d(v_j, v_i) + g(v_i, v_{j-1}, v_i, t)\}.$$

Using dynamic programming, we recursively compute $g(v_i, v_j, s, t)$ for all $n^2 + n$ subproblems and ultimately obtain $g(v_1, v_n, s, t)$ for $s \in \{v_1, v_n\}$. The optimal tour is then given by the minimum of the following two values: $g(v_1, v_n, v_1, t) + f(0) \cdot d(t, v_1)$ and $g(v_1, v_n, v_n, t) + f(0) \cdot d(t, v_n)$.

Theorem 5. *When the metric is a path graph, then the W-TSP problem with a fixed starting node t can be solved optimally in time $O(n^2)$.*

To find the optimal solution for the W-TSP problem, we consider all possible starting nodes t, compute the optimal tour for each, and select the best among them.

Theorem 6. *When the metric is a path graph, then the W-TSP problem with a freely chosen starting node can be solved optimally in time $O(n^3)$.*

5 Experiments

In our experiments, we utilize TTP benchmark instances introduced in [14] that are built on the TSP instances from TSPLIB [14]. The instances are publicly available.[1] Our benchmark set consists of Euclidean TSP benchmark instances

[1] https://cs.adelaide.edu.au/~optlog/TTP2017Comp/.

with n cities given as coordinates (x_i, y_i), $1 \le i \le n$ and Euclidean distances between the cities. For the TTP instances, items are placed at the different cities except that starting city. We use TTP instances based on the TSP instance which different numbers of items per city, namely $1, 2, 5, 6, 10, 20$ resulting in TTP instances with $m = 50, 100, 150, 250, 300, 500, 1000$ items with 51 cities. In addition, we use TTP instances based on the TSP instances which contain 101 and 575 cities. We use 1 item per city resulting in $m = 100, 574$ items, respectively. For the experiments with our approach for the path metric, we set all y_i-coordinates in the considered TSPLIB instance to 0.

For comparison, we are using the Algorithm $S5$ which iteratively executes the process defined in the Algorithm $S1$ introduced in [6] as a greedy heuristic solver for the classical TTP until a specified time bound is fulfilled. $S1$ applies the Chained Lin-Kernighan heuristic to generate high-quality TSP tours and utilizes the *PackIterative* algorithm to construct packing plans by selecting items based on their score, considering weight, profit, and distance to the last city. Note that $S5$ has achieved strong results across various TTP benchmark instances and is recognized as a top-performing algorithm for the problem [6]. We run the algorithm on each considered TTP instance for a maximum duration of 10 minutes on a MacBook Pro 2022 with a Apple M2 chip and 24GB RAM.

We start from the solution $s = (\pi, x)$ produced by the S5 heuristic, where π denotes the tour and x is the corresponding packing plan. Fixing the packing plan x, we derive a W-TSP instance in which the cost of a tour is determined by the accumulated weight induced by x along the tour. We then apply our tour-optimization algorithms to this induced W-TSP instance and evaluate whether, and by how much, they can improve upon the tour π returned by S5 under the same packing plan. Since the benchmark TTP instances specify a fixed starting node, all evaluated tours use the same starting node to ensure a fair and consistent comparison.

5.1 Performance Analysis

Tables 1 summarizes the performance of our proposed dynamic programming (DP) algorithm on instances derived from the TSPLIB dataset. Overall, our algorithm demonstrates consistent and sometimes significant improvements in the route planning objective compared to the benchmark approach. In total, we evaluated 25 problem instances. As shown in Table 1, the average improvement in route cost is 13.58%, with the maximum improvement reaching 69.71%, and no degradation observed in any instance (i.e., the worst case is 0% improvement, indicating the same with the benchmark).

To gain insight into how our DP algorithm achieves these improvements, we provide two visual comparisons of the paths produced by our method and the benchmark, as shown in Figs. 2 and 3. Regarding Figs. 2 and 3, we clarify that the Y-axis in these figures does not represent the actual geometric y-coordinates of the TSPLIB instances. For the path-metric experiments, all nodes indeed lie on a line. The Y-axis in Figs. 2 and 3 is used solely for visualization purposes: it represents an artificial level assigned according to the visiting order in the tour.

Table 1. Summary statistics of the experiment. A minimum improvement of 0.00% indicates that, in some instances, the benchmark tour is already optimal for the induced W-TSP under the fixed packing plan. Since the DP approach computes an optimal solution for the path-metric W-TSP, its solution quality is never worse than that of the benchmark.

Metric	Value
Total instances analyzed	25
Average improvement	13.58%
Best improvement	69.71%
Worst improvement	0.00%

This layout allows us to visually distinguish the traversal order of nodes along the path and to highlight differences (or similarities) between tours produced by our algorithm and the benchmark. In particular, plotting all nodes on a single horizontal line would result in heavily overlapping edges and labels, making the comparison difficult to interpret. The visualization therefore separates nodes vertically to clearly illustrate the tour structure, while preserving the correct X-coordinates that define the path metric.

Figure 2 shows a case where the path produced by our algorithm is identical to that of the benchmark. Both follow the same route pattern, indicating that the benchmark already performs optimally or near-optimally. This shows that DP algorithm pattern is able to recognize when no further improvement is possible and avoid unnecessary computation. In contrast, Fig. 3 illustrates a case where our algorithm achieves a 21.6% improvement. The benchmark path proceeds in one direction for a longer distance while collecting nodes (and their associated weights) before turning back. This strategy can lead to higher costs, as the cost function is non-decreasing with respect to the total collected weight. In comparison, our DP algorithm strategically defers collecting node weights until the return path after reaching an endpoint. This allows it to collect items more efficiently on the way back, resulting in a more cost-effective tour.

These results show that while our DP algorithm matches the benchmark in some cases, it outperforms others by leveraging an optimal item collection substructure. This insight could be used to enhance solution quality in existing TTP studies.

5.2 Time Complexity and Empirical Performance

We also conducted experiments on a synthetic dataset using a $3 \times 5 \times 5$ configuration, combining different node sizes, item sets. Specifically, we evaluated 5 node sizes ($n = 101, 501, 1001, 1501, 2001$), where one node is designated as the depot and the remaining nodes are randomly distributed. For each number of nodes, we generated 5 items per city resulting in instances with $m = 500, 2500, 5000, 7500, 10000$ items, respectively. In the DP implementation, we observed that the step-wise runtime grows quadratically with respect to the

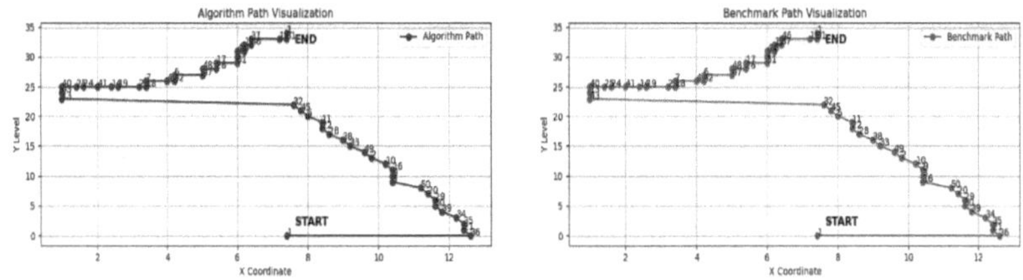

Fig. 2. Same Result as Benchmark

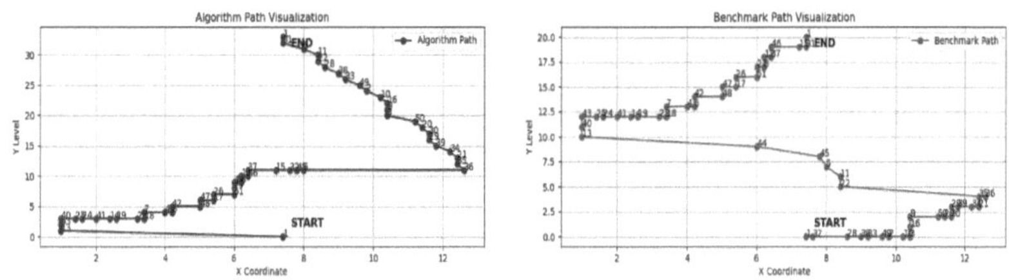

Fig. 3. 21.6% Improvement Compared to Benchmark

number of nodes n, which is consistent with the theoretical time complexity of our DP algorithm.

To improve scalability, we introduce a clustering strategy that aggregates nearby items into $\sqrt{n}$ clusters based on their spatial locations. Specifically, for each item, we use the coordinates of the node at which it is located and apply a standard k-means clustering algorithm with $k = \sqrt{n}$. Each cluster therefore consists of items whose locations are close in the underlying metric. For each cluster, we construct a single representative (clustered) item as follows. The profit and weight of the representative item are set to the sum of the profits and weights of all items in the cluster. The representative item is assigned to a center node chosen as the node of the item whose location is closest (in Euclidean distance) to the cluster centroid. We maintain an explicit mapping from each representative item to the set of original items it aggregates, which allows the clustered solution to be interpreted in terms of the original instance. After clustering, the dynamic programming algorithm is applied to the reduced instance consisting of the clustered items. This significantly reduces the problem size, allowing the DP to run independently on the clustered representation. The resulting solution is then mapped back to the original items using the maintained cluster mapping. Empirically, this approach reduces the overall runtime to be approximately linear in n (see Fig. 4), while increasing the total cost by only a small margin.

As shown in Table 2, the preprocessing step including clustering and the computation of segment weights and distances, reduced runtime by an average of −82.67%, while the DP step achieved an even greater reduction of −99.28%. The average cost increase across all instances was only 1.18%, indicating that our clustering-based approximation remains highly effective with minimal trade-offs in solution quality.

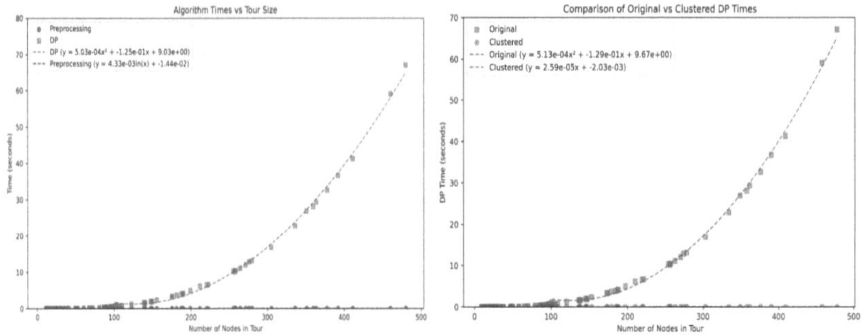

Fig. 4. Growth of Running Time with Increasing Problem Size

Table 2. Statistics Summary (%)

Metric	Mean	Median	Min	Max
Pre Time	-82.67	-93.73	-98.51	-30.14
DP Time	-99.28	-99.87	-99.98	-91.94
Cost	1.18	0.45	0.00	5.04

6 Conclusion

Motivated by multi-component problems such as the Traveling Thief Problem, we studied a generalized version of the Traveling Salesman Problem, where the unit cost of traversing an edge depends on the total weight collected from previously visited nodes, instead of 1. We analyzed this problem under various settings and present algorithms for special cases, including specific metrics (e.g., path metrics, star metrics) and special cost functions (e.g., linear functions representing speed). Our experimental results demonstrate that, for practically relevant instances, the proposed algorithm—the dynamic programming approach for the path metric—can substantially improve solution quality over existing heuristics while remaining computationally efficient.

Acknowledgments. This work was supported by the Australian Research Council through grant FT200100536, by the German Research Foundation through grant 390685813, and by the Lamarr Institute for Machine Learning and Artificial Intelligence lamarr-institute.org.

References

1. Blum, A., Chalasani, P., Coppersmith, D., Pulleyblank, W.R., Raghavan, P., Sudan, M.: The minimum latency problem. In: STOC, pp. 163–171. ACM (1994)

2. Bonyadi, M.R., Michalewicz, Z., Barone, L.: The travelling thief problem: the first step in the transition from theoretical problems to realistic problems. In: IEEE Congress on Evolutionary Computation, pp. 1037–1044 (2013). https://doi.org/10.1109/CEC.2013.6557681
3. Bonyadi, M.R., Michalewicz, Z., Wagner, M., Neumann, F.: Evolutionary computation for multicomponent problems: opportunities and future directions. In: Datta, S., Davim, J.P. (eds.) Optimization in Industry. MIE, pp. 13–30. Springer, Cham (2019). https://doi.org/10.1007/978-3-030-01641-8_2
4. Bossek, J., Casel, K., Kerschke, P., Neumann, F.: The node weight dependent traveling salesperson problem: approximation algorithms and randomized search heuristics. In: Genetic and Evolutionary Computation Conference, GECCO 2020, pp. 1286–1294. ACM (2020)
5. Epstein, L., et al.: Universal sequencing on a single machine. In: Eisenbrand, F., Shepherd, F.B. (eds.) IPCO 2010. LNCS, vol. 6080, pp. 230–243. Springer, Heidelberg (2010). https://doi.org/10.1007/978-3-642-13036-6_18
6. Faulkner, H., Polyakovskiy, S., Schultz, T., Wagner, M.: Approximate approaches to the traveling thief problem. In: Proceedings of the Genetic and Evolutionary Computation Conference, GECCO 2015, pp. 385–392. ACM (2015). https://doi.org/10.1145/2739480.2754716
7. Garey, M.R., Johnson, D.S.: Computers and Intractability: A Guide to the Theory of NP-Completeness. W. H. Freeman (1979)
8. Herring, D., Kirley, M., Yao, X.: A comparative study of evolutionary approaches to the bi-objective dynamic travelling thief problem. Swarm Evol. Comput. **84**, 101433 (2024)
9. Ibarra, O.H., Kim, C.E.: Fast approximation algorithms for the knapsack and sum of subset problems. J. ACM **22**(4), 463–468 (1975)
10. Karpinski, M., Lampis, M., Schmied, R.: New inapproximability bounds for TSP. J. Comput. Syst. Sci. **81**(8), 1665–1677 (2015)
11. Neumann, F., Polyakovskiy, S., Skutella, M., Stougie, L., Wu, J.: A fully polynomial time approximation scheme for packing while traveling. In: Disser, Y., Verykios, V.S. (eds.) ALGOCLOUD 2018. LNCS, vol. 11409, pp. 59–72. Springer, Cham (2019). https://doi.org/10.1007/978-3-030-19759-9_5
12. Nikfarjam, A., Neumann, A., Neumann, F.: On the use of quality diversity algorithms for the travelling thief problem. ACM Trans. Evol. Learn. Optim. **4**(2), 12 (2024)
13. Papadimitriou, C.H., Yannakakis, M.: The traveling salesman problem with distances one and two. Math. Oper. Res. **18**(1), 1–11 (1993)
14. Polyakovskiy, S., Bonyadi, M.R., Wagner, M., Michalewicz, Z., Neumann, F.: A comprehensive benchmark set and heuristics for the traveling thief problem. In: Genetic and Evolutionary Computation Conference, GECCO 2014, pp. 477–484. ACM (2014). https://doi.org/10.1145/2576768.2598249
15. Polyakovskiy, S., Makarowsky, A., M'Hallah, R.: Just-in-time batch scheduling problem with two-dimensional bin packing constraints. In: GECCO, pp. 321–328. ACM (2017)
16. Polyakovskiy, S., Neumann, F.: The packing while traveling problem. Eur. J. Oper. Res. **258**(2), 424–439 (2017)
17. Przybylek, M.R., Wierzbicki, A., Michalewicz, Z.: Decomposition algorithms for a multi-hard problem. Evol. Comput. **26**(3) (2018)

Efficient AB-Cycle Fusion: Boosting GPX and Assessing Its Impact on EAX

Jonathan Heins[1,2(✉)], Pascal Kerschke[1,2], and Darrell Whitley[3]

[1] TUD Dresden University of Technology, Dresden, Germany
jonathan.heins@tu-dresden.de
[2] ScaDS.AI Dresden/Leipzig, Dresden, Germany
[3] Colorado State University, Fort Collins, USA

Abstract. Many highly effective solvers have been developed for the Traveling Salesperson Problem (TSP), among which the Edge Assembly Crossover (EAX) and the Lin-Kernighan-Helsgaun (LKH) heuristic are considered state of the art. In recent years, Generalized Partition Crossover (GPX) has emerged as a promising crossover operator that preserves all parental edges. It might thus be beneficial in situations in which EAX loses important information during the search. However, existing GPX variants lack a competitive implementation and are unable to reliably determine the feasibility of fused AB-cycles. In this work, we extend a recently proposed subtour-check procedure for AB-cycles to improve the evaluation of fused AB-cycles within GPX. Based on this extension, we develop a new GPX variant and integrate it directly into EAX. Unlike previous GPX variants, which rely on restrictive feasibility criteria such as equal connections or mirror-edge tests, our approach can correctly determine the feasibility of arbitrary fused AB-cycles. This overcomes a major limitation of previous GPX implementations. We evaluate the resulting algorithm on a well-established benchmark of evolved TSP instances and show that it outperforms vanilla EAX in several settings. To better understand the impact and limitations of our proposed check, we analyze the interaction between the two stages of EAX and the distribution of optimal edges in the population. Our experiments indicate that EAX sometimes removes optimal edges from the population during the first stage, thereby limiting the potential benefit of GPX in later stages. These findings provide new insight into the behavior of EAX and suggest promising directions for future research on crossover operators and population initialization.

Keywords: TSP · Evolutionary Algorithm · EAX · GPX · AB-cycles

1 Introduction

The Traveling Salesperson Problem (TSP) is a classical NP-hard optimization problem studied for decades across disciplines and fundamental to numerous variants – such as the Vehicle Routing Problem (VRP) – that are highly relevant

G. Iacca et al. (Eds.): PPSN 2026, LNCS 16985, pp. 298–314, 2027.
https://doi.org/10.1007/978-3-032-36214-8_19

in practice [1]. Given a graph $G = (V, E)$ with a set of cities (or vertices) $V = \{v_1, \ldots, v_n\}$, a set of edges $E = \{e_k \mid e_k = (v_i, v_j) \in V \times V\}$, and a cost function $c : E \rightarrow \mathbb{R}$ which assigns a cost to each edge of the graph, the goal is to find a cost-optimal tour along all n cities (and returns to the starting city) without visiting any city twice and without traveling multiple times along the same edge. Still, despite its long history and importance in both research and practice, numerous recent works focus on a better understanding of what makes a TSP instance easy or hard for an algorithm and on designing solvers that perform efficiently across diverse instance structures and sizes [3,4,9,10,16,17,22,23,26,28,29].

For the last decades, two approaches have established themselves as highly efficient state-of-the-art heuristics for the TSP: a genetic algorithm that is based on the Edge Assembly Crossover (EAX) operator [14,15] and the Lin-Kernighan-Helsgaun (LKH) heuristic [18,19]. For each of them, a few competitive variants have been proposed in recent years [7,12,13,24].

Noticeably, despite the strong performance of EAX, only few works have investigated its inner workings. It essentially consists of two stages: a local phase (stage I), which performs small modifications to existing solutions, and a global phase (stage II), which efficiently combines large segments of high-quality tours. Recently, [13] examined stage I, whereas stage II remains largely unexplored. As indicated by its name, EAX makes use of a very specific crossover operator. Meanwhile, [31] introduced an alternative crossover operator – Generalized Partition Crossover (GPX) – which has been further refined in recent years [21,26] and shows promise for integration into other heuristics. However, to date, there exists no efficient implementation that integrates GPX within EAX, largely due to the lack of an efficient mechanism for fusing the AB-cycles produced by EAX.

This work addresses the aforementioned research gap(s) by introducing an efficient mechanism for fusing AB-cycles, combining it with GPX, and integrating the resulting approach into EAX. Moreover, the strengths and limitations of the resulting EAX_{GPX} are assessed in a benchmark study, comparing the proposed heuristic with the vanilla EAX across a diverse set of TSP instances. The remainder of this work is structured as follows. Section 2 describes the two stages of EAX, including the so-called Block2 strategy, which will be modified using our proposed approach. Section 3 then introduces GPX. In Sect. 4, we present a fast check for AB-cycles (Sect. 4.1), which is essential for realizing the boosted EAX_{GPX} (Sect. 4.2). The effectiveness of the proposed approach is analyzed through multiple experimental studies in Sect. 5. Finally, Sect. 6 summarizes the contributions and outlines directions for future research.

2 Inner Workings of EAX

EAX is a genetic algorithm that optimizes a population of TSP solutions using the eponymous Edge Assembly Crossover operator. In each generation, every individual participates in two mating operations: once as parent A and once as parent B. To perform crossover between two parent solutions, the union graph $G_{AB} = (V, E_{AB})$ is constructed, where V is the set of nodes and $E_{AB} = E_A \uplus E_B$

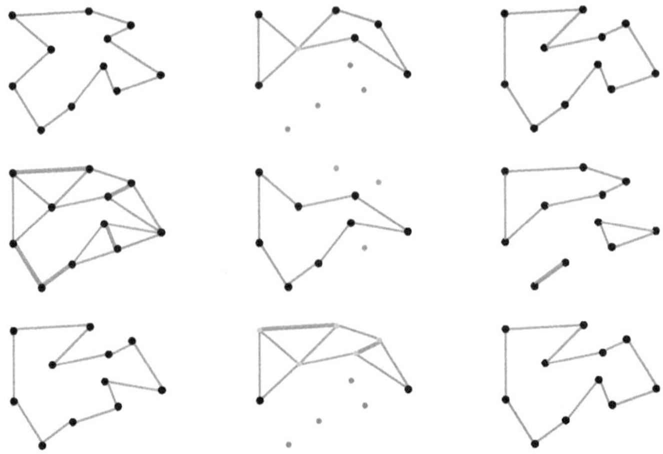

Fig. 1. Left: Parent A (blue, top) and Parent B (orange, bottom) form G_{AB}, which contains all edges (middle). Middle: Examples of AB-cycles resulting in different E-sets used by EAX. Points are distinguished into inner nodes (blue), outer nodes (orange), and portals (black) in the corresponding E-set. Right: The resulting offspring solutions after applying the E-sets. The solution in the middle requires two repairs. (Color figure online)

is the multiset sum of the edge sets of parents A and B. If an edge appears in both parents, it is treated as two distinct edges. An example is visible in Fig. 1. The crossover process relies on so-called AB-cycles, which are cycles $(v_1, v_2, \dots, v_k, v_1)$ of k vertices in G_{AB} whose edges strictly alternate between edges from E_A and E_B. That is, for any two consecutive edges in the cycle, one originates from parent A and the other from parent B, i.e.,

$$(v_i, v_{i+1}) \in E_A \Leftrightarrow (v_{i+1}, v_{i+2}) \in E_B \quad \forall i.$$

See multiple examples in Fig. 1. A crossover is performed by applying an AB-cycle C to parent A. This produces an offspring with edge set

$$E_O = (E_A \setminus E_{AC}) \cup E_{BC},$$

where E_{AC} and E_{BC} denote the subsets of edges in C originating from parents A and B, respectively. In other words, all edges of parent A that are part of the AB-cycle are removed and replaced by the edges from parent B in the AB-cycle. Thus, the connections between nodes contained in the cycle are rewired according to parent B, while all other edges remain unchanged. This viewpoint naturally induces a distinction between inner and outer components, as well as portal nodes. Edges that belong to the AB-cycle are referred to as *inner edges*, while edges in E_A and E_B that are not part of the cycle are called *outer edges*. A node is called (i) an *inner node*, if all of its incident edges in G_{AB} belong to the AB-cycle, (ii) an *outer node*, if none of its incident edges belongs to the

AB-cycle, and (iii) *portal nodes* are the remaining nodes, i.e., neither inner nor outer nodes, with degree two in the AB-cycle (see Fig. 1). The latter mark the transition points between inner and outer edges and thus correspond to the cut points of the crossover operation.

The solution obtained after applying an AB-cycle still satisfies the degree constraint of two edges per node. However, it may become infeasible due to the introduction of subtours, i.e., connected components that form cycles but do not include all nodes in V, as exemplified in Fig. 1. EAX addresses this issue using a repair mechanism that iteratively merges subtours into a single tour. The procedure starts with the smallest subtour, removes one edge from it and one edge from a nearby subtour, and reconnects them using two new edges. This repair step is the only part of the algorithm that may introduce edges not previously present in the population. If an offspring with a shorter tour length than parent A is obtained, it replaces parent A in the population. If multiple such offspring are generated, the one maximizing the ratio between tour length improvement and the potential loss in edge diversity, measured using edge entropy, is selected.

Generating AB-Cycles and E-Sets. EAX consists of two stages. Stage I is a local phase that introduces only small modifications to existing solutions. The stage II is a global phase that combines larger segments of high-quality tours while minimizing the need for repairs, under the assumption that most beneficial edges are already present in the population.

In both stages, AB-cycles are constructed in the same manner. Starting from a random vertex, a path is extended by alternately selecting edges from parent A and parent B. At each step, if multiple candidate edges are available and have not yet been used in any AB-cycle, one is chosen at random. When a previously visited node is encountered and the alternating property is preserved, a cycle is identified. This cycle is extracted as an AB-cycle, and its edges are removed from the set of available edges. The path construction then continues either from the remaining portion of the path that was not part of the extracted cycle or, if no such remainder exists, from a new starting edge. AB-cycles consisting of only two edges are termed *ineffective*, as they correspond to edges shared by both parents and thus do not modify the solution. Such cycles are discarded and not considered further. The two phases of EAX differ only in how the so-called E-set is constructed. The E-set is defined as the set of AB-cycles that are applied simultaneously to parent A in order to produce an offspring solution. In the local phase of EAX, a *single strategy* is employed in which the E-set consists of only one AB-cycle selected at random.

The Block2 Strategy. In the second stage of EAX, instead of accepting single AB-cycles, multiple cycles are merged using the *Block2 strategy*. The aim is to create an offspring solution that contains many edges from both parents while requiring as few repairs as possible. The strategy is based on the observation that the number of subtours that can arise is at most half the number of portals in the E-set, since two cuts are required to form a single subtour. Consequently, reducing the number of portals to two guarantees that the resulting offspring solution is feasible.

To efficiently search for a set of AB-cycles that satisfies this property, ineffective AB-cycles containing exactly the same edges from both parents – which were previously discarded – are incorporated by merging them with adjacent AB-cycles. If an ineffective AB-cycle connects two portals of the same AB-cycle, they can be ignored, as including the edges in the AB-cycle does not alter the resulting solution but merely converts the corresponding portals into inner nodes. If the ineffective AB-cycle connects to another AB-cycle, the two are considered adjacent, since the path between them will remain unchanged under any circumstances. The two connected portals can be merged into a single portal, since adding the ineffective AB-cycle to one of the AB-cycles would again make one portal an inner node.

After updating the AB-cycles in this way, a matrix is constructed that counts the contact points, i.e. shared portals, between pairs of AB-cycles. Using this matrix, the number of portals that remain after adding an AB-cycle to the E-set can be computed efficiently. Specifically, the number of portals in the candidate AB-cycle is added, and twice the number of contact points with AB-cycles already contained in the E-set is subtracted – once for the candidate cycle and once for the cycles already in the E-set. Analogously, the number of remaining portals after deleting an AB-cycle from the E-set can be computed by adding twice the number of contact points and subtracting the number of portals of the removed AB-cycle.

A tabu search procedure is used to identify the best AB-cycle to add or remove. If an AB-cycle is currently not part of the E-set and adding it reduces the number of portals, its edges are added to the E-set. Conversely, if an AB-cycle is already part of the E-set and removing it decreases the number of portals, its edges are removed. One pre-determined AB-cycle (the *base AB-cycle*) is never removed in order to prevent the E-set from becoming empty. If no AB-cycle can be found that decreases the number of portals, the cycle that introduces the fewest additional portals is selected. To avoid cycling in the search process, the tabu mechanism temporarily prohibits recently added or removed AB-cycles from being selected again. An AB-cycle remains tabu for at least one iteration and at most ten iterations. If no improvement is found after 20 iterations, the search terminates and returns the E-set with the smallest number of portals encountered during the search.

3 Generalized Partition Crossover

Partition Crossover (PX) is not limited to the TSP and has been applied to a variety of combinatorial optimization problems [5,6,8,20,25,27]. Consequently, improvements to this operator can have a broad impact. Although the underlying concepts were originally developed for the TSP, genetic algorithms employing GPX have not surpassed the performance of EAX in terms of time required to reach optimal solutions for TSP instances [26].

The central idea of GPX is to preserve all edges by applying one or multiple (potentially fused) AB-cycles to both parents. This requires that the selected AB-cycles yield feasible solutions when applied to both parent A and parent B. The original PX operator ensures feasibility by identifying partitions of the union graph G_{AB} using exactly two cuts at edges shared between the parents [30]. This is equivalent to applying an AB-cycle with only two portals. The first generalized version, GPX [31], extends this approach by allowing multiple such partitions to be applied simultaneously, while still restricting each partition to two cuts.

GPX2 [26] introduces additional flexibility by creating new cut possibilities. This is achieved by duplicating edges connecting four distinct nodes in G_{AB}, thereby forming virtual shared edges between the duplicated nodes. Moreover, GPX2 allows AB-cycles with more than two portals and ensures feasibility by requiring that the connections between portals induced by the AB-cycle are identical in both parent solutions. While this condition guarantees feasibility, it does not identify all feasible AB-cycles [13]. If multiple AB-cycles fail this equal-connections condition, GPX2 employs a fusion procedure that attempts to combine several AB-cycles into a single cycle that satisfies the condition.

GPX3 [21] further improves the approach by introducing a more efficient method for identifying AB-cycles in G_{AB}, as well as the so-called mirror edge test. This test detects potential subtour creation by checking whether a pair of portals is connected within the AB-cycle via edges from one parent, while simultaneously being connected via edges from the other parent that are not part of the AB-cycle. Although this test eliminates AB-cycles that would induce subtours in both offspring solutions, it does not detect all such cases [13].

4 Enhancing EAX Through GPX-Based Fusion

The limitations of GPX motivate the need for more general and efficient methods to evaluate and combine AB-cycles, which we address in the following sections.

4.1 Checking AB-Cycles for Induced Subtours

An efficient feasibility test for E-sets consisting of multiple AB-cycles is essential for making GPX practical. During the search for a feasible configuration of AB-cycles, many candidate E-sets must be evaluated, and explicitly constructing the resulting offspring solutions to count subtours would be computationally prohibitive. Existing feasibility checks, such as restricting AB-cycles to two portals, enforcing equal connections, or applying the mirror edge test, are not sufficient in this setting because they either reject feasible E-sets or admit infeasible ones. In particular, limiting E-sets to AB-cycles with only two portals is overly restrictive and excludes many promising combinations [13]. A recently proposed AB-cycle test [13] addresses this issue by enabling efficient computation of the number of subtours induced by a given single AB-cycle and has been successfully used to preselect AB-cycles in the first phase of EAX.

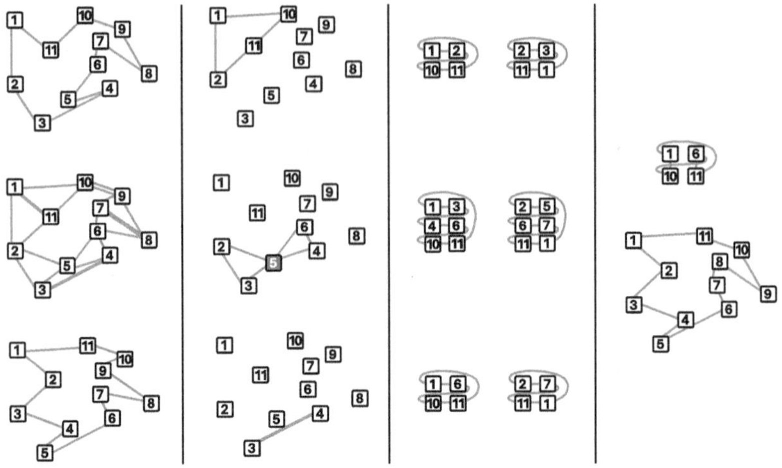

Fig. 2. Column 1: Parent A (orange) and parent B (blue), together with their node orderings and union graph G_{AB} in the center. Column 2: Three different AB-cycles. The bottom cycle is ineffective, while the middle cycle contains one inner node marked in gray. Column 3: The representation of the first AB-cycle in terms of the inner and outer portal arrays for parent A (left) and parent B (right), where inner portals are listed on the left and outer portals on the right. Inner connections are short lines. It further illustrates the iterative insertion of the second AB-cycle into the portal arrays and the effect of additionally including the ineffective AB-cycle. Column 4: Visualization of how the AB-cycle check traverses the portal arrays and the resulting tour structure. (Color figure online)

This method is based on the observation that only the connections between portal nodes determine the number of subtours, while inner and outer nodes can be ignored. To efficiently trace these connections, portals are classified as *inner* and *outer* with respect to each parent. An *inner portal* of a parent is a node at which the tour transitions from an outer edge to an inner edge, while an *outer portal* corresponds to a transition from an inner edge to an outer edge. Consequently, a portal may be an inner portal for parent A while simultaneously being an outer portal for parent B. The portals are stored in four arrays: inner and outer portal arrays (for parents A and B). Sorting these arrays according to the order of nodes in the corresponding parent tour allows efficient tracing of connections (see Fig. 2). Starting from any inner portal of parent B, the offspring solution follows the connection of parent B to the next outer portal in the tour order. Due to the sorted arrays, this corresponding portal can be found at the same index in the outer portal array of parent B. A special case occurs when the AB-cycle includes the first node of the tour as an inner node or outer portal. In this case, the index must be incremented by one modulo the array length. The next connection is then the outer connection of A incident to the current portal. To trace it, the portal is located in either the inner or outer portal array of parent A, after which the corresponding outer connection is determined. This alternating tracing process continues until the starting portal is reached. If all portals are visited, the resulting solution is feasible. Otherwise, at least two subtours would be produced. To determine the exact number of subtours, the tracing procedure is repeated starting from an unvisited portal.

4.2 Integrating a GPX-Based Fast Fusion Operator Into EAX

While the procedure for checking subtours can efficiently evaluate individual AB-cycles, GPX requires efficient evaluation of E-sets that are composed of multiple fused cycles in order to fully exploit its potential [13]. To make this practical, the Block2 strategy (see Sect. 2) must be extended so that the feasibility of dynamically changing E-sets can be evaluated efficiently during the tabu search.

In this work, we address this challenge by introducing a method that incrementally maintains consistent portal representations for combined AB-cycles. This enables efficient evaluation of arbitrary E-sets combined with the tabu search framework. To this end, the four inner and outer portal arrays (for parent A and B) used for individual AB-cycles are merged and dynamically updated as AB-cycles are added to or removed from the E-set. This process is visualized in Fig. 2. If two AB-cycles in the E-set contain the same portal, that portal must be removed because it becomes an inner node and can no longer induce cuts. To efficiently insert portals, they are processed in increasing order. For each portal, if it is, for example, an inner portal of parent A in the candidate AB-cycle, it is inserted into the appropriate position in the E-set's array of inner portals for parent A. If the same portal already appears among the outer portals of A in the E-set, both occurrences are removed. When an AB-cycle is removed from the E-set, the corresponding portals must either be deleted from the portal arrays or added if they are not already present in the respective arrays of the E-set. Ineffective AB-cycles can further reduce the number of portals per AB-cycle, as described previously, thereby decreasing the computational effort required to update the portal arrays and perform the subtour check. Therefore, before the search procedure, the portals in the AB-cycles constituting the inner and outer portal arrays are preprocessed analogously to the preprocessing of portals in the Block2 strategy. Using the updated sets of inner and outer portal arrays, the subtour check can be performed in the same way as for a single AB-cycle. In this work, the next AB-cycles to be added or removed are selected using the tabu search strategy described above. However, instead of focusing on reducing the number of portals in the E-set, a candidate AB-cycle is accepted only if it reduces the total number of subtours that would be induced in parents A and B. The procedure terminates either when an E-set is found that induces no subtours in either parent or when the search terminates without finding such a set, in which case the E-set is rejected.

In contrast to standard EAX, which focuses on improving parent A, the GPX setting requires consideration of both parents to avoid losing edges from the population. Therefore, the subtour check must be executed twice: once for parent A and once for parent B. The second evaluation is performed by swapping

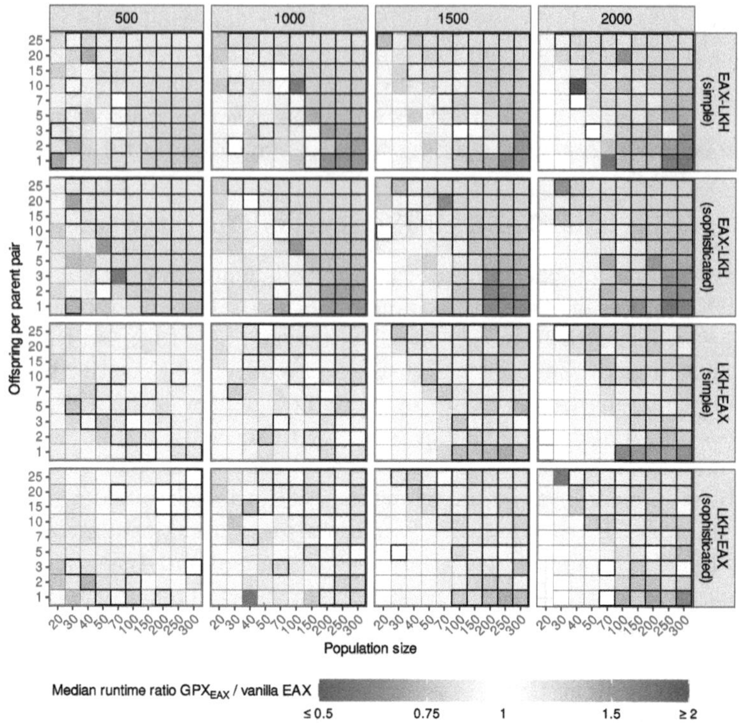

Fig. 3. Comparison of median runtime ratios between vanilla EAX and EAX_{GPX}, per instance group and for different settings of population size (x-axis) and number of produced offspring (y-axis). Non-grayed-out, bold cells indicate statistically significant differences (significance level: 5 %) according to a two-sided Wilcoxon signed-rank test.

the roles of inner and outer portals, as the second offspring will be equal to the residual of the E-set applied to parent A. During the mating procedure, each individual is modified at most once per generation: it participates either as parent A or as parent B, but not both. Parents are selected randomly, and the better individual is designated as parent A, ensuring that improvements are always applied to the higher-quality solution. Since edges are preserved in the GPX setting, the selected offspring is simply the one that yields the greatest improvement over parent A. We name the newly created fusion of EAX with the GPX crossover operator EAX_{GPX}.

5 Experiments

Although the proposed checking procedure is efficient, it still introduces additional computational overhead. In the experiments we thus investigate under which conditions this overhead is outweighed by the benefits of preserved edges. To evaluate the performance of EAX_{GPX}, we use the well-established benchmark set of evolved TSP instances that are difficult for one of the state-of-the-art solvers, EAX or LKH, while being comparatively easy for the other in terms of the time required to reach the optimal solution [3]. This makes them particularly well suited for investigating when improvements to stage II of EAX are beneficial and when they are not. Performance is measured using the penalized average runtime (PAR10), which is the average runtime with a penalty of ten times the cutoff time whenever the optimum is not reached within the allotted time budget [2]. The benchmark instances were specifically evolved to maximize the gap in PAR10 performance between the two solvers. This was achieved using both simple uniform random mutations and more sophisticated mutation operators that create distinctive structures, such as explosion, grid, and implosion operators, among others. Two directions of instance difficulty are considered: instances that are easy for LKH and difficult for EAX (LKH-EAX), and instances that are easy for EAX and difficult for LKH (EAX-LKH). For each of these instance types, we further distinguish into instances evolved using the simple or sophisticated mutation operators – resulting in a total of four instance categories. For each category, instances were generated with 500, 1 000, 1 500, and 2 000 nodes.

EAX_{GPX} Helps When EAX Frequently Enters Stage II. We compare the vanilla EAX variant with EAX_{GPX} under different settings for the population size (ranging from 20 to 300) and the number of offspring generated (ranging from 1 to 25). Experiments are conducted for all 16 settings resulting from the four instance classes times four instance sizes. For each setting, 10 independent runs are performed on 300 instances. We report the ratio between the median PAR10 score of vanilla EAX and that of EAX_{GPX} (see Fig. 3). As expected, the vanilla EAX variant performs better on most instances that are comparatively easy for EAX due to the additional computational overhead of the check described in Sect. 4.2. However, in moderately difficult settings for instances with 500 nodes, EAX_{GPX} significantly outperforms vanilla EAX despite the introduced overhead.

Table 1. Fraction of restarts terminating during the first stage of EAX within a run of five minutes by instance size and instance group.

Size	EAX-LKH (Simple)	EAX-LKH (Sophisticated)	LKH-EAX (Simple)	LKH-EAX (Sophisticated)
500	0.998	1.000	0.142	0.283
1000	0.991	0.998	0.249	0.329
1500	0.943	0.987	0.372	0.380
2000	0.856	0.921	0.418	0.417

To better understand this behavior, Table 1 reports the fraction of restarts within a total runtime budget of five minutes that terminate during the first stage of vanilla EAX. Results are shown separately for each instance group,

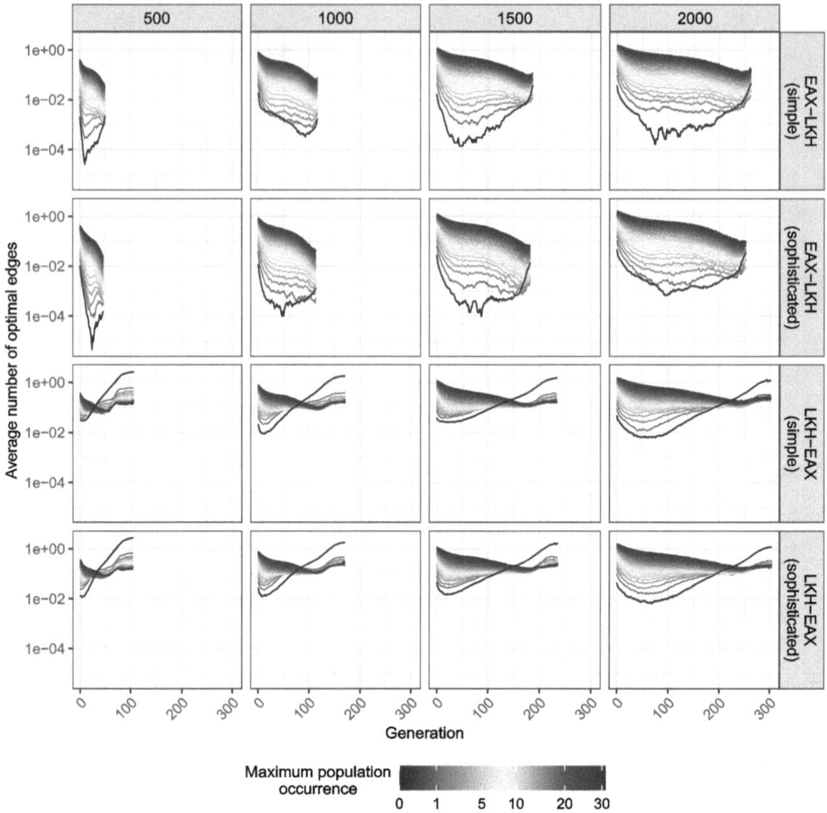

Fig. 4. Average number of optimal edges missing from the population, or present in at most 1, 2, ..., 30 individuals, over the generations of the first stage of EAX. Results are aggregated over restarts recorded for five minutes using 10 seeds on 100 instances per instance group. Only generations reached by at least 50 % of the restarts are included.

based on 100 instances and 10 random seeds. Restarts that terminate in the first stage are those for which the optimum is already found before entering the second stage of the algorithm. All remaining restarts either require the second stage or fail to find the optimum within the allotted runtime. As expected, most EAX-friendly instances (i.e., EAX-LKH) are solved already during the first stage, although this fraction decreases for larger instances. In contrast, the fraction is substantially lower for difficult instances, where more restarts are required to find the optimum. However, for these difficult instances, the fraction of first-stage solutions increases with instance size. A likely explanation is that larger instances induce longer runtimes per restart and are therefore inherently more difficult, even if a greater proportion of first-stage runs already finds the optimum. Since EAX_{GPX} is only active during the second stage, a high fraction of first-stage terminations limits its potential benefit. This is particularly relevant for large instances, where restarts are longer and the second stage is entered less

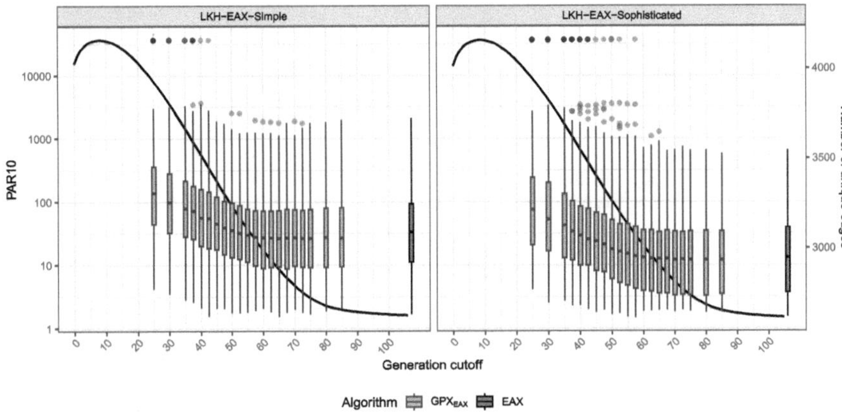

Fig. 5. Boxplots showing the runtime distributions of EAX_{GPX} for different cutoff values for the first stage of EAX compared to the runtime distributions of vanilla EAX under the default setting. Results are shown for instances of size 500 for each of the two groups that are difficult for EAX. The figure also shows the average number of unique edges in the population per generation (solid line).

frequently. Thus, EAX_{GPX} can only provide a benefit on instances for which the second stage of EAX is entered frequently enough for edge preservation to matter.

EAX Can Remove Optimal Edges from the Population. To further investigate why a medium rate of second-stage entry is not sufficient for EAX_{GPX} to improve performance on the remaining instance groups, we record the average number of optimal edges that are entirely absent from the population, or present in only one, two, and up to 30 individuals. These values are plotted over the generations of the first stage of EAX in Fig. 4.

For the EAX-friendly instances, the number of optimal edges missing entirely from the population (dark red line) decreases steadily over most generations until reaching an average value of approximately 0.001 or lower. Thus, in almost all runs, nearly all optimal edges are present in at least one individual. The same trend is visible for the lighter lines representing optimal edges that occur in only one individual, only two individuals, and so on, although naturally at higher levels. For these instances, the EAX search process successfully eliminates non-optimal edges while retaining and propagating optimal ones. For the difficult instances, the search process initially behaves similarly to that observed for EAX-friendly instances. However, after several generations, optimal edges disappear entirely from the population, reaching average values of more than two missing optimal edges per run. GPX does not delete any edges but at the same time cannot introduce new ones. Thus, the potential of EAX_{GPX} depends not only on how often stage II is entered, but also on whether optimal edges are still present in the population when that stage begins.

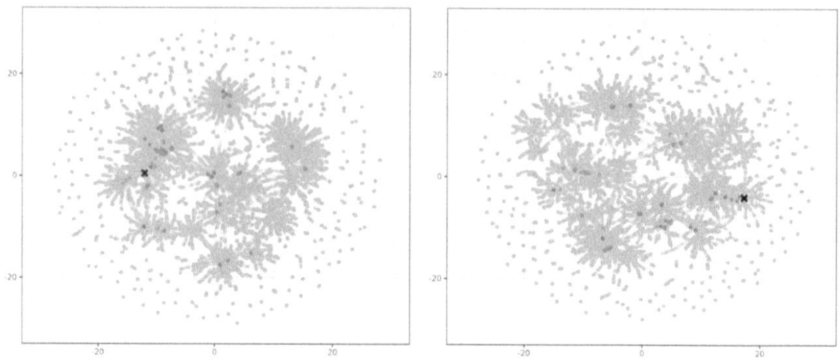

Fig. 6. Visualization of the pairwise differences between solutions (introduced in [11]), highlighting stage II of EAX (blue) and EAX_{GPX} (orange) compared to the solutions found during stage I (gray). The individual plots correspond to exemplary instances with 500 nodes from the LKH-EAX-simple group (left) and the LKH-EAX-sophisticated group (right). The optimum of each instance is marked with an x. (Color figure online)

Earlier Stage Transitions Can Improve EAX_{GPX}. Aiming to leverage this insight, we assess the impact of terminating the first stage of EAX earlier, at a point where, on average, more optimal edges are still present in the population. Note that the original transition criterion is not based on a fixed number of generations, but on the number of consecutive generations without an improvement in the best solution found so far. As shown in Fig. 5, based on the runtime distributions over 100 instances and 10 seeds, mid- to late-early cutoffs improve the runtime performance of EAX_{GPX}. In contrast, very early cutoffs, where almost all optimal edges are still present in the population, lead to worse performance. A likely explanation is that these early cutoffs result in substantially higher diversity, as indicated by the larger number of unique edges in the population. This increases the runtime of the considerably slower second stage and therefore reduces the number of restarts that can be performed. A visual comparison of the diversity in the different stages is shown in Fig. 6. The figure was created using a recent technique for visualizing algorithmic search behavior [11], which performs dimensionality reduction of the pairwise distances between all recorded solutions during ten runs of EAX and EAX_{GPX}. The second stage of EAX, in contrast to EAX_{GPX}, continues to remove optimal edges and does so even for EAX-friendly instances in runs that are not solved during the first stage (see Table 2).

These observations suggest, first, that the ability of EAX_{GPX} to preserve all edges can be highly beneficial. However, this is only the case if all optimal edges are still present in the population and have not already been removed during the first stage of EAX, since the second stage of EAX_{GPX}, like GPX in general, cannot introduce new edges. Second, the results indicate that the loss of optimal edges is not solely caused by the structure of the instance itself.

Otherwise, optimal edges would not also disappear for EAX-friendly instances. Instead, we suspect that the initial composition of the population appears to play an important role. Thus, additional research is required to develop more suitable initialization strategies for EAX_{GPX} in order to fully exploit its potential.

Overall, the results indicate that EAX_{GPX} is most beneficial in settings where the second stage is entered frequently and optimal edges are still present in the population. While the additional computational overhead limits its usefulness on EAX-friendly instances, the preservation of edges can substantially improve performance on harder instances where vanilla EAX tends to remove optimal edges from the population.

Table 2. Average increase of optimal edges missing from the entire population during the second stage of EAX until generation 50 when all unsuccessful runs terminate.

Size	EAX-LKH (Simple)	EAX-LKH (Sophisticated)	LKH-EAX (Simple)	LKH-EAX (Sophisticated)
500	1.102	1.039	0.402	0.503
1000	2.690	1.969	1.029	1.769
1500	3.343	4.189	1.803	2.186
2000	4.904	7.850	3.630	3.380

6 Conclusion

In this paper, we revisited existing GPX implementations and found them to be either too restrictive or too permissive in their E-set selection criteria. We therefore extended an existing AB-cycle feasibility check to efficiently support arbitrarily composed E-sets. This enabled the direct integration of GPX into the second stage of EAX – resulting in EAX_{GPX}, the first genetic algorithm capable of using an efficient fusion operator while guaranteeing edge preservation.

Our empirical investigations reveal that this heuristic can significantly improve runtime on small instances specifically designed to be difficult for EAX, where the original algorithm frequently enters its second stage. Further analyses showed that EAX_{GPX} is less successful on instances, where EAX enters the second stage less frequently and often removes optimal edges from the population during both the first and second stages. Thus, GPX appears to be a highly promising operator, but its effectiveness is limited by two factors: the increased runtime caused by higher diversity when earlier phase transitions are used, and the absence of optimal edges in the population once the second stage begins. The observation that optimal edges are also removed in unsolved runs on EAX-friendly instances suggests that this phenomenon is not solely caused by misleading instance structure. Instead, the initialization of the population might play an important role in determining whether optimal edges are retained or discarded during the search.

Our findings open a promising direction for future research, both in developing improved initialization strategies and in understanding which components of EAX are responsible for removing optimal edges from the population. Progress in these directions could make EAX_{GPX} an even more effective TSP heuristic.

Acknowledgments. The authors gratefully acknowledge the computing time made available to them on the high-performance computer at the NHR Center of TU Dresden. This center is jointly supported by the Federal Ministry of Education and Research and the state governments participating in the NHR (www.nhr-verein.de/unsere-partner). The authors would also like to thank Sebastian Dengel, Anna Franke, Markus Leyser and Sandy Wolf for their kind support and encouragement throughout this work.

Disclosure of Interests. The authors have no competing interests to declare that are relevant to the content of this article.

References

1. Applegate, D.L., Bixby, R.E., Chvátal, V., Cook, W. (eds.): The Traveling Salesman Problem: A Computational Study. Princeton Series in Applied Mathematics. Princeton University Press, Princeton (2011)
2. Bischl, B., et al.: Aslib: a benchmark library for algorithm selection. Artif. Intell. **237**, 41–58 (2016). https://doi.org/10.1016/j.artint.2016.04.003
3. Bossek, J., Kerschke, P., Neumann, A., Wagner, M., Neumann, F., Trautmann, H.: Evolving diverse TSP instances by means of novel and creative mutation operators. In: Proceedings of the 15th ACM/SIGEVO Conference on Foundations of Genetic Algorithms (FOGA 2019), pp. 58–71. ACM, Potsdam Germany (2019). https://doi.org/10.1145/3299904.3340307
4. Bossek, J., Kerschke, P., Trautmann, H.: Anytime behavior of inexact TSP solvers and perspectives for automated algorithm selection. In: 2020 IEEE Congress on Evolutionary Computation (CEC), pp. 1–8. IEEE, Glasgow, UK (2020). https://doi.org/10.1109/CEC48606.2020.9185613
5. Chen, W., Whitley, D., Tinós, R., Chicano, F.: Tunneling between plateaus: improving on a state-of-the-art MAXSAT solver using partition crossover. In: Proceedings of the Genetic and Evolutionary Computation Conference, pp. 921–928. ACM. https://doi.org/10.1145/3205455.3205482
6. de Godoi, G.A., Tinós, R., Sanches, D.S.: A graph-based crossover and soft-repair operators for the Steiner tree problem. In: Britto, A., Valdivia Delgado, K. (eds.) BRACIS 2021. LNCS (LNAI), vol. 13073, pp. 111–125. Springer, Cham (2021). https://doi.org/10.1007/978-3-030-91702-9_8
7. Dubois-Lacoste, J., Hoos, H.H., Stützle, T.: On the empirical scaling behaviour of state-of-the-art local search algorithms for the Euclidean TSP. In: Silva, S., Esparcia-Alcázar, A.I. (eds.) Proceedings of the Genetic and Evolutionary Computation Conference (GECCO), pp. 377–384. ACM (2015). https://doi.org/10.1145/2739480.2754747
8. Dunton, P., Whitley, D.: Reducing the cost of partition crossover on large MAXSAT problems: the PX-preprocessor. In: Proceedings of the Genetic and Evolutionary

Computation Conference, pp. 694–702. ACM. https://doi.org/10.1145/3512290.3528748
9. Heins, J., Bossek, J., Pohl, J., Seiler, M., Trautmann, H., Kerschke, P.: On the potential of normalized TSP features for automated algorithm selection. In: Proceedings of the 16th ACM/SIGEVO Conference on Foundations of Genetic Algorithms (FOGA), pp. 1–15. ACM (2021). https://doi.org/10.1145/3450218.3477308
10. Heins, J., Bossek, J., Pohl, J., Seiler, M., Trautmann, H., Kerschke, P.: A study on the effects of normalized TSP features for automated algorithm selection. Theoret. Comput. Sci. **940**, 123–145 (2023). https://doi.org/10.1016/j.tcs.2022.10.019
11. Heins, J., Dengel, S., Whitley, D., Kerschke, P.: Clearing the Combinatorial Fog: tracing the hidden paths of TSP heuristics. In: Proceedings of the 18th ACM/SIGEVO Conference on Foundations of Genetic Algorithms, pp. 73–84. ACM (2025). https://doi.org/10.1145/3729878.3746623
12. Heins, J., Schäpermeier, L., Kerschke, P., Whitley, D.: Dancing to the State of the Art? How candidate lists influence LKH for solving the Traveling Salesperson Problem. In: Affenzeller, M., et al. (eds.) Parallel Problem Solving from Nature – PPSN XVIII, pp. 100–115. Springer, Cham (2024). https://doi.org/10.1007/978-3-031-70055-2_7
13. Heins, J., Whitley, D., Kerschke, P.: To repair or not to repair? Investigating the importance of AB-cycles for the state-of-the-art TSP heuristic EAX. In: Proceedings of the Genetic and Evolutionary Computation Conference, pp. 231–239. ACM. https://doi.org/10.1145/3712256.3726436
14. Helsgaun, K.: An effective implementation of the Lin-Kernighan traveling salesman heuristic. Eur. J. Oper. Res. **126**(1), 106–130 (2000). https://doi.org/10.1016/S0377-2217(99)00284-2
15. Helsgaun, K.: General k-opt submoves for the Lin-Kernighan TSP heuristic. Math. Program. Comput. **1**(2–3), 119–163 (2009). https://doi.org/10.1007/s12532-009-0004-6
16. Kerschke, P., Bossek, J., Trautmann, H.: Parameterization of state-of-the-art performance indicators: a robustness study based on inexact TSP solvers. In: Proceedings of the Genetic and Evolutionary Computation Conference Companion (GECCO 2018), pp. 1737–1744. ACM, Kyoto, Japan (2018). https://doi.org/10.1145/3205651.3208233
17. Kerschke, P., Kotthoff, L., Bossek, J., Hoos, H.H., Trautmann, H.: Leveraging TSP solver complementarity through machine learning. Evol. Comput. (ECJ) **26**(4), 597–620 (2018). https://doi.org/10.1162/evco_a_00215
18. Nagata, Y., Kobayashi, S.: Edge assembly crossover: a high-power genetic algorithm for the travelling salesman problem. In: Bäck, T. (ed.) Proceedings of the 7th International Conference on Genetic Algorithms (ICGA), pp. 450–457. Morgan Kaufmann, East Lansing (1997)
19. Nagata, Y., Kobayashi, S.: A powerful genetic algorithm using edge assembly crossover for the traveling salesman problem. INFORMS J. Comput. **25**(2), 346–363 (2013). https://doi.org/10.1287/ijoc.1120.0506
20. Przewozniczek, M.W., Tinós, R., Frej, B., Komarnicki, M.M.: On turning black - into dark gray-optimization with the direct empirical linkage discovery and partition crossover. In: Proceedings of the Genetic and Evolutionary Computation Conference, pp. 269–277. ACM. https://doi.org/10.1145/3512290.3528734
21. Quevedo De Carvalho, O., Whitley, D.: Dramatically faster partition crossover for the traveling salesman problem. In: Proceedings of the Genetic and Evolutionary Computation Conference, pp. 818–826. ACM (2025). https://doi.org/10.1145/3712256.3726465

22. Seiler, M., Pohl, J., Bossek, J., Kerschke, P., Trautmann, H.: Deep learning as a competitive feature-free approach for automated algorithm selection on the traveling salesperson problem. In: Bäck, T., et al. (eds.) PPSN 2020. LNCS, vol. 12269, pp. 48–64. Springer, Cham (2020). https://doi.org/10.1007/978-3-030-58112-1_4
23. Seiler, M.V., Rook, J., Heins, J., Ludger Preuß, O., Bossek, J., Trautmann, H.: Using reinforcement learning for per-instance algorithm configuration on the TSP. In: 2023 IEEE Symposium Series on Computational Intelligence (SSCI), pp. 361–368 (2023). https://doi.org/10.1109/SSCI52147.2023.10372008
24. Taillard, É.D., Helsgaun, K.: POPMUSIC for the travelling salesman problem. Eur. J. Oper. Res. **272**(2), 420–429 (2019). https://doi.org/10.1016/j.ejor.2018.06.039
25. Tinós, R., Whitley, D., Chicano, F.: Partition crossover for pseudo-Boolean optimization. In: Proceedings of the 2015 ACM Conference on Foundations of Genetic Algorithms XIII, pp. 137–149. ACM. https://doi.org/10.1145/2725494.2725497
26. Tinós, R., Whitley, D., Ochoa, G.: A new generalized partition crossover for the traveling salesman problem: tunneling between local optima. Evolutionary Computation **28**(2), 255–288 (2020). https://doi.org/10.1162/evco_a_00254
27. Tlili, T., Chicano, F., Krichen, S., Alba, E.: Evolutionary algorithm based on partition crossover (EAPX) for the vehicle routing problem. In: 2015 International Conference on Intelligent Networking and Collaborative Systems, pp. 169–175. IEEE. https://doi.org/10.1109/INCoS.2015.89
28. Varadarajan, S., Whitley, D.: The massively parallel mixing genetic algorithm for the traveling salesman problem. In: Proceedings of the Genetic and Evolutionary Computation Conference, pp. 872–879. ACM. https://doi.org/10.1145/3321707.3321772
29. Varadarajan, S., Whitley, D., Ochoa, G.: Why many travelling salesman problem instances are easier than you think. In: Proceedings of the 2020 Genetic and Evolutionary Computation Conference, pp. 254–262. ACM. https://doi.org/10.1145/3377930.3390145
30. Whitley, D., Hains, D., Howe, A.: Tunneling between optima: partition crossover for the traveling salesman problem. In: Proceedings of the 11th Annual conference on Genetic and evolutionary computation, pp. 915–922. ACM (2009). https://doi.org/10.1145/1569901.1570026
31. Whitley, D., Hains, D., Howe, A.: A hybrid genetic algorithm for the traveling salesman problem using generalized partition crossover. In: Schaefer, R., Cotta, C., Kołodziej, J., Rudolph, G. (eds.) PPSN 2010. LNCS, vol. 6238, pp. 566–575. Springer, Heidelberg (2010). https://doi.org/10.1007/978-3-642-15844-5_57

Evolutionary Latent Space Search for Real-Time Dynamic Vehicle Routing Under Demand and Traffic Uncertainty

Li Li, Xun Nie, and Liang Chang(✉)

Guangxi Key Laboratory of Trusted Software, Guilin University of Electronic Technology, Guilin, China
lili@guet.edu.cn , changl@guet.edu.cn

Abstract. Dynamic Capacitated Vehicle Routing Problems (DVRPs) require real-time routing decisions under continuously changing demand and traffic conditions. Existing deep reinforcement learning (DRL) approaches generate a single solution per inference step, limiting flexibility in dynamic environments. We propose a Latent Vector-Guided Evolutionary Inference Method (LVGEIM) that maintains a population of latent vector individuals in the continuous policy space to guide parallel route decoding during inference. At each step, the population is updated via elitism, Gaussian perturbation, and random immigration, enabling gradient-free adaptation to shifting conditions without modifying the policy network. A lightweight two-stream encoding architecture separates static and dynamic features to enable efficient parallel inference. Evaluations across four problem scales and six uncertainty scenarios show that LVGEIM consistently outperforms competitive baselines including POMO, achieving lower routing cost across all tested conditions while remaining real-time.

Keywords: Dynamic Vehicle Routing · Evolutionary Inference · Latent Space Search · Deep Reinforcement Learning · Dynamic Optimization

1 Introduction

The Vehicle Routing Problem (VRP) is a fundamental NP-hard problem in combinatorial optimization. In practical logistics, real-world conditions introduce demand fluctuations, traffic accidents, and road closures that static VRP formulations cannot address. As demands and traffic evolve continuously, the optimal routing decision shifts at each time step. A routing policy trained on static instances has no built-in mechanism to adapt to such shifts and may commit to increasingly suboptimal decisions as conditions change.

Most existing DRL approaches for DVRP rely on single-shot decoding, in which the policy network generates one route per inference pass from the current state embedding. This produces a single deterministic solution with no

G. Iacca et al. (Eds.): PPSN 2026, LNCS 16985, pp. 315–331, 2027.
https://doi.org/10.1007/978-3-032-36214-8_20

mechanism to explore alternative routing strategies within the same time step. Under continuously shifting conditions, this limited exploration leaves the agent vulnerable to poor decisions. Recent work on inference-time search has shown that exploring multiple decoding trajectories during inference can substantially improve solution quality. Active Search [10] demonstrated this for static combinatorial optimization (CO) instances, and COMPASS [5] further showed that evolving a population of latent vectors via the Covariance Matrix Adaptation Evolution Strategy (CMA-ES) in the continuous policy space yields strong results on static VRPs.

However, these methods assume a fixed fitness landscape throughout the search process. In dynamic environments, the fitness landscape shifts at every time step as demands and traffic costs change, so search strategies designed for static instances require adaptation to remain effective. This challenge is closely related to the problem of tracking moving optima studied in evolutionary computation [24], where maintaining population diversity has been shown to improve the algorithm's ability to respond to environmental changes [25]. The continuous latent policy space is well suited to population-based search. Small perturbations around the current elite correspond to locally modified routing strategies, which allows gradient-free exploration of nearby policies without modifying the network at inference time. Yet existing latent-space search methods assume a static fitness landscape and cannot adapt to per-step environmental changes. We propose a Latent Vector-Guided Evolutionary Inference Method (LVGEIM) that maintains a population of latent vector individuals in the continuous policy space, updated at each inference step through elitism, Gaussian perturbation, and random immigration. The specific contributions are:

1. A conceptual connection between DVRP inference and the problem of tracking moving optima in the continuous latent space, which motivates a population-based inference design. To our knowledge, this is the first work to bring diversity maintenance from dynamic optimization into DRL-based inference for dynamic combinatorial optimization.
2. A hybrid population update operator combining elitism, Gaussian perturbation, and random immigration, which balances exploitation of the current best individual with exploration of the latent space, enabling the population to track continuously shifting environmental conditions.
3. A decoupled spatio-temporal encoding architecture that caches static graph embeddings once per episode, reducing per-step encoder complexity from $\mathcal{O}(N^3)$ to $\mathcal{O}(N^2)$ and enabling real-time population-based inference.

2 Related Work

DVRPs require algorithms that can act quickly on incomplete information while remaining effective as conditions continue to change [15,18,21]. Early approaches based on Rolling Horizon frameworks [9,20] and population-based metaheuristics such as ACO [1] require frequent re-solving upon environmental changes, resulting in inference latency that is incompatible with high-frequency dynamic dispatch.

Recent advances in DRL have established end-to-end inference as a mainstream DVRP paradigm. Pioneering works trained pointer networks [17] and Transformer-based Attention Models [13] to construct routes sequentially. To handle dynamic uncertainty, temporal difference learning [11] and multi-start parallel exploration frameworks such as POMO [14] were introduced. Kadyrov et al. [12] further examined DRL-based route construction under dual demand and traffic uncertainties. Two-stage training strategies have also been shown to improve DRL policy quality for VRP [6]. However, when applied to dynamic environments, these approaches treat each time step as an isolated static instance. Search history is discarded at every step, so the agent lacks temporal memory for tracking the shifting optimal policy. Augmented variants such as POMO with symmetry augmentation improve within-step diversity but still do not transfer information across time steps, preventing warm-starting of the next search step.

Beyond these construction methods, inference-time search further improves solution quality. COMPASS [5] maintains a population of latent vectors guided by CMA-ES, where the covariance matrix is estimated over multiple generations to direct the search toward promising regions of the latent space. This mechanism is effective when the fitness landscape remains fixed, but in dynamic environments the landscape shifts at every time step as demands and traffic evolve. The covariance estimates accumulated over previous generations become stale after each environmental change, and re-estimating them from scratch at every step would negate the warm-start benefit of population-based search. This motivates a stateless update whose operators depend only on the current population rather than on cross-step statistics.

In evolutionary computation, Dynamic Optimization Problems (DOPs) require algorithms to track a moving optimum as the environment evolves [24, 25]. A key mechanism is the elitism-based immigrants scheme [22], later extended with memory [23] and applied to dynamic routing [7,16]. At each generation, mutated copies of the previous elite replace the worst individuals, biasing exploration toward the recent optimum. LVGEIM inherits this elitism-plus-immigration structure but operates in a different space. Whereas classical schemes perturb discrete encodings (binary strings or permutation tours) via bit-flip operators, LVGEIM perturbs continuous latent vectors via Gaussian noise, producing semantically smooth changes in the decoded routing policy.

3 Problem Formulation

We formalize the DVRP under dual uncertainties as a Partially Observable Markov Decision Process (POMDP). The logistics network is defined on a complete directed graph $G = (V, E)$, where $V = \{0, 1, \ldots, N\}$ comprises the central depot (node 0) and N spatially distributed customers. A vehicle with maximum capacity C departs from the depot to fulfill customer demands and must return to the depot to reload when its remaining capacity is insufficient.

At time step t, the underlying state s_t encapsulates the vehicle's position $x_t \in V$, remaining capacity $l_t \in [0, C]$, actual customer demands $D_t = \{D_t(1), \ldots, D_t(N)\}$, and actual network traffic weights W_t. The action $a_t \in V$ selects the next node from the set of unmasked candidates, where a binary mask $M_t(j) \in \{0, 1\}$ prevents revisiting fully served customers. Since dynamic environmental shifts are hidden prior to observation, the agent acquires only a local observation slice $\omega_t = \{x_t, l_t, M_t, \hat{D}_t, \hat{W}_t\}$, where $\hat{D}_t$ and $\hat{W}_t$ are the currently revealed demand and traffic estimates.

The state transition $P(s_{t+1}|s_t, a_t)$ involves deterministic capacity updates $(l_{t+1} = l_t - \hat{D}_t(a_t))$ alongside stochastic environmental drifts. Traffic conditions evolve as an autoregressive process with temporal smoothing coefficient β, a stationary trend component W_{trend}, and Gaussian noise ξ_t^W:

$$W_{t+1} = \beta W_t + (1 - \beta) W_{\text{trend}} + \xi_t^W \tag{1}$$

where β is the temporal smoothing coefficient and ξ_t^W is Gaussian traffic noise. Customer demands drift concurrently via:

$$D_{t+1} = \max(0,\ D_t + \xi_t^D) \tag{2}$$

where $\xi_t^D \sim \mathcal{N}(0, \sigma_D^2)$ and the max operator enforces non-negative demands.

Equations (1) and (2) describe how the environment changes between consecutive time steps. A routing decision at step t incurs a cost determined by W_t, which was only partially observable at $t-1$ through $\hat{W}_{t-1}$, and unserved customer demands continue to drift throughout the episode. As a result, the cost-optimal policy at step t may differ from that at $t-1$.

The agent receives a reward $r_t = -W_t(x_t, a_t)$ reflecting the actual travel cost on edge (x_t, a_t). The objective is to find a routing policy $\pi_\theta : \omega_t \mapsto a_t$ that minimizes the expected cumulative travel cost $\min_\pi \mathbb{E}\left[\sum_{t=0}^{T} W_t(x_t, a_t)\right]$ over T steps, where the episode terminates when all N customers are served.

4 Methodology

4.1 Training Framework and Efficient Decoupling

To support high-frequency parallel evolutionary search at low per-step cost, we adopt a Decoupled Spatio-Temporal Encoding scheme, whose overall pipeline is illustrated in Fig. 1 (Phase 1). Static features (e.g., fixed node coordinates, expected demand distributions) are embedded strictly *once per episode* using a global Graph Attention Network (GATv2) [4] and a Transformer, producing a cached embedding $\boldsymbol{e}_{\text{static}}$.

In contrast, dynamic features are processed sequentially at each evolutionary step via a lightweight Multilayer Perceptron (MLP) employing pointwise mappings. Specifically, for each node j, the MLP takes a concise input vector comprising its current unfulfilled demand $D_t(j)$, the vehicle's remaining capacity l_t, and the binary mask status $M_t(j)$, directly projecting these scalars into a high-dimensional dynamic embedding $\boldsymbol{e}_{\text{dynamic}}$. Note that the

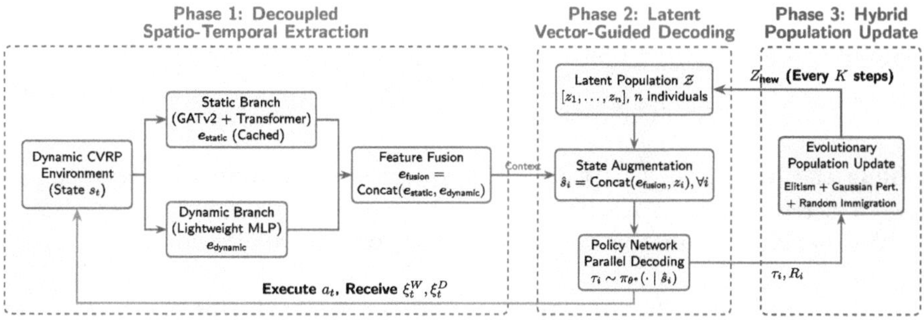

Fig. 1. Overall architecture of LVGEIM across three phases. **Phase 1** (Decoupled Spatio-Temporal Extraction): static graph features are encoded once per episode via GATv2 and a Transformer (e_{static}, cached), while dynamic features are updated at each step via a lightweight MLP (e_{dynamic}); the two streams are fused into e_{fusion}. **Phase 2** (Latent Vector-Guided Decoding): each individual $z_i \in \mathcal{Z}$ augments the fused context to form $\hat{s}_i = \text{Concat}(e_{\text{fusion}}, z_i)$, enabling n-fold parallel decoding via $\tau_i \sim \pi_{\theta^*}(\cdot \mid \hat{s}_i)$. **Phase 3** (Hybrid Population Update): the evaluated pairs (τ_i, R_i) drive elitism, Gaussian perturbation, and random immigration every K steps, producing Z_{new} that re-enters Phase 2. The orange loop closes the online-tracking cycle: execute a_t, receive ξ_t^W, ξ_t^D.

MLP processes each node independently without executing expensive graph-wide neighborhood aggregations. The two streams are then fused as $e_{\text{fusion}} = \text{Concat}(e_{\text{static}}, e_{\text{dynamic}})$. Because the static encoder runs once per episode and the dynamic encoder operates on each node independently, the per-step encoding cost is $\mathcal{O}(N)$ rather than the $\mathcal{O}(N^2)$ required by full-graph re-encoding. This separation also decouples the network's parameter dimensions from the number of customer nodes N, which supports generalization across problem scales (evaluated in Sect. 5.2).

The policy network is optimized using Proximal Policy Optimization (PPO). The reward function applies shaping (+1.0 per valid service, −0.5 per time step) to penalize capacity violations and redundant detours; these values were selected via preliminary experiments to balance service incentive against route length. During training, latent vectors are re-sampled from $\mathcal{N}(0, I)$ at each episode rather than evolved across steps, and the policy is updated via PPO on the trajectory of the best-performing individual; the population size n matches that used in inference (Algorithm 1).

4.2 Evolutionary Latent Space Inference

In dynamic vehicle routing, the routing policy that performs well under current conditions may become suboptimal as the environment changes at the next step. To allow the inference procedure to adapt to these shifts, we maintain a population of latent vector individuals $Z = [z_1, z_2, \ldots, z_n]$ in the continuous policy space, updated at each step through three lightweight operators.

As shown in Fig. 1 (Phase 2), each individual $z_i \in Z$ is concatenated with the fused context $\boldsymbol{e}_{\text{fusion}}$ to form an augmented state:

$$\hat{s}_i = \text{Concat}(\boldsymbol{e}_{\text{fusion}}, z_i), \quad \forall i \in \{1, \ldots, n\} \tag{3}$$

enabling the fixed policy network to decode n diverse routes in parallel: $\tau_i \sim \pi_{\theta^*}(\cdot \mid \hat{s}_i)$. Each route τ_i is evaluated by the environment simulator to obtain a fitness value R_i (cumulative reward over the segment since the last population update).

During inference, the population is updated through three operators. The *elitism* operator retains $z_{i^*} = \arg\max_i(R_i)$, carrying forward the latent vector that produced the best routing outcome in the current step and providing a warm start for the next step. The *Gaussian perturbation* operator generates offspring centered on the elite individual, exploring neighboring regions of the latent space that correspond to locally modified routing strategies. The *random immigration* operator injects independently sampled individuals to maintain population diversity and reduce the risk of premature convergence to a narrow region of the latent space.

The per-step evolutionary mechanism is visually detailed in Fig. 2 and formally specified in Algorithm 1.

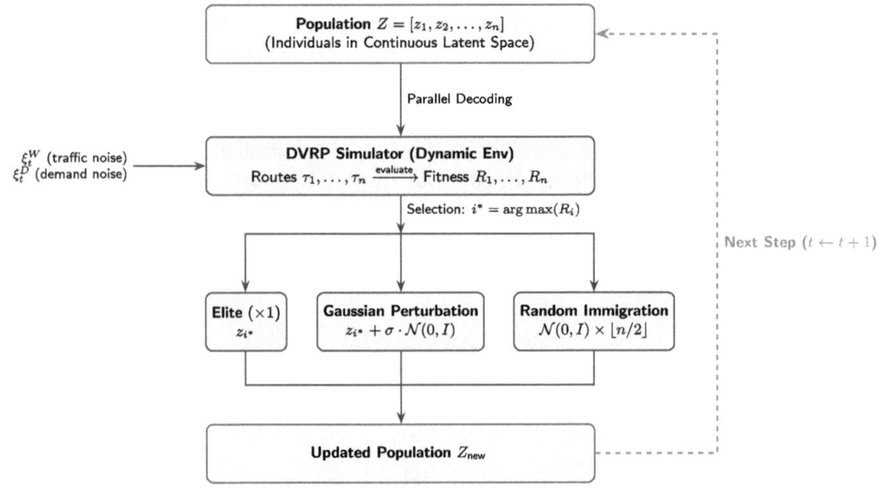

Fig. 2. One evolutionary step of LVGEIM. The population $Z = [z_1, z_2, \ldots, z_n]$ is decoded in parallel by the policy network into routes $\tau_1, \ldots, \tau_n$, which are evaluated by the DVRP Simulator under current observations ξ_t^W, ξ_t^D to obtain fitness values $R_1, \ldots, R_n$. The elite individual $i^* = \arg\max(R_i)$ seeds the next generation via three operators: elitism (retain z_{i^*}), Gaussian perturbation ($z_{i^*} + \sigma \cdot \mathcal{N}(0, I)$), and random immigration ($\mathcal{N}(0, I) \times \lfloor n/2 \rfloor$). The updated population Z_{new} feeds back into the next step $(t+1)$, enabling continuous tracking of moving optima without gradient updates.

Population Update Frequency. Our implementation uses $K = 1$, meaning the population is updated at every step. Since the fitness landscape shifts at every step as W_t and D_t change, setting $K > 1$ would allow stale latent individuals to guide decisions under outdated conditions, accumulating tracking error. This choice is consistent with findings in DOP research that high-frequency environmental change benefits from high-frequency population renewal [3].

State Synchronization and Its Implications. Prior to generating a new population, all n parallel environment states are synchronized to the elite individual's state $S_t^{i^*}$ (Algorithm 1, Line 16). This synchronization includes both the physical state (vehicle position, remaining capacity, served customers) and the decoder's GRU hidden state, which encodes the sequential decision history. As a result, the population diversity maintained by LVGEIM exists in the latent decision space (different z_i values leading to different next-step routing choices) rather than in the trajectory space (different complete paths). We adopt this design for two reasons. First, evaluating individuals from diverged physical states would make their fitness values incomparable, as differences in accumulated cost would reflect path history rather than the quality of the current latent vector. Second, synchronization to the elite state provides a consistent starting point for the next generation, analogous to restarting search from the incumbent solution after each perturbation in iterated local search.

This design trades off one form of exploration for another. By synchronizing to the elite state, the population explores alternative next actions from a shared, high-quality state rather than maintaining fully divergent trajectories. Trajectories requiring early divergence are therefore not explored within a single episode.

4.3 Complexity Analysis

We analyze the per-episode inference complexity for a population of n individuals over $T \propto N$ decoding steps.

Baseline (Coupled Encoding). Standard GNN-based DVRP approaches re-encode all node embeddings at every decoding step via full-graph attention at $\mathcal{O}(N^2)$ per step. At each step, the decoder computes attention scores between a context query and all N candidate node embeddings at $\mathcal{O}(N)$. Over T steps and n individuals, the total cost is:

$$C_{\text{baseline}} = n \cdot T \cdot \big[\underbrace{\mathcal{O}(N^2)}_{\text{encoder}} + \underbrace{\mathcal{O}(N)}_{\text{decoder}}\big] = \mathcal{O}(n \cdot N^3) \tag{4}$$

where the encoder term dominates.

LVGEIM (Decoupled Encoding). The static encoder (GATv2 + Transformer) executes once per episode at $\mathcal{O}(N^2)$. The dynamic encoder (MLP) processes each node independently at $\mathcal{O}(1)$ per node, yielding $\mathcal{O}(N)$ per step. The decoder uses

Algorithm 1. Latent Vector-Guided Evolutionary Inference (LVGEIM)

Require: Population size n; latent dimension d; pre-trained policy θ^*; update interval K; perturbation scale σ; encoders ϕ_{static}, ϕ_{dynamic}

$Z = [z_1, \ldots, z_n] \sim \mathcal{N}(0, I),\ Z \in \mathbb{R}^{n \times d}$
$\boldsymbol{e}^i_{\text{static}} \leftarrow \phi_{\text{static}}(S^i_0),\ \forall i$ ▷ Phase 1: cache once per episode
$R_{\text{seg}} \leftarrow \mathbf{0}$; $Done \leftarrow$ False
for $t = 1$ **to** $T_{\max}$ **do**
  if $Done$ **then break**
  end if
  $\boldsymbol{e}^i_{\text{dynamic}} \leftarrow \phi_{\text{dynamic}}(S^i_{t-1}),\ \forall i$ ▷ Phase 1: per-step
  $\boldsymbol{e}^i_{\text{fusion}} \leftarrow \text{Concat}(\boldsymbol{e}^i_{\text{static}}, \boldsymbol{e}^i_{\text{dynamic}}),\ \forall i$
  $\hat{S}^i_{t-1} \leftarrow \text{Concat}(\boldsymbol{e}^i_{\text{fusion}}, z_i),\ \forall i$ ▷ Phase 2: state augmentation
  $\{A^i_t\}^n_{i=1} \sim \pi_{\theta^*}(\cdot \mid \hat{S}^i_{t-1})$ ▷ Phase 2: parallel decoding
  $\{S^i_t, r^i_t, d^i_t\}^n_{i=1} \leftarrow \text{ENV.STEP}(\{A^i_t\})$
  $R_{\text{seg}} \leftarrow R_{\text{seg}} + \{r^i_t\}^n_{i=1}$
  if $t \bmod K = 0$ **and not** $Done$ **then**
   ▷ Phase 3: hybrid population update
   $i^* \leftarrow \arg\max(R_{\text{seg}})$
   $S^i_t \leftarrow S^{i^*}_t,\ \forall i$ ▷ sync all states to elite
   $Z_{\text{new}}[0] \leftarrow z_{i^*}$ ▷ **1.** Elitism
   $Z_{\text{new}}[1 : \lfloor n/2 \rfloor] \leftarrow z_{i^*} + \sigma \cdot \mathcal{N}(0, I)$ ▷ **2.** Gaussian Perturbation
   $Z_{\text{new}}[\lfloor n/2 \rfloor : n] \leftarrow \mathcal{N}(0, I)$ ▷ **3.** Random Immigration
   $Z \leftarrow Z_{\text{new}}$; $R_{\text{seg}} \leftarrow \mathbf{0}$
  end if
  $Done \leftarrow \text{all}(d^i_t = \text{True})$
end for
return OptimalPath, min(TotalCost)

a GRU-updated global context broadcast to each node, followed by a node-wise linear projection to produce action logits, at $\mathcal{O}(N)$ per step. Unlike cross-attention decoders, this design avoids query-key matrix operations, reducing the constant factor while maintaining the same asymptotic complexity. The total cost is:

$$C_{\text{LVGEIM}} = \underbrace{\mathcal{O}(N^2)}_{\text{static encoder (once)}} + n \cdot T \cdot \Big[\underbrace{\mathcal{O}(N)}_{\text{dynamic encoder}} + \underbrace{\mathcal{O}(N)}_{\text{decoder}} \Big] = \mathcal{O}(N^2 + n \cdot N^2) = \mathcal{O}(n \cdot N^2) \tag{5}$$

The asymptotic improvement stems from eliminating the dominant $\mathcal{O}(N^2)$-per-step encoder term; the decoder cost remains $\mathcal{O}(N)$ per step in both cases. This reduction from cubic to quadratic in N enables population-based search at every time step, consistent with the inference time measurements reported in Sect. 5.2.

5 Experiments

5.1 Experimental Setup

We evaluate LVGEIM against eight baselines across four problem scales ($N \in \{20, 50, 100, 200\}$) and six uncertainty scenarios designed to test different aspects of adaptation to changing conditions. The single-source scenarios (Demand-Only, Traffic-Only) isolate the contribution of each uncertainty source; the Standard Dual scenario reflects typical dynamic dispatch conditions; the Unlimited Capacity scenario removes vehicle reload constraints to isolate pure routing performance; and the High Traffic and High Demand scenarios test robustness under amplified environmental noise. Parameter settings are given in Table 1.

Table 1. Uncertainty scenario parameters. Traffic smoothing coefficient $\beta = 0.07$ and latent population hyperparameters ($K = 1$, $n = 64$, $d = 16$) are fixed across all scenarios.

Scenario	σ_D	σ_T	m	C
Standard Dual	1.0	1.0	5.0	15
Demand-Only	1.0	0.0	0.0	15
Traffic-Only	0.0	1.0	5.0	15
Unlimited Capacity	1.0	1.0	5.0	∞
High Traffic	1.0	10.0	5.0	15
High Demand	3.0	1.0	5.0	15

σ_D/σ_T: demand/traffic noise std (Eqs. 2, 1); m: traffic trend magnitude in W_{trend} (Eq. 1); C: vehicle capacity. High Traffic ($\sigma_T = 10$) and High Demand ($\sigma_D = 3$) are set to 10× and 3× standard values to stress-test robustness under amplified noise

Baselines include four DRL-based methods: Vanilla GNN, Hybrid (C-W) (GNN with Clarke-Wright [2] post-processing), POMO [14], and POMO (Aug) (POMO with symmetry augmentation [14]); and four classical heuristics: traffic-aware Tabu Search [8], traffic-aware Greedy, Clarke-Wright, and Sweep [19]. All DRL baselines are trained from scratch on the same dynamic POMDP environment using PPO and evaluated on identical test instances. Note that POMO was originally designed for static VRP instances, and its core mechanism (symmetric starting-point augmentation) does not directly address temporal shifts. Training it in the dynamic POMDP environment represents a reasonable adaptation but may not fully exploit its design strengths. This setup follows the protocol used in prior DVRP studies [12] that re-train static DRL architectures in dynamic settings.

The LVGEIM policy is trained using PPO on an NVIDIA RTX 3090 (24 GB). Models for $N \in \{20, 50\}$ (8×8 grid) are trained for 5,000 epochs; the $N = 100$ model (12×12 grid) is trained for 1,000 epochs. For $N = 200$ (15×15 grid), the $N = 100$ model is applied without fine-tuning, while all DRL baselines are trained specifically for $N = 200$. LVGEIM's advantage in this setting therefore holds against baselines with a training-scale advantage, providing a conservative estimate of cross-scale transferability. Performance is reported as mean routing cost and cost gap over 200 independent test instances per scenario-scale combination, with statistical significance assessed via paired t-test against LVGEIM.

5.2 Performance Comparison

Table 2 reports routing cost, relative cost gap, and inference time for all methods across 24 scenario-scale combinations. LVGEIM achieves the lowest routing cost in all 24 cases; all baselines incur positive cost gaps ranging from 5.5% to 158.1% relative to LVGEIM (paired t-test, $p < 0.001$; Bonferroni-corrected significance holds).

Effect of Environmental Noise Intensity. The advantage of LVGEIM over baselines varies substantially across scenarios. Under High Traffic uncertainty ($\sigma_T = 10$), LVGEIM outperforms POMO (Aug) by 76.9% at $N = 200$, compared to 24.9% under Standard Dual conditions. This widening gap is consistent with the role of elitism as a temporal memory mechanism. When the fitness landscape shifts more severely, carrying forward the best-performing latent vector provides a stronger advantage over methods that restart search from scratch at each step.

At the other extreme, under the Traffic-Only scenario at $N = 100$, the Hybrid (C-W) baseline achieves a gap of only 5.5%, the smallest among all combinations. In this setting, demand is fixed, so Clarke-Wright can be re-applied at each step using current traffic estimates as edge weights, exploiting the stable demand structure.

Scaling Behavior. As N increases from 20 to 200 under Standard Dual uncertainty, the gap of Vanilla GNN widens from 47.6% to 77.1% and POMO (Aug) from 16.6% to 24.9%, indicating that single-shot and within-step diversity become less effective at larger scales. LVGEIM's gap relative to classical heuristics remains relatively stable (e.g., Sweep: 15.3% to 19.4%), consistent with the decoupled encoding keeping per-step cost independent of N.

Inference Time. LVGEIM requires 0.53 s per episode at $N = 100$ and 1.70 s at $N = 200$, compared to 39.0 s and 236.7 s for POMO (Aug), a 139× reduction that follows from the encoder decoupling analyzed in Sect. 4.3. Classical heuristics are still faster in absolute terms, but they incur substantially higher routing cost.

Table 2. Routing cost, relative cost gap, and inference time across four problem scales and six dynamic scenarios. Gap (%) = $(C_{\text{alg}} - C_{\text{LVGEIM}})/C_{\text{LVGEIM}} \times 100\%$; a positive value indicates higher cost than LVGEIM. Bold: best result per row.

Scenario	Algorithm	$N = 20$ (8 × 8)			$N = 50$ (8 × 8)			$N = 100$ (12 × 12)			$N = 200$ (15 × 15)†		
		Cost↓	Gap§	Time/ep.	Cost↓	Gap§	Time/ep.	Cost↓	Gap§	Time/ep.	Cost↓	Gap§	Time/ep.
1. Standard	**LVGEIM (Ours)**	**207.58**	**0.0%**	**0.119 s**	**500.97**	**0.0%**	**0.254 s**	**1194.92**	**0.0%**	**0.530 s**	**2556.57**	**0.0%**	**1.704 s**
	POMO (Aug)	241.63	16.6%***	1.525 s	575.70	15.4%***	8.247 s	1401.91	17.9%***	39.013 s	3199.18	24.9%***	236.681 s
	POMO (Unaug)	245.03	18.2%***	1.396 s	581.70	16.6%***	7.488 s	1454.75	22.3%***	33.155 s	3300.58	28.9%***	195.421 s
	Vanilla GNN	306.36	47.6%***	0.068 s	741.45	48.0%***	0.160 s	1919.81	60.7%***	0.428 s	4528.74	77.1%***	1.810 s
	Hybrid (C-W)	243.76	17.4%***	0.060 s	577.93	15.4%***	0.148 s	1348.49	12.9%***	0.455 s	2928.83	14.6%***	2.651 s
	Tabu Search	264.60	27.5%***	0.021 s	631.65	26.1%***	0.049 s	1460.08	22.2%***	0.172 s	3167.47	23.9%***	1.542 s
	Greedy	264.35	27.3%***	0.021 s	631.56	26.1%***	0.049 s	1457.25	22.0%***	0.172 s	3165.99	23.8%***	1.543 s
	Sweep	239.26	15.3%***	0.018 s	581.44	16.1%***	0.041 s	1373.88	15.0%***	0.161 s	3051.79	19.4%***	1.258 s
2. Demand Only	**LVGEIM (Ours)**	**88.55**	**0.0%**	**0.122 s**	**203.88**	**0.0%**	**0.270 s**	**591.76**	**0.0%**	**0.564 s**	**1359.45**	**0.0%**	**1.810 s**
	POMO (Aug)	106.24	19.1%***	1.517 s	238.32	17.5%***	8.369 s	723.80	23.5%***	39.602 s	1838.50	35.4%***	236.427 s
	POMO (Unaug)	109.96	23.2%***	1.403 s	245.68	21.1%***	7.550 s	748.48	27.7%***	34.064 s	1964.70	44.7%***	195.845 s
	Vanilla GNN	123.16	39.1%***	0.068 s	300.79	47.5%***	0.158 s	1057.93	78.8%***	0.589 s	3075.60	126.2%***	1.841 s
	Hybrid (C-W)	105.22	18.8%***	0.061 s	226.91	11.3%***	0.150 s	638.60	7.9%***	0.472 s	1506.55	10.8%***	2.502 s
	Tabu Search	105.90	19.6%***	0.021 s	230.02	12.8%***	0.049 s	644.31	8.9%***	0.174 s	1539.07	13.2%***	1.560 s
	Greedy	105.90	19.6%***	0.021 s	230.02	12.8%***	0.049 s	644.31	8.9%***	0.176 s	1539.07	13.2%***	1.565 s
	Sweep	101.63	14.8%***	0.018 s	242.00	18.7%***	0.041 s	709.03	19.8%***	0.148 s	1713.40	26.0%***	1.344 s
3. Traffic Only	**LVGEIM (Ours)**	**205.76**	**0.0%**	**0.121 s**	**501.80**	**0.0%**	**0.278 s**	**1194.16**	**0.0%**	**0.604 s**	**2550.38**	**0.0%**	**1.745 s**
	POMO (Aug)	230.25	11.6%***	1.530 s	544.38	8.3%***	8.281 s	1326.00	11.8%***	37.544 s	3025.24	18.1%***	236.467 s
	POMO (Unaug)	235.40	14.1%***	1.408 s	558.61	11.2%***	7.509 s	1361.93	14.8%***	32.072 s	3228.94	26.1%***	195.918 s
	Vanilla GNN	294.74	43.2%***	0.068 s	697.13	38.9%***	0.158 s	1763.12	47.6%***	0.404 s	4073.01	59.7%***	1.831 s
	Hybrid (C-W)	232.81	13.1%***	0.060 s	545.80	8.8%***	0.150 s	1259.51	5.5%***	0.444 s	2741.22	7.5%***	2.651 s
	Tabu Search	259.14	25.9%***	0.022 s	621.17	23.8%***	0.049 s	1422.28	19.1%***	0.178 s	3087.50	21.1%***	1.574 s
	Greedy	259.82	26.3%***	0.021 s	620.14	23.6%***	0.049 s	1423.86	19.2%***	0.197 s	3088.54	21.1%***	1.576 s
	Sweep	239.00	16.2%***	0.018 s	580.81	15.7%***	0.041 s	1374.58	15.1%***	0.149 s	3057.19	19.9%***	1.258 s
4. Inf Capacity	**LVGEIM (Ours)**	**139.56**	**0.0%**	**0.119 s**	**317.42**	**0.0%**	**0.252 s**	**670.28**	**0.0%**	**0.512 s**	**1262.81**	**0.0%**	**1.696 s**
	POMO (Aug)	167.79	19.0%***	1.521 s	378.70	19.0%***	8.358 s	838.32	24.8%***	32.401 s	1805.86	43.1%***	235.434 s
	POMO (Unaug)	175.92	24.8%***	1.401 s	386.65	21.5%***	7.561 s	832.00	23.9%***	25.481 s	1771.02	40.3%***	196.248 s
	Vanilla GNN	270.93	94.1%***	0.068 s	603.36	90.1%***	0.160 s	1462.71	118.2%***	0.402 s	3259.34	158.1%***	1.834 s
	Hybrid (C-W)	188.71	35.2%***	0.061 s	431.95	36.1%***	0.147 s	969.49	44.6%***	0.446 s	2076.77	64.5%***	2.593 s
	Tabu Search	163.67	17.3%***	0.021 s	360.25	13.5%***	0.049 s	727.46	8.5%***	0.141 s	1425.14	12.9%***	1.488 s
	Greedy	163.70	17.3%***	0.021 s	360.81	13.7%***	0.049 s	730.10	8.9%***	0.142 s	1424.20	12.8%***	1.492 s
	Sweep	188.29	34.9%***	0.018 s	420.13	32.4%***	0.041 s	894.65	33.5%***	0.129 s	1800.06	42.5%***	1.144 s
5. High Traffic	**LVGEIM (Ours)**	**180.10**	**0.0%**	**0.121 s**	**403.13**	**0.0%**	**0.258 s**	**997.79**	**0.0%**	**0.557 s**	**2123.31**	**0.0%**	**1.776 s**
	POMO (Aug)	280.78	54.0%***	1.514 s	698.70	70.8%***	8.272 s	1651.37	65.5%***	38.387 s	3694.41	76.9%***	235.856 s
	POMO (Unaug)	299.36	64.2%***	1.401 s	719.24	75.8%***	7.488 s	1686.90	69.1%***	33.624 s	3921.93	87.8%***	195.459 s
	Vanilla GNN	360.49	100.2%***	0.068 s	883.46	119.2%***	0.158 s	2182.00	118.7%***	0.466 s	5060.65	138.3%***	1.831 s
	Hybrid (C-W)	312.58	73.6%***	0.060 s	723.63	79.5%***	0.151 s	1636.81	64.0%***	0.459 s	3498.50	64.8%***	2.651 s
	Tabu Search	326.70	81.4%***	0.021 s	775.34	92.3%***	0.049 s	1735.44	73.9%***	0.175 s	3728.40	75.6%***	1.542 s
	Greedy	327.10	81.6%***	0.021 s	774.11	92.0%***	0.049 s	1758.36	76.2%***	0.173 s	3727.89	75.6%***	1.543 s
	Sweep	372.46	106.8%***	0.018 s	917.05	127.5%***	0.041 s	2038.14	104.3%***	0.168 s	4280.82	101.6%***	1.258 s
6. High Demand	**LVGEIM (Ours)**	**209.76**	**0.0%**	**0.123 s**	**507.86**	**0.0%**	**0.259 s**	**1219.97**	**0.0%**	**0.565 s**	**2621.33**	**0.0%**	**1.745 s**
	POMO (Aug)	267.16	27.0%***	1.634 s	637.92	25.6%***	8.701 s	1594.29	30.9%***	42.719 s	3571.61	36.4%***	236.681 s
	POMO (Unaug)	270.53	28.6%***	1.488 s	648.84	27.7%***	7.846 s	1618.97	32.9%***	36.037 s	3694.27	41.0%***	195.441 s
	Vanilla GNN	338.65	61.4%***	0.068 s	874.90	72.3%***	0.160 s	2385.23	95.5%***	0.588 s	6231.23	137.7%***	1.831 s
	Hybrid (C-W)	269.55	28.5%***	0.060 s	647.33	27.5%***	0.149 s	1539.39	26.2%***	0.510 s	3400.62	29.7%***	2.651 s
	Tabu Search	289.86	38.2%***	0.021 s	692.86	36.4%***	0.049 s	1634.05	33.9%***	0.198 s	3590.62	37.0%***	1.542 s
	Greedy	288.87	37.7%***	0.021 s	692.16	36.3%***	0.049 s	1632.42	33.8%***	0.183 s	3585.44	36.8%***	1.543 s
	Sweep	237.46	13.2%***	0.018 s	579.76	14.2%***	0.041 s	1374.26	12.6%***	0.146 s	3052.94	16.5%***	1.258 s

† Zero-shot ($N = 100$ model); § Gap vs. LVGEIM (not optimum); $^{***}p < 0.001$ (t-test, 200 inst.)

5.3 Latent Space Evolution Analysis

To examine whether LVGEIM maintains population diversity during inference, we analyze latent vector evolution under the Standard Dual scenario at $N = 20$ using t-SNE visualization and mean distance tracking. We note that this analysis is conducted at the smallest problem scale for visualization clarity; quantitative results in Table 2 cover all four scales.

Figure 3 shows the t-SNE projection of all 64 latent individuals across Steps 0–24. Early steps occupy a diffuse central region; as inference proceeds, individuals migrate toward distinct clusters. Scattered individuals persist at all steps due to random immigration, suggesting persistent diversity in the projected space.

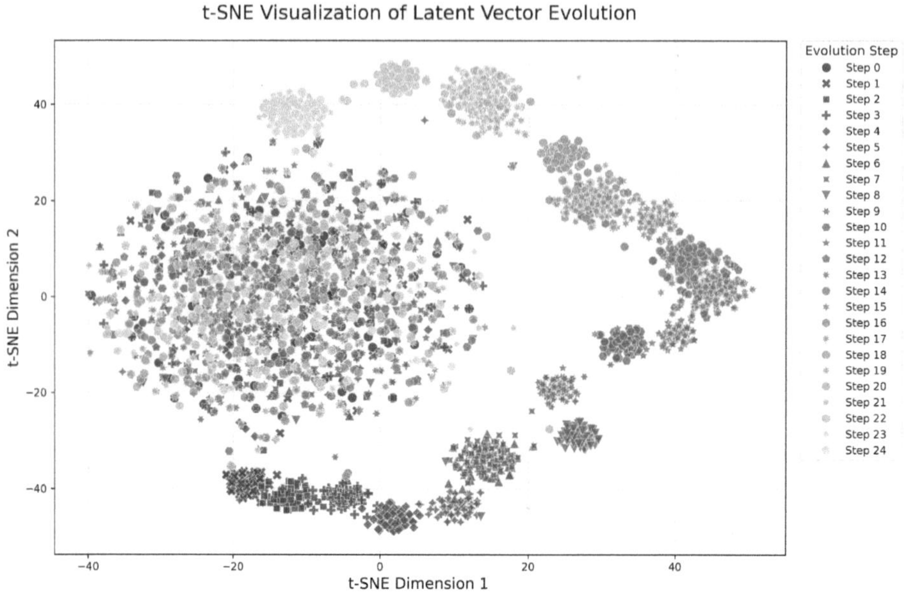

Fig. 3. t-SNE projection of the 64 latent individuals across 25 evolutionary steps (Steps 0–24) under the Standard Dual scenario ($N = 20$). Color and marker type encode the step index. Early steps (dark purple) concentrate near the arc origin; as inference proceeds, individuals migrate along a directional trajectory and form spatially distinct clusters (yellow-green). Scattered individuals persist throughout, produced by the random immigration operator. (Color figure online)

Figure 4 tracks the mean distance to centroid of the population. After a sharp drop at Step 1, the metric stabilizes within $[2.79, 3.0]$. This range reflects the population composition. Approximately half the individuals are Gaussian perturbations of the elite (clustered nearby) and half are random immigrants (spread broadly), producing a stable equilibrium between the two groups.

5.4 Ablation Study

To isolate the contribution of each proposed component, we define four ablation variants evaluated at $N = 20$:

- **Variant 1 (Heavy/Re-encode):** Replaces the two-stream decoupled encoding with a standard graph encoder that re-extracts all node features at every decoding step.

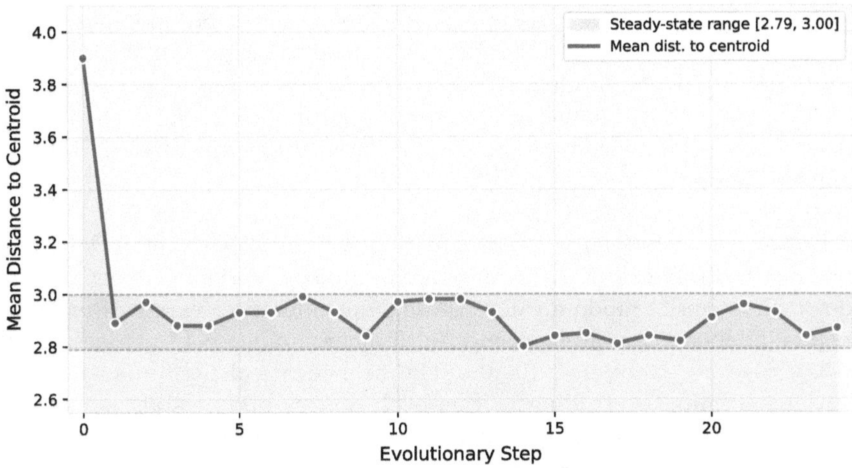

Fig. 4. Mean distance to centroid of the latent population over 25 evolutionary steps (Standard Dual, $N = 20$, $n = 64$). The drop at Step 1 reflects elitism selection pressure; the subsequent oscillation within $[2.79, 3.0]$ reflects the balance between elite-clustered perturbations and broadly spread immigrants.

- **Variant 2 (Greedy/No Latent):** Removes the latent vector mechanism entirely; the policy decodes greedily from the fused state embedding without latent conditioning.
- **Variant 3 (Static Latent/No Evo):** Retains latent vector conditioning but fixes all vectors at their initial random values throughout inference, disabling evolutionary update.
- **Variant 4 (Random Latent/No Evo):** Retains latent conditioning but replaces the directed evolutionary update with independent random resampling of all individuals at each step, removing the elite-centered structure (Table 3).

Table 3. Ablation of evolutionary operators and decoupled encoding ($N = 20$, 200 test instances).

Variant	Cost ↓	Gap (%)	Time (s)	p-value
LVGEIM (Ours)	**206.81**	**0.00%**	**0.08**	—
Variant 1 (Heavy/Re-encode)	207.71	0.44%	2.40	0.486 (ns)
Variant 2 (Greedy/No Latent)	340.08	64.45%	0.04	$1.05 \times 10^{-62***}$
Variant 3 (Static Latent/No Evo)	238.25	15.21%	0.07	$3.49 \times 10^{-50***}$
Variant 4 (Random Latent/No Evo)	238.93	15.53%	0.07	$6.75 \times 10^{-48***}$

$^{***}p < 0.001$; ns: not significant. Paired t-test vs. LVGEIM

Contribution of Latent Conditioning. Removing latent vector guidance entirely (Variant 2) increases routing cost by 64.45% ($p = 1.05 \times 10^{-62}$). Without latent conditioning, the policy produces a single deterministic output per state, eliminating the ability to evaluate multiple routing strategies within each step. This result establishes latent conditioning as the primary source of LVGEIM's performance advantage.

Contribution of Evolutionary Update. Variants 3 and 4 both retain latent conditioning but disable directed evolutionary update. Variant 3 fixes all latent vectors at their initial random values throughout inference; Variant 4 re-samples all vectors independently at each step. Both incur similar cost increases (15.21% and 15.53% respectively, $p < 10^{-47}$). The near-identical performance of these two variants admits two readings: either evolutionary update adds value beyond latent conditioning, or diversity matters more than the exact latent values.

The full LVGEIM result helps distinguish these readings. The further improvement of LVGEIM over Variants 3/4 cannot be attributed to diversity alone, since Variant 4 already has maximal per-step diversity through complete re-sampling. Instead, it stems from the temporal continuity provided by elitism. The elite vector carries information about which latent region yielded good results in the previous step, providing a directional bias that pure random sampling lacks. LVGEIM's advantage thus reflects two sources of roughly equal magnitude: latent conditioning (the gap from Variant 2 to Variants 3/4) and directed evolutionary update (the gap from Variants 3/4 to full LVGEIM).

Contribution of Decoupled Encoding. Replacing the two-stream architecture with full re-encoding (Variant 1) yields a statistically non-significant cost difference (gap 0.44%, $p = 0.486$), but increases inference time by 30× (2.40s vs. 0.08s). This indicates that the decoupled architecture achieves the computational savings analyzed in Sect. 4.3 without measurable loss in solution quality.

5.5 Path Visualization

Figure 5 shows route comparisons for all nine methods under the Standard Dual-Uncertainty scenario ($N = 50$, one depot, 8×8 grid).

LVGEIM (Panel 2, cost 481.3) produces routes that concentrate on low-congestion edges, with spatially compact sub-tours and few unnecessary cross-depot returns. In contrast, POMO variants (Panels 3–4, costs 523–535) and classical heuristics (Panels 7–9, costs 569–1145) show more red and orange edges, indicating more frequent traversal of high-congestion segments.

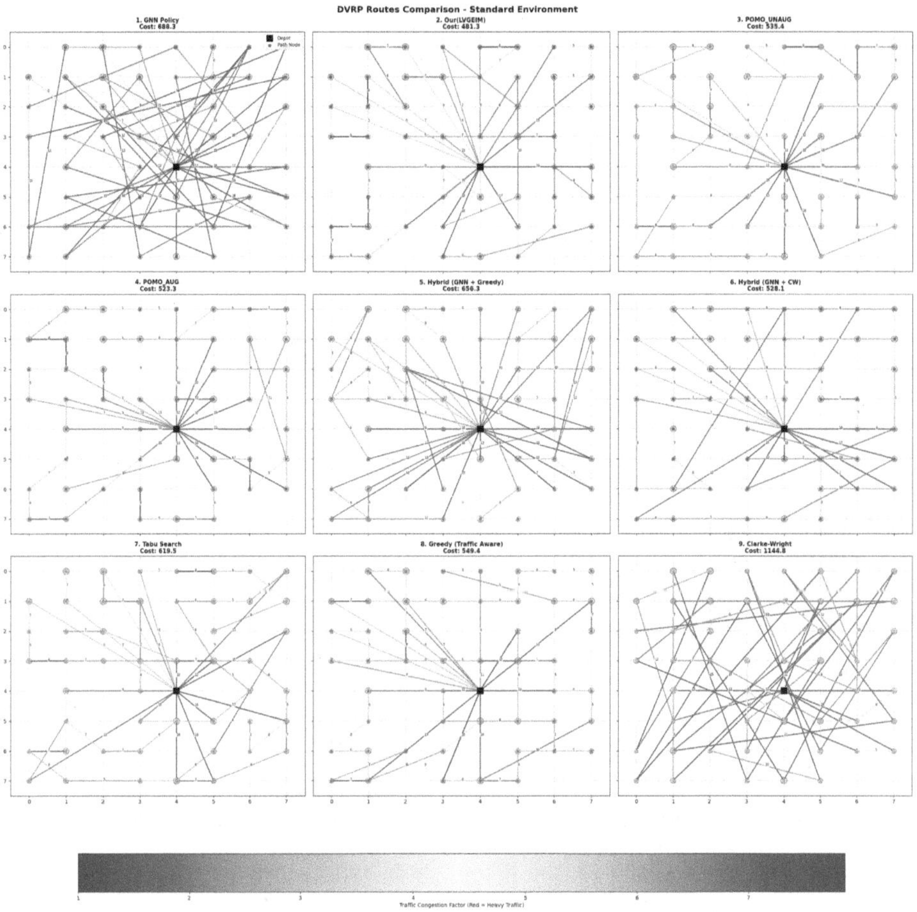

Fig. 5. Route comparison under Standard Dual-Uncertainty ($N = 50$, $8{\times}8$ grid). Edge color encodes traffic congestion at traversal time (green: low, red: high); residual red edges in Panel 2 reflect unavoidable high-traffic segments rather than suboptimal routing choices. (Color figure Online)

6 Conclusion

We presented LVGEIM, an inference-time method for dynamic vehicle routing that maintains a population of latent vector individuals updated at each step through elitism, Gaussian perturbation, and random immigration. By framing DVRP inference as a problem of tracking moving optima, the method adapts to shifting demand and traffic conditions without retraining or gradient computation at inference time. Experiments across four problem scales and six uncertainty scenarios show consistent improvements over all baselines, with larger advantages under higher environmental noise and substantial inference speedup from the decoupled encoding architecture. Several limitations remain: the population maintains diversity only in the latent decision space, and the current formulation is restricted to a single vehicle. Future work will investigate adaptive perturbation schedules and multi-vehicle extensions.

Acknowledgements. This work was supported by the National Natural Science Foundation of China under Grant 62366009.

References

1. Anastasiadou, M.N., Mavrovouniotis, M., Hadjimitsis, D.: Ant colony optimization for the dynamic electric vehicle routing problem. In: Affenzeller, M., et al. (eds.) Parallel Problem Solving from Nature - PPSN XVIII, pp. 68–84. Springer, Cham (2024)
2. Archetti, C., Coelho, L., Speranza, M., Vansteenwegen, P.: Beyond fifty years of vehicle routing: insights into the history and the future. Eur. J. Oper. Res. **330**(2), 355–372 (2026)
3. Branke, J.: Memory enhanced evolutionary algorithms for changing optimization problems. In: Proceedings of the Congress on Evolutionary Computation (CEC), pp. 1875–1882 (1999)
4. Brody, S., Alon, U., Yahav, E.: How Attentive are graph attention networks? In: International Conference on Learning Representations (2022)
5. Chalumeau, F., Surana, S., Bonnet, C., Courty, N., Vayatis, N.: Combinatorial optimization with policy adaptation using latent space search. In: Advances in Neural Information Processing Systems, vol. 36, pp. 7947–7959 (2023)
6. Chen, J., Huang, H., Zhang, Z., Wang, J.: Deep reinforcement learning with two-stage training strategy for practical electric vehicle routing problem with time windows. In: Rudolph, G., Kononova, A.V., Aguirre, H., Kerschke, P., Ochoa, G., Tušar, T. (eds.) Parallel Problem Solving from Nature - PPSN XVII, pp. 356–370. Springer, Cham (2022)
7. Cheng, H., Yang, S.: Genetic algorithms with elitism-based immigrants for dynamic shortest path problem in mobile ad hoc networks. In: IEEE Congress on Evolutionary Computation (CEC), pp. 3135–3140. IEEE (2009)
8. Gendreau, M., Hertz, A., Laporte, G.: A tabu search heuristic for the vehicle routing problem. Manage. Sci. **40**(10), 1276–1290 (1994)
9. Glomb, L., Liers, F., Rösel, F.: A rolling-horizon approach for multi-period optimization. Eur. J. Oper. Res. **300**(1), 189–206 (2022)
10. Hottung, A., Kwon, Y., Tierney, K.: Efficient active search for combinatorial optimization problems. In: The Tenth International Conference on Learning Representations, ICLR 2022, Virtual Event, 25–29 April 2022 (2022)
11. Joe, W.C., Lau, H.C.: Deep reinforcement learning approach to solve dynamic vehicle routing problem with stochastic customers. In: International Conference on Automated Planning and Scheduling, vol. 30, pp. 394–402 (2020)
12. Kadyrov, S., Azamov, A., Abdumajitov, Y., et al.: Deep reinforcement learning for dynamic vehicle routing with demand and traffic uncertainty. Oper. Res. Perspect. 100351 (2025)
13. Kool, W., van Hoof, H., Welling, M.: Attention, learn to solve routing problems! In: International Conference on Learning Representations (2019)

14. Kwon, Y.D., et al.: POMO: policy optimization with multiple optima for reinforcement learning. In: Lin, H. (ed.) Advances in Neural Information Processing Systems, vol. 33, pp. 21188–21198 (2020)
15. Li, L., Yen, G.G., Sahoo, A., Chang, L., Gu, T.: On the estimation of pareto front and dimensional similarity in many-objective evolutionary algorithm. Inf. Sci. **563**, 375–400 (2021)
16. Mavrovouniotis, M., Yang, S.: Ant colony optimization with immigrants schemes in dynamic environments. In: Schaefer, R., Cotta, C., Kołodziej, J., Rudolph, G. (eds.) PPSN 2010. LNCS, vol. 6239, pp. 371–380. Springer, Heidelberg (2010). https://doi.org/10.1007/978-3-642-15871-1_38
17. Nazari, M., Oroojlooy, A., Snyder, L.V., Takac, M.: Reinforcement learning for solving the vehicle routing problem. In: Advances in Neural Information Processing Systems, vol. 31 (2018)
18. Pillac, V., Gendreau, M., Gueret, C., Medaglia, A.L.: A review of dynamic vehicle routing problems. Eur. J. Oper. Res. **225**(1), 1–11 (2013)
19. Renaud, J., Boctor, F.F.: A sweep-based algorithm for the fleet size and mix vehicle routing problem. Eur. J. Oper. Res. **140**(3), 618–628 (2002)
20. Ritzinger, U., Puchinger, J., Hartl, R.F.: A survey on dynamic and stochastic vehicle routing problems. Int. J. Prod. Res. **54**(1), 215–231 (2016)
21. Ulmer, M.W., Goodson, J.C., Mattfeld, D.C., Doerner, K.F.: Offline-online approximate dynamic programming for dynamic vehicle routing with stochastic requests. Transp. Sci. **53**(1), 185–202 (2019)
22. Yang, S.: Genetic algorithms with elitism-based immigrants for changing optimization problems. In: Giacobini, M. (ed.) EvoWorkshops 2007. LNCS, vol. 4448, pp. 627–636. Springer, Heidelberg (2007). https://doi.org/10.1007/978-3-540-71805-5_69
23. Yang, S.: Genetic algorithms with memory- and elitism-based immigrants in dynamic environments. Evol. Comput. **16**(3), 385–416 (2008)
24. Yazdani, D., Branke, J., Omidvar, M.N., et al.: A survey of evolutionary continuous dynamic optimization over two decades, part A. IEEE Trans. Evol. Comput. **25**(4), 609–629 (2021)
25. Yazdani, D., Yazdani, D., Blanco-Davis, E., Nguyen, T.T.: A survey of multi-population optimization algorithms for tracking the moving optimum in dynamic environments. J. Membrane Comput. **7**(2), 85–107 (2024)

From Small to Large: A Heuristic Divide-and-Neural-Conquer Framework for Large-Scale Vehicle Routing Problems

Debing Wang, Junyi Luo, Zhanhong Fang, Yunfeng Xu, and Zizhen Zhang(✉)

Sun Yat-sen University, Guangzhou, China
wangdb6@mail2.sysu.edu.cn, zhangzzh7@mail.sysu.edu.cn

Abstract. Although Neural Combinatorial Optimization models have shown promise in solving Vehicle Routing Problems (VRPs), they suffer from severe performance degradation when scaling to large-scale instances beyond their training distribution. To overcome this scalability challenge, we propose Heuristic Divide-and-Neural-Conquer (HDNC), a novel framework that elegantly solves large-scale VRPs by conquering small-scale constituent sub-problems. Instead of relying on complex learning-based partitioning policies, HDNC utilizes lightweight heuristic strategies to efficiently decompose large-scale instances into manageable sub-problems. Specifically, we introduce a grid/clustering method for the Traveling Salesman Problem (TSP) and a novel Adaptive Polar-Radial Decomposition for the Capacitated VRP (CVRP). These sub-problems are solved in parallel by a pre-trained neural solver, seamlessly integrated via a heuristic merging strategy, and further optimized through a refinement mechanism to ensure global consistency. Extensive experiments on TSP and CVRP demonstrate that HDNC achieves state-of-the-art performance. It exhibits superior robustness across diverse data distributions and significantly outperforms strong baselines on the challenging TSPLIB and CVRPLIB benchmarks.

Keywords: Vehicle Routing · Large-scale Optimization · Divide-and-Conquer · Neural Combinatorial Optimization

1 Introduction

The Vehicle Routing Problem (VRP) represents a fundamental class of combinatorial optimization problems [9], with the Traveling Salesman Problem (TSP) and the Capacitated Vehicle Routing Problem (CVRP) serving as its most prominent variants. Although exact algorithms [8] and classical heuristics [17,18] are capable of yielding high-quality solutions, they typically incur prohibitive computational overhead as the problem size scales up. Recently, Neural Combinatorial Optimization (NCO) has emerged as a promising new paradigm for solving VRPs. Pioneered by the Pointer Network [16] and progressively advanced

G. Iacca et al. (Eds.): PPSN 2026, LNCS 16985, pp. 332–348, 2027.
https://doi.org/10.1007/978-3-032-36214-8_21

by the Transformer-based Attention Model (AM) [6] and the widely adopted Policy Optimization with Multiple Optima (POMO) [7], these constructive neural models can generate competitive solutions against specialized solvers [1,15] within millisecond-level inference times. However, they suffer from severe performance degradation when applied to large-scale instances (e.g., with more than 1,000 nodes), leaving a significant gap to optimality.

To tackle the scalability bottlenecks of VRPs, the recent literature has increasingly turned to divide-and-conquer strategies [13,19,20,22]. However, these approaches generally rely on a heavily parameterized decomposition model in addition to the core routing solver. This dependency not only inflates training overhead, but also struggles to match the solution quality of mature heuristic solvers. Furthermore, although several end-to-end neural models [3,4,11,12,21] exhibit some capacity to generalize to larger scales, their practical utility is often undermined by a profound sensitivity to out-of-distribution data or excessively high inference latencies. As a result, bridging the performance gap with standard heuristic baselines remains a formidable challenge.

In this work, we empirically demonstrate that complex learning-based decomposition models are not necessary for effectively handling large-scale VRPs. Instead, lightweight rule-based heuristics can rapidly partition problem instances. By coupling these heuristics with existing small-scale NCO models (e.g., POMO) through parallel inference, we establish a highly scalable and efficient computational paradigm. Accordingly, we propose **Heuristic Divide-and-Neural-Conquer (HDNC)**[1], a unified pipeline that integrates decomposition, parallel neural solving, merging, and refinement. This hybrid approach yields solutions that rival or surpass state-of-the-art traditional solvers at a fraction of the computational cost. Furthermore, by anchoring the decomposition in geometric principles rather than learned parameters, HDNC exhibits strong robustness against severe distributional shifts. In summary, our main contributions are as follows:

1. We propose HDNC, a pioneering framework that bridges lightweight heuristic rules and neural optimization models. This unified pipeline of decomposition, parallel neural solving, heuristic merging, and model-based refinement derives high-quality solutions for large-scale TSP and CVRP instances within short inference times, effectively overcoming the scalability bottlenecks of standard NCO solvers.
2. We introduce targeted heuristic decomposition strategies, specifically a grid or clustering method for TSP and an Adaptive Polar-Radial Decomposition for CVRP. These lightweight rules not only eliminate the overhead of training dedicated learning-based decomposition models, but also maintain strong robustness against severe distributional shifts.
3. Extensive experiments demonstrate that HDNC achieves state-of-the-art performance on large-scale synthetic datasets ranging from 1,000 to 10,000 nodes across four distinct distributions. Furthermore, our method outperforms all

[1] Available at https://github.com/wang-fh/HDNC-for-Large-Scale-VRP.

strong baselines on established real-world benchmarks (TSPLIB & CVRPLIB), demonstrating its practical applicability.

2 Related Work

2.1 Decomposition Methods

In recent years, decomposition-based approaches have garnered significant attention to address large-scale VRPs. A prominent line of research focuses on training deep neural networks to learn sophisticated partitioning policies. For instance, H-TSP [13] employs a hierarchical framework, where the upper level utilizes a pixel-level encoder to decompose the TSP into sub-problems, followed by solution aggregation at the lower level. Similarly, GLOP [19] proposes learning a global partitioning heatmap to decompose large-scale CVRP into multiple sub-TSP instances for separate solving. Furthermore, UDC [20] adopts a heatmap-based solver and leverages an Anisotropic Graph Neural Network (AGNN) as a partitioning strategy to generate initial solutions.

Despite these advances, existing learning-based frameworks frequently encounter severe scalability bottlenecks. Their reliance on heavily parameterized partitioning models restricts instance sizes, constraining methods such as UDC to medium-scale problems. To mitigate this training overhead, recent works have explored simpler spatial heuristics. For example, DualOpt [22] combines grid-based initialization with neural-driven iterative refinement. However, its applicability remains strictly limited to large-scale TSP, leaving a critical gap for a unified, lightweight decomposition framework capable of handling diverse routing topologies such as CVRP.

2.2 End-to-End Models

Another line of research focuses on training neural models on small-scale datasets to generalize to larger instances. To enhance scalability, LEHD [11] redesigns the Transformer into a light-encoder heavy-decoder architecture, achieving favorable performance on instances of approximately 1,000 nodes. However, its reliance on supervised learning incurs high labeling costs and limits its generalization to 10,000-node scales. To mitigate the dependency on expensive exact solvers, SIL [12] leverages self-generated, partially reconstructed solutions as pseudo-labels. It further introduces a lightweight cross-attention mechanism with linear complexity to reduce the excessive computational overhead typical of conventional self-attention.

Beyond architectural refinements, recent works prioritize robustness against distributional shifts and varying problem scales. INViT [3] proposes an Invariant Nested View Transformer, utilizing nested designs and invariant views to enhance generalization through a data-augmented policy gradient algorithm. Similarly, Omni-VRP [21] adopts a direct training strategy on 500-node instances, achieving reasonable generalization to larger scales. More recently, ELG [4] employs an ensemble strategy that merges local policies with global information to improve performance across diverse distributions. Despite these advancements, pure end-to-end models often struggle with prohibitive inference latencies or accuracy drops when confronting extreme scales, highlighting the potential of hybrid strategies that decouple decomposition from neural solving.

3 Preliminaries

3.1 Problem Definition

We define the routing problems on a graph represented by a node set $\mathcal{V} = \{0, 1, \ldots, n\}$. For TSP (excluding node 0), the objective is to determine a Hamiltonian cycle over $\mathcal{V}$ that minimizes the total Euclidean distance. Let $\boldsymbol{\pi} = (\pi_1, \ldots, \pi_k)$ denote the sequence of visited nodes, where k is the total number of nodes. The objective function is formulated as:

$$\min \sum_{t=1}^{k-1} \|\mathbf{x}_{\pi_t} - \mathbf{x}_{\pi_{t+1}}\|_2 + \|\mathbf{x}_{\pi_k} - \mathbf{x}_{\pi_1}\|_2 \tag{1}$$

where $\mathbf{x}_i$ denotes the coordinates of node i. For CVRP, node 0 is designated as the depot, while nodes 1 to n represent customers with specific demands d_i. The goal is to derive a set of routes that minimize the total travel distance, ensuring that each customer is visited exactly once and the cumulative demand of any single route does not exceed Q.

3.2 Neural Architecture

To solve the sub-problems derived in our framework, we employ the constructive POMO model as our underlying neural solver.

Encoder. The encoder aims to project input node features into robust high-dimensional embeddings. For each node i, its initial embedding is linearly projected and then updated through L attention layers. Each layer comprises a Multi-Head Attention (MHA) mechanism followed by a Feed-Forward Network (FFN). To stabilize training, Residual connections and Instance Normalization (IN) are applied after each sub-layer. The update process for node embeddings h_i in layer ℓ is formulated as:

$$\begin{aligned} \hat{h}_i &= \mathrm{IN}\left(h_i^{(\ell-1)} + \mathrm{MHA}_i^{\ell}\left(h_1^{(\ell-1)}, \ldots, h_n^{(\ell-1)}\right)\right), \\ h_i^{(\ell)} &= \mathrm{IN}\left(\hat{h}_i + \mathrm{FFN}^{\ell}\left(\hat{h}_i\right)\right). \end{aligned} \tag{2}$$

Decoder. The decoder generates the solution sequence $\boldsymbol{\pi}$ autoregressively. At each decoding step t, a context vector $h_{(c)}$ is constructed to represent the current decoding state. Specifically, it concatenates the value vectors from the encoder's specific layers:

$$h_{(c)} = [v^1, v^f], \tag{3}$$

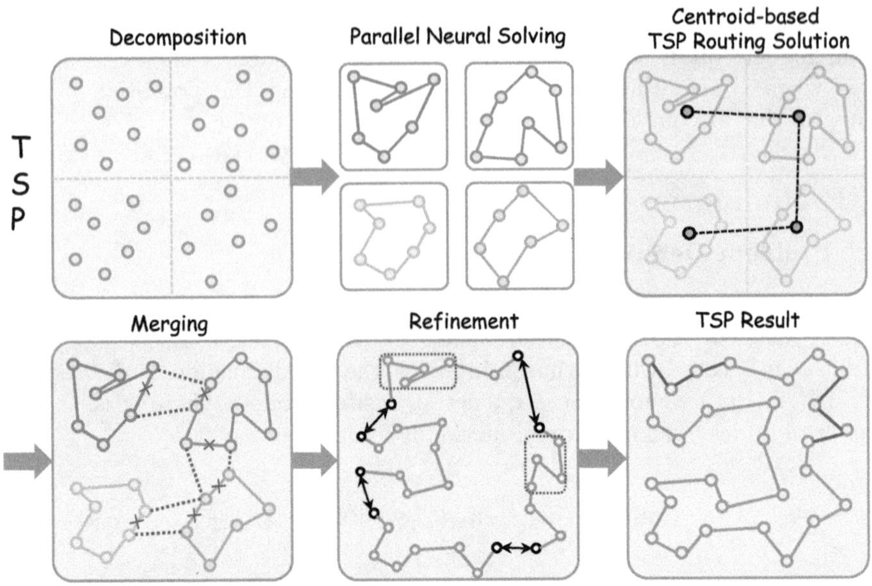

Fig. 1. Illustration of the proposed HDNC for TSP. The large-scale instance is first spatially decomposed into sub-problems and solved in parallel by the NCO model. A global routing backbone is then derived by solving a TSP over the sub-problem centroids to guide the optimal merging sequence, followed by an open-loop refinement to finalize the tour. (The legend is identical to Fig. 2).

where v^1 and v^f represent the embeddings of the starting node and the current node, respectively. Subsequently, the query vector $q_{(c)}$ for the decoder is generated through MHA over the encoded node representations:

$$q_{(c)} = \text{MHA}\left(h_{(c)}, h_1^L, \ldots, h_n^L\right). \tag{4}$$

The probability of selecting the next node j is then determined by the compatibility score $u_{(c)j}$, computed via a scaled dot-product attention mechanism with a tanh clipping scheme:

$$u_{(c)j} = \begin{cases} C \cdot \tanh\left(\frac{q_{(c)} k_j^T}{\sqrt{d}}\right) & \text{if } j \text{ is not masked,} \\ -\infty & \text{otherwise,} \end{cases} \tag{5}$$

where k_j is the key vector of node j, d is the dimension of the key vector, and C is a scaling factor. Finally, the selection probability is obtained via a softmax operation: $p_j = \text{softmax}(u_{(c)j})$.

4 Methodology

4.1 HDNC for TSP

For TSP, the absence of a depot results in a free-form topology. During the decomposition phase, uniform and non-uniform spatial distributions are addressed separately. Furthermore, the merging process conducts an optimality-preserving merge following a sequence determined by the sub-path centroids. The overall resolution process is illustrated in Fig. 1.

Decomposition. Prior research [3] suggests that in TSP, the optimal connection for a given node is predominantly influenced by its spatially nearest neighbors. Therefore, clustering nodes based on spatial proximity is an intuitive and effective decomposition strategy. We tailor our decomposition approach to the distribution characteristics of the instance to ensure robustness.

Consider a large-scale instance with N nodes, denoted as $\mathcal{V} = \{1, \dots, N\}$, where each node coordinate $(x_i, y_i) \in [0,1]^2$. For the uniformly distributed instance, we employ a grid-based decomposition. We partition the coordinate space $[0,1]^2$ into $M_1 \times M_2$ rectangular regions. Let $G_{r,c}$ denote the sub-problem located at the r-th row and c-th column (where $1 \le r \le M_1, 1 \le c \le M_2$). The subset of nodes belonging to this grid is formally defined as:

$$G_{r,c} = \left\{ i \in \mathcal{V} \mid \frac{r-1}{M_1} \le x_i < \frac{r}{M_1}, \quad \frac{c-1}{M_2} \le y_i < \frac{c}{M_2} \right\}, \tag{6}$$

where (x_i, y_i) denotes the coordinates of node i.

However, for non-uniform distributions (e.g., clustered, explosion, implosion), grid decomposition often leads to significant load imbalance, with some grids being overly dense while others are empty. To address this, we adopt the K-means clustering algorithm. We partition the node set $\mathcal{V}$ into K clusters (sub-problems) $\mathcal{S}_{kmeans} = \{C_1, \dots, C_K\}$ by minimizing the within-cluster sum of squares. With random initialization of cluster centers $\boldsymbol{\mu}_j$, the decomposition is defined as:

$$\min_{\mathcal{S}_{kmeans}} \sum_{j=1}^{K} \sum_{i \in C_j} \|\mathbf{x}_i - \boldsymbol{\mu}_j\|^2. \tag{7}$$

To solve this optimization problem efficiently, we iteratively alternate between assigning nodes to their nearest centers and updating the centers to the centroids of the formed clusters.

Parallel Neural Solving. Sub-problems decomposed from large-scale TSP instances typically vary in size, failing to satisfy the fixed-size input requirements of the POMO model for batch processing. To address this, we employ a padding strategy: for a batch of sub-problems, we pad all instances to the maximum node count N_{max} within that batch using a specific dummy coordinate.

Although padding enables parallelization, standard pre-trained models often struggle to distinguish between valid customer nodes and dummy nodes, leading to suboptimal routing. To adapt to varying sub-problem sizes, we apply rapid fine-tuning via Preference Optimization (PO) [10] using synthetic data. We minimize a pairwise preference loss $\mathcal{L}_{\text{PO}}(\theta)$, which encourages the model to assign higher probabilities to better-quality solutions (winning samples) over inferior ones (losing samples):

$$\mathcal{L}_{\text{PO}}(\theta) = -\mathbb{E}_s \left[\log \sigma \left(\alpha \left(\log p_\theta(\pi_w, s) - \log p_\theta(\pi_l, s)\right)\right)\right]. \quad (8)$$

where θ denotes the model parameters, s is the input instance, and π_w, π_l represent the winning (superior) and losing (inferior) solutions, respectively. Additionally, $p_\theta(\pi, s)$ indicates the generation probability, $\sigma(\cdot)$ is the sigmoid function, and α serves as a temperature hyperparameter.

Heuristic Merging. Once the sub-problems are solved in parallel, we reconstruct the global solution through an iterative heuristic merging process. First, to determine an optimal merging sequence, we calculate the geometric centroid for each of the T sub-problems and solve a coarse-grained TSP on these centroids using our pre-trained model. Let $\mathcal{O} = (P_1, P_2, \ldots, P_T)$ denote the resulting ordered sequence of sub-problems.

We maintain a global tour Π_{curr}, initialized as the solution of P_1. In each step ($k = 2 \ldots T$), we merge the solution of the next sub-problem $\Pi_{sub} = P_{(k)}$ into Π_{curr}. The optimal merging position is determined by identifying the best "quadruplet" of nodes that minimizes the connection cost. Let (u, u^+) be an edge in Π_{curr} and (v, v^+) be an edge in Π_{sub}. We evaluate the cost of replacing these two edges with cross-connections to fuse the two distinct loops into a single tour. We consider two topological connection directions:

$$\Delta_{cost} = \min_{\substack{(u,u^+)\in\Pi_{curr} \\ (v,v^+)\in\Pi_{sub}}} \left(\min \begin{Bmatrix} d(u,v) + d(u^+,v^+), \\ d(u,v^+) + d(u^+,v) \end{Bmatrix} - d(u,u^+) - d(v,v^+) \right), \quad (9)$$

where $d(\cdot,\cdot)$ denotes the Euclidean distance. The edges corresponding to the minimal Δ_{cost} are removed, and the new optimal connections are established.

Refinement. The solution generated in the previous stages primarily relies on decomposed sub-problems, which may limit the utilization of global structural information. To address this, we introduce a refinement phase to further optimize local path segments. We uniformly cut the current complete solution Π_{curr} into a series of fixed-length sub-sequences. Each sub-sequence is treated as an Open-loop TSP (O-TSP) instance with fixed start and end nodes. These instances are solved in parallel using a customized version of our neural model. If a re-optimized path yields a shorter distance than the original segment, it replaces the corresponding segment in the global tour.

The O-TSP model is a lightweight adaptation of the base POMO architecture described in Sect. 3.2. The key modification lies in the decoder's context query construction. We augment the context representation to explicitly include the information of the target node. The query context vector $q_{(c)}$ is reformulated as:

$$q_{(c)} = [v^1, v^f, v^n], \quad (10)$$

where v^n is the embedding of the fixed destination node. We consistently employ PO as the loss function for this phase. Leveraging the reduced search space of local segments, this open-loop model demonstrates rapid convergence capabilities.

The model typically employs a shorter sequence length configuration, which facilitates rapid convergence during the training phase. To strictly enforce the endpoint constraint, we initialize the attention mask of the destination node with a large negative value. This mask value is maintained throughout the decoding process until the destination node is selected, ensuring that the model strictly visits the fixed destination node at the end of the tour.

4.2 HDNC for CVRP

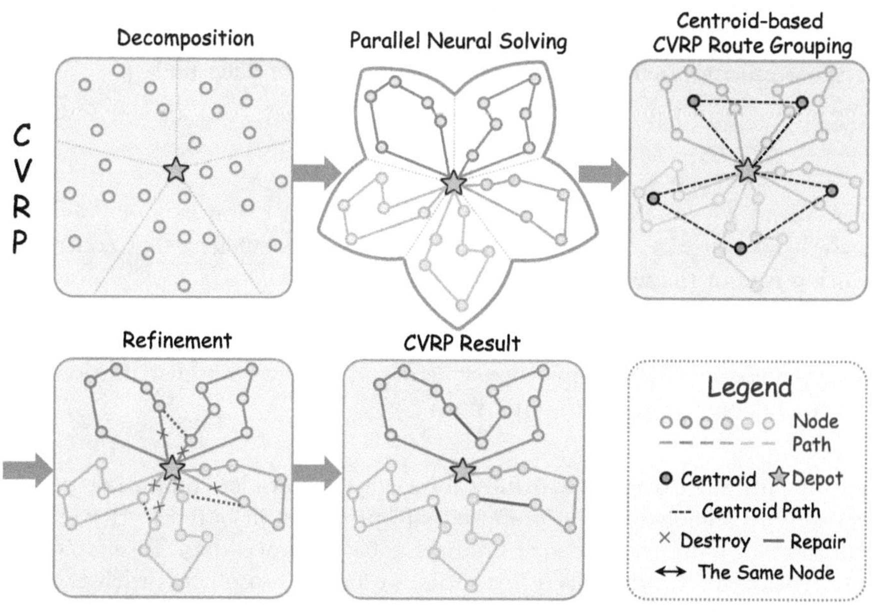

Fig. 2. Illustration of the proposed HDNC for CVRP. The process begins with the Adaptive Polar-Radial Decomposition. After parallel neural solving, a route-centric refinement strategy is applied where sub-routes are re-grouped based on their geometric centroids and optimized to yield the final solution.

For CVRP, the presence of the depot imposes strong topological constraints, inducing optimal routes to naturally form petal-shaped loops anchored at the depot. In contrast to the unconstrained topology of the TSP, such structural properties significantly simplify the solution space, thereby rendering the decomposition and solving processes highly efficient. Specifically, the pre-trained model exhibits remarkable zero-shot generalization capabilities on CVRP sub-problems, allowing for direct application without any additional training or fine-tuning. The overall resolution process is illustrated in Fig. 2.

Adaptive Polar-Radial Decomposition. Standard spatial clustering (e.g., K-means) is ill-suited for CVRP as it ignores the critical role of the depot. Empirical observations of optimal CVRP solutions reveal that routes typically span specific angular sectors. A naive angular sweep can decompose instances, but it treats nodes near the depot and those at the periphery equally, ignoring that angular proximity is less meaningful for nodes clustered closely around the depot.

To address this, we propose an **Adaptive Polar-Radial Decomposition** strategy. We first transform the Cartesian coordinates of all customer nodes into a polar coordinate system (r, α) with the depot as the origin. Premised on the rationale that the importance of the angular component grows as the customer moves further from the depot, while the radial component dominates for nodes near the center, we define a hybrid sorting score S_i for each node i:

$$\begin{aligned} w_i &= \exp\left(-k \cdot \hat{r}_i\right), \\ S_i &= (1 - w_i) \cdot \hat{\alpha}_i + w_i \cdot \hat{r}_i, \end{aligned} \tag{11}$$

where $\hat{r}_i$ and $\hat{\alpha}_i$ are the min-max normalized radial distance and polar angle of node i, respectively. The parameter k is a decay coefficient that controls the transition rate of the weighting emphasis.

Finally, all nodes are sorted based on S_i and partitioned sequentially into equal-sized sub-problems for parallel solving. Notably, this adaptive strategy is design-agnostic and remains robust across arbitrary node distributions (e.g., uniform, clustered, explosion, implosion).

Parallel Neural Solving. To fully leverage the parallel computing power of GPU, we aim to solve the decomposed sub-problems in batches. However, the POMO model requires fixed-size inputs for batch processing. To enable batch parallelization for varying-sized instances, we pad sub-problems with the depot coordinate. Unlike using arbitrary dummy nodes, padding with the depot introduces no noise since it is a structurally valid part of any route. Given that our initial decomposition yields equal-sized sub-problems, this padding strategy is employed exclusively during the refinement phase.

Merging and Refinement. Since CVRP sub-routes inherently originate from and terminate at the depot, the initial global solution is formed by simply aggregating all sub-routes.

To further optimize the global structure, we employ a **Route-Centric Refinement** strategy. We calculate the geometric centroid of each route in the initial solution and re-partition them into new, similarly-sized sub-problems by sorting these centroids using the proposed adaptive polar-radial decomposition

(Eq. 11). These new instances are re-solved in parallel (using depot-padding for alignment), and any re-optimized solution yielding a lower cost replaces the original segment. Empirical results indicate that a single refinement iteration is sufficient for high-quality convergence.

5 Experiments

5.1 Experimental Settings

Datasets. To strictly evaluate scalability and generalization, we generated synthetic datasets for both TSP and CVRP across three scales ($N \in \{1\text{k}, 5\text{k}, 10\text{k}\}$) and four distributions (*uniform, clustered, explosion, implosion*) following the protocol in [3]. This results in 12 distinct test sets per problem type. For evaluation, we generated 200 instances for each distribution at the 1k scale, and 20 instances for the 5k and 10k scales. All reported runtimes represent the average time required to solve a single instance. Furthermore, we assessed the robustness of our method on large-scale real-world instances from TSPLIB [14] and CVRPLIB [2].

Training and Test Setup. All experiments were conducted on an AMD EPYC 7702 64-core CPU and a single NVIDIA RTX 3090 GPU. We followed the training protocol of POMO [7] and reduced the training duration to only 50 epochs. We controlled the decomposed sub-problem size to be approximately 200 nodes and set the decay coefficient $k = 5$. HDNC-HGS and HDNC-LKH3 employ HGS (5 s limit) and LKH3 (default settings), respectively, to replace neural models. All results are reported without any data augmentation.

Baselines. We compare our proposed method against a comprehensive suite of competitive neural baselines, covering various paradigms: the decomposition-based framework GLOP [19]; end-to-end constructive solvers including POMO [7], LEHD [11], and INViT [3]; the meta-learning-based omni-VRP [21]; and the policy ensemble method ELG [4]. To quantify solution quality, we report the optimality gap relative to state-of-the-art heuristic solvers, employing LKH3 [5] and HGS [15] as the near-optimal reference for TSP and CVRP, respectively.

Table 1. Comparison of different methods on TSP datasets. LKH3 is the baseline. Best results (excluding LKH3) are in **bold**, second-best are underlined.

Method	Uniform			Clustered			Explosion			Implosion		
	Obj.	Gap	Time	Obj.	Gap	Time	Obj.	Gap	Time	Obj.	Gap	Time
TSP-1K												
LKH3	23.2	0.0%	1.4m	14.1	0.0%	2.3m	16.0	0.0%	1.3m	20.2	0.0%	2.9m
POMO	35.3	52.6%	1.1s	24.1	70.9%	1.1s	27.2	71.2%	1.1s	30.9	53.5%	1.1s
GLOP	24.2	4.5%	1.5s	14.8	5.2%	1.5s	16.8	4.7%	1.5s	21.0	4.3%	1.5s
omni-VRP	27.8	19.9%	0.9s	17.7	26.2%	0.8s	19.4	21.5%	0.8s	24.2	19.8%	0.8s
LEHD	23.9	3.1%	3.7s	15.3	8.7%	3.7s	17.0	6.0%	3.6s	21.0	4.3%	3.6s
INViT	25.0	8.0%	10.8s	15.5	10.5%	10.7s	18.1	13.2%	10.7s	22.2	10.0%	10.7s
ELG	27.5	18.6%	1.7s	17.8	26.3%	1.7s	19.5	21.8%	1.7s	24.2	19.8%	1.7s
HDNC	23.9	3.4%	2.7s	14.8	5.1%	2.6s	16.8	5.0%	2.6s	21.0	4.3%	2.7s
HDNC-LKH3	**23.7**	**2.2%**	4.5s	**14.6**	**4.0%**	4.3s	**16.6**	**3.6%**	4.3s	**20.9**	**3.7%**	4.4s
TSP-5K												
LKH3	51.0	0.0%	9.7m	28.8	0.0%	16.2m	32.0	0.0%	10.3m	45.0	0.0%	20.5m
POMO	91.9	80.3%	3.1s	80.8	177.7%	3.2s	78.0	152.1%	3.2s	83.6	86.6%	3.2s
GLOP	53.8	5.4%	1.9s	30.6	6.1%	1.9s	33.8	5.8%	1.9s	47.5	5.5%	1.9s
omni-VRP	76.2	49.4%	1.2s	46.3	61.5%	1.2s	48.8	52.9%	1.2s	67.9	51.0%	1.2s
LEHD	59.0	15.8%	1.7m	35.6	23.5%	1.7m	38.7	21.1%	1.7m	52.9	17.5%	1.7m
INViT	55.4	8.6%	1.3m	31.7	9.9%	1.3m	36.2	13.2%	1.3m	49.7	10.4%	1.3m
ELG	62.7	22.9%	8.8s	40.4	40.0%	8.6s	41.8	31.8%	8.8s	55.8	23.9%	8.8s
HDNC	53.2	4.3%	3.2s	30.4	5.6%	3.6s	33.7	5.5%	3.4s	47.3	4.9%	3.8s
HDNC-LKH3	**52.6**	**3.2%**	9.5s	**30.2**	**4.8%**	8.3s	**33.5**	**4.6%**	8.1s	**46.9**	**4.2%**	8.2s
TSP-10K												
LKH3	71.8	0.0%	31.6m	39.6	0.0%	39.4m	40.6	0.0%	32.2m	62.6	0.0%	60.1m
POMO	138.8	93.3%	6.4s	133.4	236.1%	6.4s	123.7	208.7%	6.4s	124.5	99.1%	6.2s
GLOP	76.0	5.8%	2.2s	41.9	6.0%	2.1s	43.1	6.1%	2.1s	66.2	5.7%	2.1s
omni-VRP	118.0	64.4%	8.5s	69.0	74.5%	8.5s	68.2	68.1%	8.5s	103.6	65.5%	8.5s
LEHD	90.8	26.4%	11.4m	54.5	37.7%	11.3m	53.9	32.7%	11.3m	80.5	28.6%	11.5m
INViT	78.0	8.6%	3.6m	43.9	10.9%	3.5m	45.5	12.2%	3.5m	68.6	9.6%	3.5m
ELG	88.6	23.5%	17.8s	68.2	71.4%	17.8s	54.9	36.3%	17.8s	77.7	24.2%	17.5s
HDNC	75.3	4.8%	4.7s	41.8	5.5%	5.2s	43.2	6.3%	5.6s	66.9	6.9%	7.1s
HDNC-LKH3	**74.4**	**3.6%**	17.2s	**41.6**	**5.0%**	13.7s	**42.6**	**4.9%**	13.7s	**66.1**	**5.6%**	17.5s

5.2 Results on Generated Datasets

Results on TSP. The experimental results on TSP datasets are detailed in Table 1. Across most evaluated scenarios, our proposed HDNC and its heuris-

Table 2. Comparison of different methods on CVRP datasets. HGS is the baseline. Best results (excluding HGS) are in **bold**, second-best are underlined.

Method	Uniform			Clustered			Explosion			Implosion		
	Obj.	Gap	Time	Obj.	Gap	Time	Obj.	Gap	Time	Obj.	Gap	Time
CVRP-1K												
HGS	131.4	0.0%	20.0m	109.8	0.0%	20.0m	90.8	0.0%	20.0m	126.1	0.0%	20.0m
POMO	233.3	82.0%	1.5s	241.8	134.6%	1.4s	240.7	177.2%	1.4s	237.8	93.9%	1.4s
GLOP	177.6	35.3%	0.6s	139.2	27.9%	0.6s	125.2	38.4%	0.6s	168.9	34.3%	0.6s
omni-VRP	144.3	10.0%	1.0s	125.0	13.0%	1.0s	101.6	12.3%	1.0s	138.5	10.0%	1.0s
LEHD	133.4	1.6%	4.6s	116.1	5.7%	4.5s	94.3	3.8%	4.6s	138.1	9.6%	4.6s
INViT	143.8	9.6%	11.5s	119.0	8.5%	11.9s	99.6	9.8%	11.9s	137.5	9.2%	2.0m
ELG	139.2	6.2%	3.5s	123.2	11.1%	3.6s	98.1	8.6%	3.4s	134.4	6.9%	3.5s
HDNC	131.7	0.3%	1.1s	110.4	0.6%	1.1s	91.5	0.9%	1.0s	127.0	0.8%	1.0s
HDNC-HGS	**129.1**	**-1.7%**	26.3s	**108.9**	**-0.3%**	26.3s	**89.4**	**-1.5%**	26.3s	**126.1**	**0.0%**	26.3s
CVRP-5K												
HGS	590.4	0.0%	4.0h	472.4	0.0%	4.0h	350.8	0.0%	4.0h	582.7	0.0%	4.0h
POMO	2165.6	281.5%	5.1s	2260.1	424.5%	5.3s	2065.3	504.6%	5.1s	2087.9	262.7%	5.3s
GLOP	756.8	29.6%	2.8s	556.4	20.3%	2.9s	464.7	33.3%	3.1s	729.8	25.9%	3.1s
omni-VRP	747.9	27.4%	5.3s	681.7	39.9%	5.3s	488.3	39.6%	5.4s	759.5	30.5%	5.2s
LEHD	637.4	8.0%	1.8m	585.2	23.9%	1.8m	410.1	16.9%	1.8m	677.6	16.3%	1.8m
INViT	638.8	8.4%	1.3m	526.2	10.5%	1.3m	382.3	9.1%	1.3m	630.0	8.2%	1.3m
ELG	638.8	8.2%	17.2s	588.7	25.8%	17.9s	410.0	16.9%	17.6s	638.5	9.6%	17.0s
HDNC	597.7	1.3%	2.4s	479.9	1.6%	2.4s	358.3	2.2%	2.4s	**591.6**	**1.5%**	2.4s
HDNC-HGS	**579.5**	**-1.8%**	2.2m	**477.6**	**0.5%**	2.2m	**347.4**	**-0.9%**	2.2m	600.1	2.7%	2.2m
CVRP-10K												
HGS	1204.2	0.0%	4.0h	849.9	0.0%	4.0h	722.4	0.0%	4.0h	1198.0	0.0%	4.0h
POMO	4567.5	294.3%	10.5s	4589.0	486.5%	10.4s	4300.4	531.8%	10.4s	5175.8	346.3%	11.3s
GLOP	1445.8	20.9%	6.1s	976.7	15.5%	6.2s	889.4	24.0%	6.7s	1425.4	20.1%	6.7s
omni-VRP	1791.5	49.1%	11.2s	1479.5	74.3%	11.2s	1189.2	63.0%	11.0s	1801.3	50.8%	10.9s
LEHD	1409.2	17.0%	11.3m	1105.1	30.0%	11.3m	862.6	19.4%	11.3m	1532.2	27.9%	11.3m
INViT	1300.1	7.9%	3.4m	915.2	7.9%	3.4m	772.1	7.0%	3.4m	1286.7	7.3%	3.4m
ELG	1340.1	11.3%	35.3s	1214.5	42.8%	36.7s	838.7	17.4%	35.4s	1392.3	16.5%	34.1s
HDNC	1220.1	1.3%	4.7s	**861.2**	**1.4%**	4.7s	734.6	1.8%	4.7s	1215.1	1.4%	4.8s
HDNC-HGS	**1178.9**	**-2.1%**	4.3m	862.0	1.7%	4.3m	**709.5**	**-1.7%**	4.3m	**1174.3**	**-2.0%**	4.3m

tic variant, HDNC-LKH3, consistently secure the top two positions in solution quality. While certain baselines exhibit competitiveness on specific configurations, such as LEHD at the 1K scale and GLOP on 10K Implosion distributions, they frequently suffer severe performance degradation when confronting distributional shifts or extreme scale expansion. In stark contrast, HDNC demonstrates exceptional robustness. It reliably controls the optimality gap at approximately

5% across all scales and distribution types, effectively circumventing the scalability bottlenecks that typically paralyze standard NCO methods.

Beyond robust solution quality, the framework exhibits remarkable computational efficiency. As the problem size scales to massive 10K nodes, where conventional neural solvers such as LEHD and INViT struggle with prohibitive latency, HDNC maintains a flat growth curve, completing inference in approximately 5 s. Furthermore, substituting the underlying neural solver with a heuristic module (HDNC-LKH3) minimizes the optimality gap even further. Although this substitution incurs a slight increase in computational cost compared to the neural solver (e.g., from roughly 5 s to 15 s on 10K instances), it remains orders of magnitude faster than the pure LKH3 baseline, which requires tens of minutes. This underscores a core advantage of our unified pipeline: it not only empowers neural solvers but also serves as a potent accelerator for traditional heuristics.

Results on CVRP. The comprehensive evaluation results for CVRP are summarized in Table 2. The strong robustness and computational efficiency observed in TSP are similarly evident and even more pronounced in CVRP, despite its complex capacity constraints. Most existing baselines, including LEHD and ELG, exhibit widened optimality gaps when moving from uniform to clustered or implosion distributions. Conversely, HDNC maintains exceptional consistency, achieving an optimality gap of strictly **less than 1%** on 1K instances and impressively holding at **approximately 2%** even when scaled to 5K and 10K instances. This consistency verifies the efficacy of our divide-and-conquer paradigm in preserving global solution quality.

In terms of computational overhead, HDNC continues to record the shortest inference times across the majority of scenarios. Mirroring the TSP results,

Table 3. Generalization Gaps (%) on TSPLIB and CVRPLIB datasets. The best results are in **bold**.

	Instance	GLOP	Omni	LEHD	INViT	ELG	HDNC (Ours)
TSPLIB	rl-5915	12.3%	64.6%	22.3%	11.6%	23.1%	**7.3%**
	rl-5934	10.7%	65.8%	35.5%	9.0%	23.9%	**7.1%**
	rl-11849	8.9%	77.2%	41.7%	9.9%	24.0%	**6.2%**
	usa-13509	5.8%	107.9%	55.8%	10.3%	32.6%	**3.6%**
	brd-14051	5.2%	78.2%	42.5%	8.8%	25.7%	**2.8%**
	d-18512	5.0%	78.0%	OOM	8.1%	25.7%	**2.9%**
	Average	8.0%	78.6%	39.6%	9.6%	25.8%	**5.0%**
CVRPLIB	L1-3000	16.9%	22.8%	14.0%	13.3%	10.8%	**4.4%**
	L2-4000	21.8%	60.4%	26.3%	20.4%	21.8%	**8.1%**
	A1-6000	20.3%	42.5%	18.9%	15.1%	10.7%	**4.9%**
	A2-7000	19.4%	48.6%	26.4%	25.0%	17.7%	**7.1%**
	Average	19.6%	43.6%	21.4%	18.5%	15.2%	**6.1%**

it smoothly solves 10K-node instances in under 5 s, empirically validating the favorable time complexity of our divide-and-conquer paradigm against standard solvers such as POMO, Omni-VRP, and INViT.

Finally, evaluating the heuristic variant HDNC-HGS further verifies our approach's capacity to push the boundaries of solution optimality. This substitution frequently achieves negative optimality gaps (e.g., -2.1% on the CVRP-10K Uniform dataset), directly outperforming the pure HGS baseline. More importantly, while the pure HGS algorithm requires roughly 4 h to solve 10K-node instances, HDNC-HGS attains these superior solutions in merely 4 min. This dramatic acceleration perfectly strikes a favorable balance between solution excellence and computational cost, proving the practical applicability of the proposed method.

5.3 Results on Real-World Benchmarks

We benchmarked our approach against competitive baselines on representative large-scale instances from TSPLIB and CVRPLIB. The results are reported in Table 3, where the numerical value in each instance name indicates its problem size, and OOM denotes Out-Of-Memory errors. For TSP, HDNC maintains an average optimality gap of only **5.0%**, leading the runner-up (GLOP) by a margin of 3.0%. Similarly, for CVRP, our approach substantially outperforms all other baselines, achieving an average gap of **6.1%**, which surpasses the second-best method (ELG) by 9.1%. This sustained effectiveness on real-world instances underscores the capability of HDNC as a robust and scalable approach for complex logistical challenges.

5.4 Sensitivity Analysis and Ablation Study

Sensitivity to Sub-problem Size. In our main experiments, the default subproblem size is configured to 200 to optimally balance local solving capability and global structure. We evaluate its specific impact through a sensitivity analysis, with detailed results presented in Table 4. As observed, an overly small size restricts the local receptive field and noticeably widens the optimality gap. For instance, reducing the size to 100 increases the gap to 5.0% for TSP and 3.3% for CVRP. Conversely, excessively large sizes incur substantial computational overhead without proportional accuracy gains. When the size reaches 500, the optimality gap for TSP deteriorates to 6.8%, while the inference time for CVRP escalates to 17.5 s. These findings validate our default setting of 200 as the optimal trade-off between solution quality and computational efficiency.

Effect of the Refinement Phase. To evaluate the efficacy of the refinement phase, we conduct an ablation study with results detailed in Table 5. As observed, omitting this module yields marginal computational savings but compromises solution quality. Specifically, the optimality gap deteriorates from 4.8% to 7.4% for TSP and from 1.3% to 2.3% for CVRP. Furthermore, the inference time breakdown in Fig. 3 illustrates that the refinement procedure consumes approximately 1.9 s for TSP and 1.4 s for CVRP. Given the substantial improvements in global accuracy, this computational overhead is highly cost-effective.

Table 4. Ablation study on sub-problem size (SubSize) using 10K Uniform instances. The gray-shaded area indicates the default experimental configuration.

Dataset	SubSize = 100			SubSize = 200			SubSize = 400			SubSize = 500		
	Obj.	Gap	Time	Obj.	Gap	Time	Obj.	Gap	Time	Obj.	Gap	Time
TSP-10K	75.4	5.0%	4.7s	75.3	4.8%	4.7s	75.9	5.7%	5.4s	76.7	6.8%	6.1s
CVRP-10K	1243.3	3.3%	2.8s	1220.1	1.3%	4.7s	1218.7	1.2%	12.3s	1221.2	1.4%	17.5s

Table 5. Ablation study on the refinement phase using 10K Uniform instances. The full HDNC is shaded gray. "w/o Ref." denotes without refinement.

Dataset	Model	Obj.	Gap	Time
TSP-10K	w/o Ref.	77.2	7.4%	**2.8s**
	HDNC	**75.3**	**4.8%**	4.7s
CVRP-10K	w/o Ref.	1232.4	2.3%	**3.2s**
	HDNC	**1220.1**	**1.3%**	4.7s

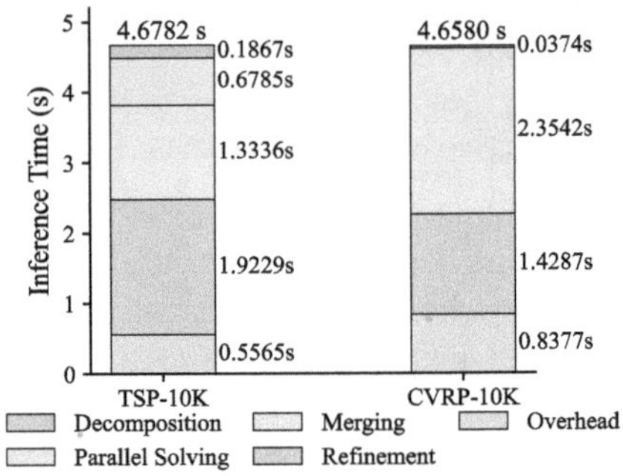

Fig. 3. Inference time breakdown of different components in HDNC.

6 Conclusion

In this paper, we propose a novel Heuristic Divide-and-Neural-Conquer (HDNC) framework, which integrates heuristic spatial decomposition with parallel neural solving, followed by heuristic merging and model-based refinement. Extensive experiments demonstrate that our method consistently achieves state-of-the-art performance across three problem scales and four distinct distributions. Furthermore, on the challenging real-world benchmarks TSPLIB and CVRPLIB, our approach significantly outperforms competitive baselines, highlighting its superior robustness and generalization capabilities. In future work, we aim to extend this scalable paradigm to tackle a broader range of large-scale combinatorial optimization problems.

Acknowledgments. This work is supported by the Guangdong Basic and Applied Basic Research Foundation (No. 2024A1515010871).

References

1. Applegate, D., Bixby, R.E., Chvátal, V., Cook, W.J.: Concorde TSP solver (2006). https://www.math.uwaterloo.ca/tsp/concorde.html
2. Arnold, F., Gendreau, M., Sörensen, K.: Efficiently solving very large-scale routing problems. Comput. Oper. Res. **107**, 32–42 (2019). https://doi.org/10.1016/j.cor.2019.03.006
3. Fang, H., Song, Z., Weng, P., Ban, Y.: INViT: a generalizable routing problem solver with invariant nested view transformer. In: Proceedings of the 41st International Conference on Machine Learning, vol. 235, pp. 12973–12992. PMLR (2024)
4. Gao, C., Shang, H., Xue, K., Li, D., Qian, C.: Towards generalizable neural solvers for vehicle routing problems via ensemble with transferrable local policy. In: Proceedings of the International Joint Conference on Artificial Intelligence (IJCAI) (2024). https://doi.org/10.24963/ijcai.2024/764
5. Helsgaun, K.: An extension of the Lin-Kernighan-Helsgaun TSP solver for constrained traveling salesman and vehicle routing problems. Technical report, Roskilde University (2017)
6. Kool, W., van Hoof, H., Welling, M.: Attention, learn to solve routing problems! In: International Conference on Learning Representations (ICLR) (2019)
7. Kwon, Y.D., Choo, J., Kim, B., Yoon, I., Gwon, Y., Min, S.: POMO: policy optimization with multiple optima for reinforcement learning. In: Advances in Neural Information Processing Systems (NeurIPS), vol. 33, pp. 21188–21198 (2020)
8. Laporte, G., Nobert, Y.: Exact algorithms for the vehicle routing problem. North-Holland Math. Stud. **132**, 147–184 (1987). https://doi.org/10.1016/S0304-0208(08)73235-3
9. Li, B., Wu, G., He, Y., Fan, M., Pedrycz, W.: An overview and experimental study of learning-based optimization algorithms for the vehicle routing problem. IEEE/CAA J. Automatica Sinica **9**(7), 1115–1138 (2022). https://doi.org/10.1109/JAS.2022.105677
10. Liao, Z., Chen, J., Wang, D., Zhang, Z., Wang, J.: BOPO: neural combinatorial optimization via best-anchored and objective-guided preference optimization. In: Proceedings of the 42nd International Conference on Machine Learning, vol. 267, pp. 37456–37475. PMLR (2025)
11. Luo, F., Lin, X., Liu, F., Zhang, Q., Wang, Z.: Neural combinatorial optimization with heavy decoder: toward large scale generalization. In: Advances in Neural Information Processing Systems (NeurIPS), vol. 36, pp. 8845–8864 (2023)
12. Luo, F., et al.: Boosting neural combinatorial optimization for large-scale vehicle routing problems. In: International Conference on Learning Representations (ICLR) (2025)
13. Pan, X., et al.: H-TSP: hierarchically solving the large-scale traveling salesman problem. In: Proceedings of the AAAI Conference on Artificial Intelligence, vol. 37, pp. 9345–9353 (2023). https://doi.org/10.1609/aaai.v37i8.26120
14. Reinelt, G.: TSPLIB–a traveling salesman problem library. ORSA J. Comput. **3**(4), 376–384 (1991). https://doi.org/10.1287/ijoc.3.4.376
15. Vidal, T.: Hybrid genetic search for the CVRP: open-source implementation and SWAP* neighborhood. Comput. Oper. Res. **140**, 105643 (2022). https://doi.org/10.1016/j.cor.2021.105643
16. Vinyals, O., Fortunato, M., Jaitly, N.: Pointer networks. In: Advances in Neural Information Processing Systems (NeurIPS), vol. 28, pp. 2692–2700 (2015)

17. Wang, D., Zhang, Z., Teng, Y.: Large language model implemented simulated annealing algorithm for traveling salesman problem. In: 2024 IEEE International Conference on Systems, Man, and Cybernetics (SMC), pp. 209–214. IEEE (2024). https://doi.org/10.1109/SMC54092.2024.10831145
18. Wang, X., Choi, T.M., Liu, H., Yue, X.: Novel ant colony optimization methods for simplifying solution construction in vehicle routing problems. IEEE Trans. Intell. Transp. Syst. **17**(11), 3132–3141 (2016). https://doi.org/10.1109/TITS.2016.2542264
19. Ye, H., Wang, J., Liang, H., Cao, Z., Li, Y., Li, F.: GLOP: learning global partition and local construction for solving large-scale routing problems in real-time. In: Proceedings of the AAAI Conference on Artificial Intelligence, vol. 38, pp. 20284–20292 (2024). https://doi.org/10.1609/aaai.v38i18.30009
20. Zheng, Z., Zhou, C., Tong, X., Yuan, M., Wang, Z.: UDC: a unified neural divide-and-conquer framework for large-scale combinatorial optimization problems. In: Advances in Neural Information Processing Systems (NeurIPS), vol. 37, pp. 6081–6125 (2024)
21. Zhou, J., Wu, Y., Song, W., Cao, Z., Zhang, J.: Towards omni-generalizable neural methods for vehicle routing problems. In: Proceedings of the 40th International Conference on Machine Learning, vol. 202, pp. 42769–42789. PMLR (2023)
22. Zhou, S., Ding, Y., Zhang, C., Cao, Z., Jin, Y.: DualOPT: a dual divide-and-optimize algorithm for the large-scale traveling salesman problem. In: Proceedings of the AAAI Conference on Artificial Intelligence, vol. 39, pp. 27178–27186 (2025). https://doi.org/10.1609/aaai.v39i25.34926

HCPS: Heuristic Candidate-Space Policy Search for Agile Earth Observation Satellite Scheduling

Pengfei Gao[1], Benkui Zhang[2], Yajing Xu[1], Jinxiao Wang[2], Zhiyuan Yu[2], Kangning Du[3], and Bo Xiao[1](✉)

[1] Beijing University of Posts and Telecommunications, Beijing 100101, China
xiaobo@bupt.edu.cn
[2] Key Lab of Target Cognition and Application Technology, Aerospace Information Research Institute, CAS, Beijing 100101, China
[3] Key Laboratory of Information and Communication Systems, Ministry of Information Industry, Beijing Information Science and Technology University, Beijing 100101, China

Abstract. Many real-world combinatorial optimization tasks are constrained constructive processes with expensive per-step fitness evaluation. Population-based methods often suffer from high evaluation cost, while single-solution methods may lack exploration. We propose Heuristic Candidate-space Policy Search (HCPS), a hybrid framework that combines heuristic candidate-space shaping with policy-based sequential selection. At each decision step, HCPS applies feasibility-aware Top-K pruning to compress the primitive action space and then uses Proximal Policy Optimization (PPO) to optimize long-horizon utility on the reduced candidate set. We provide complexity analysis showing that Top-K pruning can substantially reduce computation while keeping quality loss limited when high-value actions are retained with high probability. Experiments on agile Earth observation satellite scheduling as a representative testbed demonstrate a strong quality–time trade-off and support the framework's generalizability.

Keywords: Heuristic Candidate-space Policy Search · Hybrid Search · Agile Earth Observation Satellite Scheduling Problem · Reinforcement Learning

1 Introduction

Agile Earth Observation Satellites (AEOS) can adjust attitude along roll, pitch, and yaw axes, achieving significantly longer observation windows than traditional satellites. This flexibility introduces complex constraints (e.g., attitude transition times), making the Agile Earth Observation Satellite Scheduling Problem (AEOSSP) a challenging combinatorial optimization task [6,9,26]. A key characteristic of AEOSSP—and many real-world scheduling problems—is that evaluating a single action (assigning an observation) is computationally expensive:

G. Iacca et al. (Eds.): PPSN 2026, LNCS 16985, pp. 349–365, 2027.
https://doi.org/10.1007/978-3-032-36214-8_22

each candidate requires checking geometric visibility, attitude feasibility, resource constraints, and conflicts with scheduled tasks. This per-step cost forces a fundamental trade-off between efficiency and solution quality.

Heuristic and metaheuristic methods have long been the dominant paradigm. Constructive heuristics [13,14,22] first defined AEOSSP and proposed greedy, dynamic programming, constraint programming, and local search; later works improved them with limited backtracking [24], priority ordering [27], and complex network theory [25]. Metaheuristics include evolutionary algorithms (genetic algorithms [15,16], ant colony optimization [4,5]) and single-point search (ALNS [11,18], GRILS [20]). They escape local optima but incur high computational cost due to many feasibility checks.

Recently, reinforcement learning (RL) methods have been explored: A3C for online scheduling [23], data-driven parallel scheduling with C-NEAT [7], end-to-end pointer networks [1,2], Q-learning [12], RL-guided genetic algorithm [21], and learning-to-construct [3]. Despite advances, most RL methods either operate directly on the large primitive action space (all targets) or adopt a two-stage "sequence-then-generate" pipeline where feasibility is handled after policy decisions. Both suffer from exploration inefficiency or costly repair.

Action space reduction has succeeded elsewhere: AlphaGo [10] uses policy-guided MCTS to expand only top moves; quadruped robots use kinematic priors [8]. However, these either rely on learned networks or are tailored to continuous control, not directly transferable to combinatorial optimization. In combinatorial optimization, domain heuristics are often used only as pre-filters or reward shaping, not structurally integrated.

In this work, we treat heuristics as structural shapers of the search space through two complementary designs: a feasibility-aware Top-K heuristic pruning that compresses the primitive action space into a compact candidate set, and a structured feature extraction module that enriches each candidate. Proximal Policy Optimization (PPO) trains a policy that selects actions from this candidate space to maximize long-horizon utility. Heuristics encode feasibility and immediate reward, whereas RL focuses on sequential strategies within a structured action space.

We instantiate this framework as HCPS (Heuristic Candidate-space Policy Search) and evaluate it on AEOSSP. Our contributions are:

- A hybrid optimization paradigm combining heuristic candidate-space shaping with policy-based selection, shifting heuristics from pre-filtering to structural guidance.
- A feasibility-aware Top-K pruning mechanism with theoretical complexity analysis, showing substantial computation reduction while preserving high-quality actions with high probability.
- Extensive experiments on AEOSSP demonstrating HCPS outperforms pure heuristics, population-based methods, and existing RL-with-heuristic baselines, achieving superior quality–time trade-off.

The remainder of this paper is organized as follows. Section 2 formally defines the AEOSSP. Section 3 presents the HCPS framework in detail. Section 4 describes the experimental setup and analyzes the results. Section 5 concludes the paper and outlines future research directions.

2 Problem Description and Formulation

2.1 Description of the AEOSSP

In line with related studies [18,19], this paper introduces the following assumptions for the AEOSSP.

1. The problem assumes the availability of sufficient ground stations and relay satellites.
2. Each AEOS can observe only one target at a time.
3. The problem assumes that the satellite has sufficient energy and storage capacity.
4. Each target can be observed at most once, known as the multi-orbit uniqueness constraint.

In addition to the above assumptions, the constraints that must be satisfied are as follows:

1. Observation windows (OWs) must lie within their visible time windows (VTWs) and meet the minimum duration.
2. Sufficient transition time between consecutive OWs for attitude maneuvering.

Objective: maximize total observation profit. Table 1 lists main symbols.

2.2 Formulation of the AEOSSP

Two types of decision variables are defined:

1. Binary variable x_i^j: equals 1 if target T_i is observed in orbit O_j, otherwise 0.
2. Real variable os_i^j: start time of the observation window OW_i^j when $x_i^j = 1$.

The mathematical model is as follows:

$$\text{Maximize} \quad \sum_{i \in T} \sum_{j \in O} p_i \cdot x_i^j \tag{1}$$

$$\text{subject to} \quad \sum_{j \in O} x_i^j \leq 1, \quad \forall i \in T \tag{2}$$

$$x_i^j \in \{0,1\}, \quad os_i^j \in \mathbb{R} \tag{3}$$

$$x_i^j \cdot (os_i^j - vs_i^j) \geq 0, \quad \forall i \in T, j \in O \tag{4}$$

$$x_i^j \cdot (ve_i^j - d_i - os_i^j) \geq 0, \quad \forall i \in T, j \in O \tag{5}$$

$$\begin{aligned} & os_m^j + d_m + \text{tran}\big(\Delta g(OW_m^j, OW_n^j)\big) \leq os_n^j, \\ & \quad \forall j \in O, \ \forall m, n \in T \text{ with } x_m^j = x_n^j = 1, \\ & \quad \text{and } OW_m^j \text{ immediately precedes } OW_n^j \end{aligned} \tag{6}$$

Table 1. Main notation

Symbol	Description
T	Set of targets
T_i	Target i
p_i	Profit of target T_i
d_i	Required observation duration of target T_i
O	Set of satellite orbits
O_j	Orbit j
$[st_j, et_j]$	Start and end time of orbit O_j
VTW_i^j	Visible time window of orbit O_j for target T_i
$[vs_i^j, ve_i^j]$	Start and end time of VTW_i^j
vt_i^j	Optimal (central) time of VTW_i^j
OW_i^j	Selected observation window from VTW_i^j
os_i^j	Start time of OW_i^j
$\phi_i^j, \theta_i^j, \psi_i^j$	Roll, pitch, yaw angles of OW_i^j
$\Delta g(OW_m^j, OW_n^j)$	Total angle change between two OWs
$\mathrm{tran}(\Delta g)$	Required transition time for angle change Δg

Constraint (2) ensures each target is observed at most once. Constraints (4) and (5) enforce that the observation window lies within the visible time window and meets the minimum duration. Constraint (6) guarantees sufficient transition time between two consecutive observations on the same satellite orbit.

Transition time follows the piecewise function from [18]:

$$\mathrm{tran}(\Delta g) = \begin{cases} 11.66, & \Delta g \leq 10 \\ a_1 + \Delta g/v_1, & 10 < \Delta g \leq 30 \\ a_2 + \Delta g/v_2, & 30 < \Delta g \leq 60 \\ a_3 + \Delta g/v_3, & 60 < \Delta g \leq 90 \\ a_4 + \Delta g/v_4, & \Delta g > 90 \end{cases} \tag{7}$$

with parameters $(v, a) = (1.5, 5)$, $(2, 10)$, $(2.5, 16)$, $(3, 22)$ for the AS-01 satellite. $\Delta g(OW_m^j, OW_n^j) = |\phi_m^j - \phi_n^j| + |\theta_m^j - \theta_n^j| + |\psi_m^j - \psi_n^j|$.

3 Methodology: Heuristically-Shaped Policy Optimization

3.1 MDP Formulation and HCPS Framework

The AEOSSP can be formulated as a Markov Decision Process (MDP) defined by the tuple $(\mathcal{S}, \mathcal{A}, \mathcal{P}, \mathcal{R})$, where:

- $\mathcal{S}$: state space, each state s encoding a partial schedule.

- $\mathcal{A}$: action space, each action a corresponding to a feasible VTW insertion.
- $\mathcal{P}$: deterministic transition, $P(s_{t+1}|s_t, a_t) = 1$ if a_t leads to s_{t+1}.
- $\mathcal{R}$: immediate reward r_t, set as the profit of the completed target.

A policy π maximizes the expected cumulative reward $\mathbb{E}[\sum \gamma^t r_t]$.

The primitive action set $\mathcal{A}(s_t)$ is huge (e.g., 5800 VTWs for 1000 targets and 6 satellites), making direct optimization prohibitive. Therefore, HCPS restructures this MDP into a two-layer process: first heuristic candidate-space shaping, then policy optimization on the reduced set.

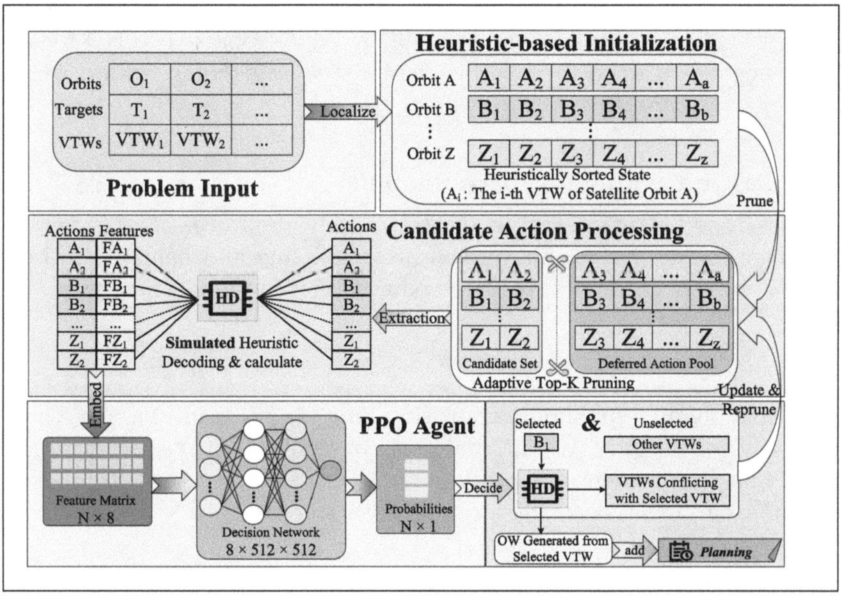

Fig. 1. Overall framework of HCPS. The heuristic component consists of three sub-modules: Heuristic-based Initialization (HBI), Candidate Action Processing (CAP), and Heuristic Decoder (HD). The policy-based selection uses Proximal Policy Optimization (PPO). HBI performs conflict-driven orbit splitting and sorting; CAP performs adaptive Top-K pruning and deep feature extraction; HD handles simulation and execution. The PPO agent selects actions from the reduced candidate set C_t.

The overall architecture of HCPS is shown in Fig. 1. The framework consists of two components: Heuristic Candidate-space Shaping and Policy-based Selection. The heuristic component comprises three sub-modules:

- **HBI**: Preprocesses the raw input via conflict-driven orbit splitting and intra orbit sorting (Earliest Deadline First [17] and start-time priority). This produces a structured initial state that respects temporal order.
- **CAP**: At each decision step, performs adaptive Top-K pruning on the candidate VTWs within each orbital group, where the number of retained candidates per orbit is dynamically adjusted based on current resource competition.

For each retained VTW, the HD is invoked in simulation mode to perform a tentative insertion, and deep features are extracted from the resulting post-decoding state. These feature vectors are then passed to the PPO agent.
- **HD**: Converts a selected VTW into an executable observation window (OW) and updates the environment. It operates in two modes: simulation returns the resulting feature vector without altering the live schedule; execution inserts the OW, computes the reward, and removes the scheduled target and conflicting VTWs.

The PPO agent receives the normalized feature vectors of finite candidates from CAP and outputs a probability distribution over them. It samples an action according to this distribution, and the chosen VTW is passed to HD for execution. The agent is trained to maximize cumulative scheduling reward using the proximal policy optimization algorithm.

3.2 Heuristic Candidate-Space Shaping

Heuristic-Based Initialization (HBI). HBI preprocesses the raw input through two operations: conflict-driven orbit splitting and heuristic intra-orbit ordering. For orbit O_j, compute the maximum viewing angle change among its VTWs:

$$\Delta g^j_{\max} = \max_{m,n \in T} \left(|\phi^j_m - \phi^j_n| + |\theta^j_m - \theta^j_n| + |\psi^j_m - \psi^j_n| \right), \tag{8}$$

and the corresponding maximum transition time $\tau^j_{\max} = \text{tran}(\Delta g^j_{\max})$. The orbit is recursively split at any gap Δt_k between consecutive VTWs satisfying $\Delta t_k > \tau^j_{\max}$, yielding independent sub-orbits with no cross-conflict.

Within each virtual orbit, assign a priority score to each VTW:

$$G^j_i = \alpha \left(1 - \frac{ve^j_i - st_j}{et_j - st_j} \right) + \beta \left(1 - \frac{vs^j_i - st_j}{et_j - st_j} \right), \tag{9}$$

The first term implements the Earliest Deadline First (EDF) principle (earlier ve^j_i gives higher score), while the second term favors VTWs with earlier start times. We set $\alpha = 1$ and $\beta = 0.2$ to prioritize deadline pressure while still considering start time. VTWs are sorted descending by G^j_i to form an initial sequence, providing a temporally coherent state for subsequent modules.

Candidate Action Processing (CAP). CAP handles the selection of promising VTWs at each decision step through two sequential operations: adaptive Top-K pruning and deep feature extraction.

Let A_t be the primitive feasible action set at state s_t. Instead of using a fixed K, we adapt the pruning ratio per orbit based on its current resource competition intensity at decision step t. For orbit O_j, let $N^j_{VTW}(t)$ be the number of remaining VTWs in that orbit at time t. Define the expected occupancy per VTW as $\bar{t}_{\text{occ}} = d_{\text{avg}} + \bar{\tau}$, where d_{avg} is the average required observation duration and $\bar{\tau}$ is the expected transition time (derived from historical data or a fixed prior). The total estimated time required to schedule all remaining VTWs is $T^j_{\text{need}}(t) = N^j_{VTW}(t) \cdot \bar{t}_{\text{occ}}$, and the remaining available time in the orbit is $T^j_{\text{avail}}(t) = et_j - st_j - t_{\text{used}}(t)$, where $t_{\text{used}}(t)$ is the total time already occupied by scheduled observations in that orbit. The resource competition ratio at step t is defined as:

$$r_j(t) = \frac{T^j_{\text{need}}(t)}{T^j_{\text{avail}}(t)}. \tag{10}$$

A larger $r_j(t)$ indicates tighter resource competition (more VTWs competing for limited time), which demands a larger candidate set to allow careful selection. We set the adaptive $K_j(t)$ for orbit O_j at step t as:

$$K_j(t) = \max\left(2,\ \lfloor r_j(t) \rfloor\right), \tag{11}$$

where $\lfloor \cdot \rfloor$ denotes the floor function. This formulation ensures that $K_j(t)$ increases with competition intensity, retaining more candidates when resources are scarce, with a minimum of 2 to maintain a minimal level of choice.

We then sort all feasible candidate VTWs in A_t (i.e., VTWs that satisfy all constraints given the current partial schedule) descending by their priority scores G^j_i computed in Sect. 3.2. For each orbit O_j, only the top $K_j(t)$ candidates are kept:

$$C_t = \bigcup_j \text{TopK}_{a \in A^j_t} G^j_i, \qquad |C_t| = \sum_j K_j(t) \ll |A_t|, \tag{12}$$

where A^j_t is the set of VTWs belonging to orbit O_j. PPO selects actions only within C_t. This shapes the search manifold using domain heuristics and allocates computation from breadth to depth, with $K_j(t)$ automatically increasing under tight resource conditions to enable more informed decisions.

After pruning, the Heuristic Decoder performs a *full decoding call* for each candidate $a \in C_t$: it converts the selected VTW into an executable OW, tentatively inserts it into a copy of the current schedule, and evaluates the consequences on all remaining VTWs in the same orbit. This process is computationally expensive because it involves OW generation, transition feasibility checks, and conflict propagation.

For each candidate, the following raw features are extracted from the simulation:

- **Transition time overhead**: the time $\text{tran}(\Delta g)$ required for the insertion, normalized by $Tran_{\max}$ (the maximum transition time in the instance).
- **Conflict count**: the number of VTWs that become infeasible due to the insertion.
- **Target kill count**: the number of targets that lose all their remaining VTWs (i.e., the insertion removes their last feasible opportunity).
- **Orbit resource pressure**: the ratio of remaining available time in the orbit to the number of remaining VTWs in that orbit.
- **Candidate rank**: the position of the VTW in the Top-$K_j(t)$ list (from 1 to $K_j(t)$), normalized by $K_j(t)$.

- **Remaining opportunities**: the number of VTWs still available for the same target across all orbits, normalized by the total number of VTWs of that target.

These raw features are then normalized within the candidate set C_t to provide relative comparisons, enabling the PPO agent to make decisions based on relative advantages rather than absolute magnitudes.

Let c_f be the cost of one full decoding call and c_h the heuristic scoring cost. In the primitive action space, naïvely evaluating all $|A_t|$ candidates would require $\mathcal{O}(|A_t|c_f)$ per step. With HCPS, the lightweight heuristic scoring first selects the $K_j(t)$ most promising VTWs per orbit, and only these $\sum_j K_j(t)$ candidates undergo the expensive decoding. Thus the per-step complexity becomes $\mathcal{O}(|A_t|c_h + (\sum_j K_j(t))c_f)$. Because $\sum_j K_j(t) \ll |A_t|$ and $c_h < c_f$, the number of costly full decoding calls is reduced substantially.

If a_t^* is the optimal primitive action and $p_t = \Pr(a_t^* \in C_t)$, the expected value loss is bounded by $\sum_{\tau=t}^{T} \gamma^{\tau-t}(1-p_\tau)\Delta_{\max}$; thus high retention probability limits quality degradation. The adaptive $K_j(t)$ ensures that p_t remains high especially when resource competition is intense, by retaining a sufficiently large candidate set for careful selection.

3.3 Policy Learning on Candidate Space

State and Action Representation. The state is implicitly encoded in the feature vectors of the candidates in C_t; the agent does not maintain an explicit global state representation. The action space is the candidate set C_t (size K), each action corresponding to selecting a specific VTW.

PPO Agent Architecture. The PPO agent comprises two neural networks with an identical architecture. The input is an 8-dimensional feature vector per candidate, which includes the six features listed in Sect. 3.2 plus two additional features: target profit p_i and required duration d_i, both normalized within the candidate set. The actor network maps these features to selection probabilities via a shared feature extraction layer followed by a softmax output over the K candidates. The critic network first transforms each candidate's features independently, then aggregates them via average pooling to produce a scalar state value estimate. Action selection follows $a_t \sim \pi_\theta(\cdot|s_t) = \text{softmax}(f_\theta(s_t))$. Training uses the standard PPO-clip objective with generalized advantage estimation (GAE).

Reward Function Design. The reward components are designed based on the features extracted in CAP, providing immediate feedback aligned with the candidate evaluation criteria. Rewards are hierarchical to provide dense learning signals while focusing on profit maximization (Table 2). Here, $Tran_{\min} = 11.6$ seconds and $Tran_{\max}$ is the maximum transition time in the current instance.

Table 2. Reward components.

Component	Formula	Weight
Target completion	$R_{comp} = \omega_1 p_i$	ω_1 (large)
Kill penalty	$R_{kill} = -\omega_2 \sum_i p_i k_i$	ω_2 (large)
Transition cost	$R_{time} = -\omega_3 \cdot \frac{tran - Tran_{\min}}{Tran_{\max} - Tran_{\min}}$	ω_3 (small)
Urgency	$R_{chance} = \omega_4 \left(1 - \frac{chance}{Chance_{\max}}\right)$	ω_4 (small)

Total reward: $R_{all} = R_{comp} + R_{kill} + R_{time} + R_{chance}$, with $\omega_1, \omega_2 \gg \omega_3, \omega_4$ to prioritize primary objectives. The specific weight values are provided in the experimental setup (Sect. 4).

3.4 Heuristic Decoder

HD converts selected VTWs into executable OWs and updates the environment. Algorithm 1 shows its dual-mode operation: *simulation* copies the schedule, inserts the candidate OW, and returns features; *execution* inserts the OW into the live schedule, computes reward, and updates state (removing scheduled target and conflicting VTWs).

Algorithm 1. Heuristic Decoder (HD)

Input: orbit index j, current schedule S, target set T, candidate VTW vtw, boolean $simulate$
$S_j \leftarrow S[j]$ {schedule of orbit j (list of scheduled OWs)}
if $simulate$ **then**
  $ow \leftarrow$ convertVTWtoOW(vtw) {generate OW from VTW, respecting attitude feasibility}
  $vtw.ow \leftarrow ow$ {store OW for potential execution}
  $S'_j \leftarrow$ copy(S_j) {create a copy of the current orbit schedule}
  $S'_j \leftarrow$ insertOW(S'_j, ow) {tentative insertion, checks transition constraints}
  $features \leftarrow$ extractFeatures(S'_j, ow) {compute feature vector for PPO}
  return $features$
else
  $ow \leftarrow vtw.ow$ {retrieve precomputed OW (must have been set during simulation)}
  $S_j \leftarrow$ insertOW(S_j, ow) {permanently update the live schedule}
  $reward \leftarrow ow.profit$ {immediate reward equals target profit}
  removeTarget($T, vtw.target$) {mark target as observed}
  removeConflictingVTWs(S_j, ow) {delete all VTWs that conflict with the new OW}
  return $reward$
end if

All helper functions (convertVTWtoOW, copy, insertOW, extractFeatures, removeTarget, removeConflictingVTWs) are standard schedule manipulation routines; their implementations are omitted for brevity.

4 Experiment

This section systematically analyzes the performance of the proposed HCPS framework. All training experiments were conducted on an NVIDIA A100 GPU, with code implemented in Python. Inference times (the solving times reported in the tables) are in seconds. Training on the 1000-target scale (six satellites) takes 3.5 h on a single A100 GPU. For testing and evaluation, all experiments were run on an Intel i7-14700K processor platform equipped with 32 GB of RAM.

4.1 Dataset

Following prior studies [18,19], the observation missions consist of point targets uniformly generated within China (3°N–53°N, 74°E–133°E). The orbital parameters of the six satellites used in the experiments are provided in the supplementary material (Table S1)[1]. Fifteen problem scales are evaluated, with the number of targets ranging from 100 to 1500 in steps of 100. Each target requires a uniformly random observation duration between 15 and 30 s, and each target is assigned a random profit uniformly drawn from [1, 10].

For each problem scale from 100 to 1500 in steps of 100, we generate 1000 independent instances for training and 20 separate instances for testing. All reported results in the following sections are statistical summaries over these 20 test instances. The primary quality metric is the target completion rate F_s, defined as:

$$F_s = \frac{\sum_{n \in N} f_n}{\sum_{n \in N} P_n}, \tag{13}$$

where f_n is the total profit obtained for instance n and P_n is the sum of profits of all targets in that instance.

4.2 Effect of Top-K Pruning

The parameter K in the Candidate Action Processing (CAP) module controls the number of candidate VTWs retained after heuristic pruning. We evaluate fixed K values $\{1, 3, 5, 10\}$ and the proposed adaptive K strategy (denoted as K_{ada}) on three representative problem scales (700, 1000, 1500 targets). Figure 2 shows the training convergence patterns for the fixed K values and the adaptive K.

Smaller fixed K values lead to faster initial convergence due to the reduced action space, which simplifies credit assignment. Larger fixed K values ($K = 5$) exhibit slower but more sustained improvement initially, yet excessively large K

[1] Available at https://github.com/Gaopf1016/hcps-supplementary.

can hinder convergence: as shown in Table 3, an overly large action space may introduce redundant candidates and complicate policy learning. The value loss patterns in Fig. 2 further confirm that smaller K yields more stable training, while larger K introduces higher variance.

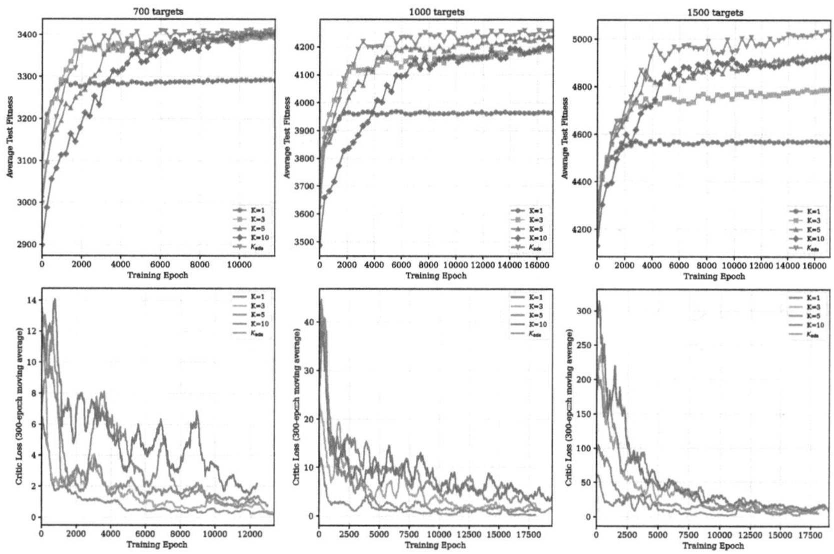

Fig. 2. Training convergence for different K values across three scales. The top row shows the average episode reward on test instances; the bottom row shows the critic loss during training. Each column corresponds to a specific problem scale (700, 1000, 1500 targets).

Table 3 quantifies the quality–time trade-off for fixed K and the adaptive K strategy. The adaptive K strategy achieves the highest completion rates on 1000 and 1500 targets, and is comparable to the best fixed K on 700 targets. More importantly, it does so with significantly lower computational cost than the best-performing fixed K values. This efficiency stems from the fact that larger K values increase computational time, while the adaptive mechanism dynamically shrinks the candidate set when resources are tight, reducing the per-step action space and accelerating decision making. Based on these results, we adopt the adaptive K strategy for all subsequent experiments.

4.3 Comparison with State-of-the-Art Algorithms

We compare HCPS with four state-of-the-art baselines: GRILS [20], A-ALNS [11], GA, and RL-GA [21]. All algorithms are evaluated on the same dataset and hardware, with parameters tuned to their optimal settings as reported in

Table 3. Performance comparison for different fixed K and adaptive K (mean ± std over 20 instances per scale).

K	700 targets		1000 targets		1500 targets	
	F_s (%)	Time (s)	F_s (%)	Time (s)	F_s (%)	Time (s)
1	85.52±2.23	**2.71**±0.32	72.10±2.36	**4.27**±0.21	55.38±2.01	**9.65**±0.65
3	88.16±1.02	5.13±0.28	76.22±2.04	9.52±0.24	58.05±1.27	17.87±0.72
5	**88.57**±1.79	8.44±0.31	77.04±1.71	13.95±0.35	59.84±1.05	28.03±0.89
10	88.49±1.08	15.92±0.44	76.35±1.62	22.23±0.58	59.93±0.93	39.76±1.23
K_{ada}	88.52±1.25	6.41±0.27	**77.39**±1.82	11.43±0.42	**60.95**±0.85	23.49±0.76

the original papers. For HCPS, we adopt the adaptive Top-K pruning strategy described in Sect. 3.2. The parameter configurations of all algorithms are summarized in Table 4.

Table 4. Parameter settings for all algorithms.

Algorithm	Key Parameters
A-ALNS [11]	Max iterations 5000, annealing coeff. 0.9975, removal rate 10%, score increments (30,20,10,1), $\phi = 50$, $\rho = 5$, $M = 100$ (Chinese area).
GRILS [20]	$StartGreed = 1$, $GreedyRange = 0.2$, $GreedDecrease = 0.02$, max non-improving iterations 300.
GA	Population size $N_p = 10$, crossover prob. 0.9, mutation prob. 0.1, $MFE = 5000$ (RL-GA without Q-learning).
RL-GA [21]	$T = 1000$, $\alpha = 0.01$, $\gamma = 0.95$, $\epsilon = 0.01$, $L = 2$, $N_p = 10$, $MFE = 5000$.
HCPS	Policy/value net hidden layers 8-512-512, activation Tanh, PPO epochs 10, batch size 256, $\gamma = 0.99$, $\lambda = 0.95$, clip=0.2, KL target 0.01.

Figure 3 presents the target completion rate (a) and runtime (b, log scale) across all 15 problem scales (100–1500 targets). HCPS achieves 100% completion for instances up to 400 targets. For scales 500 to 1500, HCPS consistently yields the highest completion rates.

In terms of computational efficiency, HCPS substantially outperforms all baseline algorithms. For scales exceeding 300, HCPS reduces runtime by an average of 84.8% compared to the second-fastest algorithm (GA) and by 91.0% compared to the algorithm with the second-best completion rate (GRILS). For smaller scales (100–300), where resources are abundant relative to the number of tasks, early termination of heuristic iterations occurs, leading to faster baseline

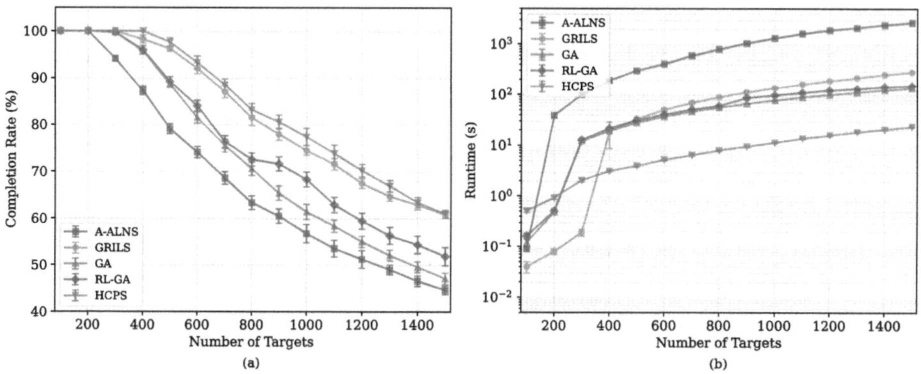

Fig. 3. Comparison of (a) target completion rate and (b) computational time. Error bars indicate standard deviation over 20 test instances per scale.

runtimes. Consequently, HCPS exhibits slightly slower solving speeds in these small-scale cases, though the absolute runtime differences remain negligible.

Statistical significance is verified using the Wilcoxon signed-rank test ($p < 0.05$), confirming that HCPS significantly outperforms all baselines in both quality and efficiency for scales ≥ 500.

Table 5 reports a condensed comparison on four scales (300,700,1000,1500). The complete numerical results (mean $\pm$ standard deviation over 20 test instances per scale) are provided in the supplementary material (Table S2).

Table 5. Selected performance comparison (mean $\pm$ std over 20 test instances per scale).

Scale	Metric	A-ALNS	GRILS	GA	RL-GA	HCPS
300	F_s (%)	94.13 ± 0.56	$\mathbf{100.0} \pm 0.0$	99.78 ± 0.34	99.75 ± 0.42	$\mathbf{100.0} \pm 0.0$
	t (s)	100.17 ± 5.23	$\mathbf{0.19} \pm 0.03$	12.20 ± 0.78	12.72 ± 0.85	$\mathbf{2.05} \pm 0.14$
700	F_s (%)	68.58 ± 1.34	$\mathbf{87.04} \pm 1.33$	75.50 ± 1.34	76.31 ± 1.28	$\mathbf{88.52} \pm 1.25$
	t (s)	593.19 ± 25.34	69.49 ± 2.91	$\mathbf{46.30} \pm 1.89$	50.62 ± 2.01	$\mathbf{6.81} \pm 0.27$
1000	F_s (%)	56.69 ± 1.97	$\mathbf{74.51} \pm 1.01$	61.38 ± 1.62	68.31 ± 1.56	$\mathbf{77.39} \pm 1.82$
	t (s)	1312.90 ± 42.23	136.72 ± 4.23	$\mathbf{77.32} \pm 2.45$	97.04 ± 2.89	$\mathbf{11.23} \pm 0.42$
1500	F_s (%)	44.65 ± 0.93	$\mathbf{60.80} \pm 0.69$	47.08 ± 1.21	51.81 ± 1.94	$\mathbf{60.95} \pm 0.85$
	t (s)	2668.85 ± 80.78	282.62 ± 6.01	$\mathbf{136.44} \pm 3.56$	149.70 ± 4.45	$\mathbf{23.28} \pm 0.76$

4.4 Ablation Studies

We conduct ablation experiments to quantify the contribution of each component in HCPS, including all engineered features (from CAP) and the hierarchical reward components. The full model (HCPS-Full) includes all features and the complete reward. We systematically remove each component and evaluate the impact on solution quality across three problem scales (700, 1000, 1500 targets). Table 6 reports the relative completion rate (F_s) normalized to the full model (set to 100%) when individual components are omitted.

The results indicate that the primary objective *target completion* has the largest impact, with its removal reducing performance by over 13%. Among the reward components, *kill penalty* also plays a critical role, causing a 6–8% drop when omitted. Among the features, *Target kill count* has the strongest influence. The relative importance of these components remains broadly consistent across problem scales, though the impact of removing components tends to be slightly smaller on larger scales.

Table 6. Ablation results: relative completion rate (F_s) when removing key components (normalized to full model = 100%).

Removed Component	700 targets	1000 targets	1500 targets
None (Full model)	100.00	100.00	100.00
Target completion (reward)	85.47	86.22	86.47
Kill penalty (reward)	92.15	93.44	93.72
Target kill count (feature)	94.60	94.24	95.44
Orbit resource pressure (feature)	97.07	96.85	97.40
Remaining opportunities (feature)	98.26	98.30	98.82
Urgency (reward)	98.46	98.71	99.15
Transition time overhead (feature)	98.65	98.83	99.61
Conflict count (feature)	99.07	99.85	99.79

5 Conclusion and Future Work

This paper proposed HCPS, a hybrid framework that treats domain heuristics as structural shapers of the search space for combinatorial optimization with expensive per-step evaluation. By combining adaptive Top-K pruning and deep feature extraction with PPO policy learning, HCPS reduces per-step complexity while preserving solution quality. Experiments on agile Earth observation satellite scheduling demonstrate that HCPS outperforms state-of-the-art baselines in both quality and efficiency, and ablation studies confirm the contribution of each component.

Future work includes incorporating uncertainty-aware features, extending temporal causality to cross-interval rescheduling, and transferring the HCPS paradigm to other large-scale engineering scheduling problems.

Acknowledgements. This work was supported by the National Natural Science Foundation of China (NSFC No.62076031 and No.62076036).

References

1. Chen, J., Chen, M., Wen, J., He, L., Liu, X.: A heuristic construction neural network method for the time-dependent agile earth observation satellite scheduling problem. Mathematics **10**(19) (2022). https://doi.org/10.3390/math10193498, https://www.mdpi.com/2227-7390/10/19/3498
2. Chen, M., Chen, Y., Chen, Y., Qi, W.: Deep reinforcement learning for agile satellite scheduling problem. In: 2019 IEEE Symposium Series on Computational Intelligence (SSCI), pp. 126–132 (2019). https://doi.org/10.1109/SSCI44817.2019.9002957
3. Chen, M., Du, Y., Tang, K., Xing, L., Chen, Y., Chen, Y.: Learning to construct a solution for the agile satellite scheduling problem with time-dependent transition times. IEEE Trans. Syst. Man, Cybern. Syst. **54**(10), 5949–5963 (2024). https://doi.org/10.1109/TSMC.2024.3411640
4. Cui, K., Xiang, J., Zhang, Y.: Mission planning optimization of video satellite for ground multi-object staring imaging. Adv. Space Res. **61**(6), 1476–1489 (2018). https://doi.org/10.1016/j.asr.2017.10.056. https://www.sciencedirect.com/science/article/pii/S0273117717307950
5. Du, B., Li, S., She, Y., Li, W., Liao, H., Fei, W.H.: Area targets observation mission planning of agile satellite considering the drift angle constraint. J. Astron. Telescopes, Instrum. Syst. **4**, 047002 – 047002 (2018). https://api.semanticscholar.org/CorpusID:173179998
6. Du, J., Jiang, C., Guo, Q., Guizani, M., Ren, Y.: Cooperative earth observation through complex space information networks. IEEE Wirel. Commun. **23**(2), 136–144 (2016). https://doi.org/10.1109/MWC.2016.7462495
7. Du, Y., Wang, T., Xin, B., Wang, L., Chen, Y., Xing, L.: A data-driven parallel scheduling approach for multiple agile earth observation satellites. IEEE Trans. Evol. Comput. **24**(4), 679–693 (2020). https://doi.org/10.1109/TEVC.2019.2934148
8. Fransen, R., Bowden, R., Hadfield, S.: Addressing dimensional scaling in reinforcement learning for symbolic locomotion policies through leveraging inductive priors. In: 2025 IEEE/RSJ International Conference on Intelligent Robots and Systems (IROS), pp. 5221–5227 (2025). https://api.semanticscholar.org/CorpusID:283358088
9. Globus, A., Crawford, J., Lohn, J., Pryor, A.: A comparison of techniques for scheduling earth observing satellites, pp. 836–843 (2004)
10. Graepel, T.: Alphago - mastering the game of go with deep neural networks and tree search. In: Machine Learning and Knowledge Discovery in Databases (2016)
11. He, L., Liu, X., Laporte, G., Chen, Y., Chen, Y.: An improved adaptive large neighborhood search algorithm for multiple agile satellites scheduling. Comput. Oper. Res. **100**, 12–25 (2018). https://doi.org/10.1016/j.cor.2018.06.020. https://www.sciencedirect.com/science/article/pii/S030505481830172
12. He, Y., Xing, L., Chen, Y., Pedrycz, W., Wang, L., Wu, G.: A generic Markov decision process model and reinforcement learning method for scheduling agile earth observation satellites. IEEE Trans. Syst. Man, Cybern. Syst. **52**(3), 1463–1474 (2022). https://doi.org/10.1109/TSMC.2020.3020732

13. Lemaître, M., Verfaillie, G., Fargier, H., Lang, J., Bataille, N., Lachiver, J.M.: Sharing the use of earth observation satellites. In: 3rd Workshop on Planning and Scheduling for Space (2002)
14. Lemaître, M., Verfaillie, G., Jouhaud, F., Lachiver, J., Bataille, N.: How to manage the new generation of agile earth observation satellites. In: Proceedings of the International Symposium on Artificial Intelligence, Robotics and Automation in Space, pp. 1–10. Citeseer (2000)
15. Lemaître, M., Verfaillie, G., Jouhaud, F., Lachiver, J.M., Bataille, N.: Selecting and scheduling observations of agile satellites. Aerosp. Sci. Technol. **6**(5), 367–381 (2002). https://doi.org/10.1016/S1270-9638(02)01173-2. https://www.sciencedirect.com/science/article/pii/S1270963802011732
16. Li, Y., Xu, M., Wang, R.: Scheduling observations of agile satellites with combined genetic algorithm. In: Third International Conference on Natural Computation (ICNC 2007), vol. 3, pp. 29–33 (2007). https://doi.org/10.1109/ICNC.2007.652
17. Liu, C.L., Layland, J.W.: Scheduling algorithms for multiprogramming in a hard-real-time environment. J. ACM **20**(1), 46–61 (1973). https://doi.org/10.1145/321738.321743
18. Liu, X., Laporte, G., Chen, Y., He, R.: An adaptive large neighborhood search metaheuristic for agile satellite scheduling with time-dependent transition time. Comput. Oper. Res. **86**, 41–53 (2017). https://doi.org/10.1016/j.cor.2017.04.006. https://www.sciencedirect.com/science/article/pii/S0305054817300977
19. Peng, G., Dewil, R., Verbeeck, C., Gunawan, A., Xing, L., Vansteenwegen, P.: Agile earth observation satellite scheduling: an orienteering problem with time-dependent profits and travel times. Comput. Oper. Res. **111**, 84–98 (2019). https://api.semanticscholar.org/CorpusID:197507578
20. Peng, G., et al.: Solving the agile earth observation satellite scheduling problem with time-dependent transition times. IEEE Trans. Syst. Man, Cybern. Syst. **52**(3), 1614–1625 (2022). https://doi.org/10.1109/TSMC.2020.3031738
21. Song, Y., Wei, L., Yang, Q., Wu, J., Xing, L., Chen, Y.: RL-GA: a reinforcement learning-based genetic algorithm for electromagnetic detection satellite scheduling problem. Swarm Evol. Comput. **77**, 101236 (2023). https://doi.org/10.1016/j.swevo.2023.101236. https://www.sciencedirect.com/science/article/pii/S221065022300010X
22. Verfaillie, G., Lemaître, M.: Selecting and scheduling observations for agile satellites: some lessons from the constraint reasoning community point of view. In: Walsh, T. (ed.) Principles and Practice of Constraint Programming – CP 2001, pp. 670–684. Springer, Heidelberg (2001)
23. Wang, H., Yang, Z., Zhou, W., Li, D.: Online scheduling of image satellites based on neural networks and deep reinforcement learning. Chin. J. Aeronaut. **32**(4), 1011–1019 (2019). https://doi.org/10.1016/j.cja.2018.12.018. https://www.sciencedirect.com/science/article/pii/S1000936119300305
24. Wang, P., Reinelt, G., Gao, P., Tan, Y.: A model, a heuristic and a decision support system to solve the scheduling problem of an earth observing satellite constellation. Comput. Ind. Eng. **61**(2), 322–335 (2011). https://doi.org/10.1016/j.cie.2011.02.015. https://www.sciencedirect.com/science/article/pii/S0360835211000702. combinatorial Optimizatiion in Industrial Engineering
25. Wang, X., Han, C., Zhang, R., Gu, Y.: Scheduling multiple agile earth observation satellites for oversubscribed targets using complex networks theory. IEEE Access **7**, 110605–110615 (2019). https://doi.org/10.1109/ACCESS.2019.2925704

26. Wang, X., Wu, G., Xing, L., Pedrycz, W.: Agile earth observation satellite scheduling over 20 years: formulations, methods, and future directions. IEEE Syst. J. **15**(3), 3881–3892 (2021). https://doi.org/10.1109/JSYST.2020.2997050
27. Xu, R., Chen, H., Liang, X., Wang, H.: Priority-based constructive algorithms for scheduling agile earth observation satellites with total priority maximization. Expert Syst. Appl. **51**, 195–206 (2016). https://doi.org/10.1016/j.eswa.2015.12.039. https://www.sciencedirect.com/science/article/pii/S0957417415008477

Insights from Multi-tasking the EAX Algorithm for the Travelling Salesperson Problem

Liam Wigney[1(✉)], Aneta Neumann[1], Yew Soon Ong[2,3], and Frank Neumann[1]

[1] Optimisation and Logistics, School of Computer Science and Information Technology, Adelaide University, Adelaide, Australia
{liam.wigney,aneta.neumann,frank.neumann}@adelaide.edu.au

[2] College of Computing and Data Science, Nanyang Technological University, Singapore, Singapore
ASYSOng@ntu.edu.sg

[3] Centre for Frontier AI Research, Institute of High Performance Computing, Agency for Science, Technology and Research, Singapore, Singapore

Abstract. Evolutionary multitasking allows several related problems to be solved in a single run of an algorithm. In this paper, we investigate integrating evolutionary multitasking with Edge Assembly Crossover (MT-EAX) to solve the classical Travelling Salesperson Problem (TSP). To fairly compare MT-EAX against standard EAX under strict compute budgets, we evaluate three scaling methods: generation scaling, population scaling, and balanced scaling. Our results show that generationally scaled MT-EAX is highly effective compute-wise in the early stages of the search, saving 60% to 90% of compute for equal or better solution quality. We observe that instance geometry has a significant impact, with clustered, normally distributed instances securing larger improvements than uniformly distributed ones. However, when scaling by population or utilising explicit solution transfer, the results are negative due to population starvation and incompatible cross-instance parent selection. We demonstrate that the advantage of MT-EAX derives from increased diversity through parallel search in early generations, which can be successfully preserved using a decoupled configuration to strictly outperform or match standard EAX performance at final convergence.

Keywords: Travelling Salesperson Problem · Multi-task Benchmark · Transfer Optimisation

1 Introduction

In recent years, there has been a significant increase in attention towards evolutionary multitasking. This is a framework where several often related problems can be solved in a single run of an algorithm using its population, rather than by

G. Iacca et al. (Eds.): PPSN 2026, LNCS 16985, pp. 366–381, 2027.
https://doi.org/10.1007/978-3-032-36214-8_23

solving the single problems individually. It has been shown to work in a broad range of areas, such as continuous and combinatorial optimisation as well as genetic programming. Key literature includes the book by [3] and the special issue edited by [6].

The travelling salesperson problem (TSP) is a classical NP-hard computer science problem, where the goal is to generate a tour where the tour length is minimised and all cities are visited exactly once before returning to the starting city. Solving the TSP can be done using both exact and heuristic methods. State-of-the-art exact methods such as Concorde [1] can solve single instances up to 10000 nodes in reasonable time with supercomputers, however practically, above a few thousand nodes, solving multiple instances is not viable. On the other hand, evolutionary state-of-the-art solvers such as EAX [11] and LKH-2 do not find exact solutions but rather approximations in a drastically shorter time frame and are thus viable for these larger problems. While EAX is stronger for large instances, LKH-2 and its extensions are competitive, sometimes beating EAX for problems fewer than around 100000 nodes. LKH-2 is based on the Lin-Kernighan variable depth search, where at each step the algorithm considers swapping edges of increasing length until no improvement is found while using a 1-tree lower bound to guide the search process. EAX works by maintaining a population of tours and iteratively combining pairs of parent solutions via edge assembly crossover, which creates offspring by selectively swapping segments of edges between parents, preserving high quality building blocks while introducing diversity. However, running EAX independently from scratch on multiple similar routing problems ignores potentially transferable structural knowledge, wasting computational effort. By integrating multitasking with EAX and creating MT-EAX, we ask whether a) MT-EAX solves problems more efficiently than EAX, and b) MT-EAX yields better solutions than EAX. As MT-EAX solves k problems in one run but for the same compute, for fairness we strictly scale MT-EAX's generation limit and population to evaluate performance under fixed computational budgets.

This paper shows that MT-EAX in the generation-scaled case is more effective compute-wise than EAX for the same solution quality. This is because the MT-EAX framework allows for parallel solving of instances, injecting increased diversity through similar-instance crossover which outperforms deeper, isolated searching in the early generations. However, when scaling is done by population, the results are more negative, as the larger global population does not improve diversity enough to offset the penalty of fewer generations. The advantage of MT-EAX is significantly extended by utilising a decoupled variation (Decoupled MT-EAX), which acts as a computational cutoff, ensuring that the early-time multitasking advantage is preserved before isolating instances to cleanly reach final convergence. These results are directly related to the limitations of effective transfer, seen in [15] which shows that even with the most optimal node matching, transferred tours are on average 1.6 times longer than instance-specific optimal solutions. It also aligns with the results in [2], confirming that the advantage of multitasking is primarily for limited computational budgets. We also observe that the structural similarity between instances dictates performance, where the geometric distribution of the instances has the largest and most significant effect. Overall our results show that MT-EAX can provide better solutions than EAX both at convergence and at fixed budgets, although the primary advantage is that for early fixed-time budgets MT-EAX finds significantly better results than EAX.

The paper is structured as follows: Sect. 2 outlines the background of the problem, in terms of the TSP, related work and the EAX algorithm. Section 3 provides a description of MT-EAX and justifications for the design. Section 4 outlines the experimental setup and the evaluation metrics. Section 5 describes the experimental results. Section 6 gives a discussion on these results, relating back to both EAX and MT-EAX.

2 Background

Traditionally, solving multiple problems, TSP or otherwise, has required solving each problem separately. However, multitasking has been shown to provide both reduced compute times for similar quality solutions and an increased quality for comparable compute times [3,5].

2.1 Problem Formulation

An instance I is given as a set $V = \{v_1, \ldots, v_n\}$ of n cities with distances $d(u, v)$ for any pair of cities $(u, v) \in V \times V$. For a given instance $I = (V, d)$, the goal is to compute a permutation $\pi = (\pi_1, \ldots, \pi_n)$ of the cities of V such that

$$c(\pi) = d(\pi_n, \pi_1) + \sum_{i=1}^{n-1} d(\pi_i, \pi_{i+1})$$

is minimal. We consider a set of TSP instances $I = \{I_1, \ldots, I_k\}$ where each instance $I_j = (V_j, d_j)$ has n cities and is defined by a set $V_j = \{v_1^j, \ldots, v_n^j\}$ with distances $d_j(u, v)$ for any pair of cities $(u, v) \in V_j \times V_j$. The goal is to find for any instance I_j, $1 \leq j \leq k$, a permutation $\pi^j = (\pi_1^j, \ldots, \pi_n^j)$ of the cities of V_j such that

$$c_j(\pi^j) = d_j(\pi_n^j, \pi_1^j) + \sum_{i=1}^{n-1} d_j(\pi_i^j, \pi_{i+1}^j)$$

is minimal. We consider sets of TSP instances I where each instance I_s, $1 \leq s \leq k$ has the same number of cities n.

2.2 Related Work

Prior work on multitasking for the TSP has focused on deep learning and reinforcement learning techniques [4,8,13] rather than evolutionary algorithms. These approaches do not explicitly transfer solutions from one instance to another, making it difficult to understand mechanistically how transfer operates. Much of the current evolutionary multitasking literature [7,9,10,14] focuses on general combinatorial problems or non-TSP problems. They lack an explicit focus on explainable techniques to compare methods, hindering the ability to understand why certain approaches succeed or fail during transfer.

More explicit transfer work exists [15], examining how node ordering can be aligned between instances to minimise distances via matching, allowing permutations generated for one instance to be more applicable to another. Multitasking has also been investigated [2] and shown to be effective in the closely related TTP, where for small time budgets it outperforms classical approaches. This suggests that the benefits of multitasking are restricted to certain stages of the optimisation process for these problems. Work on transferring solutions is not always positive, [16] showed that transferring solutions can cause negative transfer where progress degrades compared to running separately.

2.3 The EAX Algorithm

EAX works by maintaining a population of solutions and iterating over them with selection and crossover operators until termination conditions are met. It is initialised by applying 2-opt local search to random tours, producing population size P locally optimal starting solutions.

Crossover operates by taking two parent tours, p_A and p_B. Let E_A and E_B denote their edge sets. The combined multigraph is then defined as $G_{A,B} = (V, E_A \cup E_B)$. It is decomposed into AB-cycles which are alternative cycles using edges from each edge set in turn. Then an E-set is built by selecting a subset of these AB-cycles depending on the stage the algorithm is in. A temporary solution is formed from p_A by removing the E_A edges in the E-set and replacing them with the corresponding E_B edges, which creates one or potentially more subtours. These are greedily merged, minimising the added edge length to connect into a valid tour. This is done O times, where O is the number of offspring.

There are two stages, I and II, which define the strategy selected for constructing these E-sets. For stage I, a single AB-cycle is selected, which closely resembles p_A and thus maintains population diversity. This functions as a localised search. For stage II, the block2 method is used which applies a tabu search to select AB-cycles that form large adjacent segments, which more drastically changes the offspring and functions as a more global search.

Selection operates by replacing only p_A with the best offspring rather than p_B. The selection preserves diversity via preserving entropy, as offspring are evaluated not on tour length alone, but by the improvement in average tour length per unit loss of population edge entropy in an attempt to prevent premature convergence. Stage I and II both continue until no improvement in the best solution has occurred for $1500/O$ generations. Once the average tour length equals the best, the run ends.

The edge frequency matrix $F(e)$ tracks the number of individuals in the population that use each edge e. It drives diversity maintenance and is utilised by the block2 strategy. Standard EAX uses $F(e)$ to strictly guide crossover based on a single instance's population structure, evaluating population diversity via edge entropy H. Offspring selection preserves this diversity by balancing tour length improvement vs diversity loss, evaluating solutions via $\Delta L/\Delta H$. In MT-EAX, calculating $F(e)$ from a pooled subset of the population preserves rare cross-instance edges. This explicitly inflates H with low-frequency, structurally compatible edges, altering the selection criteria to permit exploratory offspring and preventing premature convergence.

3 MT-EAX

We now introduce our multi-task version of EAX, called MT-EAX. The core principle of MT-EAX is to maximise genetic sharing across instances while strictly preventing instance-specific mechanisms from introducing destructive noise. By maintaining a single, global population pooled across all instances, structural knowledge can flow between distinct routing problems during crossover. However, because edge frequencies are highly geometry-specific, merging the edge frequency matrix $F(e)$ or forcing instances to share convergence stages would influence the algorithm's local search capabilities. Therefore, the baseline MT-EAX isolates the edge frequency matrix, convergence detection, and stage transitions per-instance, while allowing for cross-instance parent selection.

MT-EAX solves multiple problems simultaneously as shown in Algorithm 1. In the baseline configuration, individuals, and thus instances, are selected uniformly at random for crossover, and there is no explicit population transfer. Outside of the baseline, several extensions were considered. These include modifying the selection operator, sharing edge frequencies, directing explicit transfer, and utilising decoupled execution. The shared edge frequency modification differs from Algorithm 1 by generating the edge frequencies for the entire global population, rather than strictly from the parents' instance. We evaluated scenarios with warm-up periods to allow the algorithm to build locally optimal structures first. Sampling from a diverse multi-instance pool explicitly increases early-stage genetic diversity forcing exploration of a broader region of the search space before converging on local structures.

In the directed parent selection modification, selection dictates crossover pairings. We applied Pearson similarity thresholds (calculated between instance distance matrices) at various r, as well a same-distribution constraint. Parent B was sampled continuously until the criteria were met or the pool of valid parents was exhausted. For explicit instance transfer, individuals were directly injected into foreign instances every M generations following a W generation warm-up. This replaced the worst s individuals in a target instance with the best from a compatible instance. To support this, we also tested kopt-supported transfer, where the best tour from Parent A was locally repaired using 2-opt under Parent B's distance evaluator prior to injection, ensuring only competitive repairs were accepted. Prior work on using tours generated from one instance and applying them to another instance [15] provided a framework for improving the ability to

Algorithm 1 MT-EAX: Multi-Task Edge Assembly Crossover

Require: k TSP instances $\mathcal{I}_1, \ldots, \mathcal{I}_k$, population size P, offspring count O
Ensure: Best tour found for each instance $\mathcal{I}_i$

Initialisation: Generate $k \times P$ individuals, assign individual j to instance j mod k via round-robin.
Improve each individual with 2-opt local search under its assigned instance evaluator.
repeat
  $r(\cdot) \leftarrow$ random permutation of $\{1, \ldots, k \times P\}$ ▷ Cross-instance mating allowed
  for $i = 1$ to $k \times P$ **do**
    $p_A \leftarrow x_{r(i)}$, $p_B \leftarrow x_{r(i+1)}$
    Compute edge frequency matrix F from individuals assigned to $\mathcal{I}_{p_A}$
    $\{c_1, \ldots, c_O\} \leftarrow \text{EAX}(p_A, p_B, F, \mathcal{I}_{p_A})$ ▷ Crossover under p_A's instance
    $x_{r(i)} \leftarrow \text{SELECTBEST}(c_1, \ldots, c_O, p_A)$ ▷ Replace p_A only if improved
    $x_{r(i)}$.instance $\leftarrow \mathcal{I}_{p_A}$ ▷ Preserve instance assignment
  end for
  for $i = 1$ to k **do** ▷ Per-instance convergence check
    if STAGNATED($\mathcal{I}_i$, stage $= 1$) **then**
      Transition instance $\mathcal{I}_i$ to Stage II (block2 EAX)
    end if
    if STAGNATED($\mathcal{I}_i$, stage $= 2$) **or** population collapsed **then**
      Mark $\mathcal{I}_i$ as converged
    end if
  end for
until all k instances converged
return best individual for each instance $\mathcal{I}_i$

generate tours for one instance but apply them to another. This would potentially allow for the effective transfer of knowledge between instances through the tours themselves.

Finally, we introduce Decoupled MT-EAX. In this configuration, we execute the baseline MT-EAX for W generations before disabling cross-instance crossover entirely. This effectively turns off multi-tasking after the initial phase, forcing the algorithm to operate as k independent EAX runs to isolate the final convergence process. This was motivated by prior work [2] suggesting that the advantages of multitasking most come in the early stages of the optimisation process.

4 Experiments

To test MT-EAX, we consider a subset of benchmark instances generated using the same methodology as in [15]. These instances are generated on an $A \times A$ grid of the Euclidean plane $[0, A]^2$ for a given number of nodes n. We considered node and axis lengths of 100, 1000, and 10000, with location distributions sampled either from a discrete uniform distribution $\sim U(0, A)$ or a normal distribution $\sim N(A/2, A/6)$. We also consider configurations mixing these distributions.

The instances are saved in the TSPLIB95 format [12] with the distance type EUC_2D, which rounds distances to the nearest integer, introducing a roughly 2% error in tour lengths. The ordering of the instances is initially lexicographical, with instances being greedily and exactly matched to the first generated instance. For mixed-distribution configurations, the reference instance for matching is the first uniformly distributed instance.

Given the parameters P for population size and O for offspring, solving k problems taking G generations in EAX consumes a computational budget of $B_e = k \cdot P \cdot O \cdot G$. The primary algorithmic overhead resides in the crossover phase. To ensure a fair comparison under fixed computational budgets, we hold the offspring parameter constant and evaluate three distinct scaling methods: **Generation Scaling:** MT-EAX utilises a global population of size P and runs for $k \cdot G$ generations. The total evaluations are equal ($B_m = k \cdot P \cdot O \cdot G$). Because the population P is distributed across instances, each instance operates with P/k individuals. **Population Scaling:** MT-EAX utilises a global population of size $P \cdot k$ and runs for G generations. The total evaluations remain equal ($B_m = k \cdot P \cdot O \cdot G$). **Balanced Scaling:** MT-EAX utilises a population of size $P \cdot \sqrt{k}$ and runs for $G \cdot \sqrt{k}$ generations, achieving an intermediary balance while maintaining the strict budget ($B_m = \sqrt{k} \cdot P \cdot O \cdot \sqrt{k} \cdot G$).

Scaling the offspring introduces complex internal interactions and obscures the underlying mechanisms, thus it is held constant. For generation and population scaling, we test 5 sets of instances ($k \in \{2, 4, 5, 9, 10\}$). For balanced scaling we use only two sets ($k \in \{4, 9\}$) to ensure clean integer scaling. For $k \in \{2, 5, 10\}$ we evaluate permutations of $P \in \{100, 200, 500, 1000\}$ with offspring $O \in \{30, 60\}$. For $k \in \{4, 9\}$ we evaluate permutations of $P \in \{100, 200, 300, 400\}$ with $O \in \{30, 60\}$ and $P \in \{100, 300, 900, 1500, 2700, 3000\}$ with $O \in \{30, 60\}$ respectively. The shared edge frequency modification had warm-up periods of $W = \{0, 20, 50, 100\}$ generations. The directed parent selection modification had Pearson similarity thresholds (calculated between instance distance matrices) at $r = \{0.2, 0.5, 0.8, 0.9\}$. For explicit instance transfer, individuals were directly injected into foreign instances every $M = \{10, 20, 50\}$ generations following a $W = \{0, 20, 50\}$ generation warm-up. This replaced the worst $s = \{1, 5, 10\}$ individuals.

We restrict analysis to instances within fixed configurations (identical axis length, node count, and matching type) to isolate factors. For single-distribution setups, the first k instances are utilised. Larger k sets are strict supersets of smaller k sets, ensuring performance is not driven entirely by instance geometry. We execute 30 independent trials per configuration, allowing runs to reach full convergence. Trials utilise deterministic seeds identical to the standard EAX implementation to ensure direct reproducibility.

4.1 Evaluation Metrics

We analyse two primary metrics: quality gap and compute saved. These are benchmarked strictly against standard EAX rather than exact Concorde solutions, as this isolates the performance delta introduced by MT-EAX operations, and exact resolution of 10000-node instances across our exhaustive benchmark set is computationally impractical.

Let c_{EAX} and c_{MT} denote the median final tour costs for EAX and MT-EAX respectively across 30 trials on the same instance. The quality gap is defined as:

$$\Delta Q = \frac{c_{\mathrm{EAX}} - c_{\mathrm{MT}}}{c_{\mathrm{EAX}}} \times 100\% \tag{1}$$

A positive ΔQ indicates MT-EAX found a shorter tour, a negative value indicates standard EAX found a shorter tour. Win rate defines the fraction of trials where MT-EAX strictly outperformed EAX. We report median tour costs, which is consistent with the non-parametric Wilcoxon signed-rank test utilised for significance testing. Let C denote the median generation at which standard EAX reaches final convergence (triggering the natural termination condition of no improvement for $1500/O$ generations). The compute saved at checkpoint G is calculated as:

$$\text{Compute Saved}(G) = \frac{C - G}{C} \times 100\% \tag{2}$$

A positive percentage signifies that terminating MT-EAX at generation G consumes less total computational overhead than executing EAX to full convergence. For reference under the primary evaluation parameters ($O = 60$), standard EAX converges at a median of $C = 137$ generations for $n = 1000$ normally distributed instances, and $C = 117$ for uniformly distributed instances. For generation scaling efficiency, the EAX-equivalent generation is calculated as G_{MT}/k, and the per-instance population is defined as P/k. In the context of population scaling analysis, we map the recovery generation, defined as the first checkpoint where the quality gap recovers to $\geq -0.01\%$. In the tabulated results, a star symbol ($\star$) indicates that MT-EAX at that specific checkpoint has matched or exceeded the final converged quality of standard EAX.

We also define the following secondary metrics to capture smaller dynamics. Peak deficit is the maximum negative ΔQ observed during a run, occurring at the peak deficit generation. For full convergence efficiency, MT conv is the median generation MT-EAX triggers termination, requiring a total MT cost (measured in tour evaluations). Sprint gap is the quality gap evaluated at an early target generation (e.g. $G = 50$), capturing initial progress prior to full convergence. Peak sprint denotes the maximum positive ΔQ observed prior to full convergence. For decoupled, the cutoff is the generation threshold where instance populations are strictly isolated, and decoupled conv gen defines the median convergence generation for this isolated phase. Base gap and decoupled gap denote final converged ΔQ without and with decoupling, respectively. Global P defines the total aggregate population across all evaluated instances. All comparative quality gaps are evaluated via the Wilcoxon signed-rank test. Significance levels are denoted as: *** $p<0.001$, ** $p<0.01$, * $p<0.05$.

5 Experimental Results

Table 1. MT-EAX generation scaling results ($n = 1000$, $O = 60$). Values represent quality gaps ΔQ (%) relative to full-convergence EAX, with win rates (%) in parentheses. The $\star$ symbol indicates MT-EAX matches or exceeds EAX final quality. Positive gaps (MT-EAX wins) are bolded.

k	Dist. Comp. saved (Normal):	EAX-equivalent generation G				
		$G = 10$ 93%	$G = 20$ 85%	$G = 50$ 64%	$G = 100$ 27%	$G = 200$ −46%
2	Normal	**+0.069** (58)	**+0.217** (70)	**+0.366*** (83)	**+0.157*** (78)	+0.000* (2)
	Uniform	−0.051 (44)	**+0.044** (54)	**+0.208*** (71)	**+0.022*** (62)	+0.000* (1)
4	Normal	**+0.042** (54)	**+0.244** (66)	**+0.476** (80)	**+0.161** (81)	—
	Uniform	−0.088 (41)	**+0.105** (56)	**+0.307** (70)	**+0.036** (67)	—
5	Normal	**+0.098** (57)	**+0.366** (70)	**+0.590** (84)	**+0.163** (79)	+0.000 (0)
	Uniform	−0.146 (38)	**+0.043** (52)	**+0.271** (69)	**+0.022*** (61)	+0.000* (0)
9	Normal	**+0.402** (70)	—	**+0.892** (88)	**+0.141** (79)	—
	Uniform	−0.035^{ns} (48)	—	**+0.399** (75)	**+0.018*** (57)	—
10	Normal	**+0.247** (62)	**+0.638** (71)	**+0.809** (83)	**+0.113** (71)	−0.055 (0)
	Uniform	−0.128 (42)	**+0.086** (53)	**+0.345** (69)	**+0.004** (51)	−0.043* (0)

For the $n = 100$ node case, there is no statistical difference between any result, as EAX converges in a median of 3 generations on these scale instances, leaving no window for parallel advantages. For the $n = 10000$ cases, we got similar results to the $n = 1000$, with larger generational lengths and thus compute limitations for the exhaustive test cases. Thus, our analysis focuses strictly on the 1000-node benchmark sets.

The baseline MT-EAX results significantly outperformed all structural extensions. The primary performance drivers were identified strictly as the scaling method, node distribution, and instance count. Consequently, for our primary results, we evaluate the baseline configuration utilising an offspring size of $O = 60$, which we later demonstrate provides a buffer against population starvation.

5.1 Scaled Generations

Table 1 details the performance of generation-scaled MT-EAX over computational time. Following the initial 10 generations, the win rate becomes substantially positive, yielding compute savings. While compute saved peaks proportionally at generation 20, the magnitude of the win rate and quality gap peak near generation 50 for virtually all parameters. The advantage is fully eroded by generation 200, at which point the median gap returns to essentially 0%.

Table 1 shows a distinct geometric bias where clustered, normally distributed instances exploit MT-EAX crossover significantly better than uniformly distributed instances, maintaining an approximately 15% higher win rate through-

Table 2. MT-EAX population scaling quality deficits relative to full-convergence EAX ($n = 1000$, $O = 60$). Recovery generation defines the checkpoint where the gap recovers to $\geq -0.01\%$. Deficits are bolded for worst-case isolation.

k	Dist.	Peak deficit ΔQ (%)	Peak def. gen	Gen 100 ΔQ (%)	Recovery gen
2	Normal	**−0.861**	80	−0.767	250
	Uniform	**−1.394**	80	−1.116	250
4	Normal	**−1.888**	100	−1.888	500
	Uniform	**−2.345**	100	−2.345	400
5	Normal	**−1.962**	100	−1.962	500
	Uniform	**−2.838**	100	−2.838	500
9	Normal	**−2.117**	100	−2.117	750
	Uniform	**−3.419**	100	−3.419	750
10	Normal	**−2.429**	100	−2.429	1000
	Uniform	**−3.401**	100	−3.401	1000

Table 3. MT-EAX full convergence efficiency ($n = 1000$, $O = 60$). Savings and gaps are relative to full-convergence standard EAX. Positive compute savings are bolded.

k	Dist.	EAX C	MT conv	EAX-equiv	Saving (%)	MT cost (evals)	Final ΔQ (%)
2	Normal	137	243	122	**+11.3**	17793	+2.381
	Uniform	117	215	108	**+8.1**	22305	+3.013
4	Normal	137	460	115	**+16.1**	18230	−0.016
	Uniform	117	384	96	**+17.8**	22998	+0.000
5	Normal	137	484	97	**+29.3**	18357	−0.713
	Uniform	117	481	96	**+17.8**	22976	+0.096
9	Normal	137	740	82	**+40.0**	18272	−0.247
	Uniform	117	700	78	**+33.5**	22998	+0.000
10	Normal	137	801	80	**+41.5**	18244	−0.093
	Uniform	117	786	79	**+32.8**	22998	+0.000

out the early search phase. Uniformly distributed instances remain highly consistent, indicating that normal distribution structures are more sensitive to cross-instance crossover parameters.

The compute efficiency displayed in Table 3 highlights the strength of the generation-scaling mechanism. At generation 50 (a 64% compute save versus full EAX convergence at $C = 137$), MT-EAX at $k = 9$ (Normal) is already +0.892% better than standard EAX at the equivalent checkpoint. More importantly, halting the algorithm at generation 100 (a 27% compute save) with $k = 9$ yields a +0.141% improvement over EAX's converged quality.

Table 4. Decoupled MT-EAX performance ($n = 1000$, $O = 60$). Max ΔQ values denote the maximum positive ΔQ, with the corresponding generation G in parentheses. Positive gaps and savings are bolded.

k	Dist.	Cutoff	Base ΔQ (%)	Dec. ΔQ (%)	Max ΔQ (%)	Dec. conv gen	Compute saved (%)
2	Normal	100	**+2.381**	**+2.381**	**+2.896** (75)	178	**+35.0**
2	Normal	200	**+2.381**	**+2.381**	**+2.560** (50)	224	**+18.2**
2	Uniform	100	**+3.013**	**+3.013**	**+3.509** (75)	162	**+30.8**
2	Uniform	200	**+3.013**	**+3.013**	**+3.233** (75)	209	**+10.7**
4	Normal	100	−0.016	−0.016	**+1.512** (50)	206	**+62.4**
4	Normal	200	−0.016	−0.016	**+0.751** (75)	276	**+49.6**
4	Uniform	100	+0.000	**+1.454**	**+2.844** (50)	180	**+61.5**
4	Uniform	200	+0.000	**+1.433**	**+1.968** (75)	250	**+46.6**
5	Normal	100	−0.713	−0.713	**+1.280** (30)	198	**+71.0**
5	Normal	200	−0.713	−0.713	**+0.821** (50)	268	**+60.9**
5	Uniform	100	**+0.096**	**+0.096**	**+1.963** (40)	187	**+68.0**
5	Uniform	200	**+0.096**	**+0.096**	**+1.260** (50)	263	**+55.0**
9	Normal	100	−0.247	−0.247	**+1.307** (50)	215	**+82.6**
9	Normal	200	−0.247	−0.247	**+1.307** (50)	295	**+76.1**
9	Uniform	100	+0.000	+0.000	**+1.410** (50)	194	**+81.6**
9	Uniform	200	+0.000	+0.000	**+1.410** (50)	276	**+73.8**
10	Normal	100	−0.093	−0.060	**+2.650** (20)	209	**+84.7**
10	Normal	200	−0.093	−0.069	**+2.274** (30)	288	**+79.0**
10	Uniform	100	+0.000	+0.000	**+2.863** (20)	197	**+83.2**
10	Uniform	200	+0.000	+0.000	**+2.369** (30)	279	**+76.2**

5.2 Scaled Population and Balanced

Conversely, when scaling the global population size, Table 2 and Fig. 1 demonstrate negative results in the early stages of optimisation. No compute is saved and the deficit is worst near generation 100, in contrast to the generation-scaled experiments which peaked in this exact window. There is a negative correlation between the number of instances k and the depth of the quality deficit. Furthermore, uniformly distributed instances are significantly worse early penalties than normally distributed instances. The balanced scaling followed identical trends, showing that partial generation scaling fails to overcome the penalty of partial population scaling.

Figure 1 shows that while the recovery trajectory is consistent, the depth of the early deficit scales with geometry and instance count. For normally distributed instances at $k = 9$, the largest gap is −2.117%, whereas uniformly distributed instances drop to −3.419%. The mechanism driving this deficit is cross-instance parent incompatibility. Because normally distributed instances possess localised geometric similarities, cross-instance parents remain more compatible. The deficit scales proportionally with $(k - 1)/k$, representing that crossovers forced to utilise an incompatible second parent. For $k = 2$, 50% of parents are

Table 5. Population starvation gradient quality gaps ($k = 10$, Normal, $n = 1000$). Gaps (ΔQ) are relative to full-convergence standard EAX. Positive sprint gaps are bolded.

Global P	Per-inst P/k	Sprint gap (%, Gen 50)		Conv gap (%, Final)	
		$O = 30$	$O = 60$	$O = 30$	$O = 60$
100	10.0	**+0.427**	**+0.793**	−0.324	−0.351
200	20.0	**+0.523**	**+0.972**	−0.088	−0.071
500	50.0	**+0.591**	**+0.992**	−0.016	−0.016
1000	100.0	**+0.589**	**+0.936**	−0.005	−0.011

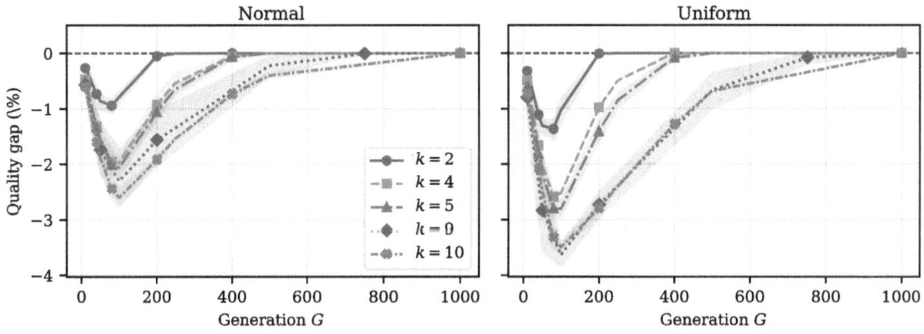

Fig. 1. Quality deficit of population-scaled MT-EAX in early generations ($n = 1000$).

incompatible, yielding a gap of −0.861%. For $k = 9$, 89% are incompatible, moving the gap to −2.117%. Notably, recovery is complete by final convergence. All configurations eventually reach parity (0.00%), demonstrating that $k \times$ pop scaling ultimately resolves into k independent EAX runs.

5.3 Decoupled MT-EAX

To address the compute inefficiency observed in the late stages of the generation-scaled baseline (Table 3), we evaluated Decoupled MT-EAX. Table 4 shows that decoupling instances at fixed generation cutoffs had minimal impact on the final global quality gap, as the gaps remain virtually identical. However, the convergence speedups achieved by Decoupled MT-EAX demonstrated a significant computational advantage. In the baseline MT-EAX, the algorithm requires more generations to converge when using incompatible cross-instance edges. Furthermore, Table 4 demonstrates that for smaller instance sets ($k \in \{2, 4, 5\}$), Decoupled MT-EAX strictly outperforms standard EAX at final convergence, yielding a positive final quality gap (up to +3.013% for $k = 2$ Uniform). It is only at highly scaled instance counts ($k = 10$) that a negligible deficit emerges. By severing cross-instance interaction, the algorithm rapidly detects convergence. For example, for $k = 10$ (Normal, Cutoff 100), Decoupled MT-EAX reduces the convergence time from 801 generations in the baseline down to 209 generations. This increases the compute saved from +41.5% to +84.7%, while preserving a peak early-stage advantage of +2.650% and yielding an identical final tour cost.

6 Discussion

The empirical results demonstrate a clear performance gap based on the available computational budget. In fixed-budget environments, generationally scaled MT-EAX is significantly more effective compute-wise than standard EAX. The mechanism driving this improvement is the injection of structural diversity via similar-instance crossover, which outperforms local search in the early generations. Conversely, when scaling is applied to the population size, the results are strongly negative, a large global population does not offset the penalty of cross-instance incompatibility.

Under generation scaling, MT-EAX maintains dominance up to an intersection point near generation 200, where parity with standard EAX is achieved. The geometry, through density, controls the performance in this window: clustered, normally distributed instances exploit the shared population more effectively than uniform distributions, indicating that structural similarities improve crossover viability. MT-EAX operates with a restricted population per instance (P/k) but cycles through genetic material k times faster than standard EAX for an equivalent compute budget. This turnover process allows the population to quickly escape poor configurations, which gives the early compute savings. However, as optimisation progresses, the lack of per-instance population depth restricts the localised search required to finalise incremental improvements. Standard EAX is slower initially, but its large, isolated population maintains the diversity necessary to construct final optimal tours.

Table 5 quantifies this population starvation dynamic. As the per-instance population P/k scales from 10.0 to 100.0, the final convergence gap moves from −0.351% to −0.011%. This demonstrates a directly proportional relationship between instance population depth and final convergence capability. With ($P/k = 10$), an offspring size of $O = 60$ provides twice as many iterations to locate a viable crossover product, nearly doubling the early-stage advantage (+0.793%) compared to $O = 30$ (+0.427%). Large k final quality gaps in the Decoupled MT-EAX validate the expectation that cross-instance noise and a lack of genetic diversity due to the starved per-instance population cause this deficit at convergence. However, in fixed-budget contexts, this late-stage deficit represents a trade-off. For state-of-the-art solvers processing 1000-node TSP instances, quality gaps are sub-1%. MT-EAX sacrifices a negligible fraction of late-stage optimality to gain large early-stage improvements in quality (up to +3.509%), while also reducing total compute by up to 84.7%.

In the early stages (0 to 100), edge frequencies operate as noise, hence blending them yields no optimisation benefits. Explicit transfer, with and without kopt, failed because redistributing converged individuals between disparate geometric spaces degrades the EAX crossover logic. Although optimal node matching yields a 1.635 length ratio against converged solutions, EAX populations surpass this within 20 generations. Transferring an optimally matched solution is unable to be competitive.

The early-stage computational advantage is linked to the preservation of edge entropy H as seen in Fig. 2. For EAX, $F(e)$ rapidly focuses on local structures, minimising H and halting exploration. Cross-instance crossover introduces compatible structural diversity, artificially inflating H with low-frequency edges. By pooling populations, MT-EAX forces H to remain higher than isolated standard EAX at equivalent average tour lengths L. This alters the $\Delta L/\Delta H$ selection criteria, allowing the algorithm to accept exploratory offspring and accelerating early-stage cost reductions. While MT-EAX accelerates early-stage cost reductions, it converges prematurely at a higher final cost compared to standard EAX. The restricted per-instance population lacks the genetic depth required for the final improvements, validating the Decoupled MT-EAX configuration that is able to reach or exceed the converged tour lengths from EAX.

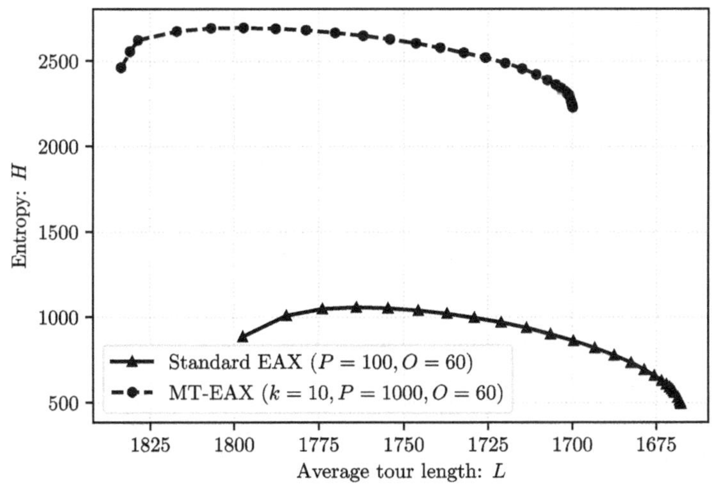

Fig. 2. Trajectory of population edge entropy H against average tour length L. MT-EAX maintains significantly higher structural diversity at equivalent tour lengths in early generations.

7 Conclusions

Parallel MT-EAX dominates isolated EAX execution during the early stages of optimisation, making it highly effective for strict, fixed-budget computational environments. Under fair generation scaling, MT-EAX gives significant early-stage quality improvements while simultaneously saving 60% to 90% of necessary compute. This computational advantage is further extended by utilising Decoupled MT-EAX. By stopping cross-instance interaction after an initial phase, the algorithm preserves the substantial early-stage speedups (saving up to 84.7% over standard EAX) while allowing individual instances to specialise, successfully matching or strictly outperforming EAX's final global convergence quality. We demonstrated that direct solution transfer is fundamentally restricted, as the native EAX populations rapidly beat the relevance of geometrically transferred solutions within narrow generational windows. The strongest determinant of MT-EAX success is instance geometry, with clustered, normally distributed geometries exploiting cross-instance structures up to 15% more effectively than uniform distributions. Population starvation dictates the bounds of this advantage, governing a trade-off between rapid early optimisation and final global convergence. Future work includes investigating mechanisms to pre-solve reference instances to seed populations for the earliest stages of the multi-task sequence.

Acknowledgments. This work was supported by the Australian Research Council through grant FT200100536.

References

1. Applegate, D.L., Bixby, R.E., Chvátal, V., Cook, W.J.: The Traveling Salesman Problem: A Computational Study. Princeton University Press, Princeton (2006). https://doi.org/10.1515/9781400841103
2. Don, T.P., Neumann, A., Neumann, F.: Evolutionary multitasking for the scenario-based travelling thief problem. In: GECCO, pp. 809–817. ACM (2025). https://doi.org/10.1145/3712256.3726468
3. Feng, L., Gupta, A., Tan, K., Ong, Y.: Evolutionary Multi-Task Optimization: Foundations and Methodologies. Machine Learning: Foundations, Methodologies, and Applications. Springer, Singapore (2023), https://doi.org/10.1007/978-981-19-5650-8
4. Gao, L.Y., Wang, R., Jia, Z.H., Liu, C.: Multi-objective optimization for traveling salesman problem: A deep reinforcement learning algorithm via transfer learning. IEEE Trans. Artif. Intell. 1–14 (2024). https://doi.org/10.1109/TAI.2024.3499946
5. Gupta, A., Ong, Y., Feng, L.: Multifactorial evolution: toward evolutionary multitasking. IEEE Trans. Evol. Comput. **20**(3), 343–357 (2016). https://doi.org/10.1109/TEVC.2015.2458037
6. Gupta, A., Ong, Y., Jong, K.A.D., Zhang, M.: Guest editorial special issue on multitask evolutionary computation. IEEE Trans. Evol. Comput. **26**(2), 202–205 (2022). https://doi.org/10.1109/TEVC.2022.3156325
7. Hu, Z., Li, Y., Sun, H., Ma, X.: Multitasking multiobjective optimization based on transfer component analysis. Inf. Sci. **605**, 182–201 (2022). https://doi.org/10.1016/j.ins.2022.05.037
8. Joshi, C.K., Cappart, Q., Rousseau, L., Laurent, T.: Learning the travelling salesperson problem requires rethinking generalization. Constraints **27**(1–2), 70–98 (2022). https://doi.org/10.1007/s10601-022-09327-y
9. Lin, J., Liu, H., Tan, K.C., Gu, F.: An effective knowledge transfer approach for multiobjective multitasking optimization. IEEE Trans. Cybern. **51**(6), 3238–3248 (2021). https://doi.org/10.1109/TCYB.2020.2969025

10. Lin, Q., Wu, Z., Ma, L., Gong, M., Li, J., Coello, C.A.C.: Multiobjective multitasking optimization with decomposition-based transfer selection. IEEE Trans. Cybern. **54**(5), 3146–3159 (2024). https://doi.org/10.1109/TCYB.2023.3266241
11. Nagata, Y., Kobayashi, S.: A powerful genetic algorithm using edge assembly crossover for the traveling salesman problem. INFORMS J. Comput. **25**(2), 346–363 (2013). https://doi.org/10.1287/ijoc.1120.0506
12. Reinelt, G.: TSPLIB - a traveling salesman problem library. INFORMS J. Comput. **3**(4), 376–384 (1991). https://doi.org/10.1287/ijoc.3.4.376
13. Souza, G.K.B., Santos, S.O.S., Ottoni, A.L.C., Oliveira, M.S., Oliveira, D.C.R., Nepomuceno, E.G.: Transfer reinforcement learning for combinatorial optimization problems. Algorithms **17**(2), 87 (2024). https://doi.org/10.3390/a17020087
14. Sun, H., Chen, P., Hu, Z., Wei, L.: Multi-objective evolutionary multitasking algorithm based on cross-task transfer solution matching strategy. ISA Trans. **138**, 504–520 (2023). https://doi.org/10.1016/j.isatra.2023.03.015
15. Wigney, L., Neumann, A., Ong, Y.S., Neumann, F.: On the use of matching algorithms to transfer solutions for the travelling salesperson problem. In: Proceedings of the Genetic and Evolutionary Computation Conference. GECCO 2025, pp. 845–853. Association for Computing Machinery, New York (2025). https://doi.org/10.1145/3712256.3726469
16. Xu, H., Qin, A.K., Xia, S.: Evolutionary multitask optimization with adaptive knowledge transfer. IEEE Trans. Evol. Comput. **26**(2), 290–303 (2022). https://doi.org/10.1109/TEVC.2021.3107435

Learning to Solve Complex Constrained Routing Problems with Feasibility-Guided Reward And Diversity-Guided Policy

Yuanxu Yang[1], Zikang Yu[1], Jiahai Wang[1](✉), Jieyi Bi[2], Jinbiao Chen[3], and Zizhen Zhang[1]

[1] School of Computer Science and Engineering, Sun Yat-sen University, Guangzhou, China
{wangjiah,zhangzzh7}@mail.sysu.edu.cn

[2] College of Computing and Data Science, Nanyang Technological University, Singapore, Singapore
jieyi001@e.ntu.edu.sg

[3] Department of Industrial Systems Engineering and Management, National University of Singapore, Singapore, Singapore
bill.cjb@nus.edu.sg

Abstract. Neural combinatorial optimization (NCO) has achieved remarkable progress in solving routing problems using deep reinforcement learning, yet its application to complex constrained routing problems is still at an early stage. Recent advances have successfully applied Lagrange multiplier variants to handle constraints, but still suffer from limitations in reward configuration and sufficient diversity in exploring, leading to high infeasibility rates and suboptimal performance. To address these challenges, we propose a feasibility-guided reward and diversity-guided policy (FRDP) framework that explicitly targets both enhancing feasibility and promoting exploration diversity. Specifically, FRDP introduces a specialized reward mechanism that differentiates feasible from infeasible solutions, ensuring correct reward ordering and guiding the model toward feasible regions. In parallel, the diversity-guided policy encourages the exploration of varied solution patterns. We evaluated FRDP on two challenging constrained routing problems: the traveling salesman problem with time windows (TSPTW) and the traveling salesman problem with draft limits (TSPDL). Experimental results demonstrate that FRDP significantly improves both solution quality and feasibility over existing neural methods, highlighting its effectiveness in handling complex constrained routing problems. Additional details and supplementary experiments are provided in the appendix, available at: https://zenodo.org/records/19162589.

Keywords: Constrained routing problems · Neural combinatorial optimization · Deep reinforcement learning · Constraint handling

G. Iacca et al. (Eds.): PPSN 2026, LNCS 16985, pp. 382–398, 2027.
https://doi.org/10.1007/978-3-032-36214-8_24

1 Introduction

Combinatorial optimization (CO) problems are fundamental challenges in computer science and operations research. Among these problems, routing problems [41], including the traveling salesman problem (TSP), vehicle routing problem (VRP) and their variants, represent a particularly important class due to their direct relevance to real-world logistics [22] and transportation [48]. Classical approaches, including exact algorithms [1] and handcrafted heuristics [17], achieve strong performance but often suffer from limited scalability and adaptability. Recently, neural combinatorial optimization (NCO) methods [24,25,39] based on deep reinforcement learning (DRL) have emerged as a promising alternative, enabling end-to-end learning without explicit domain-specific design.

Despite this progress, most existing NCO methods still focus primarily on simple constrained problems. In real-world scenarios, routing problems typically involve complex constraints whose feasibility depends on the entire solution trajectory. These constraints significantly reduce and fragment the feasible solution space, making the learning process substantially more challenging.

A common approach to handling constraints is to incorporate penalty terms into the reward function via Lagrangian multipliers. Recent works [3,8,38] have explored various improvements along this direction, such as adaptive penalty coefficients, feasibility-aware training schemes, and auxiliary supervision signals to encourage constraint satisfaction. However, we observe a fundamental limitation of this paradigm: it does not guarantee correct reward ordering between feasible and infeasible solutions. In particular, an infeasible solution with a shorter route may receive a higher reward than a feasible one under fixed penalties. This misalignment introduces biased learning signals that may systematically favor infeasible regions, resulting in high infeasibility rates and poor performance in problems with complex constraints.

Beyond this limitation, even when feasible solutions are discovered, another key challenge emerges: insufficient exploration within the feasible region. In complex constrained problems, feasible solutions are often sparsely distributed and separated by large infeasible regions, forming fragmented feasible subspaces. Standard exploration strategies fail to adequately explore these regions, causing the model to converge prematurely to suboptimal solutions.

These observations indicate that solving complex constrained routing problems requires a structured learning process with two essential capabilities: (1) reliably identifying the feasibility boundary to reach feasible regions and (2) effectively exploring the feasible solution space to approach optimality. However, existing methods do not satisfy both requirements simultaneously.

To address these challenges, we propose a **F**easibility-guided **R**eward and **D**iversity-guided **P**olicy (FRDP) framework, which explicitly differentiates feasible from infeasible solutions using a specialized reward function, while exploring diverse solutions with multiple latent embeddings to promote the policy diversity. By combining these components, FRDP guides the model toward feasible regions while enabling effective exploration within them, leading to improved performance on complex constrained routing problems.

Our main contributions can be summarized as follows:

1. We propose a feasibility-guided reward mechanism that explicitly differentiates feasible from infeasible solutions, replacing fixed-penalty designs with constraint-aware reward shaping.
2. We introduce a diversity-guided policy learning strategy that samples multiple latent embeddings from a Gaussian distribution to enhance solution diversity, thereby mitigating premature convergence to local optima.
3. We provide comprehensive experimental validation on two challenging routing problems, TSP with time windows (TSPTW) and TSP with draft limits (TSPDL), demonstrating significant improvements in both solution quality and feasibility compared to existing NCO methods.

2 Background

2.1 NCO Methods for Routing Problems

Neural combinatorial optimization (NCO) methods have recently shown strong potential in solving routing problems, and are typically categorized into neural constructive methods and neural iterative methods. Constructive methods directly generate complete solutions through sequential decisions, including autoregressive (AR) methods such as attention model (AM) [24] and policy optimization with multiple optima (POMO) [25], as well as non-autoregressive (NAR) methods [13,23,34,37] that predict edge probability heatmaps to solve routing problems. In recent years, these methods have achieved significant progress in performance [5,10,21,43], scalability [6,11,31], and generalization [2,15,26,29,45,47]. In contrast, neural iterative methods [9,20,30,32,33,42,44] refine initial solutions using learned local search operators within heuristic frameworks. In this paper, we focus on neural constructive methods for solving routing problems with complex constraints.

2.2 Constraint Handling Mechanisms for NCO Methods

Constraint handling in neural constructive methods generally follows two approaches: feasibility masking and penalty-based methods. Feasibility masking restricts the action space to valid decisions at each step [28,33,37] and is effective for simple constraints. However, it becomes computationally expensive or impractical for more complex constraints that require global feasibility checks. Penalty-based methods [7,38,46], typically using Lagrangian multipliers, incorporate constraint violations into the objective function, allowing exploration of infeasible solutions while encouraging feasibility. Recent works attempt to improve constraint handling by integrating look-ahead mechanisms [8] or proactive infeasibility masking [3] into model architectures. However, these approaches generally treat all solutions uniformly and fail to explicitly distinguish between feasible and infeasible solutions. This limitation reduces their effectiveness in handling complex constrained problems, motivating the need for more informative and structured constraint-aware reward mechanisms.

2.3 Diversity Mechanisms for NCO Methods

Diversity mechanisms have been widely explored to enhance solution quality in neural routing methods. Techniques such as multi-start strategies [25], multi-decoder architectures [16,43], and regularization-based approaches [21,27,36] aim to promote exploration of multiple solution patterns. More recent methods leverage continuous latent representations [4,18] to model solution distributions, enabling diverse solution sampling through latent space exploration. Similarly, Hottung et al. proposed PolyNet [19] to learn diverse solution strategies through a single-decoder architecture combined with strategy-specific bit vectors. Although these approaches improve performance in simple constrained settings, their integration with complex constraint handling remains underexplored.

3 Preliminaries

3.1 Problem Definition

Formally, a routing problem instance I is defined on a complete graph $G = (V, E)$ where $V = \{v_0, v_1, v_2, \dots, v_n\}$ represents the set of nodes (with node v_0 as the depot), and $e(v_i, v_j) \in E, \forall i, j \in \{1, 2, \dots, n\}$ represents the set of edges. The objective of routing problems is to plan a route that minimizes the total travel distance while visiting all customer nodes at once. This study addresses two complex constrained routing problem variants: TSPTW and TSPDL.

Definition of TSPTW. TSPTW extends the classical TSP by incorporating time window constraint [12], where each node v_i must be served within a specified time window $[l_i, u_i]$. l_i defines the lower bound of the visit time and u_i defines the upper bound. Time window constraint ensures that the arrival time t_i at each node v_i does not exceed u_i, and if arrived early ($i.e., t_i < l_i$), the vehicle must wait until l_i to start service.

Definition of TSPDL. TSPDL arises in maritime transportation [35], where ships must comply with port draft constraint. In this problem, each node v_i represents a port with a non-negative demand δ_i and a maximum allowable draft d_i. The constraint requires that the cumulative load m_i from the last i steps must not exceed d_i ($\sum_{t=1}^{i} \delta_{\tau_t} \leq d_i$).

3.2 RL-Based Constructive Method for Routing Problems

RL-based constructive methods can be formulated as a Markov decision process (MDP) defined by a state space S, an action space A, a state transition probability function P, and a reward function R. For the neural constructive methods, a solution τ is constructed sequentially in total discrete T time steps, where $T = n$ in the TSP variants. At each step $t \in (1, \dots, T)$, an action $a_t \in A$ is selected based on the probability distribution $\pi_\theta(a_t|s_t) \in P$, where $s_t \in S$ is the current state and $\pi_\theta(a_t|s_t)$ is generated by a neural network with learnable weights θ. Given an instance I in distribution $\mathcal{D}$, once a complete solution $\tau = (a_1, a_2, \ldots, a_n)$ is constructed, we can calculate its associated reward $R(\tau)$. The overall policy that the neural network generates the solution $\tau = (a_1, a_2, ..., a_n)$ can be calculated as $\pi_\theta(\tau|I) = \prod_{t=1}^{T} \pi_\theta(a_t|s_t)$.

3.3 Penalty-Based Method for Constrained Routing Problems

Penalty-based constraint handling remains the most common approach in neural constructive methods, where constraint violations are incorporated into the objective function through additional penalty terms. These approaches treat constraints as soft constraints with an associated Lagrangian multiplier variable λ. The objective function is defined as:

$$\max_{\lambda \geq 0} \min_{\tau} f(\tau|I) + \lambda c(\tau|I), \tag{1}$$

where $f(\tau|I)$ is the total travel distance, and $c(\tau|I)$ is the degree of constraint violation (time window violation in TSPTW and draft limit violation in TSPDL). Specifically, λ is a fixed parameter in proactive infeasibility prevention (PIP) [3] used to reduce computational complexity.

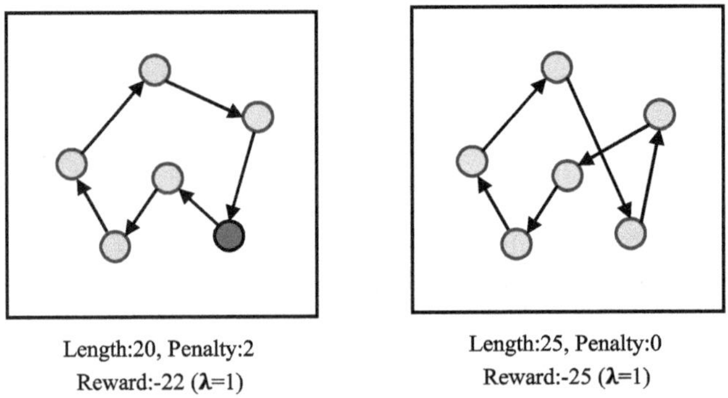

(a) Infeasible solution (b) Best feasible solution

Fig. 1. An example of the limitation in reward configuration. (a) and (b) are two solutions for the same routing problem instance. (a) is an infeasible solution which contains a red node that violates constraints, while (b) is the best feasible solution. With a fixed $\lambda = 1$, (a) achieves a higher reward than (b). (Color figure online)

4 Methodology

4.1 Motivation and General Framework

The proposed FRDP framework is motivated by two key challenges in complex constrained routing problems.

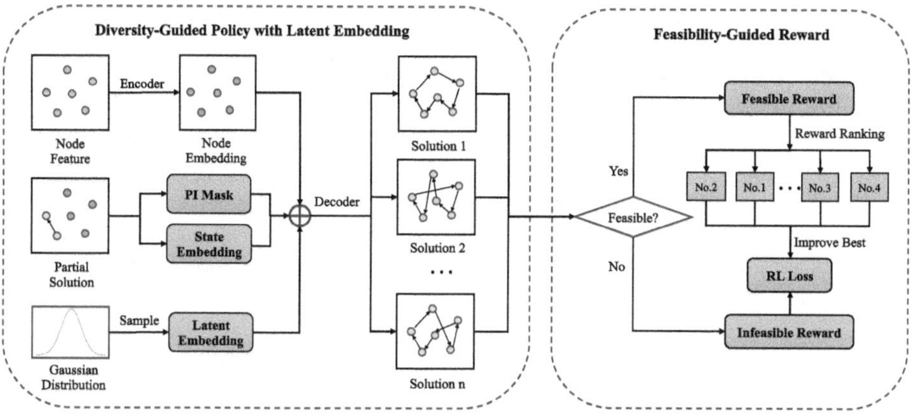

Fig. 2. The architecture of FRDP framework. The left part illustrates the incorporation of latent embeddings during decoding to promote diversity-guided policy, and the right part is the calculation of feasibility-guided reward.

Firstly, existing methods commonly rely on Lagrangian multiplier variants to handle constraints by incorporating penalty terms into the objective function. However, this paradigm suffers from a fundamental limitation in reward design: it does not guarantee correct reward ordering between feasible and infeasible solutions. In particular, under fixed penalty coefficients, infeasible solutions with shorter route lengths may receive higher rewards than feasible ones as shown in Fig. 1. This leads to biased learning signals and drives the model toward infeasible regions, especially in the context of complex constrained problems.

Secondly, exploration in constrained settings is often insufficient. Feasible solutions are often sparse and fragmented, while standard strategies fail to adequately cover these regions, resulting in premature convergence. POMO [25] artificially enhanced exploration by forcing diverse first actions during solution construction. However, in complex constrained routing problems such as TSPTW and TSPDL, first actions create cascading effects that determine the feasibility of all subsequent choices in the solution path. The recent PIP [3] implied POMO without forcing the first action, which limits the ability to explore diverse solution space and leads to suboptimal performance.

To address these issues, FRDP integrates feasibility-aware learning and diversity-driven exploration into a unified framework, as shown in Fig. 2. It introduces a feasibility-guided reward that explicitly distinguishes feasible and infeasible solutions, ensuring a correct optimization direction. Meanwhile, a diversity-guided policy leverages latent embeddings to promote exploration of diverse solutions. By combining these components, FRDP guides the model toward feasible regions while enabling effective exploration within them, thereby improving both feasibility and optimality.

4.2 Feasibility-Guided Reward

Compared with the penalty-based Lagrangian method in Eq. (1), the core of FRDP lies in a reward mechanism that explicitly differentiates between feasible and infeasible solutions, addressing the reward misalignment issue in existing methods. Instead of balancing objective value and constraint violation with fixed penalties, our design enforces a clear separation between feasible and infeasible regions, providing more reliable learning signals.

Given an instance I, we roll out N solutions $\{\tau_1, \tau_2, \cdots, \tau_N\}$ and compute the reward $R(\tau_i)$ as:

$$R(\tau_i) = \begin{cases} K - f(\tau_i|I) & \text{if } \tau_i \text{ is feasible} \\ -c(\tau_i|I) - c_{in}(\tau_i|I) & \text{if } \tau_i \text{ is infeasible} \end{cases}, \tag{2}$$

where $K = max\, f(\tau_i|I) \times 2$ over feasible solutions, $f(\tau_i|I)$ is the total travel distance, $c(\tau_i|I)$ is the constraint violation, and $c_{in}(\tau_i|I)$ denotes the number of infeasible nodes. This formulation guarantees that all feasible solutions receive higher rewards than infeasible ones, ensuring correct reward ordering and guiding the model toward feasible regions.

Importantly, this design induces two desirable properties: when all sampled solutions are infeasible, the reward depends solely on constraint violations, encouraging the model to reduce infeasibility. When feasible solutions exist, the reward of feasible solutions depends only on the path length, enabling effective optimization within the feasible region without interference from penalty terms.

To further improve solution quality, we incorporate the leader reward mechanism [40]. When feasible solutions exist, the best feasible solution is treated as the leader, while other solutions are scaled by a factor $\beta \in [0, 1]$ to maintain exploration. The advantage function is defined as:

$$\mathcal{A}_{FRi} = \begin{cases} R(\tau_i) - b & \text{if } \tau_i \text{ is the best feasible solution} \\ \beta \times (R(\tau_i) - b) & \text{else} \end{cases}, \tag{3}$$

where $b = \frac{1}{N}\sum_{i=1}^{N} R(\tau_i)$. Overall, the feasibility-guided reward decouples feasibility and optimality, providing structured guidance that first promotes reaching feasible regions and then improves solution quality within them.

4.3 Diversity-Guided Policy with Latent Embeddings

To address the insufficient exploration in constrained routing problems, FRDP incorporates a diversity-guided policy based on latent embeddings. In such problems, feasible solutions are often sparse and fragmented, and standard exploration strategies are prone to premature convergence. Therefore, enhancing diversity is essential for effectively exploring the feasible solution space.

Inspired by recent advances in latent space learning [4] for combinatorial optimization, we introduce continuous latent embeddings sampled from a Gaussian distribution. Compared with discrete strategy representations (e.g. bit vectors [19]) or uniform sampling, Gaussian sampling enables smoother transitions between solution strategies and naturally promotes exploration around promising regions.

Given an instance I, we sample N latent embeddings $\{z_1, z_2, \ldots, z_N\}$ from Gaussian distribution Z, and condition the policy on these embeddings to generate diverse solutions $\{\pi_\theta(\tau_1|I, z_1), \pi_\theta(\tau_2|I, z_2), \ldots, \pi_\theta(\tau_N|I, z_N)\}$. In this way, FRDP learns multiple complementary solution strategies implicitly, where each strategy may specialize in different regions of the solution space. During training, all sampled solutions contribute to policy optimization through the feasibility-guided reward, rather than selecting only the best solution. We use the gradient calculated by our feasibility-guided reward $\mathcal{A}_{FR}$:

$$\nabla_\theta \mathcal{L}(\theta) = \mathbb{E}_\tau[\mathcal{A}_{FR} \nabla_\theta \log \pi_\theta(\tau|I, z)]. \tag{4}$$

This design enables collaborative learning among diverse strategies while maintaining exploration efficiency.

4.4 Training Process of FRDP

The overall workflow of the training process for FRDP is presented in Algorithm 1. The training process involves generating multiple solutions per instance by sampling latent embeddings and evaluating them with a feasibility-guided reward function (lines 1–13). The diverse solutions are trained collaboratively, sharing information about promising routing patterns while maintaining distinct exploration strategies (lines 14–19). Finally, we update the model using policy gradients (lines 20–22).

4.5 Characteristics of FRDP Framework

The FRDP framework can be characterized by the following key features:

Complementary Components of FRDP Framework. FRDP integrates two complementary components to address the core challenges in complex constrained routing problems. The feasibility-guided reward explicitly distinguishes feasible and infeasible solutions, ensuring the correct order of rewards and guiding the model toward feasible regions. Meanwhile, the diversity-guided policy leverages latent embeddings to promote diverse exploration, enhancing coverage of fragmented feasible spaces and mitigating premature convergence. Together, these components enable a structured learning process that improves both feasibility and optimality.

Generality and Flexibility of FRDP Framework. FRDP demonstrates strong generality and flexibility, allowing seamless integration with existing NCO frameworks. It can be combined with methods such as POMO [25] and PIP [3] to enhance their ability to handle complex constraints, while remaining compatible with various exploration strategies, including multi-decoder architectures and latent space methods. This flexibility enables FRDP to improve constraint handling and exploration without significant computational overhead, making it a practical and extensible framework for constrained routing problems.

Algorithm 1. Training Process of FRDP

Input: Initial policy network π_θ, training step T, batch size B, number of solution N, problem distribution D, latent embedding distribution Z, learning rate α, multiplier parameter β.

Output: Trained policy network π_θ.

for $step = 1$ to T **do**
  for $i = 1$ to B **do**
    $I_i \leftarrow \text{SampleInstance}(\mathcal{D})$
    for $j = 1$ to N **do**
      $z_i^j \leftarrow \text{SampleEmbedding}(\mathcal{Z})$
      $\tau_i^j \leftarrow \text{SampleRollout}(a_i^j, I_i, z_i^j, \pi_\theta)$
      if τ_i^j is feasible **then**
        Compute feasible reward as $R(\tau_i^j)$
        $j^* \leftarrow \arg\max_j R(\tau_i^j)$
      else
        Compute infeasible reward as $R(\tau_i^j)$
      end if
    end for
    $b_i^{share} \leftarrow \frac{1}{N}\sum_{j=1}^{N} R(\tau_i^j)$
    $\mathcal{A}_{FR_i^j} \leftarrow \beta(R(\tau_i^j) - b_i^{share}), \forall j \in \{1, \dots, N\}$
    if Exist feasible τ_i^{j*} **then**
      $\mathcal{A}_{FR_i^{j^*}} \leftarrow (R(\tau_i^{j^*}) - b_i^{share})$
    end if
  end for
  $\nabla_\theta \mathcal{L}(\theta) \leftarrow \frac{1}{BN}\sum_{i=1}^{B}\sum_{j=1}^{N} \mathcal{A}_{FR_i^j} \nabla_\theta \log \pi_\theta(\tau_i^j | I_i, z_i^j)$
  $\theta \leftarrow \theta + \alpha \nabla_\theta \mathcal{L}(\theta)$
end for

5 Experiments

5.1 Experimental Settings

Implementation Details. Experiments are conducted on two kinds of complex constrained routing problem variants: TSPTW and TSPDL, each problem with different hardness levels. Problem instances are generated following the same settings as Bi et al. [3], with detailed information provided in Appendix A. The network architectures of FRDP employ an encoder-decoder framework inherited from POMO [25], enhanced with specialized components for constraint handling. Details of the network architecture are provided in Appendix B. Training takes approximately 4 days for 50-node instances and 14 days for 100-node instances, and all the experiments are conducted on a machine with Intel(R) Xeon(R) Gold 6226R CPUs @ 2.90 GHz and NVIDIA GeForce RTX 3090 GPUs.

Baselines. We compare FRDP with a diverse set of baselines across three categories: (1) For traditional solvers, we adopt LKH3 [17] (a strong solver for routing problems) and OR-Tools [14] (a flexible solver that supports various combinations of constraints). (2) For constructive heuristics, Greedy-L and Greedy-C are considered. Greedy-L selects the nearest available node, and Greedy-C chooses the earliest time window ends relative to the current time in TSPTW and the minimal draft limit in TSPDL. (3) For neural solvers, we include POMO [25] (trained with penalty of constraint violation), PIP [3], which proactively masks actions to prevent future infeasibility, and its variant PIP-D.

Metrics. Following existing works [3,8], infeasibility rate (Inf.), average tour length (Obj), average optimality gap (Gap) and inference time (Time) are utilized for evaluation. We only compute the metrics corresponding to the feasible solutions obtained. The infeasibility rate is the proportion of instances in which no feasible solution is found among all the sampled solutions (we use the 8× augmentation technique [25] in all experiments).

5.2 Experimental Results

Experiments are conducted on TSPTW with three different hardness levels (easy, medium, and hard) and TSPDL with two different hardness levels (medium and hard) across two kinds of problem size ($n = 50$ and $n = 100$). The results are presented in Table 1 and Table 2, where the best feasibility rate and Gap among neural constructive methods are highlighted in bold.

Comparison with Traditional Solvers and Constructive Heuristics. Between the two greedy heuristics, Greedy-L fails to find any feasible solutions across all problem instances. Greedy-C achieves zero infeasible rate on TSPDL problems at the expense of completely disregarding travel distance optimization. OR-Tools performs well on simple TSPTW, but its feasibility rate drops significantly in medium and hard difficulty levels and does not find feasible solutions for TSPDL. Compared with the traditional strong solver LKH3, FRDP reduces the inference time by a factor of more than 200 in many cases while maintaining the feasibility and quality of the solution. Overall, FRDP demonstrates superior overall performance on both TSPTW and TSPDL, achieving 0% infeasible rates while maintaining significantly small optimality gaps.

Comparison with Neural Constructive Solvers. FRDP is compared with neural constructive solver POMO, PIP and its variant PIP-D under the same settings. Using greedy rollout with 8× augmentation, FRDP achieves comparable or better performance while requiring only a single solution per instance in most cases. Notably, on hard TSPTW, FRDP significantly reduces the infeasibility rate (e.g., from 2.67% to 0.43% for $n = 50$ and from 6.48% to 1.68% for $n = 100$) while achieving smaller optimality gaps. FRDP further benefits from generating multiple solutions. When increasing the number of sampled solutions (sample 20), it achieves 0% infeasibility rates across all tasks and consistently reduces optimality gaps. In contrast, PIP and PIP-D show limited improvement with additional samples, indicating weaker exploration capability.

Table 1. Results on TSPTW instances. The corresponding results of / are not available due to no feasible solutions.

	Method	$n = 50$				$n = 100$			
		Inf.↓	Obj.↓	Gap↓	Time↓	Inf.↓	Obj.↓	Gap↓	Time↓
Easy	LKH3	0.00%	7.31	0.00%	4.6h	0.00%	10.21	0.00%	8.5h
	OR-Tools	0.00%	7.34	0.96%	7h	0.00%	10.41	1.97%	14h
	Greedy-L	100.00%	/	/	13.8 s	100.00%	/	/	1.3 m
	Greedy-C	0.00%	26.08	257.27%	4.5 s	0.00%	52.14	411.13%	12 s
	POMO(greedy)	0.00%	7.54	3.08%	10 s	0.00%	10.83	6.07%	20 s
	PIP(greedy)	0.00%	7.50	2.65%	11 s	0.00%	10.57	3.53%	40 s
	PIP-D(greedy)	0.00%	7.49	2.51%	11 s	0.00%	10.66	4.39%	40 s
	PIP(sample 25)	0.00%	7.46	1.97%	2 m	0.00%	10.50	2.83%	13.2 m
	PIP-D(sample 25)	0.00%	7.45	1.84%	2 m	0.00%	10.58	3.60%	13.2 m
	FRDP(greedy)	0.00%	7.48	2.32%	11 s	0.00%	10.60	3.81%	40 s
	FRDP(sample 20)	**0.00%**	7.41	**1.45%**	1.2 m	**0.00%**	10.42	**2.08%**	7.8 m
Medium	LKH3	0.00%	13.02	0.00%	7h	0.00%	18.74	0.00%	10.8h
	OR-Tools	15.77%	13.02	0.30%	5.9h	0.52%	19.34	3.23%	13.8h
	Greedy-L	100.00%	/	/	15 s	100.00%	/	/	1 m
	Greedy-C	47.52%	25.33	96.43%	4.2 s	20.34%	51.62	176.07%	11.4 s
	POMO(greedy)	3.77%	13.68	5.23%	10 s	0.12%	20.78	10.93%	20 s
	PIP(greedy)	0.90%	13.40	2.91%	11 s	0.19%	19.61	4.65%	40 s
	PIP-D(greedy)	0.65%	13.45	3.32%	11 s	0.03%	19.79	5.64%	40 s
	PIP(sample 25)	0.29%	13.29	2.09%	2 m	0.03%	19.39	3.51%	13.2 m
	PIP-D(sample 25)	0.21%	13.34	2.46%	2 m	0.00%	19.56	4.41%	13.2 m
	FRDP(greedy)	0.30%	13.35	2.54%	11 s	0.06%	19.53	4.20%	40 s
	FRDP(sample 20)	**0.00%**	13.19	**1.32%**	1.2 m	**0.00%**	19.31	**3.04%**	7.8 m
Hard	LKH3	0.12%	25.61	0.00%	7h	0.07%	51.24	0.00%	1.4d
	OR-Tools	65.72%	25.76	0.00%	2.4h	89.07%	51.61	0.00%	1.6h
	Greedy-L	100.00%	/	/	21.8 s	100.00%	/	/	1.3 m
	Greedy-C	72.55%	26.39	1.53%	4.5 s	93.38%	52.95	1.43%	11.1 s
	POMO(greedy)	35.25%	26.22	1.61%	10 s	100.00%	/	/	20 s
	PIP(greedy)	2.67%	25.66	0.18%	11 s	16.27%	51.42	0.37%	40 s
	PIP-D(greedy)	3.07%	25.69	0.28%	11 s	6.48%	51.39	0.31%	40 s
	PIP(sample 25)	1.67%	25.65	0.14%	2 m	8.04%	51.38	0.27%	13.2 m
	PIP-D(sample 25)	1.81%	25.67	0.22%	2 m	4.14%	51.35	0.25%	13.2 m
	FRDP(greedy)	0.43%	25.64	0.13%	11 s	1.68%	51.29	0.11%	40 s
	FRDP(sample 20)	**0.00%**	25.61	**0.01%**	1.2 m	**0.00%**	51.24	**0.01%**	7.8 m

5.3 Further Experiment and Analysis

Comparison with LKH3 Under Similar Time Budget. To comprehensively evaluate FRDP against the traditional strong solver LKH3, we conducted comparative experiments on hard difficulty level of TSPTW and TSPDL under constrained inference time conditions. As shown in Table 3, LKH3 exhibits a significant increase in the infeasible rate (reaching 95.92% at $n = 100$) as the inference time decreases. Crucially, FRDP demonstrates superior performance across both problem domains.

Table 2. Results on TSPDL instances

	Method	$n = 50$				$n = 100$			
		Inf.↓	Obj.↓	Gap↓	Time↓	Inf.↓	Obj.↓	Gap↓	Time↓
Medium	LKH3	0.00%	10.87	0.00%	5.1h	0.00%	16.39	0.00%	14h
	OR-Tools	100.00%	/	/	10.9 s	100.00%	/	/	56.9 s
	Greedy-L	100.00%	/	/	2.4 m	100.00%	/	/	9.5 m
	Greedy-C	0.00%	26.09	144.24%	9.1 s	0.00%	52.16	222.71%	27 s
	POMO(greedy)	12.52%	10.98	3.80%	7 s	32.19%	17.11	9.15%	16 s
	PIP(greedy)	0.43%	11.22	3.41%	8 s	0.38%	17.71	8.08%	31 s
	PIP-D(greedy)	0.37%	11.26	3.78%	8 s	0.23%	17.84	8.86%	31 s
	PIP(sample 25)	0.24%	11.14	2.61%	1.4 m	0.13%	17.50	6.78%	8.5 m
	PIP-D(sample 25)	0.18%	11.16	2.81%	1.4 m	0.07%	17.61	7.43%	8.5 m
	FRDP(greedy)	0.00%	11.22	3.29%	8 s	0.00%	17.45	6.45%	31 s
	FRDP(sample 20)	**0.00%**	11.06	**1.81%**	52 s	**0.00%**	17.17	**4.75%**	5 m
Hard	LKH3	0.00%	13.30	0.00%	6.8h	0.00%	20.70	0.00%	1.2d
	OR-Tools	100.00%	/	/	10.6 s	100.00%	/	/	56.8 s
	Greedy-L	100.00%	/	/	2.4 m	100.00%	/	/	9.4 m
	Greedy-C	0.00%	26.07	99.73%	10.9 s	0.00%	52.17	156.37%	25 s
	POMO(greedy)	29.25%	13.03	4.11%	7 s	99.85%	20.95	15.87%	16 s
	PIP(greedy)	2.10%	13.66	3.13%	8 s	20.66%	22.30	12.67%	31 s
	PIP-D(greedy)	0.82%	13.80	3.95%	8 s	7.91%	22.84	12.32%	31 s
	PIP(sample 25)	1.73%	13.57	2.37%	1.4 m	17.26%	22.04	10.86%	8.5 m
	PIP-D(sample 25)	0.32%	13.69	3.04%	1.4 m	5.95%	22.58	10.68%	8.5 m
	FRDP(greedy)	0.00%	13.64	2.67%	8 s	0.27%	22.06	6.59%	31 s
	FRDP(sample 20)	**0.00%**	13.48	**1.44%**	52 s	**0.00%**	21.48	**3.75%**	5 m

Table 3. Results on hard TSPTW and TSPDL under similar time budget

	Method	$n = 50$			$n = 100$		
		Inf.	Gap	Time	Inf.	Gap	Time
TSPTW	LKH3 (Default)	0.12%	0.00%	7h	0.07%	0.00%	1.4d
	LKH3	64.71%	0.32%	1.4 m	95.92%	0.63%	8 m
	FRDP(sample 20)	0.00%	0.01%	1.2 m	0.00%	0.01%	7.8 m
TSPDL	LKH3 (Default)	0.00%	0.00%	6.8h	0.00%	0.00%	1.2d
	LKH3	9.66%	4.52%	1 m	8.83%	6.79%	6 m
	FRDP(sample 20)	0.00%	1.44%	52 s	0.00%	3.75%	5 m

Analysis of the Solution Diversity. To evaluate the solution diversity of the FRDP framework, we conducted experiments on TSPTW instances with 50 nodes. Across different time window difficulty levels, we sampled varying numbers of solutions from the test dataset and calculated the average count of unique solutions. The results in Fig. 3 demonstrate that FRDP consistently outperforms PIP and PIP-D in solution diversity across three difficulty levels on

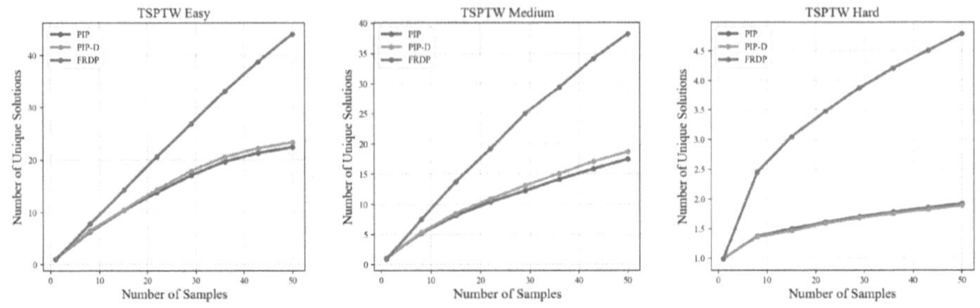

(a) Easy instance (b) Medium instance (c) Hard instance

Fig. 3. Solution diversity on TSPTW ($n = 50$). X-axis: Number of sampled solutions (1–50); Y-axis: Average count of unique solutions.

TSPTW. Notably, on hard TSPTW, PIP and PIP-D generate fewer than two unique solutions on average, while FRDP achieves approximately five unique solutions from 50 samples. This performance stems from the constrained feasible solution space in hard TSPTW instances, which naturally increases solution similarity. The FRDP framework's design enables exploration of more diverse solutions during both training and inference, effectively reducing infeasibility rates and optimality gaps.

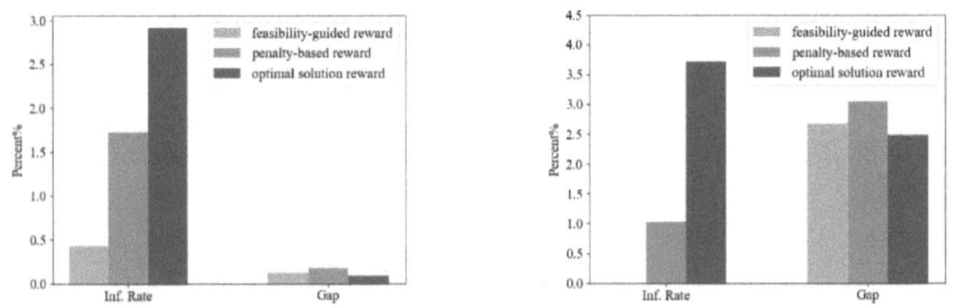

(a) TSPTW (b) TSPDL

Fig. 4. Ablation on feasibility-guided reward. The comparative experiments were conducted on hard difficulty level with $n = 50$.

5.4 Ablation Study

Ablation on Feasibility-Guided Reward. In the experiment, we replaced the feasibility-guided reward with two alternative reward mechanisms. (1) For the penalty-based reward, we retained the experimental configuration of PIP

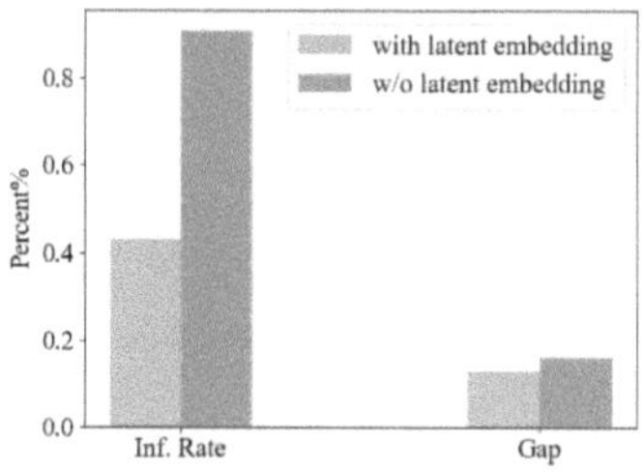

(a) TSPTW

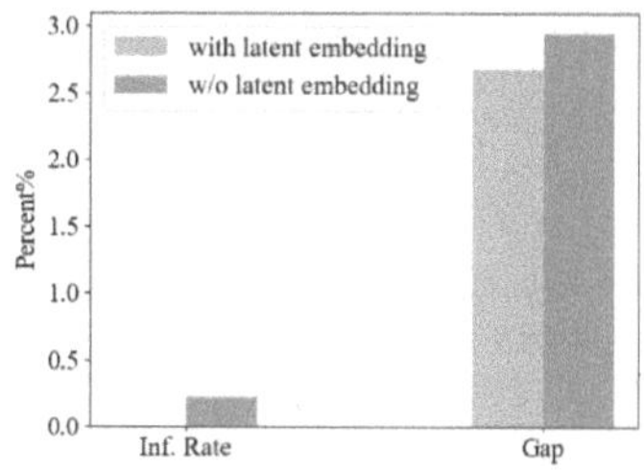

(b) TSPDL

Fig. 5. Ablation on latent embeddings. The comparative experiments were conducted on hard difficulty level with $n = 50$.

with fixed $\lambda = 1$. (2) The optimal solution reward, commonly used in diversity improvement approaches [4,16,19], calculates gradients based on the best of all sampled solutions to update the model weights. Figure 4 presents experimental results for the hard difficulty level with 50 nodes, demonstrating that the feasibility-guided reward mechanism achieves the lowest infeasibility rates across both TSPTW and TSPDL. Therefore, the feasibility-guided reward shows superior overall performance by effectively balancing solution feasibility and solution quality.

Ablation on Latent Embeddings. We conducted an ablation study on the latent embeddings component by excluding it during both training and inference, while restricting greedy rollout to a single solution. Figure 5 presents the ablation results for hard TSPTW and TSPDL, demonstrating that latent embeddings also play a role in encouraging exploration during training. In the hard TSPTW with $n = 50$, the infeasibility rate decreased from 0.92% to 0.43% (53% reduction) when latent embeddings were used. Notably, utilizing latent embeddings to generate multiple rollouts during inference is essential to fully demonstrate the effect of diversity, as shown in the previous tables.

6 Conclusion

In this paper, we propose the feasibility-guided reward and diversity-guided policy (FRDP) framework to solve complex constrained routing problems using deep reinforcement learning. FRDP introduces a feasibility-guided reward mechanism that explicitly differentiates feasible and infeasible solutions, ensuring correct optimization signals and improving constraint satisfaction. In addition, a diversity-guided policy based on latent embeddings enhances exploration, enabling effective search within fragmented feasible regions. Compared with existing traditional methods and neural constructive methods, FRDP achieves significant improvements in solution quality and constraint satisfaction.

In future work, we plan to further analyze the theoretical properties of the framework, extend it to more complex routing variants, and explore more efficient integrations with existing neural combinatorial optimization methods.

Acknowledgment. This work is supported by the National Natural Science Foundation of China (62472461), and the Guangdong Basic and Applied Basic Research Foundation (2025A1515010129).

References

1. Applegate, D.L., Bixby, R.E., Chvtal, V., Cook, W.J.: Concorde tsp solver (2006). http://www.math.uwaterloo.ca/tsp/concorde/index.html
2. Bi, J., et al.: Learning generalizable models for vehicle routing problems via knowledge distillation. In: Advances in Neural Information Processing Systems (2022)
3. Bi, J., et al.: Learning to handle complex constraints for vehicle routing problems. In: Advances in Neural Information Processing Systems (2024)
4. Chalumeau, F., et al.: Combinatorial optimization with policy adaptation using latent space search. In: Advances in Neural Information Processing Systems (2023)
5. Chen, C., Chen, R., Li, T., Ao, R., Wen, Z.: Monte Carlo policy gradient method for binary optimization. arXiv preprint arXiv:2307.00783 (2023)
6. Chen, J., et al.: Rethinking neural multi-objective combinatorial optimization via neat weight embedding. In: International Conference on Learning Representations (2025)
7. Chen, J., Huang, H., Zhang, Z., Wang, J.: Deep reinforcement learning with two-stage training strategy for practical electric vehicle routing problem with time windows. In: International Conference on Parallel Problem Solving from Nature, pp. 356–370. Springer, Cham (2022)
8. Chen, J., Gong, Z., Liu, M., Wang, J., Yu, Y., Zhang, W.: Looking ahead to avoid being late: solving hard-constrained traveling salesman problem. arXiv preprint arXiv:2403.05318 (2024)
9. Chen, X., Tian, Y.: Learning to perform local rewriting for combinatorial optimization. In: Advances in Neural Information Processing Systems, vol. 32, pp. 6281–6292 (2019)
10. Choo, J., et al.: Simulation-guided beam search for neural combinatorial optimization. In: Advances in Neural Information Processing Systems, vol. 35, pp. 8760–8772 (2022)
11. Drakulic, D., Michel, S., Mai, F., Sors, A., Andreoli, J.M.: BQ-NCO: bisimulation quotienting for generalizable neural combinatorial optimization. In: Advances in Neural Information Processing Systems (2023)
12. Dumas, Y., Desrosiers, J., Gelinas, E., Solomon, M.M.: An optimal algorithm for the traveling salesman problem with time windows. Oper. Res. **43**(2), 367–371 (1995)
13. Fu, Z.H.F., Qiu, K.B., Zha, H.: Generalize a small pre-trained model to arbitrarily large tsp instances. In: Proceedings of the AAAI Conference on Artificial Intelligence, pp. 7474–7482 (2021)
14. Furnon, V., Perron, L.: Or-tools routing library. https://developers.google.com/optimization/routing/

15. Goh, Y.L., Ma, Y., Zhou, J., Cao, Z., Dupty, M.H., Lee, W.S.: Shield: Multi-task multi-distribution vehicle routing solver with sparsity & hierarchy in efficiently layered decoder. In: International Conference on Machine Learning (2025)
16. Grinsztajn, N., Furelos-Blanco, D., Surana, S., Bonnet, C., Barrett, T.D.: Winner takes it all: training performant RL populations for combinatorial optimization. In: Advances in Neural Information Processing Systems (2023)
17. Helsgaun, K.: An extension of the Lin-Kernighan-Helsgaun TSP solver for constrained traveling salesman and vehicle routing problems. Roskilde: Roskilde Univ. **12**, 966–980 (2017)
18. Hottung, A., Kwon, Y.D., Tierney, K.: Efficient active search for combinatorial optimization problems. In: International Conference on Learning Representations (2022)
19. Hottung, A., Mahajan, M., Tierney, K.: PolyNet: learning diverse solution strategies for neural combinatorial optimization. arXiv preprint arXiv:2402.14048 (2024)
20. Hottung, A., Tierney, K.: Neural large neighborhood search for routing problems. Artif. Intell. **313**, 103786 (2022)
21. Kim, M., Park, J., Park, J.: Sym-NCO: leveraging symmetricity for neural combinatorial optimization. In: Advances in Neural Information Processing Systems (2022)
22. Konstantakopoulos, G.D., Gayialis, S.P., Kechagias, E.P.: Vehicle routing problem and related algorithms for logistics distribution: a literature review and classification. Oper. Res. Int. J. **22**(3), 2033–2062 (2022)
23. Kool, W., van Hoof, H., Gromicho, J., Welling, M.: Deep policy dynamic programming for vehicle routing problems. In: International Conference on Integration of Constraint Programming, Artificial Intelligence, and Operations Research. Springer, Cham (2022)
24. Kool, W., Van Hoof, H., Welling, M.: Attention, learn to solve routing problems. In: International Conference on Learning Representations (2018)
25. Kwon, Y.D., Choo, J., Kim, B., Yoon, I., Gwon, Y., Min, S.: POMO: policy optimization with multiple optima for reinforcement learning. In: Advances in Neural Information Processing Systems (2020)
26. Li, H., Liu, F., Zheng, Z., Zhang, Y., Wang, Z.: CADA: cross-problem routing solver with constraint-aware dual-attention. In: International Conference on Learning Representations (2025)
27. Li, Q., Cao, Z., Ma, Y., Wu, Y., Gong, Y.J.: Diversity optimization for travelling salesman problem via deep reinforcement learning. arXiv preprint arXiv:2501.00884 (2025)
28. Liao, Z., Chen, J., Wang, D., Zhang, Z., Wang, J.: BOPO: neural combinatorial optimization via best-anchored and objective-guided preference optimization. In: Proceedings of the 42nd International Conference on Machine Learning (ICML 2025), vol. 267, pp. 37456–37475 (2025)
29. Liu, F., Lin, X., Wang, Z., Zhang, Q., Xialiang, T., Yuan, M.: Multi-task learning for routing problem with cross-problem zero-shot generalization. In: Proceedings of the 30th ACM SIGKDD Conference on Knowledge Discovery and Data Mining, pp. 1898–1908 (2024)
30. Lu, H., Zhang, X., Yang, S.: A learning-based iterative method for solving vehicle routing problems. In: International Conference on Learning Representations (2020)
31. Luo, F., Lin, X., Liu, F., Zhang, Q., Wang, Z.: Neural combinatorial optimization with heavy decoder: toward large scale generalization. In: Advances in Neural Information Processing Systems (2023)

32. Ma, Y., Cao, Z., Chee, Y.M.: Learning to search feasible and infeasible regions of routing problems with flexible neural k-opt. In: Thirty-seventh Conference on Neural Information Processing Systems (2023)
33. Ma, Y., et al.: Learning to iteratively solve routing problems with dual-aspect collaborative transformer. Adv. Neural. Inf. Process. Syst. **34**, 11096–11107 (2021)
34. Qiu, R., Sun, Z., Yang, Y.: Dimes: a differentiable meta solver for combinatorial optimization problems. Adv. Neural. Inf. Process. Syst. **35**, 25531–25546 (2022)
35. Rakke, J.G., Christiansen, M., Fagerholt, K., Laporte, G.: The traveling salesman problem with draft limits. Comput. Oper. Res. **39**(9), 2161–2167 (2012)
36. Song, Y., Kingma, D.P.: How to train your energy-based models. arXiv preprint arXiv:2101.03288 (2021)
37. Sun, Z., Yang, Y.: Difusco: graph-based diffusion solvers for combinatorial optimization. In: Advances in Neural Information Processing Systems (2023)
38. Tang, Q., Kong, Y., Pan, L., Lee, C.: Learning to solve soft-constrained vehicle routing problems with lagrangian relaxation. arXiv preprint arXiv:2207.09860 (2022)
39. Vinyals, O., Fortunato, M., Jaitly, N.: Pointer networks. In: Cortes, C., Lawrence, N., Lee, D., Sugiyama, M., Garnett, R. (eds.) Advances in Neural Information Processing Systems (2015)
40. Wang, C., Cheng, P., Li, J., Sun, W.: Leader reward for pomo-based neural combinatorial optimization. arXiv preprint arXiv:2405.13947 (2024)
41. Wen, H., et al.: A survey on service route and time prediction in instant delivery: taxonomy, progress, and prospects. IEEE Trans. Knowl. Data Eng. **36**(12), 7516–7535 (2024)
42. Wu, Y., Song, W., Cao, Z., Zhang, J., Lim, A.: Learning improvement heuristics for solving routing problems. IEEE Trans. Neural Netw. Learn. Syst. **33**(9), 5057–5069 (2021)
43. Xin, L., Song, W., Cao, Z., Zhang, J.: Multi-decoder attention model with embedding glimpse for solving vehicle routing problems. In: Proceedings of the AAAI Conference on Artificial Intelligence, pp. 12042–12049 (2021)
44. Ye, T., Zhang, Z., Zhang, Q., Chen, J., Wang, J.: Solving multiobjective combinatorial optimization via learning to improve method. IEEE Trans. Emerg. Top. Comput. Intell. **9**(3), 2122–2136 (2025)
45. Yu, Z., Chen, J., Wang, J.: Combination-of-experts with knowledge sharing for cross-task vehicle routing problems. In: International Conference on Learning Representations (2026)
46. Zhang, R., Prokhorchuk, A., Dauwels, J.: Deep reinforcement learning for traveling salesman problem with time windows and rejections. In: 2020 International Joint Conference on Neural Networks (IJCNN), pp. 1–8. IEEE (2020)
47. Zhou, J., et al.: MVMoE: multi-task vehicle routing solver with mixture-of-experts. In: International Conference on Machine Learning (2024)
48. Zong, Z., Wang, J., Feng, T., Xia, T., Li, Y.: Deep reinforcement learning for demand driven services in logistics and transportation systems: a survey. ACM Trans. Knowl. Discov. Data (2025)

Linear Ordering Problem: Time for a Change

Fabrizio Fagiolo[1(✉)], Marco Baioletti[2], and Valentino Santucci[1]

[1] University for Foreigners of Perugia, 06123 Perugia, Italy
f.fagiolo@phd.unistrapg.it, valentino.santucci@unistrapg.it
[2] University of Perugia, 06123 Perugia, Italy
marco.baioletti@unipg.it

Abstract. The Linear Ordering Problem (LOP) is a fundamental combinatorial optimization problem with important applications in areas such as economics, social choice, and machine learning. Its most prominent use is the triangulation of economic input-output tables, which helps identify critical industries in an economy. Most existing algorithms have been evaluated on benchmarks derived from outdated macroeconomic data, which no longer reflect the structure of contemporary economies. Furthermore, LOP instances often exhibit many distinct global optima that can differ substantially from one another, creating challenges for applications that rely on a single solution. To address these limitations, we introduce a novel benchmark suite derived from up-to-date real-world economic data and an algorithmic scheme that leverages state-of-the-art LOP metaheuristics to generate diverse sets of high-quality solutions, together with metrics for assessing both quality and diversity. Experiments were conducted to report results on the proposed benchmark suite under both the traditional single-solution setting and the newly introduced multi-solution scenario.

Keywords: Linear Ordering Problem · Multimodal Combinatorial Optimization · Benchmarking · Permutation Diversity

1 Introduction

The Linear Ordering Problem (LOP) is a fundamental combinatorial optimization problem that has been extensively studied due to its central role in a variety of economically and computationally relevant applications [31]. Its most prominent application is the triangulation of economic input-output tables [13,26,31], a core task in input-output analysis [28], the field for which Leontief received the Nobel Prize in 1973. Beyond economics, the LOP has also found applications in diverse areas, including social choice [4], sports analytics [9], anthropology [21], natural language processing and translation [25,51], machine learning [33], and prompt engineering for generative AI [48]. Moreover, the LOP has recently attracted attention in studies examining the difficulty and structural characteristics of combinatorial problem instances [17,24,30,42] The LOP involves finding a

G. Iacca et al. (Eds.): PPSN 2026, LNCS 16985, pp. 399–416, 2027.
https://doi.org/10.1007/978-3-032-36214-8_25

permutation that reorders the rows and columns of a given matrix so as to maximize the sum of its upper-triangular part. When interpreted as the weight matrix of a digraph, the LOP is equivalent to finding a maximum-weight tournament subgraph [31]. In economic applications, the graph nodes represent industries, while the arcs correspond to inter-industry flows. The resulting triangulation provides a structured representation of these relationships and helps to identify critical industries within an economy.

The vast majority of algorithms proposed in the literature have been evaluated using the same benchmark suite, whose most recent version dates back to 2010 [32]. This suite includes a mix of artificial and real-world instances, although the latter are based on macroeconomic data that are now between 47 and 67 years old. In the meantime, financial crises, pandemics, globalization and rapid digitalization of markets have fundamentally reshaped economies, making many of these historical instances less representative of contemporary economies. Moreover, from an algorithmic perspective, although the No Free Lunch theorem does not apply to the LOP—since its objective function is not closed under permutation[1] [16]—, algorithms' performance can still vary significantly across different instances. Consequently, updated benchmarks are essential both to accurately assess algorithms' performance in contemporary settings and to ensure that algorithmic design reflects the structural characteristics of modern economic data.

To address this gap, in this work we introduce a novel benchmark suite for the LOP based on up-to-date real-world economic data. Thanks to advances in macroeconomic data collection, the proposed suite, called `EXIOBASE`, includes larger realistic instances than previous benchmarks, offering a more representative and challenging testbed for modern algorithm evaluation.

Furthermore, recent studies have shown that LOP instances often admit many distinct global optima rather than a single one [3,7,9]. These optimal solutions can differ substantially from one another, posing significant challenges for analyses that rely on a single solution and motivating the need for methods capable of identifying representative and diverse sets of global optima. Recent approaches mostly rely on exact algorithms, which are only feasible for relatively small instances due to the NP-hardness of the LOP.

To address this limitation, this work introduces an algorithmic scheme designed to leverage state-of-the-art LOP metaheuristics to generate a diverse set of high-quality solutions. We also propose appropriate metrics for evaluating the quality and diversity of the obtained solutions, providing a rigorous framework for experimental assessment.

The rest of the paper is organized as follows. Section 2 formally defines the LOP, reviews state-of-the-art algorithms, summarizes commonly adopted benchmarks, and discusses recent approaches for collecting multiple solutions. Section 3 introduces the newly proposed real-world economic benchmark suite,

---

[1] This follows immediately from the skew-symmetry property of the LOP: given an instance, the optimal solution is the reverse of the worst solution, the second-best solution is the reverse of the second-worst solution, and so on.

while Sect. 4 formally defines the Multi-Solution LOP (MS-LOP). The algorithmic scheme for generating multiple high-quality solutions is presented in Sect. 5, and experimental results are reported and analyzed in Sect. 6. Finally, Sect. 7 concludes the paper by also outlining directions for future research.

2 Linear Ordering Problem

An LOP instance is represented by a square matrix $\mathbf{B} = [b_{ij}]_{n\times n}$. Denoting the set of all permutations of $[n] = \{1, 2, \dots, n\}$ as $\mathbb{S}_n$, the goal is to find an optimal permutation $\sigma^* \in \mathbb{S}_n$ that maximizes the objective function

$$f(\sigma) = \sum_{i=1}^{n-1} \sum_{j=i+1}^{n} b_{\sigma(i),\sigma(j)}. \tag{1}$$

Algorithms for the LOP. The LOP was proven NP-hard in the seminal work of Garey and Johnson [20], and was later shown to be APX-complete [34]. Consequently, many metaheuristics have been proposed over the past decades (e.g., [2,5,6,10,12,19,27,31,40,41]).

To the best of our knowledge, the current state-of-the-art methods for the LOP are CD-RVNS [43] and MA-EDM [29].

CD-RVNS, introduced in [43], alternates between a variable neighborhood search phase and a destruction-construction phase. The variable neighborhood search identifies a local optimum under both insertion and interchange neighborhoods, using the restricted neighborhood technique for the LOP proposed in [11]. The destruction and construction operators partially disrupt the local optimum and then reconstruct a new solution by exploiting a novel representation specifically designed for the LOP. These latter procedures employ both intensification and diversification mechanisms, which appear to be the reason for the superior performance of CD-RVNS with respect to previous methods.

MA-EDM was introduced in [29] and further investigated in [36]. Its core idea is to maintain a population of local optima, under the insertion neighborhood, that is both diverse and of high quality. To this end, it employs a population selection procedure based on the permutation deviation distance, previously introduced in [46], which can be computed in linear time and is particularly well-suited for the LOP. Unlike CD-RVNS, the local search incorporated in MA-EDM does not use the restricted neighborhood, but relies on a classical acceleration technique that allows the evaluation of all the neighbors in $\Theta(n^2)$ time [45].

Benchmark Suites for the LOP. `LOLIB` is the most extensively studied benchmark suite for the LOP. It was compiled in 2010 and presented in [32]. The suite originally comprised 485 instances grouped into eight distinct sets and was later extended in [11] with the addition of the `xLOLIB2` instance set[2]. The resulting

[2] `LOLIB` and `xLOLIB2` are available at https://www.uv.es/rmarti/paper/lop.html and https://github.com/sgpceurj/EJOR2015, respectively.

suite, now consisting of nine sets and 685 instances, has become the de facto standard in LOP research. All the state-of-the-art algorithms discussed previously [29,43] have been evaluated on these nine instance sets, which are listed along with their main characteristics in Table 1. Most of the instances are artificial, while only a few are derived from real-world data.

Table 1. Main characteristics of the benchmark instance sets for the LOP.

Instance Set	#Instances	Size	Normal Form	#Optima	Type
`IO` [23]	50	44–79	Yes	50	Economic IO tables (1959–1979)
`SGB` [27]	25	75	Yes	25	Random uniform from large range
`RandA1` [10]	100	100–500	Yes	0	Random uniform from small range
`RandA2` [10]	75	100–200	Yes	0	From random permutations
`RandB` [32]	90	40–50	Yes	71	Random asymmetric
`MB` [35]	30	100–250	Yes	30	Random asymmetric + zeroed entries
`xLOLIB` [44]	78	150–250	Yes	0	Oversampled `IO` instances
`Spec` [14,22,32]	37	11–452	Yes	30	Artificial, subsampled `IO`, tennis results
`xLOLIB2` [11]	200	300–1000	No	0	Oversampled `IO` instances

The only proper real-world instances are the national input-output tables in the `IO` set, first used in the context of the LOP in [23] and based on data from Eurostat and other statistical agencies. These IO tables span the years 1959–1979 and include 50 instances: 31 with $n = 44$, 4 with $n = 50$, 11 with $n = 56$, 3 with $n = 60$, and one (`usa79`) with $n = 79$. Schiavinotto and Stützle [44] then generated the `xLOLIB` instances by oversampling 39 `IO` instances: for each, they created two larger instances with $n = 150$ and $n = 250$ by uniformly sampling entries from the corresponding IO table. They describe these `xLOLIB` instances as "real-life like" problems. The same procedure was later used by Ceberio et al. [11] to generate the `xLOLIB2` instances, 50 for each size $n \in \{300, 500, 750, 1000\}$. Finally, the `Spec` set—comprising both artificial and real-world instances—contains 15 economic-like matrices generated in [14] by subsampling the `usa79` instance from the `IO` set, as well as 8 preference counting matrices derived from the results of ATP tennis tournaments held in 1993—1994.

All remaining instances are artificially generated. The entries of `SGB` and `RandA1` instances are sampled uniformly at random from $[0, 250\,000]$ and $[0, 100]$, respectively. `RandA2` instances are preference counting matrices constructed from a set of randomly generated permutations. `RandB` and `MB` instances are generated with upper and lower triangular parts drawn from uniform distributions over intervals of different length, with `MB` instances further processed by zeroing-out a certain percentage of entries. Finally, the 14 fully artificial instances in `Spec` include 6 random binary matrices and 8 Paley graph adjacency matrices.

The number of known optima reported in Table 1 (proven via integer programming) is taken from [29]. In general, the unsolved instances are the larger ones, with a few exceptions: `MB` instances are easier due to the zeroing-out process, and `RandA1` instances appear more difficult than `SGB` instances because of their different variability. Indeed, both sparsity (the percentage of zeros) and entry variability are known to influence instance difficulty [44].

All benchmark instances have non-negative integer entries. Moreover, all instances except those in `xLOLIB2` are in normal form, i.e., $\min\{b_{ij}, b_{ji}\} = 0$ for all $i, j \in [n]$. As noted in [32], every LOP instance can be normalized without altering the ranking over the solutions induced by the objective function.

Multiple High-Quality Solutions for the LOP. Recently, there has been a growing interest in the observation that LOP instances often have multiple distinct global optima rather than a single one. Benito-Marimom et al. [7] showed, using integer programming, that: all `IO` instances have more than one global optimum, only 7 out of 50 instances had fewer than 500 optima, while 21 exhibited more than 500 000 distinct optima. Although these solutions share the same optimal objective value, they can differ substantially in their genotypic representations. In particular, [3] showed that some items may vary considerably in both their positions and relative ordering across the different optimal solutions.

Therefore, novel algorithms aimed at identifying representative sets of multiple global optima for the LOP have begun to emerge. Cameron et al. [9] proposed an integer programming model to identify two global optima that are maximally distant according to the Kendall's-τ distance [45]. Building on this idea, Anderson et al. [3] introduced a family of integer programming models to: compute a summary—referred to as the "centroid"—of item positions across all global optima, identify pairs of optima closest to and farthest from this centroid, as well as the most similar and most dissimilar pairs of optima. In all cases, the Kendall's-τ distance is used to measure (dis)similarity, as it is particularly well suited to the LOP.

More recently, [7] extended the model of [9] to generate sets of more than two global optima that are maximally distant from one another. To this end, they considered two distinct objectives: maximizing the sum of all pairwise distances and maximizing the minimum distance between any two solutions. The latter helps to prevent multiple copies of the same optimum and ensures better coverage of the entire set of global optima. The same authors also noted that the integer programming approaches are limited to small/medium-sized instances and therefore proposed a two-phase heuristic method based on a Variable Neighborhood Search (VNS) scheme using insertion neighborhoods with an increasing number of moves. In the first phase, VNS finds the best possible objective value (with no theoretical guarantee), while in the second phase, a separate VNS collects m solutions attaining this value into a representative set that tries to maximize the sum of pairwise Kendall's-τ distances. The authors explicitly note that this second-phase VNS ignores any solutions with an objective value higher than that found in the first phase.

Beyond the LOP, two related areas of research can be identified in the evolutionary computation field: multimodal optimization and evolutionary diversity optimization. Multimodal optimization [37,38] deals with the task of approximating the set of local optima whose quality is higher than a specified threshold. In this context, the diversity of solutions is primarily considered to enhance the search exploration, rather than being an explicit optimization goal in its own [15]. In contrast, evolutionary diversity optimization, introduced in [52], aims to compute a set of solutions whose quality is higher than a given threshold and that are maximally diverse from one another. While multimodal optimization is mainly applied to continuous search spaces (see the recent GECCO competition [1]), evolutionary diversity optimization has mostly been studied from a theoretical perspective (e.g., [8,15]). Neither has been applied to the LOP. Moreover, most of the works in these research lines assume that the optima or their objective values are known in advance.

3 The `EXIOBASE` Benchmark Suite

To address the limitations of existing benchmark suites for the LOP, we introduce a novel suite, `EXIOBASE`, derived from real-world data in the homonymous input-output (IO) database introduced by Stadler et al. [49] for economical and sustainability analyses. The proposed benchmarks are publicly available at https://doi.org/10.5281/zenodo.20554376.

The rest of this section outlines the structure of the original IO tables, describes the construction of our benchmark suite, and presents comparative statistics with classical benchmarks.

The Original IO Database. EXIOBASE [49] is a widely used global multi-regional input-output database that links economic transactions between countries and industries with environmental and social indicators; it is standardly applied in footprint and sustainability assessments. At the time of writing, the most recent release is version 3.9.6, available at [50], which provides annual data from 1995 to 2022. While environmental extensions are available, the present study focuses exclusively on the economic component containing the IO tables. For each year, four primary economic data structures are provided: the multi-regional transaction matrix $\mathbf{Z}$, the final demand matrix $\mathbf{Y}$, the total output vector $\mathbf{x}$, and the structural matrix $\mathbf{A}$.

Data are available in both industry-by-industry and product-by-product formats. Here, we focus on the product-by-product format. In this configuration, the rows and columns of $\mathbf{Z} = [z_{ij}]_{s \times s}$ correspond to specific region-product pairs. Each entry z_{ij} represents the monetary value of the output from region-product i used as intermediate input by region-product j, expressed in millions of euros. The order of $\mathbf{Z}$ is $s = s_\mathrm{R} \cdot s_\mathrm{P}$, with $s_\mathrm{R} = 49$ regions (44 countries and 5 rest-of-the-world regions) and $s_\mathrm{P} = 200$ products. The matrix $\mathbf{Y} \in \mathbb{R}^{s \times f}$ represents final demand, with rows corresponding to the s region-product pairs and $f = s_\mathrm{R} \cdot s_\mathrm{F}$ columns representing the $s_\mathrm{F} = 7$ standard final demand categories repeated for each of the s_R regions. The output vector $\mathbf{x} \in \mathbb{R}^s$ contains the total production of each sector and is computed as $\mathbf{x} = \mathbf{Z}\mathbf{1}_s + \mathbf{Y}\mathbf{1}_f$, where $\mathbf{1}_s$

and $\mathbf{1}_f$ are all-ones vectors of length s and f, respectively. Finally, the structural matrix $\mathbf{A} = [a_{ij}]_{s\times s}$ is a normalized version of $\mathbf{Z}$, computed as $\mathbf{A} = \mathbf{Z}\,\text{diag}(\mathbf{x})^{-1}$, where each entry $a_{ij} \geq 0$ is a *technical coefficient* representing the amount of output from region-product i required to produce one unit of region-product j. These coefficients are dimensionless ratios of monetary values, and $\mathbf{A}$, like $\mathbf{Z}$, is a 9800×9800 matrix with the following block structure:

$$\mathbf{A} = \begin{pmatrix} \mathbf{A}_{1,1} & \mathbf{A}_{1,2} & \cdots & \mathbf{A}_{1,s_\text{R}} \\ \mathbf{A}_{2,1} & \mathbf{A}_{2,2} & \cdots & \mathbf{A}_{2,s_\text{R}} \\ \vdots & \vdots & \ddots & \vdots \\ \mathbf{A}_{s_\text{R},1} & \mathbf{A}_{s_\text{R},2} & \cdots & \mathbf{A}_{s_\text{R},s_\text{R}} \end{pmatrix}, \tag{2}$$

where each block $\mathbf{A}_{kl}$ is a $s_\text{P} \times s_\text{P}$ matrix, with diagonal blocks $\mathbf{A}_{kk}$ representing inter-industry flows within region k, and the off-diagonal blocks $\mathbf{A}_{kl}$ $(k \neq l)$ describing inter-regional trade.

Construction of the Benchmark Suite. Building on the data described above, we introduce the `EXIOBASE` benchmark suite for the LOP, consisting of four distinct instance sets: `pxp`, `rxr`, `isic`, and `os300`. These sets are designed to reflect different levels of aggregation and capture a range of contemporary global economic interdependencies.

The set `pxp` (product-by-product) consists of 49 instances derived from the domestic structural matrices for the year 2022. Each instance is obtained from the corresponding diagonal block $\mathbf{A}_{kk}$ in Eq. (2), representing the inter-industry transactions for region $k \in \{1, 2, \dots, 49\}$. Formally, given a 200×200 domestic matrix $\mathbf{A}$, a corresponding LOP instance $\mathbf{B}$ is constructed in two steps. First, an intermediate matrix $\mathbf{B}'$ is computed as $\mathbf{B}' = \lfloor c\mathbf{A} - \min(c\mathbf{A}, c\mathbf{A}^T)\rceil$, where $c = 10^5$ is a scaling factor, both min and $\lfloor\cdot\rceil$ are applied element-wise, and $\lfloor\cdot\rceil$ denotes the rounding to the nearest integer. This transformation ensures that instances have integer entries within $[0, 10^5]$ and are in normal form, as defined in [32]. Second, the final LOP instance $\mathbf{B}$ is obtained from $\mathbf{B}'$ by removing all products i for which both the i-th row and i-th column are entirely zero. This filtering step is essential, since such "null products" can be placed arbitrarily within a permutation without affecting its LOP objective value; consequently, k null products would artificially increase the number of global optima by a factor of $\binom{n}{k} = \Theta(n^k)$. As a result, the instances in `pxp` have varying sizes, ranging from 89 to 176, with a median size of 149.

The set `rxr` (region-by-region) consists of 28 instances, one for each year from 1995 to 2022, obtained by first aggregating the technical coefficients into 49×49 matrices representing inter-regional trade and then applying the same scaling, rounding, and normalization steps described above. In this case, all instances have size $n = 49$.

The set `isic` comprises 49 matrices generated by aggregating the 2022 data according to the International Standard Industrial Classification (ISIC). This aggregation groups the 200 original products into the first ten ISIC categories (e.g., agriculture, manufacturing, and services). Scaling, rounding and normalization steps are applied here as well. All instances therefore have size $n = 10$ and can be exhaustively evaluated, making them particularly useful in scenarios where exact global optima are required (e.g., [30]).

Finally, the set `os300` consists of 49 real-life-like instances of size $n = 300$. Each is created by oversampling the corresponding `pxp` instance using the technique used in [44] and [11], followed by normalization of the resulting matrix.

Statistical Analysis of the Instances. Table 2 compares the main structural statistics of the `EXIOBASE` and `LOLIB` benchmark suites. In particular, we report the same measures used in [44]: sparsity (percentage of zeros), the coefficient of variation (ratio of standard deviation to mean), and the skewness index. Since the instances are in normal form, these statistics were computed only on the maximum of each pair (b_{ij}, b_{ji}) for all $1 \leq i < j \leq n$.

Table 2. Structural statistics for the `EXIOBASE` and `LOLIB` benchmark suites.

Suite	Set	Sparsity		Variation Coeff.		Skewness	
		Median	[Min,Max]	Median	[Min,Max]	Median	[Min,Max]
`EXIOBASE`	`isic`	0.00	$[0.00, 0.04]$	1.23	$[0.89, 2.86]$	2.14	$[0.74, 5.88]$
	`rxr`	0.00	$[0.00, 0.01]$	1.85	$[1.77, 2.30]$	4.67	$[4.07, 10.02]$
	`pxp`	0.17	$[0.03, 0.45]$	5.19	$[3.03, 8.26]$	16.01	$[7.88, 27.17]$
	`os300`	0.01	$[0.00, 0.02]$	3.62	$[2.20, 6.20]$	10.14	$[5.83, 16.69]$
	Overall	0.01	$[0.00, 0.45]$	2.92	$[0.89, 8.26]$	8.70	$[0.74, 27.17]$
`LOLIB`	`IO`	0.18	$[0.05, 0.71]$	3.74	$[2.59, 9.22]$	10.54	$[4.95, 23.67]$
	`SGB`	0.10	$[0.08, 0.12]$	6.48	$[6.42, 6.78]$	17.02	$[16.64, 17.65]$
	`RandA1`	0.01	$[0.01, 0.01]$	0.71	$[0.69, 0.73]$	0.56	$[0.53, 0.62]$
	`RandA2`	0.08	$[0.00, 0.12]$	0.75	$[0.73, 0.79]$	0.94	$[0.77, 1.09]$
	`RandB`	0.01	$[0.00, 0.02]$	0.57	$[0.52, 0.61]$	0.07	$[-0.05, 0.47]$
	`MB`	0.02	$[0.01, 0.06]$	0.72	$[0.70, 0.77]$	0.49	$[0.43, 0.63]$
	`xLOLIB`	0.12	$[0.05, 0.65]$	3.87	$[3.19, 11.60]$	9.81	$[7.17, 30.89]$
	`xLOLIB2`	0.34	$[0.26, 0.73]$	4.18	$[3.56, 12.35]$	9.88	$[7.53, 27.81]$
	Overall	0.10	$[0.00, 0.73]$	3.42	$[0.52, 12.35]$	7.87	$[-0.05, 30.89]$

Although these statistics are not strong predictors of instance difficulty, it is apparent that `EXIOBASE` instances exhibit lower sparsity and variation than `LOLIB` instances, while the overall skewness is roughly similar.

4 Multi-solution Linear Ordering Problem

We introduce the Multi-Solution LOP (MS-LOP) as a multimodal and diversity-aware extension of the LOP that unifies three related research directions: the

Multiple LOP of Benito-Marimom et al. [7], and the tasks of multimodal optimization (e.g., [38]) and evolutionary diversity optimization (e.g., [15]).

Let $f : \mathbb{S}_n \rightarrow \mathbb{R}$ be the LOP objective function defined as in Eq. (1) and $\mathcal{N} : \mathbb{S}_n \rightarrow 2^{\mathbb{S}_n}$ be a neighborhood relation on $\mathbb{S}_n$ (e.g., the insertion neighborhood), then $\mathcal{L} = \{\sigma \in \mathbb{S}_n : f(\sigma) \geq f(\pi),\ \forall\ \pi \in \mathcal{N}(\sigma)\}$ is the set of local optima of f with respect to $\mathcal{N}$. Given an integer $m > 0$, $\mathcal{L}_m = \{S \subseteq \mathcal{L} : |S| = m\}$ is the family of sets containing exactly m distinct local optima.

For any $S \in \mathcal{L}_m$, we define its quality $\Phi(S)$ and its diversity $\Delta(S)$. The quality $\Phi(S)$ is the average of the individual objective values in S, while the diversity $\Delta(S)$ can be instantiated by any suitable metric that quantifies the dispersion within S based on the pairwise Kendall's-τ distances of its members. The Kendall's-τ distance counts the number of pairwise disagreements between two permutations and is widely regarded as the most appropriate distance measure for the LOP [44,45]. Two reasonable choices for Δ are: the sum of nearest-neighbor distances Δ_{NN} [7,31], and the Solow-Polasky diversity Δ_{SP} [39,47]. The metrics Φ, Δ_{NN} and Δ_{SP} are formally defined as:

$$\Phi(S) = \frac{1}{m} \sum_{\sigma \in S} f(\sigma), \tag{3}$$

$$\Delta_{NN}(S) = \sum_{\sigma \in S} \min_{\pi \in S \setminus \{\sigma\}} d(\sigma, \pi), \tag{4}$$

$$\Delta_{SP}(S) = \mathbf{1}_m^T \mathbf{C}^{-1} \mathbf{1}_m, \tag{5}$$

where: $d(\cdot, \cdot)$ is the Kendall's-τ distance between permutations, $\mathbf{C} = [c_{ij}]_{m \times m}$ is the similarity matrix with entries $c_{ij} = e^{-\theta d(\sigma_i, \sigma_j)}$ for each pair $\sigma_i, \sigma_j \in S$, $\theta > 0$ is a Solow-Polasky parameter [47], and $\mathbf{1}_m$ is a (column) vector of m ones.

We compare sets in $\mathcal{L}_m$ using a lexicographic order $\succeq$ that prioritizes quality over diversity. For any $S, P \in \mathcal{L}_m$, we define $S \succeq P$ if and only if either $\Phi(S) > \Phi(P)$, or $\Phi(S) = \Phi(P)$ and $\Delta(S) \geq \Delta(P)$. Therefore, the goal of MS-LOP is to identify a set S^* of m distinct local optima that is maximal with respect to $\succeq$. More formally,

$$S^* \in \{S \in \mathcal{L}_m : S \succeq P\ \ \forall\ P \in \mathcal{L}_m\} \tag{6}$$

Interestingly, MS-LOP subsumes several established tasks: the Multiple LOP of [7], since global optima are also local optima; the objective of multimodal optimization [38], as quality is prioritized over diversity; and evolutionary diversity optimization [15], as diversity is explicitly optimized. Importantly, unlike these approaches, MS-LOP does not require the optimal objective value, which is unavailable in practice due to the NP-hardness of the LOP.

Finally, we suggest addressing the MS-LOP with small set cardinalities, such as $m \in \{5, 10, 15\}$, since in practical settings the (approximation of the) solution set S^* is ultimately analyzed by a human decision maker.

5 Algorithms for the Multi-solution LOP

As a first approach to the MS-LOP, we propose an algorithmic scheme that can be readily integrated into any single-solution LOP algorithm $\mathcal{A}$ based on local search. As discussed in Sect. 2, such methods constitute the predominant paradigm in the current state-of-the-art.

The proposed scheme maintains an archive of solutions S, initialized as the empty set. $\mathcal{A}$ is modified in such a way that, during its execution, it triggers the procedure `UpdateArchive`(σ) whenever a locally optimal solution $\sigma \in \mathbb{S}_n$ is generated. Upon termination, the archive S contains m distinct local optima that approximate the goal defined in Eq. (6).

Algorithm 1. The algorithm for updating the archive S.

procedure UPDATEARCHIVE($\sigma \in \mathbb{S}_n$)
    if $\sigma \notin S$ **then**
        if $|S| < m$ **then**
            $S \leftarrow S \cup \{\sigma\}$
        else
            $S' \leftarrow S \cup \{\sigma\}$
            for all $\pi \in S'$ **do**
                $u_0^{(\pi)} \leftarrow f(\pi)$
                Calculate the m Kendall's-τ distances $d(\pi, \rho)$ for each $\rho \in S' \setminus \{\pi\}$
                $\langle u_1^{(\pi)}, \ldots, u_m^{(\pi)} \rangle \leftarrow$ sort the distances
            $\pi \leftarrow$ lexicographic argmin of the $\mathbf{u}$ vectors
            $S \leftarrow S' \setminus \{\pi\}$

The pseudo-code of `UpdateArchive`(σ) is shown in Algorithm 1. It first checks whether $\sigma \in S$. If it is, no action is taken, since all solutions in S must be distinct. Otherwise, if $|S| < m$, σ is simply added to the archive. In the more interesting case where $\sigma \notin S$ and $|S| = m$, a temporary set $S' = S \cup \{\sigma\}$ of $m+1$ solutions is formed. To determine which solution to exclude, a $(m+1)$-dimensional utility vector $\mathbf{u}^{(\pi)} = (u_0, u_1, \ldots, u_m)$ is constructed for each $\pi \in S'$, where: $u_0 = f(\pi)$ and $u_1 \leq u_2 \leq \ldots \leq u_m$ are the Kendall's-τ distances between π and the remaining members of S', sorted in ascending order. Finally, the solution $\pi \in S'$ with the lexicographically smallest utility vector $\mathbf{u}^{(\pi)}$ is removed from S' to yield the updated archive S.

This procedure, inspired by the selection mechanism of the well-known SPEA2 algorithm [53] widely used in multi-objective optimization, enables the construction of an anytime algorithm, i.e., if interrupted at any iteration, it always returns a solution set of m elements. The underlying rationale is that u_0, i.e., the objective value, serves as the primary selection criterion. In the case of ties, u_1, i.e., the distance to the nearest neighbor, is used as a secondary criterion, and any remaining ties are resolved by recursively reapplying the nearest-neighbor principle. Consequently, an algorithm $\mathcal{A}$ equipped with `UpdateArchive` explicitly maximizes the average objective value while ensuring good diversity in the returned solution set.

Since $m \ll n$, the complexity of `UpdateArchive` is dominated by the computation of the Kendall's-τ distances, each of which requires $\Theta(n \log n)$ steps. To improve efficiency, we maintain simple bookkeeping: at each call, only the m distances between the new solution σ and those in S are computed. Hence, the worst-case asymptotic complexity of `UpdateArchive` is $\Theta(mn \log n)$.

6 Experiments

Experiments were conducted on the proposed `EXIOBASE` benchmark suite under both single-solution and multiple-solution settings using the two state-of-the-art algorithms CD-RVNS and MA-EDM. In the single-solution setting, each algorithm was executed 30 times per instance. For the multiple-solution setting (MS-LOP), the algorithms were extended with the scheme introduced in Sect. 5, resulting in the variants MS-CD-RVNS and MS-MA-EDM. Three archive sizes, $m \in \{5, 10, 15\}$, were considered, with 30 independent runs per instance and size. To ensure comparability, we adopt a unified stopping criterion, terminating each run after $100n$ local optima (not necessarily distinct) are generated. The algorithms were evaluated on the `rxr`, `pxp`, and `os300` instance sets, with the smaller `isic` instances used in a preliminary experiment to study how the number of global optima varies with instance sparsity. C++ implementations were used, and all experiments were conducted on a machine running Ubuntu 22.04, equipped with an Intel Core i9-11900F CPU (2.50 GHz) and 32 GB of RAM.

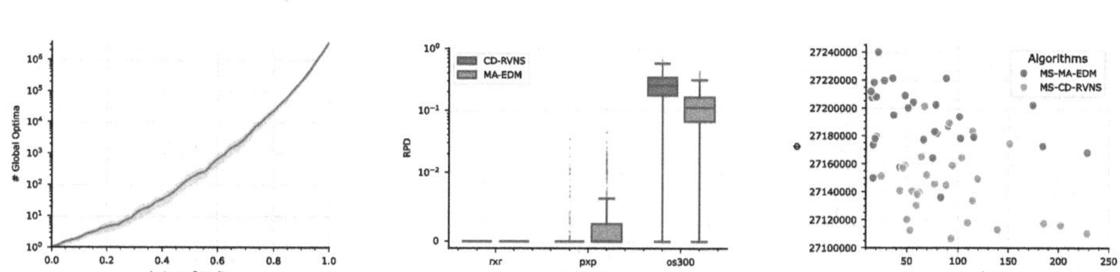

(a) (b) (c)

Fig. 1. (a) Number of global optima (log scale) as a function of sparsity for `isic` instances. (b) Relative percentage deviations (log scale) in the single-solution setting for the `rxr`, `pxp`, and `os300` instances. (c) Φ vs. Δ_{NN} for all the executions in the `os300_it_2022_n300` instance with archive size $m = 15$.

Number of Global Optima vs. Sparsity. A first experiment was conducted to investigate how the number of global optima varies with instance sparsity (percentage of zeros in the LOP matrix). The `isic` instances were used, as their size ($n = 10$) allows exhaustive evaluation of the search space. For each instance, sparsity was gradually increased by zeroing out one additional matrix entry at each iteration, generating a chain of instances ranging from no sparsity to full sparsity. Then, for every generated instance, all $n!$ solutions were evaluated and the number of global optima was recorded. The experiment was repeated for all `isic` instances, and the results were aggregated in Fig. 1a, which reports the mean number of global optima and the corresponding 95% confidence interval as a function of instance sparsity. The results clearly indicate that the number of global optima increases exponentially as sparsity grows.

Single-Solution Experiments. To evaluate the effectiveness of CD-RVNS and MA-EDM in the classical LOP setting, we use the standard Relative Percentage Deviation (RPD) metric. Let v denote the final objective value obtained from a single execution, and let *best* be the best objective value observed across all executions of all algorithms on the same instance. Then, v is converted to a RPD value as $rpd = 100 \cdot \frac{best-v}{best}$. Lower RPDs indicate better effectiveness, and this normalization allows results to be fairly aggregated across instances.

Figure 1b shows the RPDs obtained from the experiments, aggregated by instance set. On the smallest `rxr` instances, both algorithms consistently achieve near-zero RPDs, indicating that all runs converged to essentially the same value. For medium-sized `pxp` instances, CD-RVNS is more robust and effective than MA-EDM, while the latter performs better on the larger `os300` instances. These observations are supported by per-instance Mann–Whitney U tests at a 0.05 significance level and p-values adjusted via the Benjamini–Hochberg procedure: no significant differences were found on `rxr`, CD-RVNS significantly outperforms MA-EDM on 29 out of 49 `pxp` instances while being significantly outperformed on only 4, and MA-EDM is significantly better on 46 out of 49 `os300` instances.

This behavior is possibly explained by the stronger emphasis of MA-EDM on diversity management, which we believe provides an advantage on larger instances and longer runs, while the trajectory-based strategy of CD-RVNS appears better suited for medium-sized instances.

For detailed results and best known objective values in the classical LOP setting, we refer the interested reader to the supplementary material [18]

MS-LOP Experiments. To evaluate the multi-solution variant of the LOP, we assess each run in terms of both solution quality and diversity. Quality is measured using the Φ metric as defined in Eq. (3), while diversity is quantified using Δ_{NN} and Δ_{SP} as defined in Eqs. (4) and (5). For aggregation and visualization, all measures are normalized using the same RPD technique described earlier (using the best Φ value for each instance, and the maximum Δ_{NN} and Δ_{SP} values[3] for each instance-archive size pair). The resulting normalized values—which have to be minimized for better quality and diversity, due to the RPD transformation—are presented in log scale as boxplots in Fig. 2.

[3] Δ_{SP} was computed using normalized Kendall's-τ distances, with θ calibrated such that the median of the observed distances corresponds to a similarity of 0.5.

Regarding the Φ metric, for all m values, similarly to the classical setting, MS-CD-RVNS performs better on `pxp` instances, while MS-MA-EDM outperforms on the larger `os300` instances. Interestingly, the tie observed in the classical LOP for the smaller `rxr` instances does not occur in the MS-LOP case, where MS-MA-EDM is slightly more effective. To further investigate this behavior, we analyzed the archive filling patterns and found that while the archive is always filled to its maximum size m for `pxp` and `os300` instances, in the `rxr` case it is often underfilled—likely due to fewer local optima in the instance landscape—, and MS-CD-RVNS surprisingly returns a slightly larger archive on average than

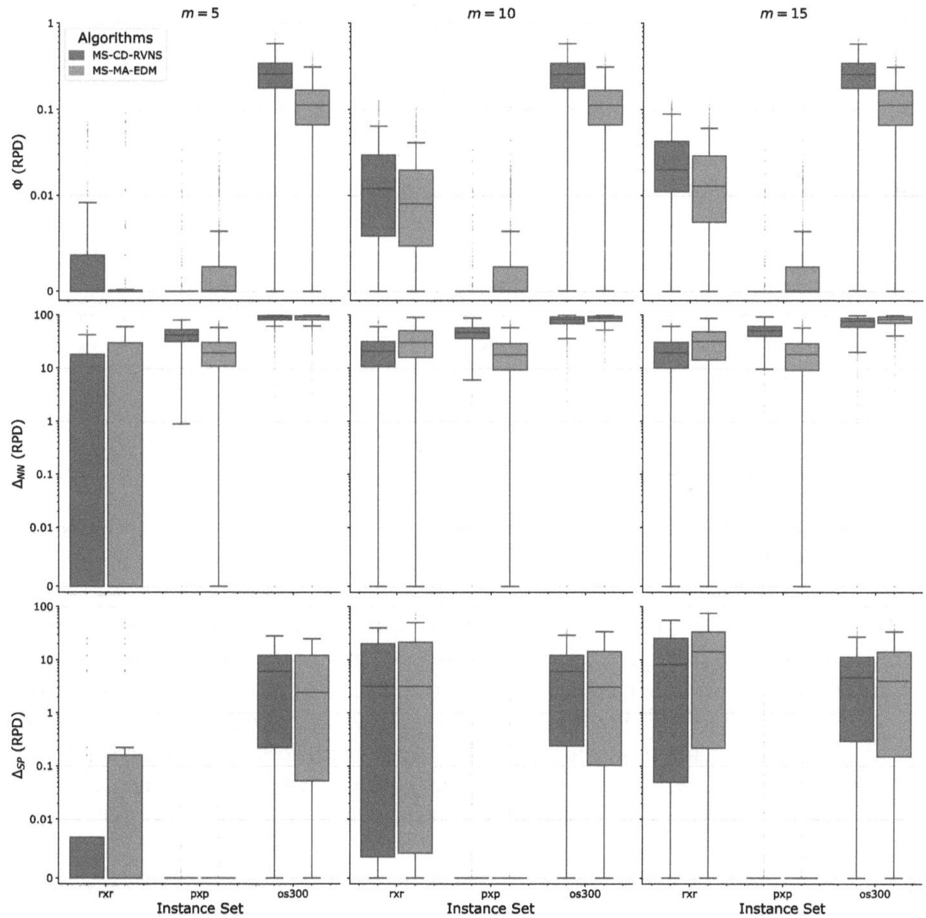

Fig. 2. Each row of boxplots shows the relative percentage deviations of the Φ, Δ_{NN}, and Δ_{SP} metrics, respectively. Columns correspond to different archive sizes (m values), and within each plot, the metrics are shown in log scale and aggregated by instance set.

MS-MA-EDM (e.g., 10.94 vs. 10.28 for $m = 15$), suggesting that its additional solutions may have worse objective values.

Regarding diversity, rxr appears to be the only instance set where diversity depends on the archive size, showing greater diversity for smaller m values. For pxp instances, the Solow-Polasky diversity remains stable and nearly identical across all runs (approximately 0 deviation) for both algorithms, while the nearest-neighbor diversity of MS-MA-EDM is higher than that of MS-CD-RVNS. In contrast, for os300 instances, there is no clear difference in diversity between the two algorithms.

Finally, in Fig. 1, for illustrative purposes, we provide a scatter plot showing the relationship between the Φ quality and the Δ_{NN} diversity achieved by each run on the os300_it_2022_n300 instance with an archive size of $m = 15$. The figure confirms the previous observation, showing that, on this illustrative instance, MS-MA-EDM achieves higher objective values (Φ) than MS-CD-RVNS, while the distribution of the average Kendall's-τ distance to the nearest neighbor within the returned solution set (Δ_{NN}) is similar for both algorithms. Furthermore, since each point in the scatter plot of Fig. 1 corresponds to the solution set obtained by a single execution of the algorithms, it becomes evident that neither method consistently returns the same solution set. This variability suggests that there is still room for improvement, particularly in terms of robustness.

For detailed results and statistical analyses in the MS-LOP setting, we refer the interested reader to the supplementary material [18], which also includes a reproducibility package.

7 Conclusion and Future Work

This work has addressed a critical gap in the study of the Linear Ordering Problem (LOP) by modernizing its benchmarking infrastructure and expanding its algorithmic objectives.

By introducing the EXIOBASE benchmark suite, derived from a widely used global multi-regional input-output database, this work provides a contemporary testbed that captures the structural complexity of modern economies from 1995 to 2022. The newly introduced instances offer a more representative evaluation environment for modern algorithms than traditional benchmarks, which are based on macroeconomic data that are now more than half a century old. Initial experimental results on the new instances are provided using two recent state-of-the-art LOP algorithms.

Furthermore, this work formalizes the Multi-Solution LOP (MS-LOP), shifting the focus from the search for a single optimum to the identification of diverse sets of high-quality solutions. The MS-LOP framework unifies three existing approaches to multi-solution optimization: one tailored to the LOP but less suitable for non-exact methods such as metaheuristics, and the two tasks of multimodal optimization and evolutionary diversity optimization that have not previously been applied to the LOP. The introduction of rigorous metrics for both quality and diversity, alongside the lexicographic prioritization of these objectives, provides a comprehensive framework for future research in multimodal and diversity-aware combinatorial optimization.

To address the MS-LOP, we proposed an algorithmic scheme that can be readily integrated into existing (single-solution) metaheuristics to identify diverse sets of high quality solutions. Specifically, our approach has been incorporated into CD-RVNS and MA-EDM, two recent state-of-the-art LOP algorithms—one trajectory-based and the other population-based. Initial experimental results for the MS-LOP are then presented using the proposed methodology and considering the `EXIOBASE` benchmark suite.

Finally, a promising avenue for future research is the development of algorithms specifically designed for the MS-LOP, rather than adaptations of existing (single-solution) LOP methods.

Acknowledgments. Valentino Santucci and Fabrizio Fagiolo have been partially supported by the research funds "Università per Stranieri di Perugia – Finanziamento Dipartimentale alla Ricerca per Progetti di Ricerca di Ateneo".

References

1. Ahrari, A., Fieldsend, J., Preuss, M., Li, X., Epitropakis, M.: Experimental setup for gecco 2024 competition on benchmarking niching methods for multimodal optimization. In: Proceedings of the Genetic and Evolutionary Computation Conference Companion, pp. 5–6 (2024)
2. Ali, A.B.: Comprehensive learning TLBO with recursive precedence-based solution construction and multilevel local search for the linear ordering problem. Expert Syst. Appl. **238**, 122315 (2024)
3. Anderson, P.E., Chartier, T.P., Langville, A.N., Pedings-Behling, K.E.: Fairness and the set of optimal rankings for the linear ordering problem. Optim. Eng. **23**(3), 1289–1317 (2022)
4. Ando, K., Sukegawa, N., Takagi, S.: Strong condorcet criterion for the linear ordering problem. J. Oper. Res. Soc. Jpn. **65**(2), 67–75 (2022)
5. Baioletti, M., Milani, A., Santucci, V.: Linear ordering optimization with a combinatorial differential evolution. In: 2015 IEEE International Conference on Systems, Man, and Cybernetics, pp. 2135–2140. IEEE (2015)
6. Baioletti, M., Milani, A., Santucci, V.: A new precedence-based ant colony optimization for permutation problems. In: Asia-Pacific Conference on Simulated Evolution and Learning, pp. 960–971. Springer, Cham (2017)
7. Benito-Marimon, M., Martí, R., Martínez-Gavara, A.: Multiple linear ordering problem. Eur. J. Oper. Res. (2025)
8. Bossek, J., Neumann, F.: Evolutionary diversity optimization and the minimum spanning tree problem. In: Proceedings of the Genetic and Evolutionary Computation Conference, pp. 198–206 (2021)
9. Cameron, T.R., Charmot, S., Pulaj, J.: On the linear ordering problem and the rankability of data. Found. Data Sci. **3**(2), 133–149 (2021)
10. Campos, V., Glover, F., Laguna, M., Martí, R.: An experimental evaluation of a scatter search for the linear ordering problem. J. Global Optim. **21**(4), 397–414 (2001)

11. Ceberio, J., Mendiburu, A., Lozano, J.A.: The linear ordering problem revisited. Eur. J. Oper. Res. **241**(3), 686–696 (2015)
12. Ceberio, J., Santucci, V.: Model-based gradient search for permutation problems. ACM Trans. Evol. Learn. Optim. **3**(4), 1–35 (2023)
13. Chiarini, B., Chaovalitwongse, W., Pardalos, P.M.: A new algorithm for the triangulation of input-output tables in economics. In: Supply chain and finance, pp. 253–272. World Scientific (2004)
14. Christof, T., Reinelt, G.: Combinatorial optimization and small polytopes. TOP **4**(1), 1–53 (1996)
15. Do, A., Guo, M., Neumann, A., Neumann, F.: Analysis of evolutionary diversity optimization for permutation problems. ACM Trans. Evol. Learn. **2**(3), 1–27 (2022)
16. Droste, S., Jansen, T., Wegener, I.: Optimization with randomized search heuristics–the (a) nfl theorem, realistic scenarios, and difficult functions. Theoret. Comput. Sci. **287**(1), 131–144 (2002)
17. Elorza, A., Hernando, L., Lozano, J.A.: Characterizing permutation-based combinatorial optimization problems in Fourier space. Evol. Comput. **31**(3), 163–199 (2023)
18. Fagiolo, F., Baioletti, M., Santucci, V.: Supplementary material for "Linear Ordering Problem: Time for a Change" (2026). https://doi.org/10.5281/zenodo.20554376
19. Garcia, C.G., Pérez-Brito, D., Campos, V., Martí, R.: Variable neighborhood search for the linear ordering problem. Comput. Oper. Res. **33**(12), 3549–3565 (2006)
20. Garey, M.R., Johnson, D.S.: Computers and Intractability: A Guide to the Theory of NP-Completeness (Series of Books in the Mathematical Sciences), W. H. Freeman. first edition edn. (1979)
21. Glover, F., Klastorin, T., Kongman, D.: Optimal weighted ancestry relationships. Manage. Sci. **20**(8), 1190–1193 (1974)
22. Goemans, M.X., Hall, L.A.: The strongest facets of the acyclic subgraph polytope are unknown. In: International Conference on Integer Programming and Combinatorial Optimization, pp. 415–429. Springer, Cham (1996)
23. Grötschel, M., Jünger, M., Reinelt, G.: A cutting plane algorithm for the linear ordering problem. Oper. Res. **32**(6), 1195–1220 (1984)
24. Hernando, L., Mendiburu, A., Lozano, J.A.: Journey to the center of the linear ordering problem. In: Proceedings of the 2020 Genetic and Evolutionary Computation Conference, pp. 201–209 (2020)
25. Kim, A., Skiena, S.: Chapter ordering in novels. In: Proceedings of the 2022 Conference on Empirical Methods in Natural Language Processing, pp. 3838–3848 (2022)
26. Kondo, Y.: Triangulation of input-output tables based on mixed integer programs for inter-temporal and inter-regional comparison of production structures. J. Econ. Struct. **3**(1), 2 (2014)
27. Laguna, M., Marti, R., Campos, V.: Intensification and diversification with elite tabu search solutions for the linear ordering problem. Comput. Oper. Res. **26**(12), 1217–1230 (1999)
28. Leontief, W.: Input-Output Economics. Oxford University Press, Oxford (1986)
29. Lugo, L., Segura, C., Miranda, G.: A diversity-aware memetic algorithm for the linear ordering problem. Memetic Comput. **14**(4), 395–409 (2022)
30. Ma, Y., Santucci, V., Witt, C.: Theoretical and empirical analysis of lehmer codes to search permutation spaces with evolutionary algorithms. In: Proceedings of the AAAI Conference on Artificial Intelligence, vol. 40, no. 43, pp. 37099–37107 (2026). https://doi.org/10.1609/aaai.v40i43.41039

31. Marti, R., Reinelt, G.: Exact and Heuristic Methods in Combinatorial Optimization, vol. 175, Springer, Cham (2022)
32. Martí, R., Reinelt, G., Duarte, A.: A benchmark library and a comparison of heuristic methods for the linear ordering problem. Comput. Optim. Appl. **51**(3), 1297–1317 (2012)
33. Mishra, N.K., Singh, P.K.: Linear ordering problem based classifier chain using genetic algorithm for multi-label classification. Appl. Soft Comput. **117**, 108395 (2022)
34. Mishra, S., Sikdar, K.: On approximability of linear ordering and related np-optimization problems on graphs. Discret. Appl. Math. **136**(2–3), 249–269 (2004)
35. Mitchell, J.E., Borchers, B.: Solving linear ordering problems with a combined interior point/simplex cutting plane algorithm. In: High Performance Optimization, pp. 349–366. Springer, Cham (2000)
36. Pérez Lugo, L.J., Segura, C., Miranda, G.: Future trends in the design of memetic algorithms: the case of the linear ordering problem. Neural Comput. Appl. **37**(18), 12471–12485 (2025)
37. Preuss, M.: Multimodal Optimization by Means Of Evolutionary Algorithms. Springer, Cham (2015)
38. Preuss, M., Epitropakis, M., Li, X., Fieldsend, J.E.: Multimodal optimization: Formulation, heuristics, and a decade of advances. In: Metaheuristics for Finding Multiple Solutions, pp. 1–26. Springer, Cham (2021)
39. Preuss, M., Wessing, S.: Measuring multimodal optimization solution sets with a view to multiobjective techniques. In: Emmerich, M., et al. (eds.) EVOLVE–A Bridge between Probability, Set Oriented Numerics, and Evolutionary Computation IV, vol. 227, pp. 123–137. Springer, Cham (2013)
40. Qian, Y., Lin, J., Li, D., Hu, H.: Block-insertion-based algorithms for the linear ordering problem. Comput. Oper. Res. **115**, 104861 (2020)
41. Sakuraba, C.S., Yagiura, M.: Efficient local search algorithms for the linear ordering problem. Int. Trans. Oper. Res. **17**(6), 711–737 (2010)
42. Santucci, V., Baioletti, M., Tomassini, M.: Optimization through iterative smooth morphological transformations. In: Proceedings of the Genetic and Evolutionary Computation Conference, pp. 241–248 (2024)
43. Santucci, V., Ceberio, J.: Using pairwise precedences for solving the linear ordering problem. Appl. Soft Comput. **87**, 105998 (2020)
44. Schiavinotto, T., Stützle, T.: The linear ordering problem: instances, search space analysis and algorithms. J. Math. Model. Algorithms **3**(4), 367–402 (2004)
45. Schiavinotto, T., Stützle, T.: A review of metrics on permutations for search landscape analysis. Comput. Oper. Res. **34**(10), 3143–3153 (2007)
46. Sevaux, M., Sörensen, K., et al.: Permutation distance measures for memetic algorithms with population management. In: Proceedings of 6th Metaheuristics International Conference (MIC'05), pp. 832–838 (2005)
47. Solow, A.R., Polasky, S.: Measuring biological diversity. Environ. Ecol. Stat. **1**(2), 95–103 (1994)
48. Song, W., Ahn, C.W.: Order matters: permutation-based prompt optimization for personalized image generation. IEEE Access **13**, 206155–206164 (2025). https://doi.org/10.1109/ACCESS.2025.3638839
49. Stadler, K., et al.: Exiobase 3: developing a time series of detailed environmentally extended multi-regional input-output tables. J. Ind. Ecol. **22**(3), 502–515 (2018)
50. Stadler, K., et al.: Exiobase 3 (2025). https://doi.org/10.5281/zenodo.15689391

51. Tromble, R., Eisner, J.: Learning linear ordering problems for better translation. In: Proceedings of the 2009 Conference on Empirical Methods in Natural Language Processing, pp. 1007–1016 (2009)
52. Ulrich, T., Thiele, L.: Maximizing population diversity in single-objective optimization. In: Proceedings of the 13th Annual Conference on Genetic and Evolutionary Computation, pp. 641–648 (2011)
53. Zitzler, E., Laumanns, M., Thiele, L.: Spea 2: improving the strength pareto evolutionary algorithm. TIK Rep. **103** (2001)

Parallel Local Search for MaxSAT with Solutions and Score Functions Co-evolving

Mengchuan Zou[1], Peng Lin[1,2], Yi Chu[1], and Shaowei Cai[1,2](✉)

[1] Key Laboratory of System Software, Institute of Software, Chinese Academy of Sciences, Beijing, China
{zoumc,linpeng,caisw}@ios.ac.cn, chuyi2020@iscas.ac.cn
[2] School of Computer Science and Technology, University of Chinese Academy of Sciences, Beijing, China

Abstract. The Weighted Partial MaxSAT (WPMS) problem is a classical problem that generalizes the fundamental Satisfiability (SAT) problem and has many applications. Local search is an important class of algorithms for computing high-quality solutions to combinatorial optimization problems. Although there has been much progress in serial local search algorithms for the WPMS problem, research on parallel local search algorithms for WPMS remains to be explored. We develop a parallel local search framework for WPMS to leverage the power of parallelism with co-evolving solutions and score functions to improve search efficiency. Based on this parallel framework, we develop the first parallel local search WPMS solver, ParMaxSAT, by plugging in the efficient serial WPMS local search algorithm NuWLS. Experiments demonstrate the efficiency of our solver and the effectiveness of our techniques.

Keywords: Maximum Satisfiability · Local Search · Evolution Process

1 Introduction

The Weighted Partial MaxSAT (WPMS) problem is a generalization of the Boolean Satisfiability (SAT) problem, a fundamental problem in computer science, by differentiating hard and soft clauses and incorporating an optimization objective. The classic SAT problem takes a set of Boolean variables and a propositional formula in conjunctive normal form (CNF) as inputs and decides whether there is an assignment of truth values (true or false) to the Boolean variables that makes the CNF formula evaluate to true.

As an extension, the WPMS problem involves soft and hard clauses in a CNF formula and has weights associated with clauses. The hard clauses are those required to be satisfied, while the soft clauses represent clauses expected to be satisfied. The aim of the WPMS problem is to maximize the weights of all satisfied soft clauses while keeping all hard clauses satisfied.

G. Iacca et al. (Eds.): PPSN 2026, LNCS 16985, pp. 417–432, 2027.
https://doi.org/10.1007/978-3-032-36214-8_26

Like other types of general-form optimization problems, such as SAT, Integer Linear Programming (ILP), and Pseudo-Boolean Optimization (PBO), the WPMS problem is usually used to describe and solve concrete combinatorial optimization problems. Applications of WPMS arise in different domains, ranging from classical combinatorial optimization problems, such as time-tabling [2], staff scheduling [13], planning [15], to recent problems of data analysis [5], machine learning [3,14], and so on. In the above applications, the application problems could be encoded into WPMS format and then solved by an WPMS solver.

The algorithm design of WPMS, like other general-form optimization problems, is thus different and presents distinctive challenges compared to specific optimization problems: (1) the solving algorithm must address general-form constraints and variables, rather than specific meaning variables; (2) the problem structure and physical meaning are usually unknown to the solving algorithm, which brings hardness to the algorithm design for general-form optimization problems, compared to concrete optimization problems.

Algorithms for solving the WPMS can be categorized as complete or incomplete methods. Complete methods compute the optimal assignment and prove its optimality, whereas incomplete algorithms aim to find high-quality assignments without proving optimality. Since complete algorithms often encounter hardness when dealing with large-scale instances, incomplete algorithms play an important role in solving WPMS. The local search is an important class of incomplete algorithms. Due to its execution efficiency, local search usually yields high-quality solutions quickly and has become a fundamental method for solving WPMS, serving as an end-to-end solver or a component of other complex solvers.

In recent years, multi-core structures have made remarkable progress in the basic infrastructure of computing. It is an important topic to study what parallel computing could bring to solving WPMS. However, there are not many works on this topic; the state-of-the-art parallel WPMS solver is PWBO [20,21], which is a complete solver and a hybrid of different algorithms. For parallel local search for other general-form optimization problems, there have been works on parallel local search for the ILP [18] and on the PBO [7] problem. As far as we know, there is no prior work on parallel local search for WPMS.

In this work, we focus on the algorithmic design of parallel local search for WPMS, aiming to develop an algorithm that leverages parallelism to enhance search efficiency while addressing key design challenges, particularly the reduction of overlapping searches—a common issue in parallel local search algorithms.

1.1 Related Works

Over the years, important progress has been made in the study of classical complete and incomplete methods. Representative complete methods include core-guided search [1,4], SAT-based methods [12,16,19,22,23], and a recent branch-and-bound method [17]. Incomplete algorithms aim to find high-quality assignments in a short time, which does not have to be optimal. Recent years have witnessed significant progress in Stochastic Local Search (SLS) incomplete algorithms for WPMS [6,10,11,25,26]. One of the state-of-the-art SLS algorithms is NuWLS [10]. Moreover, hybrid algorithms that combine the local search and the complete solvers are usually the most performant in MaxSAT Evaluation, with NuWLS-c [8,9] being the champion of the incomplete track in MaxSAT Evaluations 2022 and 2023, and an efficient incomplete solver DT-HyWalk [24], etc.

As for parallel algorithms for WPMS, the state-of-the-art parallel WPMS solver is PWBO [20,21]. It is a complete solver based on a portfolio strategy that combines unsatisfiability-based algorithms and linear search algorithms with other strategies, such as learned-clause sharing, space splitting, and heuristics. As for parallel local search, we have not found any studies on WPMS, but there are works on parallel local search for ILP [18] and the PBO [7].

The most relevant work of our paper is the one on parallel local search for Pseudo-Boolean Optimization (PBO) [7] since the PBO problem also handles Boolean variables. However, the nature of constraints and optimization objectives of WPMS is distinct from PBO: while the optimization objective and constraints are independent in PBO, they are relevant in the WPMS problem since, in a WPMS problem, the objective is defined to maximize the weighted sum of satisfied clauses. Thus, the parallel local search for WPMS should take into account the nature of the WPMS problem.

In our work, we propose a parallel local search algorithm for WPMS with co-evolving solutions and score functions. The idea of the algorithm is to execute threads with different score functions and make solutions and score functions co-evolving, meaning that solutions and score functions are considered together to generate new solutions and score functions to continue the search.

1.2 Our Contributions

We propose a WPMS solver, ParMaxSAT, with a co-evolving solution-and-score-function scheme, implemented via a pool that collects high-quality solutions and information about the score functions that produce them. We define a new family of ratio-based score functions for adjusting the form of a score function and generating new ones. A mechanism is designed to evaluate the solutions and score functions, where score functions are evaluated by the quality of the solutions they produce. Based on the mechanism, selected solutions and score ratios are stored and used to generate new solutions and score functions, which are then used to help threads trapped in local optima redirect their search while maintaining good diversity of search paths among parallel search threads. We implement our ParMaxSAT solver based on the parallel framework with NuWLS, an efficient local search algorithm for WPMS, as a subroutine. Experiments are conducted to evaluate our ParMaxSAT and our proposed techniques.

Our main contributions are as follows:

1. We propose a new parallel local search method for WPMS by co-evolving the solutions and score functions among threads.

2. For adjusting the form of score functions, we define a new form of score functions, i.e. the ratio-based score function family, that could generate new score functions while keeping the form of score functions of an existing serial local search algorithm.
3. For evolving solutions and score functions, we design a new evaluation and selection mechanism that considers solutions and score functions together, to generate new solutions and score functions for WPMS.
4. Based on the above ideas, we develop the ParMaxSAT solver, based on a serial local search algorithm NuWLS. As far as we know, our ParMaxSAT solver is the first parallel local search solver for WPMS. Experimental results show that our algorithm outperforms the state-of-the-art parallel WPMS solver PWBO, the state-of-the-art parallel local search PBO solver ParLS-PBO, and the diversity-based parallel local search WPMS algorithm.

2 Preliminaries

2.1 The Weighted Partial MaxSAT Problem

We first introduce the formal description of the WPMS problem. Consider a set of n boolean variables $V = \{x_1, ..., x_n\}$ and a set of $2n$ **literals** $L = \{x_1, \neg x_1, ..., x_n, \neg x_n\}$ corresponding to these n variables, a **clause** is defined as a disjunction of some literals, e.g. $c_i = l_{i_1} \vee ... \vee l_{i_k}$, where k is the length of the clause. The CNF formula F is defined as a conjunction of clauses, e.g. $F = c_1 \wedge ... \wedge c_m$. A **complete assignment** (or a solution) is a mapping that maps every variable of a CNF formula to a truth value (true or false). For an assignment α, we say a clause is satisfied if its value is evaluated to be true (at least one of its literals is true), otherwise, we say the clause is unsatisfied.

We call a CNF formula F **partial** if its clauses are classified into **hard** ones and **soft** ones. When each soft clause is associated with a positive integer as a weight, the objective could be defined as maximizing the total weights of satisfied soft clauses while keeping all hard clauses satisfied. This is called the Weighted Partial MaxSAT problem and is denoted as WPMS for short. For a partial CNF formula F, and α an assignment of F, we say that α is feasible if α makes all hard clauses in F satisfied, and we call α a feasible assignment (or feasible solution). Here, we adopt the convention of treating the problem as a minimization of the total weight of unsatisfied clauses, following the descriptions of MaxSAT Evaluations and MaxSAT solvers. It is easy to check the equivalence of the maximization description.

In this paper, for simplicity, we denote the input formula as F, and we use C to denote the set of all clauses in F. The weight of each clause in the input is denoted as $w_{cost}(c)$. A WPMS solver takes an input of the CNF and computes the optimal solution with its corresponding *cost* value.

2.2 The Local Search Algorithm

The local search framework is of great importance in solving combinatorial optimization problems. The local search algorithms start with an initial solution, and then iteratively apply modifications to the current solution to obtain a neighboring solution, recording the best-known solution. For a given current solution, an **operator** defines how a variable is modified. When an operator is specified with a variable to be modified, we get an **operation**. An operation generates a new solution from a given solution.

In WPMS, for a given assignment, numerous operations can be candidate operations. The **score function** is a measure to evaluate the impacts of operations, based on which the local search algorithm will select an operation to apply for obtaining a new assignment. Thus, the score function decides the strategy for selecting a variable to be modified and is a crucial component of a local search algorithm.

In local search algorithms of WPMS, a typical form of score function is to compute a weighted sum of changes of satisfaction states of clauses: when an operation *op* turns an unsatisfied clause c into satisfied, it incurs a reward, and if *op* turns a satisfied clause c into unsatisfied, it incurs a penalty. The score function is the sum of all rewards and penalties of an operator *op* over all clauses. In our work, we construct a family of score functions based on the typical form, to realize our parallel collaborative score function.

3 Basics of Our Algorithm

3.1 Main Ideas: Co-evolution of Solutions and Score Functions

In parallel local search, the question of how to make use of information from different threads while increasing the diversity of their search trajectories is crucial to algorithm design.

Collecting high-quality solutions of different threads and generating new solutions may help the parallel local search. Since solutions found by different threads are heterogeneous, providing high-quality solutions to local search procedures trapped in local optimums may help them change the search direction and jump to different search regions, increasing the chance of finding a better solution. However, considering only elite solutions in parallel local search could bring problems with the diversity of parallel local searches: if only solutions with good objective values are recorded, the solutions in the pool could tend to be similar, increasing the possibility that different threads search in similar regions, reducing the efficiency of parallel search.

The score function is another important element of a local search algorithm. The score function is used to evaluate a candidate operation and decide whether to apply it to obtain a new neighboring solution. Assuming the current assignment is α, let *op* be a neighbor operator and $op(\alpha)$ be an operation on α. We denote $f_{op,c}(\alpha)$ as the score associated with clause c induced by applying *op* with current assignment α. Then the score function of a serial local search algorithm with operator *op*, for which the score with a clause is provided by $f_{op,c}(\alpha)$, could be denoted as $f_{op}(\alpha)$.

In a serial local search algorithm, it is usual that a fixed form of score function is used, while the parallel setting offers the possibility to use different forms of score functions in different threads. Compared to serial settings, using different score functions across threads can increase the diversity of search strategies, thereby enlarging the search space within the same time period, and can make score functions adaptive to the collective behavior of different threads, increasing the algorithm's adaptability.

From a given serial local search algorithm, we could extend its score function into a general family and generate new score functions. For this purpose, we define a family of ratio-based score functions that can take a score function for a WPMS local search algorithm and generate new ones.

Definition 1. *Given a score function $f_{op}(\alpha)$, the* ***ratio-based score function family*** *is defined as $\mathcal{F} = \{\sum_{c \in C} r_c \cdot f_{op,c}(\alpha) | \boldsymbol{r} \in \mathbb{R}^m\}$. Here r_c is called the* ***score ratio*** *(or ratio for short) of clause c, and $\boldsymbol{r}$ is the score ratio vector.*

It is easy to see that the new score function family generalizes the score function f by adding a coefficient factor to the original score of each clause. The intuition behind this setting is that different score ratios lead to different priorities of clauses in the local search process.

Consider three different settings of score ratios as examples:

1. If all score ratios are set to be 1, then the new score function is the same as the one in the original algorithm, and the algorithm with the new score function is equivalent to executing the original algorithm.
2. If all ratios of soft clauses are set to 0, then the score function does not count the influence of operations on soft clauses, and the local search algorithm with the new score function is targeted at finding feasible solutions.
3. If the ratio of a soft clause is set to be greater than 1 while keeping others as 1, then the priority of satisfaction of this soft clause is increased, and the search direction is guided to find an assignment satisfying it.

In the examples above, different score-ratio configurations yield different effects on search directions. From Definition 1, we can also observe that, taking any serial local search algorithm for WPMS of which scores are computed for each clause, we can build the ratio-based score function family and generate new score functions by modifying ratios.

During the execution, high-quality solutions and the score function that generates the high-quality solutions are collected, then an evaluate-and-update mechanism is used to realize the evolution of solutions and score functions. The new solutions are generated based on the objective quality and distances of solutions collected, while the new score functions are generated based on the solutions found by executed score functions. In this way, we intend to use information from successful search threads while promoting diversity of searches across different threads.

3.2 The Parallel Framework of ParMaxSAT Solver

As shown in Fig. 1, our ParMaxSAT solver takes a WPMS problem instance as the input and will output the highest-quality feasible solution found through the search process. Our ParMaxSAT solver adopts a portfolio-based framework that contains a **master** thread and some **worker** threads. The master thread reads the instance's input, performs initialization, and then distributes information to initiate and launch worker threads. Each worker thread executes the serial local search algorithm using the given parameters and the initial score function. A **shared pool** is used to store shared data as solutions and ratios, depending on which the solution sharing and score function updating mechanisms are developed. When the predefined time limit is reached, the master thread outputs the best solution found by all worker threads.

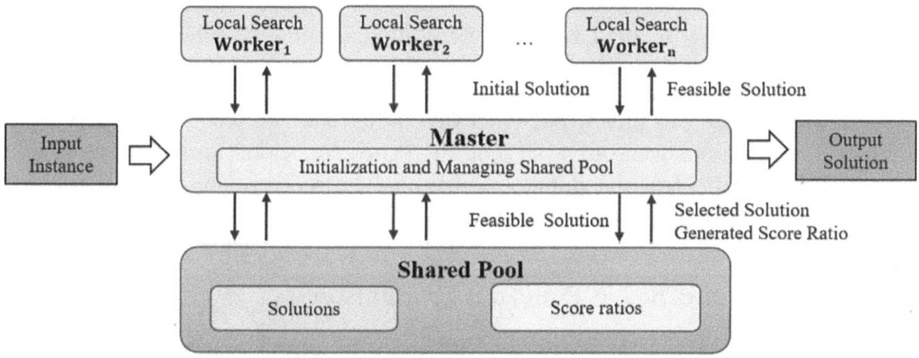

Fig. 1. Parallel Framework for Solving WPMS.

During execution, a worker thread can send and require data from the shared pool. When a worker thread finds a new feasible solution, it sends the solution and the score ratios used to find it to the shared pool. The master thread performs an evaluation mechanism to decide whether to accept the data. If it accepts, the shared pool will be updated. When a worker thread cannot find a better solution for some given time (we say the thread is blocked), it could request a solution or a vector of score ratios from the shared pool to update its search process. The initialization, evaluation, and updating processes are the core components of our algorithm. We present in the next section the detailed algorithmic components of our solver.

4 Full Algorithm with Co-evolving Solutions and Score Functions

We present the implementation of our algorithm here, including the initialization method, evaluation rules, and the updating process. Through these routines, our main ideas are realized.

4.1 Initialization

At the beginning of the algorithm, the initialization is performed by the master thread and includes two parts:

1. Random diversification: we set different random seeds and use random perturbations to the internal parameters of the base solver (NuWLS).
2. Initialization of score ratios: (1) in 1 thread, the score ratios of all clauses are set to 1, i.e. running the original serial algorithm (NuWLS); (2) in 1 thread, the score ratios of all soft clauses are set to 0 and ratios of all hard clauses are set to be 1, i.e. targeting at searching for feasible solutions; (3) for other threads, all score ratios are initialized to 1 and then perturbed with a probability of p_p to be a random value in $[1-\varepsilon, 1+\varepsilon]$, where p_p and ε are parameters of our algorithm

The score ratios are initialized with diversified values, with small perturbations to maintain both the diversity of the score functions and their closeness to the original form. In subsequent executions, the score ratios will be changed and will differ from the initial configuration.

4.2 Solutions and Score Function Evaluation

The rules for evaluating the solutions and score functions are important to our algorithm. Since the performance of a score function is strongly related to how good the solutions it can produce are, we take the solution and the corresponding score ratio to represent each solution as a pair, and define evaluation measures on these pairs.

The data that a worker sends to the shared pool can be represented by a pair $Q = (\boldsymbol{s}, \boldsymbol{r})$, where $\boldsymbol{s}$ is a solution, $\boldsymbol{r}$ is the score ratio vector to find $\boldsymbol{s}$. The worker threads and the master threads follow the rules below to send data to the shared pool:

- In a worker thread, when a new feasible solution $\boldsymbol{s}$ is found with score ratios $\boldsymbol{r}$, the worker thread will send the pair $Q = (\boldsymbol{s}, \boldsymbol{r})$ to the master thread.
- When data is received, the master thread evaluates and decides whether to accept it and update the shared pool.

The evaluation needs to consider two factors: (1) the quality of objective value and (2) the diversity among all threads. They need to be balanced: if we only measure the objective value of the solution, it may appear that best-found solutions are very similar, and the diversity of data in the shared pool is reduced. Since worker threads use data in the shared pool to guide the search, reducing diversity could restrict the search space and negatively affect the overall performance of parallel searches. Thus, the objective quality and diversity of solutions should be considered. This idea has also been adopted in [7], but our diversity measure for WPMS is new and based on the satisfiability state of clauses, unlike that in [7]. Our measurements are described as follows.

Let $\boldsymbol{Q}_{pool}$ be the multi-set of solution-ratio pairs in current solution pool, when a pair $Q' = (\boldsymbol{s}', \boldsymbol{r}')$ is received, we consider the multi-set $\boldsymbol{Q}_{temp}$ that contains all pairs in the pool and the new received pair: $\boldsymbol{Q}_{temp} = Q' \cup \boldsymbol{Q}_{pool}$. Then we define the quality and diversity ranks to evaluate the newly received pair.

Definition 2. *For every $Q = (\boldsymbol{s}, \boldsymbol{r}) \in \boldsymbol{Q}_{temp}$, we define $R_{obj}(Q)$ the* ***quality rank*** *of Q to be the rank of $cost(\boldsymbol{s})$ among all pairs in $\boldsymbol{Q}_{temp}$, by increasing order.*

It is easy to see that the smaller $R_{obj}(Q)$ is, the better the objective value of Q is among the solutions in the shared pool. For measuring the diversity, we denote $z_c(Q)$ as the satisfiability state of clause c by solution $\boldsymbol{s}$, i.e. $z_c(Q) = 1$ if clause c is satisfied with solution $\boldsymbol{s}$, and $z_c(Q) = 0$ otherwise. Let the function $d(Q)$ to be the sum of hamming distances of $\boldsymbol{z}$ to the others, i.e. $d(Q) = \sum_{Q'' \in \boldsymbol{Q}_{temp}} \sum_{c \in C} \mathbf{1}_{\{z_c(Q) \neq z_c(Q'')\}}$, then we measure the diversity by:

Definition 3. *For every $Q = (\boldsymbol{s}, \boldsymbol{r}) \in \boldsymbol{Q}_{temp}$, we define the* ***diversity rank*** *of Q as the rank of $d(Q)$ among all pairs in $\boldsymbol{Q}_{temp}$ by decreasing order.*

Similarly, the smaller $R_d(Q)$ is, the clauses satisfied by Q are more different than other solutions in $\boldsymbol{Q}_{temp}$, then the Q is more diversified with $\boldsymbol{Q}_{pool}$. Taking both of the two factors into account, we define the fitness rank of Q:

Definition 4. *For every $Q = (\boldsymbol{s}, \boldsymbol{r}) \in \boldsymbol{Q}_{temp}$, we define the* ***fitness rank*** *of Q as $R(Q) = R_{obj}(Q) + \eta \cdot R_d(Q)$ where η is a parameter that controls the trade-off of the two ranks in the evaluation.*

The fitness rank is used in the accepting rules to define whether a received pair Q' could be accepted to the pool, depending on $R(Q)$:

Definition 5. *Let n_{pool} be a parameter as the limit of the number of pairs to be stored in the shared pool. The* ***accepting rules*** *of our algorithm are as follows:*

- *When a pair $Q' = (\boldsymbol{s}', \boldsymbol{r}')$ is received, it is directly admitted if the capacity limit will not be exceeded.*
- *When the number of solutions reaches n_{pool}, we compute $R(Q)$ for all $Q \in \boldsymbol{Q}_{temp}$, if $R(Q')$ is better than some Q'' in $\boldsymbol{Q}_{pool}$, we then add Q to the pool and replacing the worst Q''; otherwise Q' will not be accepted.*

Every time a solution-ratio pair is received, the master thread checks the pool's capacity limit and performs the evaluation. Since low-fitness pairs will be deleted from the pool, the acceptance rules keep solutions and ratio vectors in the pool, ensuring high quality and diversity.

4.3 Solutions and Score Functions Updating

The update process is launched when a worker thread cannot make improvements for enough steps. According to the situation, its current solution or score function will be updated:

- When the best solution of a worker thread has not been improved for T_{sol} steps, the thread requests a solution, and the master thread returns a randomly selected solution from the pool. When acquiring such a sol, the worker thread continues the local search with sol. Here, the T_{sol} is a parameter.
- When the best solution of a worker thread has not been improved for T_{ratio} steps, where the T_{ratio} is a parameter, the thread requests and the master thread returns a randomly selected vector of ratios $\boldsymbol{r}$. When such a $\boldsymbol{r}$ is acquired, a modification on $\boldsymbol{r}$ will be performed, and our algorithm will update the score function in the worker thread, then the worker thread continues the local search.

The reason not to take a ratio vector in the shared pool directly to update the score function is that the score ratio vector in the pool must be ratios already utilized by some thread, taking the same ratio vector brings repetitiveness and reduces diversity. Thus, we add a small modification for the diversity.

Let $\boldsymbol{s}$ be the solution associated with $\boldsymbol{r}$, let $N_{unsat}(c)$ be the number of times that a solution enters the pool that makes c unsatisfied. Combining $N_{unsat}(c)$ with the cost weight in the input, we define a function $H(c) = w_{cost}(c) \cdot N_{unsat}(c)$. We select the clause(s) with the largest $H(c)$, then increase its ratio in $\boldsymbol{r}''$ by $r''_c = r''_c + \varepsilon$, where ε is a parameter. If there are more than 1 clauses with the same largest $H()$ value, we modify all their ratios in $\boldsymbol{r}''$. With this modification, the algorithm maintains diversity and increases the priority of clauses that are hard to satisfy in high-quality solutions across different threads. After the modification, the worker updates its score function by replacing its score ratios with the modified ratio vector. Moreover, in our framework, the T_{ratio} is set to be much greater than T_{sol} because we would like score ratios to be updated based on longer-term performance.

During the above process, the solutions are evolved: the shared pool maintains a set of high-quality solutions, and when a worker thread cannot find a better solution for a long time period, it requests a solution from the pool and continues the search. Since the shared solution could be obtained from another thread, the update helps the worker thread escape the current local optimum and explore other search regions, expanding its search space.

During the above process, score functions are evolved with solutions found: different score functions are executed, and those that yield good solutions are recorded, helping threads obtain a good score function. By modifying score functions, we can generate new ones, allowing our algorithm to run with more score functions and learn good ones.

By combining the components, we leverage parallelism to expand the search space and adaptively adjust search strategies collectively.

5 Experiments

We conduct experiments to assess the performance of our solver and proposed techniques. Firstly, to evaluate the efficiency of our algorithm, we compare our ParMaxSAT solver with the state-of-the-art parallel WPMS solver PWBO. Further, since our solver uses the serial NuWLS solver as a subroutine, we compare our algorithm with the diversity-based parallel version of NuWLS to validate the effectiveness of our new method. We also present a comparison between our solver and the state-of-the-art parallel local solver for the PBO problem, i.e., the ParLS-PBO solver; the results are in the Supplementary Material[1]. For all competing solvers, we evaluate their ability to find feasible solutions and their ability to find high-quality solutions within the time limit.

5.1 Solvers in Evaluation

For the performance evaluations, we compare three solvers:

(1) Our **ParMaxSAT** solver, which is implemented by the parallel framework presented in this paper with the serial NuWLS algorithm as a sub-solver on each thread. Our solver is implemented in C++ and compiled with g++. The source code and detailed results are included in the Supplementary Material.
(2) **PWBO** [20,21]: The state-of-the-art parallel Weighted Partial MaxSAT solver, version 2.2 [2].
(3) **Div-NuWLS**: A diversified parallel variant of the SOTA serial WPMS local search solver NuWLS [10]. It runs NuWLS on different threads with different random seeds and returns the best solution found by all threads, keeping all other implementations unchanged.

5.2 Experimental Configurations

The configuration of the experiments is as follows, including the running environments, benchmarks, execution settings, and evaluation metrics.

Running Environments. All experiments are carried out on a server with two AMD EPYC 7763 CPUs @ 2.45Ghz of 128 physical cores and 1TB memory running the operating system Ubuntu 20.04 LTS (64bit), each solver is executed with the cutoff time of 300 s.

Benchmarks. We take the instances from the MaxSAT Evaluations (MSEs) competition[3] of years 2021 to 2023 to evaluate our algorithm. There are six benchmarks in total: the PMS (unweighted) and WPMS (weighted) benchmarks, derived from incomplete tracks of the MaxSAT Evaluations (MSEs) for the years 2021, 2022, and 2023, for a total of 1021 instances.

[1] https://github.com/imzmc/ParMaxSAT.
[2] https://sat.inesc-id.pt/pwbo/.
[3] https://maxsat-evaluations.github.io/.

Parallel Setting. All solvers are tested with 4 and 8 thread settings, since the maximum number of threads that PWBO allows is 8. It is worth noting that our algorithm does not have this limit and could be run with a larger number of thread settings.

Parameter Setting. There are 6 parameters in our algorithm: n_{pool} for the pool size, η for the coefficient of the rank of difference, p_p for the probability of perturbation, ε for the distance limit of perturbations used in our algorithm, T_{ratio} for the step of ratio updates, and T_{sol} for the step of solution sharing. The parameters are tuned according to the empirical literature and our preliminary experiments on 20% sampled instances by grid search, and are set as follows: $n_{pool} = 10$, $\eta = 10$, $p_p = 1\%$, $\varepsilon = 0.00001$, $T_{ratio} = 10000000$, and $T_{sol} = 750000$ for all benchmarks.

Evaluation Metrics. The evaluation metrics we adopted are listed below:

- *#feas*. A measure of the ability to find a feasible solution for given instances, we denote *#feas* as the number of instances where a solver successfully finds a feasible solution.
- *#win*. A measure of the ability to find a high-quality feasible solution, meaning the number of instances where a solver finds the best solution among all tested solvers. If multiple solvers find the same best solution, all of those solvers are credited with a *#win* count. If none of the solvers find a solution, no solver counts a #win for that instance.
- *Score*. A well-established measure of average performance, used in the Incomplete Track of MaxSAT competition. For a solver and an instance set, *Score* represents the average value of ratios between the output of this solver and the best-known solution of all solvers. The greater the value of the *Score* is, the better the overall performance of a solver is.

5.3 Comparison with the SOTA Parallel MaxSAT Solver PWBO

Table 1 compares our ParMaxSAT solver with the SOTA parallel WPMS solver PWBO. We measured the *#feas*, *#win*, and *Score* in 4-thread and 8-thread settings.

From Table 1, our ParMaxSAT outperforms PWBO on the *#win* measure in all six test sets under both thread settings, and on the *Score* measure in all test sets except PMS_2021 in the 4-thread setting. For *#feas*, our solver outperforms PWBO in 5/6 test sets. In summary, our solver shows better overall performance than the SOTA parallel WPMS solver PWBO on these benchmarks.

5.4 Comparison with the Diversity-Based Parallel Local Search

We compare our ParMaxSAT against a diversity-based parallelization of the base solver NuWLS to demonstrate the effectiveness of our parallel framework. We

Table 1. Comparison with SOTA solver PWBO

Benchmark	#ins	PWBO			ParMaxSAT		
		$\#feas$	$\#win$	$Score$	$\#feas$	$\#win$	$Score$
		Thread Number = 4					
PMS_2021	155	**133**	90	**0.755**	129	**101**	0.746
PMS_2022	179	152	97	0.733	**155**	**116**	**0.757**
PMS_2023	180	150	99	0.733	**151**	**118**	**0.740**
WPMS_2021	151	127	92	0.721	**128**	**100**	**0.767**
WPMS_2022	197	156	110	0.689	**167**	**133**	**0.766**
WPMS_2023	160	128	84	0.671	**136**	**108**	**0.760**
ALL	1021	846	572	0.717	**866**	**676**	**0.757**
		Thread Number = 8					
PMS_2021	155	**131**	87	0.738	129	**99**	**0.743**
PMS_2022	179	148	91	0.706	**155**	**122**	**0.776**
PMS_2023	180	142	95	0.697	**151**	**122**	**0.761**
WPMS_2021	151	127	90	0.721	**128**	**104**	**0.774**
WPMS_2022	197	152	107	0.669	**167**	**131**	**0.760**
WPMS_2023	160	128	85	0.666	**136**	**110**	**0.767**
ALL	1021	828	555	0.699	**866**	**688**	**0.764**

can see from Table 2 that in all test sets, our ParMaxSAT solver performs better in the $\#feas$, $\#win$, and $Score$ measures. This result shows that our algorithm is more effective compared to the directly diversified parallel version of NuWLS.

5.5 Experiment with Different Number of Threads

To measure the scalability of our solver, we perform tests on PWBO and ParMaxSAT in 4-thread and 8-thread settings. From Fig. 2 we can see that the performance of our solver improves in both $\#win$ and $Score$, while the PWBO solver shows little improvement in $\#win$ and worse $Score$ results when the number of threads increases from 4 to 8.

Further, we tested our algorithm with 16 and 32 threads; since PWBO cannot support more than 8 threads, we did not compare it with PWBO. The results show that our solver's performance continues to improve. From Fig. 3, we can see that the $\#win$ metric steadily increases as the number of threads rises, although the $\#feas$ metric has small fluctuations at 16 and 32 threads. Considering the obvious gap in the $\#win$ metric, our solver still shows good scalability from 4 to 32 threads.

Table 2. Comparison with Div-NuWLS

Benchmark	#ins	Div-NuWLS			ParMaxSAT		
		#feas	#win	*Score*	#feas	#win	*Score*
Thread Number = 4							
PMS_2021	155	121	115	0.775	**129**	**120**	**0.8014**
PMS_2022	179	144	126	0.7841	**155**	**151**	**0.8521**
PMS_2023	180	150	123	0.8019	**151**	**143**	**0.8156**
WPMS_2021	151	125	107	0.7955	**128**	**120**	**0.8177**
WPMS_2022	197	163	140	0.7921	**168**	**161**	**0.8304**
WPMS_2023	160	132	113	0.7829	**136**	**129**	**0.8249**
ALL	1021	835	724	0.7897	**867**	**824**	**0.8253**
Thread Number = 8							
PMS_2021	155	121	115	0.7726	**129**	**123**	**0.8037**
PMS_2022	179	144	128	0.7854	**155**	**148**	**0.8374**
PMS_2023	180	150	129	0.8116	**151**	**142**	**0.8083**
WPMS_2021	151	125	110	0.7976	**128**	**119**	**0.8075**
WPMS_2022	197	163	144	0.7954	**167**	**159**	**0.822**
WPMS_2023	160	132	118	0.7903	**136**	**127**	**0.8184**
ALL	1021	835	744	0.7934	**866**	**818**	**0.8176**

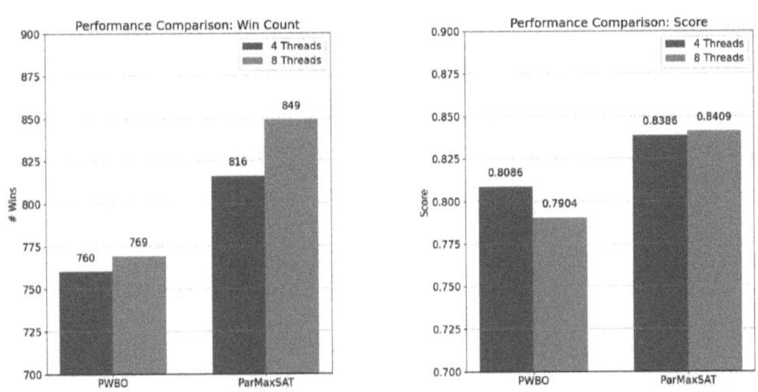

Fig. 2. Scalability: comparison of the *#win* and *Score* of 4-thread and 8-thread versions of PWBO and ParMaxSAT

Benchmark	#ins	Threads	#feas	#win
ALL(1021)	1021	4	866	756
ALL(1021)	1021	8	866	783
ALL(1021)	1021	16	865	801
ALL(1021)	1021	32	867	813

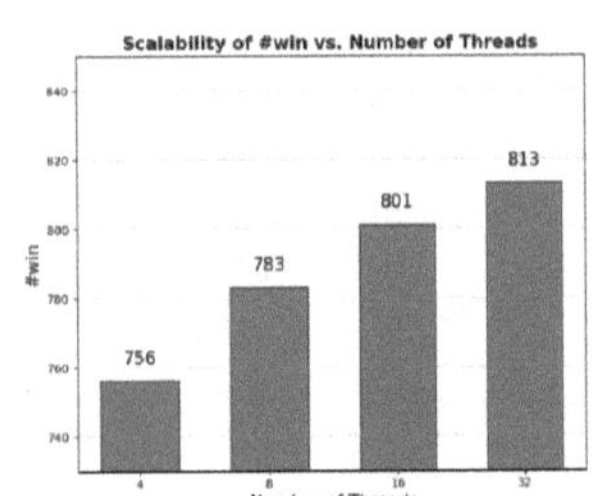

Fig. 3. Scalability: the results of our ParMaxSAT solver from 4 threads to 32 threads

6 Conclusions and Future Work

We develop the first parallel local search solver for Weighted Partial MaxSAT with co-evolving solutions and score functions. The shared pool maintains a set of solutions and score ratios, with an evaluation mechanism that selects high-quality, diverse solutions and score ratios to produce elite solutions. Solutions and score ratios are then modified and updated to generate new solutions and score functions, helping threads change the search routine. Our parallel framework is implemented with the state-of-the-art serial local search algorithm NuWLS as a subroutine. Experimental results confirmed the performance of our solver. In the future, we expect to explore the ways to apply our framework to other serial algorithms and related problems.

Acknowledgements. This research was supported by National Cyber Security-National Science and Technology Major Project 2026ZD1500700, the National Natural Science Foundation of China under Grant 62302492, and the National Natural Science Foundation of China under Grant 62502509.

References

1. Abelson, H., Sussman, G.J., Sussman, J.: Structure and Interpretation of Computer Programs. MIT Press, Cambridge, Massachusetts (1985)
2. Achá, R., Nieuwenhuis, R.: Curriculum-based course timetabling with SAT and MaxSAT. Ann. Oper. Res. **218**, 71–91 (2012)
3. Alos, J., Ansótegui, C., Torres, E.: Interpretable decision trees through MaxSAT. Artif. Intell. Rev. **56**, 8303–8323 (2022)
4. Ansótegui, C., Bonet, M.L., Levy, J.: Solving (weighted) partial MaxSAT through satisfiability testing. In: International Conference on Theory and Applications of Satisfiability Testing, pp. 427–440. Springer (2009)
5. Berg, J., Hyttinen, A., Järvisalo, M.: Applications of MaxSAT in data analysis. In: Pragmatics of SAT (2015)
6. Cai, S., Lei, Z.: Old techniques in new ways: clause weighting, unit propagation and hybridization for maximum satisfiability. Artif. Intell. **287**, 103354 (2020)
7. Chen, Z., Lin, P., Hu, H., Cai, S.: ParLS-PBO: a parallel local search solver for pseudo Boolean optimization. In: 30th International Conference on Principles and Practice of Constraint Programming (CP 2024), pp. 5–1. Schloss Dagstuhl–Leibniz-Zentrum für Informatik (2024)
8. Chu, Y., Cai, S., Lei, Z., He, X.: NuWLS-c: solver description. MaxSAT Eval. **2022**, 28 (2022)
9. Chu, Y., Cai, S., Luo, C.: NuWLS-c-2023: solver description. MaxSAT Eval. **23–24**, 2023 (2023)
10. Chu, Y., Cai, S., Luo, C.: NuWLS: improving local search for (weighted) partial MaxSAT by new weighting techniques. In: Proceedings of the AAAI Conference on Artificial Intelligence, vol. 37, pp. 3915–3923 (2023)
11. Chu, Y., Li, C.-M., Ye, F., Cai, S.: Enhancing MaxSAT local search via a unified soft clause weighting scheme. In: Proceedings of SAT 2024, vol. 305, pp. 8:1–8:18 (2024)

12. Davies, J., Bacchus, F.: Solving MaxSAT by solving a sequence of simpler sat instances. In: International Conference on Principles and Practice of Constraint Programming, pp. 225–239. Springer (2011)
13. Demirović, E., Musliu, N., Winter, F.: modeling and solving staff scheduling with partial weighted MaxSAT. Ann. Oper. Res. **275**, 79–99 (2017)
14. Hu, H., Siala, M., Hebrard, E., Huguet, M.: Learning optimal decision trees with MaxSAT and its integration in AdaBoost (2020)
15. Juma, F., Hsu, E., McIlraith, S.: Preference-based planning via MaxSAT, pp. 109–120 (2012)
16. Koshimura, M., Zhang, T., Fujita, H., Hasegawa, R.: QMaxSAT: a partial max-sat solver. J. Satisfiability Boolean Model. Comput. **8**(1–2), 95–100 (2012)
17. Li, C.-M., Xu, Z., Coll, J., Manyà, F., Habet, D., He, K.: Combining clause learning and branch and bound for MaxSAT. In: 27th International Conference on Principles and Practice of Constraint Programming (CP 2021). Schloss Dagstuhl-Leibniz-Zentrum für Informatik (2021)
18. Lin, P., Cai, S., Zou, M., Lin, J.: New characterizations and efficient local search for general integer linear programming (2023)
19. Martins, R., Joshi, S., Manquinho, V., Lynce, I.: Incremental cardinality constraints for MaxSAT. In: Principles and Practice of Constraint Programming: 20th International Conference, CP 2014, Lyon, France, September 8-12, 2014. Proceedings 20, pp. 531–548. Springer (2014)
20. Martins, R., Manquinho, V., Lynce, I.: Clause sharing in parallel MaxSAT. In: Learning and Intelligent Optimization: 6th International Conference, LION 6, Paris, France, January 16-20, 2012, Revised Selected Papers, pp. 455–460. Springer (2012)
21. Martins, R., Manquinho, V., Lynce, I.: Parallel search for maximum satisfiability. AI Commun. **25**(2), 75–95 (2012)
22. Nadel, A.: Anytime weighted MaxSAT with improved polarity selection and bit-vector optimization. In: 2019 Formal Methods in Computer Aided Design (FMCAD), pp. 193–202. IEEE (2019)
23. Nadel, A.: TT-Open-WBO-Inc-20: an anytime MaxSAT solver entering MSE'20. MaxSAT Eval. **32** (2020)
24. Zheng, J., He, K., Chen, Z., Zhou, J., Li, C.-M.: Decision tree based hybrid walking strategies. MaxSAT Evaluation 2022, pp. 24 (2022)
25. Zheng, J., He, K., Zhou, J.: Farsighted probabilistic sampling based local search for (weighted) partial MaxSAT. In: Proceedings of AAAI, vol. 2023, pp. 4132–4139 (2023)
26. Zheng, J., He, K., Zhou, J., Jin, Y., Li, C.-M., Manya, F.: BandMaxSAT: a local search MaxSAT solver with multi-armed bandit. In: Proceedings of IJCAI, vol. 2022, pp. 1901–1907 (2022)

SAT-Based Search for Minwise Independent Families

Enrico Iurlano(✉) and Günther R. Raidl

Algorithms and Complexity Group, TU Wien, Austria
{eiurlano,raidl}@ac.tuwien.ac.at

Abstract. Optimal minwise independent families of permutations are combinatorial structures designed to realize a large set of highly symmetric monotonicity patterns while using as few permutations as possible. Such families play a central role in efficient Jaccard similarity estimation and related applications in document analysis and machine learning. While practical variants often tolerate slight relaxations of their attributes, we focus on the computational construction of minwise independent families in their pure form. We introduce a constraint-based framework that (i) provides a direct SAT encoding of the underlying combinatorial conditions and (ii) extends a heuristic by Mathon and van Trung to guide the search within this constrained space. Our results show that this hybrid approach is effective in discovering new, provably optimal families and yields substantial speed-ups compared to previous methods. The framework also offers structural insights into these combinatorial objects and supports them with certified optimality proofs.

Keywords: Minwise independence · SAT solving · Heuristic search · Symmetry breaking · Certified optimality

1 Introduction

A family $\mathcal{F}$ of permutations on a finite universe U is (k-restricted) minwise independent, if it ensures that a random permutation π drawn from $\mathcal{F}$ and applied to any $X \subseteq U$ (of cardinality at most k) exhibits perfect fairness in the sense that each element $x^* \in X$ has the same chance of being mapped to the minimum of the values in $\{\pi(x) : x \in X\}$. These families are a crucial ingredient of the well-known `MinHash` algorithm [4], which was proposed in the 1990s to rapidly estimate the similarity between documents in large World Wide Web page collections. Since then, these structures have inspired thousands of subsequent studies in the broader field of big data and machine learning [35]. More precisely, minwise independence is strongly linked to established techniques such as b-bit minwise hashing [22], one-permutation hashing [23], and locality-sensitive hashing [14],

This work was partially supported by the program VGSCO of the Austrian Science Fund (FWF) [10.55776/W1260-N35].

G. Iacca et al. (Eds.): PPSN 2026, LNCS 16985, pp. 433–449, 2027.
https://doi.org/10.1007/978-3-032-36214-8_27

which are used in applications including spam detection [22], dimensionality reduction [35], derandomization [28], and computational geometry [28].

At a technical level, minwise independence is used to accurately approximate the pairwise Jaccard similarity between high-dimensional data based on precomputed *sketches* [5] associated with the data points, i.e., subsets of a large, finite universe U. The Jaccard similarity $r(A, B) := |A \cap B|/|A \cup B|$ of two sets $A, B \subseteq U$ can be estimated by repeatedly subjecting A and B to random permutations of U and computing a simple statistic on the results [5].

The main challenge in this sampling approach is that, for large-scale universes U, it is difficult to draw a permutation of U (even approximately) uniformly at random: The combinatorial explosion of the symmetric permutation group on n elements makes it impossible even for state-of-the-art pseudo random number generators (PRNG) to guarantee full sampling of its elements before reaching the PRNG's period-length [26]; this applies in particular to the PRNG *Mersenne Twister* [26] being the standard[1] of the `C++` programming language. Therefore, Broder et al. [5] introduced minwise independent families of permutations that can be generated with far fewer members than the full symmetric group on U, yet any random permutation drawn from such a smaller family serves as an equivalent replacement for a random permutation. The impact in the field of big data is largely influenced by relaxed and approximate forms of such families, which accept a certain degree of inaccuracy for practical purposes. Often, certain families of practical hash functions are employed as a compromise; see Broder et al. [5] and subsequent work.

Optimality-achieving algorithms or constructions for these families are known only for specific parameter choices [17]. Therefore, providing a general framework for finding and studying provably optimal representatives (at computationally tractable scales) adds valuable insight into their nature. Indeed, constructive approaches for generating minwise independent families [32] and related structures often benefit from knowledge of optimal behavior on smaller scales as they are designed in a recursive [19,32], product-type [7], replicative [34], or extensible [13] manner. Such a tool may also help to disprove or empirically support conjectures on minwise independence that might be inferable by analogy from open conjectures in related fields [21,29].

For related structures [13,24,29] the development of exact computational approaches has been initiated. Notably, in computational (non-)existence proofs, their proposed algorithms exhibit long runtimes even for small parameter values [13,29]. This raises questions about the existence of better performing alternatives. On the other hand, the aforementioned works have the disadvantage that no verification across diverse and independent software platforms for their claims of optimality is available, which would be important to enhance trust.

Our Contribution. We propose a framework to address questions on the existence and structural decomposability of minwise independent families of a postulated size, thereby providing insights on (optimal) such representatives with

[1] https://cplusplus.com/reference/random/mt19937/ (accessed 2026-02-12).

specific characteristics. The framework relies on a reduction to the Boolean satisfiability problem, i.e., a SAT encoding, which is then solved with a modern SAT solver. To manage the high-dimensional search space, our encoding employs problem-specific symmetry breaking whereas to accelerate heuristic solution finding, we adapt a group-theoretic decomposition strategy to our setting; originally, it was proposed for a similar structure [24,29] and implemented as a backtracking algorithm. For parameter instances that allow direct comparison with [29], we observe a dramatic speed-up in terms of computation time by a factor of up to 2.09×10^5. Using our SAT framework, we can also detect non-existence and subsequently certify it with independent software. Furthermore, the framework supports flexible customization and the integration of additional assumptions for further studies. As a byproduct of potentially broader interest, we provide pure-SAT evaluations (without resorting to bitvector arithmetic) of certain permutation attributes arising from symbolic permutation emulation.

The paper is structured as follows. Relevant terminology and related work are reported in Sect. 1.1–1.2. The SAT encoding is derived in Sect. 2 and extended to subgroup-parametrized encodings in Sect. 3. Experimental evaluations are reported in Sect. 4 and concluding remarks in Sect. 5.

1.1 Preliminaries

In the following, we provide the most relevant definitions and properties concerning minwise independence and the methodology we develop later.

Let $[n] := \{1, \dots, n\}$ and denote by S_n the symmetric group of all $n!$ permutations of $[n]$. Let us introduce sets of *semiordered patterns* $\mathsf{SOP}(n, j) := \{(s_1, \dots, s_j) \in [n]^j : s_2 < s_3 < \dots < s_j \text{ and } s_1 \notin \{s_2, \dots, s_j\}\}$ for $1 \leq j \leq n$. In the following we always assume that appearing families are indexed by $[d]$ for some $d \in \mathbb{N}$. We refer to the number d of (not necessarily distinct) members of a family as its size.

Following [5], we say that a finite non-empty family $\mathcal{F} = (\pi_1, \dots, \pi_d)$ of permutations $\pi_i \in S_n$, $i = 1, \dots, d$ is *k-restricted minwise independent* if for each $j \in [k]$, for each $X \subseteq [n]$ with $|X| = j$, and each x^* in X,

$$\Pr\left[\pi(x^*) = \min_{x \in X} \pi(x) \,\middle|\, \pi \text{ is drawn uniformly at random from } \mathcal{F}\right] = \frac{1}{j}. \quad (1)$$

It will be convenient to consider an equivalent replacement for (1) given by

$$|\{i \in [d] : \pi_i(s_1) = \min\{\pi_i(s_1), \pi_i(s_2), \dots, \pi_i(s_j)\}\}| = \frac{d}{j}, \quad (2)$$

for each $j \in [k]$ and each $(s_1, s_2, \dots, s_j) \in \mathsf{SOP}(n, j)$. This implies that d must be a multiple of lcm $([k])$, the least common multiple of the numbers $1, 2, \dots, k$. We call a k-restricted minwise independent family $\mathcal{F} = (\pi_1, \dots, \pi_d) \subseteq S_n$ *optimal* or *minimally-sized* if no k-restricted minwise independent family $\mathcal{G} \subseteq S_n$ has fewer than d members.

Example 1. Let $\mathcal{F} := (\pi_1, \ldots, \pi_6) \subseteq S_4$ be the family formed by the following six permutations in Fig. 1. No matter which $d \times j$ submatrix of $j \leq 3$ columns we select together with an attribute of "speciality" for one of its columns, the number of rows of this submatrix in which this special column entry is row-wise the smallest is invariant of the submatrix-choice and subsequent speciality-choice. More precisely, notice that for $(2, 1, 4) \in \mathsf{SOP}(4, 3)$, we have $2 = |\{\pi_3, \pi_5\}| = 6/3$ and $\pi_3(2) = 1 = \min\{1, 4, 3\} = \min\{\pi_3(2), \pi_3(1), \pi_3(4)\}$ as well as $\pi_5(2) = 1 = \min\{1, 2, 3\} = \min\{\pi_5(2), \pi_5(1), \pi_5(4)\}$. A respective observation can be made for all 12 elements in $\mathsf{SOP}(4, 3)$, all 12 elements in $\mathsf{SOP}(4, 2)$, and all four elements in $\mathsf{SOP}(4, 1)$. Optimality follows from the fact that the family size must be a multiple of $\operatorname{lcm}(1, 2, 3) = 6$, and 6 itself already suffices.

$\pi_1 =$	2	3	1	4
$\pi_2 =$	1	4	2	3
$\pi_3 =$	4	1	2	3
$\pi_4 =$	2	3	4	1
$\pi_5 =$	2	1	4	3
$\pi_6 =$	4	3	2	1

Fig. 1. A 3-restricted minwise independent and optimal family. Each row corresponds to a permutation: a (throughout the row) unique value is assigned to each column-position.

Let us denote by $S_{n,k}$ the set of injective functions $\sigma : [k] \rightarrow [n]$, henceforth called *k-subpermutations*. The latter are used to strengthen the requirements of k-restricted minwise independent families towards *k-rankwise independent* families [32] satisfying for each subpermutation $\sigma \in S_{n,k}$

$$|\{i \in [d] : \pi_i(\sigma(1)) < \cdots < \pi_i(\sigma(k))\}| = \frac{d}{k!}. \tag{3}$$

In the sequel, we recall some concepts from group theory. Let G be a (not necessarily commutative) group with operation $\circ : G \times G \rightarrow G$, neutral element $e \in G$, and inversion $(\cdot)^{-1} : G \rightarrow G$. For each subgroup H of G—a subset of G closed under the group operations—a *left-coset* is a set of the form $gH := \{g \circ h : h \in H\}$ for an arbitrary $g \in G$. Similarly, *right-cosets* $Hg := \{h \circ g : h \in H\}$ can be considered for any $g \in G$. We focus on subgroups of the symmetric group S_n, where $\circ$ is function composition. We use the standard function composition convention $f \circ g := f(g(\cdot))$, i.e., f is applied to the output of g.

1.2 Related Work

In their foundational work, Broder et al. [5] proposed an exponential-size construction of k-restricted minwise independent families for $k = n$. Subsequently,

Itoh et al. [17] gave a provably optimal constant-factor improvement for this case, and Tarui et al. [32] derived polynomial-size asymptotic upper bounds for the cases $k \in \{3, 4\}$ via finite-geometric methods. Error-tolerant approximate versions of biased-distribution settings were also examined in [5]. The work [17] also introduced the concept of rankwise independence, identified it as a proper special case of minwise independence, and provided the subsequent useful observations. In contrast, Matoušek and Stojaković [25] studied slight relaxations.

Remark 1. Tarui et al. [32] observed that (2) for $j \leq 3$ is equivalent to (3) with $k = 3$, and consequently 3-restricted minwise independence coincides with 3-rankwise independence. Itoh et al. [17, p. 141] give a transformation that turns a k-restricted min- (or rankwise) independent family $\mathcal{F} \subseteq S_n$ into one in $S_{\tilde{n}}$ of the same size, for any $\tilde{n} \leq n$. By contrapositive, non-existence of k-restricted minwise independent families $\widetilde{\mathcal{F}} \subseteq S_{\tilde{n}}$ of d members implies non-existence for such families $\mathcal{F} \subseteq S_n$ with larger $n \geq \tilde{n}$, provided $|\mathcal{F}| = d$.

Asymptotic bounds on the size of optimal k-restricted minwise independent families appear in [2,17,18]. Upper bounds for optimal rankwise independent families have also been derived [15,20] as well as lower bounds [2].

It was recently recognized [19] that requirement (3) "up to isomorphy" exactly characterizes perfect sequence covering arrays [34]. For the latter, several construction [12,21,34] and search approaches [13,24,29] have been proposed as alternatives to the ones addressing mainly the asymptotics [2,32] of rankwise independence. The enumerative strategies of Mathon and van Trung [24] and Na et al. [29] are based on the assumption that families are unions of cosets in S_n; both works motivate our SAT-based adaptation of their group-decomposition approach. For minwise independence, the following theorem seems to be the first result proving the existence of a minwise independent family by a particular decomposition into (non-trivial) *right-cosets*. Although originally stated without group-theoretic language, the result is rephrased here to match our framework; see Sect. 3.

Theorem 1 (adapted from *[3]***).** *Let $k \geq 3$ be odd and $\mathcal{F} = (\theta_1, \ldots, \theta_d) \subseteq S_n$ be k-restricted minwise independent. With $G = \{\gamma_1, \gamma_2\} := \{\mathrm{id}, \sigma\}$ denoting the subgroup of S_n containing the identity permutation and the order-reversing permutation $\sigma : [n] \to [n]$, $i \mapsto n + 1 - i$, we have that $(\gamma_\ell \circ \theta_m : \ell \in [2], m \in [d])$ is a $(k+1)$-restricted minwise independent family of size* $2d$.

Another connection of minwise independence to group theory has been theoretically studied within a more restrictive setting in [6].

Remark 2. While deriving the apparently best lower bound on the size of rankwise independent families, Bargachev [2] asked whether there is a "natural bijection" between the class of permutations having k so-called waste indices and the class of k-partial derangements. In passing we highlight that such a bijection follows from a slight adaptation and generalization of [36, Proposition 2.1] which bijects on top of the cycle notation of permutations [9]; more details can be found in the electronic appendix. [2]

2 A SAT Approach

In this section we provide an initial SAT encoding. In the upcoming Sect. 3 we then refine the encoding into two more sophisticated ones, each constituting a generalization.

The main idea relies on the natural approach to represent permutations via their order-theoretic incidence structure already used, e.g., in [1]. In fact, if $\pi \in S_n$ is a permutation, it is uniquely determined by full knowledge on $X^\pi = (x^\pi_{i,j})^n_{i,j=1} \in \{0,1\}^{n\times n}$ capturing by $x^\pi_{i,j}$ the information if $\pi(i) < \pi(j)$ (1 in the affirmative case, otherwise 0). Note that X^π can be seen as the incidence matrix of a strict total order on $[n]$ representing the permutation. We can, in fact, recover the j-th entry of the permutation via $\pi(j) = 1 + \sum_{i=1}^n x^\pi_{i,j}$. All solutions for $x^\pi_{i,j}$ subject to the following constraints (irreflexivity, asymmetry, totality, and transitivity) correspond to a permutation:

$$\neg x^{\pi_\ell}_{i,i} \quad \text{for } \ell \in [d]\,,\ i \in [n]\,; \tag{4}$$

$$\neg x^{\pi_\ell}_{i,j} \vee \neg x^{\pi_\ell}_{j,i} \quad \text{for } \ell \in [d]\,,\ i \in [n]\,,\ j \in [n]\,; \tag{5}$$

$$x^{\pi_\ell}_{i,j} \vee x^{\pi_\ell}_{j,i} \quad \text{for } \ell \in [d]\,,\ i \in [n]\,,\ j \in [n]\,; \tag{6}$$

$$\neg x^{\pi_\ell}_{i,j} \vee \neg x^{\pi_\ell}_{j,h} \vee x^{\pi_\ell}_{i,h} \quad \text{for } \ell \in [d]\,,\ i \in [n]\,,\ j \in [i+1\,{:}\,n],\ h \in [n] \setminus \{i,j\}. \tag{7}$$

These constraints are already stated in conjunctive normal form (CNF). Here for two integers p and q, we denote by $[p:q] := \{p, p+1, \ldots, q\}$. Constraints (5)–(6) in fact are equivalent to $x_{j,i} \leftrightarrow \neg x_{i,j}$ for $i < j$; by substitution, it is hence possible to rely exclusively on the strict upper-diagonal entries of X^π thereby allowing us to omit (4)–(6).

To encode property (2) in the definition of minwise independence, we need to impose specific cardinality constraints which are stated in (8)—note that $j \leq 3$ is intentionally excluded due to a more favorable replacement in (9):

$$\sum_{\ell=1}^{d} \left[\bigwedge_{h=2}^{j} x^{\pi_\ell}_{s_1,s_h} \right] \leq d/j \quad \text{for } j \in [4\,{:}\,k],\ s \in \mathsf{SOP}(n,j); \tag{8}$$

$$\sum_{\ell=1}^{d} \left[\bigwedge_{h=2}^{3} x^{\pi_\ell}_{\sigma(h-1),\sigma(h)} \right] \leq d/3! \quad \text{for } \sigma \in S_{n,3}. \tag{9}$$

In fact, we avoid the higher upper bound $d/3$ (which in (8) would result for $j = 3$) and we can also relinquish the case $j = 2$—this is justified by the aforementioned particularity that 3-restricted minwise independence is precisely 3-rankwise independence (see Remark 1).

[2] https://www.ac.tuwien.ac.at/files/resources/instances/minw-idp/.

Note that upper bounds instead of equalities appear in (8)–(9), which provide a considerable simplification in view of a conversion to CNF. These seemingly weaker conditions are, however, equivalent to the ones with equalities: Any not attained upper bound in (8)–(9) would imply a strict excess of the upper bound for some other $s' \in \mathsf{SOP}(n, j)$ respectively $\sigma' \in S_{n,3}$—infeasibility would be a consequence; in fact, the number of patterns/subpermutation conditions met by a family $\mathcal{F}$ is fixed. How the cardinality constraints are technically translated into CNF is deferred to Sect. 4.

Remark 3. If the interest is in modeling k-rankwise independence, we can simply replace (8)–(9) with inequalities $\sum_{\ell=1}^{d} \left[\bigwedge_{h=2}^{k} x^{\pi_\ell}_{\sigma(h-1),\sigma(h)}\right] \leq d/k!$ for $\sigma \in S_{n,k}$.

For symmetry breaking, for a permutation π, we consider the binary string resulting from $X^\pi = (x^\pi_{i,j})^n_{i,j=1}$ by concatenating all side-diagonals of X^π lying above the main diagonal to form a string, denoted by $z_{\text{cat}}[X^\pi] := (x^\pi_{i+1,i+j} : j = 2, \ldots n, i = 0, \ldots, n-j)$.

It is then natural, without loss of generality, to enforce an ordering on X^{π_ℓ}, $\ell = 1, \ldots, d$, defined through an ordering of their associated strings $z_{\text{cat}}[X^{\pi_\ell}]$. With $\preceq$ denoting the lexicographical ordering relation on equal-length binary strings, i.e., $s_1 s_2 \cdots, s_M \preceq t_1 t_2 \cdots t_M$ iff $\sum_{h=1}^{M} (t_h - s_h) 2^{M-h} \geq 0$, we thus require

$$z_{\text{cat}}[X^{\pi_{\ell+1}}] \preceq z_{\text{cat}}[X^{\pi_\ell}] \quad \text{for } \ell \in [d-1]\,. \tag{10}$$

The identity permutation, whose upper diagonal part of the incidence matrix consists of exclusively 1-entries, is hence the maximum element with respect to this ordering. For $\pi \in S_n$, the string $z_{\text{cat}}[X^\pi]$ is of length $n(n-1)/2$. Therefore, we might prefer the following weakened constraint depending on a user-defined "accuracy" parameter $L \in [0 : n(n-1)/2]$ (denoting for a string s the substring consisting of the first L entries of s by $s_{[1..L]}$):

$$z_{\text{cat}}[X^{\pi_{\ell+1}}]_{[1..L]} \preceq z_{\text{cat}}[X^{\pi_\ell}]_{[1..L]} \quad \text{for } \ell \in [d-1]\,. \tag{11}$$

In [30, entry A036604] one can look up $\mathsf{A}(n)$, the minimum number of (dynamically choosable) pairwise comparisons needed to uniquely determine a strict total order on $[n]$; see also the follow-up paper [31]. In the following we regard the choice $L := \mathsf{A}(n)$ as a meaningful guideline but we could also choose a different value for L, if an alternative trade-off between symmetry breaking and lower complexity of the encoding is preferred. Finally, in this setting we can without loss of generality add the following constraint (12) as we preserve minwise independence by composing each permutation in the family with π_1^{-1} and afterwards bringing the transformed family into lexicographic order.

$$\pi_1 = \text{id}\,;\ \text{i.e., unit clauses} \quad x^{\pi_1}_{i,j} \quad \text{for } i \in [n]\,,\ j \in [i+1 : n]. \tag{12}$$

3 Incorporating a Group-Theoretic Decomposition

This section emulates in SAT the group-theoretic heuristic of Mathon and van Trung [24] (see also [29]) on top of the encoding in Sect. 2.

As the approaches in the latter two works are—up to isomorphy, see [19]—applied to rankwise independence it is a natural question if they are a useful approach when rankwise independence is now relaxed towards minwise independence: Clearly, this decomposition property is inherited by the relaxation, but according to asymptotic results a minwise independent family obtained this way could in general be far from optimal [17,18]. Therefore, it is a priori unclear whether this approach can be successful for the search of (near-)optimal minwise independent representatives, too.

Moreover, let us anticipate that it is currently unknown whether solutions always carry the underlying structure which will be postulated in what follows. Hence, this section provides a heuristic; a negative answer here does not entail non-existence in the full search space.

Let us address the required steps for answering this question: In the spirit of [24,29] we will hence assume that our families $(\pi_1, \ldots, \pi_d)$ with $\pi_i \in S_n$ such that $\operatorname{lcm}([k])$ is a divisor of d have the following underlying structure: There is a subgroup $G = \{\gamma_1, \ldots, \gamma_q\}$ of S_n whose order $q = |G|$ is a divisor of d and for which there are "offsets" $\theta_1, \ldots, \theta_{d/|G|} \in S_n$ such that

$$\mathcal{F} = (\pi_1, \ldots, \pi_d) = (\theta_\ell \circ \gamma_m)_{(\ell,m) \in [d/|G|] \times [|G|]}. \tag{13}$$

Condition (13) models the coincidence of $(\pi_1, \ldots, \pi_d)$ with a union of $d/|G|$ (not necessarily distinct) left-cosets of a subgroup G of S_n. By fixing a non-trivial subgroup G of S_n, the number of decision variables is reduced, as every choice for an offset θ_i automatically fixes a larger set of size $|G|$ of permutations in $\mathcal{F}$. Note that we will always assume that by default $\gamma_1 := \mathrm{id}$ in our enumeration of G.

An encoding in our current setting consists of the following parts: Variables $X^{\theta_\ell} = (x_{i,r}^{\theta_\ell})_{(i,r) \in [n] \times [n]}$, for $\ell = 1, \ldots, d/|G|$, with permutation constraints precisely as in (4)–(7), cardinality constraints

$$\sum_{\ell=1}^{d/|G|} \sum_{m=1}^{|G|} \left[\bigwedge_{h=2}^{j} x_{\gamma_m(s_1), \gamma_m(s_h)}^{\theta_\ell} \right] \leq d/j \quad \text{for } j \in [4\,{:}\,k],\ s \in \mathsf{SOP}(n,j); \tag{14}$$

$$\sum_{\ell=1}^{d/|G|} \sum_{m=1}^{|G|} \left[\bigwedge_{h=2}^{3} x_{\gamma_m(\sigma(h-1)), \gamma_m(\sigma(h))}^{\theta_\ell} \right] \leq d/3! \quad \text{for } \sigma \in S_{n,3}; \tag{15}$$

and lexicographical symmetry breaking as in (11),

$$z_{\mathrm{cat}}[\theta_{\ell+1}]_{[1..L]} \preceq z_{\mathrm{cat}}[\theta_\ell]_{[1..L]} \quad \text{for } \ell \in [d/|G| - 1]\,. \tag{16}$$

Fixing the trivial subgroup solely consisting of the identity permutation $\{\gamma_1\} = \{\mathrm{id}\}$ of S_n, we recover precisely the original, non-heuristic model (4)–(9), (11). Thus, we derived a generalization parameterized by the group G.

Apart from left-cosets, one can just as well look at *right-cosets* based on the same motivation. This was also analyzed in [29] and the underlying assumption here translates to

$$\mathcal{F} = (\pi_1, \ldots, \pi_d) = (\gamma_m \circ \theta_\ell)_{(\ell,m) \in [d/|G|] \times [|G|]}. \tag{17}$$

In the framework of SAT, this needs a more elaborate modeling approach: Although we notice that $\gamma_m \circ \theta_\ell = \left(\theta_\ell^{-1} \circ \gamma_m^{-1}\right)^{-1}$, leading to the idea of using θ_ℓ^{-1} directly as a new decision variable, there is then no direct way to recover via an encoding in CNF the incidence matrix of the inverse of a permutation (when the latter is itself specified by its incidence matrix). In fact we now face the difficulty of applying a fixed permutation given as parameter (via G) to a symbolically represented permutation (specified by decision variables) and afterwards accessing its incidence matrix. This considerably differs from the left-coset approach where the need is to access the incidence matrix of a symbolic permutation in entries pre-permuted by fixed permutations passed as parameters.

We circumvent this issue by encoding the offsets θ_m as permutation matrices $T^{\theta_\ell} = (t_{ij}^{\theta_\ell})_{i,j=1}^n \in \{0,1\}^{n\times n}$ whose bi-stochasticity we enforce via

$$\bigvee_{j=1}^{n} t_{ij}^{\theta_\ell} \quad \text{for } \ell \in [d/q]\,,\ i \in [n]\,; \tag{18}$$

$$\neg t_{i,j}^{\theta_\ell} \vee \neg t_{i,r}^{\theta_\ell} \quad \text{for } \ell \in [d/q]\,,\ i \in [n]\,,\ \{j,r\} \in \binom{[n]}{2}; \tag{19}$$

$$\bigvee_{i=1}^{n} t_{ij}^{\theta_\ell} \quad \text{for } \ell \in [d/q]\,,\ j \in [n]\,. \tag{20}$$

Cardinality-2 subsets of $[n]$ are denoted as $\binom{[n]}{2}$. The first two clause families together enforce exactly one 1-entry per row, hence exactly n such entries in total; demanding in addition at least one 1-entry per column then already forces T^{θ_ℓ} to be a permutation matrix keeping the encoding more compact. The subsequent constraint (21) encodes the incidence matrix of $\gamma_m \circ \theta_\ell$ for any fixed γ_m and a symbolic θ_ℓ inside the fresh variables $x_{i,r}^{\gamma_m\circ\theta_\ell}$, $(i,r) \in [n]^2$. The formulation depends on the permutation matrix T^{θ_ℓ} of θ_ℓ:

$$x_{i,r}^{\gamma_m\circ\theta_\ell} \leftrightarrow \left((t_{\gamma_m(r),j}^{\theta_\ell})_{j=1}^n \preceq (t_{\gamma_m(i),j}^{\theta_\ell})_{j=1}^n \right) \quad \text{for } \ell \in [d/q]\,,\ m \in [q]\,,\ (i,r) \in [n]^2\,. \tag{21}$$

Regarding (21), note that firstly, row i is a lexicographical successor of row r of the permutation matrix of an arbitrary permutation π iff $\pi(i) < \pi(r)$; secondly, the permutation matrix of $\gamma_m \circ \theta_\ell$ can be recovered from T^{θ_ℓ} by applying the permutation γ_m to its rows.

The main task to be addressed here is accessing the truth value of $x_{i,r}^{\gamma_m\circ\theta_\ell}$, storing as a Boolean whether the i-th and the r-th row are in ascending lexicographic ordering. A case distinction for the value of $x_{i,r}^{\gamma_m\circ\theta_\ell}$ shows that imposing the subsequent lexicographical ordering on the tuples that suitably link $x^{\gamma_m\circ\theta_\ell}$ to T^{θ_ℓ}-entries precisely ensures such a behavior for $x_{i,r}^{\gamma_m\circ\theta_\ell}$:

$$\text{Relations(23)} - -\text{(24)are satisfied} \quad \text{for } \ell \in [d/q]\,,\ m \in [q]\,,\ (i,r) \in [n]^2\,; \tag{22}$$

$$(x_{i,r}^{\gamma_m\circ\theta_\ell}, t_{\gamma_m(r),1}^{\theta_\ell}, \ldots, t_{\gamma_m(r),n}^{\theta_\ell}) \preceq (1, \quad t_{\gamma_m(i),1}^{\theta_\ell}, \ldots, t_{\gamma_m(i),n}^{\theta_\ell}), \tag{23}$$

$$(0, \neg t_{\gamma_m(r),1}^{\theta_\ell}, \ldots, \neg t_{\gamma_m(r),n}^{\theta_\ell}) \preceq (x_{i,r}^{\gamma_m\circ\theta_\ell}, \neg t_{\gamma_m(i),1}^{\theta_\ell}, \ldots, \neg t_{\gamma_m(i),n}^{\theta_\ell}). \tag{24}$$

The implementation of $\preceq$ in SAT is deferred to Sect. 4. What remains are the analogous cardinality constraints, now stated as

$$\sum_{\ell=1}^{d/|G|} \sum_{m=1}^{|G|} \left[\bigwedge_{h=2}^{j} x_{s_1,s_h}^{\gamma_m \circ \theta_\ell} \right] \leq d/j \quad \text{for } j \in [4:k],\ s \in \mathsf{SOP}(n,j); \tag{25}$$

$$\sum_{\ell=1}^{d/|G|} \sum_{m=1}^{|G|} \left[\bigwedge_{h=2}^{3} x_{\sigma(h-1),\sigma(h)}^{\gamma_m \circ \theta_\ell} \right] \leq d/3! \quad \text{for } \sigma \in S_{n,3}, \tag{26}$$

without loss of generality, we require

$$z_{\text{cat}}[X^{\gamma_1 \circ \theta_{\ell+1}}]_{[1:L]} \preceq z_{\text{cat}}[X^{\gamma_1 \circ \theta_\ell}]_{[1:L]} \quad \text{for } \ell \in [d/q-1]\,. \tag{27}$$

The question is now, which subgroups are promising choices, and how many of them should be tested. Potentially, for each q dividing $n!$ and d, there can exist a subgroup of S_n of order q that is fruitful for the coset approach. In [29] it is shown for their problem setting that for the left-coset approach, a single representative per *conjugacy class* $\text{Cl}(G) := \{(\psi G)\psi^{-1} : \psi \in S_n\}$ is sufficient to be considered, as it reflects the behavior of all groups serving as representatives of this class. For the right-coset approach in [29], all subgroups of a given order have to be examined, where, however, without loss of generality one can assume the first offset θ_1 to be the identity permutation. We incorporate these insights from [29] into our experimental setup too, for reducing the number of inspected groups, respectively tightening the search space.

4 Computational Experiments

We use `Julia` in version 1.11.1 for generating strict total order constraints. The cardinality constraints are obtained by calling the library `PySAT` in version 0.1.7.-dev15 [16]. The latter not only provides an interface to different SAT solvers, but also contains conversion routines to bring the encountered cardinality constraints (14),(15),(25),(26) into pure CNF: Seven encodings are available and we pick the one recommended by the default strategy of `PySAT`. For accessing the set of all (conjugacy classes of) subgroups of a given symmetric group S_n, we fall back on the `Julia`-package `Oscar` in version 1.2.0-dev [8], having an interface to `GAP` [11]—the access time is negligible and therefore not reported. In the literature, formulations for enforcing that for two binary vectors $a = (a_1, \ldots, a_r)$ and $b = (b_1, \ldots, b_r)$ the lexicographic ordering $a \preceq b$ applies (needed for the symmetry breaking in Sect. 2), have been proposed. However, as these are unsupported by `PySAT`, we opted for a custom implementation of the "AND Encoding using Common Subexpression Elimination" from [10], which depends on fresh variables x_i, $i = 1, \ldots, r-1$, and less than $6(r-1)$ clauses.

We choose for our experimental evaluation the SAT solver `Glucose` in version 4.2.1 which is frequently used for combinatorial problems. The time limit of the solver is set to 3600 s per instance if not stated otherwise. Selected non-existence results are independently verified by running the SAT solver in proof-logging modality such that a proof in the DRAT (Deletion Resolution Asymmetric Tautology [33]) file format is generated which is subsequently certified by the software `drat-trim` [33]. The experiments have been run on a cluster with an Intel® Xeon® E5-2640 v4 CPU with 2.40GHz and 160GB RAM on a single thread.

To first motivate competitiveness of our approach via a SAT encoding, we carried out the initial experiment (see Table 1) for $k = 3$ which is the only parameter allowing a comparison with existing literature results—this, given the particular isomorphy mentioned in Sect. 1.2. It can be noted that our proposed encoding can be solved in a fraction of time: The solving time on the smallest instance is smaller by three orders of magnitude, the medium instance by five orders (more precisely, the latter, by a factor of 2.09×10^5). Most remarkably, the largest instance $(12, 8)$ which requires to return non-existence as answer can be solved by the SAT solver in just two seconds while the computation was inconclusive apparently even after hours (although the authors do not explicitly report their maximum runtime; presumably, it would require days or weeks of computation time given the fact that it is a non-existence result). These observations demonstrate the SAT solver's excellent comparative performance, apparently forming a preferable method for such a task. We next transfer to this SAT setting the heuristic decomposition strategy into cosets of S_n.

Table 1. Runtimes in seconds for both approaches where $k = 3$. Both approaches were run without postulated decomposition into cosets. A slightly faster CPU (Intel® Xeon® E5-2680 with 2.70GHz) is used by the authors of [29]. The symbol "/" indicates an inconclusive experiment (with unreported time measurement in [29]).

(d, n)	Solver run on (7)–(10),(12)	Algorithm 1 of [29]	Representative exists?
$(12, 6)$	0.00591 s	3.0 s	yes
$(12, 7)$	0.01146 s	2400.0 s	yes
$(12, 8)$	2.03096 s	/	no

For the subsequent, we decided not to examine the case $k = 3$ whose properties can be deduced from the considerations in the works [12, 29] for $n \leq 8$. The search for optimality when $n = 9$—the smallest open case unanswerable by those works—was inconclusive in our setting despite a week of computation time. Therefore, we begin a systematic series of experiments starting with $k = 4$ relying on almost a month of cumulative computation time: Table 2 presents computational results as follows.

While the first two columns indicate the choice of parameters d, n, and $|G|$, the remainder contains results for the left- (shaded background) and the right-coset approach, respectively. Column `sg` provides information about the number of *conjugacy classes of subgroups of* S_n *of order* $|G|$ (`#exst L`). It also shows how many *subgroups of* S_n *of order* $|G|$ exist (`#exst R`). The experiment has been conducted for every existing subgroup, i.e., the number of considered subgroups always coincides with **#exst**. The count of subgroups for which the left- respectively right-coset leads to a feasible solution is displayed in `sg-feas #L/R`. This information is accompanied by the average time (in seconds), which the solver needed to find such a feasible solution (`avg_t[s] L/R`). The same structure applies for the group of columns `sg-infeas`, which reports these counts and time measurements for subgroups resulting in infeasibility. The right-coset approach is never run for $|G| = 1$, symbolized by the entry "/", as this represents the non-heuristic approach where we only employ the left-coset approach.

From the experiments in Table 2 we can derive the following observation.

Observation 1. *For* $n \in \{4, 5\}$, *respectively* $n \in \{6, 7\}$, *there are optimal* 4*-restricted minwise independent families of* 12 *members, respectively* 24 *members, which are all coincident with a union of non-trivial cosets of* S_n. *For* $n = 4$, *respectively* $n = 6$, *a family being a coset of cardinality* 12, *respectively* 24, *can be found.*

Computer-Aided Proof. We have $\text{lcm}([4]) = 12$. Therefore the representatives found in Table 2 are minimally-sized for $n \in \{4, 5\}$. In 0.04 s the solver showed non-existence of a minwise independent family with 12 members for $n = 6$; an accompanying solver run producing a proof-file of size 334 KB (in binary encoded DRAT format) in 0.21 s is certified in 0.13 s by `drat-trim`. The next smallest cardinality which is a multiple of 12, namely $d = 24$, is therefore reported. For $d = 24$ and $n \in \{6, 7\}$ such representatives, again being decomposable into cosets, indeed exist; see Table 2. □

It is noteworthy that, consistently across Table 2, for every subgroup order considered, infeasibility on the left-coset side implies infeasibility on the right-coset side. In contrast, for several subgroup orders, for which the left-coset approach was fruitful, the right-coset approach was not successful. Moreover, the more complex SAT encoding for the right-coset approach increases the runtime by one to two orders of magnitude in comparison to the left-coset approach for feasible but also infeasible instances. Due to these two observations we conclude that in particular for higher values of d and n one should rather focus on the left-coset approach: If one faces a fixed and limited margin of computation time, this allows examining more instances (which also seem the more promising ones).

Observation 2. *There is no* 4*-restricted minwise independent family* $\mathcal{F} \subseteq S_8$ *consisting of* 24 *members.*

Computer-Aided Proof. The associated SAT instance turned out to be unsatisfiable after 64.7 hours of computation time; since unsatisfiability was already suspected, full lexicographical accuracy $L := n(n-1)/2 - 1$ was chosen here. Additionally, we ran the solver in proof-logging modality, which after 184.5 hours returned a 28.3 GB long proof (in binary encoded DRAT format). It took `drat-trim` 59.3 hours to validate the latter. □

Proposition 1. k*-restricted minwise independent families* $\mathcal{F} \subseteq S_n$ *with* $d = |\mathcal{F}|$, *which coincide with a union of left-cosets of* S_n, *exist for constellations*

$$(d, n, k; |G|) \in \{(60, 6, 6; 6), (60, 6, 5; 6), (60, 5, 5; 60), (48, 8, 4; 24)\}.$$

Table 2. Results for $k = 4$; 3600 s time limit given for each run of the SAT solver. Asterisks followed by numbers indicate the count of inconclusive experiments, i.e., the solver hit the time limit, for the right-coset approach.

		sg		sg-feas				sg-infeas			
		#exst		#		avg_t[s]		#		avg_t[s]	
(d, n)	$\|G\|$	L	R	L	R	L	R	L	R	L	R
(12, 4)	12	1	1	1	1	$2.1 \cdot 10^{-4}$	$2.3 \cdot 10^{-4}$	0	0	/	/
	6	1	4	0	0	/	/	1	4	$3.7 \cdot 10^{-4}$	$2.4 \cdot 10^{-4}$
	4	3	7	2	2	$2.2 \cdot 10^{-4}$	$2.9 \cdot 10^{-3}$	1	5	$1.2 \cdot 10^{-3}$	$3.6 \cdot 10^{-3}$
	3	1	4	1	4	$3.3 \cdot 10^{-4}$	$1.3 \cdot 10^{-3}$	0	0	/	/
	2	2	9	1	3	$5.7 \cdot 10^{-4}$	$1.2 \cdot 10^{-3}$	1	6	$1.1 \cdot 10^{-3}$	$6.6 \cdot 10^{-3}$
	1	1	/	1	/	$1.2 \cdot 10^{-3}$	/	0	/	/	/
(12, 5)	12	2	15	0	0	/	/	2	15	$8.9 \cdot 10^{-4}$	$1.8 \cdot 10^{-5}$
	6	3	30	1	0	$2.1 \cdot 10^{-3}$	/	2	30	$2.8 \cdot 10^{-3}$	$1.2 \cdot 10^{-3}$
	4	3	35	0	0	/	/	3	35	$1.0 \cdot 10^{-2}$	$4.8 \cdot 10^{-3}$
	3	1	10	1	0	$2.7 \cdot 10^{-3}$	/	0	10	/	$2.8 \cdot 10^{0}$
	2	2	25	1	2	$2.8 \cdot 10^{-2}$	$6.1 \cdot 10^{-3}$	1	23	$1.5 \cdot 10^{-3}$	$2.1 \cdot 10^{-1}$
	1	1	/	1	/	$4.1 \cdot 10^{-3}$	/	0	/	/	/
(24, 6)	24	6	90	2	8	$2.9 \cdot 10^{-3}$	$2.6 \cdot 10^{-3}$	4	82	$1.2 \cdot 10^{-2}$	$1.5 \cdot 10^{-5}$
*2	12	4	150	2	4	$4.0 \cdot 10^{-3}$	$1.2 \cdot 10^{-2}$	2	144	$3.2 \cdot 10^{-3}$	$2.4 \cdot 10^{1}$
	8	7	255	2	32	$1.4 \cdot 10^{-2}$	$6.7 \cdot 10^{-2}$	5	223	$8.9 \cdot 10^{-1}$	$1.1 \cdot 10^{-1}$
*27	6	6	280	3	30	$2.6 \cdot 10^{-2}$	$2.5 \cdot 10^{0}$	3	223	$6.0 \cdot 10^{-2}$	$8.5 \cdot 10^{1}$
*38	4	7	255	5	72	$3.6 \cdot 10^{-2}$	$6.8 \cdot 10^{0}$	2	145	$2.0 \cdot 10^{1}$	$2.5 \cdot 10^{2}$
*15	3	2	40	1	14	$9.1 \cdot 10^{-2}$	$4.6 \cdot 10^{2}$	1	11	$6.2 \cdot 10^{-1}$	$4.4 \cdot 10^{2}$
*38	2	3	75	2	35	$9.8 \cdot 10^{-1}$	$9.0 \cdot 10^{2}$	1	2	$3.3 \cdot 10^{0}$	$2.0 \cdot 10^{3}$
	1	1	/	1	/	$8.5 \cdot 10^{-1}$	/	0	/	/	/
(24, 7)	24	14	1435	0	0	/	/	14	1435	$1.9 \cdot 10^{-2}$	$1.5 \cdot 10^{-5}$
	12	13	1715	2	0	$2.6 \cdot 10^{-2}$	/	11	1715	$1.7 \cdot 10^{-1}$	$1.2 \cdot 10^{0}$
*3	8	7	1575	0	0	/	/	7	1572	$1.4 \cdot 10^{-1}$	$3.8 \cdot 10^{-1}$
*30	6	8	1645	3	0	$5.4 \cdot 10^{-2}$	/	5	1615	$8.6 \cdot 10^{-2}$	$2.8 \cdot 10^{1}$
*532	4	7	1295	3	3	$2.7 \cdot 10^{-1}$	$4.0 \cdot 10^{1}$	4	760	$2.4 \cdot 10^{1}$	$4.1 \cdot 10^{2}$
*151	3	2	175	1	0	$6.1 \cdot 10^{0}$	/	1	24	$2.0 \cdot 10^{0}$	$3.5 \cdot 10^{2}$
*230	2	3	231	2	1	$1.4 \cdot 10^{1}$	$4.1 \cdot 10^{2}$	1	0	$5.9 \cdot 10^{1}$	/
	1	1	/	1	/	$3.5 \cdot 10^{1}$	/	0	/	/	/

Proof. Certificates for this assertion can be found in the supplementary material of the aforementioned electronic appendix, where some `Julia` routines are also deposited.

We also remark that for $k \in \{5,6\}$ we have $\text{lcm}([5]) = \text{lcm}([6]) = 60$ and thus the respective instances in Proposition 1 are even optimal. We highlight suboptimality for $(d,n,k) = (48,8,4)$, as we can construct an instance with constellation $(d,n,k) = (36,8,4)$ by applying Theorem 1 to the 3-restricted minwise independent family deducible from $\mathcal{F} \subseteq S_8$ of 18 members in [29, Proposition 4.6]—due to Observation 2 the end product is even optimal. However, although we can obtain this insight, even with runtimes up to 48 hours, the solver was not able to find any respective family counting 36 members via decomposition into left-cosets. Finally, we point out the particular situation occurring for $(d,k) = (12,4)$, see Table 2, where for $n = 4$ no decomposition into left-cosets of cardinality 6 is possible, while this is the case for $n = 5$, which intuitively is a stronger constraint setting. This, apparently, is explainable by the structurally different order-6 subgroups present in S_5 compared with those in S_4. At the same time it is notable that for $|G| = 12$ the left-coset approach worked for $n = 4$ but not for $n = 5$.

5 Conclusion

We have developed SAT encodings that are well suited for examining the existence and structural decomposability of (near-)optimal k-restricted minwise independent families $\mathcal{F} \subseteq S_n$ for computationally tractable values of n and k. In addition to discovering several previously unknown minwise independent families, our results reveal the following key insight: All computed (near-)optimal representatives of these families indeed carry the aforementioned group-theoretic structure. A further benefit of the decomposition is that it dramatically reduces running times, making the search for feasible representatives possible where a purely non-heuristic approach would be prohibitively expensive. Collecting further empirical evidence, or even obtaining a generally valid result, affirming the decomposability of optima into unions of non-trivial cosets seems an interesting challenge.

The following aspects offer interesting directions for further work. In Observations 1 and 2, the partial replacement of cardinality constraints with parity constraints (using a solver featuring exclusive-or) or other weakened forms might offer a useful strategy for finding shorter computational non-existence proofs. They might be convertible into considerably shorter Lean proofs [27]. Further experimentation with the cardinality constraints, e.g., by using solvers with built-in cardinality propagators or by employing integer linear programming, seems a worthwhile alternative to study.

Disclosure of Interests. The authors have no competing interests.

References

1. Banbara, M., Tamura, N., Inoue, K.: Generating event-sequence test cases by answer set programming with the incidence matrix. In: Technical Communications of the 28th International Conference on Logic Programming. Schloss Dagstuhl-Leibniz-Zentrum für Informatik (2012). https://doi.org/10.4230/lipics.iclp.2012.86
2. Bargachev, V.: An improved lower bound on the size of k-rankwise independent families of permutations. Tech. rep., St. Petersburg Department of Steklov Institute of Mathematics (2004). https://www.pdmi.ras.ru/preprint/2004/04-13.html. Accessed 30 May 2026
3. Bargachev, V.: On some properties of min-wise independent families and groups of permutations. J. Math. Sci. **134**(5), 2340–2345 (2006). https://doi.org/10.1007/s10958-006-0110-1
4. Broder, A.Z.: On the resemblance and containment of documents. In: Proceedings of the Compression and Complexity of SEQUENCES, pp. 21–29. IEEE (1997)
5. Broder, A.Z., Charikar, M., Frieze, A.M., Mitzenmacher, M.: Min-wise independent permutations. J. Comput. Syst. Sci. **60**(3), 630–659 (2000). https://doi.org/10.1145/276698.276781
6. Cameron, P.J., Spiga, P.: Min-wise independent families with respect to any linear order. Comm. Algebra **35**(10), 3026–3033 (2007). https://doi.org/10.1080/00927870701404812
7. Chee, Y.M., Colbourn, C.J., Horsley, D., Zhou, J.: Sequence covering arrays. SIAM J. Discret. Math. **27**(4), 1844–1861 (2013). https://doi.org/10.1137/120894099
8. Decker, W., Eder, C., Fieker, C., Horn, M., Joswig, M. (eds.): The Computer Algebra System OSCAR: Algorithms and Examples, Algorithms and Computation in Mathematics, vol. 32. Springer, 1 edn. (2024). https://doi.org/10.1007/978-3-031-62127-7
9. Désarménien, J.: Une autre interprétation du nombre de dérangements. Séminaire Lotharingien de Combinatoire **8**, 1–6 (1984). https://www.emis.de/journals/SLC/opapers/s08desar.html
10. Elgabou, H.: Encoding The Lexicographic Ordering Constraint in Satisfiability Modulo Theories. Ph.D. thesis, University of York (2015). https://etheses.whiterose.ac.uk/10387/
11. The GAP Group: GAP—Groups, Algorithms, and Programming, Version 4.13.1 (2024). https://www.gap-system.org
12. Gentle, A.R.: A polynomial construction of perfect sequence covering arrays. Algebraic Comb. **6**(5), 1383–1394 (2023). https://doi.org/10.5802/alco.308
13. Gentle, A.R., Wanless, I.M.: On perfect sequence covering arrays. Ann. Comb. **27**(3), 539–564 (2023). https://doi.org/10.1007/s00026-022-00610-6
14. Gionis, A., Indyk, P., Motwani, R.: Similarity search in high dimensions via hashing. In: Atkinson, M.P., Orlowska, M.E., Valduriez, P., Zdonik, S.B., Brodie, M.L. (eds.) Proceedings of 25th International Conference on Very Large Data Bases, pp. 518–529. Morgan Kaufmann (1999). http://www.vldb.org/conf/1999/P49.pdf
15. Harvey, N., Sahami, A.: Explicit and near-optimal construction of t-rankwise independent permutations. In: Kumar, A., Ron-Zewi, N. (eds.) Approximation, Randomization, and Combinatorial Optimization. LIPIcs, vol. 317, pp. 67:1–67:12. Schloss Dagstuhl - Leibniz-Zentrum für Informatik (2024). https://doi.org/10.4230/lipics.approx/random.2024.67

16. Ignatiev, A., Morgado, A., Marques-Silva, J.: PySAT: a Python toolkit for prototyping with SAT oracles. In: International Conference on Theory and Applications of Satisfiability Testing, pp. 428–437. Springer (2018). https://doi.org/10.1007/978-3-319-94144-8_26
17. Itoh, T., Takei, Y., Tarui, J.: On permutations with limited independence. In: Proceedings of the Eleventh Annual ACM-SIAM Symposium on Discrete Algorithms, pp. 137–146 (2000). https://doi.org/10.5555/338219.338245
18. Itoh, T., Takei, Y., Tarui, J.: On the sample size of k-restricted min-wise independent permutations and other k-wise distributions. In: Proceedings of the Thirty-Fifth Annual ACM Symposium on Theory of Computing, pp. 710–719 (2003)
19. Iurlano, E.: Growth of the perfect sequence covering array number. Des. Codes Crypt. **91**(4), 1487–1494 (2023). https://doi.org/10.1007/s10623-022-01168-3
20. Kuperberg, G., Lovett, S., Peled, R.: Probabilistic existence of regular combinatorial structures. Geom. Funct. Anal. **27**(4), 919–972 (2017). https://doi.org/10.1007/s00039-017-0416-9
21. Levenshtein, V.I.: On perfect codes in deletion and insertion metric. Discret. Math. Appl. **2**(3), 241–258 (1992). https://doi.org/10.1515/dma.1992.2.3.241
22. Li, P., König, A.C.: Theory and applications of b-bit minwise hashing. Commun. ACM **54**(8), 101–109 (2011). https://doi.org/10.1145/1978542.1978566
23. Li, P., Owen, A., Zhang, C.H.: One permutation hashing. Adv. Neural. Inf. Process. Syst. **25** (2012)
24. Mathon, R., van Trung, T.: Directed t-packings and directed t-Steiner systems. Des. Codes Crypt. **18**(1), 187–198 (1999). https://doi.org/10.1023/a:1008353723204
25. Matoušek, J., Stojaković, M.: On restricted min-wise independence of permutations. Random Struct. Algorithms **23**(4), 397–408 (2003)
26. Matsumoto, M., Nishimura, T.: Mersenne twister: a 623-dimensionally equidistributed uniform pseudo-random number generator. ACM Trans. Model. Comput. Simul. **8**(1), 3–30 (1998). https://doi.org/10.1145/272991.272995
27. de Moura, L., Ullrich, S.: The Lean 4 theorem prover and programming language. In: Platzer, A., Sutcliffe, G. (eds.) Automated Deduction - CADE 28 - 28th International Conference on Automated Deduction, Virtual Event, July 12-15, 2021, Proceedings, pp. 625–635. LNCS, Springer (2021). https://doi.org/10.1007/978-3-030-79876-5_37
28. Mulmuley, K.: Computational Geometry: An Introduction Through Randomized Algorithms. Prentice-Hall, Englewood Cliffs, N.J. (1994)
29. Na, J., Jedwab, J., Li, S.: A group-based structure for perfect sequence covering arrays. Des. Codes Crypt. **91**(3), 951–970 (2023). https://doi.org/10.1007/s10623-022-01132-1
30. OEIS Foundation Inc.: The On-Line Encyclopedia of Integer Sequences (2024). published electronically at https://oeis.org
31. Peczarski, M.: New results in minimum-comparison sorting. Algorithmica **40**, 133–145 (2004). https://doi.org/10.1007/s00453-004-1100-7
32. Tarui, J., Itoh, T., Takei, Y.: A nearly linear size 4-min-wise independent permutation family by finite geometries. In: Arora, S., Jansen, K., Rolim, J.D.P., Sahai, A. (eds.) Approximation, Randomization, and Combinatorial Optimization. Algorithms and Techniques. LNCS, vol. 2764, pp. 396–408. Springer, Berlin, Heidelberg (2003). https://doi.org/10.1007/978-3-540-45198-3_33
33. Wetzler, N., Heule, M., Hunt Jr., W.A.: DRAT-trim: efficient checking and trimming using expressive clausal proofs. In: Theory and Applications of Satisfiability Testing—17th International Conference. LNCS, vol. 8561, pp. 422–429. Springer (2014). https://doi.org/10.1007/978-3-319-09284-3_31

34. Yuster, R.: Perfect sequence covering arrays. Des. Codes Crypt. **88**(3), 585–593 (2020). https://doi.org/10.1007/s10623-019-00698-7
35. Zamora, J., Mendoza, M., Allende, H.: Hashing-based clustering in high dimensional data. Expert Syst. Appl. **62**, 202–211 (2016). https://doi.org/10.1016/j.eswa.2016.06.008
36. Zhang, J., Gray, D., Wang, H., Zhang, X.D.: On the combinatorics of derangements and related permutations. Appl. Math. Comput. **431**, 127341 (2022). https://doi.org/10.1016/j.amc.2022.127341

Stein Variational Black-Box Combinatorial Optimization

Thomas Landais, Olivier Goudet, Adrien Goëffon(✉), Frédéric Saubion, and Sylvain Lamprier

LERIA, Université d'Angers, 2 Boulevard Lavoisier, 49045 Angers, France
{thomas.landais,olivier.goudet,adrien.goeffon,frederic.saubion, sylvain.lamprier}@univ-angers.fr

Abstract. Combinatorial black-box optimization in complex and rugged settings demands a careful trade-off between exploiting promising regions of the search space and preserving sufficient exploration to identify multiple optima. Although Estimation-of-Distribution Algorithms (EDAs) provide a powerful model-based framework, they often concentrate on a single region of interest, which may result in premature convergence when facing complex or multimodal objective landscapes. In this work, we incorporate the Stein operator to introduce a repulsive mechanism among particles in the parameter space, thereby encouraging the population to disperse and jointly explore several modes of the fitness landscape. Empirical evaluations across diverse benchmark problems show that the proposed method achieves performance competitive with, and in several cases superior to, leading state-of-the-art approaches, particularly on the larger evaluated instances. These findings highlight the potential of Stein variational gradient descent as a promising direction for addressing large, computationally expensive, discrete black-box optimization problems.

Keywords: Black-box Optimization · Estimation-of-Distribution Algorithm · Stein Variational Gradient Descent

1 Introduction

Black-box optimization (BBO) deals with the challenge of optimizing an objective function whose internal structure or analytical form remains unknown. We focus here on the single-objective optimization setting, where the goal is to maximize a function $f : \mathcal{X} \rightarrow \mathbb{R}$ over a discrete search space $\mathcal{X}$. In most practical scenarios, f is computationally expensive to evaluate. Whether an evaluation involves running a high-fidelity simulation, conducting a physical experiment, or executing a complex algorithm, the high cost per query makes sample efficiency a key requirement [3,7].

To solve BBO, two predominant families of approaches have emerged. On the one hand, Bayesian optimization [11,12] treats the unknown objective function f as the outcome of a stochastic process, typically a Gaussian process. On

G. Iacca et al. (Eds.): PPSN 2026, LNCS 16985, pp. 450–465, 2027.
https://doi.org/10.1007/978-3-032-36214-8_28

the other hand, Evolutionary Algorithms (EAs) evolve a population of candidate solutions through variation operators (mutation, crossover) and selection mechanisms based on fitness [4,10]. Estimation-of-Distribution Algorithms (EDAs) are a subcategory of EAs [16]. These stochastic methods guide the search for optima by explicitly learning and sampling from a probabilistic model of promising candidate solutions.

Despite their adaptive capabilities, EDAs often suffer from premature collapse of the search distribution. Since the probabilistic model is updated iteratively based on the current population's performance, the search tends to concentrate rapidly around a single region of the search space, reducing diversity and potentially overlooking alternative optima.

To address this limitation, particle-based methods have given rise to numerous approaches, often expressed through bio-inspired metaphors. While these metaphors have facilitated intuition, they have also drawn attention to the reuse of similar underlying mechanisms, which have been carefully examined [2,19].

While these heuristics can be effective in practice, they generally lack the theoretical guarantees required for principled exploration of the fitness landscape. In this work, we propose to adapt the particle-based dynamics of Stein Variational Gradient Descent (`SVGD`) [18] for black-box optimization. `SVGD` transports a collection of particles toward a target distribution by combining functional-gradient updates with kernel-induced repulsion. This mechanism allows the algorithm to ascend efficiently toward high-quality regions of the search space while preserving population diversity through a mathematically grounded repulsive term.

Recently, the integration of `SVGD` with `CMA-ES` for continuous black-box optimization has been investigated [6]. In our work, we demonstrate the feasibility of combining EDAs with `SVGD` for combinatorial black-box optimization, where applying direct gradient descent techniques is not possible.

Our contributions in this paper are as follows:

- We introduce a general framework that enables a principled extension of any EDA into a multi-EDA scheme that seeks to approximate directly a target Boltzmann distribution.
- We propose a new `SVGD` objective that preserves invariance under monotonic transformations of the objective function. From this formulation, we derive a general rank-based update rule suitable for any multi-agent EDA.
- We provide a proof of concept of our framework that extends univariate EDAs in a multi-agent EDA with `SVGD`. Empirical evidence indicates that repulsion arising from the kernel plays a role in maintaining population diversity, facilitating exploration in deceptive discrete spaces where traditional EDAs commonly experience premature convergence. These early results suggest strong potential for convergence toward high-quality solutions in challenging combinatorial landscapes.

The remainder of this paper is organized as follows. First, we describe the background and position our work (Sect. 2). We then introduce our `SVGD-EDA` algorithm (Sect. 3). Next, we provide an empirical evaluation through a series of experiments on benchmark tasks (Sect. 4). Finally, we discuss future research directions in the conclusion.

2 Background and Motivations

We consider a discrete search space $\mathcal{X} = \mathcal{X}_1 \times \cdots \times \mathcal{X}_n$ and aim to solve the maximization problem $\max_{x \in \mathcal{X}} f(x)$, where $x = (x_1, \ldots, x_n)$ denotes a candidate solution in $\mathcal{X}$.

In this section, we begin by summarizing the core principles of Estimation-of-Distribution Algorithms (EDAs), before explaining why we shift toward a broader perspective centered on approximating a target distribution. We then present the Stein Variational Gradient Descent method [18], which we will use in the next section to leverage our probabilistic formulation of combinatorial optimization as a variational inference problem, free from the constraints imposed by a restricted parametric family of distributions.

2.1 From Estimation-of-Distribution Algorithms (EDAs) to Variational Inference

EDAs are population-based algorithms that frame optimization as an iterative density estimation process, where the objective is to learn a probability distribution that increasingly concentrates its mass on high-quality regions of the search space. Instead of relying on genetic operators such as crossover or mutation, an EDA evolves a distribution π_θ from a parametric family Π, with parameters $\theta \in \Theta$.

Thus, rather than directly dealing with the former optimization problem over $\mathcal{X}$, EDAs consider an equivalent reformulation of the search over the parameter space Θ of a probabilistic model π_θ: $\max_{\theta \in \Theta} \mathbb{E}_{x \sim \pi_\theta}\left[f(x)\right]$. At each iteration t, the algorithm performs three steps: (i) a population of λ samples $x^1, \ldots, x^\lambda$ is generated from the current model π_{θ^t}; (ii) the objective function is evaluated on these samples; (iii) the parameter vector is updated to θ^{t+1} so as to increase the likelihood of generating high-fitness solutions. This cycle continues until a predefined stopping criterion is reached.

From an information-theoretic viewpoint, EDAs can be embedded in a more general variational framework in which we consider a target Boltzmann distribution over the parameter space Θ, expressed as:

$$p(\theta) \propto \exp\left(\frac{1}{\gamma}\, \mathbb{E}_{x \sim \pi_\theta}[f(x)]\right). \tag{1}$$

Following the maximum-entropy principle, the distribution $q \in \mathcal{Q}$ that minimizes the Kullback–Leibler divergence from p is

$$q^* = \arg\min_{q \in \mathcal{Q}} D_{\mathrm{KL}}(q \| p) = \arg\max_{q \in \mathcal{Q}} \mathbb{E}_{\theta \sim q} \mathbb{E}_{x \sim \pi_\theta}[f(x)] + \gamma \mathcal{H}(q), \tag{2}$$

where $\mathcal{H}(q)$ denotes the entropy of q. This formulation highlights the role of the temperature parameter γ: as $\gamma \to 0$, p becomes sharply peaked, and the optimal q collapses to a Dirac distribution centered at the maximizer θ^*. In this regime, corresponding to classical EDAs, exploration over $\mathcal{X}$ arises solely from the epistemic uncertainty encoded in the generator π_θ and the diversity of high-quality sampled solutions. Because traditional EDAs typically rely on unimodal generator models, such as multivariate Gaussians (CMA-ES) [14] or factorized Bernoulli models (PBIL) [5], the resulting search is constrained to represent only simple, single-mode structures, often leading to premature convergence on multimodal landscapes

We argue that optimizing (2) with $\gamma > 0$ can mitigate such collapse by explicitly regularizing the search distribution through its entropy, thereby preserving diversity at the level of $q(\theta)$, even when the underlying generator π_θ remains unimodal. Under this perspective, the variational inference problem acts as a landscape exploration mechanism: it aims to discover and characterize the dominant basins of attraction that structure the fitness landscape, enabling subsequent intensification within these basins to obtain high-quality solutions.

To address this variational inference problem without restricting the search to a predefined parametric family $\mathcal{Q}$, we propose to rely on Stein Variational Gradient Descent (SVGD). The next subsection introduces SVGD in its original formulation, while Sect. 3 adapts it to our combinatorial optimization setting, leveraging a set of interacting particles to capture complex, potentially multimodal distributions over Θ.

2.2 Stein Variational Gradient Descent (SVGD)

Given a target distribution p on a space Θ, the SVGD algorithm [18] considers a tractable family $\mathcal{Q}$ of probability measures on Θ, where each $q \in \mathcal{Q}$ is represented by the empirical distribution of particles $\{\theta_i\}_{i=1}^m$. The method updates an initial empirical measure q_0 through transport maps

$$T_t(\theta) = \theta + \epsilon_t\, \phi^\star_{q_t,p}(\theta), \qquad q_{t+1} = q_{t[T_t]},$$

so as to decrease $\mathrm{KL}(q_t \| p)$.

The vector field $\phi^\star_{q_t,p}$ is given by the kernelized Stein operator

$$\phi^\star_{q_t,p}(\theta) = \mathbb{E}_{\theta' \sim q_t}\left[k(\theta',\theta)\, \nabla_{\theta'} \log p(\theta') + \nabla_{\theta'} k(\theta',\theta)\right].$$

In practice, with particles (or samples) $\{\theta_i^t\}_{i=1}^m \sim q_t$, the updates become $\theta_i^{t+1} \leftarrow \theta_i^t + \epsilon_t\, \widehat{\phi}^\star(\theta_i^t)$ with

$$\widehat{\phi}^\star(\theta_i^t) = \frac{1}{m}\sum_{j=1}^m \left[k(\theta_j^t, \theta_i^t)\, \nabla_{\theta_j} \log p(\theta_j)\Big|_{\theta_j=\theta_j^t} + \nabla_{\theta_j} k(\theta_j, \theta_i^t)\Big|_{\theta_j=\theta_j^t} \right]. \qquad (3)$$

The kernel term involving $\nabla_{\theta_j} k()$ induces repulsion, whereas the term weighted by $\nabla_{\theta_j} \log p()$ transports particles toward regions of high target density (See Fig. 1(a)). These interactions prevent collapse and allow SVGD to capture multimodal targets, providing a flexible nonparametric alternative to parametric inference.

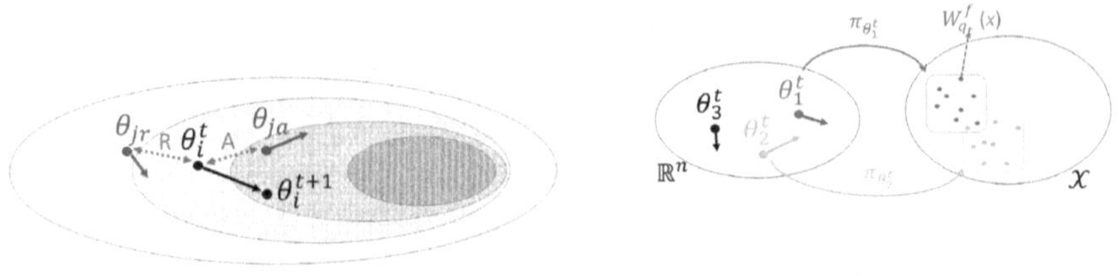

(a) SVGD particle update from θ_i^t to θ_i^{t+1}, using (3). Green particle θ_{ja} induces attraction via $k(\theta_{ja}, \theta_i^t)$; red particle θ_{jr} induces repulsion.

(b) SVGD-EDA process. At step t, particles θ_i^t sample solutions in $\mathcal{X}$ via $\pi_{\theta_i^t}$. Solutions estimate $W_{q_t}^f(x)$ (Section 3.2).

Fig. 1. Illustration of SVGD mechanisms: (a) SVGD-EDA at the particle-level, (b) SVGD-EDA process, including solution sampling.

3 Stein Variational Combinatorial Optimization

This section begins by presenting the proposed methodology, which leverages the SVGD framework to directly approximate the target Boltzmann distribution. We then introduce a variant featuring a modified objective function that remains invariant under any monotonic transformation of the original objective. Finally, we derive practical update rules that enable a multi-agent EDA to tackle discrete black-box optimization problems. In this formulation, each agent's parameter vector is interpreted as a particle within the SVGD algorithm.

3.1 Derivation of a Multi-agent Estimation of Distribution Algorithm with SVGD

We formalize a multi-agent EDA as a collection of m parameter vectors $\{\theta_i\}_{i=1}^m$, where each θ_i parameterizes a policy acting as the underlying probabilistic model of an agent. At each iteration t, every agent i generates λ solutions. Within the SVGD framework presented in Sect. 2.2, each parameter vector θ_i^t is treated as a sample from a time-varying empirical distribution q_t. Our goal is to make q evolve so that it converges toward the target Boltzmann distribution $p(\theta)$, defined by (1).

First, leveraging the log-derivative trick on the Boltzmann target (1), for $j = 1, \ldots, m$, we estimate $\nabla_{\theta_j} \log p(\theta_j)\Big|_{\theta_j=\theta_j^t}$ using the λ solutions $\{x^{j(\ell)}\}_{\ell=1}^{\lambda}$ generated by agent j:

$$\nabla_{\theta_j} \log p(\theta_j)\Big|_{\theta_j=\theta_j^t} = \frac{1}{\gamma} \int_{x\in\mathcal{X}} f(x) \nabla_{\theta_j} \log \pi_{\theta_j}(x)\Big|_{\theta_j=\theta_j^t} \pi_{\theta_j^t}(x) dx \tag{4}$$

$$\approx \frac{1}{\lambda\gamma} \sum_{\ell=1}^{\lambda} f(x^{j(\ell)}) \nabla_{\theta_j} \log \pi_{\theta_j}(x^{j(\ell)})\Big|_{\theta_j=\theta_j^t}. \tag{5}$$

Then, by substituting this approximation into the empirical Stein operator (3), we obtain the following update rule for each agent i at iteration t, which is influenced by all parameter vectors $\{\theta_1^t, \dots, \theta_m^t\}$ through

$$\theta_i^{t+1} \leftarrow \theta_i^t + \epsilon \frac{1}{m} \sum_{j=1}^{m} \left[\frac{k(\theta_i^t, \theta_j^t)}{\lambda\gamma} \sum_{\ell=1}^{\lambda} f(x^{j(\ell)}) \nabla_{\theta_j} \log \pi_{\theta_j}(x^{j(\ell)}) \Big|_{\theta_j=\theta_j^t} + \nabla_{\theta_j} k(\theta_i^t, \theta_j) \Big|_{\theta_j=\theta_j^t} \right]. \quad (6)$$

It gives us a first version of multi-agent EDA with SVGD allowing the maximization of $\mathbb{E}_{x\sim\pi_\theta}[f(x)]$ through the variational approximation of the target distribution (1).

However, this update rule is highly sensitive to extreme values of the objective function f. As a consequence, the parameters involved in (6), such as the learning rate ϵ, must be carefully tuned to the scale and variability of the problem-specific fitness landscape. By contrast, a desirable property exhibited by many widely used EDA algorithms, such as CMA-ES [14] and PBIL [5], is the invariance of the search dynamics under monotonic transformations of the objective function. This ensures that algorithmic behavior does not depend on the absolute scaling of f, thereby improving robustness across heterogeneous optimization tasks. This motivates the development, presented in the next section, of an alternative formulation whose objective inherently preserves such invariance.

3.2 Invariance-Preserving Objective for Black-Box Optimization with SVGD

Drawing on the Information-Geometric Optimization (IGO) framework [20], the raw objective f can be replaced by a monotone, quantile-based transformation $W^f_{\theta^t}$ that depends on the current parameter vector θ^t of the generator. In [20], the optimization problem $\max_{\theta\in\Theta} \mathbb{E}_{x\sim\pi_\theta}[f(x)]$ (see Sect. 2.1) is accordingly reformulated at each iteration t as $\max_{\theta\in\Theta} \mathbb{E}_{x\sim\pi_\theta}\left[W^f_{\theta^t}(x)\right]$, with $W^f_{\theta^t}(x) := w\big(\mathbb{P}_{x'\sim\pi_{\theta^t}}(f(x') > f(x))\big)$ where w is a non-increasing function.

However, in our context, since we aim to evolve a distribution q over parameters θ with SVGD, the objective cannot depend on a single θ^t. Instead, we propose to optimize $J_t^f(\theta) = \mathbb{E}_{x\sim\pi_\theta}[W^f_{q_t}(x)]$, with $W^f_{q_t}(x) = w\Big(\mathbb{E}_{\theta^t\sim q_t}\Big[\mathbb{P}_{x'\sim\pi_{\theta^t}}(f(x') > f(x))\Big]\Big)$ and $w : [0,1] \to \mathbb{R}$ a non-increasing function.

At iteration t, this objective can be interpreted as searching for parameters θ that maximize, up to the monotone transformation w, the probability that samples drawn from π_θ outperform on average those generated by the previous generation of EDA agents π_{θ^t} with $\theta^t \sim q_t$.

This transformation promotes consistent policy improvement throughout the search process, and guarantees invariance with respect to any monotonic transformation of the objective function f.

The Boltzmann distribution that we seek to approximate then becomes, at iteration t, $p_t(\theta) \propto \exp\Big(\frac{1}{\gamma}\mathbb{E}_{x\sim\pi_\theta}\left[W^f_{q_t}(x)\right]\Big)$.

When approximating this target distribution with SVGD, the update direction for each agent at each iteration t becomes then

$$\theta_i^{t+1} \leftarrow \theta_i^t + \epsilon \frac{1}{m} \sum_{j=1}^{m} \left[\frac{k(\theta_i^t, \theta_j^t)}{\lambda\gamma} \sum_{\ell=1}^{\lambda} W_{q_t}^f(x^{j(\ell)}) \nabla_{\theta_j} \log \pi_{\theta_j}(x^{j(\ell)}) \Big|_{\theta_j=\theta_j^t} + \nabla_{\theta_j} k(\theta_i^t, \theta_j) \Big|_{\theta_j=\theta_j^t} \right], \tag{7}$$

where an unbiased estimator of $W_{q_t}^f(x^{j(\ell)})$ using sampled solutions at iteration t is given by
$w\left(\frac{1}{m}(\sum_{i=1,i\neq j}^{m} \frac{1}{\lambda} \left|\{k : f(x^{i(k)}) > f(x^{j(\ell)})\}\right| + \frac{1}{\lambda-1} \left|\{k \neq \ell : f(x^{j(k)}) > f(x^{j(\ell)})\}\right|)\right)$.

The choice of the weighting function w used to compute the score $W_{q_t}^f(x^{j(\ell)})$ for each candidate solution plays a central role in shaping the selection pressure exerted during the update of agent parameters. By transforming the ordinal ranks of individuals into utility values, the function w determines the relative contribution of each sampled point when updating the model. This principle is fundamental to rank-based optimization methods such as CMA-ES, where careful tuning of utility weights has been shown to substantially influence performance across diverse classes of optimization problems [1].

In this work, we adopt a simplified form of utility weighting by introducing the function $w : [0,1] \to \mathbb{R}$ defined as $w(x) = 1 - 2x$. When this function is used and all sampled points $x^{j(\ell)}$ are distinct, $W_{q_t}^f(x^{j(\ell)})$ can be approximated by $\hat{S}(x^{j(\ell)}) = 1 - 2\,\frac{\text{rk}(x^{j(\ell)}, \Gamma^t, f)}{m\lambda - 1}$, where $\Gamma^t = \{x^{i(k)}\}_{i=1..m,\ k=1..\lambda}$ denotes the complete population at iteration t, and $\text{rk}(x, \Gamma, f) = \left|\{x' \in \Gamma : f(x') > f(x)\}\right|$ is the number of individuals in Γ with strictly higher fitness (with ties broken at random). Thus, under this very simple scoring scheme, the best solution of the complete population sampled at iteration t attains rank 0 and obtains score +1, while the worst receives rank $m\lambda - 1$ and score -1.

3.3 Practical Implementation for Black-Box Discrete Optimization

In this section, we implement the general framework of the multi-agent EDA presented in the previous section to solve discrete black-box optimization problems (See Fig. 1(b)). This algorithm is called SVGD-EDA. The version of SVGD-EDA designed to solve pseudo-Boolean problems is summarized in Algorithm 1.

For pseudo-Boolean problems, the probabilistic model associated with each agent is chosen to be a univariate Bernoulli distribution.[1] In this case, for each agent $i \in [1..m]$, the parameter vector $\theta_i = (\theta_{i1}, \dots, \theta_{in}) \in \mathbb{R}^n$ specifies the log-odds of the binary variables. The probability of sampling a solution $x \in \mathcal{X}$ from π_{θ_i} is $\pi_{\theta_i}(x) = \prod_{l=1}^{n} \sigma(\theta_{il})^{x_l}(1 - \sigma(\theta_{il}))^{1-x_l}$, with $\sigma(z) = (1 + e^{-z})^{-1}$ the sigmoid function.

[1] More expressive multivariate discrete probabilistic models, such as those employed in MIMIC [8], BOA [21], or more recently in RL-EDA [13] could also be incorporated to capture dependencies between variables, but exploring such extensions is left for future work.

When the search space is categorical, $\mathcal{X} = [1..D]^n$. For each agent i and variable l, we maintain a vector of logits $\theta_{il} = (\theta_{il1}, \dots, \theta_{ilD}) \in \mathbb{R}^D$. The softmax function converts these into the categorical distribution $\mathbb{P}(x_l = d \mid \theta_{il}) = \frac{\exp(\theta_{ild})}{\sum_{d'=1}^{D} \exp(\theta_{ild'})}$. Assuming independence across variables, the probability of sampling a solution $x \in \mathcal{X}$ from the distribution π_{θ_i} is given by $\pi_{\theta_i}(x) = \prod_{l=1}^{n} \frac{\exp(\theta_{il,x_k})}{\sum_{d=1}^{D} \exp(\theta_{ild})}$.

Regarding the choice of the kernel function k, we propose to employ the standard Radial Basis Function (RBF) defined as $k(\theta_i, \theta_j) = \exp\left(-\frac{\|\theta_i - \theta_j\|^2}{2h^2}\right)$, where $h > 0$ is the bandwidth. To ensure that the kernel scale remains appropriate throughout optimization, the bandwidth h is adapted using the median trick of [18].

Algorithm 1. SVGD-EDA for pseudo-Boolean problems with parameters $m, \lambda \in \mathbb{N}^*$, $\epsilon, \gamma \in \mathbb{R}$

Input: instance $(\mathcal{X}, f)$, with $\mathcal{X} = \{0,1\}^n$, objective $f : \mathcal{X} \to \mathbb{R}$ and number of iterations T.
# Initialization
for $i = 1$ **to** m **do**
  Randomly initialize the parameter vector $\theta_i^0 = (\theta_{i1}^0, \dots, \theta_{in}^0)$
end for
$f^* \leftarrow -\infty$
for $t = 0$ **to** $T - 1$ **do**
  # Generation and evaluation of solutions
  for $j = 1$ **to** m **do**
    for $\ell = 1$ **to** λ **do**
      $x^{j(\ell)} \sim \pi_{\theta_j^t}$
      $f' \leftarrow f(x^{j(\ell)})$
      if $f' > f^*$ **then**
        $x^* \leftarrow x^{j(\ell)}$ and $f^* \leftarrow f'$
      end if
    end for
  end for
  # Update EDAs parameters using SVGD
  for $i = 1$ **to** m **do**
    $\theta_i^{t+1} \leftarrow \theta_i^t + \epsilon \frac{1}{m} \sum_{j=1}^{m} \left[\frac{k(\theta_i^t, \theta_j^t)}{\lambda\gamma} \sum_{\ell=1}^{\lambda} \hat{S}(x^{j(\ell)}) \nabla_{\theta_j} \log \pi_{\theta_j}(x^{j(\ell)}) + \nabla_{\theta_j} k(\theta_i^t, \theta_j) \right]$
  end for
end for
Output: best solution x^*

4 Experiments

The objectives of this section are twofold. First, we benchmark the performance of the proposed multi-agent EDA against established state-of-the-art methods for pseudo-Boolean black-box optimization. Second, we conduct an ablation study to identify the critical algorithmic components that facilitate effective exploration of the search space, providing insight into the mechanism underlying the algorithm's diversity-preserving properties.

The source code, the instances and the complete results for all algorithms are available at the URL https://github.com/LandOS973/Supplementary-material.

4.1 Experimental Protocol and Parameters Setting

We focus on the NKD model, which is a natural extension of the NK model of Kauffman [15] to cases where variables can take more than two categorical values. This is a framework for describing fitness landscapes whose problem size and ruggedness are both parameterizable. The NKD function is defined as $f_{\text{NKD}}(x) = \frac{1}{n}\sum_{i=1}^{n} \gamma_i(x_i, x_{l_{i1}}, \ldots, x_{l_{iK}})$, where each subfunction $\gamma_i : \{0, 1, \ldots, D_i - 1\}^{K+1} \rightarrow [0, 1[$ is defined over categorical variables with D_i possible values for each variable. We generate instances with $D_i = 2$ for every variable, which corresponds to the original pseudo-Boolean NK problem, but we also generate instances of a categorical problem called NK3 with $D_i = 3$ for every variable.

For each problem variant (NK or NK3), we generated four instance distributions with $K \in \{1, 2, 4, 8\}$ are built. Increasing K expands the size of each epistatic neighborhood, which in turn increases the ruggedness of the landscape and typically leads to a larger number of local optima. For each combinatorial problem, we conducted evaluations across three problem sizes $n \in \{64, 128, 256\}$ and we generated 10 distinct instances for each configuration (n, K). Following a black-box optimization setting, we assume no prior knowledge of the objective function or of the interaction structure between variables.

The behavior of the multi-agent `SVGD-EDA` depends on several key hyperparameters. Across all experiments, we use a fixed configuration with $m = 7$, $\lambda = 13$, and $\gamma = 0.015$, selected via a grid search over a representative set of problem instances. This choice is not critical: the performance remains stable and qualitatively similar for a range of nearby parameter settings.

Thanks to a fully tensorial PyTorch implementation, `SVGD-EDA` efficiently exploits hardware acceleration for large-scale evaluations. Empirical measurements on an Intel Core i5-12400F CPU and an NVIDIA RTX 3060 GPU show that solving a single NK instance ($n = 256$) with a budget of 50,000 iterations takes respectively 7 s and 3 s (including fitness evaluations). By leveraging GPU vectorization, an entire batch of 100 independent runs with 50,000 iterations is processed simultaneously in about 20 s on the GPU, compared to 300 s on the CPU for instances of this size.

4.2 Experimental Validation on Discrete Black-Box Benchmarks

We evaluate the performance of our multi-agent `SVGD-EDA` against a comprehensive set of 80 baseline algorithms from the `Nevergrad` library (v1.0.12) [23], which includes classic metaheuristics (Evolutionary Strategies, Differential Evolution) and machine-learning-driven adaptive portfolios, often represented by multiple variants of the same algorithm with different hyperparameter configurations. We enriched this pool with three classical EDAs: `PBIL` [5], `MIMIC` [8], and `BOA` [22], using their default hyperparameter settings and the source code available at https://github.com/e5120/EDAs. This yields a total of 84 evaluated algorithms (our method and 83 baselines) for the pseudo-Boolean NK instances. However, because `PBIL` is strictly limited to binary search spaces, it was excluded from the experiments on the categorical NK3 instances, reducing the total ranking pool to 83 algorithms for that specific problem class. We impose a budget of 50,000 objective function evaluations per run, consistent with recent evolutionary algorithm benchmarking protocols with this number of function evaluations for these instance sizes [24,25]. To ensure statistical robustness, each algorithm is evaluated over 100 independent runs per (n, K) problem configuration (10 runs on 10 distinct instances).

Global Experimental Results. Table 1 presents a selection of these results, comparing `SVGD-EDA` to three other EDAs of the same category: `PBIL` [5], `MIMIC` [8], and `BOA` [22]. The final columns report the performance of the best algorithm among all remaining competitors, including the 80 Nevergrad discrete algorithms [23]. Based on this average score, the algorithms are ranked, and their position among all competitors is indicated.

As seen in Table 1, `SVGD-EDA` reaches the 1st rank in 17 out of the 24 tested distributions of instances (and obtains the best average rank of 5.375 across (n, K) configurations). When not ranked 1st, the algorithm is very frequently 2nd. `SVGD-EDA` consistently shows superior performance compared to other Estimation-of-Distribution Algorithms such as `PBIL`, `MIMIC`, or `BOA`. On pseudo-Boolean NK landscapes, it demonstrates strong scalability as the instance size increases with $n = 128$ and $n = 256$, where it systematically leads the 84 algorithms across all ruggedness levels ($K = 1, 2, 4, 8$). On NK3 instances, `SVGD-EDA` dominates the rankings for $n = 128$ and $n = 256$, with the exception of the most constrained cases ($K = 8$) where specific `OnePlusOne` variants maintain an edge.

This trend suggests that the cooperative mechanism driven by the `SVGD` kernel is effective on the largest instances evaluated in our study. By maintaining population diversity through agent interaction, the method appears to mitigate the premature convergence and stagnation that affect standard independent agents or simpler EDAs as the problem size increases.

Evolution of the Average Scores During the Search. Figure 2 on the left illustrates the evolution of the average best score from the start of the search on a categorical NK3 landscape ($n = 128$, $K = 2$, with three categories per variable). The green curve corresponds to `SVGD-EDA` compared to the five best-performing competing methods (over 83) with a budget of 50,000 evaluations. From this plot, we observe that `DiscreteLengler3OnePlusOne` (orange curve), an elitist (1+1) evolutionary strategy with a decreasing mutation scope, exhibits faster

Table 1. Global rankings and average scores at 50,000 evaluations (100 runs). **Bold** values indicate the best mean score. An asterisk (*) denotes a statistically significant improvement over the second-best method (Wilcoxon signed-rank test on instance-wise means, $p < 0.05$). The final row summarizes the global mean ranks and mean scores across all instances.

Instances			SVGD-EDA		PBIL		MIMIC		BOA		Best Method (Others)		
Pb	n	t	Rank	Score	Rank	Score	Rank	Score	Rank	Score	Name	Rank	Score
NK	64	1	2/84	0.7124	36/84	0.7097	54/84	0.7052	56/84	0.7008	DiscreteDE	**1/84**	**0.7126**
		2	2/84	0.7475	10/84	0.7412	53/84	0.7312	56/84	0.7262	DiscreteDE	**1/84**	**0.7497**
		4	2/84	0.7609	4/84	0.7492	54/84	0.7330	55/84	0.7316	DiscreteDE	**1/84**	**0.7658**
		8	**1/84**	**0.7538**	8/84	0.7368	54/84	0.7176	55/84	0.7174	DiscreteDE	2/84	0.7535
	128	1	**1/84**	**0.7118***	3/84	0.7085	54/84	0.6956	56/84	0.6928	DiscreteDE	2/84	0.7094
		2	**1/84**	**0.7406***	3/84	0.7334	55/84	0.7147	56/84	0.7135	DiscreteDE	2/84	0.7364
		4	**1/84**	**0.7646***	3/84	0.7514	54/84	0.7229	53/84	0.7285	DiscreteDE	2/84	0.7535
		8	**1/84**	**0.7454***	3/84	0.7369	55/84	0.7034	56/84	0.6681	DiscreteDE	2/84	0.7373
	256	1	**1/84**	**0.7091***	2/84	0.7048	56/84	0.6802	50/84	0.6863	DiscreteDE	3/84	0.7029
		2	**1/84**	**0.7395***	2/84	0.7314	54/84	0.6989	48/84	0.7093	HugeLognormalDiscreteOnePlusOne	3/84	0.7269
		4	**1/84**	**0.7598***	2/84	0.7429	55/84	0.7019	47/84	0.7162	HugeLognormalDiscreteOnePlusOne	3/84	0.7352
		8	**1/84**	**0.7334***	2/84	0.7315	55/84	0.6805	63/84	0.5852	OnePtRecombiningDiscreteLenglerOnePlusOne	3/84	0.7242
NK3	64	1	2/83	0.7894	—	—	54/83	0.7659	55/83	0.7651	DiscreteDE	**1/83**	**0.7918**
		2	**1/83**	**0.8204***	—	—	51/83	0.7825	54/83	0.7793	DiscreteDE	2/83	0.8154
		4	**1/83**	**0.8066***	—	—	55/83	0.7645	58/83	0.7548	DiscreteDE	2/83	0.8033
		8	14/83	0.7627	—	—	59/83	0.6583	57/83	0.7000	DiscreteDE	**1/83**	**0.7779***
	128	1	**1/83**	**0.7975***	—	—	48/83	0.7600	50/83	0.7530	DiscreteDE	2/83	0.7918
		2	**1/83**	**0.8123***	—	—	48/83	0.7621	55/83	0.7523	DiscreteDE	2/83	0.7959
		4	**1/83**	**0.7989***	—	—	54/83	0.7469	56/83	0.7321	RandRecombiningDiscreteLenglerOnePlusOne	2/83	0.7829
		8	44/83	0.7491	—	—	63/83	0.6098	58/83	0.6297	OnePtRecombiningDiscreteLenglerOnePlusOne	**1/83**	**0.7618***
	256	1	**1/83**	**0.7889***	—	—	47/83	0.7363	49/83	0.7238	DiscreteLengler3OnePlusOne	2/83	0.7741
		2	**1/83**	**0.8035***	—	—	47/83	0.7393	49/83	0.7249	DiscreteLengler3OnePlusOne	2/83	0.7839
		4	**1/83**	**0.7823***	—	—	47/83	0.7265	54/83	0.7072	DiscreteLengler3OnePlusOne	2/83	0.7783
		8	46/83	0.7285	—	—	63/83	0.5784	62/83	0.5790	SmallLognormalDiscreteOnePlusOne	**1/83**	**0.7552***
SVGD-EDA			5.375	0.763							**DiscreteDE**	10.625	0.756

initial progress than SVGD-EDA. Its favorable performance for small optimization budgets stems from its large initial mutation neighborhood, which allows for fast progress by combining multiple elementary moves [9,17]. However, the progressive reduction of this neighborhood turns the algorithm into a 1-flip random hill-climber, leading to early stagnation once a local optimum is reached. In contrast, SVGD-EDA exhibits sustained improvement throughout the optimization process and ultimately attains better final results. Figure 2 on the right compares the distribution of final scores with those of the five best competing algorithms. SVGD-EDA not only attains a higher mean performance but also exhibits strong reliability, as evidenced by a tight interquartile range across 100 independent runs, indicating consistent solution quality.

4.3 Sensitivity Analysis: Impact of the Number of EDA Agents m

To assess the influence of the multi-agent architecture and the repulsive Stein force on optimization performance, we conducted a sensitivity analysis on the number of EDA agents, denoted by m.

We evaluated the framework across different sizes of instances ($n \in \{64, 128, 256\}$) on NK and NK3. We tested 16 configurations for the number of agents: $m \in [1..16]$. The other hyperparameters remain at their baseline values as described

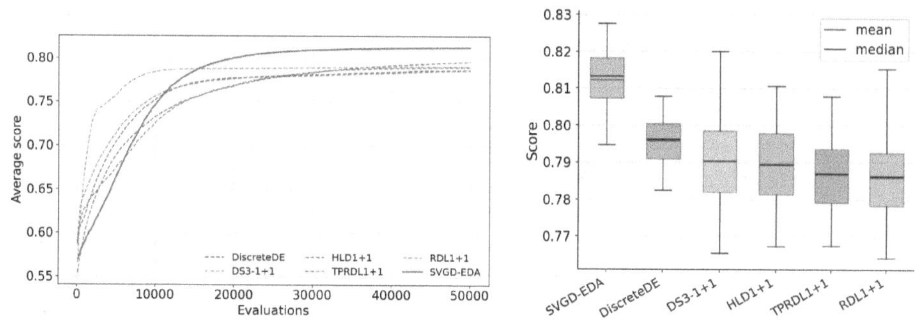

Fig. 2. Performance analysis of `SVGD-EDA` with top-five competitors on an NK3 instance distribution ($n = 128, K = 2$). Left: Evolution of the average score. Right: Final score distribution after 50,000 evaluations.
Acronyms in legends:
`DS3-1+1` $\rightarrow$ `DiscreteLengler3OnePlusOne`,
`HLD1+1` $\rightarrow$ `HugeLognormalDiscreteOnePlusOne`,
`TPRDL1+1` $\rightarrow$ `TwoPtRecombiningDiscreteLenglerOnePlusOne`,
`RDL1+1` $\rightarrow$ `RecombiningDiscreteLenglerOnePlusOne`.

in Sect. 4.1. These experiments were conducted under a strict fixed budget of 50,000 evaluations. Consequently, an increase in m mechanically reduces the total number of generations available for the EDA agents to converge, as the budget per EDA agent is approximately $50,000/m$. This setup tests the trade-off between broad parallel exploration and deep local exploitation. Finally, to ensure statistical robustness, each reported result represents the aggregate of 100 independent runs.

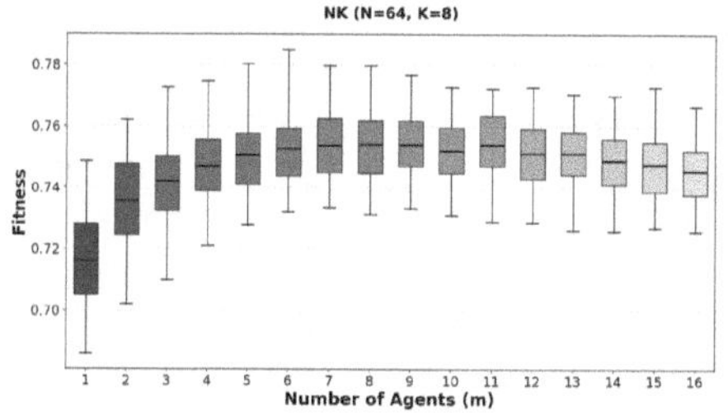

(a) NK $n = 64, K = 8$

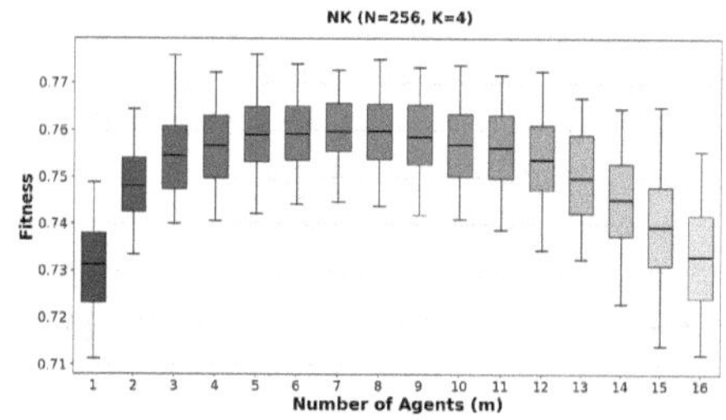

(b) NK $n = 256, K = 4$

Fig. 3. Sensitivity analysis on NK landscapes. The boxplots illustrate the distribution of final fitness scores across 100 independent runs for varying numbers of EDA agents m (x-axis). Higher scores indicate better performance.

Figure 3 displays the distribution of the final fitness scores achieved at the end of the 50,000 evaluations for different EDA agent counts $m \in [1..16]$. These box-plots reveal a distinct behavioral shift governed by the interaction between dimensionality and ruggedness.

On moderate-dimensional problems ($n = 64$), deploying a larger number of EDA agents ($m \in [10..14]$) is generally beneficial to maximize broad exploration. However, as the dimensionality increases to $n = 256$ or as the landscape becomes more rugged (e.g., $K = 8$), the optimal configuration systematically shifts towards a more compact EDA agent count ($m \approx 5$ to 7). This phenomenon suggests that the optimal m is not solely a function of dimensionality, but rather a consequence of the combined complexity: both higher dimensionality and increased ruggedness amplify the need for a larger individual evaluation budget per EDA agent to ensure convergence.

Importantly, performance degrades noticeably across all problem types for $m > 14$. This is directly attributed to *budget dilution*: with too many EDA agents under a strictly fixed evaluation budget, the number of generations available per EDA becomes insufficient to perform the necessary model refinements. In conclusion, for expensive black-box optimization, a balanced configuration (such as $m = 7$) appears to offer the most robust compromise between collective landscape coverage and individual search depth.

4.4 Ablation Study on the Impact of Stein Attraction/Repulsion Mechanisms in SVGD-EDA

To isolate the specific contribution of the Stein repulsive force, we conducted an ablation study comparing the standard multi-agent SVGD-EDA ($m = 7$) against a baseline of non-interacting EDA agents. In the "No Interaction" setting, the RBF kernel in (7) is replaced by the identity matrix (i.e., $k(\theta_i, \theta_j) = \delta_{ij}$) and the gradient of the kernel term $\nabla_{\theta_j} k(\theta_i, \theta_j)$ is forced to zero. This effectively decouples the EDA agents, reducing the algorithm to m independent EDAs running in parallel.

While detailed tabular results over all configurations are omitted for brevity, aggregated data reveal a clear correlation between problem complexity and the utility of cooperation. On simple, low-dimensional binary spaces (e.g., NK $n = 64, K \leq 4$), independent agents perform similarly to the interacting ones. However, as the dimensionality increases, the landscape becomes more rugged, or the search space shifts to a categorical domain (NK3), the interacting approach systematically outperforms the independent baseline. For instance, performance gaps reach up to +15 % on large, highly rugged NK landscapes ($n = 256, K = 8$) and consistently exceed +10 % across all categorical NK3 instances.

To illustrate the mechanics behind this performance gap, Fig. 4 highlights a distinct behavioral advantage provided by the multi-agent interaction on a specific NK3 instance ($n = 64, K = 2$). Although both methods initially progress and appear to plateau around 20,000 evaluations, the independent baseline gets permanently trapped in a sub-optimal region. In contrast, the kernel-induced repulsion maintains enough population diversity for the interacting EDA agents to eventually escape this deceptive local optimum and locate a superior global solution ($\approx$ +10 % performance gap).

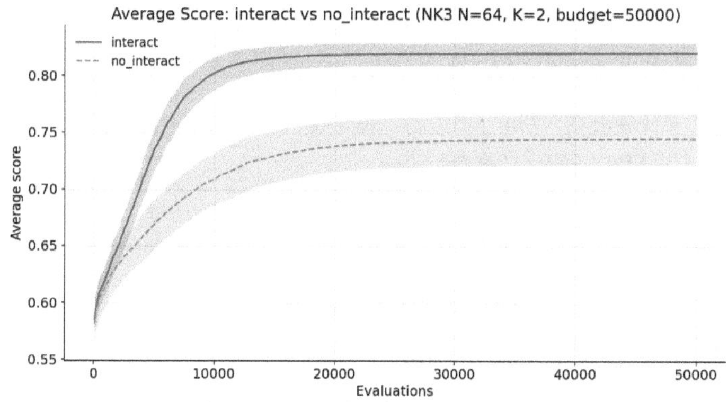

Fig. 4. Impact of Interaction on NK3 with $n = 64$ and $K = 2$. Comparison of average score evolution between cooperative (blue) and independent (orange) agent EDAs. The repulsive force prevents early stagnation in deceptive local optima. (Color figure online)

5 Conclusion

In this paper, we introduced a principled multi-agent Estimation-of-Distribution Algorithm (`SVGD-EDA`) designed for complex black-box combinatorial optimization. By explicitly mapping the Stein Variational Gradient Descent framework to the parameter space of independent Bernoulli and categorical distributions, our approach structurally prevents the premature variance collapse that typically afflicts standard EDAs and local search heuristics.

Our empirical validation on instances of up to $n = 256$ variables of binary NK and categorical NK3 landscapes demonstrates that `SVGD-EDA` effectively avoids early stagnation and sustains continuous improvement throughout the search. The ablation study further isolates and quantifies the contribution of kernel-induced repulsion, revealing performance improvements of up to 15 % over non-interacting parallel EDAs by enabling the population to bypass deceptive local optima.

Future work will focus on two concrete technical extensions. First, given the limitations of the Euclidean RBF kernel applied to logit parameters, we will investigate information-geometric kernels (e.g., Fisher-Rao or Jensen-Shannon) defined directly on the statistical manifold, so that repulsive forces reflect meaningful distributional divergences. Second, to mitigate the budget dilution effect while preserving the exploratory benefits of a large population, we plan to study sparse or asynchronous update schemes, in which only a subset of agents is updated at each iteration.

Acknowledgments. The authors acknowledge ANR – FRANCE (French National Research Agency) for its financial support of the COMBO project (PRC - AAPG 2023 - Axe E.2 - CE23). This work was granted access to the HPC resources of IDRIS (Grant No. AD010617253 and AD010611887R3) from GENCI. Some computations were also performed using the computer clusters and data storage resources of the GLiCID (Groupement Ligérien pour le calcul Intensif Distribué, www.glicid.fr).

Disclosure of Interests. The authors have no competing interests to declare that are relevant to the content of this article.

References

1. Andersson, M., Bandaru, S., Ng, A.H., Syberfeldt, A.: Parameter tuned CMA-ES on the CEC'15 expensive problems. In: 2015 IEEE Congress on Evolutionary Computation (CEC), pp. 1950–1957. IEEE (2015)
2. Aranha, C., et al.: Metaphor-based metaheuristics, a call for action: the elephant in the room. Swarm Intell. **16**(1), 1–6 (2022)
3. Audet, C., Kokkolaras, M.: Blackbox and derivative-free optimization: theory, algorithms and applications. Optim. Eng. **17**(1), 1–2 (2016)
4. Back, T.: Evolutionary algorithms in theory and practice: evolution strategies, evolutionary programming, genetic algorithms, Oxford University Press (1996)
5. Baluja, S.: Population-based incremental learning. a method for integrating genetic search based function optimization and competitive learning, Carnegie Mellon University (1994). Tech. rep
6. Braun, C.V., Lange, R.T., Toussaint, M.: Stein variational evolution strategies. arXiv preprint arXiv:2410.10390 (2024)
7. Brochu, E., Cora, V.M., de Freitas, N.: A tutorial on bayesian optimization of expensive cost functions, with application to active user modeling and hierarchical reinforcement learning. arXiv preprint arXiv:1012.2599 (2010)
8. Bonet, D., Isbell, J., Viola, C., P.: MIMIC: finding optima by estimating probability densities. In: Advances in Neural Information Processing Systems, vol. 9 (1996)
9. Doerr, B., Doerr, C., Lengler, J.: Self-adjusting mutation rates with provably optimal success rules. In: Proceedings of the Genetic and Evolutionary Computation Conference, pp. 1479–1487 (2019)
10. Eiben, A.E., Smith, J.E.: Introduction to evolutionary computing, Springer (2015)
11. Forrester, A.I., Keane, A.J.: Recent advances in surrogate-based optimization. Prog. Aerosp. Sci. **45**(1–3), 50–79 (2009)
12. Frazier, P.I.: Bayesian optimization. In: Recent Advances in Optimization and Modeling of Contemporary Problems, pp. 255–278. Informs (2018)
13. Goudet, O., Suire, Q., Goëffon, A., Saubion, F., Lamprier, S.: Black-box combinatorial optimization with order-invariant reinforcement learning. arXiv preprint arXiv:2510.01824 (2025)
14. Hansen, N., Ostermeier, A.: Completely derandomized self-adaptation in evolution strategies. Evol. Comput. **9**(2), 159–195 (2001)
15. Kauffman, S.A., Weinberger, E.D.: The NK model of rugged fitness landscapes and its application to maturation of the immune response. J. Theor. Biol. **141**(2), 211–245 (1989)
16. Larranaga, P., Bielza, C.: Estimation of distribution algorithms. In: Handbook of Heuristics, pp. 583–598. Springer (2025)
17. Lengler, J., Spooner, N.: Fixed budget performance of the (1+1) EA on linear functions. In: Proceedings of the 2015 ACM Conference on Foundations of Genetic Algorithms XIII, pp. 52–61 (2015)

18. Liu, Q., Wang, D.: Stein variational gradient descent: a general purpose bayesian inference algorithm. Adv. Neural. Inf. Process. Syst. **29** (2016)
19. Molina, D., Del Ser, J., Poyatos, J., Herrera, F.: The paradox of success in evolutionary and bioinspired optimization: revisiting critical issues, key studies, and methodological pathways. Swarm Evol. Comput. **98**, 102063 (2025)
20. Ollivier, Y., Arnold, L., Auger, A., Hansen, N.: Information-geometric optimization algorithms: a unifying picture via invariance principles. J. Mach. Learn. Res. **18**(18), 1–65 (2017)
21. Pelikan, M.: Bayesian optimization algorithm: From single level to hierarchy, University of Illinois at Urbana-Champaign (2002)
22. Pelikan, M., Goldberg, D., Cantu-Paz, E.: BOA: The Bayesian Optimization Algorithm, pp. 525–532 (1999)
23. Rapin, J., Teytaud, O.: Nevergrad - A gradient-free optimization platform. https://GitHub.com/FacebookResearch/Nevergrad (2018)
24. Ye, F., Doerr, C., Wang, H., Bäck, T.: Automated configuration of genetic algorithms by tuning for anytime performance. IEEE Trans. Evol. Comput. **26**(6), 1526–1538 (2022)
25. Ye, F., Wang, H., Doerr, C., Bäck, T.: Benchmarking a genetic algorithm with configurable crossover probability. In: International Conference on Parallel Problem Solving from Nature, pp. 699–713. Springer (2020)

The Suffering Salesperson

Measuring the Difficulty of TSP Instances and Generating Instances with Weighted Anytime Runtime Performance

Jonathan Heins[1,2(✉)], Anna Franke[1,2], Markus Leyser[1], Sebastian Dengel[1], and Pascal Kerschke[1,2]

[1] TUD Dresden University of Technology, Dresden, Germany
jonathan.heins@tu-dresden.de
[2] ScaDS.AI Dresden/Leipzig, Dresden, Germany

Abstract. The Traveling Salesperson Problem (TSP) is a fundamental problem with widespread relevance in modern applications such as parcel delivery. Still, the factors that determine instance difficulty for state-of-the-art TSP solvers, such as LKH and EAX, remain insufficiently understood. Prior work has focused on constructing instances that are challenging for either of the two solvers based on the runtime required by LKH or EAX, respectively, for finding the optimal tour of the instance. However, such measures do not adequately capture the anytime behavior of heuristic algorithms. To address this limitation, we introduce *weighted Anytime Runtime Performance (wARP)*, a novel measure which explicitly accounts for the progression of solution quality over time. An analysis of instances considered difficult by established criteria reveals their diverse anytime behaviors, ranging from instances where solvers quickly find near-optimal tours to instances where near-optimal tours are not found. These patterns can be captured by the new *wARP* measure. Leveraging the sensitivity of wARP, we propose a gradual instance generation process that constructs instances that are consistently challenging for both LKH and EAX. This provides an alternative basis for evaluating algorithmic variants and enables a more nuanced analysis of their strengths and weaknesses. Furthermore, the proposed concept is transferable to other domains, providing a foundation for more informative benchmarking and targeted instance generation beyond the TSP.

Keywords: Traveling Salesperson Problem · LKH · EAX · Anytime Performance · Instance Generation · Benchmarking

1 Introduction

Although it was first formulated almost a century ago [4,26], the *Traveling Salesperson Problem (TSP)* remains a cornerstone of teaching and research in various disciplines, ranging from computer science and operations research via logistics and transportation science to physics and biology [2]. Essentially, the TSP seeks a minimum-cost Hamiltonian cycle through a set of N vertices $V = \{v_1, \ldots, v_N\}$.

G. Iacca et al. (Eds.): PPSN 2026, LNCS 16985, pp. 466–482, 2027.
https://doi.org/10.1007/978-3-032-36214-8_29

That is, the goal is to find a cyclic tour that visits every vertex exactly once and returns to the starting vertex. Given edge costs $c : E \rightarrow \mathbb{R}$, the objective is to identify the Hamiltonian cycle of minimum total cost.

Over the past two decades, (variants of) two heuristics have emerged as the dominant methods for efficiently tackling the TSP: the Lin–Kernighan-Helsgaun (LKH) [15,19,20] heuristic and the Edge Assembly Crossover (EAX) [16,28,29]. Both algorithms have demonstrated remarkable performance across a wide range of benchmark instances and are widely regarded as state of the art in practical TSP solving. Understanding the respective strengths and limitations of these methods, as well as how their performance is influenced by structural properties of the instances, requires systematic benchmark studies. Ideally, such studies would rely on extensive collections of real-world instances that vary in size and properties. In practice, however, the availability of such instances is limited, and their ability to represent real-world characteristics remains uncertain.

Consequently, researchers started to generate artificial TSP instances that are easy or hard for a given solver [5,37]. Within their evolutionary mechanisms, simple and sophisticated mutation operators have been considered, enabling the systematic generation of TSP instances with diverse structures – ranging from nearly unstructured, random configurations to instances exhibiting pronounced structural patterns. These instance sets formed the basis for several benchmarking and algorithm selection studies [12,13,35]. Nevertheless, despite the substantial amount of experimental studies, the increased usage of feature-based analyses and novel visualization techniques [9,13,14,21,27,32], our understanding of what determines the difficulty of a given TSP instance remains very limited. It thus is still largely unclear which structural properties render an instance easy or challenging for state-of-the-art solvers such as EAX and LKH, and, consequently, how instance generation mechanisms influence algorithmic performance.

An essential component of any benchmarking study is the performance assessment, which fundamentally depends on the choice of an appropriate metric. In TSP, performance is typically assessed under the following paradigms: a fixed-target setting (e.g., reaching an optimal solution or achieving a predefined approximation ratio α), a fixed-budget setting (e.g., setting a maximum runtime), or a hybrid approach that terminates once one of the two conditions is met. Since TSP solvers often incorporate stochastic elements, benchmark studies commonly involve multiple independent runs per instance, necessitating performance measures that aggregate outcomes across both successful and unsuccessful trials [3]. The most common performance metric in this domain is PAR10, but other approaches such as PQR10 or ERT – which is a widely accepted performance metric in continuous optimization [3,11] – have also been considered in past studies [22]. However, these metrics share an inherent limitation: they rely on fixed budgets or targets, and thus enable only a very limited view at the performance. In particular, they fail to capture the algorithm's anytime behavior, i.e., how solution quality evolves over time. In other domains, such as continuous optimization, algorithmic runtime profiles are increasingly employed, enabling more nuanced comparisons [3,10,11,41]. In TSP, however, such assessments are still the exception [6]. As a consequence, it is usually difficult to assess whether an instance is intrinsically challenging or whether a solver simply requires most of its computational budget for identifying the final few improving edge exchanges.

2 Related Work

Generating problem instances that are diverse, challenging for specific solvers (or ideally both) is a promising approach for understanding the relationship between instance characteristics and solver performance, which can ultimately support the design of more effective algorithms [38]. Consequently, a substantial body of work has investigated methods for generating problem instances across various combinatorial optimization domains, including the Knapsack Problem [24,25,33], graph coloring [36], NK landscapes [34], or bin packing [1]. In the context of the TSP, early work proposed a method based on an elitist evolutionary loop to evolve random uniform euclidean (RUE) instances with 100 cities each, with the objective of making them either easier or harder for two variants of the Lin-Kernighan heuristic [37]. Employing a related evolutionary approach, [5] evolved TSP instances that are challenging for either EAX or LKH. Extending the work of [37], they designed and employed more sophisticated mutation operators and generated larger instances (between 500 and 2,000 cities per instance), resulting in a greater diversity of structural patterns. The results indicate that sophisticated mutation strategies can induce substantial diversity in instance features, even without explicitly optimizing for diversity. Other approaches focus primarily on promoting diversity, either exclusively [36] or by incorporating explicit diversity measures into the evolutionary instance generation process. These include novelty search [24,25], discrepancy measures [30], multi-objective indicators [31], and quality-diversity approaches [8]. However, existing TSP instance generation methods exhibit two key limitations. First, only [5] considers state-of-the-art solvers. Second, the performance metrics commonly used are sensitive to whether the exact optimal solution is found. In practical settings, however, the focus might shift from exact optimality to the time-dependent quality gap between the found solution and the optimum. Furthermore, performance metrics must account for unsuccessful runs, which is typically achieved by applying penalty factors or adapting bootstrap concepts. To investigate the impact of these aggregation strategies, [22] compared the PAR10 score, one of the most widely used metrics in the TSP, with alternative measures such as PQR and ERT, demonstrating that PQR tends to favor more robust solvers, which is often desirable. Additionally, it has been proposed to combine the number of failed runs with the PQR score using a hypervolume-based indicator [7]. Nevertheless, these metrics still primarily emphasize whether and when the optimal solution is reached. Small investigations of the anytime behavior of EAX and LKH on specifically evolved hard instances suggest that even slight relaxations of the target solution quality can drastically reduce the number of failed runs [5]. However, the authors mention that a general measure for capturing the anytime performance of solvers remains an open challenge.

3 Background

Since the proposed difficulty measure and instance generation process are evaluated with respect to the heuristics LKH and EAX, the two algorithms considered in this work are briefly introduced in the following.

LKH. [19,20,39] is Helsgaun's efficient extension of the classical Lin-Kernighan (LK) heuristic [23]. The LK heuristic aims to efficiently search for k-opt moves, i.e., removing k edges from a solution and add k new edges to form a valid, improved tour. Rather than enumerating arbitrary k-opt exchanges, LK constructs such moves as chains of sequential 2-opt and potentially one 3-opt step, so that each removed edge is incident to an added edge, and vice versa, while the partial exchange remains extendable to a feasible tour. Consequently, not all improving k-opt moves are explored, since non-sequential exchanges are excluded. Further, instead of considering all possible edges when adding edges in the sequential exchange, LK restricts the search to a candidate set, typically consisting of the five nearest neighbors of each node. This reduces the search space and improves computational efficiency. However, it also introduces the risk that edges belonging to the optimal tour are not included in the candidate sets.

Based on the LK framework, Helsgaun proposed several improvements in the LKH heuristic [19]. In particular, candidate sets are constructed using minimum 1-trees and α-nearness [17,18]. A minimum 1-tree captures important global structures of a good tour. It is obtained by constructing a minimum spanning tree (MST) on all but one node and then connects the excluded node by its two shortest incident edges. The α-value of an edge indicates how much the cost of the minimum 1-tree increases if that edge is forced into the tree. Helsgaun [19] shows that edges of an optimal tour receive, on average, a lower rank when the incident edges of each node in the complete TSP graph are ranked by α-nearness rather than by Euclidean distance, which makes it a more effective criterion for selecting candidate edges in the subsequent LK search. To further improve candidate selection, LKH employs subgradient optimization [19]. This procedure iteratively transforms the distance matrix so that the minimum 1-tree becomes more tour-like while leaving the optimal tour unchanged. To further reduce runtime, Helsgaun [19] replaces random initialization with a simple construction heuristic by mainly using edges with small α-values and which are already part of the best currently found tour.

EAX. [28,29] is a powerful genetic algorithm named for its distinctive crossover operator, Edge Assembly Crossover. In this paper, however, EAX refers only to the heuristic, not to the crossover operator itself. EAX is initialized with a population of tours that have already been locally optimized using 2-opt. Even a relatively small collection of 2-opt tours is generally sufficient to contain most of the edges necessary for constructing optimal solutions [15,40].

The main algorithm consists of a local phase and a global phase, both of which generate offspring solutions. The key concepts for this construction are

AB-cycles and *E-sets*. Two parent tours, denoted by A and B, are combined by forming the union graph G_{AB} of their edge sets. If an edge occurs in both parent tours, it appears twice in G_{AB}. AB-cycles are obtained by traversing G_{AB} while alternating between edges from A (E_A) and edges from B (E_B). E-sets are then constructed as collections of AB-cycles according to a selection strategy, which constitutes the main distinction between the local and the global phase. An intermediate solution is obtained by removing the edges E_A and adding the edges E_B of the E-set. Since this application of an E-set to a parent may produce several subtours, a repair procedure is employed to obtain a valid offspring. In the local phase, the E-set consists of a single randomly selected AB-cycle, so that the offspring solution remains close to the parent A. In the global phase, the E-set is constructed by combining multiple AB-cycles such that the number of required repair operations is reduced. In both phases, the construction of an E-set is repeated 30 times using the default parameter setting. For each resulting offspring, a fitness value is computed on the basis of tour cost and diversity. Diversity is measured by the change in edge entropy relative to the current population, that is, by the increase or decrease in edge diversity that would result if the offspring were accepted. The best offspring replaces parent A only if it yields a tour of lower cost.

4 Methodology

Both EAX and LKH generate intermediate solutions that can be either retrieved when the time budget is exhausted or accepted once a predefined quality criterion is satisfied. However, existing performance measures typically focus solely on the time required to obtain an optimal solution, which might not reflect the practical objective guiding the selection of TSP heuristics. Instead, observing the change of performance over the runtime (referred to as the *anytime performance* as part of the *anytime behavior* of the algorithm) can capture a more complete picture. In the following, we introduce an alternative measure to summarize the anytime performance named *weighted anytime runtime performance (wARP)* to capture these characteristics more accurately.

Consider a set of instances $\mathcal{I}$ of a single-objective optimization problem, and set of algorithms $\mathcal{A}$ to solve that problem. For each independent run $\rho \in \mathcal{R}$ of algorithm $\alpha \in \mathcal{A}$ on instance $\iota \in \mathcal{I}$, observe the anytime performance $\tilde{p}_{\iota,\alpha,\rho}(t) \in \mathbb{R}$ over the finite time budget T. The anytime performance $\tilde{p}_{\iota,\alpha,\rho}(t)$ records the length of the best tour found by algorithm α on instance ι in run ρ up to time t. For improved readability, the ι, α, ρ-indices of $p_{\iota,\alpha,\rho}(t)$ are omitted in the following. In the context of the TSP, the anytime performance can be represented as a right-continuous, non-increasing step function:

$$\tilde{p}(t) = m\,\tilde{p}(t_1)\,\mathbb{1}_{[t_0,t_1)}(t) + \sum_{k=1}^{K-1}\left(\tilde{p}(t_k)\,\mathbb{1}_{[t_k,t_{k+1})}(t)\right) + \tilde{p}(t_K)\,\mathbb{1}_{[t_K,T]}(t) \quad (1)$$

where t_0 is the start time, t_k is the time when the kth next-best solution is found and T is the stop time. $\mathbb{1}$ denotes the indicator function and $m \in \mathbb{R}$ is a

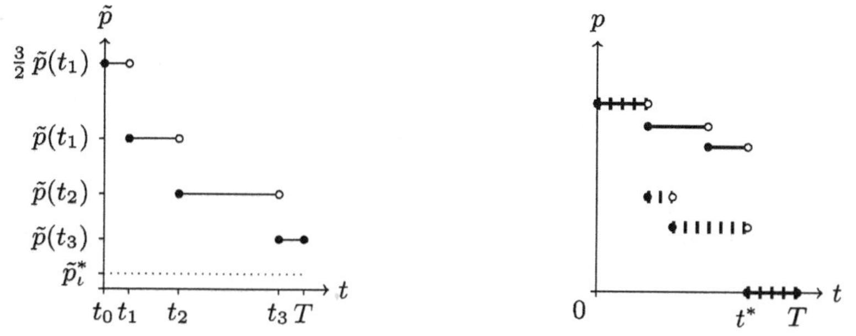

(a) Example anytime performance as described in Eq. (1) for $K = 3$ and $m = \frac{3}{2}$.

(b) Normalized anytime performance with early (hatched) / late (solid) improvements.

Fig. 1. Illustrations of (a) example and (b) normalized anytime performances.

multiplier. The interval $[t_0, t_1)$ represents the time in which no solution has been found, yet, and a multiple of the first found solution's performance value $m\, p(t_1)$ is defined as a substitute anytime performance there. This multiplier $m > 1$ penalizes a delayed discovery of the first feasible solution by assigning a worse-than-observed performance prior to t_1. Figure 1a illustrates $\tilde{p}(t)$ for $m = \frac{3}{2}$. In the experiments (see Sect. 5), we set $m = 1$ as the two algorithms considered in this paper (see Sect. 3) have very small initialization times compared to their runtime, making the choice of m negligible here.

To enable meaningful comparisons across problem instances, $p_{\iota,\alpha,\rho}(t)$ (written without tilde) denotes a normalized optimality gap rather than raw objective values $\tilde{p}_{\iota,\alpha,\rho}(t)$:

$$p_{\iota,\alpha,\rho}(t) = \frac{\tilde{p}_{\iota,\alpha,\rho}(t) - p_\iota^*}{p_\iota^*}, \tag{2}$$

where p_ι^* is the optimal tour length for instance ι, which is independent from the run ρ and the choice of the algorithm α. It should be noted that – while the TSP (i.e., a *minimization* problem) is considered here – the general concepts can be transferred to other problems as well.

4.1 Capturing Anytime Performance

Our goal is to summarize the observed normalized anytime performance $\{p_{\iota,\alpha,\rho}(t),\ \ t \in [0, T]\}$ into a scalar, facilitating statistical analysis as well as ranking of algorithms and instances, and enabling integration into instance generation or selection procedures. One simple scalarization option is $p_{\iota,\alpha,\rho}(T)$, i.e., the performance measured after the complete time budget has passed. But then, the algorithm's anytime performance is ignored, and just the end result is taken into account. Current approaches instead use t^*, i.e., the time required for the algorithm to reach the optimal tour length p_ι^*. Similarly, the algorithm's anytime performance is ignored, and just the time required to find the optimal solution is taken into account. Figure 1b illustrates two runs that would be considered equal by both aforementioned approaches, but neither approach yields a sufficiently informative description: The run shown as hatched line exhibits early improvements while the solid line represents a run with later performance improvements. In other words, they show different anytime performance. However, both runs find the optimal solution $\tilde{p}_\iota^*$ at the same time $t^* < T$ such that neither of the two above-described approaches differentiates between the two runs.

To overcome this shortcoming, we argue that the scalar performance measure should favor early performance improvements over later performance improvements: In many practical settings, optimization must be interrupted prematurely due to limited computational budgets. Early improvements are more valuable, as they increase the probability that high-quality solutions are available at interruption time. Weighting later time steps more penalizes algorithms for deferring improvements, and favors (ι, α, ρ)-combinations that are robust to unknown stopping times. From an instance perspective, instances that require sustained computation to achieve a meaningful performance receive higher scores than instances that achieve a meaningful performance faster. Yet, the early phase often reflects the quality of the random initial solutions rather than the search dynamics, and should therefore receive less attention. In the following, we propose a scalar performance measure that takes this design goal into account.

4.2 The wARP Measure

Based on the optimality gap representation from Eq. (2), we consider a scalar *anytime runtime performance (ARP)* measure:

$$ARP_{\iota,\alpha,\rho} = \int_{t_0=0}^{T} p_{\iota,\alpha,\rho}(t)dt \tag{3}$$

$ARP_{\iota,\alpha,\rho}$ measures the area under the runtime-normalized performance-curve and allows the behavior of the two runs shown in Fig. 1b to be distinguished, as ARP is smaller for runs with faster convergence, while all else being equal. However, a late improvement of the solution quality by a certain amount is still weighted the same as an early improvement by the same amount, contradicting the above-described design goal. Therefore, we introduce the *weighted anytime runtime performance* $wARP$ measure:

$$wARP_{\iota,\alpha,\rho} = \int_{0}^{T} w(t)\, p_{\iota,\alpha,\rho}(t)dt \tag{4}$$

$wARP$ can be interpreted as a weighted area-under-the-curve measure tailored to the requirements of anytime optimization. Lower $wARP$ values indicate faster convergence to high-quality solutions. To satisfy the design goal, $w(t)$ must fulfill certain requirements. First, it should be *monotonically non-decreasing* in order to emphasize later stages of the run, such that delayed improvements are penalized compared to early progress. Second, $w(t)$ must be *Riemann-integrable* to ensure that the weighted integral exists and yields a finite value over $[0, T]$. In the

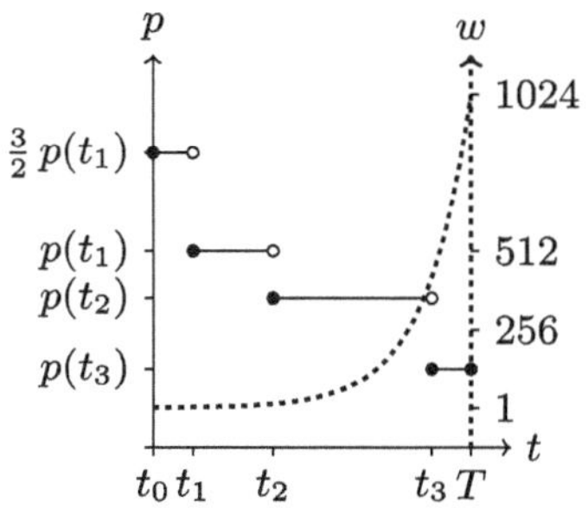

Fig. 2. Example step function and weight function for $K = 3$ and $m = \frac{3}{2}$.

following, a specific weight function $w(t)$ is chosen to instantiate the proposed framework. In particular, an exponentially increasing weighting scheme is used, such that the weight function doubles every tenth of the total runtime T.

$$w(t) = 2^{\frac{10t}{T}} \tag{5}$$

This choice emphasizes late-stage performance differences while still accounting for the anytime behavior during the complete runtime and matching the typically observed exponentially decreasing performance improvements. This is illustrated by the dashed lines in Fig. 2 where the solid lines show a normalized version of the example anytime performance from Fig. 1a. This choice fulfills the mentioned constraints: Since $w'(t) = 2^{\frac{10t}{T}} \cdot \ln(2) \cdot \frac{10}{T} > 0$ for all $t \in \mathbb{R}$, $w(t)$ is monotonous increasing. Also since $w(t)$ is a continuous function in a set interval $[0, T]$ as a composition of continuous functions and is therefore Riemann integrable. It should be noted that this choice is not unique. The proposed *wARP* formulation is intentionally flexible, and alternative weight functions may be used to reflect different preferences regarding the relative importance of early versus late improvements. Here the focus lies on introducing the performance measure and demonstrating its usefulness for analyzing and constructing challenging problem instances. By substituting $w(t)$ and $p_{\iota,\alpha,\rho}(t)$ with Eqs. (1) and (5), the indicator function induces a decomposition of the integral over $[t_0 = 0, T]$ into subintegrals over the weight function. With $\int_a^b w(t)\,dt \overset{eq.(5)}{=} \frac{T}{10ln(2)} 2^{\frac{10t}{T}}|_a^b =: W|_a^b$, *wARP* is obtained as follows:

$$\begin{aligned} wARP &= \int_{t_0=0}^{T} p(t)\,w(t)\,dt \overset{eq.(2)}{=} \int_{t_0=0}^{T} \frac{\tilde{p}(t) - \tilde{p}^*}{\tilde{p}^*}\,w(t)\,dt \\ &= \frac{1}{\tilde{p}^*} \int_{t_0=0}^{T} \tilde{p}(t) w(t)\,dt - \int_{t_0=0}^{T} w(t)\,dt \\ &= \frac{1}{\tilde{p}^*} \left(m\,\tilde{p}(t_1) W|_0^{t_1} + \sum_{k=1}^{K-1} \left(\tilde{p}(t_k) W|_{t_k}^{t_{k+1}} \right) + \tilde{p}(t_K) W|_{t_K}^{T} \right) - W|_0^T \end{aligned} \tag{6}$$

In this formulation, *wARP* is the sum of (from left to right): (i) an expression that accounts for the time until the first solution is found, (ii) weighted contributions by the first $K-1$ found next-best solutions, (iii) the weighted contribution of the best found solution within the time budget, and (iv) a negative constant reflecting the weighted performance of a run where the optimum is found right at the beginning. In other words, $wARP$ compares a given run against the ideal run. An important advantage of this formulation is that it can be computed incrementally during the optimization process. In particular, no caching of past performance values is required, as the measure only depends on the current performance value $p(t_k)$ and the elapsed times t_k and t_{k+1}. Additionally, it is not necessary to calculate the optimal tour length $\tilde{p}^*$ in advance as this is just a constant in (6). This makes wARP especially suitable for online evaluation.

5 Results

To evaluate the proposed measure not only theoretically but also empirically, we compare it against the PAR10 measure on existing instances and further investigate its capabilities through a newly designed instance generation process.

5.1 Evaluation of the wARP

Using the proposed metric, the performance of both algorithms can be reevaluated on the benchmark instances evolved by [5]. Figure 3 shows the penalized average runtime with factor 10 (PAR10) alongside the proposed $wARP$ measure for EAX (left) and LKH (right) for ten individual runs per instance. Point colors indicate the normalized optimality gap $p_{\iota,\alpha,\rho}(T)$ at the end of the run.

As described in Sect. 4, $wARP$ is designed to capture differences between early and late improvements during the optimization process. Several runs accumulate at the PAR10-penalized cutoff time, yet remain clearly separated by $wARP$. Moreover, the final relative gap increases consistently with $wARP$ in this cluster, confirming that it effectively reflects anytime performance for both algorithms. PAR10 and $wARP$ further differ in their assessment of low- and medium-difficulty instances: $wARP$ indicates lower difficulty than PAR10 for EAX, but higher difficulty for LKH. This aligns with the algorithms' behavior, EAX exhibits a longer initial setup phase due to the generation of a 2-opt-optimized initial population, resulting in a larger t^*. Since wARP places less emphasis on the initialization, the influence of this setup overhead is reduced. Interestingly, $wARP$ also identifies instances in which the optimum is reached only at the very end of the run but the best solution found before was far from the optimal one (encircled in Fig. 3). These cases might attain higher $wARP$ values than instances that do not reach the optimum, a distinction not captured by PAR10. This further supports that $wARP$ captures not only final solution quality but also anytime performance, as intended in Sect. 4.

5.2 Instance Generation Process

Furthermore, the new measure enables the generation of novel instances that are smaller in size yet more difficult for both *EAX* and *LKH*. To this end, we initialize 100 instances per algorithm, each consisting of 100 nodes sampled uniformly

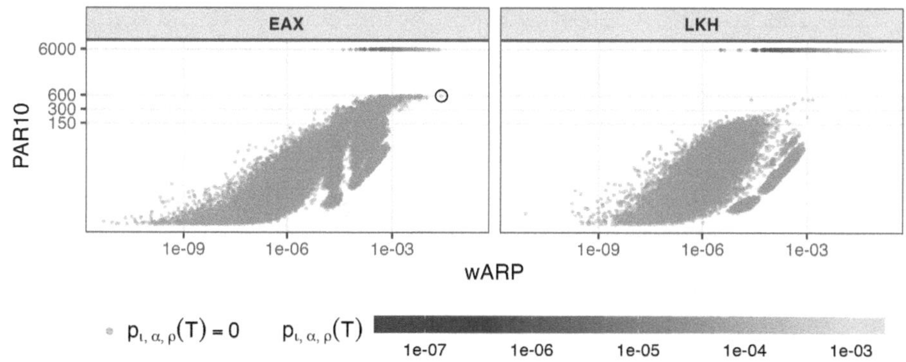

Fig. 3. wARP vs. PAR10 for EAX and LKH on original benchmark instances evolved by [5]. Point colors indicate $p_{\iota,\alpha,\rho}(T)$. Circled point: optimum found only at the cutoff, yet higher wARP than all unsolved runs.

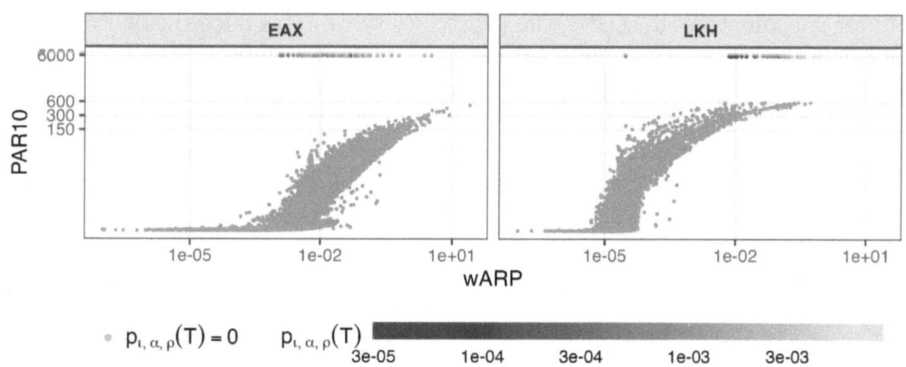

Fig. 4. wARP vs. PAR10 for EAX and LKH on instance set generated with the method described in Sect. 5.2. Point colors indicate $p_{\iota,\alpha,\rho}(T)$.

at random within an area of 1,000,000 × 1,000,000. Since $wARP$ captures gradual improvements in anytime performance, these instances can be iteratively modified in small steps to increase their difficulty for a given solver. At each iteration, a single node is added at the location that maximizes the increase in instance difficulty. To determine this position, a grid of size 100 × 100 is overlaid on the instance area, yielding 10,000 candidate locations. For each candidate, the $wARP$ value is evaluated using two different random seeds and a time budget of three seconds, which is sufficient to allow multiple algorithm restarts given the small instance size. Subsequently, successive halving with a reduction rate of 30% is applied, such that only the 7,000 most difficult candidates are evaluated using three additional seeds. The resulting $wARP$ values are aggregated by computing the mean of $\log(x + 1)$ across seeds. This aggregation emphasizes consistently

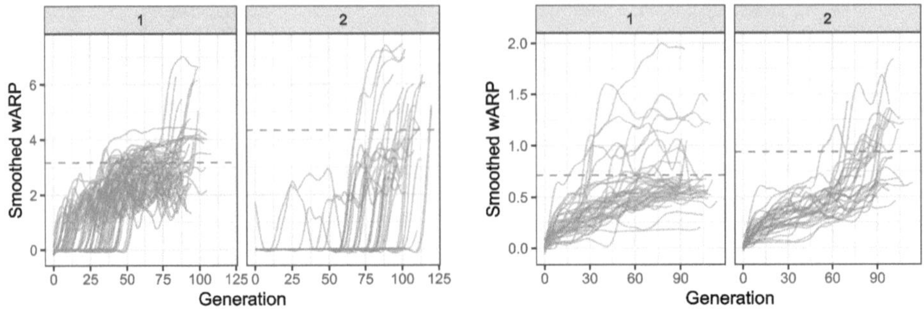

(a) EAX (b) LKH

Fig. 5. Hierarchical clustering results for smoothed *wARP*-curves visualizing the difficulty increase throughout the evolving of instances.

difficult instances across multiple runs rather than instances that are challenging only for specific random seeds. The procedure terminates once the 300 most difficult instances have been evaluated using 50 seeds. The most difficult instance is then selected as the new base instance for the next generation, where the process is repeated by adding another node. Figure 4 compares the PAR10 measure and *wARP* for a runtime limit of ten minutes across all instances generated during the instance generation process. Compared to the previously used benchmark set, a similar range of PAR10 values is obtained even though the generated instances contain only approximately 100 to 200 nodes. At the same time, the *wARP* scores are substantially higher (note the logarithmic scale on the x-axis), and, most notably, the optimality gaps are considerably larger (note the range of the color scale). We therefore conclude that the instance generation process is highly effective and produces TSP instances that are genuinely difficult.

Figure 5 illustrates the instance generation procedure for all instances whose difficulty increased substantially relative to their initialization. Each curve, representing the evolving of instances based on their associated wARP value is first smoothed using a cubic smoothing spline, before Euclidean distances are computed and agglomerative hierarchical clustering with Ward linkage is applied. The results suggest that instance difficulty is not determined solely by instance size. The different clusters vary substantially in the generation at which the instances become difficult for the solver. Some instances become challenging almost immediately, whereas others develop difficult structures only at much later stages of the generation process. In the latter case, earlier generations exhibit only modest increases in difficulty. This difference is most noticeable for EAX (Fig. 5a), but it also appears for LKH (Fig. 5b). In both cases, the concave-down shape (cluster 1) and the concave-up shape (cluster 2) are visible, though less clearly for LKH. Since the concave-up curves attain higher final mean values (dotted line), Fig. 6 provides a more detailed view of the generation process for two concave-up shaped instances, with LKH shown in the upper row and the EAX in the lower row.

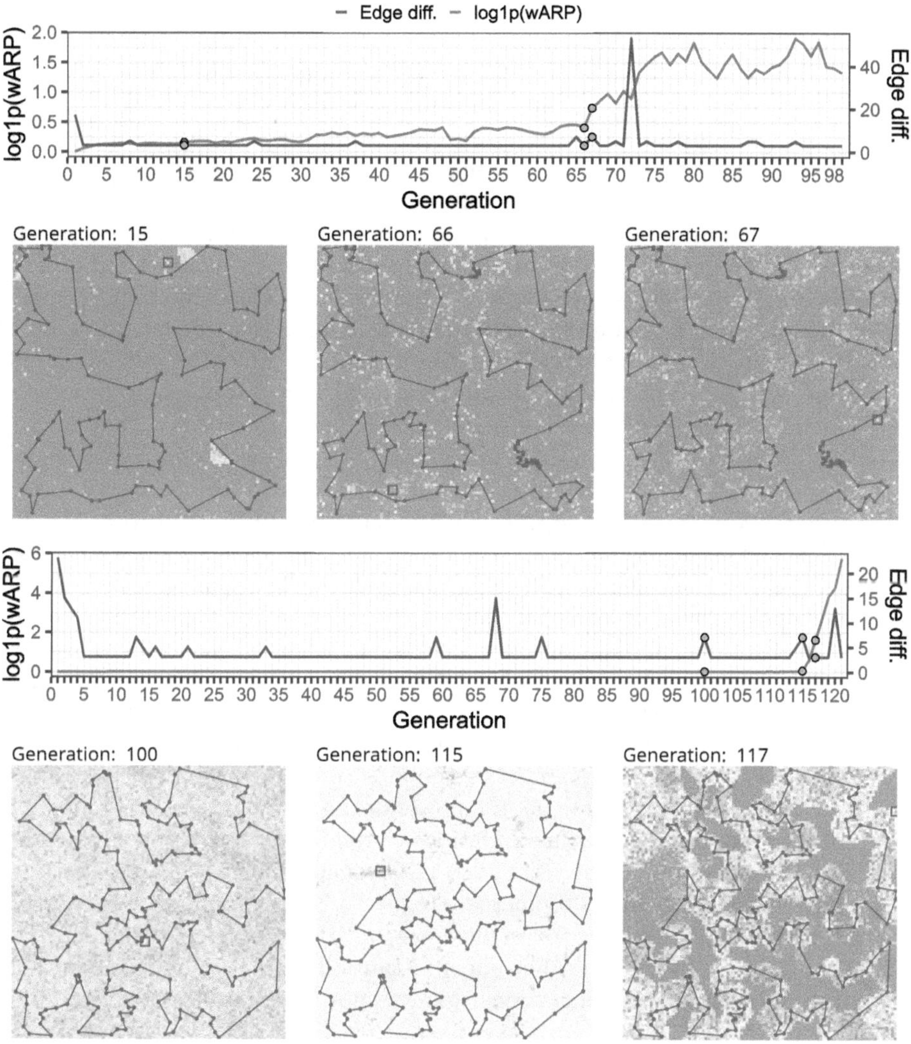

Fig. 6. Examples of evolved instances for LKH (top) and EAX (bottom). Colors show, if adding a point at the respective coordinate would increase (red), does not affect (white) or decrease (blue) instance difficulty, respectively. Red rectangles highlight the points added in the shown generation. (Color figure online)

The *wARP* curve is highlighted together with three representative generations, for which heatmaps of the difficulty landscape over possible insertion positions are shown. For each generation, the heatmap uses a continuous color gradient centered symmetrically around the *wARP* value of the point added in the previous generation, which is marked in white. Colors shifting toward red indicate regions that increase instance difficulty, while colors shifting toward blue indicate regions that decrease it.

The LKH instance begins at a slightly higher difficulty level than EAX and appears to increase more gradually, showing a pattern that is characteristic of evolved LKH instances. Comparing the heatmaps for generations 16 and 66 reveals a feature which might cause this gradual growth phase: once a region associated with a slight increase in difficulty is identified, it tends to attract repeated point insertions, resulting in the formation of small clusters of points along the tour. At generation 66, the difficulty then rises sharply. The heatmap indicates that this increase can be attributed to a single specific difficult point rather than a whole region and is accompanied by only a minor modification of the optimal tour. Consequently, whether an instance is difficult may depend not only on structural properties of the instance itself but also on the relationship between the optimal tour and the search behavior of the solver. This demonstrates the strength of the process in finding difficult structures which would have been difficult to find otherwise.

For the EAX instance, the $wARP$ value remains low until generation 114. Up to this point, the difficulty landscape resembles the one shown for generation 100: differences between cells are small overall, while the white regions tend to lie close to the current solution. In each generation, the selected cell yields only a slight increase in difficulty, and the number of edge differences between consecutive solutions also remains low. A notable change occurs at generation 115, when several dark red regions suddenly emerge, leading to the first substantial increase in difficulty. From that point onward, inserting additional nodes along the edges of the optimal tour often appears to be the most promising strategy for further increasing difficulty. This highlights the power of the instance generation process: after 114 comparatively minor modifications to the tour, typically involving only three edge differences per generation, the 115th small change reveals a structure from which the instance difficulty increases rapidly. Notably, this structure would have been difficult to anticipate in advance, as there is no obvious intuitive indication that the tour should become substantially more difficult at precisely this point and in previous generations this was not an area of interest at all.

6 Conclusion

In our study we introduced the new performance measure $wARP$ to assess the difficulty of TSP instances for a given algorithm by incorporating anytime runtime behavior. In contrast to measures based solely on runtime or the final normalized optimality gap, $wARP$ applies a weighting function that rewards early improvements and penalizes late ones. This enables a more informative quantification of instance difficulty across different algorithms. Our study shows that $wARP$ differentiates runs with different normalized optimality gap $p_{\iota,\alpha,\rho}(T)$ for both LKH and EAX on an established benchmark set with both easy and difficult instances. This is not possible for measures based only on runtime. $wARP$ also identifies difficult instances for which the optimum is reached only shortly before the cutoff time. In addition, we generated 100 difficult instances per algorithm using $wARP$ as the optimization criterion. Despite their comparatively few nodes, the generated instances also yield large PAR10 scores and substantially larger optimality gaps than instances from the literature that were considered to be challenging for EAX or LKH, respectively.

Furthermore, the generation process successfully identifies structural modifications that create pronounced challenges for both algorithms, even when the specific cause of the increased difficulty is not immediately apparent. These observations may help explain the limited success of algorithm selection studies for the TSP. First, existing performance measures may fail to capture the aspects of solver behavior that are most relevant in practice. Second, instead of using features that characterize an instance's structural node layout by summary statistics, features capturing the structure of the *optimal tour* may be required to accurately predict whether an instance will be difficult for a given solver. More broadly, the proposed generation process enables the controlled and gradual modification of instances toward increased difficulty. This opens a promising avenue for future research on instance understanding and, ultimately, the development of more effective solvers.

Acknowledgments. The authors gratefully acknowledge the computing time made available to them on the high-performance computer at the NHR Center of TU Dresden. This center is jointly supported by the Federal Ministry of Education and Research and the state governments participating in the NHR (www.nhr-verein.de/unsere-partner). The authors would also like to thank Sandy Wolf for her kind support and encouragement throughout this work.

Instances. The instances generated for this work are publicly available at https://github.com/jonathan-h1/TSP-Instances.

Disclosure of Interests. The authors have no competing interests to declare that are relevant to the content of this article.

Author Contributions. Sebastian Dengel: Investigation; Software; Validation; Visualization; Writing – original draft; Writing – review & editing. Anna Franke: Formal analysis; Methodology; Software; Validation; Visualization; Writing – original draft; Writing – review & editing. Jonathan Heins: Conceptualization; Data curation; Formal analysis; Investigation; Methodology; Project administration; Software; Supervision; Validation; Visualization; Writing – original draft; Writing – review & editing. Pascal Kerschke: Funding acquisition; Resources; Validation; Writing – original draft; Writing – review & editing. Markus Leyser: Formal analysis; Methodology; Software; Validation; Visualization; Writing – original draft; Writing – review & editing.

References

1. Alissa, M., Sim, K., Hart, E.: Algorithm selection using deep learning without feature extraction. In: Proceedings of the Genetic and Evolutionary Computation Conference (GECCO '19), pp. 198–206. ACM, Prague Czech Republic (2019). https://doi.org/10.1145/3321707.3321845

2. Applegate, D.L., Bixby, R.E., Chvátal, V., Cook, W. (eds.): The Traveling Salesman Problem: A Computational Study. Princeton Series in Applied Mathematics, Princeton University Press, Princeton (2011)
3. Bartz-Beielstein, T., et al.: Benchmarking in optimization: Best practice and open issues. arXiv:2007.03488v2 (2020) cs.NE
4. Bock, F.: Mathematical programming solution of traveling salesman examples. Presented at the (1963) Recent Advances in Mathematical Programming
5. Bossek, J., Kerschke, P., Neumann, A., Wagner, M., Neumann, F., Trautmann, H.: Evolving diverse TSP instances by means of novel and creative mutation operators. In: Proceedings of the 15th ACM/SIGEVO Conference on Foundations of Genetic Algorithms (FOGA '19), pp. 58–71. ACM, Potsdam Germany (2019). https://doi.org/10.1145/3299904.3340307
6. Bossek, J., Kerschke, P., Trautmann, H.: Anytime behavior of inexact TSP solvers and perspectives for automated algorithm selection. In: 2020 IEEE Congress on Evolutionary Computation (CEC), pp. 1–8. IEEE, Glasgow, United Kingdom (2020). https://doi.org/10.1109/CEC48606.2020.9185613
7. Bossek, J., Kerschke, P., Trautmann, H.: A multi-objective perspective on performance assessment and automated selection of single-objective optimization algorithms. Appl. Soft Comput. **88**, 105901 (2020). https://doi.org/10.1016/j.asoc.2019.105901
8. Bossek, J., Neumann, F.: Exploring the feature space of TSP instances using quality diversity. In: Proceedings of the Genetic and Evolutionary Computation Conference (GECCO '22), pp. 186–194. ACM, Boston Massachusetts (2022). https://doi.org/10.1145/3512290.3528851
9. Elorza, A., Benavides, X., Ceberio, J., Hernando, L., Lozano, J.A.: Transforming combinatorial optimization problems in fourier space: consequences and uses. IEEE Trans. Evol. Comput. **29**(4), 977–989 (2024)
10. Hansen, N., Auger, A., Brockhoff, D., Tusar, T.: Anytime performance assessment in blackbox optimization benchmarking. IEEE Trans. Evol. Comput. **26**(6), 1293–1305 (2022). https://doi.org/10.1109/TEVC.2022.3210897
11. Hansen, N., Auger, A., Ros, R., Finck, S., Pošík, P.: Comparing results of 31 algorithms from the black-box optimization benchmarking BBoB-2009. In: Proceedings of the 12th Annual Conference Companion on Genetic and Evolutionary Computation, pp. 1689–1696 (2010)
12. Heins, J., Bossek, J., Pohl, J., Seiler, M., Trautmann, H., Kerschke, P.: On the Potential of Normalized TSP Features for Automated Algorithm Selection. In: Proceedings of the 16th ACM/SIGEVO Conference on Foundations of Genetic Algorithms (FOGA), pp. 1–15. ACM (2021). https://doi.org/10.1145/3450218.3477308
13. Heins, J., Bossek, J., Pohl, J., Seiler, M., Trautmann, H., Kerschke, P.: A study on the effects of normalized TSP features for automated algorithm selection. Theoret. Comput. Sci. **940**, 123–145 (2023). https://doi.org/10.1016/j.tcs.2022.10.019
14. Heins, J., Dengel, S., Whitley, D., Kerschke, P.: Clearing the combinatorial fog: tracing the hidden paths of TSP heuristics. In: Proceedings of the 18th ACM/SIGEVO Conference on Foundations of Genetic Algorithms, pp. 73–84. ACM (2025). https://doi.org/10.1145/3729878.3746623
15. Heins, J., Schäpermeier, L., Kerschke, P., Whitley, D.: Dancing to the state of the art? how candidate lists influence LKH for solving the traveling salesperson problem. In: Affenzeller, M., Winkler, S.M., Kononova, A.V., Trautmann, H., Tušar, T., Machado, P., Bäck, T. (eds.) Parallel Problem Solving from Nature – PPSN XVIII, pp. 100–115. Springer Nature Switzerland, Cham (2024). https://doi.org/10.1007/978-3-031-70055-2_7

16. Heins, J., Whitley, D., Kerschke, P.: To repair or not to repair? Investigating the importance of AB-cycles for the state-of-the-art TSP heuristic EAX. In: Proceedings of the Genetic and Evolutionary Computation Conference, Malaga, Spain, pp. 231–239. ACM. https://doi.org/10.1145/3712256.3726436
17. Held, M., Karp, R.M.: The traveling-salesman problem and minimum spanning trees. Oper. Res. **18**(6), 1138–1162 (1970)
18. Held, M., Karp, R.M.: The traveling-salesman problem and minimum spanning trees: part II. Math. Program. **1**(1), 6–25 (1971). https://doi.org/10.1007/BF01584070
19. Helsgaun, K.: An effective implementation of the Lin-Kernighan traveling salesman heuristic. Eur. J. Oper. Res. **126**(1), 106–130 (2000). https://doi.org/10.1016/S0377-2217(99)00284-2
20. Helsgaun, K.: General k-opt submoves for the Lin-Kernighan TSP heuristic. Math. Program. Comput. **1**(2–3), 119–163 (2009). https://doi.org/10.1007/s12532-009-0004-6
21. Hutter, F., Xu, L., Hoos, H.H., Leyton-Brown, K.: Algorithm runtime prediction: methods & evaluation. Artif. Intell. (AIJ) **206**, 79–111 (2014). http://www.sciencedirect.com/science/article/pii/S0004370213001082
22. Kerschke, P., Bossek, J., Trautmann, H.: Parameterization of state-of-the-art performance indicators: a robustness study based on inexact TSP solvers. In: Proceedings of the Genetic and Evolutionary Computation Conference Companion (GECCO '18), pp. 1737–1744. ACM, Kyoto Japan (2018). https://doi.org/10.1145/3205651.3208233
23. Lin, S., Kernighan, B.W.: An effective heuristic algorithm for the traveling-salesman problem. Oper. Res. **21**(2), 498–516 (1973). https://doi.org/10.1287/opre.21.2.498
24. Marrero, A., Segredo, E., Hart, E., Bossek, J., Neumann, A.: Generating diverse and discriminatory knapsack instances by searching for novelty in variable dimensions of feature-space. In: Proceedings of the Genetic and Evolutionary Computation Conference (GECCO '23), pp. 312–320. ACM, Lisbon Portugal (2023). https://doi.org/10.1145/3583131.3590504
25. Marrero, A., Segredo, E., León, C., Hart, E.: A novelty-search approach to filling an instance-space with diverse and discriminatory instances for the knapsack problem. In: Rudolph, G., Kononova, A.V., Aguirre, H., Kerschke, P., Ochoa, G., Tušar, T. (eds.) Parallel Problem Solving from Nature – PPSN XVII. vol. 13398, pp. 223–236. Springer International Publishing, Cham (2022). https://doi.org/10.1007/978-3-031-14714-2_16
26. Menger, K.: Das Botenproblem. In: Menger, K. (ed.) Ergebnisse Eines Mathematischen Kolloquiums, vol. 2. Teubner, Leipzig (1932)
27. Mersmann, O., Bischl, B., Trautmann, H., Wagner, M., Bossek, J., Neumann, F.: A novel feature-based approach to characterize algorithm performance for the traveling salesperson problem. Ann. Math. Artif. Intell. **69**, 151–182 (2013). https://doi.org/10.1007/s10472-013-9341-2
28. Nagata, Y., Kobayashi, S.: Edge assembly crossover: a high-power genetic algorithm for the travelling salesman problem. In: Bäck, T. (ed.) Proceedings of the 7th International Conference on Genetic Algorithms (ICGA), pp. 450–457. Morgan Kaufmann, East Lansing, MI, USA (1997)
29. Nagata, Y., Kobayashi, S.: A powerful genetic algorithm using edge assembly crossover for the traveling salesman problem. INFORMS J. Comput. **25**(2), 346–363 (2013). https://doi.org/10.1287/ijoc.1120.0506

30. Neumann, A., Gao, W., Doerr, C., Neumann, F., Wagner, M.: Discrepancy-based evolutionary diversity optimization. In: Proceedings of the Genetic and Evolutionary Computation Conference (GECCO '18), pp. 991–998. ACM, Kyoto Japan (2018). https://doi.org/10.1145/3205455.3205532
31. Neumann, A., Gao, W., Wagner, M., Neumann, F.: Evolutionary diversity optimization using multi-objective indicators. In: Proceedings of the Genetic and Evolutionary Computation Conference (GECCO '19), pp. 837–845. ACM, Prague Czech Republic (2019). https://doi.org/10.1145/3321707.3321796
32. Pihera, J., Musliu, N.: Application of machine learning to algorithm selection for TSP. In: Proceedings of the IEEE 26th International Conference on Tools with Artificial Intelligence (ICTAI), pp. 47–54. IEEE (2014). https://doi.org/10.1109/ICTAI.2014.18, https://www.dbai.tuwien.ac.at/staff/musliu/art_ictai_cam.pdf
33. Plata-González, L.F., Amaya, I., Ortiz-Bayliss, J.C., Conant-Pablos, S.E., Terashima-Marín, H., Coello Coello, C.A.: Evolutionary-based tailoring of synthetic instances for the Knapsack problem. Soft. Comput. **23**(23), 12711–12728 (2019). https://doi.org/10.1007/s00500-019-03822-w
34. Santana, R., Mendiburu, A., Lozano, J.: Using network measures to test evolved NK-landscapes. Tech. rep., Department of Computer Science and Artificial Intelligence, University of the Basque Country (2012). technical Report EHU-KZAA-TR: 2012-03
35. Seiler, M., Pohl, J., Bossek, J., Kerschke, P., Trautmann, H.: Deep learning as a competitive feature-free approach for automated algorithm selection on the traveling salesperson problem. In: Bäck, T., Preuss, M., Deutz, A., Wang, H., Doerr, C., Emmerich, M., Trautmann, H. (eds.) Parallel Problem Solving from Nature – PPSN XVI, vol. 12269, pp. 48–64. Springer International Publishing. https://doi.org/10.1007/978-3-030-58112-1_4
36. Smith-Miles, K., Bowly, S.: Generating new test instances by evolving in instance space. Comput. Oper. Res. **63**, 102–113 (2015). https://doi.org/10.1016/j.cor.2015.04.022
37. Smith-Miles, K., Van Hemert, J., Lim, X.Y.: Understanding TSP Difficulty by Learning from Evolved Instances. In: Hutchison, D., et al. (eds.) Learning and Intelligent Optimization (LION 4). vol. 6073, pp. 266–280. Springer Berlin Heidelberg, Berlin, Heidelberg (2010). https://doi.org/10.1007/978-3-642-13800-3_29
38. Smith-Miles, K.A.: Cross-disciplinary perspectives on meta-learning for algorithm selection. ACM Comput. Surv. **41**(1), 1–25 (2009). https://doi.org/10.1145/1456650.1456656
39. Taillard, É.D., Helsgaun, K.: POPMUSIC for the travelling salesman problem. Eur. J. Oper. Res. **272**(2), 420–429 (2019). https://doi.org/10.1016/j.ejor.2018.06.039
40. Varadarajan, S., Whitley, D.: The massively parallel mixing genetic algorithm for the traveling salesman problem. In: Proceedings of the Genetic and Evolutionary Computation Conference (GECCO '19), pp. 872–879. Association for Computing Machinery, New York, NY, USA (2019). https://doi.org/10.1145/3321707.3321772
41. Wang, H., Vermetten, D., Ye, F., Doerr, C., Bäck, T.: IOHanalyzer: Detailed performance analyses for iterative optimization heuristics **2**(1) (2022). https://doi.org/10.1145/3510426

Landscape Analysis and Benchmarking

A Generator for Creating Streaming Continuous Optimisation Benchmarks

Mate Botond Nemeth[1(✉)], Emma Hart[1], Kevin Sim[1], and Quentin Renau[1,2]

[1] Edinburgh Napier University, Edinburgh, UK
{mate.nemeth,e.hart,k.sim,q.renau2}@napier.ac.uk
[2] University of Stirling, Stirling, UK

Abstract. In many real-world scenarios, instances arrive in a stream whose characteristics change over time, i.e. the stream is subject to data-drift. The ability to detect and respond to drift is crucial to optimise performance over the stream. However, there is currently no suitable stream-generator on which to evaluate the aforementioned methods. We address this by proposing *SCOBench*: a fully configurable generator that can generate streams of instances with different drift patterns specified by the user. In contrast to existing stream generators that define drift as a shift in the location of the optima over time of a single instance, *SCOBench* generates a stream of new instances from multiple functions, while drift between consecutive instances can be defined according to a change in three different metrics (landscape features, probing-trajectories and optima location). We verify that *SCOBench* correctly generates the intended drift patterns, then demonstrate that algorithms perform differently over streams, highlighting the need for drift detection and algorithm-selection per instance. *SCOBench* is fully integrated with IOHexperimenter for ease of use. All code and documentation are available as open-source.

Keywords: Benchmarking · Streaming Data · Continuous Optimisation

1 Introduction

In many real-world situations, instances of an optimisation problem arrive in a continual stream and must be solved as they arrive—in most scenarios, the characteristics of those instances will change over time, often in an unpredictable manner. Real-world examples of this can be seen in multiple domains, such as traffic management, managing smart energy grids, and in manufacturing environments [1,7,25].

This change in the data distribution of instances in a stream over time with respect to the instance characteristics is generically termed *data-drift*[1]. For clar-

[1] Note that this is different from *concept drift* commonly addressed in machine-learning, which refers to a drift in the input/output relationship between variables.

G. Iacca et al. (Eds.): PPSN 2026, LNCS 16985, pp. 485–500, 2027.
https://doi.org/10.1007/978-3-032-36214-8_30

ity, as this article concerns generating streams of instances whose features change over time, from herein we refer to this as *instance-drift.*

To optimise performance over a stream of instances, it is crucial to be able to detect instance-drift and react accordingly, e.g., by reconfiguring or switching a solver, potentially aided by some form of transfer-learning [22]. To facilitate this, streaming benchmark datasets are required to rigorously compare approaches for both detecting and reacting to instance-drift over a stream. However, unlike in the machine-learning community, where there are several well-known benchmarks for detecting concept-drift [4], there are currently very few benchmarks designed specifically for benchmarking algorithms over a stream in an optimisation context. Those that exist are often similar to or modified versions of *dynamic optimisation* benchmarks in which the *objective function of a single instance* changes over time, e.g., SDDOBench [24], proposed to aid the development of streaming data-driven evolutionary algorithms: these algorithms require the construction of new surrogate models of the changing objective function over the stream. In contrast, the purpose of *SCOBench* is to provide a benchmark that generates controllable drift in the *properties* of *each instance* generated in the stream, where *properties* can refer to any combination of the features, optimum location, and associated probing trajectory [15] of an instance. This can facilitate the development of online algorithm-selection or configuration methods; promote the development of drift-detection approaches; and underpin the development of novel transfer-learning approaches.

There are a number of challenges in creating a benchmark for streaming optimisation [23,24]. The first lies in the manner in which drift is defined. Existing solutions, such as [23,24], define drift to have occurred if the distance between the location of the optima of an instance of a function between two time points is greater than a pre-defined threshold. However, there are alternative ways of defining instance-drift between two consecutive instances, e.g. the difference between the Exploratory Landscape Analysis (ELA) feature-vectors [6] of the instance, or between probing-trajectories [15] obtained from an algorithm, recently shown to provide a reliable 'signature' of an instance [15]. Our proposed stream generator offers considerable flexibility in enabling similarity to be defined by any of these three metrics, or by a weighted combination of them. Secondly, a useful benchmark generator should enable diverse instance streams to be formed from a variety of functions *and* instances for maximum flexibility. Existing generators, such as SDDO, created to benchmark dynamic optimisation methods, generate streams from a *single* instance of a chosen function (by shifting the optimum of the instance). In contrast, we enable the generation of streams in which each new instance is generated *independently* from a function that can also change over the stream. This is realised by generating from MA-BBOB functions [19] which exploits affine combinations of two or more functions in which the weighting between each function can be adjusted to generate new functions. Hence, in *SCOBench*, the user can generate streams where the *function* also shifts over the stream. As in SDDObench [24], a straightforward user-interface is provided.

The major contributions of this work are as follows:

- Introduction of **SCOBench**—an open-source generator of streams of independent instances generated from multiple functions in which each instance generated in the stream can be subject to any combination of four different types of instance-drift, with respect to the previous instance (a minor, major or recurrent version of these drifts) simulating changes in the data distribution between consecutive instances that occur in streams of real-world problems.
- A customisable generator that: enables instance-drift to be defined using three different metrics; facilitates user-configuration of drift magnitude; generates streams from affine function pairs, enabling functions to drift over the stream rather than simply a single instance of a single function; gives full control of all hyper-parameters (e.g., sample size for generating features, instance search budget, similarity error weight); is integrated with IOHexperimenter [11] to provide strong scalability.
- Experimental evidence that verifies the streams exhibit the properties intended by the user; demonstrates that different algorithms exhibit different performance over the streams; demonstrates the need for algorithm-selection on a per-instance basis.

We believe that *SCOBench* will facilitate future research in transfer-learning and algorithm-selection over streams via a standardised benchmark that provides a common ground for experimentation and is built on well-understood functions. All code, documentation, experiments, and supplementary materials are uploaded to a repository [9].

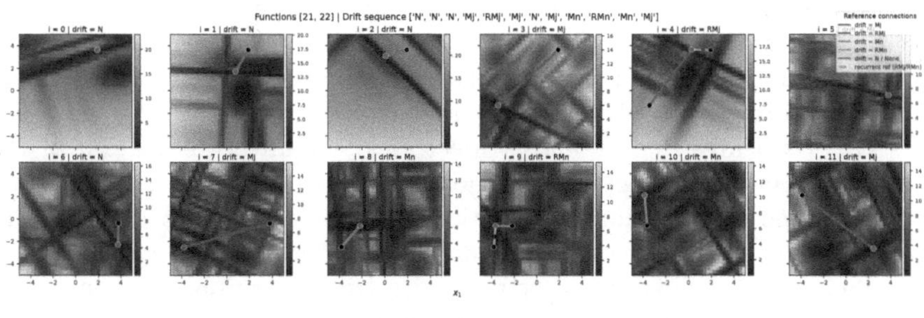

Fig. 1. A stream of MA-BBOB problems using functions 21 and 22 visualising the movement of the optimum of an instance due to instance-drift over 10 steps. At each step, a new instance is generated based on the specified instance-drift pattern. Drift is relative to a reference instance, which updates following any type of minor or major drift. The type of drift is indicated at each step; the optimum of the reference instance is shown in black and the new optimum in red. (Color figure online)

2 Related Work

As previously noted, the literature lacks benchmarks for conducting research in streaming optimisation problems. However, we have identified two recent benchmarks that share some characteristics of our proposed generator.

SDDObench [24] is a recent benchmark created to compare *data-driven evolutionary optimisation algorithms* (DDEAs) over a stream of data in which the data distribution is non-stationary due to changes in the environment. DDEAs build surrogate models as a substitute for the objective function; hence, detecting drift in such a scenario is critical to update the surrogates.

It supports stream generation from two distinct sets of functions for the objective formulation. The first group relies on a multi-peak function, which can generate a landscape containing multiple peaks via user-controlled parameters. The second group uses what the authors refer to as a 'DDEA Common Function' that aligns with the established norms in the field of data-driven optimisation (DDO), using five frequently employed problems for evaluating DDEAs. However, the paper defines drift exclusively in terms of a shift in the optima location of a single function, which limits the types of streams that can be generated.

The Sequential Transfer Optimisation Problem generator (STOP generator) [23] was created for benchmarking *transfer optimisation* techniques. However, this generator does not create a true optimisation stream: it first generates a target task and then generates a set of source tasks within a pre-defined similarity category. As with SDDOBench, it utilises the optimum location to define drift. The generator defines three similarity categories (low, mixed, and high) and offers support for 8 functions, allowing the user to benchmark inter-function transfer optimisation scenarios. However, since the functions are pre-assigned to each category, the scalability of the benchmark is limited.

Our proposed benchmark *SCOBench* builds a stream of instances in a similar manner to SDDOBench but has some key differences. Rather than relying only on optimum similarity to define drift, similarity can be measured in two other spaces (ELA features and probing-trajectories), or any combination of these. Secondly, we enable streams to be generated from more than one function by leveraging affine combinations of functions, enabling a smooth transition from one function to another over the stream. Finally, it facilitates multiple types of benchmarking, such as transfer-learning, drift detection, and per-instance algorithm-selection, as well as benchmarking algorithms over the stream.

3 Streaming Continuous Optimisation Benchmark

We now describe the design of the benchmark, focusing on how similarity metrics are used to define drift, the generation of instances, and the definition of drift types.

Stream: A stream consists of a list of instances, generated from one or two functions. Compared to SDDObench [24], our proposed benchmark generates

Table 1. Benchmark configuration settings

Config parameter	Description
Seed	The random seed
Verbose	The verbosity level (0: no info printed, 1: print at end of stream generation, 2: details printed), default is 2
Optimum Similarity	Normalised optimum similarity distance, default is 0.2.
Feature Similarity	Normalised feature similarity, default is 0.2.
Trajectory Similarity	Normalised trajectory similarity, default is 0.2.
Max Instance Generation	Number of instances to choose from for next instance, default is 100.
Use Opt Sim	Whether to use optimum similarity, default is True.
Optimum Transform	How the next instance is generated with the set optimum similarity, default is 'random'. 'shift' applies a transformation to the function so the instance optimum lies within the target similarity region. 'random' iterates through different instances generated by IOH until the right instance is found.
Features	Features used for feature similarity, default is 'ELA'.
Num Feature Calculations	Number of times to compute features for an instance, default is 5.
Feature Sample Size	Feature sample size, default is 50 * problem_dimension.
Feature Similarity Method	Feature similarity method in ['cosine', 'pearson', 'spearman'], default is 'cosine'.
Trajectory Method	What objective value is used for the trajectory time series 'best' (so far) or 'current', default is 'current'.
Trajectory Similarity Method	Trajectory similarity method in ['cosine', 'pearson', 'spearman'], default is 'cosine'.
Similarity Weight	Weight of feature and trajectory similarities in [0.0, 1.0], default is 0.5. Feature similarity weight is similarity_weight, trajectory similarity is 1-similarity_weight.
Recurrent Drift Type	0: major or minor drift is applied from the recurrent reference instance. 1: major or minor drift is applied from the previous drifted instance towards the recurrent reference. Default: 1.
Major Drift Opt Severity	Major drift optimum similarity, default is 0.8.
Major Drift Feature Severity	Major drift feature similarity, default is 0.8.
Major Drift Trajectory Severity	Major drift trajectory similarity, default is 0.8.
Re-Major Drift Opt Severity	Re-Major drift optimum similarity, default is 0.8.
Re-Major Drift Feature Severity	Re-Major drift feature similarity, default is 0.8.
Re-Major Drift Trajectory Severity	Re-Major drift trajectory similarity, default is 0.8.
Minor Drift Opt Severity	Minor drift optimum similarity, default is 0.2.
Minor Drift Feature Severity	Minor drift feature similarity, default is 0.2.
Minor Drift Trajectory Severity	Minor drift trajectory similarity, default is 0.2.
Re-Minor Drift Opt Severity	Re-Minor drift optimum similarity, default is 0.2.
Re-Minor Drift Feature Severity	Re-Minor drift feature similarity, default is 0.2.
Re-Minor Drift Trajectory Severity	Re-Minor drift trajectory similarity, default is 0.2.
Major Drift MA Weight Change	Automatic MA weight change during (re)major drift, default is 0.4.
Minor Drift MA Weight Change	Automatic MA weight change during (re)minor drift, default is 0.05.

a genuine stream, where each instance is independently produced from potentially different underlying functions rather than being a simple transformation of the previous instance. This contrasts with dynamic optimisation benchmarks, where a single instance of a single function is modified over time. Adopting the drift terminology introduced in [24], given a start (reference) instance, the next instance is generated according to the type of instance-drift specified by the user. The magnitude of the drift between two instances is defined with respect to the extent to which the instances are *similar* in a space defined by the user.

SCOBench defines drift between two instances via three similarity metrics, either individually or in combination (the latter is intended to provide a more informative description of the underlying problem than any single metric compared to previous benchmarks such as [23,24]). At an *instance level*, drift is specified as major (indicating instances are very different according to similarity metric(s)), or minor (instances are more similar). By specifying a pattern of drift between *instances*, the user can create drift patterns at a *stream* level: e.g. a repeated sequence of minor drifts can result in a stream exhibiting *(noisy) incremental* drift, while applying a major drift followed by a period of no drifting results in a stream characterised by a *sudden* drift. *Recurrent* drift within the streams can be produced using a drift type followed by a reverse version of the original drift.

Similarity Metrics: Previous benchmarks rely on *optimum similarity* as a simple mechanism to define drift within streams; however, this metric provides little insight into the difference between the landscapes of instances. We try to better approximate landscape similarities by additionally providing the option of using vectors composed of Exploratory Landscape Analysis features (ELA) [6] and/or probing trajectories [15]. ELA features are commonly used to characterise a landscape [6]. However, given the well-known limitations of these features [12,13,17], we also use the recently introduced idea of a probing-trajectory obtained from a short run of an algorithm [15] to characterise each instance from an algorithmic perspective. We leverage recent literature, which showed that *probing-trajectories* obtained via a simulated annealing algorithm (SA) tuned to produce trajectories that maximally discriminate between instances provide a reliable *signature* of an instance [14][2]. Note, however, that the generator design allows the user to use any form of probing-trajectory of their choice.

Functions: The benchmark generates streams of instances using IOHexperimenter [11] allowing for integration with popular IOHprofiler [3] framework for analysis. For single-function stream generation, any function can be used in principle as long as the range of possible locations for the optimum spans a continuous part of the search space[3]. *SCOBench* provides two options for selecting

[2] The SA algorithm is used to provide an instance-signature, not to solve an instance.
[3] Strictly speaking, the range of possible optimum locations is only important if the user wants to include the random optimum transform option for optimum similarity in the generation of new instances.

the function(s) to be used to generate a stream. The first generates a stream of instances from a single function: a subset of BBOB functions [5] is provided for this purpose. We omit functions (5, 9, 19, 20, 24) as they do not match the description above[4]. With IOHexperimenter, any function with similar properties can be used. Additionally, *SCOBench* supports the generation of a stream of instances from multiple functions: streams are generated from BBOB function pairs via an affine combination from MA-BBOB [20], in which modifying the weighting of the affine combination enables controllable drift from one of the functions to the other.

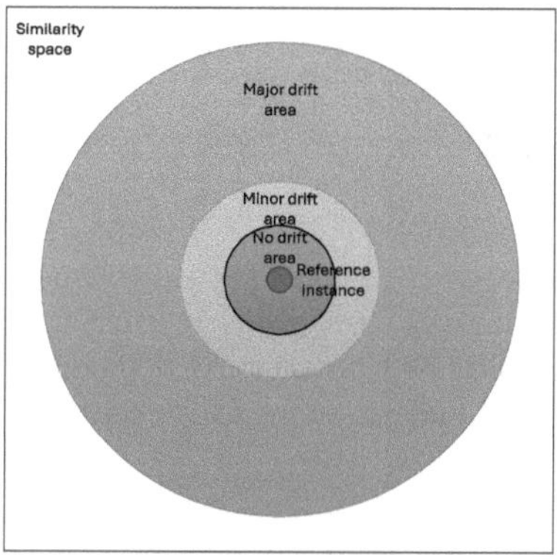

Fig. 2. Instance-drift occurs relative to a reference instance. Depending on the magnitude of the drift w.r.t. a similarity metric, it is categorised as minor or major. If a newly generated instance falls outside of the intended band, the distance to the intended band is calculated, and the instance that minimises this metric is used. The relative sizes of each band are customisable.

Instance Drift: The benchmark supports multiple types of instance-drift. *Minor* drift corresponds to a small shift of the new instance in the similarity space of choice with respect to a reference instance, while *major* drift generates a new instance with a larger shift (see Fig. 2). Note that the thresholds for major/minor drift are configurable by the user, denoted as the *severity value* (default values were chosen based on the resulting changes in algorithm performance). The first instance in a stream is denoted as the *reference* instance. As subsequent instances are generated, if no drift is indicated, the reference instance does not change, and a new instance is generated that is similar to the reference. This process avoids a random walk within the similarity spaces. Following a minor or major drift, the current reference instance is updated to the new drifted instance.

For both major and minor drift, the benchmark supports a recurrent version of a previous drift, denoted as *re*-major or *re*-minor. These drifts are applied

[4] Since the optima in these functions do not span a continuous space, they cannot be used with the random optimum transform setting.

from a recurrent reference instance, which is the reference instance of the last drifted instance (these are updated recursively as shown in Fig. 3). There are two options here. The default version reverses the effect of the previous drift by searching for a new instance close to the recurrent reference instance. The alternate version applies the same drift from the recurrent reference instance. Figure 1 shows how the optima move and the reference is updated for a specific drift pattern.

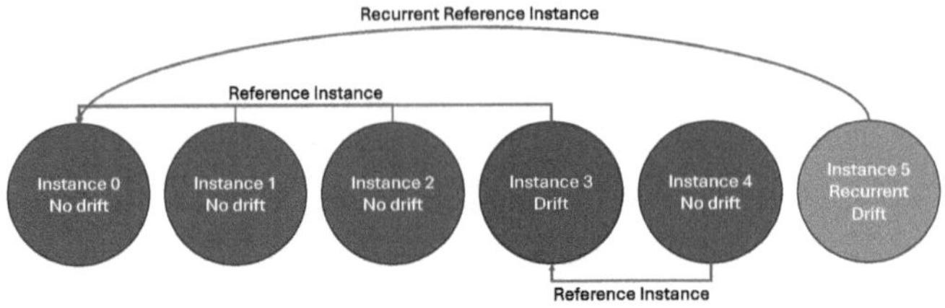

Fig. 3. Visualisation of reference instances for different drift types, in which the reference instance is always the most recently drifted instance. Under recurrent drift, the reference instance is updated recursively: the reference of the last drifted instance becomes the new reference instance.

Instance Generation: Instances are generated following a drift pattern provided by the user, e.g., [N, N, Mj, RMj, Mn], where N corresponds to no drift; Mj to major drift; RMj to re-major drift; Mn to minor drift; RMn to re-minor drift. The generator searches for a new instance that matches the desired drift pattern using a random-search process (see Algorithm 1, with more details on parameters in Table 1). If optimum similarity is enabled, the user has two options—shift and random. For '*shift*', after generating a new candidate instance, we shift its optimum such that it falls within the target similarity band around the optimum of the reference instance (see Fig. 2). The objective function is then translated along this vector. For *'random'*, a new random candidate instance is generated with a user-defined budget until the optimum of the candidate instance falls within the target similarity band. If the budget is reached, the candidate instance with the smallest deviation from the target region is selected[5]. Stream generation is deterministic based on the seed provided.

4 Methodology

To verify that the generator works as intended and to illustrate its use, we conduct experiments using the methods described below.

[5] Verification tests provided in the Supplementary material show that depending on budget, the success rate of generating an exactly matching instance is 72.5% while the median normalised feature similarity/trajectory deviation is 0.0698/0.1649 respectively, hence in practice generating an instance slightly out of bounds has little effect.

Algorithm 1. Next instance generation

```
Require: params, config
if no reference exists then
    generate first (no-drift) instance, store as reference; return
end if
pick reference (normal vs recurrent) based on drift type
for k = 1 to K do                                   ▷ K = max_instance_gen
    generate candidate
    if optimum similarity is enabled then
        if config.optimum_transform is shift then
            shift candidate optimum into the band
        else if config.optimum_transform is random then
            resample up to B times
                                        ▷ B = opt_search_multiplier * K
        end if
    end if
    compute similarity; accept if valid, else save to list
end for
if none valid: use best-so-far (with smallest weighted deviation from target simi-
    larity thresholds)
update references; return
```

4.1 Verification of Stream Generation Process

Similarity Metrics: In order to demonstrate that the two newly introduced similarity metrics (ELA features and probing-trajectories) when used in conjunction with the optimum similarity used in recent work [24] realise the desired drift pattern, we run a well known solver modular CMA-ES (modCMA) [10] over 20 streams in 10 dimensions (each with 5 seeds) containing the pairwise combinations of the last functions in each BBOB function group, with the drift pattern of [N, N, Mj, RMj, Mn, RMn, Mn, N, Mj, RMj, Mj, N, Mn, RMn, Mn, Mn, Mn, Mj, N, N]. We evaluate the change in the anytime performance of the algorithm compared to the reference instance, using area over convergence curve (AOCC) as an anytime metric. Using the Wilcoxon test and Fisher's method, we determine if the anytime performance change over all runs is significantly different.

Drift Detection: To demonstrate that drift can be detected by standard detection methods, we apply an off-the-shelf detection method used in recent literature [16] called NannyML [8] to the previously used streams with a drift pattern of 15 non-drifted instances, then a minor drift for 15 instances, then a major drift for 15 instances. This method detects multivariate feature drift via a PCA-based data reconstruction method, which detects subtle changes in the data structure, triggering a drift alert when the reconstruction error exceeds a threshold. This is a proof-of-concept experiment over a short stream, simply to confirm that drift detection is possible[6].

4.2 Algorithm Performance over Streams

In a well-designed streaming optimisation benchmark, drift within the problem stream would be expected to influence algorithm performance. We therefore benchmark the performance of five algorithms over multiple streams, specifically, RCobyla, DiagonalCMA, DifferentialEvolution from the Nevergrad framework [2], as well as modCMA [10] and modular differential evolution (modDE) [18]. Algorithm runs were repeated 30 times on each instance, recording the anytime performance using IOHanalyser [21].

We evaluate anytime performance on streams created from 72 MA-BBOB function pairs. Since the pairwise combination of all BBOB functions would produce an impractically large dataset, we formed unordered pairs of the 19 BBOB functions (as stated above, the optima of these functions continuously span the search space enabling the use of optimum similarity), requiring each pair to span two different groups (BBOB function categories) and taking every second pair to result in 72 pairs. Each stream has 20 instances generated from a single randomly assigned random drift pattern containing all drift types [N, N, Mj, RMj, Mn, RMn, Mn, N, Mj, RMj, Mj, N, Mn, RMn, Mn, Mn, Mn, Mj, N, N]. To show that streams generalise well, we have conducted additional experiments on a subset of these functions using two different drift patterns: [N, N, N, Mn, Mn, Mn, Mj, Mj, Mj] and [N, Mn, RMn, Mn, RMn, Mj, RMj, Mj, RMj]. Each stream generation was repeated over 5 seeds, measuring the difference in the anytime performance of the solvers per instance compared to the reference instance.

4.3 Algorithm Selection

Optimising performance over a drifting stream may require algorithm-selection, i.e., selecting a different algorithm per instance. To illustrate the potential for developing an algorithm-selection method, we plot the performance of the 5 algorithms described in Sect. 4.2 for two different streams. Median convergence curves are plotted per instance, aggregated over 5 seeds and 30 runs per instance.

5 Results

Similarity Metrics. In Fig. 4, we isolate the effects of the two newly introduced similarity metrics. Using either feature or trajectory similarity (along with optimum similarity) yields broadly comparable outcomes in terms of changes in algorithm performance in the example given. Combining all three metrics potentially results in a more robust metric as seen in Fig. 6.

[6] Note that any drift detection technique could have been chosen as a proof-of-concept.

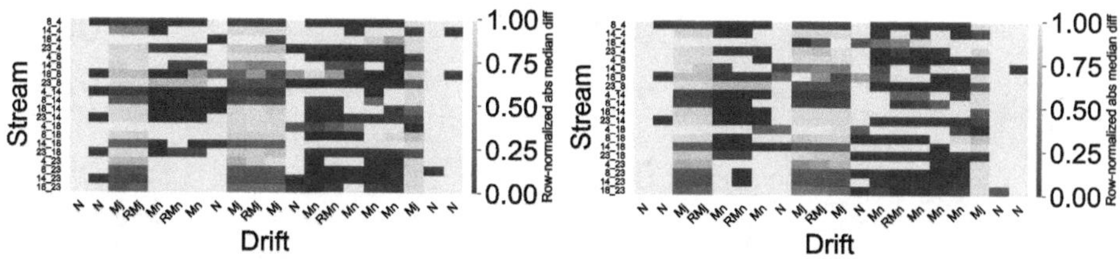

(a) Feature and optimum similarity (b) Trajectory and optimum similarity

Fig. 4. Visualisations of the differences in median absolute AOCC compared to the reference instance, where only cells with statistically significant differences are coloured. Impact on performance of modCMA (yellow indicates greater difference c.f. reference instance) of selecting different similarity metrics (evaluated on two identical streams). (Color figure online)

Drift Detection. To show that *SCOBench* could be used to benchmark drift detection methods, we provide an illustration that an off-the-shelf detection method (see Sect. 4.1) can detect drift over a single short stream (generated from an MA-BBOB function pair 4_8). Figure 5 plots the 'alerts' generated that indicate if the data-distribution of a block of instances has drifted w.r.t the initial reference block.

The figure shows that instances in the minor drift block generate a series of alerts, while the instances in the major drift block generate higher alerts, indicating that drift has occurred. Note there is significant variance in each alert due to the small amount of data in the stream. While, of course, we cannot generalise from a single stream, the example provides an indication that more sophisticated drift-detection methods could be developed and tested using *SCOBench* in future.

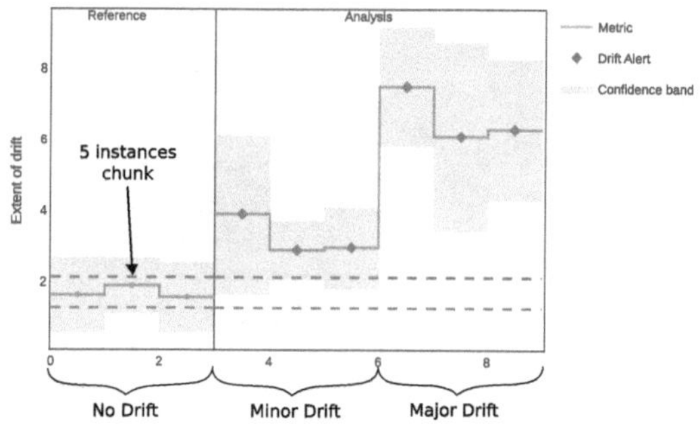

Fig. 5. Drift detection using pattern (no drift; minor drift; major drift), with 15 instances per drift-type. Dotted red lines define regions with no drift in the data-distribution, red circles indicate alerts denoting that drift was detected, with the extent of the drift indicated on the y-axis. (Color figure online)

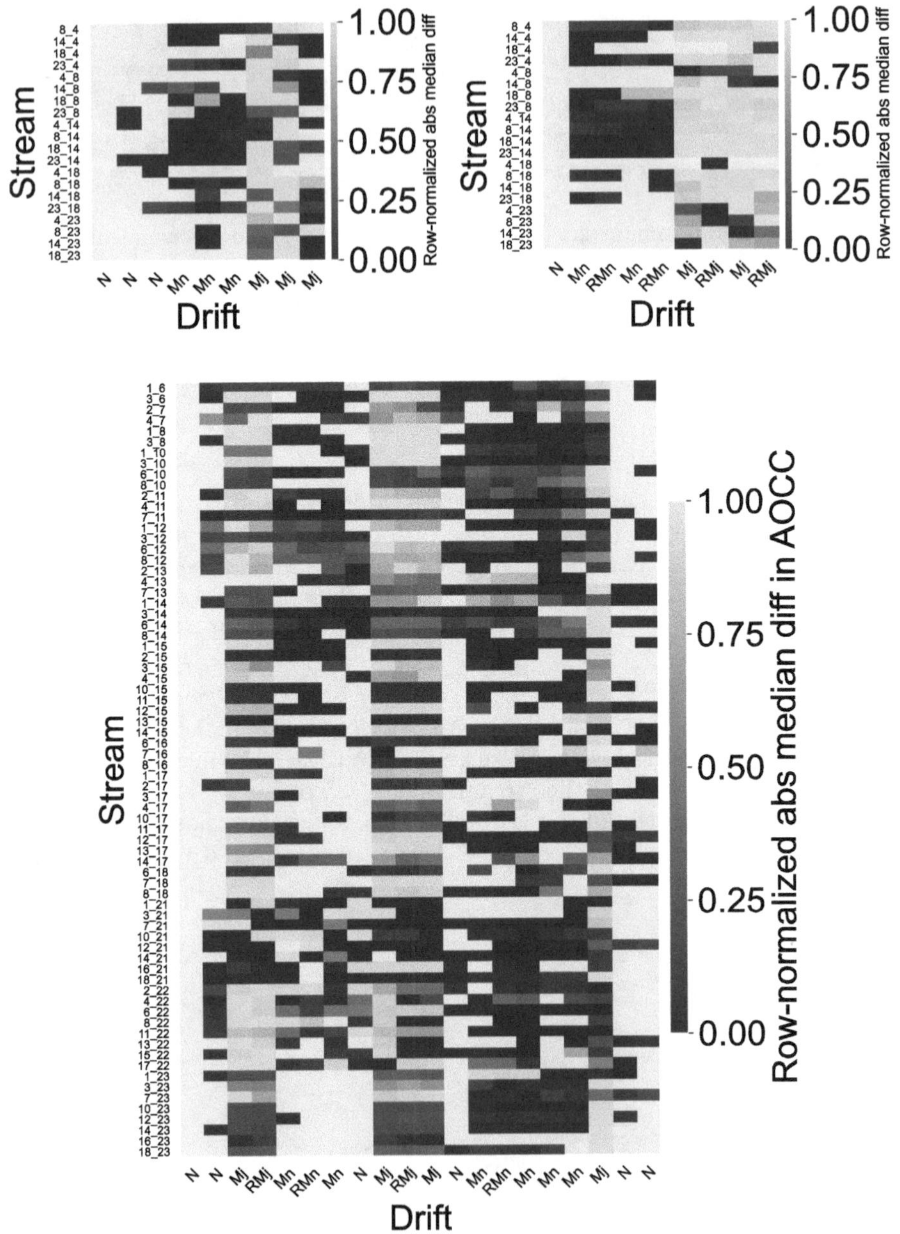

Fig. 6. Visualisations of differences in median absolute AOCC compared to the reference instance across all *modCMA* runs and seeds: rows represent streams; columns represent drift type. Cell values correspond to the algorithm behaviour change normalised across the streams (lower values indicate similar behaviour to the reference instance, higher values indicate different behaviour). Cells are coloured if the difference in anytime performance is statistically significant. Plots of other algorithms are in the Supplementary materials.

Algorithm Performance. Figure 6 illustrates how algorithm behaviour changes within streams under different types of drift using a single algorithm as an exemplar (modCMA).

We provide results aggregated over all streams with figures for the other four algorithms described in Sect. 4.2 in the Supplementary material. In all these plots, behavioural change is quantified relative to the reference instance as the absolute median difference in AOCC, computed across all seeds and algorithm repetitions. In a ‘perfect’ stream, the absence of drift should result in behaviour closely matching the reference instance; minor drift should induce a small change; and major drift should lead to a larger change.

Overall, the figures exhibit this expected pattern: the biggest changes in algorithm performance correlate with major (and re-major) drift. For example, in Fig. 6, the major drift columns show that the changes in algorithm performance are almost always statistically significant. This trend holds across most functions with the notable exception of streams whose initial instances are mostly from functions 21 and 22 (x_21, x_22). In these cases, algorithms display differences in performance unrelated to drift.

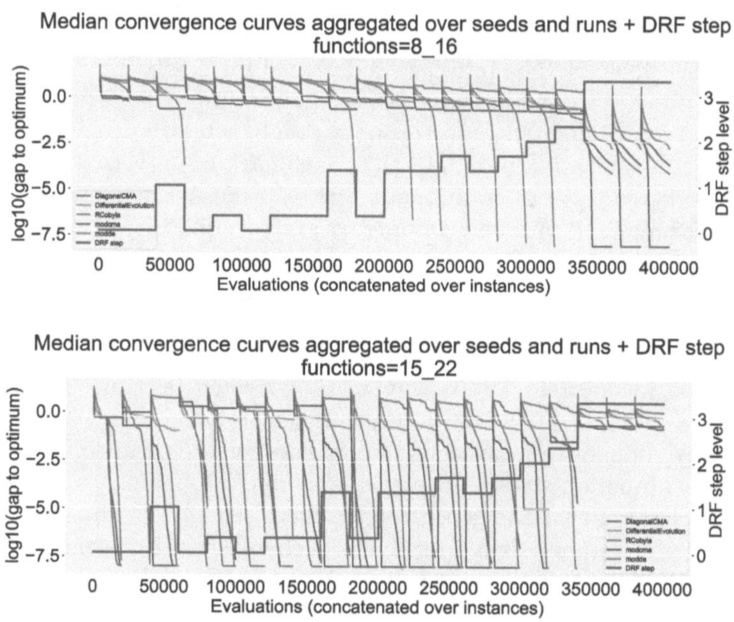

Fig. 7. Performance curve for 5 solvers per instance in the stream. The red line shows the drift pattern (DRF step level). The best performing algorithm is the one with the smallest gap to the optimum of the problem instance. If multiple algorithms reached the same best solutions, the best algorithm is the one with the steepest descent. (Color figure online)

Algorithm Selection. Figure 7 indicates that under drift, the best-performing algorithm within the portfolio can change per-instance over the course of a stream. This highlights the importance of developing drift detection and adaptable algorithm selection/configuration models to maintain performance throughout the stream. In the upper figure, *modCMA* wins on 8 instances, *DiagonalCMA* on 3 and *modDE* on 10. In the lower figure, *RCobyla* wins on 5 instances, *modDE* on 3, *DiagonalCMA* on 4 and *modCMA* on 8. The change in the best-performing algorithm in most cases corresponds to drift.

6 Conclusion

We have created *SCOBench* as a tool for conducting research in streaming optimisation domains in which instances arriving in a stream are subject to instance-drift. The streams generated by *SCOBench* can be used in multiple experimental scenarios, e.g., to develop novel drift detection methods; to develop an online per-instance algorithm selection or configuration method, or to facilitate transfer-learning. The tool is flexible in enabling multiple types of drift to be realised at the *stream level* (e.g., noisy incremental drift, sudden drift, recurrent drift) by defining sequences of drift per-instance. As the benchmark is fully configurable and is integrated with IOHexperimenter, it also enables integration with new continuous optimisation suites. Code and documentation are fully open-source to enable extensions [9]. The interface is designed for flexibility and ease of use, e.g., enabling a user to add new functions representing real-world problems.

Compared to existing streaming optimisation benchmarks (mainly developed with dynamic optimisation in mind), the benchmark has several innovations. In particular, it introduces two additional metrics for measuring similarity between instances (feature similarity and probing trajectory similarity) and enables similarity to be defined as a weighted sum of these two metrics and optimum similarity. The combination of these metrics allows for more accurate comparison of problem landscapes than methods that rely on optimum similarity alone. *SCOBench* also introduces the use of affine-recombinations of functions to generate streams, in contrast to existing methods where streams are generated from a single function. This allows streams to be generated where the function gradually changes from one function to another. *SCOBench* currently only supports multi-function stream generation for function pairs. A series of experiments demonstrated that the generator produces streams as intended. We then showed that per-instance drift introduces a change in algorithm performance over a stream. These empirical results highlight the need to develop reliable methods for drift detection and to progress research relating to the development of adaptive algorithm-selection and configuration to optimise performance over the stream. *SCOBench* will underpin future research in this direction.

Future Work. The benchmark currently only features a noisy version of incremental drift, where instances randomly move around the similarity space when multiple incremental drifts occur in the stream. We plan to add a more directional version of minor drift to enable the simulation of non-noisy incremental drift. Future versions of the benchmark will include more extended MA-BBOB support, where affine function components will not be limited.

References

1. Azeem, A., Ismail, I., Mohani, S.S., Danyaro, K.U., Hussain, U., Shabbir, S., Bin Jusoh, R.Z.: Mitigating concept drift challenges in evolving smart grids: an adaptive ensemble lstm for enhanced load forecasting. Energy Rep. **13**, 1369–1383 (2025). https://doi.org/10.1016/j.egyr.2024.12.078
2. Bennet, P., Doerr, C., Moreau, A., Rapin, J., Teytaud, F., Teytaud, O.: Nevergrad: black-box optimization platform. SIGEVOlution **14**(1), 8–15 (2021). https://doi.org/10.1145/3460310.3460312
3. Doerr, C., Wang, H., Ye, F., van Rijn, S., Bäck, T.: Iohprofiler: A benchmarking and profiling tool for iterative optimization heuristics. arXiv e-prints:1810.05281 (oct 2018). https://arxiv.org/abs/1810.05281
4. Gunasekara, N., Pfahringer, B., Gomes, H.M., Bifet, A., Koh, Y.S.: Recurrent concept drifts on data streams. In: Proceedings of the Thirty-Third International Joint Conference on Artificial Intelligence. IJCAI '24 (2024). https://doi.org/10.24963/ijcai.2024/888
5. Hansen, N., Auger, A., Ros, R., Mersmann, O., Tusar, T., Brockhoff, D.: COCO: a platform for comparing continuous optimizers in a black-box setting. Optim. Methods Softw. **36**(1), 114–144 (2021). https://doi.org/10.1080/10556788.2020.1808977
6. Mersmann, O., Bischl, B., Trautmann, H., Preuss, M., Weihs, C., Rudolph, G.: Exploratory Landscape Analysis. In: Proceedings of the Genetic and Evolutionary Computation Conference, GECCO '11, pp. 829–836. ACM (2011). https://doi.org/10.1145/2001576.2001690
7. Nallaperuma, D., et al.: Online incremental machine learning platform for big data-driven smart traffic management. IEEE Trans. Intell. Transp. Syst. **20**(12), 4679–4690 (2019). https://doi.org/10.1109/TITS.2019.2924883
8. NannyML (version 0.8.3). https://github.com/NannyML/nannyml (Mar 2023), nannyML, Belgium, OHL
9. Nemeth, M.B., Hart, E., Sim, K., Renau, Q.: A generator for creating streaming continuous optimisation benchmarks - supplementary materials (2026). https://doi.org/10.5281/zenodo.19629227
10. de Nobel, J., Vermetten, D., Wang, H., Doerr, C., Bäck, T.: Tuning as a means of assessing the benefits of new ideas in interplay with existing algorithmic modules. In: Proceedings of the Genetic and Evolutionary Computation Conference Companion, pp. 1375–1384. GECCO '21. Association for Computing Machinery, New York (2021). https://doi.org/10.1145/3449726.3463167
11. de Nobel, J., Ye, F., Vermetten, D., Wang, H., Doerr, C., Bäck, T.: Iohexperimenter: Benchmarking platform for iterative optimization heuristics. arXiv e-prints:2111.04077 (nov 2021). https://arxiv.org/abs/2111.04077
12. Renau, Q.: Landscape-Aware Selection of Metaheuristics for the Optimization of Radar Networks. Ph.D. thesis, Polytechnic Institute of Paris, Palaiseau, France (2022). https://tel.archives-ouvertes.fr/tel-03593606
13. Renau, Q., Dréo, J., Doerr, C., Doerr, B.: Towards explainable exploratory landscape analysis: Extreme feature selection for classifying BBOB functions. In: Applications of Evolutionary Computation - 24th International Conference, EvoApplications 2021, Held as Part of EvoStar 2021, Proceedings. LNCS, vol. 12694, pp. 17–33. Springer (2021). https://doi.org/10.1007/978-3-030-72699-7_2

14. Renau, Q., Hart, E.: Improving algorithm-selectors and performance-predictors via learning discriminating training samples. In: Proceedings of the Genetic and Evolutionary Computation Conference, pp. 1026–1035. ACM, Melbourne VIC Australia (2024). https://doi.org/10.1145/3638529.3654025
15. Renau, Q., Hart, E.: On the utility of probing trajectories for algorithm-selection. In: Smith, S., Correia, J., Cintrano, C. (eds.) Applications of Evolutionary Computation, vol. 14634, pp. 98–114. Springer, Cham (2024). https://doi.org/10.1007/978-3-031-56852-7_7
16. Rook, J., Renau, Q., Trautmann, H., Hart, E.: Efficient online automated algorithm selection in the face of data-drift in optimisation problem instances. In: Proceedings of the 18th ACM/SIGEVO Conference on Foundations of Genetic Algorithms, pp. 262–272 (2025). https://doi.org/10.1145/3729878.3746615
17. Skvorc, U., Eftimov, T., Korosec, P.: A comprehensive analysis of the invariance of exploratory landscape analysis features to function transformations. In: IEEE Congress on Evolutionary Computation, CEC 2022, Padua, Italy, July 18-23, 2022, pp. 1–8. IEEE (2022). https://doi.org/10.1109/CEC55065.2022.9870313
18. Vermetten, D., Caraffini, F., Kononova, A.V., Bäck, T.: Modular differential evolution, pp. 864–872. GECCO '23. Association for Computing Machinery, New York (2023). https://doi.org/10.1145/3583131.3590417
19. Vermetten, D., Ye, F., Bäck, T., Doerr, C.: MA-BBOB: a problem generator for black-box optimization using affine combinations and shifts. ACM Trans. Evol. Learn. Optim. **5**(1), 1–19 (2025). https://doi.org/10.1145/3673908
20. Vermetten, D., Ye, F., Bäck, T., Doerr, C.: Ma-bbob: A problem generator for black-box optimization using affine combinations and shifts. ACM Trans. Evol. Learn. Optim **5**(1), (2025). https://doi.org/10.1145/3673908
21. Wang, H., Vermetten, D., Ye, F., Doerr, C., Bäck, T.: Iohanalyzer: detailed performance analyses for iterative optimization heuristics. ACM Trans. Evolutionary Learn. Optim. **2**(1) (2022). https://doi.org/10.1145/3510426
22. Xue, X., Yang, C., Feng, L., Zhang, K., Song, L., Tan, K.C.: Solution transfer in evolutionary optimization: an empirical study on sequential transfer. IEEE Trans. Evol. Comput. **28**(6), 1776–1793 (2024). https://doi.org/10.1109/tevc.2023.3339506
23. Xue, X., Yang, C., Feng, L., Zhang, K., Song, L., Tan, K.C.: A scalable test problem generator for sequential transfer optimization. IEEE Trans. Cybern. **55**(5), 2110–2123 (2025). https://doi.org/10.1109/TCYB.2025.3547565
24. Zhong, Y., Wang, X., Sun, Y., Gong, Y.J.: SDDObench: a benchmark for streaming data-driven optimization with concept drift. In: Proceedings of the Genetic and Evolutionary Computation Conference, pp. 59–67. ACM, Melbourne VIC Australia (2024). https://doi.org/10.1145/3638529.3654063
25. Zhou, H., Yin, H., Zhao, D., Cai, L.: Incremental learning and conditional drift adaptation for nonstationary industrial process fault diagnosis. IEEE Trans. Industr. Inf. **19**(4), 5935–5944 (2023). https://doi.org/10.1109/TII.2022.3179423

Combinatorial Landscape Analysis for Dominating Set and Vertex Coloring

Johanna Gasse(✉), Antonia Heinen, Felix Knöfel, Timo Kötzing, and Maxim Stanko

Hasso Plattner Institute, Potsdam, Germany
Johanna.Gasse@student.hpi.uni-potsdam.de

Abstract. We analyze the two combinatorial problems of Dominating Set and Vertex Coloring regarding what kind of local optima are present for various instances. For a variety of graph classes each, we determine whether the induced landscapes are unimodal, plateau-unimodal (all optima are just one plateau), equimodal (all local optima are global) or truly multimodal. We do this for two different neighborhood operators, one based on making only a single change and one also allowing swaps (interchanging two parts of the solution).

1 Introduction

A lot of theory on randomized search heuristics is focused on deriving run time bounds for given algorithms on given problems. These analyses come to a limit when the search space is too complex to allow such a careful analysis, particularly when local optima are present. Many interesting areas of application for randomized search heuristics are for problems with such complex characteristics, leading naturally to a divide between what problems can get a rigorous run time analysis and what problems search heuristics are actually applied to.

With this paper, we extend our knowledge to larger problem classes by relaxing the goal. Instead of finding bounds on the run time of a given algorithm, we explore problem characteristics in terms of local optima for two combinatorial problems. This approach was taken experimentally for the number partitioning problem, the binary knapsack problem and the quadratic binary knapsack problem in [1] (see also [9,10] for similar experimental analyses). In this paper we derive mathematical results for the combinatorial problems of Dominating Set and Vertex Coloring.

Concretely, we are interested in the existence of a non-global local optimum (NGLO). Note that the definition of a local optimum is tricky when neighboring search points can have equal fitness, leading to plateaus. A plateau where all search points neighboring the plateau have worse fitness we call a local optimum plateau, giving then the notion of a non-global local optimum plateau (NGLOP). We distinguish the following four cases for landscapes: (a) unimodal (only one global optimum, no NGLOPs); (b) plateau-unimodal (there can be a single plateau of local optima); (c) equimodal (all local optimum plateaus have

G. Iacca et al. (Eds.): PPSN 2026, LNCS 16985, pp. 501–515, 2027.
https://doi.org/10.1007/978-3-032-36214-8_31

equal fitness, equivalent to there being no NGLOPs); and (d) multimodal (there are NGLOPs). See Sect. 2.1 for precise definitions.

The absence of NGLOPs informs us about the fact that a simple hill climbing algorithm (eventually) finds a global optimum. It is well known that even unimodal search spaces can be hard to optimize, so our results do not inform directly about problem difficulty in terms of run time. However, the results illustrate

- which problem classes suffer from local optima and thus require non-local operators;
- which problem classes admit hill climbing in principle, and could thus be targeted by further analysis.

For example, on instances where we show that even a complex set of instances does not have NGLOPs, a run time analysis could now continue and make a finer analysis of these instances, knowing that there are no fundamental road blocks, just potentially large detours.

Finally, our analyses have the advantage that they clearly expose where problem difficulty comes from, while large run times may be attributed to local optima, long paths, large plateaus or other reasons.

We consider the two combinatorial problems of Dominating Set and Vertex Coloring. Both are problems on simple graphs $G = (V, E)$ with $n = |V|$ vertices. Note that all graphs in this paper are assumed to be connected and simple. For the remainder of this introduction, we give an overview of our results for these two structurally rather different problems. In Sect. 2, we define landscapes and all related terms. Sections 3 and 4 give our results for Dominating Set and Vertex Coloring respectively. Note that some results are derived from *reconfiguration theory* (see, for example, [7] for an introduction).

Due to space constraints, the proofs are omitted. For version containing all proofs, see [5].

1.1 Overview of Results for Dominating Set

The optimization task for Dominating Set is to find a cardinality-minimal set $S \subseteq V$ such that every vertex is either in S or has a neighbor in S. We consider the following two neighborhoods, corresponding to local search operators. The Flip neighborhood either adds or removes a vertex from a given set; we denote this neighborhood by $\mathcal{N}_{\text{FLIP}}$. The Flip/Swap neighborhood either removes a vertex, adds a vertex, or does both (swap); we denote this neighborhood by $\mathcal{N}_{f/s}$.

We consider different classes of graphs and determine, for each graph in the given class, what the induced landscape is in terms of local optima. An overview of our results is given in Table 2. As we can see, for the Flip neighborhood, even very simple graph classes, such as complete bipartite graphs, induce multimodal landscapes (except for finitely many such graphs). We identify only a small graph class, containing certain trees, that induce an equimodal landscape (in fact, for trees we give a full characterization of when they induce an equimodal or multimodal landscape, see Theorem 20). This shows that Dominating Set is very hard for hill climbers, when only single bit flips are permitted.

From Table 2 we see that a lot of graph classes induce plateau-unimodal landscapes when using the Flip/Swap neighborhood; we complement these positive findings with an example graph class which, while simple, induces multimodal landscapes.

1.2 Overview of Results for Vertex Coloring

The optimization task for Vertex Coloring is to find a coloring of all vertices of the given graph using as few as possible colors while maintaining that, for each edge, the colors of the endpoints of the edge are colored differently. We first consider the neighborhood of recoloring a single vertex ($\mathcal{N}_{\text{FLIP}}$). As a second operator we either recolor a single vertex or interchange the colors of two vertices ($\mathcal{N}_{f/s}$). Note that we use the same names for the operators as for Dominating Set, since we make a general definition which covers both cases simultaneously (see Sect. 2).

We again consider different classes of graphs. An overview of our results is given in Table 1. As we can see, certain types of bipartite graphs induce unimodal landscapes (up to renaming colors) even under the Flip neighborhood. Furthermore, the large class of chordal graphs is also amenable to analysis, giving an equimodal landscape.

Turning to the Flip/Swap neighborhood, we see that here also crown graphs and certain cycles induce unimodal landscapes. We construct a simple example graph class based on cycles with certain chords added to showcase which (simple) graphs induce multimodal landscapes.

Table 1. Classification of the landscapes of various graph classes for the Vertex Coloring problem into plateau-unimodal (there exists a single local optimum plateau), equimodal (all local optimum plateaus are globally optimal) and multimodal (there exist NGLOPs, see Definition 7) for the Flip and Flip/Swap neighborhood respectively (see Definitions 3 and 5). See Definition 8 for the definition of modality.

Vertex Coloring			
Graph Class	Neighborhood	Optima	Theorem
Universal Bipartite Graphs	$\mathcal{N}_{\text{FLIP}}$	unimodal	36
C_{6n}	$\mathcal{N}_{\text{FLIP}}$	multimodal	39
Crown Graphs	$\mathcal{N}_{\text{FLIP}}$	multimodal	42
Chordal Graphs	$\mathcal{N}_{\text{FLIP}}$	equim., not plateau-unim.	46
Chordal Bipartite Graphs	$\mathcal{N}_{\text{FLIP}}$	unimodal	48
C_{6n}	$\mathcal{N}_{f/s}$	unimodal	50
Crown Graphs	$\mathcal{N}_{f/s}$	unimodal	51
Spoked C_{12n}	$\mathcal{N}_{f/s}$	multimodal	53

2 Introduction to Landscapes

In this section we formally introduce the notion of a landscape. For all $n \in \mathbb{N}$ we use the notation $[n] = \{1, \ldots, n\}$. Just as in [1] we model fitness landscapes as triples as follows. Note that these definitions assume minimization.

Definition 1. *A directed graph* $(\mathcal{X}, \mathcal{N})$ *consisting of set* $\mathcal{X}$ *of* individuals, *a set* $\mathcal{N}$ *of directed edges is called a* neighborhood graph, *with* $\mathcal{N}$ *being the* neighborhood. *For* $x \in \mathcal{X}$ *we write* $\mathcal{N}(x) = \{y \in \mathcal{X} \mid (x, y) \in \mathcal{N})\}$ *for the neighbors of* x. *For a* fitness function $f \colon \mathcal{X} \to \mathbb{R}$, *we call* $(\mathcal{X}, \mathcal{N}, f)$ *a* landscape.

For a given landscape $(\mathcal{X}, \mathcal{N}, f)$ *and* $x, y \in \mathcal{X}$, *we write* $x \longrightarrow y$ *if* $(x, y) \in \mathcal{N}$; *we write* $x \Longrightarrow y$ *if it is possible to reach* y *from* x *in any number of such steps, i.e. if there is a path from* x *to* y *in the graph* $(\mathcal{X}, \mathcal{N})$. *We write* $x \xrightarrow{+} y$ *if* $(x, y) \in \mathcal{N}$ *and* $f(x) \geq f(y)$; *we write* $x \stackrel{+}{\Longrightarrow} y$ *if it is possible to reach* y *from* x *in any number of such steps, i.e. if there are* $z_1, \ldots, z_k$ *such that*

$$x \xrightarrow{+} z_1 \xrightarrow{+} z_2 \xrightarrow{+} \ldots \xrightarrow{+} z_k \xrightarrow{+} y.$$

We refer to such a sequence $x, z_1, \ldots, z_k, y$ *as a* positive path.

Note that both $\Longrightarrow$ and $\stackrel{+}{\Longrightarrow}$ are transitive relations.

For this work we focus on only individuals which are fixed-length strings of a bounded set of natural numbers; most common are fixed-length bit strings.

Definition 2. *Let* $n \in \mathbb{N}$ *and* $r \in \mathbb{N}_{\geq 2}$. *We define* $\mathcal{X}_r = \{0, \ldots, r-1\}^n$.

Note that we can encode many problems with this search space. For example, to denote any subset of n given options (vertices of a graph, ...), we can use bit strings ($\mathcal{X}_2$), where a 1 indicates that the corresponding option is chosen, and a 0 indicates that it is not chosen.

For this specific set of individuals we define the following classic neighborhoods

Definition 3. *Let* $r \in \mathbb{N}_{\geq 2}$. *We define the neighborhood* $\mathcal{N}_{\text{FLIP}}$ *on* $\mathcal{X}_r$ *such that, for all* $u, v \in \mathcal{X}_r$, $(u, v) \in \mathcal{N}_{\text{FLIP}}$ *if and only if* u *can be transformed into* v *by changing a single position, i.e. iff* $u \neq v$ *and there is* $i \in [n]$ *such that for all* $j \in [n] \setminus \{i\}$ *we have* $u(j) = v(j)$. *We call* $\mathcal{N}_{\text{FLIP}}$ *the* Flip neighborhood.

Definition 4. *Let* $r \in \mathbb{N}_{\geq 2}$. *We define the neighborhood* $\mathcal{N}_{\text{SWAP}}$ *on* $\mathcal{X}_r$ *such that, for all vertices* $u, v \in \mathcal{X}_r$, $(u, v) \in \mathcal{N}_{\text{SWAP}}$ *if and only if* u *can be transformed into* v *by swapping two positions, i.e. iff there are* $i, j \in [n]$ *such that for all* $k \in [n] \setminus \{i, j\}$ *we have* $u(k) = v(k)$, $u(i) = v(j)$ *and* $u(j) = v(i)$. *We call* $\mathcal{N}_{\text{SWAP}}$ *the* Swap neighborhood.

Definition 5. *Let* $r \in \mathbb{N}_{\geq 2}$. *We define the neighborhood* $\mathcal{N}_{f/s}$ *on* $\mathcal{X}_r$ *such that,* $\mathcal{N}_{f/s} = \mathcal{N}_{\text{FLIP}} \cup \mathcal{N}_{\text{SWAP}}$, *so if either a flip or a swap transformation is allowed. We call* $\mathcal{N}_{f/s}$ *the* Flip/Swap neighborhood.

2.1 Local Optima

In this paper, we consider minimization. Note that all definitions carry over to maximization.

We seek to analyze the landscape for the existence of local optima, meaning, intuitively, individuals where all neighboring individuals have a worse fitness. We are especially interested in non-global local optima. We now give a formal definition for the different types of local optima.

Definition 6 (Local and Global Optimum). *Let x be an individual and $(\mathcal{X}, \mathcal{N}, f)$ a landscape. We call x a* local optimum *if, for all individuals y with $(x, y) \in \mathcal{N}$, we have $f(x) < f(y)$. Further, we call x a* global optimum *if, for all individuals $z \in \mathcal{X}$, $f(z) \geq f(x)$. A local optimum which is not a global optimum we call an* NGLO *(non-global local optimum).*

Now we define the term *plateau* to refer to a set of neighboring individuals such that all of them have the same fitness.

Definition 7 (Plateau). *Let $(\mathcal{X}, \mathcal{N}, f)$ be a landscape and let $P \subset \mathcal{X}$ be a set of individuals such that, for all $x, y \in P$, $f(x) = f(y)$ and P is connected in the graph $(\mathcal{X}, \mathcal{N})$. Then we call P a* plateau. *We call a plateau P a* local optimum plateau *if, for all individuals $y \notin P$ reachable in one step from an individual $x \in P$, it holds that $f(x) < f(y)$. A local optimum plateau P where the elements are not global optima is called an* NGLOP *(non-global local optimum plateau).*

Note that any local optimum, taken as a singleton set, is also a local optimum plateau. This means that in proving no NGLOPs exist, we also show that no NGLOs exist.

Definition 8 (Modality). *Let $\mathcal{L} = (\mathcal{X}, \mathcal{N}, f)$ be a landscape. We call $\mathcal{L}$*

- unimodal *if there is exactly one local optimum plateau, and it has size* 1*;*
- plateau-unimodal *if there is exactly one local optimum plateau;*
- equimodal *if the elements of all local optimum plateaus have the same fitness;*
- multimodal *if there are two or more elements of local optimum plateaus with different fitness.*

Note that any unimodal landscape is plateau-unimodal (but not vice versa) and any plateau-unimodal landscape is equimodal (but not vice versa). Furthermore, a landscape is multimodal if and only if it is not equimodal.

Definition 9 (Modality and Instances). *We apply the terms unimodal, plateau-unimodal and equimodal to a set of instances if all landscapes induced by an instance of the given set have that property. The term multimodal we apply to a set of instances if all but finitely many*[1] *of these instances induce a multimodal landscape.*

[1] We exclude finitely many instances to allow for small instances in the set, which are typically too small to be multimodal.

3 Dominating Set

In this section we regard the Flip neighborhood and the Flip/Swap neighborhood in turn for the Dominating Set problem. For this, we define the fitness as follows.

Definition 10. *Given a graph $G = (V, E)$ on $n = |V|$ vertices, we fix an ordering of the vertices and identify each vertex with its index from $[n]$. Furthermore, we identify bit strings $x \in \mathcal{X}_2$ with the corresponding sets $S = \{i \in [n] \mid x_i = 1\}$. We let f_G^{DS} be the fitness function induced by G regarding the dominating set problem, defined with penalty terms for each undominated vertex as follows.*

$$f_G^{\mathrm{DS}} \colon \mathcal{X}_2 \to \mathbb{R}_{\geq 0},$$
$$x \mapsto \sum_{i=1}^{n} x_i + (n+1) \cdot |\{i \in [n] \mid x_i = 0 \wedge \forall j \in [n] \colon \{i,j\} \in E \to x_j = 0\}|.$$

3.1 Dominating Set with the Flip Neighborhood

We start by considering different graph classes and analyzing, for given graphs G from these classes, the induced landscape for the Flip neighborhood $(\mathcal{X}_2, \mathcal{N}_{\mathrm{FLIP}}, f_G^{\mathrm{DS}})$.

First, we give the following proposition, establishing that local optima in that landscape are exactly the $\subseteq$-minimal dominating sets. We will use this in the following without further mention.

Proposition 11. *Let $G = (V, E)$ be a graph and $S \subseteq V$. Then S is an $\subseteq$-minimal dominating set if and only if S is a local optimum in $(\mathcal{X}_2, \mathcal{N}_{\mathrm{FLIP}}, f_G^{\mathrm{DS}})$.*

Now, we start our analysis with some general considerations about landscapes from Dominating Set instances.

Lemma 12. *Let G be a graph. In the fitness landscape $(\mathcal{X}_2, \mathcal{N}_{\mathrm{FLIP}}, f_G^{\mathrm{DS}})$, every plateau $P \subseteq \mathcal{X}$ is a singleton set.*

From this, we immediately get the following corollary.

Corollary 13. *Let G be a graph and consider landscape $(\mathcal{X}_2, \mathcal{N}_{\mathrm{FLIP}}, f_G^{\mathrm{DS}})$. The property of plateau-unimodality is equivalent to unimodality. Consequently, a landscape with multiple local optima cannot be plateau-unimodal; it is* equimodal *if all local optima share the same fitness value, and* multimodal *otherwise.*

The first graph class we consider are bipartite graphs. The following lemma is central to our analyses.

Lemma 14. *Let $G = (U \uplus V, E)$ be a connected bipartite graph. Then the sets U and V are local optima for the landscape $(\mathcal{X}_2, \mathcal{N}_{\mathrm{FLIP}}, f_G^{\mathrm{DS}})$.*

Note that it is possible that neither of the partitions is a global optimum; for example, in the path with six vertices P_6 there is a dominating set of size two, while both partitions have size three. We can conclude the following two theorems.

Theorem 15. *Let $G = (U \uplus V, E)$ be an unbalanced bipartite graph ($|U| \neq |V|$, assume w.l.o.g. that $|U| < |V|$). Then the fitness landscape $(\mathcal{X}_2, \mathcal{N}_{\text{FLIP}}, f_G^{\text{DS}})$ is multimodal.*

Theorem 16. *The graph class of all complete bipartite graphs induces multimodal landscapes.*

The next graph class we consider is the class of all trees. We start by proving a result on more general graphs before using this to characterize the landscape induced by trees.

Definition 17 (Corona Graphs). *A connected graph G is called a* Corona graph *if every vertex of G is either a leaf (that is, has degree of 1) or is adjacent to exactly one leaf. A* Corona tree *is a Corona graph that is also a tree.*

The terminology derives from the *Corona product*, introduced by [4]. Here, the *Corona product* $G \circ H$ of two graphs G and H is obtained by taking one copy of G and $|V(G)|$ copies of H, and then joining the i-th vertex of G to every vertex in the i-th copy of H. This specific class of trees has been studied under this name by [2,8]. Specifically, a Corona tree is isomorphic to the graph $H \circ K_1$, where H is a tree.

We generalize this definition of Corona graphs as follows.

Definition 18 (Simplicial-Cluster Graphs). *A graph G is a* ***simplicial-cluster graph*** *if its vertex set can be partitioned into n cliques $L_1, \ldots, L_k$ of size $|L_i| \geq 2$, such that each clique L_i contains a vertex q_i (the* ***simplicial anchor****) that is adjacent to no vertex outside of L_i.*

Remark. *The simplicial anchors act as "private" leaves for each cluster. This class generalizes the Corona graph, which corresponds to $|L_i| = 2$ for all i. In particular, Corona trees are obtained when the clusters are connected only via bridges.*

Theorem 19. *Let G be a simplicial-cluster graph. The landscape $(\mathcal{X}_2, \mathcal{N}_{\text{FLIP}}, f_G^{\text{DS}})$ is equimodal, but not plateau-unimodal.*

We use this result to get a full characterization of the landscapes induced by arbitrary trees.

Theorem 20. *Let G be a tree. The landscape $\mathcal{L} = (\mathcal{X}_2, \mathcal{N}_{\text{FLIP}}, f_G^{\text{DS}})$ is equimodal if G is a Corona tree, and multimodal if G is* not *a Corona tree.*

3.2 Dominating Set and the Flip/Swap neighborhood

We will use results from *reconfiguration theory* in order to find results for the Flip/Swap neighborhood for some elaborate graph classes. For this we need the following definition.

Definition 21 (Reconfiguration Sequence). *Let G be a graph on n vertices. Let A, B be two dominating sets of G and $k \in \mathbb{N}$. Then we write $A \rightsquigarrow_k B$ if there is a sequence of dominating sets transforming A to B such that two successive sets differ by either adding or removing a vertex and such that the cardinality of any set in the sequence is bounded from above by k. We call the corresponding sequence of dominating sets a* reconfiguration sequence.

We give the connection to landscapes in the following lemma.

Lemma 22. *Let G be a graph on n vertices and $k \in \mathbb{N}$. Let A, B be two dominating sets of G with $|A| \geq |B|$ such that $A \rightsquigarrow_k B$.*

If a reconfiguration sequence between A and B exists such that no two token additions happen in direct succession, it holds that $A \overset{+}{\Longrightarrow} B$ for $(\mathcal{X}_2, \mathcal{N}_{f/s}, f_G^{\mathrm{DS}})$.

Haddadan et al. [6] investigate the properties of reconfiguration on dominating sets, introducing the concept of *canonical dominating sets.*

Definition 23 (Canonical Dominating Set). *A minimum dominating set C for a graph G is called a* canonical dominating set *if, for every dominating set D, it holds that $D \rightsquigarrow_k C$ with $k = |D| + 1$.*

They then prove the existence of canonical dominating sets for trees, cographs and interval graphs. We argue that this implies plateau-unimodality of the landscapes induced by such graphs.

Theorem 24. *Let G be a graph and suppose G contains a canonical dominating set C. Regarding the landscape $(\mathcal{X}_2, \mathcal{N}_{f/s}, f_G^{\mathrm{DS}})$, it holds for all feasible dominating sets D that $D \overset{+}{\Longrightarrow} C$.*

Using this theorem, we can now conclude that cographs, trees and interval graphs are plateau-unimodal.

Definition 25 (Cograph). *A* cograph, *or complement-reducible graph, or P_4-free graph, is a graph that can be generated from the single-vertex graph K_1 by complementation and disjoint union.*

Examples of cographs are complete graphs, complete bipartite graphs and threshold graphs.

Theorem 26. *Let G be a cograph. Then the fitness landscape $(\mathcal{X}_2, \mathcal{N}_{f/s}, f_G^{\mathrm{DS}})$ is plateau-unimodal.*

Theorem 27. *Let G be a tree. Then the fitness landscape $(\mathcal{X}_2, \mathcal{N}_{f/s}, f_G^{\mathrm{DS}})$ is plateau-unimodal.*

Definition 28 (Interval Graph). *A graph* $G = (V, E)$ *with* $V = \{v_1, v_2, \ldots, v_n\}$ *is an* interval graph *if there exists a set* I *of (closed) intervals* $I_1, I_2, \ldots, I_n \subseteq \mathbb{R}$ *such that, for all* $i, j \in [n]$, $\{v_i, v_j\} \in E$ *if and only if* $I_i \cap I_j \neq \emptyset$.

Theorem 29. *If* G *is an interval graph, then the fitness landscape* $(\mathcal{X}_2, \mathcal{N}_{f/s}, f_G^{\mathrm{DS}})$ *is plateau-unimodal.*

We now turn to negative results for the Flip/Swap neighborhood; note that these are not connected to reconfiguration theory.

We have seen that the class of all cographs is plateau-unimodal for the Flip/Swap neighborhood. In the following, we show that it is not unimodal, as well as for trees and interval graphs.

Theorem 30. *Let* $k \in \mathbb{N}_+$. *For the graph* $G = P_{3k+2}$, *which is a tree, a cograph, and an interval graph, we have that* $(\mathcal{X}_2, \mathcal{N}_{f/s}, f_G^{\mathrm{DS}})$ *is not unimodal.*

Finally, we give an example graph class which is multimodal for the Flip/Swap neighborhood.

Definition 31 (n-bouquet). *An* n-bouquet of 6-cycles *is a graph constructed by joining* $n \geq 2$ *copies of the cycle graph* C_6 *at a single shared vertex* v_0, *i.e., by performing a 1-clique-sum of these cycles. For the* i*-th cycle, we denote the remaining vertices by* $v_{i,1}, v_{i,2}, v_{i,3}, v_{i,4}, v_{i,5}$, *where* $v_{i,1}$ *and* $v_{i,5}$ *are adjacent to the shared vertex* v_0, *and consecutive vertices are adjacent in the natural order of the cycle.*

Theorem 32. *Let* $G = (V, E)$ *be an* n-bouquet of 6-cycles. *Then the landscape* $\mathcal{L} = (\mathcal{X}, \mathcal{N}_{f/s}, f_G^{\mathrm{DS}})$ *is multimodal.*

4 Vertex Coloring

In this section we consider the vertex coloring problem with the Flip neighborhood, then with the Flip/Swap neighborhood. For this, we define the fitness as follows.

Definition 33. *Given a graph* $G = (V, E)$ *on* $n = |V|$ *vertices, we fix an ordering of the vertices and identify each vertex with its index from* $[n]$. *Any string* $x \in \mathcal{X}_n$ *is a coloring of the vertices of* G. *We call a coloring proper if, for all* $i, j \in V$ *with* $\{i, j\} \in E$ *we have* $x_i \neq x_j$. *We let* f_G^{Col} *be the fitness function induced by* G *regarding the vertex coloring problem, defined as the number of colors used with a penalty terms for each monochromatic edge as follows.*

$$f_G^{\mathrm{Col}} \colon \mathcal{X}_n \to \mathbb{R}_{\geq 0}, x \mapsto |\{x_i \mid i \in [n]\}| + (n+1) \cdot |\{\{i,j\} \in E \mid x_i = x_j\}|.$$

Note that, for any infeasible coloring, recoloring any vertex which is part of a monochromatic edge with a color that does not currently appear in the coloring will improve fitness. Thus, there are no local optima among infeasible solutions, and we will focus on feasible solutions.

4.1 Flip neighborhood

We first make the following observation.

Proposition 34. *On connected bipartite graphs, any fitness landscape that is equimodal is also unimodal, assuming that two optimal colorings are considered the same if they differ only in the naming of colors.*

This follows from the fact that for connected bipartite graphs, there is exactly one feasible 2-coloring, namely each partition colored in a unique color. While the names of those colors might differ between colors (one coloring might use red and green, the other blue and red), they are still essentially the same coloring and we therefore consider them the same for the sake of unimodality.

We now seek to further characterize the fitness landscape for coloring, focusing on the family of bipartite graphs. Here, we identify one unimodal case as well as two multimodal cases. Additionally, we derive further results from reconfiguration theory. We start with a definition.

Definition 35 (Universal Bipartite Graphs). *Let $G = (V, E)$ be a bipartite graph with partitions $U \uplus W = V$. We call $u \in U$* universal to the opposite partition *if, for all $w \in W$, it holds that $\{u, w\} \in E$. Analogously we define universality of $w \in W$. A bipartite graph with at least one vertex universal to the opposite partition is called a* universal bipartite graph.

Next we state that the existence of a vertex that is universal to the opposite partition implies that the fitness landscape $(\mathcal{X}_n, \mathcal{N}_{\text{FLIP}}, f_G^{\text{Col}})$ is unimodal.

Theorem 36. *For a universal bipartite graph G, the landscape $(\mathcal{X}_n, \mathcal{N}_{\text{FLIP}}, f_G^{\text{Col}})$ is unimodal.*

Now, we examine structural properties that allow the existence of NGLOPs, meaning the fitness landscape $(\mathcal{X}_n, \mathcal{N}_{\text{FLIP}}, f_G^{\text{Col}})$ is multimodal.

First, we regard $G = C_{6k}$. We find the fitness landscape $(\mathcal{X}_n, \mathcal{N}_{\text{FLIP}}, f_G^{\text{Col}})$ to contain NGLOs, making it multimodal. Further, we define conditions for a graph G containing the C_{6k} as a subgraph that lead to $(\mathcal{X}_n, \mathcal{N}_{\text{FLIP}}, f_G^{\text{Col}})$ containing NGLOPs, making it likewise multimodal.

We examine a situation in which an NGLOP exists at three colors. We find that the C_{6k} contains NGLOPs at three colors. In order to describe a graph configuration with the C_{6k} as an induced subgraph that likewise contains NGLOPs, we first give necessary definitions.

We refer to vertices in the C_{6k} as $v_1, \ldots, v_{6k}$, with the numbering corresponding to their order in the circle.

Definition 37 (3-chordless C_{6k}-subgraph). *Let $G = (V, E)$ be a graph and $v_1, v_2, \ldots v_{6k} \in V$ a C_{6k}-subgraph in G. We call $v_1, v_2, \ldots v_{6k} \in V$* 3-chordless *if, for all $j \leq 6k$, there exists no $l \leq 6k$ such that $l \equiv j \pmod 3$ and $\{v_j, v_l\} \in E$.*

Further, we define the term *d-opposing vertices* to describe a way vertices outside of the C_{6k} can be connected to vertices within.

Definition 38 (d-opposing vertices). *Let $G = (V, E)$ be a graph and let $v_1, \ldots v_{6k} \in V$ be a C_{6k}-subgraph. Let $v \in V$ be a vertex that is not part of the C_{6k} and let $d \in \{0, 1, 2\}$. We call v* d-opposing *if, for all $j \leq 6k$ such that $j \equiv d \pmod 3$, no edge exists between v and v_j.*

Next we prove that a bipartite graph contains an NGLOP at three colors provided that it contains a 3-chordless C_{6k}-subgraph and certain criteria for the other vertices are met. To prove this, we first show that the C_{6k} is multimodal with an NGLO at three colors for the fitness landscape $(\mathcal{X}_n, \mathcal{N}_{\text{FLIP}}, f_G^{\text{Col}})$ and then prove that such a coloring of the C_{6k}-subgraph is likewise possible in a graph meeting the given criteria.

Theorem 39. *Let $G = (V, E)$ be a bipartite graph such that*

1. *it contains a 3-chordless C_{6k}-subgraph,*
2. *for all $v \in V$ outside of the C_{6k}-subgraph, there exists a $d \in \{0, 1, 2\}$ such that v is d-opposing and*
3. *if two vertices $u, v \in V$ are both d-opposing, then $\{u, v\} \notin E$.*

Then the fitness landscape $(\mathcal{X}_n, \mathcal{N}_{\text{FLIP}}, f_G^{\text{Col}})$ for G is multimodal with NGLOPs at three colors.

We now generalize those findings to a class of graphs whose fitness landscape is multimodal with NGLOPs at k colors.

Definition 40 (k-Crown Graph). *We refer to a bipartite graph where both partitions have size k and each vertex has a degree of $k - 1$ as a* k-crown graph.

Table 2. Classification of the landscapes of various graph classes for the Vertex Coloring problem into plateau-unimodal (there exists a single local optimum plateau), equimodal (all local optimum plateaus are globally optimal) and multimodal (there exist NGLOPs, see Definition 7) for the Flip and Flip/Swap neighborhood respectively (see Definitions 3 and 5). See Definition 8 for the definition of modality.

Dominating Set			
Graph Class	Neighborhood	Optima	Theorem
Unbalanced Bipartite Graphs	$\mathcal{N}_{\text{FLIP}}$	multimodal	15
Complete Bipartite Graphs	$\mathcal{N}_{\text{FLIP}}$	multimodal	16
Simplicial-Cluster Graphs	$\mathcal{N}_{\text{FLIP}}$	equim., not plateau-unim.	19
Non-Corona Trees	$\mathcal{N}_{\text{FLIP}}$	multimodal	20
Cographs	$\mathcal{N}_{f/s}$	plateau-unim., not unim.	26 & 30
Trees	$\mathcal{N}_{f/s}$	plateau-unim., not unim.	27 & 30
Interval Graphs	$\mathcal{N}_{f/s}$	plateau-unim., not unim.	29 & 30
n-Bouquet of 6-Cycles	$\mathcal{N}_{f/s}$	multim.	32

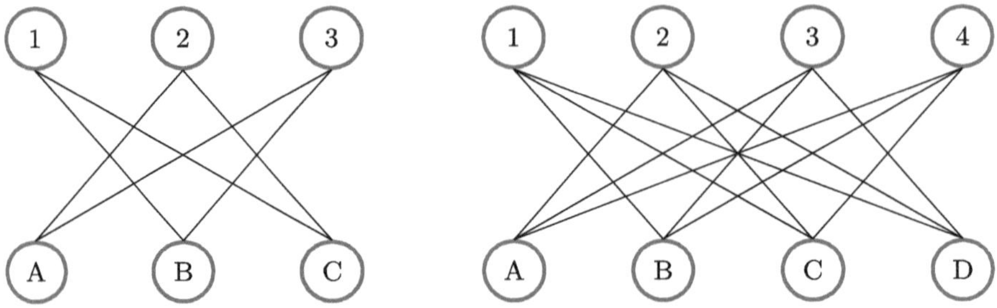

(a) $k = 3$ (b) $k = 4$

Fig. 1. Two example instances of a k-crown graph for $k = 3$ and $k = 4$.

Figure 1 shows two example instances of a k-crown graph for $k = 3$ and $k = 4$. Note that the 3-crown graph is isomorphic to the C_6.

We further introduce the term *opposing pair* to refer to two vertices in opposite partitions that are not connected by an edge.

Definition 41 (Opposing Pair). *Let $G = (V, E)$ be a bipartite graph with partitions $U \uplus W$. Let $u \in U$ and $w \in W$. Then, we call u and v an* opposing pair *if $\{u, v\} \notin E$.*

We now prove that the fitness landscape for any bipartite graph $G = (V, E)$ contains NGLOPs at k colors with $k > 2$ and is therefore multimodal, provided that G contains a k-crown graph as an induced subgraph and certain criteria regarding the structure of the rest of the graph are met. Intuitively, the fitness landscape for the k-crown graph contains an NGLO (which we prove in Lemma 43), which leads to an NGLOP in the fitness landscape for G unless a part of G's structure makes the necessary coloring of the k-crown graph impossible.

Theorem 42. *Let $k > 2$. Let $G = (V, E)$ be a bipartite graph such that*

1. *G contains a k-crown graph as an induced subgraph,*
2. *no vertex outside of the k-crown graph is connected to all vertices in one of the k-crown graph's partitions and*
3. *for all $u, v \in V$ and a, b as vertices in the k-crown graph, it holds that if $(u, a), (v, b)$ and (a, b) are opposing pairs, then so is (u, v).*

Then, the fitness landscape $(\mathcal{X}_n, \mathcal{N}_{\text{FLIP}}, f_G^{\text{Col}})$ is multimodal with NGLOPs at k colors.

To prove this, we first show that the fitness landscape for the k-crown graph contains an NGLO at k colors and then conclude that this causes an NGLOP in the fitness landscape for G.

Lemma 43. *The fitness landscape $(\mathcal{X}_n, \mathcal{N}_{\text{FLIP}}, f_G^{\text{Col}})$ for the k-crown graph contains an NGLO at k colors.*

Just as for the Dominating Set problem, we derive further results by turning to results from reconfiguration theory. In the following, for a given graph G, we use $\mathcal{F}_G^\ell$ to denote the set of all proper ℓ-colorings of G.

Bonamy et al. [3] consider changing the color of a single vertex as a modification rule; this corresponds to the Flip neighborhood. They give a general graph class, called k-color-dense graphs, and show that, for any such graph G and $\ell \geq k+1$, $(\mathcal{F}_G^\ell, \mathcal{N}_{\text{FLIP}})$ is connected under this modification rule. As example subclasses of this general graph class they give the class of all chordal graphs and all chordal bipartite graphs.

We give the following general theorem to turn such connectedness results into landscape results.

Theorem 44. *Let G be a graph with chromatic number k such that for every $\ell \geq k+1$, $(\mathcal{F}_G^\ell, \mathcal{N}_{\text{FLIP}})$ is connected. Then the fitness landscape $(\mathcal{X}_n, \mathcal{N}_{\text{FLIP}}, f_G^{\text{Col}})$ is equimodal.*

Using this as well as results given by Bonamy et al. [3], we can now derive the following theorems for chordal graphs and chordal bipartite graphs; we give the well-known definitions for completeness.

Definition 45 (Chordal Graph). *A* chordal graph *is a graph with no induced cycle of length more than* 3.

Theorem 46. *Let G be a chordal graph with chromatic number k. Then the fitness landscape $(\mathcal{X}_n, \mathcal{N}_{\text{FLIP}}, f_G^{\text{Col}})$ is equimodal.*

Definition 47 (Chordal Bipartite Graph). *A* chordal bipartite graph *is a bipartite graph with no induced cycle of length greater than* 4.

Theorem 48. *Let G be a chordal bipartite graph. Then the fitness landscape $(\mathcal{X}_n, \mathcal{N}_{\text{FLIP}}, f_G^{\text{Col}})$ is equimodal, with all global optima being isomorphic under permuting the colors.*

We further demonstrate that there are chordal graphs that have multiple distinct colorings, even taking into account relabeling of colors and graph automorphisms. Therefore, this fitness landscape is not plateau-unimodal.

Theorem 49. *There is a chordal graph G such that the landscape $(\mathcal{X}_n, \mathcal{N}_{\text{FLIP}}, f_G^{\text{Col}})$ is not plateau-unimodal.*

4.2 Flip/Swap neighborhood

We now turn to the Flip/Swap neighborhood. We show that $(\mathcal{X}_n, \mathcal{N}_{f/s}, f_G^{\text{Col}})$ is unimodal for both the C_{6k} and crown graphs, but multimodal for a graph class we introduce called *spoked* C_{12k}, which is a variant of the C_{12k} with certain chords added.

Theorem 50. *Let $G = (V, E)$ be a C_{6k}. Then the landscape $(\mathcal{X}_n, \mathcal{N}_{f/s}, f_G^{\text{Col}})$ is unimodal.*

Theorem 51. *Let $G = (V, E)$ be a k-crown graph. Then, the fitness landscape $(\mathcal{X}_n, \mathcal{N}_{f/s}, f_G^{\text{Col}})$ is unimodal.*

Finally, we give an example of a class of graphs that induces a multimodal landscape on the Flip/Swap neighborhood to demonstrate that flipping and swapping cannot color even some simple graphs.

Definition 52 (Spoked C_{12k}). *Let G be the cycle graph C_{12k} with additional edges as follows. Let $V = \{0, \dots 12k - 1\}$ be the vertices of G in order around the cycle. For any vertex i we call $(i + 6k) \bmod 12k$ its opposite vertex. For any vertex v we then add an edge from v to both of its opposite vertex's neighbors on the cycle.*

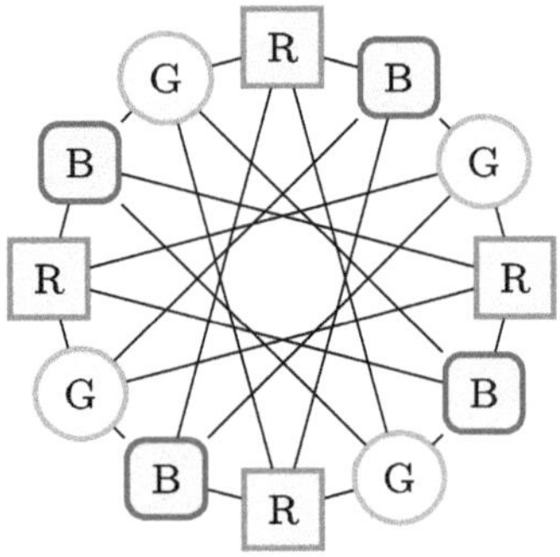

Fig. 2. The spoked C_{12}.

Note that the spoked C_{12k} is bipartite as C_{12k} is bipartite and we only added edges between different partitions. Figure 2 shows the spoked C_{12}.

We now state that any spoked C_{12k} induces a multimodal landscape on the Flip/Swap neighborhood.

Theorem 53. *Let G be a spoked C_{12k}. Then $(\mathcal{X}_n, \mathcal{N}_{f/s}, f_G^{\text{Col}})$ is multimodal.*

References

1. Alyahya, K., Rowe, J.E.: Landscape analysis of a class of NP-hard binary packing problems. Evol. Comput. **27**(1), 47–73 (2019). issn: 1063-6560. https://doi.org/10.1162/evco_a_00237
2. Barik, S., Pati, S., Sarma, B.K.: The spectrum of the corona of two graphs. SIAM J. Discrete Math. **21**(1), 47–56 (2007). issn: 0895–4801. https://doi.org/10.1137/050624029
3. Bonamy, M., et al.: Reconfiguration graphs for vertex colourings of chordal and chordal bipartite graphs. J. Comb. Optim. **27**(1), 132–143 (2014). issn: 1573-2886. https://doi.org/10.1007/s10878-012-9490-y

4. Frucht, R., Harary, F.: On the corona of two graphs'. aequationes mathematicae **4**(3), 322–325 (1970). issn: 1420-8903. https://doi.org/10.1007/BF01844162
5. Gasse, J., et al.: Combinatorial landscape analysis for dominating set and vertex coloring (2026). https://doi.org/10.48550/arXiv.2606.07361
6. Haddadan, A., et al.: The complexity of dominating set reconfiguration. In: Algorithms and Data Structures. In: Dehne, F., Sack, J.-R., Stege, U.: Springer, pp. 398–409 (2015). isbn: 978-3-319-21840-3. https://doi.org/10.1007/978-3-319-21840-3_33
7. Nishimura, N.: Introduction to reconfiguration. Algorithms **11**(4), 52 (2018). issn: 1999-4893. https://doi.org/10.3390/a11040052
8. Panda, S.K.: Unicyclic graphs with bicyclic inverses. Czechoslovak Math. J. **67**(4), 1133–1143 (2017). issn: 1572-9141. https://doi.org/10.21136/CMJ.2017.0429-16
9. Prügel-Bennett, A., Tayarani-Najaran, M.-H.: Maximum satisfiability: anatomy of the fitness landscape for a hard combinatorial optimization problem. IEEE Trans. Evol. Comput. **16**(3), 319–338 (2012). https://doi.org/10.1109/TEVC.2011.2163638
10. Tayarani-Najaran, M.-H., Prügel-Bennett, A.: Anatomy of the fitness landscape for dense graph-colouring problem. Swarm Evol. Comput. **22**, 47–65 (2015). https://doi.org/10.1016/J.SWEVO.2015.01.005

Decomposing Anytime Algorithm Performance with Bayesian Ranking ANOVA

Jonathan Wurth(✉), Helena Stegherr, Michael Heider, and Jörg Hähner

Organic Computing Group, Universität Augsburg, Augsburg, Germany
{jonathan.wurth,helena.stegherr,michael.heider,joerg.haehner}@uni-a.de

Abstract. Comparing multiple optimization algorithms across problem instances involves understanding not only which algorithm is best on average, but how instance characteristics shift relative performance over time. The PolarBear framework identifies Pareto-optimal anytime algorithms via scale-free Bayesian ranking models, but pools all instances into a single marginal comparison, discarding structure that could inform algorithm selection at deployment. We extend its temporal Plackett-Luce model with an ANOVA decomposition of latent utilities: algorithm performance over time is decomposed into a global baseline, main effects of instance characteristics, and their interactions, all modeled as time-varying processes. Factor levels are explicitly modeled independently, making the decomposition an analytical tool rather than a predictive model. We develop a diagnostic toolkit for the fitted posterior: Shapley information values quantify the decision-theoretic gain from learning each factor, log-lifts reveal how each factor level shifts performance ratings relative to the global baseline, and conditional anytime Pareto sets identify specialist algorithms that emerge under specific conditions. We demonstrate the framework on the BBOB benchmark suite, decomposing anytime performance of a diverse algorithm portfolio into effects of function identity and dimensionality. Clustering the level-specific deviations from the global baseline produces a data-driven problem taxonomy grounded in algorithmic performance rather than landscape properties.

Keywords: Anytime optimization · Algorithm benchmarking · Algorithm selection · Plackett-Luce · ANOVA

1 Introduction

Choosing which optimization algorithm to deploy on a given problem requires comparing candidates across a representative set of instances. When the computational budget is unknown at benchmarking time, anytime comparison is necessary: algorithms must be evaluated not at a single final budget but throughout execution. PolarBear [18] addresses this by treating each budget as a separate objective, identifying the Pareto set of non-dominated algorithms over

G. Iacca et al. (Eds.): PPSN 2026, LNCS 16985, pp. 516–531, 2027.
https://doi.org/10.1007/978-3-032-36214-8_32

time through sequential experiment design based on a Bayesian Plackett–Luce ranking model. This provides scale-free comparison over heterogeneous instances, without requiring normalization or known optima. However, PolarBear treats this instance distribution as monolithic: it estimates marginal performance ratings pooled across all instances, discarding any known structure. When instance characteristics that the practitioner knows at deployment modulate relative algorithm performance, the marginal Pareto set underutilizes available information. An algorithm inferior globally may be the best choice for a specific problem class; an algorithm that appears competitive on average may be dominated in every individual condition, surviving only because it is mediocre everywhere.

The natural response is to condition on known instance characteristics. The simplest approach runs the comparison separately for each condition, but this discards shared structure across conditions, requires sufficient data per condition, and provides no framework for asking whether the conditioning matters at all. At the other extreme, algorithm selection methods predict per-instance algorithm performance from instance features [6], typically requiring feature extraction (e.g., exploratory landscape analysis [8]). These methods assume that nearby points in feature space induce similar algorithm performance, which is a smoothness assumption that enables generalization but also extrapolation to regions where it may not hold.

Between these approaches, analysis of variance (ANOVA) provides a principled framework for decomposing variation into attributable sources. We propose a temporal Plackett–Luce (PL) ANOVA that decomposes algorithm ratings into a global baseline, main effects of categorical instance characteristics, and their interactions, each modeled as a time-varying process with Bayesian priors (cf. Sect. 2). Crucially, factor levels are modeled independently, i.e., no similarity between levels is assumed, and the model does not share strength across factor levels; the value of the joint model is therefore not online algorithm selection (for new levels) but post hoc decomposition. It provides a structured diagnostic framework that attributes variation in algorithm rankings to specific factors, quantifies whether conditioning matters, and identifies where the global picture breaks down, without making assumptions about the relationship between factor levels. We then develop a diagnostic toolkit that extracts progressively finer information from the fitted posterior. Together, this guides a practitioner from "does conditioning matter?" through "on what factors?" to "which algorithm for a specific observed combination of levels?" (cf. Sect. 3). We demonstrate the framework in an exemplary case study on the BBOB benchmark suite [5], decomposing anytime performance of six algorithms into effects of function identity and dimensionality (cf. Sect. 4).

Prior work has addressed structured algorithm comparison from several directions. Chiarandini and Goegebeur [2] treat benchmarking as a linear mixed-effects ANOVA problem on scalar performance, with algorithms as fixed effects, instances as random effects, and instance-by-algorithm interactions capturing heterogeneity across sampled problems. Bang-Jensen et al. [1] use the same mixed-model logic in a factorial component-analysis setting, decomposing scalar heuristic performance into effects of algorithmic design choices, instance features, and their interactions while nesting sampled instances within instance classes. Mattos et al. [7] use Bayesian item response theory to evaluate benchmark suites themselves by judging suite quality through item/test information rather than direct comparative performance modeling. Smith-Miles et al. [15] propose an instance-space methodology that maps problem instances into a low-dimensional feature space, and defines and uses algorithm "footprints" to visualize and predict where algorithms are strong or weak, with the emphasis on benchmark diversity and sample-bias reduction. Nikolikj and collaborators pursue two closely related empirical lines of work on structured algorithm assessment. One develops *footprint*-based, explainable-analysis methods that learn fixed-budget performance from landscape features and then use SHAP-based meta-representations and clustering to identify and compare easy and hard problem regions for single algorithms, benchmark portfolios, and modular meta-heuristic frameworks [9,12,13]. The second studies modular optimizers through random-forest functional ANOVA, decomposing scalar performance variance into individual, pairwise, and triplet module effects, first across benchmark suites and then at the level of problem classes [10,11]. PL trees [17] address heterogeneity by recursively partitioning rankings into covariate-defined subgroups and fitting a separate static PL model in each leaf, so the rating parameters vary across hard splits rather than through a structured effect decomposition.

2 Temporal Plackett–Luce ANOVA

We observe rankings over a set of n algorithms $\mathbf{A} = \{A_1, \ldots, A_n\}$, recorded at a discrete grid of timepoints $\mathbf{T} = (t_1, \ldots, t_T)$. Each ranking is annotated with the levels of K categorical factors, whose cardinalities we denote $G_1, \ldots, G_K$.

Formally, each ranking $r \in \mathcal{R}$ consists of:

- a permutation π_r over $n_r \leq n$ algorithms,
- a timepoint $t_r \in \mathbf{T}$,
- factor levels $\boldsymbol{g}_r = (g_r^{(1)}, \ldots, g_r^{(K)}) \in \prod_{k=1}^{K}\{1, \ldots, G_k\}$,
- a non-negative weight $w_r \geq 0$.

A factor may represent any known (categorical) property of the instance or experimental condition, such as problem class or dimensionality (band). Factor levels are assumed known for each ranking. Rankings within a factor-level combination are exchangeable draws from the corresponding conditional distribution.

We decompose the PL log-ratings (or latent *utilities*) μ of algorithm A at timepoint t under factor configuration $\boldsymbol{g} = (g^{(1)}, \ldots, g^{(K)})$ as

$$\mu_A(t, \boldsymbol{g}) = \underbrace{\mu_A(t)}_{\text{global}} + \sum_{k=1}^{K} \underbrace{\delta_A^{(k)}(t, g^{(k)})}_{\text{main effects}} + \sum_{j<k} \underbrace{\delta_A^{(jk)}(t, g^{(j)}, g^{(k)})}_{\text{two-way interactions}} + \cdots . \quad (1)$$

The global term $\mu_A(t)$ captures marginal algorithm performance across all factor levels. Each $\delta_A^{(\tau)}$ is a deviation from this baseline, depending on time and on the factor levels indexed by τ only.

More compactly, let $\mathcal{T} \subseteq \mathcal{P}(\{1,\ldots,K\})$ be the collection of *active terms*, where each $\tau \in \mathcal{T}$ is a subset of factor indices. The empty set $\tau = \varnothing$ represents the global effect $\mu_A(t)$. Singletons $\tau = \{k\}$ represent main effects of factor k. Pairs $\tau = \{j,k\}$ represent two-way interactions, and so on. Writing $\delta_A^{(\varnothing)}(t) \equiv \mu_A(t)$ for uniformity, the decomposition is

$$\mu_A(t,\, \boldsymbol{g}) \;=\; \sum_{\tau \in \mathcal{T}} \delta_A^{(\tau)}\big(t,\, \boldsymbol{g}^{(\tau)}\big)\,, \tag{2}$$

where $\boldsymbol{g}^{(\tau)} = (g^{(k)})_{k\in\tau}$ collects the factor levels involved in term τ. When $\mathcal{T} = \{\varnothing\}$, the model reduces to the temporal PL model described in [18].

The conditional rating of algorithm A_i at time t under factor configuration $\boldsymbol{g}$ is

$$\theta_i(t,\boldsymbol{g}) \;=\; \operatorname{softmax}\big(\mu(t,\boldsymbol{g})\big)_i \;=\; \frac{\exp \mu_i(t,\boldsymbol{g})}{\sum_{j=1}^{n} \exp \mu_j(t,\boldsymbol{g})}\,, \tag{3}$$

extending the unconditional rating $\theta_i(t) = \operatorname{softmax}(\mu(t))_i$. Under PL assumptions, $\theta_i(t)$ can be interpreted as the probability of A_i being ranked first (i.e., its *win probability*) at time t. The pairwise Bradley–Terry structure is preserved:

$$P(A \overset{t}{\succ} B \mid \boldsymbol{g}) = \frac{\theta_A(t,\boldsymbol{g})}{(\theta_A(t,\boldsymbol{g}) + \theta_B(t,\boldsymbol{g}))}\,. \tag{4}$$

Identifiability Constraints. Two families of constraints ensure a unique decomposition. The PL likelihood is invariant to additive shifts across algorithms. We therefore impose, for every term τ:

$$\sum_{i=1}^{n} \delta_i^{(\tau)}(t,\, \boldsymbol{g}^{(\tau)}) = 0 \qquad \forall\, t,\ \boldsymbol{g}^{(\tau)}\,. \tag{5}$$

To separate main effects from interactions and from the global mean, we require that each non-global term sums to zero over every factor it involves:

$$\sum_{g^{(k)}=1}^{G_k} \delta_i^{(\tau)}(t,\, \boldsymbol{g}^{(\tau)}) = 0 \qquad \forall\, k \in \tau,\ \forall\, t,\ \forall\, i,\ \forall\, \boldsymbol{g}^{(\tau\setminus\{k\})}\,. \tag{6}$$

These are the standard functional ANOVA marginality constraints, applied pointwise in t. Together they ensure that $\mu_i(t)$ is the marginal utility obtained by averaging $\mu_i(t,\boldsymbol{g})$ uniformly over all factor-level combinations, and that each $\delta^{(\tau)}$ captures only the variation attributable to its specific factor combination, orthogonal to all lower-order terms.

Helmert Parametrization. Let $Q_m \in \mathbb{R}^{m\times(m-1)}$ denote the Helmert contrast matrix for an m-level factor, satisfying $Q_m^\top \mathbf{1}_m = \mathbf{0}$ and $Q_m^\top Q_m = I_{m-1}$. The orthogonality of the columns guarantees that the zero-sum constraints (5)–(6) are automatically satisfied when the constrained effects are obtained by contraction with the appropriate Helmert matrices.

For a term $\tau = \{k_1, \ldots, k_m\}$ with $m = |\tau|$ factors, the free parameters are

$$\tilde{\delta}^{(\tau)} \in \mathbb{R}^{(G_{k_1}-1)\times\cdots\times(G_{k_m}-1)\times(n-1)\times T}, \tag{7}$$

comprising $\prod_{k\in\tau}(G_k - 1)\cdot(n-1)$ independent temporal functions. The constrained effects are recovered by contracting along each reduced axis:

$$\delta^{(\tau)} = \Big(\bigotimes_{k\in\tau} Q_{G_k}\Big) \otimes Q_n \cdot \tilde{\delta}^{(\tau)}, \tag{8}$$

where $\otimes$ denotes the sequential tensor contraction applied axis by axis. The result has shape $G_{k_1} \times \cdots \times G_{k_m} \times n \times T$.

For the global term $\tau = \varnothing$, the free parameters reduce to $(n-1)$ temporal functions $(\tilde{\delta})^{\varnothing} \in \mathbb{R}^{(n-1)\times T}$, contracted through Q_n alone. This recovers exactly the Helmert parametrization of [18].

Priors on Temporal Functions. Each element of $\tilde{\delta}^{(\tau)}$ is an independent temporal function $\eta_q^{(\tau)} : \mathbb{R} \to \mathbb{R}$, and all functions within the same term share a common set of hyperparameters $\psi^{(\tau)}$:

$$\eta_q^{(\tau)} \sim \mathcal{F}(\psi^{(\tau)}), \qquad q = 1, \ldots, \prod_{k\in\tau}(G_k - 1)\cdot(n-1). \tag{9}$$

The distribution $\mathcal{F}(\psi^{(\tau)})$ is a stochastic process prior over functions of time. We give the Gaussian Process (GP) instantiation from [18] here; others (e.g., B-splines) are equally applicable. Each function is drawn from a zero-mean GP:

$$\eta_q^{(\tau)} \sim \mathcal{GP}\big(0,\, (\sigma^{(\tau)})^2\, k_{\psi^{(\tau)}}\big), \tag{10}$$

where $k_{\psi^{(\tau)}}$ is a stationary kernel (e.g., Matérn-3/2) with lengthscale $\ell^{(\tau)}$. The shared hyperparameters are $\psi^{(\tau)} = (\sigma^{(\tau)}, \ell^{(\tau)}, \ldots)$. Inducing-point approximations (DTC) or Hilbert-space methods (HSGP) apply when the number of timepoints or outputs is large.

Hyperprior Hierarchy. Each term τ receives its own hyperparameters $\psi^{(\tau)}$, drawn from priors $p(\psi^{(\tau)})$. In a well-identified ANOVA decomposition, the bulk of the variation is typically captured by the global term and the main effects, with progressively less variation attributable to higher-order interactions. Encoding this belief through the hyperpriors—for instance by placing tighter priors on $\sigma^{(\tau)}$ for higher-order terms—provides regularization and improves identifiability, particularly when some cells of the factorial design are sparsely observed.

The decomposed utilities (2) plug directly into the weighted PL likelihood of [18], with $\mu_i(t_r, \boldsymbol{g}_r)$ replacing the unconditional $\mu_i(t_r)$. Inference proceeds as described there; the factor structure adds parameters but does not change the likelihood form.

3 Diagnostics for Structured Anytime Comparison

The fitted posterior provides the full conditional utility $\mu_i(t, \boldsymbol{g})$ for every algorithm, timepoint, and factor-level combination. This section defines a sequence of diagnostics that extract progressively finer information from this posterior.

Utility-Space Variance: *Does the factor structure contain enough signal to justify its inclusion?*

The ANOVA orthogonality constraints (5)–(6) ensure that cross-terms vanish, so the total cross-algorithm sum of squares in utility space decomposes exactly into per-term contributions. For a given timepoint t, the sum of squares of τ is

$$\mathrm{SS}^{(\tau)}(t) = \sum_{\boldsymbol{g}^{(\tau)}} \sum_{i=1}^{n} \left[\delta_i^{(\tau)}(t, \boldsymbol{g}^{(\tau)})\right]^2, \tag{11}$$

with $\delta_i^{(\varnothing)} \equiv \mu_i$ for the global term. Two normalizations are useful. The *variance* $\mathrm{Var}^{(\tau)}(t) = \mathrm{SS}^{(\tau)}(t)/(\prod_{k\in\tau} G_k \cdot n)$ divides by the number of cells and algorithms. Variance shares $\mathrm{Var}^{(\tau)}(t)/\sum_{\tau'} \mathrm{Var}^{(\tau')}(t)$ sum to one by orthogonality and show how total variation is allocated across terms. The *mean square* $\mathrm{MS}^{(\tau)}(t) = \mathrm{SS}^{(\tau)}(t)/(\prod_{k\in\tau}(G_k - 1) \cdot (n-1))$ divides by degrees of freedom instead, giving the average signal per free parameter. Mean squares do not sum to the total but are comparable across terms with different cardinalities. However, the softmax map from utilities in log-space to ratings on the simplex is nonlinear: when base utilities are well-separated, even large offsets barely change ratings; when utilities are close, small offsets can flip rankings. Consequently, near-zero mean squares reliably indicate that a term is negligible, but nonzero mean squares are necessary, not sufficient, for decision relevance. The probability-space diagnostics that follow do not share this limitation. Since each posterior sample yields a mean square, the posterior distribution of mean squares provides a Bayesian screening criterion: a term whose credible interval is concentrated near zero is a candidate for removal.

Global Ratings: *What is the marginal performance rating of each algorithm?* The global rating $\theta_i(t)$ summarizes the marginal performance of each algorithm over time, pooling across all factor levels, and can be interpreted as the expected win probability on a random instance under PL. This is the output of the temporal PL model of [18]: it allows identifying crossover points where the leading algorithm changes and determines the global anytime Pareto set $\mathcal{P}$. The global ratings serve as the reference against which all subsequent diagnostics are measured.

Log-Lift: *How does learning a factor level shift each algorithm's ratings relative to the global baseline?*

The log-lift of factor level $g^{(k)}$ on algorithm A at time t is

$$\Lambda_A^{(k)}(t, g^{(k)}) = \log P(A \mid t, g^{(k)}) - \log \theta_A(t), \tag{12}$$

where $P(A \mid t, g^{(k)})$ conditions on factor k while handling the remaining factors in one of two ways.

Partial Log-Lift. Set all terms involving unobserved factors to zero:

$$P_{\text{partial}}(A \mid t, g^{(k)}) \;=\; \text{softmax}\big(\mu(t) + \delta^{(k)}(t, g^{(k)})\big)_A . \tag{13}$$

This isolates the main effect of factor k: it shows what $\delta_A^{(k)}$ alone does to the global ratings, before interactions modify it.

Marginalized Log-Lift. Average the full conditional model over the unobserved factors:

$$P_{\text{marg}}(A \mid t, g^{(k)}) \;=\; \frac{1}{|\mathcal{G}_{-k}|} \sum_{g_{-k}} \theta_A(t, \boldsymbol{g}) , \tag{14}$$

where $\mathcal{G}_{-k}$ denotes the set of all level combinations of the factors other than k. This is the rating a practitioner faces when they know only factor k but nothing else.

The two variants coincide when all interaction terms involving factor k are zero. Otherwise, the marginalized variant incorporates the additional variation created by interactions and is the decision-relevant quantity. The gap between partial and marginalized log-lifts is itself an indicator of interaction strength in probability space.

Shapley Information Values: *How much does learning each factor improve algorithm selection, and are the factors complementary?*

A practitioner faces a random instance with factor levels $\boldsymbol{g} \sim \mathcal{G}$ and must choose an algorithm. Payoff at time t is the rating $\theta_i(t, \boldsymbol{g})$ of the chosen algorithm in the realized cell. For any information set $I \subseteq \{1, \ldots, K\}$ revealing the levels of the corresponding factors, the optimal policy selects

$$A^*(\boldsymbol{g}_I, t) \;=\; \arg\max_{A_i} \; \mathbb{E}_{\boldsymbol{g}_{-I}}\big[\theta_i(t, \boldsymbol{g})\big] , \tag{15}$$

and the expected value of information set I at time t is

$$V(I, t) \;=\; \mathbb{E}_{\boldsymbol{g}_I}\Big[\mathbb{E}_{\boldsymbol{g}_{-I}}\big[\theta_{A^*(\boldsymbol{g}_I, t)}(t, \boldsymbol{g})\big]\Big] . \tag{16}$$

The information value of learning factors I is $\text{IV}(I, t) = V(I, t) - V(\varnothing, t)$, where $V(\varnothing, t)$ is the payoff from selecting the globally best algorithm without any factor information. Note that for a singleton $I = \{k\}$, the inner expectation $\mathbb{E}_{\boldsymbol{g}_{-k}}[\theta_i(t, \boldsymbol{g})]$ is exactly the marginalized log lift; the Shapley framework evaluates each factor by the same conditional distribution.

The Shapley value of factor k at time t is

$$\phi_k(t) \;=\; \sum_{I \subseteq \{1, \ldots, K\} \setminus \{k\}} \frac{|I|!\,(K - |I| - 1)!}{K!} \big[V(I \cup \{k\}, t) - V(I, t)\big] . \tag{17}$$

The Shapley values partition the total information value exactly: $\sum_k \phi_k(t) = \text{IV}(\{1, \ldots, K\}, t)$ for each t. For K factors, computation requires 2^K information sets, which is tractable up to moderate K; for large K, sampling-based Shapley approximations apply.

The complementarity between factors j and k is

$$C_{jk}(t) = V(\{j,k\},t) - V(\{j\},t) - V(\{k\},t) + V(\varnothing,t). \tag{18}$$

Positive complementarity means knowing one factor increases the value of learning the other, which occurs when interactions are large.

Note that for a specific time preference $w(t)$, the preference-weighted information value $V^w(I) = \mathbb{E}[\max_{A_i} \int w(t)\, \theta_i(t, \boldsymbol{g})\, dt]$ is *not* the weighted average $\int w(t)\, V(I,t)\, dt$ of per-budget values, because the arg max over algorithms does not commute with the time integration. Under a time preference, the practitioner picks one algorithm for the whole weighted objective, and that choice depends on the joint profile across t, not on the pointwise optima. A practical consequence is that properties of the per-budget curves (such as complementarity being positive at every t) need not hold for the preference-weighted quantity. Computing exact Shapley values for a specific preference requires re-solving the arg max under that preference. Therefore, visually integrating the curves to infer results for a specific non-degenerate preference gives only a rough guide, not an exact answer.

Factor-Level Clustering: *Which levels of a factor induce similar algorithm rankings?*
For each level $g^{(k)}$ of factor k, the posterior provides a signature: the factor term $\delta^{(k)}(\cdot, g^{(k)}) \in \mathbb{R}^{n \times T}$, being the effect on algorithm ratings across the full time horizon under that level. Hierarchical clustering on these signatures (using some distance metric, e.g., Wasserstein distance between posteriors) yields a data-driven grouping of factor levels based on their effect on algorithm performance.

Conditional Anytime Pareto Sets: *Which algorithms survive for deployment under a specific factor-level combination?*

We write that Algorithm A anytime α-dominates B in context $\boldsymbol{g}$ if

$$A \underset{\alpha}{\overset{\mathbf{T}}{\succ}} B \mid \boldsymbol{g} \iff P\big(\mu_A(t,\boldsymbol{g}) > \mu_B(t,\boldsymbol{g})\; \forall t \in \mathbf{T} \mid \mathcal{R}\big) \;\geq\; \alpha\,, \tag{19}$$

where $\alpha \in (0,1]$ is a decision threshold, typically 0.99. The conditional anytime Pareto set in context $\boldsymbol{g}$ is

$$\mathcal{P}(\boldsymbol{g}) \;=\; \big\{A \in \mathbf{A} : \nexists B \in \mathbf{A} \text{ s.t. } B \underset{\alpha}{\overset{\mathbf{T}}{\succ}} A \mid \boldsymbol{g}\big\}. \tag{20}$$

This is computed at three levels of conditioning: global (using $\mu_i(t)$ only, recovering the global $\mathcal{P}$), per factor level (conditioning on one factor, marginalizing over others), and per cell (full conditioning on $\boldsymbol{g}$). An algorithm absent from the global Pareto set may appear in a conditional one, indicating it is a specialist for specific conditions. Conversely, a globally non-dominated algorithm may be conditionally dominated.

4 Case Study

We apply the temporal PL ANOVA to the BBOB benchmark suite [5], comparing several algorithms across the BBOB functions at multiple dimensionalities[1]. The two factors are:

- **Function identity** corresponding to the 24 BBOB functions $f_1, \ldots, f_{24}$.
- **Dimensionality** with 5 levels, $d \in \{5, 10, 20, 40, 60\}$.

For each combination of function and dimensionality, instances are generated by the standard BBOB transformations (e.g., random translations and rotations), providing exchangeable rankings within each cell. We sample a total of $24 \cdot 5 \cdot 10 = 1200$ instances uniformly over factor levels. The budget axis is function evaluations, discretized to 30 log-spaced timepoints from 10^2 to 10^5.

The active terms are $\mathcal{T} = \{\varnothing, \{f\}, \{d\}, \{f, d\}\}$, yielding the decomposition

$$\mu_i(t, g^{(f)}, g^{(d)}) = \underbrace{\mu_i(t)}_{\text{global}} + \underbrace{\delta_i^{(f)}(t, g^{(f)})}_{\text{function effect}} + \underbrace{\delta_i^{(d)}(t, g^{(d)})}_{\text{dimension effect}} + \underbrace{\delta_i^{(f,d)}(t, g^{(f)}, g^{(d)})}_{\text{interaction}}, \quad (21)$$

where each component satisfies the identifiability constraints of (5)–(6). The global term captures the marginal ratings across all functions and dimensionalities. The function main effect captures how each BBOB function shifts the ranking relative to the global average, uniformly across dimensionalities. The dimension main effect captures how dimensionality shifts the ranking, uniformly across functions. The interaction captures residual effects specific to particular function–dimensionality combinations that are not explained by either main effect alone.

We use a set of 6 algorithms in this case study, chosen to be diverse:

- A Covariance Matrix Adaptation Evolution Strategy (CMA-ES) [4] variant with full covariance, cumulative step size adaptation and random restarts.
- A Differential Evolution (L-SHADE) [16] variant with linear population size reduction (LPSR), that additionally resets stagnant individuals.
- A Particle Swarm Optimization (PSO) [14] variant with a ring topology, random social/cognitive coefficients, LPSR, and reset of stagnant individuals.
- Limited-memory BFGS (L-BFGS) [19] and Nelder–Mead (NM) [3], embedded into a dual annealing loop, which restarts the method when converged.
- Random search (RS) as sanity-check baseline.

All terms use an independent Matérn-3/2 GP prior with lengthscale prior $\ell^{(\tau)} \sim \text{InvGamma}(2, 5)$. The amplitude priors reflect the expectation that interactions are smaller than main effects: $\sigma^{(\tau)} \sim \text{LogNormal}(0, 0.5)$ for the global, function, and dimension terms, shrunk to $\sigma^{(f,d)} \sim \text{LogNormal}(-1, 0.5)$ for the

[1] Code and supplementary discussion: https://doi.org/10.5281/zenodo.20593205.

interaction. Posterior samples are obtained via Markov chain Monte Carlo (4 chains, 2000 warmup, 8000 samples per chain).

Figure 1 shows the marginal ratings across the full budget range. Three distinct regimes emerge. At low budgets, L-BFGS dominates with a rating approaching 0.5, benefiting from rapid initial progress of its quasi-Newton steps. CMA-ES overtakes at around 350 evaluations and maintains the highest rating throughout the mid-budget regime, peaking near 0.85 at ∼5000 evaluations. L-SHADE picks up around ∼20000, and surpasses CMA-ES at approximately 50 000 evaluations, reaching 0.8 at the final budget. The remaining algorithms—PSO, NM, and RS—are globally uncompetitive, with rating below 0.05 beyond 1000 evaluations. The credible bands are tight in the mid-budget regime (the CMA-ES dominance is well-determined) but slightly widen near the crossover with L-SHADE, reflecting uncertainty about the exact transition.

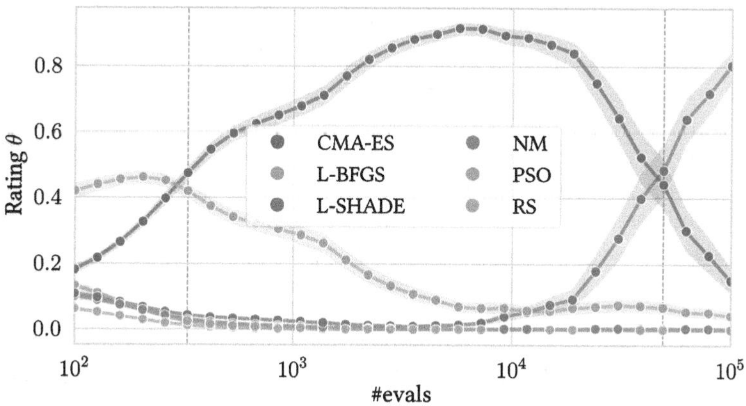

Fig. 1. Global ratings. Posterior mean (points) and 95% credible band of $\theta_i(t)$ for each algorithm, marginalized uniformly over all functions and dimensionalities. Under PL, $\theta_i(t)$ is the probability that algorithm i achieves the best objective value among all competitors at budget t on a randomly drawn problem instance.

Figure 2 shows how the total cross-algorithm variance in utility space is distributed across ANOVA terms. The global term accounts for 15% of variance at the lowest budget but grows steadily to nearly 70% at 100 000 evaluations, reflecting the increasing separation between algorithms as budget grows. Function identity is the largest factor effect, contributing 30–40% of variance at low budgets and declining to ∼20% at high budgets. Dimensionality accounts for 5–15%, and the interaction term for 5–10%. The mean squares reveal a similar picture. The global term has only 5°C of freedom but dominates throughout. The function, dimension, and interaction terms have tight mean squares well over zero, indicating that all three carry nontrivial signal per degree of freedom. In particular, the interaction term has mean squares comparable to or even greater than the dimension main effect, indicating that the data contains sufficient signal to overcome the tighter prior on the interaction amplitude.

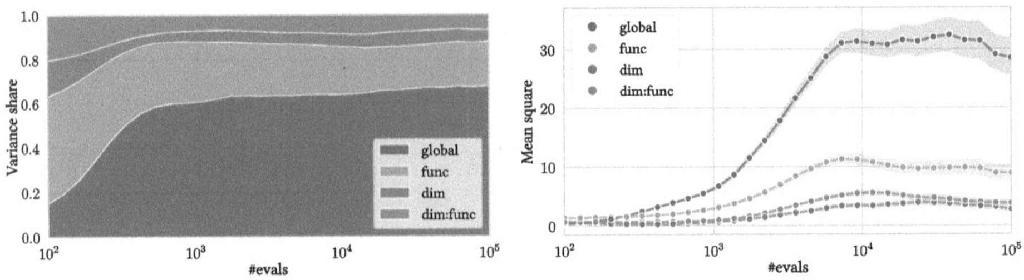

(a) Variance share. (b) Mean squares.

Fig. 2. Utility-space variance decomposition. (a) Posterior mean variance shares over budget. Shares sum to one by orthogonality of the ANOVA constraints. (b) Mean squares (variance divided by degrees of freedom) with 95% credible bands. Mean squares are comparable across terms with different numbers of levels.

Figure 3 shows the log-lift of individual levels per factor. For dimensionality, several patterns emerge. CMA-ES is lifted at late budgets and $d \in \{40, 60\}$ and suppressed at $d \in \{5, 10\}$, while L-SHADE benefits from low dimensions in the mid-regime (log-lift reaching 4 at $d = 5$), and is suppressed at late budgets for $d \in \{40, 60\}$. NM and RS show large positive log-lifts at $d = 5$, reflecting that the competitive landscape compresses at low dimensions: the globally dominant algorithms lose their edge and the weaker methods become relatively more viable. For function identity, the highlighted extremes reveal each algorithm's niches: CMA-ES is most suppressed on f2 (separable) and f7 (step-like plateaus). L-BFGS's deepest negative log-lifts occur on functions f17, f18, and f19, all high-conditioning or multimodal problems where gradient-based search stagnates. Conversely, NM, PSO, and RS achieve their highest positive log-lifts on f23 and f16, which are weakly structured multi-modal landscapes where the globally dominant algorithms lose their structural advantage. PSO and RS also strongly benefit on f11, however only in the mid-regime. Natural for RS and surprising for PSO, both struggle on f1, the unimodal sphere function. Some functions show log-lifts that change sign over the budget axis, indicating that the algorithm best suited to that function depends on the budget regime.

Figure 4 quantifies the payoff from conditioning: the expected rating of the selected algorithm rises from ~0.45 (knowing nothing) to ~0.80 (knowing both factors) at mid-budget. Knowing function identity alone captures most of this gain, while dimensionality alone provides a smaller but nontrivial improvement. The Shapley shares attribute 65–85% of the total information value to function identity at most budgets. Dimensionality's share grows at higher budgets, reaching 40–50% around 30 000–50 000 evaluations, which is the regime where the CMA-ES vs. L-SHADE crossover occurs and dimensionality determines which algorithm leads. The alone-vs-conditional decompositions reveal positive complementarity throughout the budget range: for both factors, the conditional value exceeds the standalone value, meaning that knowing one factor increases the value of learning the other. Complementarity peaks at 3 000–10 000 evaluations,

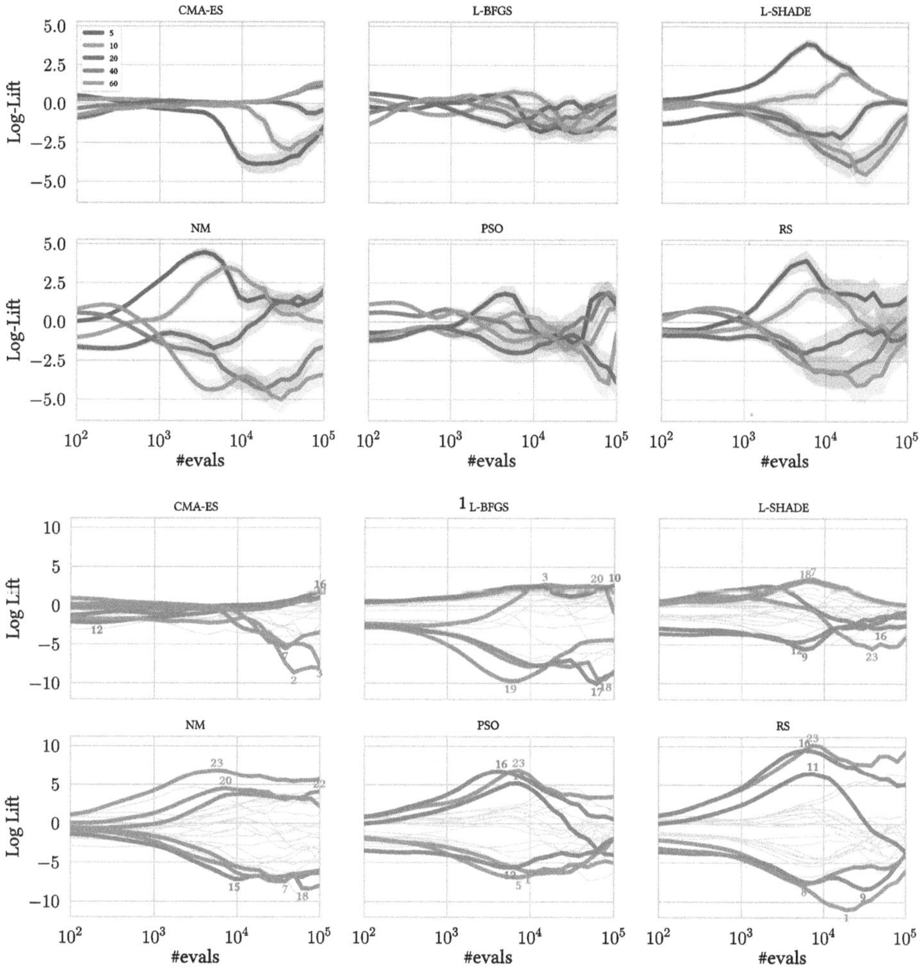

Fig. 3. Partial log-lift of dimensionality (upper) and function identity (lower), showing how each factor level shifts each algorithm's rating relative to the global baseline. In the lower figure, the most extreme functions per algorithm are highlighted in color (ranked by mean absolute log-lift across budgets); remaining functions are shown in gray. A log-lift of $\Lambda = 2$ means the rating is multiplied by $\exp(2) \approx 7.4$ relative to the global baseline; $\Lambda = -2$ means division by the same factor.

where performance is most sensitive to the joint function–dimensionality configuration. Dimensionality benefits proportionally more from the interaction: its standalone value is close to zero around 8000 evaluations while the conditional value is 0.2. This reflects that the effect of dimensionality is substantially mediated by function identity, i.e., knowing $d = 60$ alone is much less informative than knowing $d = 60$ on a specific function.

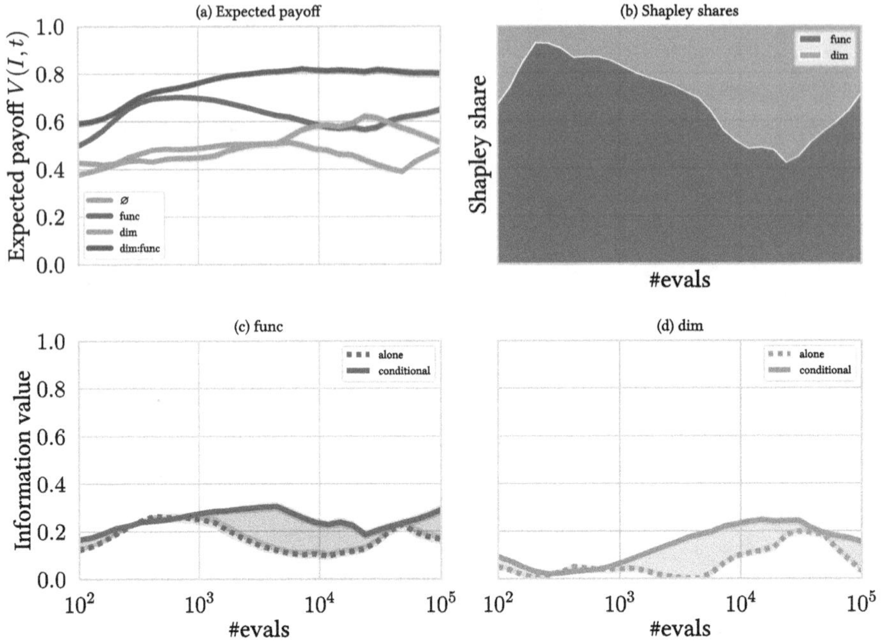

Fig. 4. Shapley information values for function identity and dimensionality. (a) Expected rating of the selected algorithm under each information set: knowing nothing (∅), knowing one factor, or knowing both. (b) Shapley share of each factor as a fraction of total information value. (c, d) Decomposition of each factor's Shapley value (shaded band) into its standalone information value (dotted) and its conditional value given the other factor is known (solid). The gap between the two lines is the complementarity.

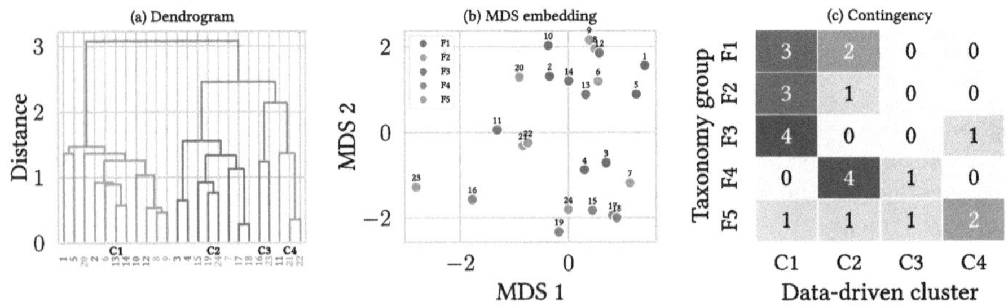

1

Fig. 5. Hierarchical clustering of BBOB functions by the Wasserstein distance between posteriors of its main effect $\delta_i^{(f)}(t, g^{(f)})$ across algorithms and budgets. (a) Dendrogram with average linkage. Leaf colors indicate the BBOB function group; branch colors and C-labels indicate the data-driven clusters. (b) Two-dimensional MDS embedding of the distance matrix, colored by BBOB function group. Spatial proximity reflects similarity in how functions perturb the global algorithm ranking. (c) Contingency matrix comparing the data-driven clusters against the BBOB function groups.

Four clusters emerge from the largest gap in merge distances (cf. Fig. 5). The largest (C1, 11 functions) contains all unimodal functions plus f20, unified by following the global ranking pattern closely. C2 (6 functions) collects multimodal functions with exploitable global structure, where L-SHADE and PSO gain at the expense of L-BFGS. C3 (f16, f23) isolates the most rugged landscapes, with f23 being the only function where NM dominates across most budgets. C4 (f11, f21, f22) groups functions where PSO and RS are boosted. Overall agreement with the canonical BBOB groups is low (ARI = 0.156, AMI = 0.236): functions that differ structurally may produce similar rankings, and functions within the same group may diverge.

CMA-ES, L-SHADE, and L-BFGS are globally non-dominated, and members of the conditional Pareto set in most cells (97%, 83%, and 74%, respectively). However, the three globally dominated algorithms (NM, PSO, RS) still enter the conditional Pareto set in 50%, 33%, and 10% of cells. They are therefore not useless; they are specialists whose value is unlocked by conditioning on problem characteristics. The "practical takeaway" for this artificial case study is therefore to always condition on "function identity" when choosing an algorithm. If budget exceeds ~10000 evaluations, also condition on "dimension", since it increasingly determines whether CMA-ES or L-SHADE is optimal. These findings are specific to the BBOB suite and the chosen algorithm portfolio; the diagnostic workflow that produced them is not, and is applicable to arbitrary instances and factors.

5 Conclusion

We introduced a temporal Plackett–Luce ANOVA that decomposes anytime algorithm performance into global, factor-specific, and interaction components. Rather than predictive algorithm selection, the diagnostic toolkit built on this decomposition provides a structured post hoc workflow: screen for signal via variance attribution, quantify the value of conditioning via Shapley information values, cluster factor levels by their effect on performance ratings, and identify where the global picture breaks down. Our BBOB case study demonstrates this workflow on analytically defined function classes and dimensionalities; however, the framework's scope is not limited to synthetic benchmarks. Any categorical property of the instance known at deployment can serve as a factor. Future work includes testing the robustness of conclusions in real-world case studies and against comparable methods. Additionally, extending the discrete factor structure to continuous covariates where sensible is a natural next step towards joint anytime- and instance-based algorithm selection.

References

1. Bang-Jensen, J., Chiarandini, M., Goegebeur, Y., Jørgensen, B.: Mixed models for the analysis of local search components. In: Stützle, T., Birattari, M., Hoos, H. (eds.) Engineering Stochastic Local Search Algorithms. Designing, Implementing and Analyzing Effective Heuristics, pp. 91–105. Springer, Berlin, Heidelberg (2007). https://doi.org/10.1007/978-3-540-74446-7_7

2. Chiarandini, M., Goegebeur, Y.: Mixed models for the analysis of optimization algorithms. In: Bartz-Beielstein, T., Chiarandini, M., Paquete, L., Preuss, M. (eds.) Experimental Methods for the Analysis of Optimization Algorithms, pp. 225–264. Springer, Berlin, Heidelberg (2010). https://doi.org/10.1007/978-3-642-02538-9_10
3. Gao, F., Han, L.: Implementing the Nelder-Mead simplex algorithm with adaptive parameters. Comput. Optim. Appl. **51**(1), 259–277 (2012). https://doi.org/10.1007/s10589-010-9329-3
4. Hansen, N.: The CMA Evolution Strategy: A Tutorial (2023). https://doi.org/10.48550/arXiv.1604.00772
5. Hansen, N., Finck, S., Ros, R., Auger, A.: Real-Parameter Black-Box Optimization Benchmarking 2009: Noiseless Functions Definitions. Report, INRIA (2009)
6. Kerschke, P., Hoos, H.H., Neumann, F., Trautmann, H.: Automated algorithm selection: survey and perspectives. Evol. Comput. **27**(1), 3–45 (2019). https://doi.org/10.1162/evco_a_00242
7. Mattos, D.I., Ruud, L., Bosch, J., Olsson, H.H.: On the assessment of benchmark suites for algorithm comparison (2021). https://doi.org/10.48550/arXiv.2104.07381
8. Mersmann, O., Bischl, B., Trautmann, H., Preuss, M., Weihs, C., Rudolph, G.: Exploratory landscape analysis. In: Proceedings of the 13th Annual Conference on Genetic and Evolutionary Computation, GECCO 2011, pp. 829–836. ACM, New York, NY, USA (2011). https://doi.org/10.1145/2001576.2001690
9. Nikolikj, A., Džeroski, S., Muñoz, M.A., Doerr, C., Korošec, P., Eftimov, T.: Algorithm instance footprint: separating easily solvable and challenging problem instances. In: Proceedings of the Genetic and Evolutionary Computation Conference, GECCO 2023, pp. 529–537. ACM, New York, NY, USA (2023). https://doi.org/10.1145/3583131.3590424
10. Nikolikj, A., Eftimov, T.: Exploring module interactions in modular CMA-ES across problem classes. Swarm Evol. Comput. **98**, 102116 (2025). https://doi.org/10.1016/j.swevo.2025.102116
11. Nikolikj, A., Kostovska, A., Vermetten, D., Doerr, C., Eftimov, T.: Quantifying individual and joint module impact in modular optimization frameworks. In: 2024 IEEE Congress on Evolutionary Computation (CEC), pp. 1–8 (2024). https://doi.org/10.1109/CEC60901.2024.10611779
12. Nikolikj, A., Muñoz, M.A., Eftimov, T.: Benchmarking footprints of continuous black-box optimization algorithms: explainable insights into algorithm success and failure. Swarm Evol. Comput. **94**, 101895 (2025). https://doi.org/10.1016/j.swevo.2025.101895
13. Nikolikj, A., Muñoz, M.A., Tuba, E., Eftimov, T.: Tracing the interactions of modular CMA-ES configurations across problem landscapes. In: 2025 IEEE Congress on Evolutionary Computation (CEC), pp. 1–8 (2025). https://doi.org/10.1109/CEC65147.2025.11042974
14. Poli, R., Kennedy, J., Blackwell, T.: Particle swarm optimization. Swarm Intell. **1**(1), 33–57 (2007). https://doi.org/10.1007/s11721-007-0002-0
15. Smith-Miles, K., Baatar, D., Wreford, B., Lewis, R.: Towards objective measures of algorithm performance across instance space. Comput. Oper. Res. **45**, 12–24 (2014). https://doi.org/10.1016/j.cor.2013.11.015
16. Tanabe, R., Fukunaga, A.S.: Improving the search performance of SHADE using linear population size reduction. In: 2014 IEEE Congress on Evolutionary Computation (CEC), pp. 1658–1665 (2014). https://doi.org/10.1109/CEC.2014.6900380

17. Turner, H.L., van Etten, J., Firth, D., Kosmidis, I.: Modelling rankings in R: The PlackettLuce package. Comput. Statistics **35**(3), 1027–1057 (2020). https://doi.org/10.1007/s00180-020-00959-3
18. Wurth, J., Stegherr, H., Kemper, N., Heider, M., Hähner, J.: Pareto-optimal anytime algorithms via bayesian racing (2026). https://doi.org/10.48550/arXiv.2603.08493
19. Zhu, C., Byrd, R.H., Lu, P., Nocedal, J.: Algorithm 778: L-BFGS-B: fortran subroutines for large-scale bound-constrained optimization. ACM Trans. Math. Softw. (TOMS) **23**(4), 550–560 (1997). https://doi.org/10.1145/279232.279236

From Networks to Landscapes: Sampling and Topographic Visualisation of Continuous LONs

Quentin Renau[1(✉)], Gabriela Ochoa[1], Arnaud Liefooghe[2], and Jonathan E. Fieldsend[3]

[1] University of Stirling, Stirling, UK
quentin.renau@stir.ac.uk, gabriela.ochoa@stir.ac.uk
[2] Univ. Littoral Côte d'Opale, UR 4491, LISIC, 62100 Calais, France
arnaud.liefooghe@univ-littoral.fr
[3] University of Exeter, Exeter, UK
J.E.Fieldsend@exeter.ac.uk

Abstract. Local Optima Networks (LONs) are established graph-based models for landscape analysis, yet their methodological maturity in continuous optimisation lags behind their combinatorial counterparts. This study investigates the challenges of constructing LONs for the Black-Box Optimisation Benchmark (BBOB) suite using a basin-hopping framework. We first identify a critical sampling anomaly—spurious local optima on theoretically unimodal functions—and suggest that vanishing gradients are the primary cause, proposing specific parameters to mitigate this behaviour. To improve high-dimensional interpretability, we introduce an integrated topographic visualisation that superimposes the LON graph onto a reconstructed fitness surface. This approach extends the intuition of 2d heatmaps to higher dimensions, capturing both connectivity and topography. Finally, we characterise the 5d BBOB suite using an expanded set of LON features. Comparative analysis reveals that LON-derived metrics outperform traditional landscape features for 3 of the 4 optimisers tested, suggesting that graph-based models capture structural nuances distinct from traditional metrics. These results contribute to establish the LON framework as a complementary quantitative and visualisation tool to characterise continuous optimisation landscapes.

Keywords: Landscape analysis · Local optima networks · Continuous optimisation

1 Introduction

Local Optima Networks (LONs) [27,43] are graph-based models of fitness landscapes, where nodes represent local optima and edges denote transitions between them. While LONs have been predominantly applied to combinatorial landscapes [27,29,38,43], their conceptual roots lie in computational chemistry,

G. Iacca et al. (Eds.): PPSN 2026, LNCS 16985, pp. 532–549, 2027.
https://doi.org/10.1007/978-3-032-36214-8_33

where the topology of continuous energy surfaces was modelled as a transition network [8]. Only recently, these graph-based models have been extended to continuous optimisation problems beyond molecular energy minimisation, including kinetic transition models for machine learning landscapes [5] and *basin-hopping* models for both synthetic benchmarks [1,24,41,44] and real-world problems [7,23]. The basin-hopping algorithm [46], originally designed for non-linear multimodal problems in computational chemistry, operates by jumping between basins of attraction through a sequence of perturbations followed by local minimisation. This process is conceptually analogous to *iterated local search* (ILS) in combinatorial optimisation [21], a metaheuristic that has been instrumental in the sampling and construction of LONs for discrete problems [28,38].

While LONs are powerful tools for landscape analysis, our preliminary explorations using basin-hopping sampling on the Black-Box Optimisation Benchmark (BBOB) [14] produced counter-intuitive results. For example, the sampling identified multiple nodes (local optima) in half of the functions classified as unimodal. By construction, these functions possess a single basin of attraction and should yield only a single global optimum. Beyond addressing these sampling anomalies, we identify a gap in the current literature regarding the limited scope of available LON features, their lack of empirical validation against the performance of optimisers, and the lack of comparison against traditional features. This study addresses these limitations through the following contributions:

1. **Investigation of sampling anomalies:** We suggest that vanishing gradients are responsible for the spurious nodes produced by basin-hopping on unimodal functions and propose specific sampling parameters to mitigate this phenomenon.
2. **Integrated topographic visualisation:** We propose a visualisation framework that superimposes the LON graph structure onto a reconstructed fitness surface. By extending the concept of 2d fitness heatmaps to higher-dimensional landscapes via spatial embedding and surface approximation, we provide a consistent visual representation that captures both the landscape's connectivity and topography.
3. **Quantitative characterisation:** We model instances of all 5d BBOB functions and compute an expanded set of LON features to provide a more thorough landscape characterisation.
4. **Performance analysis:** We conduct a correlation analysis to quantify the relationship between LON features and the performance of state-of-the-art heuristic optimisers, and contrast them with other feature sets from the literature. This establishes the usefulness of LON's features for characterising continuous landscapes.

The work now proceeds as follows: Sect. 2 briefly outlines the fundamentals of fitness landscapes and prior work on LONs in continuous spaces. Section 3 details the experimental work, followed by a presentation and analysis of results in Sect. 4. The work concludes with Sect. 5.

2 Background and Definitions

This section introduces fitness landscapes, local optima networks, and their extensions and limitations for continuous spaces.

2.1 Fitness Landscape

The search space of optimisation problems can be regarded as a spatial structure where each candidate solution has a height (objective or fitness value), collectively forming an optimisation surface called a fitness landscape. Formally [36], a fitness landscape in the context of continuous optimisation is a triplet $(\mathbf{X}, N, f)$ where $\mathbf{X} \in \mathbb{R}^d$ is the set of all real-valued solutions of d dimensions, *i.e.*, the search space; N is a function that assigns to every solution $\mathbf{x} \in \mathbf{X}$ a set of neighbours $N(\mathbf{x})$; and $f : \mathbb{R}^d \to \mathbb{R}$ is the fitness function. A solution $\mathbf{x}$ is denoted as a vector $\mathbf{x} = (x_1, x_2, \ldots, x_d)$, the fitness function is a box-constrained optimisation problem, and the neighbourhood is commonly based on hyper-rectangles/spheres.

2.2 Local Optima Network

A Local Optima Network is the directed graph $LON = (L, E)$, where the nodes are local optima, and the edges represent transitions between two local optima. L and E denote the set of nodes and edges, respectively. The edges between two local optima are directed, and their types depend on the algorithm used to sample the search space. We use edges composed of a random perturbation followed by a local minimisation process.

Definition 1 (Local Optimum). *A local optimum, which is a local minimum in this work, is a solution* $\mathbf{x}^* \in \mathbf{X}$ *such that* $\forall \mathbf{x} \in N(\mathbf{x}^*)$, $f(\mathbf{x}^*) \leqslant f(\mathbf{x})$.

We use the Compressed Monotonic LON model (CMLON), first introduced in [29] as a coarser model to capture the global structure of fitness landscapes with neutrality in combinatorial optimisation, and later extended to continuous optimisation [1]. We first define the monotonic LON model, and then the compressed monotonic LON model.

Definition 2 (Monotonic LON). *A monotonic LON is the directed graph* $MLON = (L, ME)$ *where the nodes* L *are local optima and the edges* $ME \subset E$ *are monotonic perturbation edges.*

Definition 3 (Monotonic Perturbation Edge). *There is a monotonic perturbation edge from a local optimum* $\mathbf{l}_1$ *to a local optimum* $\mathbf{l}_2$, *if* $\mathbf{l}_2$ *can be obtained from a random perturbation of* $\mathbf{l}_1$ *followed by a local minimisation process, and* $f(\mathbf{l}_2) \leqslant f(\mathbf{l}_1)$. *The edge is called monotonic because the transition between two local optima is non-deteriorating. In this model, the edges are weighted by the number of transitions between the two local optima.*

Definition 4 (Compressed Monotonic LON). *A compressed monotonic LON is the directed graph $CMLON = (CL, CE)$ where the nodes are compressed local optima CL, and the edges $CE \subset ME$ are aggregated from the monotonic edge set ME by summing up the edge weights.*

Definition 5 (Compressed local optimum). *A compressed local optimum is a single node that represents a set of connected nodes in the MLON model with the same fitness value.*

2.3 LONs for Continuous Landscapes

The development of LONs in continuous space has necessitated answering three key questions: (1) How should we define the concept of a neighbourhood, $N(\mathbf{x})$, for a continuous space? (2) How should we define the local search operation? (3) How should we sample from the search space $\mathbf{X}$?

There are two main branches in the literature—the first discretises the search space and employs hill-climbers, and the second instead utilises local search optimisers directly for the continuous space, along with basin-hopping [46]. We outline some of the properties and limitations of these approaches below.

Discretisation and Hill-Climbers. Early continuous LON discretisation methods relied on variable-space *gridding* to define neighbourhoods [4,9,10], a method that scales exponentially poorly with dimensionality. To overcome this, more recent approaches utilise *scalable sampling* via Sobol' sequences [35], defining neighbourhoods based on spatial proximity—such as a d-ball radius—to maintain computational feasibility in higher-dimensional spaces [11].

Employing Continuous Space Local Search Methods. A more widely used approach considers basin-hopping with derivative-based local search—frequently L-BFGS [20]—to construct LONs [1,7,41,44]. This process iteratively perturbs a current optimum within a d-ball radius to "hop" into adjacent basins, where move acceptance criteria range from strict monotonic improvement [1,7] to unbiased acceptance of all transitions [41]. These approaches primarily define edges via escape transitions; however, recent work [16] has also considered gradient-free alternatives such as the Nelder-Mead simplex algorithm [25].

Limitations of LONs for Continuous Landscapes. The two methodologies represent a fundamental trade-off between topographic coverage and point precision. Discretisation-based approaches use every sample to approximate both basin volumes and optima; however, the accuracy of the resulting optima is strictly limited by the resolution of the grid or sampling density. Conversely, local search-based construction methods overwhelmingly spend their computational budget on the search process, which is necessarily biased. While this yields accurate coordinates for local optima, the infrequent restarts and finite "hopping" attempts provide a comparatively sparse approximation of basin sizes and global connectivity.

3 Methodology

This section presents the methods used to construct the LON, and the metrics extracted from the network. We also present the benchmark functions, and the established optimisation algorithms experimented.

3.1 Sampling Methods

We use four sampling methods to construct the monotonic LONs in dimension $d = 2$ and choose one of these four to extract LONs in dimension $d = 5$ (discussed in Sect. 4.1). Unless stated otherwise, every method is repeated for 100 runs. The four extraction methods are the following:

1. We use the same procedure as described in Sect. 2, *i.e.*, the monotonic basin-hopping (MBH) algorithm with L-BFGS-B as local search solver.[1] We allocate 100 iterations for each L-BFGS-B run, a step size of $p = 2$ for the perturbation, and 1,000 random perturbations. We coin this method $\texttt{LON}_{\texttt{sota}}$.
2. We use the same procedure as in [16], *i.e.*, the MBH algorithm with Nelder-Mead as local search solver. As for $\texttt{LON}_{\texttt{sota}}$, we use 100 iterations for each Nelder-Mead run, a step size of $p = 2$ for the perturbation, and 1,000 random perturbations. We coin this method $\texttt{LON}_{\texttt{nm}}$.
3. We modify the parameters of the L-BFGS-B in the $\texttt{LON}_{\texttt{sota}}$ to decrease the chances of early-stopping of the algorithm. We set both the stopping criteria on fitness tolerance and gradient tolerance to 10^{-20}. We coin this method $\texttt{LON}_{\texttt{tol}}$ and $\texttt{LON}_{\texttt{tol30}}$ when 30 runs are used.
4. We compare results obtained using the Python package `pflacco` [30]. It uses the same extraction method as $\texttt{LON}_{\texttt{sota}}$. We directly use data collected in [40] available at [39]. They use 30 runs instead of 100. We coin this method $\texttt{LON}_{\texttt{pflacco}}$.

For all sampling methods, except $\texttt{LON}_{\texttt{pflacco}}$, we set a parameter for the MBH to record a node if there is a fitness improvement greater than 10^{-8}. This choice is motivated by the target fitness value to reach in BBOB: $f_{opt} + 10^{-8}$, with f_{opt} being the fitness of the global optimum [12].

3.2 LON Construction and Network Metrics

The local optima sampled with the methods described in the previous section are used to construct the LONs. For each run, we log the full edge list of a basin-hopping algorithm that accepts improving moves. From this full edge list, we extract the monotonic edges corresponding to a MBH. To construct the graph model, unique local optima and (monotonic) transitions are stored as nodes

[1] We note many prior works use L-BFGS, which requires modifying the optimisation function to penalise boundary violations, whereas L-BFGS-B is designed specifically for box-constrained problems [6].

and edges, with counters tracking their respective sampling frequencies. Additionally, to produce a compressed MLON model, we aggregate connected local optima with identical fitness values into *supernodes* (plateau nodes), employing a rounding factor of 10^{-2} to determine fitness equivalence. On highly mulit-modal functions, distant nodes can occasionally share identical rounded fitness values by coincidence. To prevent the compression method from inadvertently merging nodes from different basins of attraction, we introduced a distance-aware constraint. Under this refinement, an edge connecting equal-fitness nodes is only compressed if their spatial Euclidean distance falls within a threshold of 0.1, ensuring only adjacent states are aggregated. Note that LONs constructed with `pflacco` use $\varepsilon = 10^{-5}$ and no rounding factor for the fitness.

Once the LON is constructed, it can both be visualised and used to derive informative metrics. Building on previous studies of combinatorial and continuous LONs [1,24,29,38], we consider an extended set of network metrics reported in Table 1. These include the number of nodes and edges, the graph density, and the average node degree (incoming, outgoing, and total). In addition, we compute the number of best nodes (*i.e.*, nodes matching global optima, or best found solution), the number of sinks (nodes with no outgoing edges), and the number of connected components. The following set of metrics refer to the global optimum or best-found local optimum, without loss of generality we add a suffix to them (best): the proportion of nodes from which a sequence of monotonic edges leads to a best-found local optimum, the proportion of best solutions found (counting individual solutions within compressed nodes), the average distance (path length) to a best-found local optimum, the fraction of paths that reach a best-found local optimum node, the incoming degree of best-found local optima, and the average deviation from the best fitness across nodes.

3.3 Benchmark Functions and Algorithms

We use the 24 BBOB functions from the COCO platform [14] as a test-bed. We consider functions in dimensions $d \in \{2, 5\}$. Both dimensions are computed on the first instance: dimension $d = 2$ functions are used to visually check the correctness of LON extraction, while dimension $d = 5$ functions are used to measure the correlations with algorithm performance.

To validate the LON model and to assess how well metrics capture valuable information, we record the performance of four state-of-the-art algorithms: (1) Covariance Matrix Adaptation Evolution Strategy (CMA-ES) [15], (2) Dual Annealing algorithm, a simulated annealing [42] coupled with L-BFGS-B (DA) [48], (3) Differential Evolution [37] (DE), and (4) Particle Swarm Optimisation [17] (PSO). We ran CMA-ES using the `modcma` package [26] with a population size of 10, mirrored sampling, saturate bound correction, CSA step size adaptation, and an initial $\sigma = 2$, and Dual Annealing algorithm from `scipy` package [45], both using 15 runs of 5,000 function evaluations. We obtained data for DE-best and PSO from the IOHAnalyzer platform [47], which contains 15 runs of 5,000 function evaluations for both algorithms. The performance metric

Table 1. Considered network features.

metric	description
n_nodes	number of (compressed) nodes
n_edges	number of edges
density	density of edges
avg_deg	average node degree
max_in_deg	maximum in-degree
max_out_deg	maximum out-degree
n_best	number of best nodes
n_sinks	number of sink nodes
n_components	number of weak connected components
prop_reach_best	proportion of nodes reaching a best node
count_best	proportion of best solutions
avg_path_best	average path length to a best node
avg_in_deg_best	average in-degree of best nodes
avg_fit_dev_best	average fitness deviation from the best fitness

is the median precision of 15 runs, *i.e.*, the median fitness deviation to the global optimum fitness.

4 Results

We first study the four models introduced in Sect. 3 on 2-dimensional unimodal functions in order to validate them on simple functions and visualise LONs obtained on 2-dimensional functions. We then apply the $\texttt{LON}_{\texttt{tol}}$ model in $d = 5$ dimensions and study how much the model captures information about the landscape and correlates with algorithm performance.

4.1 2d Functions

Unimodal Functions. The rationale for focusing on unimodal functions in dimension $d = 2$ is to have a small and deliberately chosen setting where we know what result should be expected. By definition, a unimodal function has a single basin of attraction and thus, exactly one global optimum with no other local optima in the landscape. BBOB is composed of 12 unimodal functions in $d = 2$ [12]: F1 (Sphere), F2 (Ellipsoid), F5 (Linear Slope), F6 (Attractive Sector), F7 (Step Ellipsoid), F8 (Rosenbrock), F9 (Rosenbrock Rotated), F10 (Ellipsoid Rotated), F11 (Discus), F12 (Bent Cigar), F13 (Sharp Ridge), and F14 (Different Powers). Given that these functions are unimodal as defined in COCO, no edges and a single node are expected in the LONs.

Table 2 displays the number of nodes found by the four LON sampling methods on the BBOB unimodal functions. Note that one of these nodes for all

Table 2. Number of nodes in the LONs in dimension $d = 2$ for the different sampling methods on unimodal functions. 100 runs are used to construct LONs, except for $\texttt{LON}_{\texttt{pflacco}}$ and $\texttt{LON}_{\texttt{tol30}}$ where 30 runs are used.

	F1	F2	F5	F6	F7	F8	F9	F10	F11	F12	F13	F14
$\texttt{LON}_{\texttt{sota}}$	1	17	1	16	1 006	1	1	4	2	2	175	4
$\texttt{LON}_{\texttt{nm}}$	1	3	2	12	824	19	11	10	4	6	6	12
$\texttt{LON}_{\texttt{tol}}$	1	1	1	12	1 015	1	1	1	1	1	186	1
$\texttt{LON}_{\texttt{pflacco}}$	1	54	1	20	316	2	1	32	12	5	147	25
$\texttt{LON}_{\texttt{tol30}}$	1	1	1	3	313	1	1	1	1	1	72	1

sampling methods is the global optimum. We observe that only one of the four methods is consistently composed of a single node on 9 out of 12 unimodal functions: $\texttt{LON}_{\texttt{tol}}$. Other methods report more nodes than what is expected on most unimodal functions, *i.e.*, $\texttt{LON}_{\texttt{sota}}$ reports more than one node on 7 out of 12 functions, $\texttt{LON}_{\texttt{nm}}$ 11 out of 12, and $\texttt{LON}_{\texttt{pflacco}}$ on 9 out of 12 functions. $\texttt{LON}_{\texttt{tol}}$ is the only MBH-based method where the tolerance parameters on fitness and gradients values has been modified. Hence, vanishing gradients or points with very similar fitness values could be reasons why more than one node is recorded for $\texttt{LON}_{\texttt{sota}}$ and $\texttt{LON}_{\texttt{pflacco}}$. Given these results, we will use the $\texttt{LON}_{\texttt{tol}}$ method to extract LONs in the rest of the paper.

From Table 2, there is a set of three functions where none of the methods manages to record only one node in the LON: F6, F7, and F13. Even though these functions are coined unimodal in BBOB [12], they all exhibit secondary landscape properties that explain the additional number of nodes obtained. F6 exhibits a small ridge on one side of the global optimum. The twelve nodes of $\texttt{LON}_{\texttt{tol}}$ are located on one line on this ridge. Given that L-BFGS-B estimates gradients, any time the local optimiser reaches that non-differentiable ridge, the algorithm stops, and a local optimum is recorded in the LON. The same behaviour can be observed for function F13, but the ridge for this function traverses the search space, explaining the large number of nodes recorded. Note that for F13, $\texttt{LON}_{\texttt{nm}}$ records only 5 nodes. As the Nelder-Mead algorithm does not rely on the estimation of gradients, it is less affected on this function. The number of nodes in F7 is not explained by a ridge but is still a product of non-differentiable areas: plateaus. F7 is a step function, and the fitness on each step is equal, which triggers an L-BFGS-B early-stopping. According to Definition 1, F7 has an infinite number of local optima, and since we use compressed LON for an exhaustive sampling, the number of nodes in the LON should be equal to the number of neutral regions.

Multimodal Functions. We observe that three out of four sampling methods for the LONs are showing an oversampled number of local optima for unimodal functions. In this section, we compare the number of nodes obtained by the four sampling methods on multimodal functions. Table 3 reports the number of nodes found using the four sampling methods. Interestingly, while $\texttt{LON}_{\texttt{sota}}$ was oversampling the number of local optima on unimodal functions, the number of nodes obtained for multimodal functions is comparable to those found using $\texttt{LON}_{\texttt{tol}}$ except for two functions: $\texttt{LON}_{\texttt{sota}}$ finds 126 more local optima on F19 and 142 on F22.

Table 3. Number of nodes in the LONs in dimension $d = 2$ for the different sampling methods on multimodal functions. 100 runs are used to construct the LONs, except for $\texttt{LON}_{\texttt{pflacco}}$ and $\texttt{LON}_{\texttt{tol30}}$ where 30 runs are used.

	F3	F4	F15	F16	F17	F18	F19	F20	F21	F22	F23	F24
$\texttt{LON}_{\texttt{sota}}$	116	130	124	626	523	607	687	43	42	152	935	346
$\texttt{LON}_{\texttt{nm}}$	157	154	135	342	934	844	552	36	27	26	3590	539
$\texttt{LON}_{\texttt{tol}}$	124	133	116	596	495	620	561	38	30	10	1008	344
$\texttt{LON}_{\texttt{pflacco}}$	68	119	92	277	249	269	204	36	67	92	215	139
$\texttt{LON}_{\texttt{tol30}}$	54	69	67	213	167	238	176	29	18	9	390	147

The number of local optima found by $\texttt{LON}_{\texttt{nm}}$ is comparable to $\texttt{LON}_{\texttt{tol}}$ on 6 out of 12 functions. Some functions, such as F16, F17, F18, F23, or F24 have a discrepancy of over 200 nodes between the two methods, either in undersampling or oversampling the nodes count. The variation in number of nodes can be explained by the difference in the neighbourhood definition between Nelder-Mead and L-BFGS-B. Moreover, we observed on the unimodal functions that Nelder-Mead was coping better than L-BFGS-B with non-differentiable regions of the search space, which can also be a factor explaining the different number of nodes.

Finally, the number of nodes obtained with $\texttt{LON}_{\texttt{pflacco}}$ is higher than that of $\texttt{LON}_{\texttt{tol30}}$ for all multimodal functions, except F23 and F24. The larger number of nodes obtained with the $\texttt{LON}_{\texttt{pflacco}}$ sampling is due to two factors: the early-stopping tolerance on fitness and gradients in L-BFGS-B, and the different epsilon used to merge points into the same node. While $\texttt{LON}_{\texttt{tol30}}$ uses $\varepsilon = 10^{-2}$, $\texttt{LON}_{\texttt{pflacco}}$ uses $\varepsilon = 10^{-5}$.

2D Visualisations. Having presented the results of the LON construction on 2d functions, we now visualise the $\texttt{LON}_{\texttt{tol}}$ superimposed on the normalised rank heatmaps of these functions. Figure 1 visualises eight functions (plots for the remaining functions can be found in the supplementary material at [34]). The four functions in Fig. 1, top row, are unimodal: one function with one node (F1) and the three unimodal functions that have more than one node (F6, F7, F13). The four functions in Fig. 1, bottom row, are multimodal (F3, F20, F21, F24).

As previously discussed, the LON from F1 is composed of a single node: the global optimum. This point can be seen in Fig. 1 in the centre of the Sphere. The LON from F6 contains 12 nodes, including the global optimum. We can observe that all nodes in the figure are located on the same line at the landscape's ridge. The ridge in function F13 is also the reason why $\texttt{LON}_{\texttt{tol}}$ records more nodes than what could be expected on unimodal functions. All nodes for these functions can be seen following the ridge going through the landscape. As for F7, we observe that all nodes are located on the plateaus of the step function.

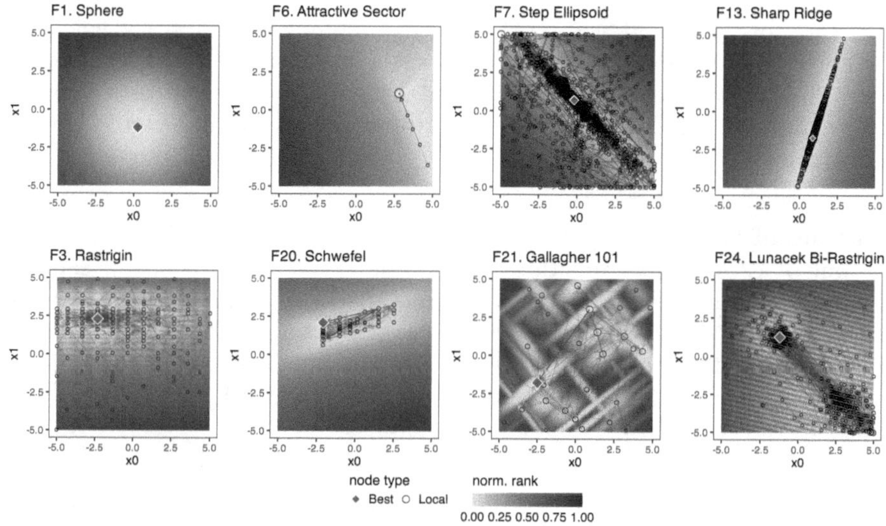

Fig. 1. Normalised rank heatmap with $\texttt{LON}_{\texttt{tol}}$ visualisation for unimodal functions F1, F6, F7, and F13 (top row) and multimodal functions F3, F20, F21, and F24 (bottom row). Lower ranks indicate better fitness values.

On F3 (Fig. 1, bottom row), we observe that the majority of $\texttt{LON}_{\texttt{tol}}$ nodes are located at the centre of basins, most of them having a low rank, *i.e.*, better fitness. This is consistent with what could be expected for the Rastrigin function. The same behaviour is observed for functions F20, F21, and F24. Interestingly, the nodes for F24 are clustered into two groups; this is to be expected for a Bi-Rastrigin function, which by construction has two funnels (cluster of local optima). Regarding F20, all nodes are concentrated in the same region of good fitness in the search space. In F20, areas of worse fitness form a smooth penalisation landscape, whereas solutions with better fitness values are gathered in the region where the Schwefel function is defined. This explains why no local optima are recorded outside the area of higher fitness. For all functions in the bottom row of Fig. 1, we observe long edges in the LONs. Some of these edges can be as long as half of the search space, which was unexpected. We hypothesise that these long edges may be due to the perturbation size of $p = 2$, and/or L-BFGS-B being able to reach points far away from its starting point, and thus "tunnel" across local optima basins. Further work is required to investigate this phenomenon.

For all functions, the nodes are located in regions of good fitness, which can be explained by the use of MBH as a sampling method. MBH is an optimisation method with a greedy acceptance criterion, and thus favours higher-quality local optima.

4.2 5D Functions

As shown above, our modification of the LON sampling with $\texttt{LON}_{\texttt{tol}}$ yields results consistent with our expectations with respect to the number of local optima for $d = 2$. We now increase the dimensionality to $d = 5$ to investigate whether these results still hold and whether the network metrics correlate with algorithm performance.

Evolution of the Number of Nodes with Dimension. In dimension $d = 2$, for 9 out of 12 unimodal functions, the LONs extracted with $\texttt{LON}_{\texttt{tol}}$ sampling show only one node in the network. Table 4 displays the number of nodes for both dimensions $d = 2$ and $d = 5$. We observe that for 5-dimensional functions, 5 out of 12 functions have only one node in the network. The reasons for the extra nodes for F6, F7, and F13 are the same as in Sect. 4.1, *i.e.*, the presence of steps or ridges in the corresponding functions.

Table 4. Number of nodes in the LONs in dimension $d = 2$ and $d = 5$.

	F1	F2	F5	F6	F7	F8	F9	F10	F11	F12	F13	F14
$d = 2$	1	1	1	12	1 015	1	1	1	1	1	186	1
$d = 5$	1	1	1	3	923	2	3	33	9	1	148	1

Four more functions exhibit more than one node in higher dimensions: F8, F9, F10, and F11. F8 and F9 are Rosenbrock functions, and they have an additional local optimum in higher dimensions with an attraction volume of about 25% [12], explaining the second node for F8 and F9. Moreover, the node count across 100 runs is 17 for F8, and 23 for F9, representing an attraction close to 25%. The third node present in F9 is a result of numerical approximations, as it has the same fitness as the additional local optima, and the distance between these two nodes in the search space is 10^{-3}. 33 nodes are found for F10 and 12 nodes are found for F11: the global optimum and 32/11 nodes reached only once in the 100 runs. Moreover, the LONs for these functions contain no edges. These nodes may be coming from local irregularities introduced in some functions.

5D Visualisations. To visualise the 5d fitness landscape, our proposal is to map the LON nodes into a 2d surface using classical multidimensional scaling (MDS) [13]. We project the 5d coordinates of each node onto a plane that preserves as much as possible their original Euclidean distances, where proximity on the plot represents actual proximity in the search space. We then generate a continuous fitness surface using bivariate Akima interpolation [2], which treats sampled local optima as survey points to estimate the surrounding landscape topography. To approximate the surface, the visualisation always uses all nodes in the LON. This core set is extended with additional points from the larger sample of local optima via uniform random sampling[2]. To maintain computational efficiency, the final combined sample is capped at a maximum of 5,000 total points; if the entire available pool of unique local optima is smaller than this threshold, all points are included. This approach creates a visualisation consistent with the standard 2d function heatmaps: the global optimum or best local optimum found is highlighted as a red diamond with a white outline to ensure visibility, while gray semi-transparent local optima and edges reveal the density and connectivity of the search space.

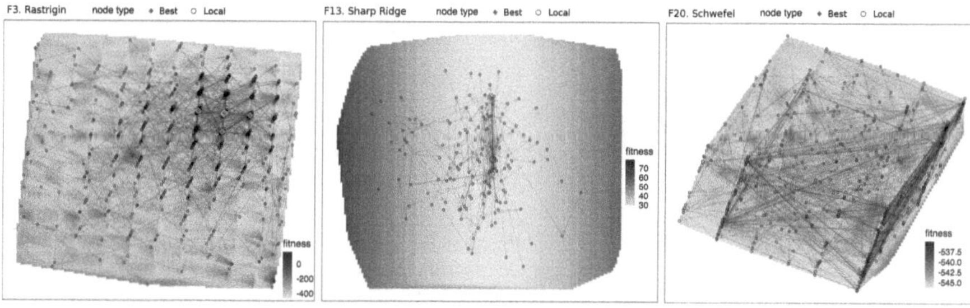

Fig. 2. 5d $\texttt{LON}_{\texttt{tol}}$ visualisations using MDS and Akima interpolation to generate an approximated fitness surface for F3, F13 and F20.

Figure 2 displays the 5-dimensional LONs for F3, F13 and F20. Plots for functions that have more than 30 nodes are provided in the supplementary materials (a threshold in the number of nodes is required to ensure a sufficient density of survey points for statistically representative spatial embedding and to prevent artifacts in the bivariate interpolation).

As in 2d LONs, nodes are concentrated in areas of better fitness. We also observe the separability of the function through parallel lines formed of nodes lying in the funnels or the Rastrigin function.

F13 is a unimodal function with a ridge that runs through the landscape and contains solutions with the best fitness values. We observe that most local optima lie in the same region of good fitness corresponding to this ridge.

Regarding F20, unlike in 2d where there is a unique area of good fitness, the landscape appears to contain multiple such regions. Most nodes are concentrated in the same area, consistent with the concentration of nodes observed on the 2d plot (Fig. 1). As in the 2d plot, the global optimum is located close to a corner of the unpenalised area.

[2] The LON models are monotonic, so the deteriorating edges are removed and with them the isolated local optima.

Correlation with Algorithm Performance. We evaluate the LON model by computing the Spearman correlation of the metrics extracted from the LON with the performance of CMA-ES, DA, DE, and PSO. We also compare these correlations with those of Exploratory Landscape Analysis (ELA) features [18,22]. We select 10 features from the cheap feature sets coming from [32]: these features have been carefully selected for their expressiveness, stability, and invariance to transformations. We get the feature data from [33] and use the median feature values of 100 runs sampled using $50 \times d = 250$ points [19] from a Sobol' low-discrepancy sequence [31]. Figure 3 shows the Spearman correlation of the LON and ELA features with algorithm performance. `avg_bet_best` is not plotted because of N/A values (all feature values are 0). Feature distribution plots can be found in the supplementary materials at [34].

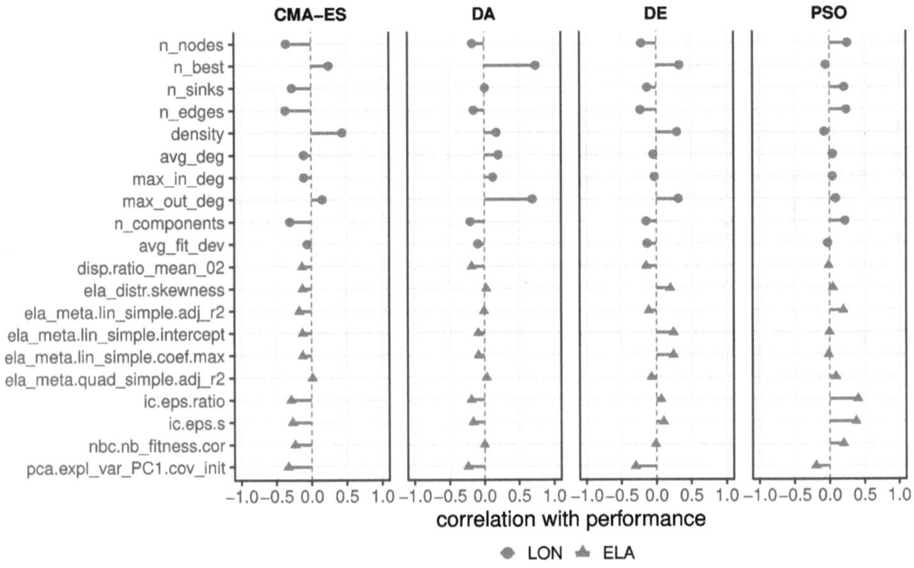

Fig. 3. Spearman correlation between LON metrics, ELA features and the performance of four algorithms.

We recall that algorithm performance is measured in terms of median fitness deviation, such that a lower value is better. Therefore, a positive correlation suggests that the corresponding metric negatively affects search performance, and vice versa. Unexpectedly, the performance of CMA-ES, DA and DE tends to increase as the number of LON nodes increases, whereas the effects differs for PSO. For three algorithms, CMA-ES, DA, and DE, LON metrics show higher correlations with performance than ELA features. The highest correlations are achieved by `density` with a correlation of 0.44 for CMA-ES, and `n_best` with a correlation of 0.74 for DA and 0.33 for DE. Consequently, highly connected LONs tend to be more challenging for CMA-ES. For CMA-ES, DA, and DE,

LON metrics have an overall better correlation with performance than ELA features. LON metrics have a median absolute Spearman correlation of 0.26, while ELA features median absolute Spearman correlation is 0.15 for CMA-ES. The same tendency can be observed for correlations with DA and DE performances with a median absolute correlation of 0.16 for both DA and DE using LON metrics and 0.09 for DA, and 0.13 for DE using ELA features. When examining correlations with PSO performance, the most strongly correlated features are ELA features: `ic.eps_ratio` with 0.4. Overall, ELA features correlate more with the performance of PSO, with a median absolute Spearman correlation of 0.19 against 0.09 for LON metrics.

Interestingly, LON metrics correlate more strongly with CMA-ES and DA performance than with DE or PSO. Even though LON metrics correlates more than ELA features using DE performance, these correlations tend to be lower than for CMA-ES and DA. This may come from the similarity of the CMA-ES and DA algorithms with the LON sampling. LON sampling is based on the basin-hopping algorithm, relying on repeated runs of L-BFGS-B, a method that estimates the gradient. DA may be the closest algorithm to the LON sampling in its design, as it is a hybrid algorithm combining simulated annealing with local search. The final solution produced by simulated annealing is refined using a local search algorithm, in this case L-BFGS-B. Because DA and LON sampling share this L-BFGS-B component, this may explain the higher correlations we observe. CMA-ES is also similar to the LON sampling as it follows the natural gradient [3]. Since L-BFGS-B estimates the gradient, this may also explain why LON metrics correlate more strongly with CMA-ES than with DE or PSO, which are population-based algorithms that do not estimate the gradient.

5 Conclusion

Our study revealed that state-of-the-art methods for sampling LONs in continuous domains can produce counter-intuitive results, *i.e.*, multiple nodes (local optima) on unimodal functions. To address this, we introduced a principled parametrisation of the LON sampling procedure that substantially reduces its early stopping. We experimented this new parametrisation, compared it with state-of-the-art methods, and confirmed that it does not produce multiple nodes for most unimodal functions. When it does, the corresponding functions exhibit non-differentiable areas, which explain the early stopping.

We introduced a landscape topographic visualisation that unifies the structural connectivity of LONs with reconstructed fitness surfaces. By employing spatial embedding to project 5d landscapes into a 2d coordinate system, we extend the traditional fitness heatmap into higher dimensions. This provides a consistent visual paradigm that allows for simultaneously capturing landscape topography and the underlying local optima connectivity. On 5-dimensional functions, we extracted new LON metrics and showed that they correlate with algorithm performance about as strongly as, and sometimes more strongly than, ELA features. We observed that these correlations are higher for algorithms that estimate gradients or follow the natural gradient.

Next steps include a deeper analysis of sampling biases, and the study of the long edges that can appear in the LONs. Another next step is to extend LONs to continuous multi-objective optimisation problems.

Acknowledgments. This work was supported by the Leverhulme Trust [grant number RPG-2025-185].

Disclosure of Interests. The authors have no competing interests to declare that are relevant to the content of this article.

References

1. Adair, J., Ochoa, G., Malan, K.M.: Local optima networks for continuous fitness landscapes. In: Genetic and Evolutionary Computation Conference Companion, GECCO (Companion), pp. 1407–1414. ACM (2019)
2. Akima, H.: A method of bivariate interpolation and smooth surface fitting for irregularly distributed data points. ACM Trans. Math. Software (TOMS) **4**(2), 148–159 (1978)
3. Akimoto, Y., Nagata, Y., Ono, I., Kobayashi, S.: Bidirectional relation between CMA evolution strategies and natural evolution strategies. In: Schaefer, R., Cotta, C., Kołodziej, J., Rudolph, G. (eds.) PPSN 2010, Part I. LNCS, vol. 6238, pp. 154–163. Springer, Heidelberg (2010). https://doi.org/10.1007/978-3-642-15844-5_16
4. Alyahya, K., Doherty, K., Akman, O.E., Fieldsend, J.E.: Reduced models of gene regulatory networks: visualising multi-modal landscapes. In: Preuss, M., Epitropakis, M.G., Li, X., Fieldsend, J.E. (eds.) Metaheuristics for Finding Multiple Solutions. NCS, pp. 229–258. Springer, Cham (2021). https://doi.org/10.1007/978-3-030-79553-5_10
5. Ballard, A.J., et al.: Energy landscapes for machine learning. Phys. Chem. Chem. Phys. **19**(20), 12585–12603 (2017)
6. Byrd, R.H., Lu, P., Nocedal, J., Zhu, C.: A limited memory algorithm for bound constrained optimization. SIAM J. Sci. Comput. **16**(5), 1190–1208 (1995)
7. Contreras-Cruz, M.A., Ochoa, G., Ramirez-Paredes, J.P.: Synthetic vs. real-world continuous landscapes: a local optima networks view. In: Filipič, B., Minisci, E., Vasile, M. (eds.) BIOMA 2020. LNCS, vol. 12438, pp. 3–16. Springer, Cham (2020). https://doi.org/10.1007/978-3-030-63710-1_1
8. Doye, J.P.: Network topology of a potential energy landscape: a static scale-free network. Phys. Rev. Lett. **88**(23), 238701 (2002)
9. Fieldsend, J.E., Alyahya, K.: Visualising the landscape of multi-objective problems using local optima networks. In: Proceedings of the Genetic and Evolutionary Computation Conference Companion, pp. 1421–1429 (2019)
10. Fieldsend, J.E.: Representation, resolution, and visualization in multimodal optimization. In: Preuss, M., Epitropakis, M.G., Li, X., Fieldsend, J.E. (eds.) Metaheuristics for Finding Multiple Solutions. NCS, pp. 27–55. Springer, Cham (2021). https://doi.org/10.1007/978-3-030-79553-5_2
11. Fieldsend, J.E.: Scalable local optima networks for continuous search spaces. In: Legrand, P., et al. (eds.) Artificial Evolution. LNCS, vol. 15926, pp. 77–90. Springer, Cham (2026). https://doi.org/10.1007/978-3-032-07998-5_6

12. Finck, S., Hansen, N., Ros, R., Auger, A.: Real-parameter black-box optimization benchmarking 2010: presentation of the noiseless functions (2010). http://coco.gforge.inria.fr/downloads/download16.00/bbobdocfunctions.pdf
13. Gower, J.C.: Some distance properties of latent root and vector methods used in multivariate analysis. Biometrika **53**(3–4), 325–338 (1966)
14. Hansen, N., Auger, A., Ros, R., Mersmann, O., Tusar, T., Brockhoff, D.: COCO: a platform for comparing continuous optimizers in a black-box setting. Optim. Methods Software **36**(1), 114–144 (2021). https://doi.org/10.1080/10556788.2020.1808977
15. Hansen, N., Ostermeier, A.: Completely derandomized self-adaptation in evolution strategies. Evol. Comput. **9**(2), 159–195 (2001). https://doi.org/10.1162/106365601750190398
16. Karatas, M.D., Akman, O.E., Fieldsend, J.E.: Towards population-based fitness landscape analysis using local optima networks. In: Proceedings of the Genetic and Evolutionary Computation Conference Companion, pp. 1674–1682 (2021)
17. Kennedy, J., Eberhart, R.: Particle swarm optimization. In: Proceedings of ICNN'95 - International Conference on Neural Networks. vol. 4, pp. 1942–1948 (1995). https://doi.org/10.1109/ICNN.1995.488968
18. Kerschke, P., Hoos, H., Neumann, F., Trautmann, H.: Automated algorithm selection: survey and perspectives. Evol. Comput. **27**(1), 3–45 (2019)
19. Kerschke, P., Preuss, M., Wessing, S., Trautmann, H.: Low-budget exploratory landscape analysis on multiple peaks models. In: Proceedings of the Genetic and Evolutionary Computation Conference, GECCO '16. pp. 229–236. ACM (2016). https://doi.org/10.1145/2908812.2908845
20. Liu, D.C., Nocedal, J.: On the limited memory BFGS method for large scale optimization. Math. Program. **45**(1), 503–528 (1989)
21. Lourenço, H.R., Martin, O.C., Stützle, T.: Iterated local search. In: Glover, F., Kochenberger, G.A. (eds.) Handbook of Metaheuristics. pp. 320–353. Springer, Boston (2003). https://doi.org/10.1007/0-306-48056-5_11
22. Mersmann, O., Bischl, B., Trautmann, H., Preuss, M., Weihs, C., Rudolph, G.: Exploratory landscape analysis. In: Proceedings of the Genetic and Evolutionary Computation Conference, GECCO '11, pp. 829–836. ACM (2011). https://doi.org/10.1145/2001576.2001690
23. Mitchell, P., Ochoa, G., Chassagne, R.: Local optima networks of the black box optimisation benchmark functions. In: Genetic and Evolutionary Computation Conference (Companion). GECCO '23 Companion, New York, NY, USA, pp. 2072–2080. ACM (2023). https://doi.org/10.1145/3583133.3596311
24. Mitchell, P., Ochoa, G., Lavinas, Y., Chassagne, R.: Local optima networks for assisted seismic history matching problems. In: Correia, J., Smith, S., Qaddoura, R. (eds.) EvoApplications 2023. LNCS, vol. 13989, pp. 86–101. Springer, Cham (2023). https://doi.org/10.1007/978-3-031-30229-9_6
25. Nelder, J.A., Mead, R.: A simplex method for function minimization. Comput. J. **7**(4), 308–313 (1965)
26. de Nobel, J., Vermetten, D., Wang, H., Doerr, C., Bäck, T.: Tuning as a means of assessing the benefits of new ideas in interplay with existing algorithmic modules. In: Krawiec, K. (ed.) GECCO '21: Genetic and Evolutionary Computation Conference, Companion Volume, Lille, France, July 10-14, 2021. pp. 1375–1384. ACM (2021). https://doi.org/10.1145/3449726.3463167
27. Ochoa, G., Tomassini, M., Vérel, S., Darabos, C.: A study of NK landscapes, basins and local optima networks. In: Conference on Genetic and Evolutionary Computation, GECCO, pp. 555–562. ACM (2008)

28. Ochoa, G., Veerapen, N.: Mapping the global structure of TSP fitness landscapes. J. Heuristics **24**(3), 265–294 (2018). https://doi.org/10.1007/S10732-017-9334-0
29. Ochoa, G., Veerapen, N., Daolio, F., Tomassini, M.: Understanding phase transitions with local optima networks: number partitioning as a case study. In: Hu, B., López-Ibáñez, M. (eds.) EvoCOP 2017. LNCS, vol. 10197, pp. 233–248. Springer, Cham (2017). https://doi.org/10.1007/978-3-319-55453-2_16
30. Prager, R.P., Trautmann, H.: Pflacco: feature-based landscape analysis of continuous and constrained optimization problems in python. Evol. Comput. **32**(3), 211–216 (2024). https://doi.org/10.1162/EVCO_A_00341
31. Renau, Q., Doerr, C., Dreo, J., Doerr, B.: Exploratory landscape analysis is strongly sensitive to the sampling strategy. In: Bäck, T., Preuss, M., Deutz, A., Wang, H., Doerr, C., Emmerich, M., Trautmann, H. (eds.) PPSN 2020, Part II. LNCS, vol. 12270, pp. 139–153. Springer, Cham (2020). https://doi.org/10.1007/978-3-030-58115-2_10
32. Renau, Q., Dreo, J., Doerr, C., Doerr, B.: Towards explainable exploratory landscape analysis: extreme feature selection for classifying BBOB functions. In: Castillo, P.A., Jiménez Laredo, J.L. (eds.) EvoApplications 2021. LNCS, vol. 12694, pp. 17–33. Springer, Cham (2021). https://doi.org/10.1007/978-3-030-72699-7_2
33. Renau, Q., Dreo, J., Doerr, C., Doerr, B.: Exploratory Landscape Analysis Feature Values for the 24 Noiseless BBOB Functions (2021). https://doi.org/10.5281/zenodo.4449934
34. Renau, Q., Ochoa, G., Lifooghe, A., Fieldsend, J.: From networks to landscapes: sampling and topographic visualisation of continuous lons - code and data (2026). https://doi.org/10.5281/zenodo.20665791
35. Sobol', I.: On the distribution of points in a cube and the approximate evaluation of integrals. USSR Comput. Math. Math. Phys. **7**(4), 86–112 (1967). https://doi.org/10.1016/0041-5553(67)90144-9, https://www.sciencedirect.com/science/article/pii/0041555367901449
36. Stadler, P.F.: Fitness landscapes. In: Lässig, M., Valleriani, A. (eds.) Biological Evolution and Statistical Physics. LNP, vol. 585, pp. 183–204. Springer, Heidelberg (2002). https://doi.org/10.1007/3-540-45692-9_10
37. Storn, R., Price, K.: Differential evolution - a simple and efficient heuristic for global optimization over continuous spaces. J. Global Optim. **11**(4), 341–359 (1997). https://doi.org/10.1023/A:1008202821328
38. Thomson, S.L., Ochoa, G., Vérel, S., Veerapen, N.: Inferring future landscapes: sampling the local optima level. Evol. Comput. **28**(4), 621–641 (2020)
39. Thomson, S.L., Renau, Q., Vermetten, D., Hart, E., van Stein, N., Kononova, A.V.: Code, data, and plots for the paper: "stalling in space: Attractor analysis for any algorithm" (2024). https://doi.org/10.5281/zenodo.14170241
40. Thomson, S.L., Renau, Q., Vermetten, D., Hart, E., van Stein, N., Kononova, A.V.: Stalling in space: attractor analysis for any algorithm. In: García-Sánchez, P., Hart, E., Thomson, S.L. (eds.) EvoApplications 2025, Part I. LNCS, pp. 510–526. Springer, Cham (2025). https://doi.org/10.1007/978-3-031-90062-4_32
41. Tomassini, M.: A local optima network view of real function fitness landscapes. Entropy **24**(5), 703 (2022). https://doi.org/10.3390/e24050703
42. Tsallis, C., Stariolo, D.A.: Generalized simulated annealing. XXPhys. A **233**(1), 395–406 (1996). https://doi.org/10.1016/S0378-4371(96)00271-3
43. Verel, S., Ochoa, G., Tomassini, M.: Local optima networks of NK landscapes with neutrality. IEEE Trans. Evol. Comput. **15**(6), 783–797 (2010)
44. Vinkó, T., Gelle, K.: Basin-hopping networks of continuous global optimization problems. CEJOR **25**(4), 985–1006 (2017)

45. Virtanen, P., et al.: SciPy 1.0 contributors: SciPy 1.0: fundamental algorithms for scientific computing in Python. Nat. Meth. **17**, 261–272 (2020). https://doi.org/10.1038/s41592-019-0686-2
46. Wales, D.J., Doye, J.P.: Global optimization by basin-hopping and the lowest energy structures of Lennard-Jones clusters containing up to 110 atoms. J. Phys. Chem. A **101**(28), 5111–5116 (1997)
47. Wang, H., Vermetten, D., Ye, F., Doerr, C., Bäck, T.: Iohanalyzer: detailed performance analyses for iterative optimization heuristics. ACM Trans. Evol. Learn. Optim. **2**(1) (2022). https://doi.org/10.1145/3510426
48. Xiang, Y., Gubian, S., Suomela, B., Hoeng, J.: Generalized simulated annealing for global optimization: The gensa package. R J. **5**(1), 13 (2013). https://doi.org/10.32614/RJ-2013-002

Generating Point Sets of Low Star Discrepancy by Optimizing Kronecker Constructions

Imène Ait Abderrahim[1,2(✉)], Carola Doerr[1], and Martin Durand[1]

[1] Sorbonne University, CNRS, LIP6, 75005 Paris, France
imene.ait-abderrahim@lip6.fr
[2] Khemis Miliana University, Ain Defla, Algeria

Abstract. The L_∞ star discrepancy of a point set is a measure for how uniformly a point set is distributed in a given space. Point sets of low star discrepancy are used as designs of experiments, as initial designs for Bayesian optimization algorithms, for quasi-Monte Carlo integration methods, and many other applications. Recent work has shown that classical constructions such as Sobol', Halton, or Hammersley sequences can be outperformed by large margins when considering point sets of fixed sizes rather than their asymptotic behavior when the number of points tends to infinity. These results, highly relevant to the aforementioned applications, raise the question of how much existing constructions can be improved through size-specific optimization. In this work, we study this question for the 3-dimensional setting, focusing on how well the so-called Kronecker construction performs for various set sizes n. We show that, for settings with at least 500 points, optimizing the two configurable parameters of its construction using evolutionary computation approaches such as CMA-ES yields point sets whose discrepancy values outperform those of the previously best-known sets. Using the algorithm configuration technique Irace, we then derive parameters that yield new state-of-the-art discrepancy values for whole ranges of set sizes. An investigation of the problem landscapes using exploratory landscape analysis reveals that the problem is highly multimodal and that some of its characteristics differ substantially from the 24 BBOB functions.

Keywords: Low-discrepancy point sets · uniformity of distribution · Evolutionary computation

1 Introduction

Discrepancy measures are designed to quantify how well-distributed a set of points is in a given space. Among the various discrepancy measures, the L_∞ star discrepancy is one of the most important. It measures how well spread the points in a point set P are. It is defined as the maximum absolute difference between the volume of a box anchored in the origin and the proportion of points from the

G. Iacca et al. (Eds.): PPSN 2026, LNCS 16985, pp. 550–565, 2027.
https://doi.org/10.1007/978-3-032-36214-8_34

P falling inside it. Indeed, the Koksma-Hlawka inequalities [18,19] show that it is possible to bound the error made on the approximation of an integral by the average of a finite number of local evaluations: The smaller the discrepancy of the point set, the smaller the error guarantee for the quasi-Monte Carlo approximation. Although its primary use is in quasi-Monte Carlo integration [9], point sets with minimal L_∞ star discrepancy are also employed in one-shot optimization [2], experiment design [29], computer vision [25], and financial mathematics [13]. Optimizing L_∞ star discrepancy is a computationally challenging problem. Indeed, optimal point sets are known only for up to 21 points in dimension 2 and for up to 8 points in dimension 3 [3]. For the 2-dimensional hypercube, the Fibonacci set gives consistently good point sets, regardless of the size [3]. It builds the coordinates in one dimension using the golden ratio and in the other using the inverse of the number of points. This ensures that the points are spread very uniformly in both dimensions. No such method exists for higher dimensions. The construction of low L_∞ star discrepancy point sets has traditionally relied on number-theoretic sequences such as Halton [14], Hammersley [15], or Sobol' [31], which offer strong asymptotic guarantees but were recently shown to be less suitable for sets of fixed size [6]. To address this limitation, research on the optimization-driven construction of low-star-discrepancy point sets tailored to specific dimensions and sample sizes has received increased attention in recent years.

1.1 Related Work

To obtain point sets of low discrepancy, a broad range of different techniques have been used: exact linear programming techniques (in isolation [3] or as components of heuristic search [4]), graph neural networks [27], large language models [28], heuristic search approaches [6–8,11].

More precisely, one of the first approaches to tackle the optimization of low discrepancy is the work of Doerr et al. [11], who propose a genetic algorithm that optimizes the generation of points from generalized Halton sequences to directly produce low-star-discrepancy point sets. Their hybrid evolutionary framework integrates efficient discrepancy estimation with stochastic search. In addition, Patel and Education [24] revisit designs based on lattices and sequences using parameter optimization instead of structural redesign. Specifically, optimized Kronecker sets show that uniformity can be significantly increased while maintaining simplicity, scalability, and progressive sampling features through careful empirical selection of irrational parameters. Another approach of Clément et al. [5,6] investigates exact and heuristic strategies for selecting subsets of points that minimize star discrepancy. The study formulates the construction problem as a combinatorial optimization task and evaluates greedy, local search, and evolutionary approaches.

In more recent work, point-set generation has been reframed as a geometric learning problem using learning-based techniques. In Rusch et al. [27], they present Message-Passing Monte Carlo (MPMC), a graph neural network architecture that achieves contemporary performance for small to medium dimensions by learning point set transformations to minimize discrepancy-related objectives. Clément et al. [4] propose a permutation-based decomposition that separates relative point ordering from coordinate optimization, achieving state-of-the-art discrepancy reductions compared to classical sets. This decomposition yields a post-processing procedure, which we use in our experiments, to lower the discrepancy of a point set. Intuitively, this procedure maintains the relative position of points in all dimensions. It finds an optimal point set among the ones which preserve these relative positions. Clément et al. [7] propose a projected gradient descent algorithm for optimizing L_2 discrepancies in low dimensions. The main contribution is demonstrating that this computationally inexpensive method can achieve results comparable to or better than state-of-the-art techniques, even for L_∞ star discrepancy, when starting with a reasonably uniform point set like the Fibonacci lattice. Sadikov [28] explores the emerging use of large language models (LLMs) as generative and guidance mechanisms for constructing low-discrepancy point sets.

1.2 Contributions and Outline of the Paper

In this paper, we investigate Kronecker sets for the construction of point sets of low star discrepancy. Kronecker constructions which generalize the 2-dimensional Fibonacci set to larger dimensions. Kronecker sets have $D-1$ degrees of freedom, i.e., parameters that need to be set to obtain a concrete set, one for each dimension. With the two-dimensional setting largely explored in previous works [4,27,28], we focus our work on constructing low-discrepancy point sets in dimension 3. That is, we are dealing with a 2-dimensional optimization problem to identify the two parameters that yield good 3-dimensional point sets.

We use evolutionary computational methods such as CMA-ES and local search variants to identify parameters that minimize the discrepancy of point sets of a fixed size, and we use the Irace configurator to obtain parameters that yield constructions for broad ranges of point set sizes. Even though this generation method is quite generic, it remains competitive with state-of-the-art solutions for small sizes and outperforms the best-known methods for larger sizes starting from 500 points.

We also investigate the landscape of the 2D-optimization problem using exploratory landscape analysis [22] and compare it to the well-studied BBOB functions from the COCO environment [16].

We then briefly study the 4-dimensional Kronecker constructions, but we were unable to identify parameter values that would yield constructions outperforming the current state of the art, indicating that the Kronecker constructions may not be the right approach for dimension 4 (and possibly higher).

Outline of the Work. In Sect. 2, we provide relevant background information such as the formal definitions of the L_∞ star discrepancy and the Kronecker sets. In Sect. 3, we describe the methods used to tune the parameters of Kronecker sets. Our experimental results are detailed in Sect. 4. 5 provides an analysis of the problems landscape. Finally, we complete the paper with a discussion and a conclusion in Sects. 6 and 7.

2 Preliminaries

2.1 L_∞ Star Discrepancy

The L_∞ star discrepancy of a point set P measures how well spread the points in P are in the d-dimensional $[0, 1)$ cube. It is defined as the maximum absolute difference between the volume of a box anchored in the origin and the proportion of points from P falling inside it. Formally, given a point set P of n points in $[0, 1)^d$, the L_∞ star discrepancy of P, denoted by $d^*_\infty(P)$, is defined as:

$$d^*_\infty(P) = \sup_{q\in[0,1)^d} \left| \frac{D(q, P)}{|P|} - \lambda(q) \right|, \tag{1}$$

where $D(q, P)$ is the number of points of P in the box $[0, q)$ and $\lambda(q)$ is the Lebesgue measure of the d-dimensional box $[0, q)$.

Computing the L_∞ star discrepancy of a point set is a discrete problem as only points on a specific grid can reach the maximal value [23]. Intuitively, if one draws a grid using the coordinates of the points in a point set P, then the maximum discrepancy can only be reached by boxes having their corner on that grid. Importantly, the evaluation of a point set in low dimensions can be done very efficiently with the DEM algorithm [10] and its parallelizable implementation available from [8].

2.2 Kronecker Point Sets

There are various methods to generate point sets with low discrepancy. In 2 dimensions, the Fibonacci set provides good solutions. For a given n, the Fibonacci point set F_n is obtained as follows:

$$F_n = \{(i/n, i\phi\%1)\big| i \in \{0, \dots, n-1\}\} \tag{2}$$

where $\phi = (1 + \sqrt{5})/2 \approx 1.618$ is the golden ratio and the notation %1 means that we remove the integer part.

Kronecker sets are a generalization of Fibonacci. One can replace ϕ with other values and generate a point set in the same way. Given a set of parameters $\{p_2, \dots, p_d\}$, p_j determining the coordinates in dimension j, one can then define a Kronecker point set as follows:

$$K_n^{(p_2,\dots,p_d)} = \{(i/n\%1, i/p_2\%1, \dots, i/p_d\%1)|\forall i \in \{0, \dots, n-1\}\} \tag{3}$$

In 2 dimensions, one can slightly improve the performance of the Fibonacci point set by shifting the set and starting with $i = 1$. We therefore define the shifted Kronecker set as follows.

$$^{+1}K_n^{(p_2,\dots,p_d)} = \{(i/n\%1, ip_2\%1, \dots, ip_d\%1)|\forall i \in \{1, \dots, n\}\} \tag{4}$$

From now on, we use "Kronecker point set" instead of "shifted Kronecker point set" as experiments are only run for shifted sets. Additionally, since most of our experiments are conducted for point sets in 3 dimensions, we will focus on the 2-dimensional problem of finding values for parameters p_2 and p_3. Figure 1 displays heatmaps showing the quality of Kronecker point sets obtained with parameters (p_2, p_3) for $n = 100$. The x and y axis represent the parameters p_2 and p_3, respectively. Each dot on a heatmap represent the discrepancy of the point set generated with parameters (p_2, p_3). On the left, we display all the points, in the center only points corresponding to point sets of discrepancy lower than 0.055 and on the right only points corresponding to point sets with discrepancy lower than 0.045.

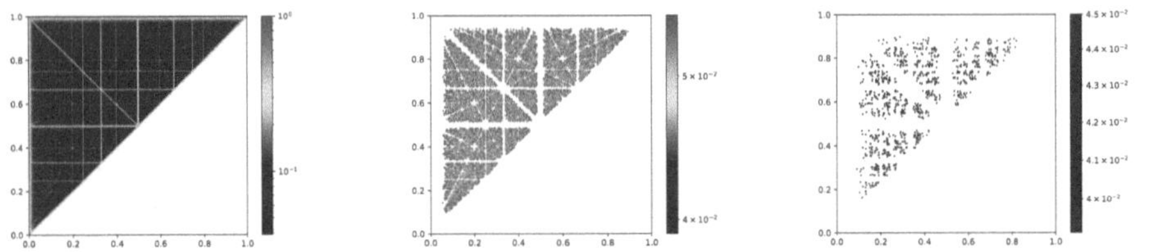

Fig. 1. Heatmaps. The X (resp. Y) coordinate corresponds to the p_2 (resp. p_3) value, the color indicates the L_∞ star discrepancy of the 3D point set generated with the $(p_2, p_3) \in [0,1]^2$ parameters, for $n = 100$. The second and last heatmaps contain only points for which the corresponding Kronecker point set has a discrepancy below 0.055 and 0.045, respectively.

These heatmaps reveal that the optimization landscape is intricate to explore, as the problem appears highly multimodal. Very good values for parameters p_2 and p_3 are scattered, and there seem to be many local optima. Additionally, this landscape depends on the value of n and a good set of parameters for $n = 100$ is not guaranteed to be good for other values. Indeed, Fig. 2 displays the discrepancies of the point sets obtained for $n = 100$ and $n = 1000$ using randomly sampled pairs of (p_2, p_3).

Interestingly, while many very good pairs for $n = 100$ seem to be also good for $n = 1000$. There are also several pairs that provide a very low discrepancy point set for either $n = 100$ or $n = 1000$ but not both. In particular, one can see that the best pair found for $n = 100$, represented by the red dot, does not yield a satisfactory point set for $n = 1000$.

We are facing two challenges: (1) *given a value* n *, how can we find good parameter values* (p_2, p_3) *that optimize the discrepancy of the set* ${}^{+1}K_n^{(1/n, p_2, p_3)}$ *?*, (2) *is it possible to find parameter values that return good point sets, for broad ranges of* n *?*

3 Optimizing Kronecker Point Set Generation

In this section, we present the methods that we use to find Kronecker sets that minimize the discrepancy. We suggest using two optimization strategies: The first

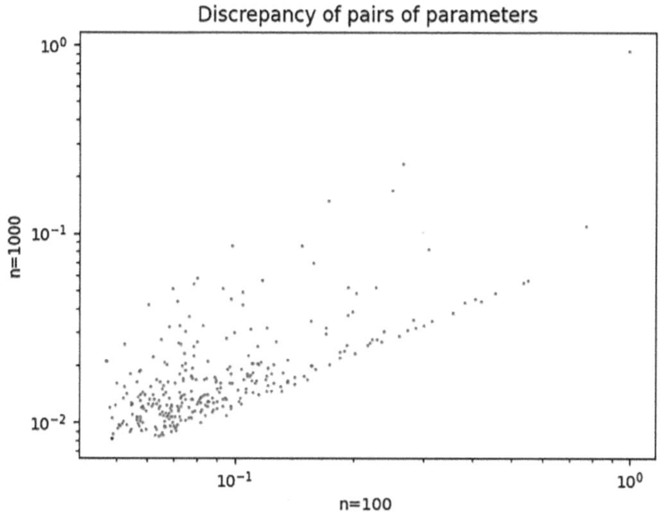

Fig. 2. L_∞ star discrepancies of point sets obtained with randomly sampled (p_2, p_3). The X (resp. Y) coordinate of a point corresponds to the discrepancy of the point set obtained for $n = 100$ (resp. $n = 1000$). Red (resp. Black) point is the best found for $n = 100$ (resp.$n = 1000$). (Color figure online)

uses heuristics and black-box approaches to find specific values of the Kronecker parameters $(p_2, p_3) \in [0, 1]^2$ for each number of points n, where we consider $0 \leq p_2 \leq p_3 \leq 1$. Three heuristics are selected for the optimization of point generation: Covariance Matrix Adaptation Evolution Strategy (CMA-ES) [17], Iterated Local Search (ILS) [21] with snowball embedding local search, and Reactive Tabu Search (RTS) [1], however, in lack of a clear difference in performance, we decide to continue only with CMA-ES for the later experiments for higher n values, as this method is better established in the community. The second strategy consists of applying the Automatic Algorithm Configuration (AAC) technique to tune Kronecker's parameters for points generation for selected intervals of n points, e.g., we find one configuration $(p_2, p_3) \in [0, 1]^2$ that yields good point sets for all $n \in [5, 100]$. Algorithm configuration (AC) refers to the systematic process of tuning algorithm parameters and has been extensively studied in the literature [30, 32]. AAC, also called parameter tuning, aims to identify high-performing parameter settings without extensive manual intervention by framing parameter configuration as an optimization problem guided by empirical performance. This data-driven exploration improves robustness, reproducibility, and performance across diverse problem instances. Among state-of-the-art methods, Irace [20] is widely adopted due to its statistically grounded racing mechanism, which efficiently focuses computational effort on the most promising configurations.

4 Experimental Results

To evaluate the discrepancy values of our optimization techniques, we run several experiments to investigate the following research questions regarding the heuristics and automatic algorithm configuration.

- (RQ1) How efficient are CMA-ES and parameter tuning for optimizing 3D Kronecker for star discrepancy?
- (RQ2) Is post-processing improving the Kronecker point sets performance?
- (RQ3) Using parameter tuning, can we find general (p_2, p_3) pairs that are good for ranges $[n_1, n_2]$ of n?
- (RQ4) How efficient can the Kronecker method be if we increase the problem's dimension?

4.1 Assessing the Effectiveness of CMA-ES and Parameter Tuning for Low Discrepancy

As the heuristic's computation time is expensive for a larger number of points, we limit experiments with CMA-ES to constructions of sets with at most $n = 1000$ points. We run the algorithm five times with 10,000 evaluations per run. The reported result is the best found value among all evaluations. For the tuning of Kronecker's parameter settings, three scenarios were selected. In the first scenario, we tune IRACE on a separate subselection for identifying good (p_2, p_3) for ranges of n. Then, we compare our results in terms of quality of the returned discrepancy to the Sobol' sequence [31], for which we use the implementation by Fox [12], and other state-of-the-art methods, namely KRONECKER21 [24], L2_Subset [7]. The results are reported in Table 1 and Figs. 3a and 3b.

The parameter pair values (p_2, p_3) obtained from the parameter tuning for I_200, I_1500, and I_2500 are (0.5494, 0.7867), (0.6193, 0.7830), (0.71810558, 0.81422429), respectively.

Table 1 displays L_∞ star discrepancy values for chosen n points in dimension $D = 3$. The results show that our CMA-ES is competitive for small n and sometimes better than the state of the art. For sizes 25, 60, 80, and $n > 300$, we outperform L2_subset and found new low discrepancy values. Additionally, for higher values of $n \geq 500$, CMA-ES and the parameter tuning I_1500 and I_2500 using IRACE have outperformed the L2_subset and Sobol and achieved new state-of-the-art discrepancy values.

Furthermore, we compare the values returned by IRACE in the different experiments. In Table 1 we highlight cells with the lowest discrepancy value among the IRACE experiments. The tuning is generally adapted to the size of n on which it is tuned, but gets slightly worse for bigger n values that were not seen in the training. This is particularly relevant as one can see in Fig. 3b that the parameters obtained with I_200 seem to produce significantly worse point sets for higher values of n. The opposite does not seem to hold as I_1500 and I_2500 produce acceptable point sets for smaller sizes, although they were trained to

Table 1. Best found L_∞ star discrepancy values for $d = 3$

n	Sobol	CMA-ES	I_200	I_1500	I_2500	L2_Subset
20	0.17742	0.12221	0.15029	0.18194	0.16935	**0.1202**
25	0.14742	**0.10066***	0.12738	0.14555	0.14618	0.1012
32	0.14917	0.08813	0.10094	0.12083	0.11420	**0.07931**
40	0.11167	0.07292	0.08273	0.09666	0.09221	**0.06392**
50	0.11276	0.06371	0.06986	0.09356	0.08682	**0.05337**
60	0.08052	**0.05799**	0.06166	0.08021	0.07639	0.0606
80	0.07862	**0.04709**	0.05099	0.06199	0.07525	0.0547
100	0.06057	0.04017	0.04325	0.04996	0.06020	**0.0374**
150	0.04062	0.03009	0.03489	0.03331	0.04014	**0.02499**
200	0.03654	0.02474	0.02762	0.02592	0.03010	**0.02181**
250	0.02645	0.02023	0.02210	0.02158	0.02408	**0.01837**
300	0.02686	**0.01826**	0.01849	0.01969	0.02241	
500	0.01524	**0.01115**	0.01402	0.01252	0.01344	0.01125
750	0.01236	**0.00826**	0.01014	0.00904	0.00915	
1000	0.00836	**0.00657**	0.00804	0.00736	0.00698	0.0081
1250	0.00853		0.00770	0.00650	**0.00592**	
1500	0.00713		0.00695	**0.00548**	0.00572	
1750	0.00649		0.00677	0.00545	**0.00474**	
2000	0.00480		0.00624	0.00477	**0.00439**	0.00503
2500	0.00501		0.00593	0.00382	**0.00305**	
3000	0.00435		0.00537	0.00328	**0.00304**	
3500	0.00383		0.00520	0.00314	**0.00268**	
4000	0.00314		0.00501	0.00286	**0.00262**	
5000	0.00264		0.00485	0.00241	**0.00210**	

Note: The value with * is the smallest value found by any of the three evolutionary heuristics. We highlight in gray the cell containing the best value among the three IRACE experiments.

also fit much larger values of n. Finally, one can see that I_2500 keeps outperforming the Sobol sequences even outside of its training space for sizes 3000, 3500, 4000, and 5000.

Figure 4 displays the inverse L_∞ star discrepancy, i.e., the number of points required to achieve a target discrepancy. For readability, the x axis shows the inverse of the discrepancy. This figure can be read as follows: "Given an objective on the discrepancy, how many points are needed?". Overall, the results show that our optimized approaches outperform classical constructions. In particular, the Sobol sequence exhibits higher variability and generally requires more points, highlighting its limitations in finite-sample regimes despite strong asymptotic guarantees. The L2_Subset method performs competitively for moderate discrepancy levels but scales less effectively as stricter targets are imposed. In contrast, CMA-ES and IRACE-based configurations consistently achieve lower sample requirements across a wide range of discrepancy values. CMA-ES is particularly effective in the low-to-medium regime, while IRACE demonstrates superior scalability and robustness, especially for tight discrepancy thresholds. These results emphasize the advantage of adaptive, black-box optimization methods over static constructions, enabling more efficient generation of low-discrepancy

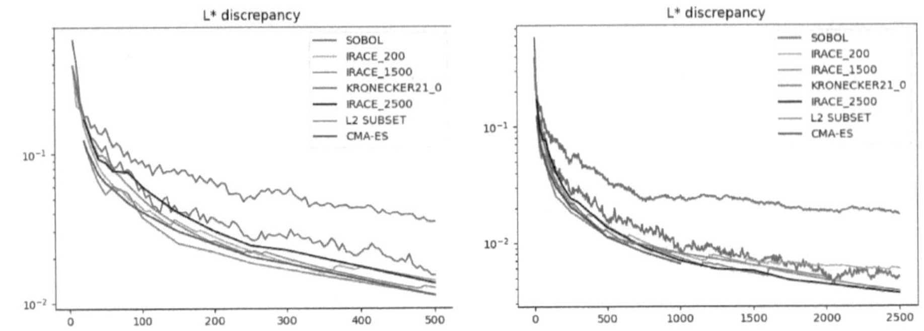

(a) $n \in [5, 500]$ (b) $n \in [5, 2500]$

Fig. 3. Plot of the L_∞ discrepancy for n points.

point sets for practical applications. Finally, this shows a great improvement if it were applied for real-world applications that require big sample for their experiments, where with CMA-ES and IRACE-based results, we can minimize the size of the sample to achieve a certain target discrepancy, e.g.;, for an inverse discrepancy of 10^2, with the L2_Subset method, we need a sample of size of about 700 points to reach this discrepancy value, where with CMA-ES we need only 500 points, and this difference can result in a big gain of time process for the real world application.

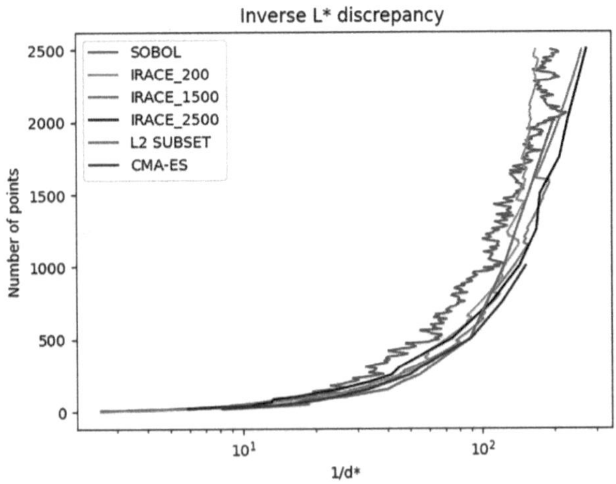

Fig. 4. Plot of the inverse L_∞ star discrepancy for n between 5 and 2500.

4.2 Using Post-processing on Kronecker Point Sets

Clément et al. [4] propose a method to improve the discrepancy starting from any given point set. Their method maintains the relative position of the points, i.e., for two points $x^{(i)}$ and $x^{(j)}$, if $x^{(i)}$ has a smaller coordinate in any given dimension k than $x^{(j)}$, then this will still be the case after the post-processing step. It returns the best possible point set such that this relative position is maintained using mathematical programming. This post-processing is, however, rather costly and therefore can only be applied for small values of n. Results can be found in Table 2.

Table 2. L_∞ star discrepancy with post-processing. Notation o(name) means after post-processing.

n	I_200	o(I_200)	L2_Subset	o(L2_Subset)	MPMC	o(MPMC)
25	0.12738	0.11029	0.1012	0.08519	0.10664	**0.08335**
32	0.10094	0.08956	0.07931	0.07309	0.08234	**0.07085**
40	0.08273	0.06082	0.06392	**0.05988**	0.08139	0.06242
50	0.06986	0.05420	0.05337	**0.04979**	0.05828	0.05067

Unsurprisingly, point sets that are better before post-processing remain better afterwards. However, one can note that the Kronecker point sets seem to benefit more from the post-processing. This may be due to the fact that coordinates for Kronecker point sets are generated in a very rigid and generic way and are more likely to benefit from small modifications than point sets obtained through optimization techniques.

4.3 Assessing Parameter Tuning on Different n Ranges

We compare the evolution of the L_∞ discrepancy for multiple interval configurations (INTERVAL1 - INTERVAL10) of $n \in [5, 1000]$ and the obtained configuration IRACE _1500 in Table 1. INTERVAL1 means $n \in [5, 100]$, INTERVAL2 has $n \in [101, 200]$ and so on until INTERVAL10 that has $n \in [901, 1000]$. The main objective of this study is to find good pairs $(p_2, p_3) \in [0, 1]^2$ that are good for a specific interval but that could also be good as well for all n. Then we rerun the algorithm for the whole $n \in [5, 2500]$. To have a clear vision on the results, we keep only the results for odd-number intervals as shown in the plots in Fig. 5.

Figure 5 illustrates how the lines are closely packed when n is small, suggesting that most configurations perform similarly for smaller sizes. Performance differences become more noticeable as the number of points increases. For example, INTERVAL_5 and INTERVAL_9 consistently outperform the majority of other interval configurations and are competitive with IRACE _1500 up to about 1000–1200 points; beyond this range, a clear improvement emerges. We do not have a full understanding of why this happens. This suggests that there exists

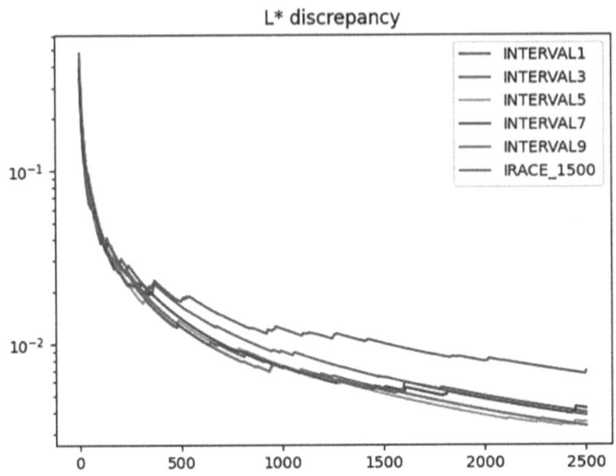

Fig. 5. L_∞ discrepancy (y-axis) with Parameters tuning for n small intervals (problem size on x-axis).

some space to further improve the Kronecker sets beyond the results we already obtained with IRACE.

However, in addition to demonstrating the benefit of data-driven parameter tuning in navigating the highly nonconvex discrepancy landscape and preventing premature performance saturation, this suggests that automatic algorithm configuration is more effective at identifying parameter settings that generalize across increasing sample sizes.

4.4 Assessing the Kronecker Method for Dimension $D = 4$

For further study, we evaluate the Kronecker method and the optimized approaches for dimension $D = 4$. This is the only section for which we optimize three parameters (p_2, p_3, p_4) instead of just (p_2, p_3). Results are reported in Table 3. We observe that even the CMA-ES and RTS algorithms that tune the parameters specifically for one given n cannot outperform the discrepancy values of the truncated Sobol' sequence in 4 dimensions. One question that remains is whether this is due to the nature of Kronecker sets or to the ability of the algorithms to find very good parameters. The former would mean that Kronecker sets would be unable to provide good point sets for dimension $D = 4$, and possibly beyond.

5 Analysis of the Problem Landscape

We have already observed in Fig. 1 that the optimization landscape is highly multimodal. To further probe the characteristics of the optimization problem, we investigate the Kronecker configuration problem using exploratory landscape analysis (ELA) [22]. We compare the ELA features using the pflacco package

Table 3. L_∞ discrepancy values of the best-found Kronecker constructions in dimension 4, for different values of n (3 columns on the right), compared to those obtained from the Sobol' sequence

n	Sobol'	CMA-ES	RTS	IRACE _512
5	0.28281	0.34054	0.34184	0.43494263
8	0.23438	0.29388	0.27840	0.32895087
16	0.13672	0.17685	0.19424	0.24601968
20	0.11172	0.15750	0.16216	0.22276414
32	0.08984	0.12989	0.12888	0.1486705
50	0.07994	0.08886	0.09975	0.09884006
64	0.05371	0.07782	0.08338	0.10571434

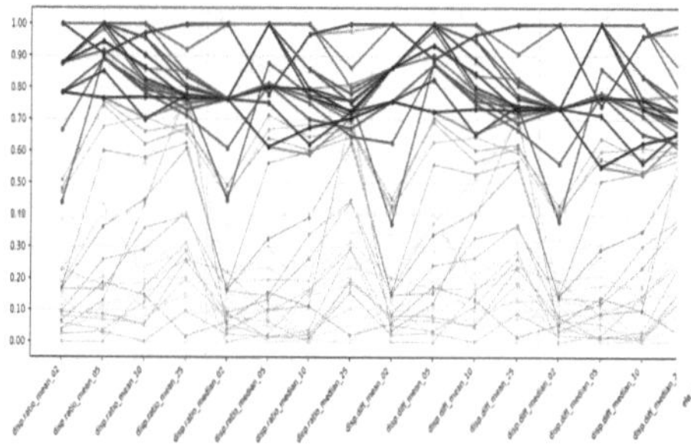

(a) dispersion features

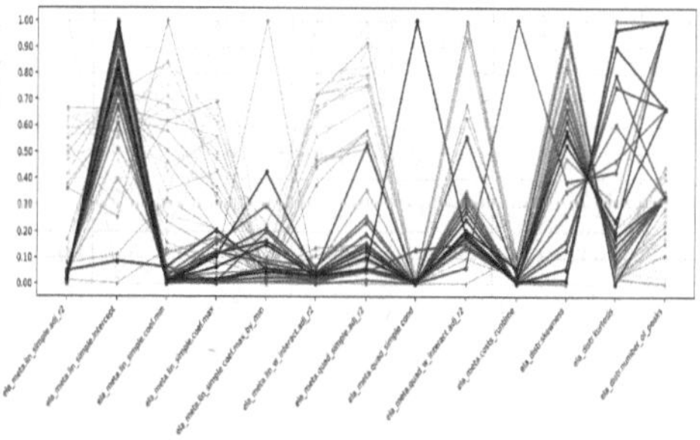

(b) ela features

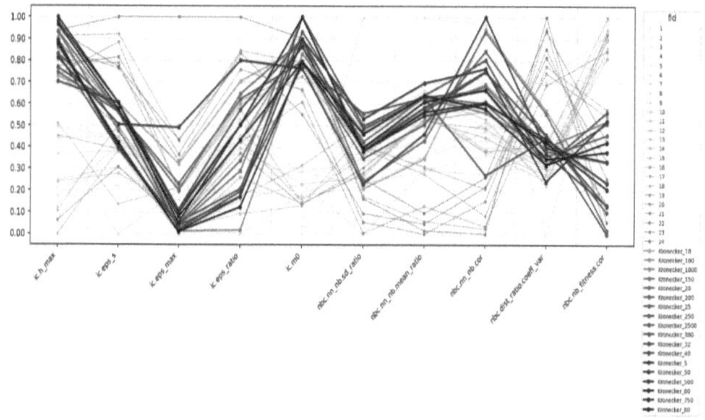

(c) information content, and nearest better clustering features

Fig. 6. Comparison of selected feature values of the 3D kronecker problem (blue lines) with those of the 24 BBOB functions (slim red lines). All values have been normalized to the interval [0, 1]. (Color figure online)

[26] with its default setting, which samples $50 \times D = 100$ points uniformly at 'LHS'. We evaluate 18 samples and compare the averages of the feature values of 15 evaluations with those of the 24 BBOB functions of the COCO framework

[16], computed using the exact same protocol. The results of the comparison are illustrated in Fig. 6. We see that the 3D Kronecker problem has high dispersion feature values compared to all the 24 BBOB functions (see Fig. 6a), additionally, we see higher values for the distribution kurtosis feature, and the number of peaks in the ELA features (Fig. 6b), which gives insights of the high multimodality landscape of the problem, and for the other feature sets displayed, we see that no BBOB function could be identified for a good fit in Fig. 6b and 6c. Additionally, we can also see from the figures that the landscape for the Kronecker problem is closely similar for the different instances of the same problem.

6 Discussion

Using optimization strategies based on the CMA-ES heuristic and automatic algorithm configuration, we have optimized the parameters of Kronecker sets for the generation of sets of low star discrepancy. Recent state-of-the-art approaches, such as the L2_Subset method [4], are particularly effective for constructing low-discrepancy point sets with a small number of points. Our results show that the CMA-ES heuristic is similarly competitive in the small to medium regime of point-set sizes n compared to L2_subset. In addition, automatic algorithm configuration using IRACE proves more effective for larger point sets, where systematic exploration of the parameter space yields superior performance. However, while we obtain good results in 3 dimensions, the Kronecker set seems to perform worse in 4 dimensions.

Figure 7 shows the distribution of (α, β) parameter pairs obtained by CMA-ES and IRACE. CMA-ES explores a wide range of configurations, with dense clusters around $\alpha \in [0.2, 0.4]$ and varying β, indicating convergence to multiple local optima and highlighting the multimodal nature of the problem. Frequently repeated points suggest the presence of several competitive parameter settings.

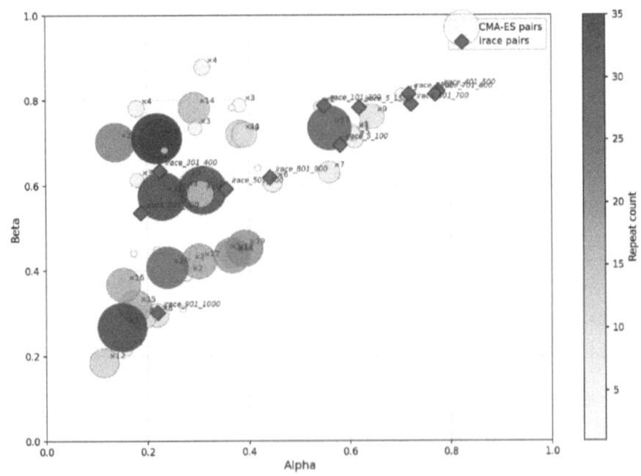

Fig. 7. Distribution of (α, β) parameter pairs.

In contrast, IRACE exhibits a more structured trajectory, progressively favoring larger α and β values as the problem size increases. While both methods overlap in some robust regions (e.g., $\alpha \approx 0.25$, $\beta \approx 0.6$), IRACE identifies additional high-value configurations that CMA-ES does not consistently reach.

Overall, this study highlights the complexity of the parameter optimization problem and underscores the value of combining different optimization paradigms.

7 Conclusion

We have used algorithm configuration techniques to find good parameters for Kronecker sets in 3 dimensions. The point sets generated are competitive with state-of-the-art methods for small set sizes and consistently outperform the previously best known constructions for point sets containing at least 300 points.

Our experiments reveal that there may be some space for further improvements when using Kronecker sets in dimension $D = 3$. In dimension $D = 4$ (and probably beyond), however, Kronecker sets do not seem to be able to outperform standard constructions like truncated Sobol' sequences.

Our results call for a better understanding of the parameter to discrepancy mapping for Kronecker sets in $D = 3$. We believe that further improvements might be possible. We also believe that a finer understanding of the trade-off between size-specific constructions and parameters that yield good sets across broader ranges of set sizes would be of great practical relevance, as in real-world situations, it may not be uncommon that an existing design of experiment or an initial sample can be extended by additional points.

Finally, our results also indicate a lack of high-performing constructions for dimensions $D \geq 4$. Here, the subset selection approaches proposed in [5,6] seem to be state of the art. However, with these being dependent on the sets that the subsets are taken from, we expect significant potential in the optimization of low-discrepancy point sets for these settings.

Last but not least, from an evolutionary computation perspective, we have demonstrated that the construction of low-discrepancy point sets offers challenging optimization problems of high practical relevance. We expect to see further contributions of our community to this important application.

Acknowledgments. This research was supported in part by the French PEPR integrated project HQI (ANR-22-PNCQ-0002), by the French National Research Agency (ANR-23-CE23-0035) and the German Research Foundation (DFG; LI 2801/7-1), through project OPT4DAC, and by the European Union (ERC, "dynaBBO", grant no. 101125586). Views and opinions expressed are, however, those of the author(s) only and do not necessarily reflect those of the European Union or the European Research Council Executive Agency. Neither the European Union nor the granting authority can be held responsible for them.

Availability of Code and Data. Our code and data are publicly available at https://gitlab.lip6.fr/durandm/low_discrepancy.

References

1. Battiti, R., Tecchiolli, G.: The reactive tabu search. ORSA J. Comput. **6**(2), 126–140 (1994)
2. Cauwet, M.L., et al.: Fully parallel hyperparameter search: Reshaped space-filling. In: International Conference on Machine Learning, pp. 1338–1348. PMLR (2020)
3. Clément, F., Doerr, C., Klamroth, K., Paquete, L.: Constructing optimal star discrepancy sets. Proc. Am. Math. Soc. Ser. B **12**(07), 78–90 (2025)
4. Clément, F., Doerr, C., Klamroth, K., Paquete, L.: Searching permutations for constructing uniformly distributed point sets. Proc. Natl. Acad. Sci. **122**(14), e2424464122 (2025)
5. Clément, F., Doerr, C., Paquete, L.: Star discrepancy subset selection: problem formulation and efficient approaches for low dimensions. J. Complex. **70**, 101645 (2022)
6. Clément, F., Doerr, C., Paquete, L.: Heuristic approaches to obtain low-discrepancy point sets via subset selection. J. Complex. **83**, 101852 (2024)
7. Clément, F., Huang, L., Lee, W., Smidt, C., Sodt, B., Zhang, X.: Low-discrepancy set post-processing via gradient descent. arXiv preprint arXiv:2511.10496 (2025)
8. Clément, F., Vermetten, D., De Nobel, J., Jesus, A.D., Paquete, L., Doerr, C.: Computing star discrepancies with numerical black-box optimization algorithms. In: Proceedings of the Genetic and Evolutionary Computation Conference, pp. 1330–1338 (2023)
9. Dick, J., Pillichshammer, F.: Digital Nets and Sequences: Discrepancy Theory and Quasi-Monte Carlo Integration, Cambridge University Press (2010)
10. Dobkin, D.P., Eppstein, D., Mitchell, D.P.: Computing the discrepancy with applications to supersampling patterns. ACM Trans. Graphics (TOG) **15**(4), 354–376 (1996)
11. Doerr, C., De Rainville, F.M.: Constructing low star discrepancy point sets with genetic algorithms. In: Proceedings of the 15th Annual Conference on Genetic and Evolutionary Computation, pp. 789–796 (2013)
12. Fox, B.L.: Algorithm 647: implementation and relative efficiency of quasirandom sequence generators. ACM Trans. Math. Software (TOMS) **12**(4), 362–376 (1986)
13. Galanti, S., Jung, A.: Low-discrepancy sequences: Monte Carlo simulation of option prices. J. Deriv. **5**(1), 63–83 (1997)
14. Halton, J.H.: On the efficiency of certain quasi-random sequences of points in evaluating multi-dimensional integrals. Numer. Math. **2**, 84–90 (1960)
15. Hammersley, J.: Monte Carlo methods for solving multivariable problems. Ann. N. Y. Acad. Sci. 86, (1960)
16. Hansen, N., Auger, A., Ros, R., Mersmann, O., Tušar, T., Brockhoff, D.: Coco: a platform for comparing continuous optimizers in a black-box setting. Optim. Meth. Software **36**(1), 114–144 (2021)
17. Hansen, N., Müller, S.D., Koumoutsakos, P.: Reducing the time complexity of the derandomized evolution strategy with covariance matrix adaptation (CMA-ES). Evol. Comput. **11**(1), 1–18 (2003)
18. Hlawka, E.: Funktionen von beschränkter variatiou in der theorie der gleichverteilung. Annali di Matematica **54**(1), 325–333 (1961)
19. Koksma, J.F.: A general theorem from the theory of uniform distribution modulo 1. Mathematica B (Zutphen) **11**, 7–11 (1942)
20. López-Ibáñez, M., Dubois-Lacoste, J., Cáceres, L.P., Birattari, M., Stützle, T.: The irace package: Iterated racing for automatic algorithm configuration. Oper. Res. Perspect. **3**, 43–58 (2016)

21. Lourenço, H.R., Martin, O.C., Stützle, T.: Iterated local search: framework and applications. In: Gendreau, M., Potvin, J.-Y. (eds.) Handbook of Metaheuristics. ISORMS, vol. 272, pp. 129–168. Springer, Cham (2019). https://doi.org/10.1007/978-3-319-91086-4_5
22. Mersmann, O., Bischl, B., Trautmann, H., Preuss, M., Weihs, C., Rudolph, G.: Exploratory landscape analysis. In: Proceedings of the 13th annual conference on Genetic and evolutionary computation, pp. 829–836 (2011)
23. Niederreiter, H.: Methods for estimating discrepancy. In: Applications of Number Theory to Numerical Analysis, pp. 203–236. Elsevier (1972)
24. Patel, M., Education, P.: Optimizing kronecker sequences for multidimensional sampling. J. Comput. Graphics Techniques **11**(1) (2022)
25. Paulin, L., Bonneel, N., Coeurjolly, D., Iehl, J.C., Keller, A., Ostromoukhov, V.: Matbuilder: mastering sampling uniformity over projections. ACM Trans. Graphics (TOG) **41**(4), 1–13 (2022)
26. Prager, R.P., Trautmann, H.: PFLACCO: feature-based landscape analysis of continuous and constrained optimization problems in python. Evol. Comput. 1–25 (2023)
27. Rusch, T.K., Kirk, N., Bronstein, M.M., Lemieux, C., Rus, D.: Message-passing Monte Carlo: generating low-discrepancy point sets via graph neural networks. Proc. Natl. Acad. Sci. **121**(40), e2409913121 (2024)
28. Sadikov, A.: Generation of low-discrepancy point sets and sequences via LLM evolutionary search (2026)
29. Santner, T.J., Williams, B.J., Notz, W.I., Williams, B.J.: The Design and Analysis of Computer Experiments, vol. 1. Springer, Heidelberg (2003). https://doi.org/10.1007/978-1-4757-3799-8
30. Schede, E., et al.: A survey of methods for automated algorithm configuration. J. Artif. Intell. Res. **75**, 425–487 (2022)
31. Sobol, I.M.: On the distribution of points in a cube and the approximate evaluation of integrals. USSR Comput. Math. Math. Phys. **7**(4), 86–112 (1967)
32. Stützle, T., López-Ibáñez, M.: Automated design of metaheuristic algorithms. In: Handbook of Metaheuristics, pp. 541–579 (2019)

Reshaping the Fitness Landscape of Classification Rule Mining for Clinical Natural Language Processing

Mathilde Lepers(✉), Clarisse Dhaenens, Olivier Caron, and Julie Jacques

Univ. Lille, CNRS, Centrale Lille, UMR 9189 CRIStAL, 59000 Lille, France
mathilde.lepers@univ-lille.fr

Abstract. Clinical texts are valuable sources of patient information, but they are hard to process automatically due to synonyms, abbreviations, and terminological variability. Classification rule mining can extract readable "IF-THEN" rules that clinicians can understand and verify. Standard evolutionary rule miners rely on exact word matching, which fails when the same concept is expressed with different words (e.g., "renal failure" vs. "kidney failure"). This fragments the fitness landscape into vast neutral plateaus where standard operators cannot progress.

We propose a multi-objective evolutionary framework that integrates word embeddings into binary classification rule mining at two levels: soft matching, which allows semantically similar terms to contribute to rule coverage during fitness evaluation, and a semantic mutation operator, which replaces terms with embedding-space neighbors during search. Together, these components restore a continuous optimization gradient in place of the flat landscape caused by terminological variability.

Experiments on MIMIC-IV-Note across five diagnostic targets show that our framework more than doubles the mean hypervolume (+116%) and substantially improves the F1-score (+80%) compared to exact-matching baseline NSGA-II and bitflip mutation, while maintaining fully interpretable rule structures. Fitness landscape analysis confirms that the combined approach reduces neutrality and increases information content, explaining its superior search dynamics.

Keywords: Multi-objective optimization · Classification rule mining · Word embeddings · Soft matching · Semantic mutation · Clinical NLP

1 Introduction

Medical letters contain valuable patient information, but their unstructured format makes automatic processing difficult. They often contain errors, acronyms, abbreviations, and high variability in medical terminology [10].

Moreover, in medical applications, algorithmic transparency is not optional: clinicians must be able to understand, challenge, and trust the reasoning behind

G. Iacca et al. (Eds.): PPSN 2026, LNCS 16985, pp. 566–581, 2027.
https://doi.org/10.1007/978-3-032-36214-8_35

any automated decision support system. This requirement is now formally recognized at the regulatory level through the European Artificial Intelligence Act [11], which classifies medical AI as high-risk and mandates explainability.

Classification rule mining (CRM) directly addresses this need by discovering explicit logical rules of the form *"IF symptom A AND symptom B THEN disease C"*. Metaheuristic algorithms, particularly multi-objective evolutionary approaches, have proven effective for CRM on structured (tabular) clinical data, balancing predictive performance and rule simplicity [26]. However, these methods have not been designed for unstructured text and cannot be applied directly.

The core difficulty lies in terminological variability. Clinical texts are rich in synonyms, abbreviations, and paraphrases. For instance "renal failure", "kidney failure", and "CKD" refer to the same concept but appear as entirely distinct tokens. This fragments semantically identical information, resulting in high landscape neutrality and low Information Content (IC). Consequently, exact-matching rule miners are trapped in vast neutral plateaus or face sudden fitness drops when random bit-flip mutations break fragile medical patterns. Without a semantic guide, these blind jumps effectively block any steady improvement, forcing the discovery of sparse, over-specific rules with poor generalization.

Word embeddings [17] offer a principled solution to this variability. By representing each term as a vector in a continuous semantic space, they capture the fact that semantically related terms appear in similar contexts. In clinical NLP, embeddings have shown strong performance and have been shown to capture meaningful medical analogies on biomedical corpora [5,10]. However, they are almost exclusively used within black-box predictive models, sacrificing interpretability that medical applications require. To our knowledge, no prior work has integrated word embeddings into a metaheuristic classification rule mining framework.

In this work, we propose to bridge this gap by combining the interpretability of evolutionary CRM with the semantic richness of word embeddings. We integrate embedding-based similarity into multi-objective evolutionary algorithm (MOEA) at two levels:

- soft matching [21] integrated evaluation, which allows semantically related terms to contribute to rule coverage during fitness evaluation
- semantic mutation operator, which replaces terms with embedding-space neighbors during search, enabling informed local exploration and reducing synonym-induced plateaus

This framework retains the explicit structure of logical rules while overcoming the vocabulary fragmentation that limits classical rule miners on clinical texts.

This work is motivated by a clinical application in partnership with the Lille University Hospital, targeting systemic sclerosis, a rare disease with strong clinical heterogeneity [28]. Although the ultimate goal is to apply this framework to the FHU PRECISE cohort [12], privacy constraints currently prevent its use in this study. We evaluate our method on five distinct diseases from *MIMIC-IV-Note* [14], a large deidentified collection of discharge summaries. This multi-disease evaluation ensures robustness before deployment on the hospital cohort.

The paper is structured as follows. Section 2 reviews related work. Section 3 and 4 formalize the problem and detail our framework. Section 5 presents experimental results and fitness landscape analysis on *MIMIC-IV-Note*. Finally, Sect. 6 concludes and discusses perspectives.

2 Related Work

The proposed framework sits at the intersection of four research areas that have mostly been treated independently: semantic representations for clinical NLP, interpretable rule mining, soft matching and MOEA.

2.1 Semantic Representations in Clinical NLP

Word embeddings models such as Word2Vec [17] and GloVe [23] represent lexical items in continuous vector spaces where geometric proximity reflects semantic similarity, most commonly measured using cosine similarity [15]. Let $\mathbf{u}$ and $\mathbf{v}$ denote the vector representations of two words:

$$\cos(\mathbf{u}, \mathbf{v}) = \frac{\mathbf{u} \cdot \mathbf{v}}{\|\mathbf{u}\| \|\mathbf{v}\|} \tag{1}$$

Embedding representations capture linguistic variability: synonyms, paraphrases, and abbreviations that refer to the same concept tend to have close vector representations. These methods have demonstrated strong performance in clinical NLP tasks [10] where they capture meaningful medical analogies. However, they are typically incorporated into black-box predictive models, limiting their applicability in medical context where interpretability is required.

2.2 Classification Rule Mining and Interpretability

Classification rule mining (CRM) aims to discover explicit logical rules of the form "IF condition THEN class" that can be directly read and validated by domain experts. It is a specific case of association rule mining, where the prediction is the class to predict. Association rule mining was formalized by the Apriori algorithm [1], which introduced support and confidence as core measures to filter relevant patterns from large itemsets. In medical applications, rule-based systems are particularly valuable because they provide auditable decisions [11]. Neuro-fuzzy approaches such as NEFCLASS [22] further showed that compact, readable rules can be learned directly from medical data.

CRM can be formulated as a multi-objective optimization problem, since rule quality cannot be evaluated by a single criterion: a rule covering many positive documents tends to also cover negative ones, while a highly specific rule may miss relevant cases. Balancing these competing objectives requires finding a set of trade-off solutions rather than a single optimum [13]. Metaheuristic algorithms, particularly multi-objective evolutionary approaches, have proven effective in this setting. MOEA are widely used to approximate Pareto fronts in discrete rule spaces, producing a diverse set of trade-off solutions rather than a single rule. More recent work such as EVOTER [26] has further demonstrated the ability of evolutionary CRM to generate high-quality interpretable classifiers on structured tabular data. However, these approaches rely on exact matching between items and have been designed exclusively for structured data. Applying them to unstructured clinical text raises a fundamental challenge: terminological variability.

2.3 Soft Rule Evaluation

Classical rule miners rely on exact matching, treating semantically identical concepts such as *"renal failure"* and *"kidney failure"* as distinct tokens, which artificially lowers the support of meaningful patterns. Nahm and Mooney [21] addressed this with Soft-Apriori, replacing binary item presence with a graded similarity metric. However, their approach targets association rule mining on general text and does not extend to multi-objective evolutionary frameworks or embedding-based similarity.

2.4 Multi-objective Evolutionary Rule Generation

Metaheuristics have been widely applied to classification rule mining [9], but a fundamental limitation of standard evolutionary operators such as bit-flip mutation or uniform crossover is that they operate strictly on the syntax of the rules rather than their semantics. As highlighted by Moraglio et al. in the context of Geometric Semantic Genetic Programming (GSGP) [19], purely syntactic search results in a rugged fitness landscape where small representational changes lead to unpredictable jumps in performance. By defining operators that move directly within a semantic space, they showed that the fitness landscape can be transformed into a more easily optimizable structure. While GSGP was designed for program trees, its core principle, using semantic proximity to guide variation, is directly relevant for combinatorial rule mining over high-dimensional, synonymous vocabularies such as those found in clinical texts.

2.5 Positioning of the Present Work

Existing approaches typically treat semantic embeddings, interpretable rule systems, soft matching, and evolutionary optimization as separate concerns. In clinical NLP, embeddings are generally confined to black-box models, sacrificing the transparency required for medical applications. Evolutionary rule miners often rely on purely syntactic operators that remain "blind" to semantic relationships, a critical limitation in medical vocabularies where synonymy is pervasive. Despite a decade of rule mining applications in health informatics [2], these approaches have been applied almost exclusively to structured data, leaving unstructured clinical text largely unexplored.

This disconnect has a direct impact on search dynamics: synonyms create neutral plateaus in the fitness landscape [29,31], where standard bit-flip mutations fail to escape. The lack of semantic awareness in these operators often leads to sudden fitness drops, as random mutations break the fragile logical structure of exact-matching rules. As our Fitness Landscape Analysis (FLA) (Sect. 5.4) demonstrates, this results in a "flat desert" topology where the statistical signal is too diluted for traditional greedy or heuristic learners (e.g., RIPPER [6]). Moreover, as single-objective learners, such methods cannot expose the trade-off front required for clinician-driven decision making.

Embedding-based cosine similarity provides a natural solution to this issue: by quantifying semantic relatedness automatically, without manual annotation, it enables soft matching that bridges synonymous terms such as *"pneumonia"* and *"pulmonary infection"*, restoring a continuous optimization gradient in place of the flat desert. This makes it a principled and scalable foundation for extending the original framework, particularly well-suited to the high terminological variability of clinical texts.

While state-of-the-art contextual embeddings can navigate this terminological variability, they lack the auditable decision structure necessary for clinical support. Our proposed framework bridges this gap. By grounding both fitness evaluation and variation operators in embedding geometry, we "activate" the search space and restore a continuous optimization gradient. This approach enables the discovery of high-quality rules while maintaining the explicit "IF-THEN" structure essential for clinical validation. To our knowledge, this is the first framework to integrate semantic similarity into both the evaluation and the variation phases of a multi-objective evolutionary classification rule learner.

3 Problem Formalization

We now formalize the problem and detail the two core components of the proposed framework: soft matching evaluation and semantic mutation operator.

3.1 Definitions

We consider a corpus of N clinical documents $\mathcal{D} = \{d_1, \ldots, d_N\}$, where each document corresponds to a patient medical letter. Each document d is associated with a set of diagnostic labels extracted from structured fields of the electronic health record. We focus on binary classification: given a target diagnosis (e.g., pneumonia), each document receives a label $y_d \in \{0, 1\}$, where $y_d = 1$ indicates that the target diagnosis is mentioned in the document, and $y_d = 0$ otherwise. A document with label $y_d = 1$ is called a *positive document*, and a document with label $y_d = 0$ is called a *negative document*.

Let $V = \{v_1, \ldots, v_{|V|}\}$ denotes the vocabulary (set of terms) extracted from $\mathcal{D}$. Each document d is represented as a binary vector in $\{0,1\}^{|V|}$, giving rise to the document-term matrix $X \in \{0,1\}^{N \times |V|}$, where $X_{d,i} = 1$ if term v_i appears in document d, and $X_{d,i} = 0$ otherwise.

Word embeddings [17] are used to capture semantic relationships between vocabulary terms. Each term $v_i \in V$ is associated with a dense vector $\boldsymbol{w}_i \in \mathbb{R}^{100}$, learned from a large text corpus such that terms appearing in similar contexts have geometrically close representations. These vectors are used at two levels of the framework: in the fitness evaluation via soft matching (Sect. 3.4), and in the variation operators via semantic mutation (Sect. 4.4).

3.2 Solution Representation

A candidate rule is represented as a binary vector $r \in \{0,1\}^{|V|}$, where $r_i = 1$ indicates that term v_i is selected. Semantically, a rule corresponds to a conjunction of terms: a document is said to be covered by the rule if all selected terms appear in it. This conjunctive structure ensures that rules remain readable. To improve interpretability and reduce noise, a set of banned terms (negations, stop words, overly common medical terms) is defined a priori. Rules are further restricted to a maximum of 9 selected terms to ensure readability and limit combinatorial complexity, in line with known cognitive limits on human information processing (7 ± 2 elements) [18,25].

The search space is $\{0,1\}^{|V|}$, which grows exponentially with the vocabulary size. Even for moderate vocabularies, exhaustive exploration is infeasible, which motivates the use of metaheuristic search.

3.3 Standard Fitness Evaluation

Classical measures for evaluating classification models compare the predicted coverage of a rule with the truth labels, and are typically derived from the confusion matrix, which records the number of true positives (TP), true negatives (TN), false positives (FP), and false negatives (FN). Following [24], we propose two complementary measures to capture different types of errors:

- **F1-score (2 * precision * recall/(precision + recall))**: this is a combined measure that is the harmonic mean of precision (TP/(TP+FP)) and recall (TP/(TP+FN)). It penalizes rules that miss positive documents (low recall) as well as rules that over-generalize and cover too many negative documents (low precision). It is robust to class imbalance and does not favor trivial strategies that systematically predict the majority class.
- **Specificity (TN/(TN + FP))**: this is the proportion of negative documents correctly rejected by the rule. It ensures that the rule remains clinically precise and does not flag too many negative documents.

This gives a bi-objective setup, maximizing both measures at once and producing a Pareto front of rules r showing different trade-offs. Optimizing both F1-score and specificity simultaneously prevents the two extreme solutions: covering all documents (high F1-score, zero specificity) and covering no document (perfect specificity, zero F1-score).

3.4 Fitness Evaluation in Soft Matching Context

In the standard setting, a rule covers a document only if all selected terms appear exactly in that document. Missing even one term means the document is not covered. This strictness can be problematic: two documents may be semantically equivalent but use different words, leading to sparse coverage and fragmented rules. This is denoted by *hard matching.*

To handle vocabulary variations, we use the notion of *soft matching* and our contribution extends it using embeddings and cosine similarity. We say a term x is similar to a term y if their cosine similarity exceeds a threshold $\tau \in [0, 1]$. Formally we define a function called $\sim$:

$$x \sim y \iff \cos(w_x, w_y) \geq \tau$$

Using this, a rule covers a document if, for every selected term x, the document contains either x itself or at least one term y such that $x \sim y$. This keeps the "AND" logic of the original rule while allowing flexibility at the term level. The hard matching case is simply the special case where only exact matches are allowed ($x \sim y \iff x = y$) (Fig. 1).

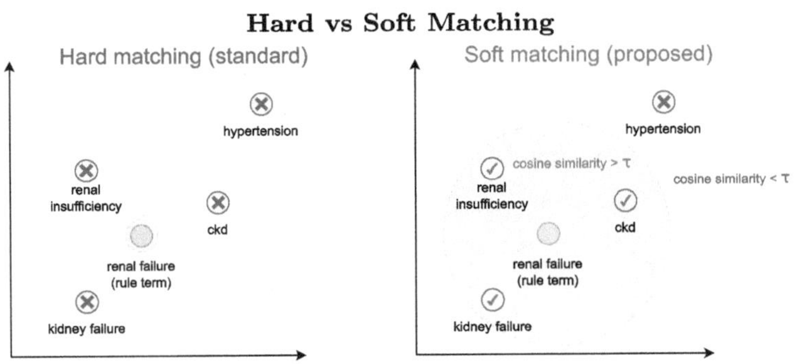

Fig. 1. Illustration of the proposed soft matching mechanism compared to standard hard matching. Semantically related terms activate the rule when their cosine similarity with the rule term is above a threshold.

4 Resolution with MOEA

In order to deal with this new introduced modeling of the problem, we propose to integrate word embedding information into a multi-objective evolutionary algorithm (MOEA) at two levels: fitness evaluation and mutation operators. Starting from a corpus of clinical documents, word embeddings are first trained to capture semantic relationships between vocabulary terms. The evolutionary search then proceeds iteratively: candidate rules are initialized from positive documents, evaluated using F1-score and specificity with optional soft matching, and improved through bit-flip mutation or semantically-informed variation operators. The following sections detail each component of this pipeline.

4.1 Training of the Word Embedding Model

To ensure high-quality semantic representations, the raw clinical corpus underwent a standard preprocessing pipeline before training [20]. First, all texts were converted to lowercase and special characters were removed to reduce noise and ensure terminological consistency. Non-informative words, such as "the" or "and" (*stop words*), were filtered out to focus the model on clinical concepts.

Finally, we applied lemmatization, a linguistic process that reduces different inflected forms of a word to their common base or dictionary form. For example, medical terms like *"treating"*, *"treated"*, and *"treats"* are all mapped to the single lemma *"treat"*. This step is crucial in clinical NLP as it helps to group semantically identical terms that only differ in their grammatical conjugation [20]. Following this cleaning phase, word embeddings were trained on the processed corpus. Each term $v_i \in V$ is mapped to a dense vector $\boldsymbol{w}_i \in \mathbb{R}^{100}$ using Word2Vec [17].

We chose to use Word2Vec because contextual models such as BERT [8] are unsuitable here for two reasons: their subword tokenization fragments medical terms into unreadable pieces, and their context-dependent vectors prevent the construction of a stable similarity matrix required by our threshold-based soft matching.

4.2 Population Initialization

Candidate rules are initialized from positive documents to improve convergence and produce plausible initial solutions. A positive document is randomly selected from the training set, and a random number between 1 and 3 terms present in that document (excluding banned words) are included in the rule vector. All other terms are set to zero. This strategy ensures that initial rules are clinically plausible while preserving population diversity. Keeping the initial rules small also leaves room for the algorithm to expand them during evolution, in line with cognitive constraints such as the Miller limit (7 ± 2 elements) [18] and the imposed maximum of 9 terms.

4.3 Soft Matching Evaluation

In the standard evaluation, each rule is assessed using hard matching. To overcome the limitations of hard matching in clinical texts, we replace this evaluation with the soft matching evaluation, where semantically similar terms are allowed to contribute to rule coverage. This reshapes the fitness landscape by reducing synonym-induced plateaus: two rules that differ only by similar semantic terms now receive similar fitness values, allowing the search to escape neutral regions that trap standard evolutionary operators.

4.4 Semantic Mutation Operator

Standard bit-flip mutation activates or deactivates terms uniformly at random, with no knowledge of semantic relationships between terms. We propose a semantic mutation operator that replaces an active term with one of its semantically similar terms in the embedding space. This method is described in Algorithm 1

Algorithm 1. Semantic Mutation Operator

```
1: Input: rule r, threshold τ, banned word set B
2: Output: mutated rule r'
3: r' ← r
4: Select a random active term v_i such that r_i = 1
5: N(v_i) ← {v_j ∈ V | cos(v_i, v_j) ≥ τ ∧ v_j ∉ B} ∧ i ≠ j ∧ r_j ≠ 1
6: if N(v_i) ≠ ∅ then
7:     Select a random v_j from N(v_i)
8:     r'_i ← 0                                   ▷ Remove v_i
9:     r'_j ← 1                                   ▷ Insert v_j
10: else
11:     Apply standard bit-flip mutation to r'
12: end if
13: return r'
```

If no similar term is available for the selected term, the operator falls back to standard bit-flip mutation. This semantic mutation ensures that variations correspond to meaningful semantic changes, improving search efficiency.

5 Experiments

This section evaluates the proposed clinical rule mining approach. We first describe the experimental setup and data sampling, followed by a quantitative and qualitative analysis of the results and a Fitness Landscape Analysis (FLA).

5.1 Experimental Setup

While black-box ensemble methods (e.g., Random Forests [3], XGBoost [4]) or deep learning models often achieve high predictive accuracy, they are intentionally excluded from this study. In a clinical context, interpretability is more important than predictive performance, and post-hoc explanations cannot fully replace it [25]. Our objective is not only to predict a label, but also to discover explicit logical rules that characterize a disease. Such rules allow clinicians to audit, challenge, and trust the decision process. Consequently, we focus exclusively on rule-based learners that provide a direct mapping between clinical concepts and diagnostic outcomes.

We perform systematic experiments to evaluate the proposed approach across multiple target classes, random seeds, and algorithmic configurations. Specifically:

- **Target classes:** 5 primary diagnoses are considered (e.g., diabetes, hepatitis, pneumonia, hypertension, sepsis).

- **Random seeds:** 10 seeds are used to ensure statistical robustness.
- **Data splitting:** To ensure robust evaluation, the dataset is partitioned using stratified k-fold cross-validation ($k = 3$), where stratification preserves the class imbalance ratio across each fold. For each target class, the NSGA-II algorithm is trained on the training fold. At the end of the optimization process, the obtained solutions are evaluated on the test fold.
- **Algorithm selection:** NSGA-II [7] is chosen as the multi-objective evolutionary algorithm due to its effectiveness in exploring Pareto fronts in combinatorial spaces. Selection uses binary tournament, and recombination uses SPX crossover. Population size and offspring size are both set to 50, with a total of 100,000 evaluations.
- **Mutation operator:** We use either a semantic mutation, guided by a similarity matrix, or a standard bit-flip mutation with probability $1/|V|$ when semantic mutation is disabled.
- **Modes:** four algorithmic modes are evaluated: (1) baseline, (2) soft matching ($\tau = 0.5$), (3) semantic mutation, and (4) soft matching + semantic mutation (combined, $\tau = 0.5$).

For each combination of target class, seed, fold and mode, the NSGA-II algorithm is executed independently. Intermediate non-dominated fronts are recorded at fixed evaluation intervals to monitor convergence.

All experiments are implemented in Python using `jMetalPy 1.7.0`. Word embeddings are obtained with Word2Vec using the `Gensim 4.4.0` library [32]. Sparse matrices (`csr_matrix`) are employed to efficiently handle high-dimensional document-term matrices. Rules are intrinsically sparse (few active terms) but stored as dense binary vectors for compatibility with evolutionary operators.

5.2 Data Sampling

We evaluate our method on MIMIC-IV-Note [14], a large collection of deidentified clinical notes derived from MIMIC-IV, which includes 331,794 discharge summaries from 145,915 patients. A discharge summary is a document written when a patient leaves the hospital, summarizing diagnoses, treatments, test results, and follow-up recommendations. For preprocessing purpose, we randomly sample 5,000 discharge summaries, from which preprocessing yields a vocabulary of 14,513 unique terms.

The dataset exhibits strong class imbalance: among the 5,000 sampled discharge summaries, positive instances represent a small fraction of the corpus for each target class (see Table 1). To reduce this imbalance, we apply random undersampling on negative documents: all positive instances are retained, while a fixed subset of 1000 negative instances is randomly sampled independently for each target class and each seed.

Hypertension is among the most frequent diagnoses in the corpus (imbalance ratio of 17:1, meaning 17 negative documents for every positive one), whereas hepatitis and sepsis correspond to rarer conditions (see Table 1). This setup

Table 1. Target classes and their occurrence count in the dataset

Disease	Number of occurrences	Imbalance ratio
Hepatitis	39	127
Sepsis	51	97
Diabetes	177	27
Pneumonia	208	23
Hypertension	280	17

allows us to evaluate the robustness of the proposed approach across both relatively common and rare diseases.

5.3 Results

Table 2 reports the hypervolume and F1-score test (from cross-validation) for each mode, averaged over 10 runs × 3 folds for a total of 30 runs per dataset and mode. As specificity reaches near-perfect values across all configurations and is not the primary differentiator, we therefore focus on hypervolume and F1-score test, which are more sensitive to the effects of the proposed operators.

The hypervolume captures the overall quality and diversity of the Pareto front, providing a global measure of multi-objective optimization performance, while the F1-score specifically monitors the convergence on the positive class, which is challenging in our setting. The hypervolume was calculated using (0, 0) as the reference point, representing the theoretical worst possible performance for both F1-score and specificity.

Soft matching alone provides a substantial improvement, while semantic mutation alone has no effect. Their combination, however, produces the best results, demonstrating that the synergy between the two components is essential to reshape the fitness landscape and enhance search dynamics.

Statistical significance of the differences between methods was assessed using the Wilcoxon signed-rank test for each disease. Prior to pairwise comparisons, a Friedman test on the ranks confirmed that there were overall significant differences among the methods. Results show that semantic mutation alone has no significant effect ($p > 0.05$ for all diseases), while soft matching alone consistently improves performance. The combination of both components achieves the best results, significantly outperforming the baseline and individual methods in terms of both hypervolume and F1-score test ($p < 0.05$ for all pairwise comparisons where applicable).

The Pareto fronts shown in Figs. 2a–2b represent the union of all non-dominated fronts obtained across runs and folds for each mode; solutions from different runs may therefore dominate each other, explaining the presence of apparently dominated points in the figures. As a reminder, hypertension is the less imbalanced class in the dataset (1/17), whereas sepsis is the second least represented (1/97), which explains the differences in Pareto fronts.

Table 2. Comparison of methods using hypervolume and F1-score test (mean ± std.). Bold values indicate statistical significance at the $\alpha = 0.05$ level (Wilcoxon signed-rank test vs. baseline, Holm-Bonferroni correction on Friedman ranks).

Disease	Metric	Baseline	Semantic	Soft	Combined
Hepatitis	Hypervolume	0.27 ± 0.09	0.35 ± 0.13	0.39 ± 0.10	0.63 ± 0.10
	F1-score	0.13 ± 0.14	0.17 ± 0.16	0.11 ± 0.12	0.20 ± 0.13
Sepsis	Hypervolume	0.24 ± 0.08	0.25 ± 0.07	0.39 ± 0.09	0.55 ± 0.07
	F1-score	0.12 ± 0.11	0.08 ± 0.09	0.16 ± 0.11	0.24 ± 0.11
Diabetes	Hypervolume	0.30 ± 0.14	0.30 ± 0.11	0.50 ± 0.06	0.62 ± 0.06
	F1-score	0.28 ± 0.16	0.25 ± 0.15	0.46 ± 0.09	0.50 ± 0.10
Pneumonia	Hypervolume	0.33 ± 0.12	0.36 ± 0.14	0.59 ± 0.07	0.76 ± 0.03
	F1-score	0.32 ± 0.16	0.33 ± 0.16	0.53 ± 0.08	0.67 ± 0.05
Hypertension	Hypervolume	0.29 ± 0.09	0.27 ± 0.07	0.46 ± 0.03	0.53 ± 0.02
	F1-score	0.28 ± 0.11	0.23 ± 0.09	0.42 ± 0.06	0.42 ± 0.05
Rank	Hypervolume	3.45	3.35	**2.19**	**1.01**
	F1-score	3.07	3.14	**2.17**	**1.62**

The baseline method (red dots) struggles to escape local optima, resulting in clustered and sub-optimal solutions. In contrast, the proposed method (dark purple dots) consistently pushes the Pareto front towards the top-right corner (maximizing both F1-score and Specificity). Furthermore, our approach generates a much wider and denser spread of non-dominated solutions. This diversity is crucial in clinical settings, as it provides medical practitioners with a rich set of interpretable and complementary rules.

In summary, the synergy between soft matching and semantic mutation effectively overcomes the limitations of exact-matching rule mining. The combination of both components is necessary to achieve the best predictive quality and diversity of interpretable rules.

To further illustrate the qualitative impact of our approach, we present a representative rule extracted for the pneumonia diagnostic target:

bacteroides AND bal AND cardioembolic AND cefuroxime AND daptomycin AND hcap AND imipenem AND lobar AND pneumonitis

On the independent test set, this rule yields an F1-score of 0.48 and a specificity of 0.97. It combines:

- **Bacterial agents**: bacteroides
- **Procedures**: bal (bronchoalveolar lavage)
- **Treatments**: cefuroxime, daptomycin, imipenem
- **Pneumonia subtypes**: hcap, lobar, pneumonitis

This example illustrates that the performance gains reported in Table 2 reflect a qualitative shift: our approach discovers clinically coherent rules grounded in

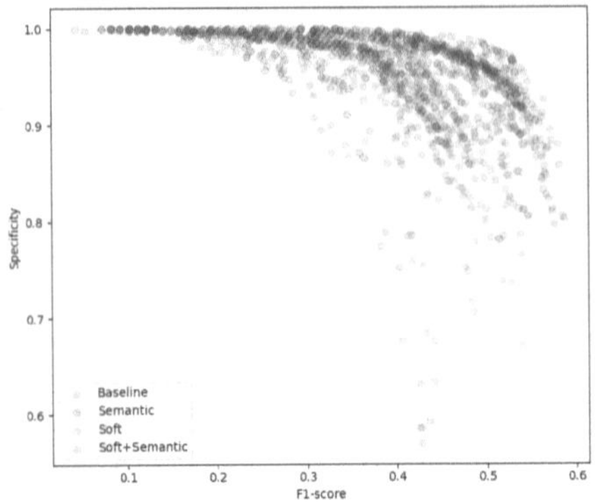

(a) Hypertension

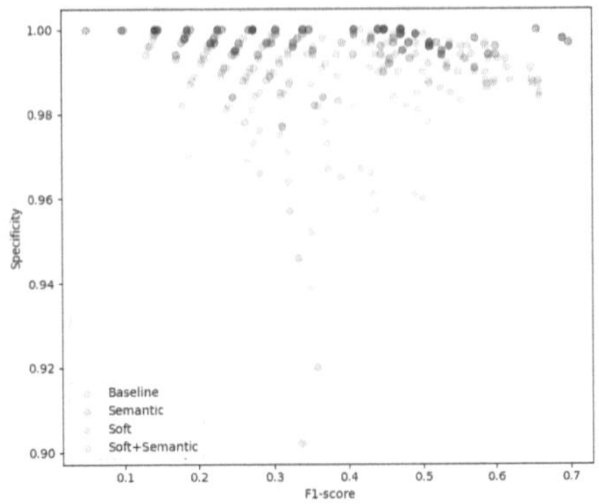

(b) Sepsis

Fig. 2. Pareto fronts for hypertension and sepsis.

the semantic structure of medical vocabulary. The clinician at Lille University Hospital (CHU Lille) further noted that the terms captured by these rules predominantly refer to severe forms of pneumonia, which is consistent with the nature of our dataset, collected in an emergency department setting where critical presentations are overrepresented.

5.4 Fitness Landscape Analysis

To characterize the difficulty of the rule mining task, we employ *Fitness Landscape Analysis* (FLA). According to a recent survey by Zou et al. [31], FLA is an essential tool to understand the relationship between problem-specific features (such as terminological variability) and the behavior of evolutionary algorithms. Specifically, we measure the *neutrality* to detect the presence of plateaus and the *Information Content* (IC) [30] to quantify the ruggedness of the landscape, following the taxonomy established in [31]. IC estimates the entropy of a fitness sequence $(f_1, \ldots, f_n)$ generated during a random walk. While a near-zero IC characterizes a "flat" uninformative landscape, a higher IC indicates a navigable gradient. We use this metric to demonstrate how our proposed operators reshape the search space, restoring the evolutionary signal lost to terminological variability. We analyze the search space using a 1,000-step random walk starting from a valid initial rule.

As shown in Table 3, the results reveal a structural transformation of the optimization problem.

For all diseases, the baseline (exact matching + bit-flip) exhibits an extreme neutrality rate ($n \approx 1.00$) and near-zero Information Content ($IC \approx 0.00$). This describes a landscape where the fitness signal is virtually absent. In clinical NLP, synonyms and terminological variations fragment semantically identical concepts. Consequently, a syntactic change (bit-flip) often results in a rule with zero document coverage, trapping the algorithm in vast, sterile plateaus.

Table 3. Fitness Landscape Analysis (mean ± std over 10 runs). Bold values indicate the best performance for each metric.

Target	Metric	Baseline	Soft	Semantic	Combined
Hepatitis	Neutrality	0.99 ± 0.00	0.99 ± 0.00	0.94 ± 0.07	$\mathbf{0.02 \pm 0.03}$
	IC	0.00 ± 0.00	0.00 ± 0.00	0.18 ± 0.21	$\mathbf{0.84 \pm 0.09}$
Sepsis	Neutrality	0.99 ± 0.00	0.99 ± 0.00	0.88 ± 0.20	$\mathbf{0.06 \pm 0.04}$
	IC	0.00 ± 0.00	0.00 ± 0.00	0.26 ± 0.37	$\mathbf{0.94 \pm 0.11}$
Diabetes	Neutrality	0.99 ± 0.00	0.99 ± 0.00	0.89 ± 0.18	$\mathbf{0.04 \pm 0.04}$
	IC	0.00 ± 0.00	0.00 ± 0.00	0.26 ± 0.38	$\mathbf{0.89 \pm 0.10}$
Pneumonia	Neutrality	0.99 ± 0.00	0.99 ± 0.00	0.80 ± 0.26	$\mathbf{0.04 \pm 0.10}$
	IC	0.00 ± 0.00	0.00 ± 0.00	0.41 ± 0.40	$\mathbf{0.84 \pm 0.13}$
Hypertension	Neutrality	0.99 ± 0.00	0.99 ± 0.00	0.75 ± 0.31	$\mathbf{0.02 \pm 0.03}$
	IC	0.00 ± 0.00	0.00 ± 0.00	0.41 ± 0.47	$\mathbf{0.84 \pm 0.09}$

In contrast, our framework dramatically reshapes this landscape. Neutrality drops significantly (from ≈ 1.00 to < 0.10), while the IC rises above 0.80. This indicates that our proposed operators successfully "activate" the landscape. By allowing semantically related terms to contribute to rule coverage and guiding mutations through the embedding space, we restore a continuous gradient. This transformation turns a combinatorial search for a rare solution search into a navigable optimization task, explaining superior quality of the extracted rules.

6 Conclusion and Future Work

This work presents a multi-objective evolutionary framework for classification rule mining in clinical texts, integrating soft matching and semantic mutation into NSGA-II. By leveraging word embeddings, the framework reshapes the fitness landscape, reducing neutral plateaus caused by terminological variability and restoring a continuous optimization gradient. Experiments on five diagnostic targets demonstrate significant improvements in hypervolume and F1-score over exact-matching baselines, while maintaining fully interpretable rule structures suitable for clinical validation.

Several promising directions remain for future work. First, the framework will be applied to the FHU PRECISE cohort, targeting systemic sclerosis on French clinical texts, which also serves as a direct evaluation of its adaptability to French medical vocabulary. Second, a systematic parameter tuning study using irace [16] will be conducted to optimize key algorithmic settings such as population size, offspring size, and similarity threshold τ, which are currently based on preliminary experiments. Third, the size of the word embedding vectors will be investigated. Finally, the framework will be extended to a Pittsburgh [27] representation, where each individual encodes a complete rule set rather than a single rule. This extension will also allow disjunctions between rules, simulating OR conditions at the rule-set level.

References

1. Agrawal, R., Srikant, R.: Fast algorithms for mining association rules. In: Proceedings of the 20th International Conference on Very Large Data Bases (VLDB), pp. 487–499. Morgan Kaufmann (1994)
2. Altaf, W., Shahbaz, M., Guergachi, A.: Applications of association rule mining in health informatics: a survey. Artif. Intell. Rev. **47**(3), 313–340 (2017)
3. Breiman, L.: Random forests. Mach. Learn. **45**(1), 5–32 (2001)
4. Chen, T., Guestrin, C.: Xgboost: a scalable tree boosting system. In: Proceedings of the 22nd ACM SIGKDD International Conference on Knowledge Discovery and Data Mining, pp. 785–794 (2016)
5. Chiu, B., Crichton, G., Korhonen, A., Pyysalo, S.: How to train good word embeddings for biomedical NLP. In: Proceedings of the 15th Workshop on Biomedical Natural Language Processing, pp. 166–174 (2016)
6. Cohen, W.W.: Fast effective rule induction. In: Machine Learning Proceedings, pp. 115–123. Elsevier (1995)
7. Deb, K., Pratap, A., Agarwal, S., Meyarivan, T.: A fast and elitist multiobjective genetic algorithm: NSGA-II. IEEE Trans. Evol. Comput. **6**(2), 182–197 (2002)
8. Devlin, J., Chang, M.W., Lee, K., Toutanova, K.: Bert: pre-training of deep bidirectional transformers for language understanding. In: Proceedings of the 2019 Conference of the North American Chapter of the Association for Computational Linguistics: Human Language Technologies, vol. 1 (long and short papers), pp. 4171–4186 (2019)
9. Dhaenens, C., Jourdan, L.: Metaheuristics for data mining: survey and opportunities for big data. Ann. Oper. Res. **314**(1), 117–140 (2022)
10. Dynomant, E., et al.: Word embedding for the French natural language in health care: comparative study. JMIR Med. Inform. **7**(3), e12310 (2019)
11. European Commission: Proposal for a regulation of the European parliament and of the council laying down harmonised rules on artificial intelligence (artificial intelligence act) (2021). https://eur-lex.europa.eu/legal-content/EN/TXT/?uri=CELEX:52021PC0206
12. FHU PRECISE: Fhu precise – médecine de précision dans les maladies rares autoimmunes (2024). https://www.fhu-precise.fr/
13. Ishibuchi, H., Yamamoto, T.: Fuzzy rule selection by multi-objective genetic local search algorithms and rule evaluation measures in data mining. Fuzzy Sets Syst. **141**(1), 59–88 (2004)
14. Johnson, A.E., et al.: Mimic-iv, a freely accessible electronic health record dataset. Sci. Data **10**(1), 1 (2023)
15. Levy, O., Goldberg, Y., Dagan, I.: Improving distributional similarity with lessons learned from word embeddings. Trans. Assoc. Comput. Linguist. **3**, 211–225 (2015)
16. López-Ibáñez, M., Dubois-Lacoste, J., Cáceres, L.P., Birattari, M., Stützle, T.: The irace package: iterated racing for automatic algorithm configuration. Oper. Res. Perspect. **3**, 43–58 (2016)
17. Mikolov, T., Sutskever, I., Chen, K., Corrado, G.S., Dean, J.: Distributed representations of words and phrases and their compositionality. Adv. Neural. Inf. Process. Syst. **26** (2013)

18. Miller, G.A.: The magical number seven, plus or minus two: some limits on our capacity for processing information. Psychol. Rev. **63**(2), 81 (1956)
19. Moraglio, A., Krawiec, K., Johnson, C.G.: Geometric semantic genetic programming. In: Coello, C.A.C., Cutello, V., Deb, K., Forrest, S., Nicosia, G., Pavone, M. (eds.) PPSN 2012. LNCS, vol. 7491, pp. 21–31. Springer, Heidelberg (2012). https://doi.org/10.1007/978-3-642-32937-1_3
20. Mujtaba, G., et al.: Clinical text classification research trends: systematic literature review and open issues. Expert Syst. Appl. **116**, 494–520 (2019)
21. Nahm, U.Y., Mooney, R.J.: Using soft-matching mined rules to improve information extraction. In: Proceedings of the AAAI-2004 Workshop on Adaptive Text Extraction and Mining (2004)
22. Nauck, D., Kruse, R.: Obtaining interpretable fuzzy classification rules from medical data. Artif. Intell. Med. **16**(2), 149–169 (1999)
23. Pennington, J., Socher, R., Manning, C.D.: Glove: global vectors for word representation. In: Proceedings of the 2014 Conference on Empirical Methods in Natural Language Processing (EMNLP), pp. 1532–1543 (2014)
24. Powers, D.M.: Evaluation: from precision, recall and f-measure to roc, informedness, markedness and correlation. arXiv preprint arXiv:2010.16061 (2020)
25. Rudin, C., Chen, C., Chen, Z., Huang, H., Semenova, L., Zhong, C.: Interpretable machine learning: fundamental principles and 10 grand challenges. Stat. Surv. **16**, 1–85 (2022)
26. Shahrzad, H., Hodjat, B., Miikkulainen, R.: Evoter: evolution of transparent explainable rule-sets. ACM Trans. Evol. Learn. **5**(2), 1–30 (2025)
27. Smith, S.F.: A learning system based on genetic adaptive algorithms. University of Pittsburgh (1980)
28. Sobanski, V., Lescoat, A., Launay, D.: Novel classifications for systemic sclerosis: challenging historical subsets to unlock new doors. Curr. Opin. Rheumatol. **32**(6), 463–471 (2020). https://doi.org/10.1097/BOR.0000000000000747
29. Vanneschi, L., Pirola, Y., Collard, P.: A quantitative study of neutrality in GP Boolean landscapes. In: Proceedings of the 8th Annual Conference on Genetic and Evolutionary Computation, pp. 895–902 (2006)
30. Vassilev, V.K., Fogarty, T.C., Miller, J.F.: Information characteristics and the structure of landscapes. Evol. Comput. **8**(1), 31–60 (2000)
31. Zou, F., Chen, D., Liu, H., Cao, S., Ji, X., Zhang, Y.: A survey of fitness landscape analysis for optimization. Neurocomputing **503**, 129–139 (2022)
32. Řehůřek, R., Sojka, P.: Gensim: statistical semantics in python (2011). https://radimrehurek.com/gensim

Sampling on Random Subspaces Under Limited Data in the Context of Exploratory Landscape Analysis

Iván Olarte Rodríguez[1(✉)], Anja Jankovic[2], Thomas Bäck[1], and Elena Raponi[1]

[1] Leiden Institute of Advanced Computer Science (LIACS), Leiden University, Leiden, The Netherlands
{i.olarte.rodriguez,t.h.w.baeck,e.raponi}@liacs.leidenuniv.nl
[2] RWTH Aachen University, Aachen, Germany
jankovic@aim.rwth-aachen.de

Abstract. Classical space-filling designs often fail to provide reliable statistical results for Exploratory Landscape Analysis (ELA) when only limited evaluation budgets are available, as commonly occurs in high-dimensional problems or other resource-constrained settings, resulting in noisy and unstable landscape descriptors.

To address this challenge, we propose an alternative sampling strategy for ELA based on random linear embeddings. Rather than sampling uniformly in the full decision space, we allocate the budget to randomly oriented low-dimensional subspaces and investigate whether this improves the robustness of the resulting landscape descriptors.

We compare full-space and embedding-based sampling strategies across several classical ELA feature sets on the noiseless Black-Box Optimization Benchmarking (BBOB) test suite from the COmparing Continuous Optimizers (COCO) environment, in a 20-dimensional setting. Our results suggest that random linear embeddings constitute a promising alternative for budget-constrained ELA, although their effectiveness remains dependent on the feature class and the underlying problem.

Keywords: Exploratory Landscape Analysis · Linear Random Embeddings · Sampling under Limited Budgets · Problem Characterization · Continuous Black-Box Optimization

1 Introduction

Deriving informative and robust descriptors of optimization problems is fundamental to many downstream tasks in numerical optimization, such as landscape similarity assessment, problem classification, and automated algorithm selection and configuration [7,10,12,14,20]. In this context, Exploratory Landscape Analysis (ELA) [18] has established itself as a leading data-driven methodology. ELA transforms a collection of sampled decision vectors and their corresponding objective function values into a structured numerical feature space, thereby

G. Iacca et al. (Eds.): PPSN 2026, LNCS 16985, pp. 582–599, 2027.
https://doi.org/10.1007/978-3-032-36214-8_36

enabling the systematic characterization of black-box problems without relying on explicit analytical expressions.

While ELA has demonstrated considerable promise, its applicability in expensive settings remains limited [24,35]. A central challenge is its reliance on relatively large sample sizes to obtain stable and reliable feature estimates [21,34], which is an assumption that becomes impractical when function evaluations are costly [35]. This situation frequently arises in real-world applications, such as engineering design, where evaluating a single candidate solution may require computationally intensive simulations (*e.g.,* finite element or computational fluid dynamics analyses) [30,32], or costly experimental procedures involving significant time and resources [5]. As a result, optimization is typically performed under strict evaluation budgets, where only a limited number of function evaluations are available to explore and characterize the search landscape. Under these conditions, commonly used ELA features exhibit substantial sampling variance, with values varying noticeably across independent samples of identical size [33].

As a result, the behavior of ELA features under limited budgets is inherently shaped by the sampling process. The sampling design not only determines the coverage of the decision space but also induces the neighborhood structure on which many feature classes rely. While space-filling designs promote global coverage, they lead to sparse representations in higher-dimensional settings, which reduces the stability and discriminative power of features based on local relationships [11,34]. Conversely, sampling within defined regions of the landscape can improve the reliability of such features but introduces a bias toward specific regions of the search space. This highlights that the effectiveness of ELA features is governed by a trade-off between global representativeness and locally meaningful structure.

To address this trade-off, we consider a sampling strategy based on random linear embeddings. Rather than projecting an already sampled set of points onto random subspaces, as in previous work [24], we consider the setting in which a fixed sampling budget is available and samples are generated directly within randomly oriented low-dimensional affine subspaces defined by linear embedding maps. This allows global variation to be explored along low-dimensional directions while maintaining higher sampling density within each subspace. Moreover, it avoids the artificial landscape distortions introduced by projection [24]. As a result, neighborhood relationships are defined intrinsically within these subspaces, leading to more stable local feature estimates while still capturing variability in the ambient space, *i.e.,* in the original representation.

Our Contribution. We study the stability of ELA features under limited sampling budgets by comparing features computed from samples in embedded subspaces and the full ambient space. The main contributions are:

- We **propose a sampling strategy** based on randomly generated affine subspaces that reallocates a fixed evaluation budget to increase effective sampling density.

- We **conduct an empirical analysis across benchmark functions and feature sets**, distinguishing between feature computation in the ambient and reduced spaces, with the aim to examine how the effectiveness of subspace sampling depends on the objective function and feature class.
- We **analyze the impact of compression ratios**, highlighting the trade-off between sampling density and information loss and its effect on feature stability.
- We **assess the preservation of ELA features** relative to a large-budget baseline and provide practical insights into the reliability of structured sampling under scarce evaluations.

Our results show that sampling in randomly embedded subspaces can improve the stability of subsets of ELA features under tight budgets. This provides a practical alternative to conventional space-filling designs, enabling more reliable landscape characterization in resource-constrained optimization problems.

Code and Data Availability. All code used to run the experiments and generate the results presented in this paper is publicly available at https://github.com/olarterodriguezivan/Random_embeddings_BBOB. A permanent archive of the datasets, results, and additional plots is available on Zenodo at https://zenodo.org/records/19633768.

2 Background

2.1 Sampling Challenges in Exploratory Landscape Analysis

ELA [18] provides a modular collection of feature sets designed to capture structural properties of optimization landscapes based on sampled data. These feature classes target complementary characteristics, such as global distributional properties, local neighborhood structure, meta-model behavior, and information-theoretic measures.

Several ELA feature classes explicitly rely on local geometric or structural information. For example, *information content* (`ic`) features [19] estimate ruggedness and modality from symbol sequences derived from ordered function values along random walks. Similarly, *nearest better clustering* (`nbc`) features [11,28] quantify funnel structures via directed graphs linking each point to its nearest improving neighbor. ELA relies on sufficiently dense, well-distributed samples to capture landscape characteristics, particularly local trends and neighborhood structure. Under sparse sampling (*e.g.,* limited budgets or high dimensionality), feature estimates become unstable, exhibiting high variance across independent samples [34]. Consequently, the sampling strategy is critical: feature values depend strongly on how well samples represent the landscape and can vary significantly across sampling schemes [33].

Recent work has explored trajectory-based (dynamic) sampling, where features are computed from points generated during an optimizer run [8,14]. This avoids additional sampling cost and is effective for online algorithm selection. However, such samples are biased toward visited regions and are not space-filling, so the resulting features capture local, algorithm-dependent properties rather than global landscape characteristics. Consequently, they are not directly comparable to features obtained from uniform or quasi-random designs.

2.2 Random Linear Embeddings for Budget-Constrained Sampling

Random linear embedding techniques [25,37] provide a structured way to restrict sampling to predefined low-dimensional subspaces of the ambient space.

The idea of randomized linear mappings comes from compressed sensing [2], where signals can be recovered from a limited number of measurements when they admit sparse representations. In optimization, we use this idea in a broader sense: randomized linear mappings define transformed design variables that combine the original variables, allowing the objective function to be explored through a reduced parametrization without assuming sparsity in a prescribed basis.

This principle already underlies several high-dimensional Bayesian optimization methods, including REMBO [37], ALEBO [15], and PCA-based approaches [1,31]. The same reasoning applies to ELA, where the goal is to characterize landscape structure through compact numerical descriptors. Under a limited sampling budget, samples generated from such transformed variables can provide a denser and more reliable characterization of relevant landscape features than samples distributed uniformly across the full ambient domain.

Building on this principle, the present work systematically investigates the stability of ELA features when sampling is performed on random linear embeddings. We compare feature estimates obtained *(i)* from embedded subspaces and *(ii)* from uniform sampling in the ambient space. This comparison allows us to assess whether embedding-based sampling can improve the robustness of landscape descriptors under strict evaluation budgets. In contrast to uniformly distributing points across the entire decision domain, embedding-based sampling emphasizes structured coverage of linear subspaces. From the perspective of landscape analysis, this may provide more reliable neighborhood relations and graph structures when evaluation budgets are limited, thereby potentially stabilizing feature classes that rely on local structural information.

3 Methodology

To address the limitations of space-filling sampling under restricted evaluation budgets, in this work we adopt a sampling strategy for ELA based on random linear embeddings. Rather than distributing samples across the full search space, we allocate them within low-dimensional subspaces. This increases the effective sampling density and improves local resolution, which is critical for estimating neighborhood relationships and local structural properties. As a result, embedding-based sampling can provide more reliable geometric and graph-based

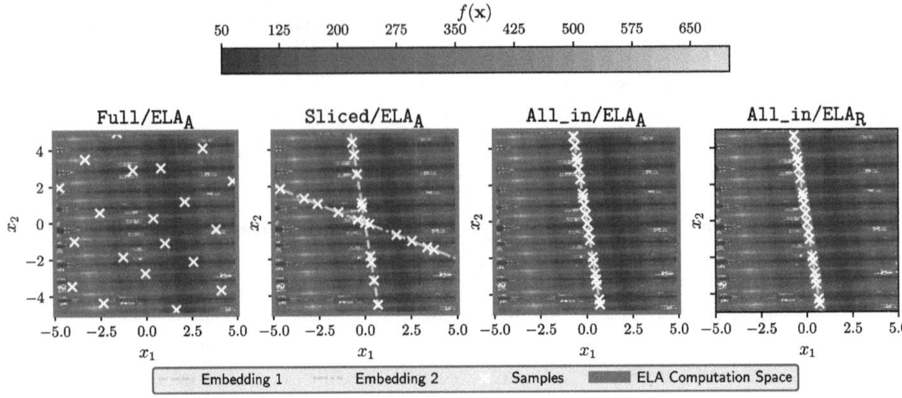

Fig. 1. Representation of the proposed sampling strategies for ELA. All the panels show the original 2D landscape of the Weierstrass (f16) function with points sampled on the landscape. Each panel depicts different sampling strategies. The red lines represent the space wherein ELA features are computed. (Color figure online)

structures under limited budgets, thereby improving the stability of ELA features that depend on local information.

To generate samples and compute the ELA features afterwards, we consider four sampling strategies (graphical representation in Fig. 1):

(1) **Full-space sampling:** Samples are drawn uniformly in the full D-dimensional decision space $\mathcal{X} = [l, u]^D$, and ELA features are computed directly in this space. We denote this setting by $\texttt{Full/ELA}_{\texttt{A}}$, where A indicates evaluation in the ambient space.
(2) **Multiple random embeddings (distributed budget):** We generate K random linear embeddings of effective dimension d with compression ratio $r = d/D$. The total sampling budget N_s is evenly distributed, yielding N_s/K samples per embedding, where $K = \lfloor 1/r \rfloor$. All sampled points are mapped back to the D-dimensional space, and ELA features are computed on their union. This approach is inspired by random projection techniques from compressed sensing [3]. We refer to it as $\texttt{Sliced/ELA}_{\texttt{A}}$, reflecting the use of multiple embeddings ("slices") with evaluation in the ambient space.
(3) **Single embedding (ambient evaluation):** The full sampling budget N_s is allocated to a single d-dimensional random embedding. Samples are mapped to the D-dimensional space, where ELA features are computed. By analogy to poker, we denote this strategy as $\texttt{All\_in/ELA}_{\texttt{A}}$, indicating that the entire budget is committed to one embedding with evaluation in the ambient space.
(4) **Single embedding (reduced evaluation):** As above, all samples are generated within a single d-dimensional embedding, but ELA features are computed directly in the reduced space. This variant is denoted $\texttt{All\_in/ELA}_{\texttt{R}}$, where R indicates evaluation in the reduced space.

The distinction between strategies (3) and (4) is important because linear embeddings can introduce strong linear dependencies between variables when represented in the ambient space. This can negatively affect feature sets that rely on regression or classification models (*e.g., meta-model* or *level-set* features), potentially leading to ill-conditioned estimation problems.

Algorithm 1 summarizes the general procedure for the computation of ELA features based on embedding-based sampling. The method consists of three main components: *(i)* an **embedding generation step** (Algorithm 2); *(ii)* a **sampling step** performed in the embedded space, followed by **evaluation of the objective function** in the ambient space (Algorithm 3); and *(iii)* a **feature computation step**, controlled by a flag indicating whether features should be computed in the embedded space (Algorithm 4). Symbols are introduced progressively within the corresponding pseudocode blocks, where they become relevant.

In Algorithm 2, the main component is the construction of random linear embeddings using the matrix $\mathbf{A} \in \mathbb{R}^{D\times d}$. The entries of $\mathbf{A}$ are sampled independently from a standard normal distribution, *i.e.*, $\mathcal{N}(0,1)$. Optionally, a normalization step, controlled by the flag `normalize`, is applied, in which each column of $\mathbf{A}$ is scaled to unit norm. This step helps preserve Euclidean distances between the embedded and ambient spaces, which is important since ELA features are not invariant to such distortions.

Next, the pseudo-inverse of $\mathbf{A}$, denoted by $\mathbf{A}^{\dagger}$, is computed. This step is required because points generated in the embedded space may be mapped, through the inverse transformation, to points that lie outside the feasible domain of the original ambient problem. To address this issue, Algorithm 3 first draws samples in the embedded space according to a sampling strategy $\mathcal{S}$, such as Sobol, Latin hypercube, or Halton sampling. Since these sampling strategies generate points in the unit hypercube $[0,1]^d$, the samples must be rescaled to the search region defined in the embedded space. In this work, the embedded search box is scaled as $\frac{1}{\sqrt{r}}[l,u]^d$. This scaling accounts for the relative change in the length of the maximum diagonal of the unit hypercube when moving from the ambient dimension D to the embedding dimension d. After the embedded samples are mapped back to the ambient space, the resulting points are clipped onto the feasible domain using the operator $\Pi_{\mathcal{X}}(\cdot)$, as shown in line 6. This step

Algorithm 1. Embedded Sampling and Feature Computation Pipeline

Require: Problem dimension D, embedding dimension d, number of subspaces K, sampling budget N_s, sampling strategy $\mathcal{S}$, search domain $\mathcal{X}$, objective function f, and flags: `normalize`, `compute_in_reduced_space`

Ensure: Sampled points $\{\mathbf{x}_i, \mathbf{z}_i, y_i\}_{i=1}^{N_s}$ and ELA feature vectors $\mathbf{t}$ (computed in ambient space), $\mathbf{t}'$ (computed in embedded space)

1: $\{\mathbf{A}^{(k)}, \mathbf{A}^{\dagger(k)}\}_{k=1}^{K} \leftarrow$ GenerateEmbeddings($d, D, K,$ `normalize`)

2: $\{(\mathbf{x}_i, \mathbf{z}_i, y_i)\}_{i=1}^{N_s} \leftarrow$ SampleAndEvaluate($\mathbf{A}^{(k)}, \mathbf{A}^{\dagger(k)}, N_s, \mathcal{S}, \mathcal{X}, f$)

3: $(\mathbf{t}, \mathbf{t}') \leftarrow$ ComputeFeatures($\{\mathbf{x}_i, \mathbf{z}_i, y_i\}$, `compute_in_reduced_space`)

ensures that the final candidate points satisfy the bounds of the ambient space $\mathcal{X} = [l, u]^D$.

However, since some feature computations require points to be expressed in the embedded space, an additional step is necessary. After correction, the feasible points in the ambient space are mapped back to the embedded space using the pseudo-inverse $\mathbf{A}^\dagger$. This ensures consistency between the domain in which feasibility is enforced and the domain in which features are evaluated. Note that clipping in the previous step may move points away from the embedded subspace and concentrate them near the domain boundary. Since our aim is to characterize the landscape from feasible samples, rather than to model and optimize the objective directly in the embedded space, we treat this distortion as negligible for feature computation.

Algorithm 2 GenerateEmbeddings

Require: Problem dimension D, embedding dimension d, number of subspaces K, and a flag: `normalize`

Ensure: Embedding matrices $\{\mathbf{A}^{(k)}, \mathbf{A}^{\dagger(k)}\}_{k=1}^{K}$

```
1: for k = 1,...,K do
2:     Sample A^(k) ∈ R^(D×d)            ▷ Elements are sampled according to N(0,1)
3:     if normalize then
4:         for j = 1,...,d do
5:             A^(k)_{:,j} ← A^(k)_{:,j} / ||A^(k)_{:,j}||_2      ▷ Column-wise normalization
6:         end for
7:     end if
8:     A^†(k) ← pinv(A^(k))              ▷ Compute Moore-Penrose pseudo-inverse [26]
9: end for
```

Algorithm 3 SampleAndEvaluate

Require: Random projection matrices and their pseudo-inverses $\{\mathbf{A}^{(k)}, \mathbf{A}^{\dagger(k)}\}$, sampling budget N_s, sampling strategy $\mathcal{S}$, search domain $\mathcal{X}$, objective function f

Ensure: Sampled points $\{(\mathbf{x}_i, \mathbf{z}_i, y_i)\}_{i=1}^{N_s}$

```
1: N_k ← ⌊N_s/K⌋                      ▷ Determine number of samples per embedding
2: for k = 1,...,K do
3:     for i = 1,...,N_k do
4:         z_i ← S(d)                  ▷ Sample in 1/√r · [l,u]^d
5:         x_i ← A^(k) z_i             ▷ Express sample in ambient space coordinates
6:         x_i ← Π_X(x_i)              ▷ Clip onto box X = [l,u]^D
7:         y_i ← f(x_i)                ▷ Evaluate the objective function
8:         z_i ← A^†(k) x_i            ▷ Map clipped point back to reduced-space coordinates
9:     end for
10: end for
```

Algorithm 4 ComputeFeatures

Require: Sampled points $\{(\mathbf{x}_i, \mathbf{z}_i, y_i)\}_{i=1}^{N_s}$, with a flag: `compute_in_reduced_space`
Ensure: ELA feature vectors $\mathbf{t}$ (computed in ambient space), $\mathbf{t}'$ (computed in embedded space)

if `compute_in_reduced_space` **then**
  $\mathbf{t}' \leftarrow \mathcal{F}\left(\{(\mathbf{z}_i, y_i)\}_{i=1}^{N_s}\right)$
end if
$\mathbf{t} \leftarrow \mathcal{F}\left(\{(\mathbf{x}_i, y_i)\}_{i=1}^{N_s}\right)$

Finally, ELA features are computed once the sample points are represented in the ambient space and their corresponding function values have been evaluated. Algorithm 4 outlines the standard procedure for computing the feature vector $\mathbf{t}$ via the operator $\mathcal{F}[\cdot]$. An additional flag (`compute_in_reduced_space`) enables the computation of an alternative feature vector $\mathbf{t}'$ in the reduced space when required.

4 Experimental Setup

Benchmark Problems. We evaluate the proposed sampling strategies on noiseless functions from the Black-Box Optimization Benchmarking (BBOB) test suite from the COmparing Continuous Optimizers (COCO) environment [6]. This suite comprises 24 continuous optimization problems that cover a wide range of landscape characteristics, including separability, conditioning, and multimodality. The search domain of the BBOB functions is defined over $[-5, 5]^D$. Function evaluations and instance definitions are obtained using `IOHexperimenter` [23], for which the first 15 instances of each function are used. Experiments are conducted only for $D = 20$, representing a moderate-dimensional setting [35].

Sampling Budget and Strategies. For each problem, the total sampling budget is fixed to $N_s = 10 \cdot$ D. Samples are generated using the four strategies described in Sect. 3.

For embedding-based methods, the reduced dimension is determined by the compression ratio $r = d/D$, $r \in \{0.1, 0.25, 0.5\}$. In the multi-embedding setting, the number of embeddings is set to $K = \lfloor 1/r \rfloor$, with the budget uniformly allocated across embeddings. This yields $K \in \{10, 4, 2\}$ embeddings with $\{20, 40, 100\}$ samples per embedding for the respective values of r.

Sampling is performed via Latin Hypercube Sampling (LHS) in both ambient and embedded spaces. The search domain in the embedded space is expanded to $1/\sqrt{r} \cdot [-5, 5]^d$. This was performed with the `scipy.stats.qmc` package [36].

Each configuration is repeated 40 times with different random seeds. For the `slice` approach, each embedding is generated using distinct seeds to avoid repetition.

Considered Features. We consider six classical ELA feature sets: *y-distribution* (`ela_distr`), *level-set* (`ela_level`), *meta-model* (`ela_meta`), *nearest-better clustering* (`nbc`), *dispersion* (`disp`), and *information content* (`ic`). In addition, *fitness-distance correlation* (`fitness_distance`) is included solely as a baseline, as its computation requires knowledge of the global optimum, which is typically unavailable in practice.

We exclude the `pca` feature set [13] as it primarily captures the geometric distribution of sampled points in the search space and is thus strongly influenced by the sampling strategy rather than intrinsic landscape properties [16].

The selected feature sets are computationally inexpensive [7,14] and can be computed from a single sample without requiring repeated evaluations or structured sampling. An overview is given in Table 1, with further details available in the cited literature.

Table 1. Feature sets considered in this study.

Feature set	Name	Num. features
`ela_distr`	distribution-based features [18]	3
`ela_level`	level-set features [18]	9
`ela_meta`	meta-model features [18]	9
`nbc`	nearest-better clustering [11,28]	5
`disp`	dispersion features [17]	16
`ic`	information content features [19]	5
`fitness-distance`	fitness distance correlation [9]	6

Feature Computation. ELA features are computed using the `pflacco` package [27]. Depending on the strategy, features are computed either in the ambient space or directly in the reduced subspace as explained in Sect. 3.

Evaluation Protocol. We evaluate sampling strategies by measuring how well feature distributions obtained with a reduced budget approximate a high-fidelity reference. For each ELA feature, function, and instance, a reference distribution is computed using 2000 ($100 \cdot D$) LHS samples. In line with Renau et al. [34], these values are not interpreted as exact feature quantities, but rather as high-budget approximations within a fixed sampling scheme. The resulting distribution serves as an operational baseline.

Distance Measure. The discrepancy between reference and approximated distributions is quantified using the Wasserstein-1 (Earth Mover's) distance [29], computed per feature–function–instance triplet. This metric compares empirical distributions directly and is sensitive to differences across their full support.

In contrast to divergence-based measures such as the Kullback-Leibler divergence, it remains well-defined under limited or non-overlapping support, providing a stable notion of distributional discrepancy without relying on parametric or moment-based assumptions.

Aggregation and Comparison. For each feature and function, the Wasserstein distances are aggregated over instances and used to rank the sampling strategies based on the median. Rankings are averaged across *(i)* ELA features and *(ii)* functions to analyze overall performance. Results are reported using average ranking plots and Critical Difference (CD) diagrams [4], where statistical significance is assessed via the post-hoc Nemenyi test [22].

5 Results

Following the experimental setup choices described in Sect. 4, we summarize the ranking and distance measures in a compact form in Fig. 2. In particular, we empirically evaluate ten different methods, constructed based on the four strategies outlined in Sect. 3; we use $\texttt{Full/ELA}_\texttt{A}$ as a baseline, and we assess the performance of the combinations of three embedding-based strategies ($\texttt{Sliced/ELA}_\texttt{A}$, $\texttt{All\_in/ELA}_\texttt{A}$, $\texttt{All\_in/ELA}_\texttt{R}$) with three different values of the compression ratio r. Figure 2 presents a heatmap of method rankings based on the Wasserstein distance between estimated and reference distributions of ELA features across the BBOB test suite. Rows correspond to functions and columns to features; colors indicate the best-performing method, while symbols denote the second-best.

Overall, the heatmap does not reveal strong global trends. Nevertheless, two consistent observations can be made. First, $\texttt{Full/ELA}_\texttt{A}$ tends to outperform subspace-based approaches for the subclass `ela_meta.lin_simple`. This is expected, as meta-model-based features aim to capture global landscape structure, which may not be preserved under projection to lower-dimensional subspaces. Second, $\texttt{Sliced/ELA}_\texttt{A}$ with $r = 0.5$ frequently achieves the best ranking for the `disp` features. This indicates that these features can often be estimated reliably from subspace samples, suggesting a certain robustness to localized sampling. However, this behavior is not consistent across all functions. In particular, for functions exhibiting strong anisotropy or ill-conditioning, such as the Discus function (f11), as well as highly irregular landscapes such as the Weierstrass (f16) and Katsuura (f23) functions, subspace sampling appears insufficient to capture relevant structural properties. In these cases, full-space sampling provides more accurate approximations of the feature distributions.

For the `ela_level` features, the best-performing strategies are consistently of the $\texttt{All\_in/ELA}_\texttt{R}$ type, with $r = 0.25$ often achieving the top ranking. While no single configuration dominates uniformly, concentrating samples within a single subspace is preferred over full-space sampling. In fact, $\texttt{Full/ELA}_\texttt{A}$ is rarely among the top strategies, even as a second-best choice, reinforcing that it is not well suited for these features. The variability observed across functions reflects sensitivity to the compression ratio rather than to the sampling paradigm itself.

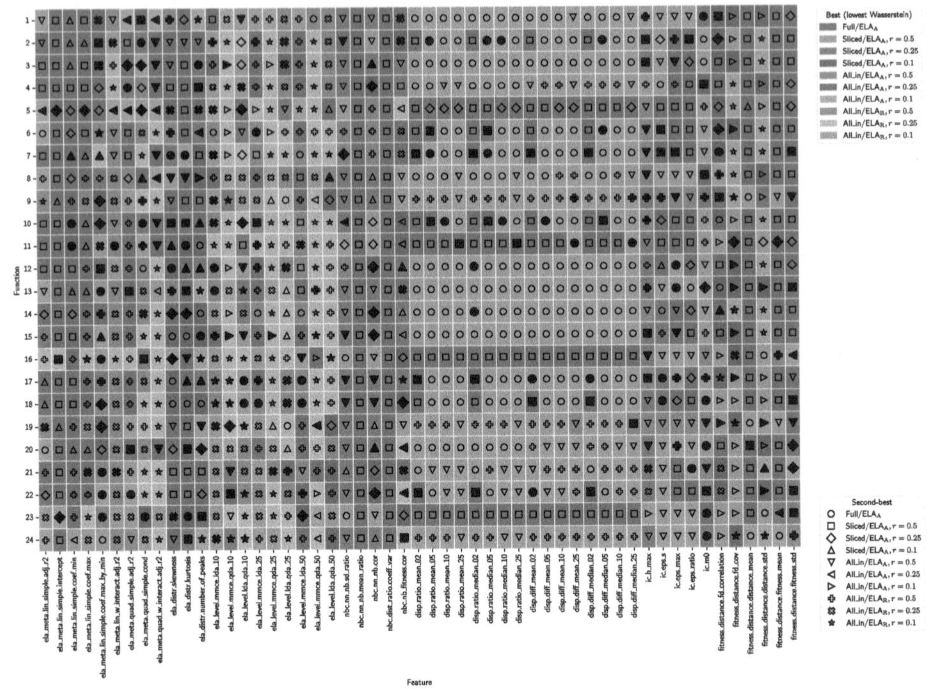

Fig. 2. Rankings of ELA feature sampling methods based on the Wasserstein-1 distance relative to LHS sampling with 2000 ($100 \cdot D$) samples. Colored blocks indicate the method that achieved the lowest median distance for each function–feature pair, while the symbol overlaid on each block denotes the second-lowest distance. A **filled** symbol indicates that the difference between the methods with the lowest and second-lowest distances is not statistically significant with 95% confidence according to the signed Wilcoxon test.

The remaining question is how far the compression ratio can be reduced without losing essential structure.

To illustrate how Wasserstein distances vary depending on both the function and the analyzed feature, we present Fig. 3, which shows distributions of the Wasserstein distances across instances for the Rosenbrock (f8) and Rotated Rosenbrock (f9) functions for the `ic.eps_s` feature, which defines the sensitivity of the landscape to variable scaling and, intuitively, quantifies how much of the landscape structure is preserved when the resolution is changed from fine- to coarse-grained. This example highlights that, despite the strong similarity between these functions, noticeable differences in Wasserstein distances can still arise. This observation motivates the use of aggregate statistics such as medians and rankings.

Because Wasserstein distances are on different scales for different features and functions, *i.e.*, reflecting differences in feature magnitude and variability rather than in method quality, direct comparisons of raw Wasserstein distance values

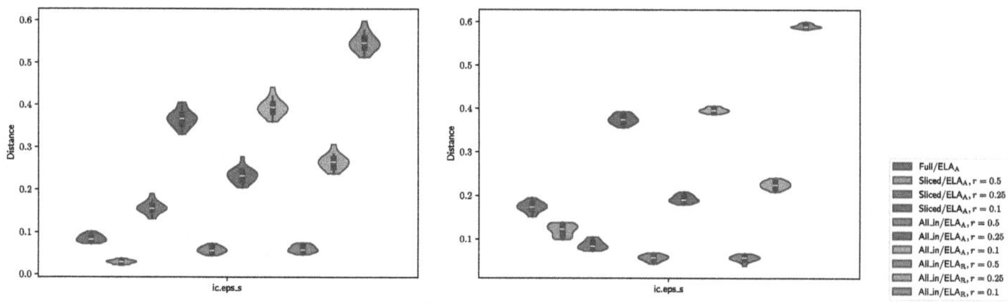

(a) Rosenbrock (f8) (b) Rotated Rosenbrock (f9)

Fig. 3. Per instance Wasserstein distance distribution of the `ic.eps_s`.

across features or functions would be misleading. We therefore conduct a ranking-based analysis to identify relative strengths and weaknesses of the sampling strategies. Figure 4 reports the average rankings per function (aggregated over all features), while Fig. 5 reports the average rankings per feature (aggregated over all functions). From Fig. 4, the methods Full/ELA$_A$ and Sliced/ELA$_A$ with $r = 0.5$ consistently achieve top rankings across functions. In contrast, methods with $r = 0.1$ systematically rank lowest, indicating that aggressive compression degrades performance. Thus, for the considered feature sets, retaining a moderate level of landscape information is necessary to preserve descriptive power.

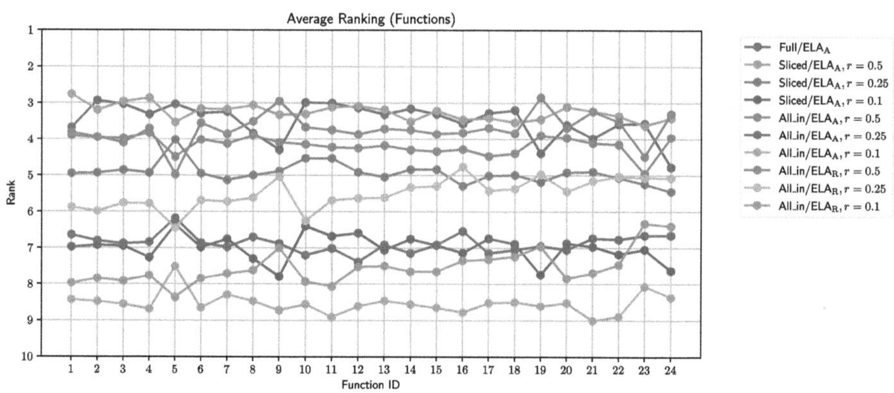

Fig. 4. Average method rankings based on the median Wasserstein-1 distance across features per function.

Figure 5 shows consistent trends across feature classes. For `ela_meta.lin_simple`, Full/ELA$_A$ performs best, as these features depend on global coverage for fitting linear meta-models. In contrast, for `ela_distr`, `ic`, and `disp`, embedding-based sampling with moderate compression ($r = 0.5$) is competitive with or outperforms full-space sampling, indicating that subspace

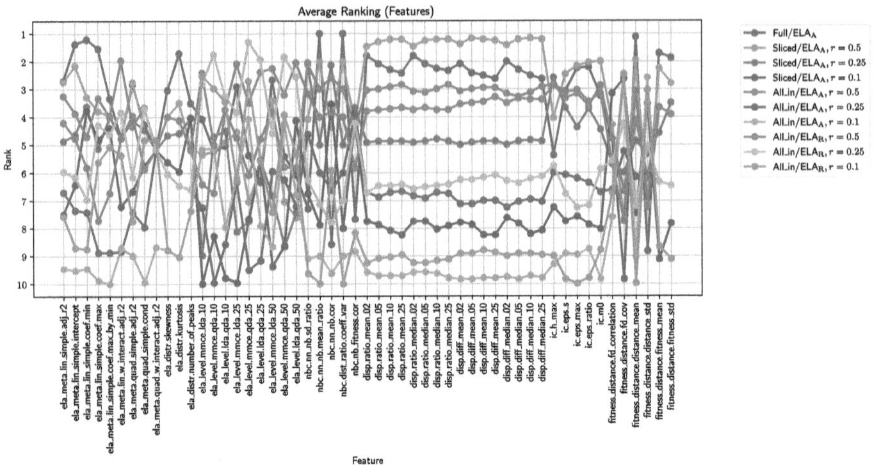

Fig. 5. Average method rankings based on the median Wasserstein-1 distance across functions per feature.

sampling can match or improve upon the default approach. For `ela_level`, single-embedding strategies combined with feature computation in the reduced space (*e.g.*, $\texttt{All\_in/ELA}_\texttt{R}$) achieve the best rankings, suggesting that level-set features benefit from higher sampling density within a single subspace. Overall, the effectiveness of a sampling strategy is strongly feature-dependent. However, all these results underscore a key practical implication: the common practice of applying default full-space sampling ($\texttt{Full/ELA}_\texttt{A}$) with a limited budget to characterize problem landscapes is often suboptimal. Targeted use of structured subspace sampling can meaningfully improve the quality of landscape characterization for specific feature sets.

Overall Comparison Across Features and Functions. The CD diagram (Fig. 6) summarizes the performance of the sampling strategies by aggregating results across both ELA features and problem landscapes. While the overall trends are consistent, the grouping structure reflects the combined variability across descriptors and functions, motivating a cautious interpretation of the aggregated results.

To obtain an overall comparison, we define a consensus ranking by averaging the mean ranks obtained in the feature-wise and function-wise analyses. This criterion assigns equal importance to robustness across features and across functions, and can be interpreted as a compromise between the two views. In addition, we consider the two-dimensional rank space induced by both criteria, where non-dominated methods represent jointly competitive approaches.

According to this consensus criterion, $\texttt{Sliced/ELA}_\texttt{A}$ with $r = 0.5$ achieves the best overall performance, as it consistently ranks among the top methods in both views. $\texttt{Full/ELA}_\texttt{A}$ and $\texttt{All\_in/ELA}_\texttt{R}$ with $r = 0.5$ follow closely, forming a group of competitive strategies with only minor differences in average rank.

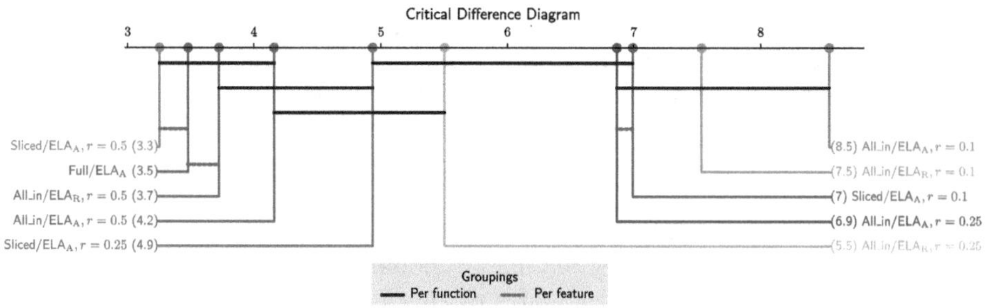

Fig. 6. Critical difference diagram with groupings obtained by averaging over all functions and features.

In contrast, methods using lower sampling ratios ($r = 0.25$ and $r = 0.1$) are consistently dominated, confirming that aggressive reduction of the compression ratio significantly degrades the approximation of feature distributions.

Overall, the results indicate that, despite no single method attains statistically superior results across all considered settings, moderate reduction ratios, when combined with structured sampling strategies, most notably slicing, achieves a favorable trade-off between computational cost and approximation fidelity, leading to consistently strong performance across both ELA features and BBOB functions.

6 Outlook and Conclusion

In this paper, we investigated sampling in Exploratory Landscape Analysis (ELA) under limited evaluation budgets. Using *random linear embeddings*, we introduced sampling strategies that allocate samples to randomly generated low-dimensional subspaces, increasing sampling density and improving the resolution of local geometric relationships.

Across BBOB functions and feature sets, these strategies are consistently competitive with, and often outperform, full-space sampling under the same budget. The advantage is most pronounced for the feature sets capturing *dispersion* and *information content*. Moderate compression ($r = 0.5$) yields the best trade-off, suggesting that concentrating samples within informative subspaces is more effective than distributing them uniformly across the full search space. These results show that, under limited budgets, sampling is a key design choice in ELA rather than a secondary detail. The standard practice of sampling in the full decision space can therefore be suboptimal and should be reconsidered.

Limitations. Our study is restricted to a 20-dimensional benchmark setting and a specific family of random linear embeddings. The effectiveness of the approach depends on both the problem structure and the choice of subspace dimension: overly low-dimensional slices may fail to capture complex or global landscape properties, and randomly sampled subspaces are not guaranteed to intersect informative regions.

In addition, not all ELA feature classes benefit equally from subspace-based sampling. In particular, features relying on global structure, such as those from the *meta-model*, *y-distribution*, or parts of the *fitness-distance correlation* sets, remain difficult to estimate reliably under localized sampling.

These limitations, however, do not alter the main conclusion: under limited budgets, sampling uniformly in the full space should not be taken as the default, and structured subspace sampling provides a viable alternative.

Future Work. Several directions follow naturally from this work. Our study focuses on BBOB functions, which are not designed to exhibit low intrinsic dimensionality, meaning that relevant variation of the objective function is concentrated in a lower-dimensional subspace. Given that sampling on random embeddings naturally aligns with this setting, future work will consider benchmarks where this property holds, including real-world problems. Another important direction is to assess the impact of these sampling strategies on downstream tasks such as algorithm selection and configuration.

Disclosure of Interests. The authors have no competing interests to declare that are relevant to the content of this article

References

1. Antonov, K., Raponi, E., Wang, H., Doerr, C.: High dimensional bayesian optimization with kernel principal component analysis. In: Proceedings of the 17th International Conference on Parallel Problem Solving from Nature (PPSN'22), pp. 118–131 (2022). https://doi.org/10.1007/978-3-031-14714-2_9
2. Candès, E., Tao, T.: Near-optimal signal recovery from random projections: universal encoding strategies? IEEE Trans. Inf. Theory **52**(12), 5406–5425 (2006). https://doi.org/10.1109/TIT.2006.885507
3. Candès, E., Tao, T.: The dantzig selector: statistical estimation when p is much larger than n. Ann. Stat. **35**(6), 2313–2351 (2007). https://www.jstor.org/stable/25464587
4. Demšar, J.: Statistical comparisons of classifiers over multiple data sets. J. Mach. Learn. Res. **7**, 1–30 (2006). https://dl.acm.org/doi/10.5555/1248547.1248548
5. Falaschetti, M.P., Semprucci, F., Birnie Hernández, J., Troiani, E.: Experimental and numerical assessment of crashworthiness properties of composite materials: a review. Aerospace **12**(2), 122 (2025). https://doi.org/10.3390/aerospace12020122
6. Hansen, N., Auger, A., Ros, R., Mersmann, O., Tušar, T., Brockhoff, D.: COCO: a platform for comparing continuous optimizers in a black-box setting. Optim. Methods Softw. **36**(1), 114–144 (2021). https://doi.org/10.1080/10556788.2020.1808977

7. Jankovic, A., Doerr, C.: Landscape-aware fixed-budget performance regression and algorithm selection for modular CMA-ES variants. In: Proceedings of the Genetic and Evolutionary Computation Conference (GECCO'20), pp. 841–849 (2020). https://doi.org/10.1145/3377930.3390183
8. Jankovic, A., Eftimov, T., Doerr, C.: Towards feature-based performance regression using trajectory data. In: Castillo, P.A., Jiménez Laredo, J.L. (eds.) EvoApplications 2021. LNCS, vol. 12694, pp. 601–617. Springer, Cham (2021). https://doi.org/10.1007/978-3-030-72699-7_38
9. Jones, T., Forrest, S.: Fitness distance correlation as a measure of problem difficulty for genetic algorithms. In: Proceedings of the 6th International Conference on Genetic Algorithms, pp. 184–192. Morgan Kaufmann Publishers Inc., San Francisco (1995)
10. Kerschke, P., Hoos, H.H., Neumann, F., Trautmann, H.: Automated algorithm selection: survey and perspectives. Evol. Comput. **27**, 3–45 (2019). https://doi.org/10.1162/evco_a_00242
11. Kerschke, P., Preuss, M., Wessing, S., Trautmann, H.: Detecting funnel structures by means of exploratory landscape analysis. In: Proceedings of the 2015 Annual Conference on Genetic and Evolutionary Computation, pp. 265–272. ACM, Madrid (2015). https://dl.acm.org/doi/10.1145/2739480.2754642
12. Kerschke, P., Trautmann, H.: Automated algorithm selection on continuous black-box problems by combining exploratory landscape analysis and machine learning. Evol. Comput. **27**, 99–127 (2019). https://doi.org/10.1162/evco_a_00236
13. Kerschke, P., Trautmann, H.: Comprehensive feature-based landscape analysis of continuous and constrained optimization problems using the R-package flacco. In: Bauer, N., Ickstadt, K., Lübke, K., Szepannek, G., Trautmann, H., Vichi, M. (eds.) Applications in Statistical Computing. SCDAKO, pp. 93–123. Springer, Cham (2019). https://doi.org/10.1007/978-3-030-25147-5_7
14. Kostovska, A., et al.: Per-run algorithm selection with warm-starting using trajectory-based features. In: Proceedings of the International Conference on Parallel Problem Solving from Nature (PPSN'22), pp. 46–60 (2022). https://doi.org/10.1007/978-3-031-14714-2_4
15. Letham, B., Calandra, R., Rai, A., Bakshy, E.: Re-examining linear embeddings for high-dimensional Bayesian optimization. In: Proceedings of the 34th International Conference on Neural Information Processing Systems (NeurIPS'20), pp. 1546–1558 (2020). https://papers.neurips.cc/paper_files/paper/2020/file/10fb6cfa4c990d2bad5ddef4f70e8ba2-Paper.pdf
16. Long, F.X., Vermetten, D., van Stein, N., Kononova, A.: BBOB instance analysis: landscape properties and algorithm performance across problem instances. In: Proceedings of the 26th European Conference on Applications in Evolutionary Computation (EvoApplications'23), pp. 380–395 (2023). https://doi.org/10.1007/978-3-031-30229-9_25
17. Lunacek, M., Whitley, D.: The dispersion metric and the CMA evolution strategy. In: Proceedings of the Conference on Genetic and Evolutionary Computation (GECCO'06), pp. 477–484. ACM, Seattle (2006). https://dl.acm.org/doi/10.1145/1143997.1144085
18. Mersmann, O., Bischl, B., Trautmann, H., Preuss, M., Weihs, C., Rudolph, G.: Exploratory landscape analysis. In: Proceedings of the Conference on Genetic and Evolutionary Computation (GECCO'11), pp. 829–836. ACM, Dublin (2011). https://dl.acm.org/doi/10.1145/2001576.2001690

19. Muñoz, M.A., Kirley, M., Halgamuge, S.: Exploratory landscape analysis of continuous space optimization problems using information content. IEEE Trans. Evol. Comput. **19**(1), 74–87 (2015). http://ieeexplore.ieee.org/document/6719480/
20. Muñoz, M.A., Sun, Y., Kirley, M., Halgamuge, S.K.: Algorithm selection for black-box continuous optimization problems: a survey on methods and challenges. Inf. Sci. **317**, 224–245 (2015). https://doi.org/10.1016/j.ins.2015.05.010
21. Muñoz, M.A., Kirley, M., Smith-Miles, K.: Analyzing randomness effects on the reliability of exploratory landscape analysis. Natural Comput. **21**(2), 131–154 (2022). https://link.springer.com/10.1007/s11047-021-09847-1
22. Nemenyi, P.: Distribution-Free Multiple Comparisons. Princeton University (1979)
23. de Nobel, J., Ye, F., Vermetten, D., Wang, H., Doerr, C., Bäck, T.: IOHexperimenter: benchmarking platform for iterative optimization heuristics. Evol. Comput. **32**, 205–210 (2024). https://doi.org/10.1162/evco_a_00342
24. Olarte Rodríguez, I., Jankovic, A., Bäck, T., Raponi, E.: Does Dimensionality Reduction via Random Projections Preserve Landscape Features? (2026). http://arxiv.org/abs/2604.13230
25. Owen, A.B.: Monte Carlo theory, methods and examples (2013). https://artowen.su.domains/mc/
26. Penrose, R.: On best approximate solutions of linear matrix equations. Math. Proc. Cambridge Phil. Soc. **52**(1), 17–19 (1956). https://www.cambridge.org/core/product/identifier/S0305004100030929/type/journal_article
27. Prager, R.P., Trautmann, H.: Pflacco: feature-based landscape analysis of continuous and constrained optimization problems in python. Evol. Comput. **32**(3), 211–216 (2024). https://doi.org/10.1162/evco_a_00341
28. Preuss, M.: Improved topological niching for real-valued global optimization. In: Di Chio, C., et al. (eds.) EvoApplications 2012. LNCS, vol. 7248, pp. 386–395. Springer, Heidelberg (2012). https://doi.org/10.1007/978-3-642-29178-4_39
29. Ramdas, A., Garcia Trillos, N., Cuturi, M.: On wasserstein two sample testing and related families of nonparametric tests. Entropy **19** (2017). https://doi.org/10.3390/e19020047
30. Raponi, E., Bujny, M., Olhofer, M., Aulig, N., Boria, S., Duddeck, F.: Kriging-assisted topology optimization of crash structures. Comput. Methods Appl. Mech. Eng. **348**, 730–752 (2019). https://doi.org/10.1016/j.cma.2019.02.002
31. Raponi, E., Wang, H., Bujny, M., Boria, S., Doerr, C.: High dimensional bayesian optimization assisted by principal component analysis. In: Proceedings of the 16th International Conference on Parallel Problem Solving from Nature (PPSN'20), pp. 169–183 (2020). https://doi.org/10.1007/978-3-030-58112-1_12
32. Ren, C., Meng, Z.: A survey on expensive optimization problems using differential evolution. Appl. Soft Comput. **170**, 112727 (2025). https://doi.org/10.1016/j.asoc.2025.112727
33. Renau, Q., Doerr, C., Dreo, J., Doerr, B.: Exploratory landscape analysis is strongly sensitive to the sampling strategy. In: Proceedings of the 16th International Conference on Parallel Problem Solving from Nature (PPSN'20), pp. 139–153 (2020). https://doi.org/10.1007/978-3-030-58115-2_10
34. Renau, Q., Dreo, J., Doerr, C., Doerr, B.: Expressiveness and robustness of landscape features. In: Proceedings of the Genetic and Evolutionary Computation Conference Companion, GECCO '19, pp. 2048–2051. Association for Computing Machinery, New York (2019). https://doi.org/10.1145/3319619.3326913

35. Tanabe, R.: Towards exploratory landscape analysis for large-scale optimization: a dimensionality reduction framework. In: Proceedings of the Genetic and Evolutionary Computation Conference, pp. 546–555. ACM, Lille (2021). https://dl.acm.org/doi/10.1145/3449639.3459300
36. Virtanen, P., et al.: SciPy 1.0: fundamental algorithms for scientific computing in python. Nat. Methods **17**, 261–272 (2020). https://doi.org/10.1038/s41592-019-0686-2
37. Wang, Z., Hutter, F., Zoghi, M., Matheson, D., De Freitas, N.: Bayesian optimization in a billion dimensions via random embeddings. J. Artif. Intell. Res. **55**, 361–387 (2016). https://jair.org/index.php/jair/article/view/10983

The Traveling Thief Problem with Time Windows: Benchmarks and Heuristics

Helen Yuliana Angmalisang(✉) and Frank Neumann

Optimisation and Logistics, School of Computer Science and Information Technology, Adelaide University, Adelaide, Australia
{helen.angmalisang,frank.neumann}@adelaide.edu.au

Abstract. While traditional optimization problems were often studied in isolation, many real-world problems today require interdependence among multiple optimization components. The traveling thief problem (TTP) is a multi-component problem whose components are interdependent and has been widely studied in the literature. In this paper, we introduce and investigate the TTP with time window constraints which provides a TTP variant highly relevant to real-world situations where good can only be collected at given time intervals. We examine adaptions of existing approaches for TTP and the Traveling Salesperson Problem (TSP) with time windows to this new problem and evaluate their performance. Furthermore, we provide a new heuristic approach for the TTP with time windows. To evaluate algorithms for TTP with time windows, we introduce new TTP benchmark instances with time windows based on TTP instances existing in the literature. Our experimental investigations evaluate the different approaches and show that the newly designed algorithm outperforms the other approaches on a wide range of benchmark instances.

Keywords: benchmarking · constraints · evolutionary algorithms · heuristic search · time windows · traveling thief problem

1 Introduction

There are many real-world applications that require simultaneous and inseparable routing and resource selection optimization, while being subject to time limitations and deadlines. Emergency services [16,28], home health care [20], and municipal waste management [3,15] are some examples of such applications. In these problems, routing optimization and resource selection optimization cannot be addressed separately. An important benchmark problem that shares the interacting characteristics of routing and resource selection optimization is the Traveling Thief Problem (TTP), which is an interdependent combination of the Traveling Salesperson Problem (TSP) and the Knapsack Problem (KP). The TTP has gained significant attention in the evolutionary computation literature [6,7,27,34]. However, to the best of our knowledge, there is no study that investigates the time window constraint in TTP, which is highly relevant in many real-world situations.

Nevertheless, with time window constraints, TTP becomes more complex as the travel speed in TTP depends on the current cumulative weight of the stolen items carried

G. Iacca et al. (Eds.): PPSN 2026, LNCS 16985, pp. 600–615, 2027.
https://doi.org/10.1007/978-3-032-36214-8_37

in the knapsack. Taking an item from a city may lead to either profit or loss, or cause late arrival at subsequent cities. However, the thief cannot make any profit if no item is taken. Moreover, with time windows, the feasible space becomes drastically narrower in a wide search space. Therefore, it is more challenging to find a feasible solution.

1.1 Related Work

Bonyadi et al. [6] introduced and defined TTP as an optimization problem that is more similar to real-world problems for its combination and interdependence of the KP and TSP components. Since then, numerous studies have been conducted to find the best algorithms for solving TTP. Most of them used the Chained Lin-Kernighan (CLK) heuristic to create the initial tour [10, 11, 19, 23, 36]. The tour is then optimized using various algorithms. Operators like crossover and insertion are popular for optimizing the tour [2, 21, 22, 24–26, 30]. Faulkner et al. [13] proposed Pack and PackIterative algorithms as item scoring strategies for the packing plan. They and some studies afterwards [23, 31, 35, 36] found that repetition of CLK and PackIterative algorithm (S5) shows particularly strong results. Compared to population-based algorithms, S5 which is a single-solution algorithm also performs better on the TTP [34]. Moreover, it is found that the initial tour really affects the final objective value [13]. BitFlip, Insertion, Pack, and PackIterative have become popular operators that have been used in many investigations afterwards [1, 2, 8, 19, 30, 36]. Wuijits and Thierens [33] found that the TTP fitness landscape contains numerous local optima, while the global optimum lies within a narrow region of attraction. These findings underlie the development of our proposed algorithms that use more perturbation to improve the chances of discovering the global optimum and to improve the accuracy of fitness estimates.

Regarding time windows, there have been many studies in decades on the Traveling Salesman Problem with Time Windows (TSPTW), which is a component of TTPTW. Helsgaun [14] improved the classic Lin-Kernighan algorithm [17], for constrained TSP and Vehicle Routing Problems (VRP), named LKH-3. The LKH-3 algorithm can solve different variants of those problems, including TSPTW. Subsequently, Zheng et al. [37] proposed a Variable Strategy Reinforced LKH (VSR-LKH) and VSR-LKH-3 algorithms, and combined them with three reinforcement learning methods, namely Q-learning, Sarsa, and Monte Carlo, for the TSP. VSR-LKH-3 was reported to have better performance in solving TSPTW compared to LKH-3. However, to the best of our knowledge, no study or algorithm has been specifically proposed for TTP with time windows.

1.2 Our Contribution

This study extends the Traveling Thief Problem with time window constraints (TTPTW). The time windows constraint adds a condition in which the thief has to wait if he is early. This condition changes the objective function, as the total duration of the trip may increase and affect the rental fee.

For the evaluation, we construct the benchmark with varying level of time window tightness. Furthermore, we propose a set of evolutionary algorithms, named Dual Search Evolutionary Algorithm (DSEA), that utilize a new tour initialization algorithm, modified operators, and different packing plan repair approaches. We compared the performance of the packing plan repairs and analyzed them to study which repair works best in this problem. For comparison, we adapt the state-of-the-art TTP and TSPTW algorithms and evaluate their performance on TTPTW instances. We also implement our new tour initialization algorithm to the TTP heuristics and evaluate their performance on TTPTW. We compare and analyze the effect of using the tour initialization algorithm and the variation of operators on these TTP optimizers. Finally, we compare the performance of DSEA, S4, S5, C5, LKH-3, and VSR-LKH-3 algorithms.

The rest of the paper is structured as follows. In Sect. 2, we provide an overview of general TTP. Section 3 introduces the Traveling Thief Problem with Time Windows. Section 4 explains how we adapt state-of-the-art TTP and TSPTW algorithms for TTPTW. Section 5 describes the proposed algorithms, their components, and the difference among the variants. Section 6 explains the benchmark problem used in this study. Section 7 presents the experimental settings and results. Finally, Sect. 8 provides the conclusion.

2 Traveling Thief Problem

Given are a set of cities $N = \{1, 2, \ldots, n\}$ and a set of items $M = \{1, 2, \ldots, m\}$. The items are distributed among the cities, except city 1, and each item $j \in M$ has a profit p_j and a weight w_j. The distance between any pair of cities $i, j \in N$, d_{ij}, is known. The thief has to start from city 1 and return there after visiting all the other cities once. The thief can steal some items as long as the total weight of the stolen items does not exceed the knapsack capacity W. The knapsack rental rate R is charged for each time unit used during the tour. The minimal and maximal speed of the thief, $v_{\min}$ and $v_{\max}$, are known. The objective of this problem is to find a tour $\Pi = (x_1, x_2, ..., x_n), x_i \in N$ and a packing plan $P = (y_1, y_2, ..., y_m) \in \{0,1\}^m$ to maximize profit. The formulation of the objective functions is as follows.

$$\max_{\Pi,\, P} \quad Z(\Pi, P) = \sum_{j=1}^{m} p_j y_j - R \left(\frac{d_{x_n x_1}}{v_{\max} - vW_{x_n}} + \sum_{i=1}^{n-1} \frac{d_{x_i x_{i+1}}}{v_{\max} - vW_{x_i}} \right) \tag{1}$$

where $v = \frac{v_{\max} - v_{\min}}{W}$. If $y_j = 0$, then the item $j \in M$ is not stolen. Otherwise, it is stolen. W_x denotes the total weight of the knapsack when the thief leaves the city $x \in N$.

3 Traveling Thief Problem with Time Windows

In TTPTW, the city can only be visited in certain time windows. Given a set of time windows in each city $T = \{q_1, \ldots, q_n\}$, where each $q_i = (L_i, U_i)$. L_i and U_i are the lower and upper bound of the time window in city $i \in N$, respectively. The planned arrival time in city x_i is given as

$$t^*_{x_i}(\Pi, P) = t_{x_{i-1}}(\Pi, P) + \frac{d_{x_{i-1} x_i}}{v_{\max} - vW_{x_{i-1}}}. \tag{2}$$

where $v = \frac{v_{\max}-v_{\min}}{W}$ and represents a component of the travel speed update. We have $t_1 = t_1^* = 0$ and t_{x_i} is the exact arrival time in city x_i given by

$$t_{x_i}(\Pi, P) = \max(t^*_{x_i}(\Pi, P), L_{x_i}). \tag{3}$$

The planned and exact arrival time may be different as the thief has to wait during the trip if the planned arrival time $t^*_{x_i}$ is earlier than the lower bound of the time window in the city ($t^*_{x_i} < L_{x_i}$). If the thief has to wait, the waiting duration may affect the arrival times in the subsequent cities also. Therefore, the formulation of objective function and constraints for TTPTW is the following:

$$\begin{aligned}
\max_{\Pi, P} \quad & Z(\Pi, P) = \sum_{j=1}^{m} p_j y_j - R\left(\frac{d_{x_n x_1}}{v_{\max} - vW_{x_n}} + t_{x_n}\right) \\
\text{s.t.} \quad & \sum_{i=1}^{n}\sum_{j=1}^{m} w_j y_j \leq W \\
& x_1 = x_{n+1} = 1 \\
& x_a \neq x_b, \quad \forall x_a, x_b \in N\backslash\{1\} \\
& t_{x_i} \leq U_{x_i}
\end{aligned} \tag{4}$$

All constraints are deterministic. In order to handle infeasibility due to time window constraints, the value of time window constraints violation (cv) is:

$$cv(\Pi, P) = \sum_{i=1}^{n} \max(0, t_{x_i}(\Pi, P) - U_{x_i}) \tag{5}$$

This study uses the constraint-objective value separation technique [9], where the violations of the time-window constraints are calculated separately and not combined with the objective function value. A solution with a lower constraint violation is considered better. If two solutions have the same constraint violation value, then the one with the higher objective function value is considered better.

We do not employ a penalty-based method due to the highly variable scale of the objective function, which can take extremely negative values due to prolonged travel times even for feasible solutions. Under such condition, the penalty-based method may fail to properly balance constraint violation and objective value. Since this study considers hard time windows, it is very important to ensure that infeasible solutions are not preferred over feasible ones regardless of their objective values.

4 Adaptation of Existing Approaches

At first, we adapt the state-of-the-art TTP and TSPTW heuristics to solve TTPTW.

4.1 Adaptation of TTP Heuristics

We adapt S4, S5, C5 algorithms [13] to TTPTW by adding the time windows and the procedure to calculate the violation of the time window constraints in the evaluation function. We also modified all the solution comparison procedures using the constraint-objective value separation technique [9]. In the preliminary study using the same experimental setup in Sect. 7, we found that these algorithms cannot find any feasible solution in all datasets, except S4 in 51-A with positive l, as shown in Table 1.

In this study, we applied our proposed tour initialization algorithm (Sect. 5.2) to these algorithms (S4-Initial, S5-Initial, and C5-Initial) to create the initial tour at the beginning of the algorithms.

4.2 Adaptation of TSPTW Heuristics

We also adapt LKH-3 and VSR-LKH-3 for TTPTW, because they are state-of-the-art algorithms for TSP, including TSPTW. After finding the tour for TSPTW, the process continues to generate the packing plan based on the tour using PackIterative [13]. With the same experimental setup as in Sect. 7, we conducted the evaluation of these algorithms. Based on the results as shown in Table 1, VSR-LKH-3 barely finds a feasible solution, while LKH-3 can find feasible solutions, but not in all runs in the instances with more than 100 cities. The feasible rate decreases as the instance gets bigger.

Table 1. Feasible Rate of S4, S5, C5, LKH-3, and VSR-LKH-3 on TTPTW Benchmark

Instance	l	S4	S5	C5	LKH-3	VSR-LKH-3
51-A	100	40%	0%	0%	100%	13%
	1000	100%	0%	0%	100%	50%
	−100	0%	0%	0%	100%	0%
	−1000	0%	0%	0%	0%	0%
100-A	100	0%	0%	0%	97%	0%
	1000	0%	0%	0%	100%	0%
	−100	0%	0%	0%	97%	0%
	−1000	0%	0%	0%	97%	0%
150-A	100	0%	0%	0%	80%	0%
	1000	0%	0%	0%	97%	0%
	−100	0%	0%	0%	40%	0%
	−1000	0%	0%	0%	17%	0%
225-A	100	0%	0%	0%	67%	0%
	1000	0%	0%	0%	0%	0%
	−100	0%	0%	0%	0%	0%
	−1000	0%	0%	0%	0%	0%

5 Dual Search Evolutionary Algorithm (DSEA)

Because the adapted TTP and TSPTW algorithms do not perform well in solving TTPTW, we develop a set of algorithms for TTPTW, named Dual-Search Evolutionary Algorithm (DSEA). In this section, we discuss the main algorithm of DSEA, its variants, their components, and the packing plan repair approaches employed in this study.

5.1 Outline of the Algorithm and Its Variants

The pseudo-code of Dual Search Evolutionary Algorithm (DSEA) is shown in Algorithm 1. It begins with Tour Initialization (Sect. 5.2) to generate an initial tour (line 1). The process then continues to initialize the packing plan based on the tour using PackIterative algorithm [13] (line 2). The generated tour and packing plan are saved as the initial best solution (line 3). The current best solution is then optimized using Operator 1 (O1) and Operator 2 (O2) (line 10). After O2 ends, the best solution is updated if the new solution is better than the current best one (lines 11–15). Subsequently, a new packing plan is created using PackIterative based on the new tour (line 16). If the new tour and newly generated packing plan are better than the current best solution, the best solution is updated (lines 16–18). If in ρ consecutive main loops, no better solution is found, the tour mutation is conducted to enhance exploration before the tour is optimized again (lines 7–9). Mutation, O1, O2, PackIterative, and best solution update are repeated until the stopping criterion.

The mutation procedure is only a simple swap of two random cities in the current best tour for a random number of times between 1 and h. There is also an adaptation for the Pack algorithm by Faulkner et al. [13]. In this study, the algorithm performs an objective function and a constraint violation value evaluation every time it tries to add an item to the knapsack. Our preliminary study shows that this setting generates a significantly better packing plan. Three variants of DSEA are investigated in this study. Each of them has a different packing plan repair approach. Table 2 shows the differences between the DSEA variants.

Table 2. Variants of DSEA

Variant	O1	O2	Packing Plan Repair
$DSEA_1$	Topo	Rain	No Repair
$DSEA_2$	Topo	Rain	Repack
$DSEA_3$	ITP	IIP	Integration of Repair and Optimizer

5.2 Tour Initialization

As Faulkner et al. [13] stated that the initial tour really affects the final objective value, tour initialization is an important part that can help the main algorithm in finding a

Algorithm 1. Dual Search Evolutonary Algorithm (DSEA)

$\Pi_B \leftarrow$ TourInitialization()
$P_B \leftarrow$ PackIterative(Π_B, c, δ, q, e)
$B \leftarrow (\Pi_B, P_B)$
$k \leftarrow 0$
while not meet stopping criterion **do**
    $\Pi_x \leftarrow \Pi_B$
    if $k > \rho$ **then**
        $\Pi_x \leftarrow$ Mutate(Π_x)
    end if
    $(\Pi_x, P_x) \leftarrow$ O2(O1(Π_x, P_B))
    $X_1 \leftarrow (\Pi_x, P_x)$
    if X_1 is better than B **then**
        $B \leftarrow X_1$
        $k = 0$
    end if
    $X_2 \leftarrow (\Pi_x,$ PackIterative$(\Pi_x, c, \delta, q, e))$
    if X_2 is better than B **then**
        $X_1 \leftarrow X_2$
        $k = 0$
    end if
    $k{+}=1$
end while

feasible solution from the beginning. Instead of using CLK like majority studies on TTP, we proposed a tour initialization algorithm. This is because the shortest route, which CLK tries to find, is not always feasible due to time-window constraints. Moreover, the time windows significantly reduce the feasible area of a very large search space. It makes the search for feasible tours using CLK computationally very expensive. The tour initialization adapts the nearest neighborhood approach by adding memory and scoring procedure to generate a tour. The scoring procedure calculates the arrival time from the previous city x_i to all unvisited cities $j \in N$.

$$score_j(\Pi, P) = \begin{cases} travelTime_{x_i j}, & L_j \le t_j^* \le U_j, \\ -(travelTime_{x_i j} + M(t_j^* - U_j)), & \text{otherwise} \end{cases}$$

where $travelTime_{x_i j} = \frac{d_{x_i j}}{v_{\max}}$ and M is a large positive value to distinguish between the cities predicted to be reached on time, early, or late. If the arrival time in the next city, j, satisfies the time window in the city ($L_j \le t_j^* \le U_j$), the score of that city is the travel time from x_i to j. In contrast, if the time window is violated, then the score will be penalized. If the thief is late, the score will be negative, so it can be prioritized, whereas if the thief is too early, the city will be considered less important to visit next. This scoring will create a sequence of unvisited cities.

When a city is predicted to be visited late, the algorithm does not attempt to explore alternative sequences. It only avoids the constraint violation from increasing. This approach minimizes the computational effort as an initialization algorithm, because the solution will be refined in the optimization process later. It also avoids unnecessary searching when the instance has no any feasible solution.

It starts in city 1 and the scoring is conducted. The cities, except city 1, are then sorted according to their scores. This sequence and the reverse one are memorized as the first and second tours. After that, the city with the lowest score will be visited next. For example, if city 4 gets the lowest score, then the visited cities are $(1, 4)$, while city $2, 3, 5, 6, \ldots$ will be scored again from city 4. After all the cities are visited, a new tour is created based on the visit sequence and is memorized as the third tour. Its reverse is also memorized as the fourth one. The tour with the least constraint violation and the highest objective function value among all memorized tours is selected as the result.

5.3 Search Operators

The DSEA uses two different search operators in the same run (line 10 in Algorithm 1). One operator is 2-opt-based, while the other is insertion-based, as our preliminary study shows that the combination of these two operators helps the algorithm perform better and both of them complement each other.

For the 2-opt-based operator, we proposed Two-Opt with Perturbation Operator (Topo) and Integration of Two-opt and Pack (ITP). DSEA uses 2-opt as the basic move instead of the k-opt operator like Lin-Kernighan and LKH-3 for efficiency reasons and because [18] show that the 2-opt operator makes the most improvement compared to the others, e.g., 3-opt and 4-opt. For the insertion-based operator, we proposed Random Insertion (Rain) and Integration of Insertion and Pack (IIP). Insertion is used as the basic move, because S4 [13] which uses only insertion outperforms S5 and C5 according to Table 1 and Table 4.

Topo is used in $DSEA_1$ and $DSEA_2$ to optimize the current tour. It tries to reverse the subtour at random. After reverse, the new tour is evaluated with a probability of μ. If the new tour is better than the current tour, the current tour is updated with the new one. Otherwise, there will be packing plan repair if the algorithm uses the repair procedure. It takes the probability of μ as the perturbation to prevent the algorithm from being trapped in local optima. If not evaluated, the algorithm continues to reverse the next subtour. In other words, the probability of μ provides the possibility of creating a different tour, rather than conducting an evaluation after every reverse to enhance the exploration. This process repeats θ times.

Random Insertion (Rain) is used in $DSEA_1$ and $DSEA_2$ to optimize the tour by randomly inserting a city into another index in the tour. If the insertion gives a better result, the current tour will be updated with the new one. Otherwise, there will be packing plan repair if the algorithm uses the repair scheme. The operator also repeats θ times. Randomness is used to give different generated tours that might be a better tour, while without randomness, the insertion operator might give the same tour because it always does insertion in the exactly same procedure in each iteration.

Integration of Two-opt and Pack (ITP) is only used in $DSEA_3$. In the beginning, the item scores of all items are calculated. Then, the tour is generated by 2-opt randomly. After that, the item scores in the affected cities are updated. If a random number is greater than μ, the newly found solution will be evaluated and the current best solution updated. When the repetition counter is divisible by ρ, the packing plan is updated using

Pack; otherwise, the evaluated packing plan is the current one. If the new solution is better than the current best one, the best solution is updated. Like *Topo*, this entire procedure is repeated θ times.

Integration of Insertion and Pack (IIP) is only used in DSEA_3. It is similar to ITP, but the new tour is generated using random insertion and only the items in the inserted city are updated. We do not update the scores of the items in the rest city that move slightly because of the insertion to make it efficient. In addition, the evaluation is performed every time there is a new tour.

5.4 Packing Plan Repair

For packing plan repair, we propose two approaches, namely Repack and Integration of Repair and Optimizer. Repack approach is only used in DSEA_2. It is a modification of Pack algorithm, designed specially for time windows problem. The modification is based on the understanding that in TTP there are items that are not worth taking as it can make a loss, because the profit is too small, but the weight is too heavy. Firstly, it calculates the contributions of each item that was taken before the thief was late. If the solution after deleting the item is better than taking it, the item is deleted. After that, we use the similar pack algorithm, but only for the items that are previously not in the list to be taken, for the additional items until the knapsack is full. This repair approach is conducted if the counter in *Topo* or *Rain* is divisible by β.

Meanwhile, integration of repair and optimizer approach is only used in DSEA_3. It is proposed based on the consideration that after Topo or Rain operator, the tour changes, but only in some part. ITP and IIP fall under this approach. Instead of using PackIterative every time a new tour is generated, we use Pack algorithm as in the preliminary study, we found that Pack and PackIterative algorithms mostly give the same packing plan. Moreover, Pack is much faster than PackIterative. Similarly to the Repack approach, we do not repair the packing plan every time in order to save time. Thus, after updating the item scores, if the counter is divisible by β, the packing plan is repaired using Pack.

6 Benchmark Instances for the the Traveling Thief Problem with Time Windows

To investigate the performance of the proposed algorithm in solving the Traveling Thief Problem with Time Windows, we conducted an algorithm evaluation. In the following, we describe the new benchmark instances that we generated for TTPTW.

We use classical TTP benchmark instances introduced in [27][1] and add time windows to obtain TTPTW instances. The time windows are generated using the same method as in [12] and [5], specifically following the hard setting. The lower bound of the time windows in each city i, $L_i \sim \mathcal{U}(\max(0, t - l), t)$ and the upper bound $U_i \sim \mathcal{U}(t', (t' + l))$, where $l \in \{100, 1000, -100, -1000\}$; t_i and t'_i denote the arrival time of the thief in each city i, traveling at maximum speed $v_{\max}$ and minimum speed

[1] Instances available at https://cs.adelaide.edu.au/~optlog/CEC2014Comp/.

$v_{\min}$, respectively. Note that l denotes a parameter controlling the tightness of the time windows and does not represent the actual width of the time window. Considering that there are negative l, to ensure that there is a feasible solution that satisfies all the time windows constraints, we modified them into the more general formula: $L_i = \max(0, \min(t_i - \mathcal{U}(0,l), t_i'))$ and $U_i = \max(t_i' + \mathcal{U}(0,l), t_i)$. To ensure that the time windows in an instance with smaller l is smaller than the greater one, we use more conditional that if $l_a > l_b$, $L_i^{(l_a)} \leq L_i^{(l_b)}$ and $U_i^{(l_a)} \geq U_i^{(l_b)}$ where $L_i^{(l_c)}$ and $U_i^{(l_c)}$ denote the lower bound and upper bound of the time window in city i in the same TTP instance where $\exists l_c \in \{100, 1000, -100, -1000\}$.

We used various l values to study the performance of algorithms in different time window widths. We also consider very tight time windows as it naturally arises in several real-world applications such as emergency response services [28] and food delivery [32]. It also serves for the robustness evaluation of the proposed algorithms.

There are two types of instances: Type A which uses the optimal route in the corresponding TSP instance for generating the time windows, and Type B which uses the other tour than the TSP optimal tour to generate the time windows. Type A instances are used for the result comparison to the basic TTP as the optimal solution in the instances are already provided. Thus, the performance of the algorithms is measurable, as it contains the optimal solution in the basic TTP for instances with positive l.

Meanwhile, Type B is created as in the real-world, the time windows pattern is not always in line with the shortest route. The TSP optimal tour is not always the optimal tour in TTP, including in the problem with time windows. The tour for generating the time windows is generated randomly.

The instances consist of various numbers of cities, ranging from 51 to 1000, with different numbers of items, while the number of items ranges from 50 to 999. In addition, we also examined different knapsack instance types. We denote the instance name using the format x-KP where x is the number of cities and KP is the type of knapsack types. x-u denotes the instances where profit and weight are uncorrelated, while x-usw denotes the instance that has similar weights for all items but different profits [27]. x-A, x-B, and x-2 instances have bounded strongly correlated settings from [27], x-2 instances have the type A time windows and larger knapsack capacity and renting ratio (R) than the x-A and x-B. Meanwhile, x-A and x-B share the same knapsack component, knapsack capacity and R, but have different time windows. Table 3 in the extended version of this paper [4] presents the list of instances, along with the aliases assigned to each for ease of reference.

7 Experimental Investigations

We implemented the proposed algorithms, DSEA, using Python language. For comparison, we use the original code of S4, S5, C5, LKH-3[2] and VSR-LKH-3[3]. S4, S5, and C5 were written in Java, whereas LKH-3 and VSR-LKH-3 were written in C. At the end, the final objective and constraint violation values were calculated using the same

[2] available at http://webhotel4.ruc.dk/~keld/research/LKH-3/.

[3] available at https://github.com/JHL-HUST/VSR-LKH-V2.

evaluation function code to calibrate the results. For LKH-3 and VSR-LKH-3, after the final tour is generated, we call $PackIterative$ to find the packing plan based on the tour using the Python language. We used the custom parameters reported in [14] and [38]. For the parameters in DSEA, we set $\mu = 0.5, \theta = 1000, \beta = 10, \rho = 5, M = 10000, h = 10$ which were selected based on preliminary experiments that several candidate values were evaluated and those that provided the best performance were chosen, and $c = 5, \delta = 2.5, q = 20, e = 0.1$ as suggested in [13].

All algorithms are evaluated in 30 independent runs, where in each run, all algorithms stop after 1,000,000 Function Evaluations (FE). Each evaluation in any S4, S5, C5 and DSEA component, including the repair procedures, counts as one FE. Meanwhile, following their original implementations, LKH-3 and VSR-LKH-3 increment FE counter only after a complete-tour evaluation. We do not use CPU time as the stopping criterion, because the algorithms are not written in the same programming language. According to Ravber et al. [29], when CPU time cannot be applied, max FE is the best stopping criterion. We run all algorithms on an Ubuntu system with an Intel(R) Core(TM) i7-1255U (1.70 GHz CPU, 10 cores) and 16 GB of RAM. The results were analyzed using the Kruskal-Wallis test for the significance test and The Dunn-Bonferroni post-hoc test for the pair-wise test.

7.1 Impact of Packing Plan Repair Mechanisms

The evaluation results of the proposed algorithms with different packing plan mechanisms, $DSEA_1$-$DSEA_3$, are presented in Table 3. mean OB, std OB, FR, and stat denote the mean of the objective function values, the standard deviation of the objective function values, feasible rate, and the statistical test results, respectively. For example, $2(+), 3(-)$ means that $DSEA_1$ is significantly better than $DSEA_2$ and significantly worse than $DSEA_3$. $1(*), 3(-)$ means that there is no significant difference between $DSEA_1$ and $DSEA_2$, while $DSEA_2$ is significantly worse than $DSEA_3$. The Feasible Rate (FR) is the ratio of the number of runs that find a feasible solution to the total number of runs.

To provide a clearer overall comparison, we rank the algorithms according to their mean values of the objective function for each instance. The result is that $DSEA_1$, $DSEA_2$, and $DSEA_3$ have an average rank of 1.42, 2.33, and 2.00, respectively. In other words, $DSEA_1$, the proposed algorithm without packing plan repair, outperforms the other variants that use the packing plan repair mechanism, $DSEA_2$ and $DSEA_3$. Furthermore, Table 3 shows that all DSEA variants obtain $FR = 100\%$ in the majority instances, except when $l = -1000$ which means that they are consistent in finding feasible solution when the time windows are not too thigh. Although all DSEA variants have the same feasibility rate in almost all instances, there is a difference in 225-A with $l = -1000$. $DSEA_1$ consistently finds feasible solutions, whereas the other variants fail in some runs. This is because the repair procedure requires an additional computation cost.

In-depth analysis reveals that out of 1,000,000 FEs, the repair mechanism consumes substantial bugdet across instances (up to 260,000 FEs in $DSEA_2$ and up to 85,000 FEs in $DSEA_3$), but its success rate in finding better solutions is lower than 1%, even often drops to 0%. Taking into account that the difference between $DSEA_1$ and the others is only the use of the packing plan repair, it can be inferred that the absence of the repair mechanism benefits the algorithm. Rather than focusing on repair, $DSEA_1$ uses the full FE budget to explore more tours and generates the packing plan only in the last part of the main loop. This explains the superior performance of $DSEA_1$ in many instances.

However, $DSEA_1$ is stuck in the local optima in 1000-A and 1000-B instances due to the absence of packing plan repair. The proposed algorithm that employs the integration of the repair and optimizer, $DSEA_3$, on the other hand, shows better results in these instances because it explores better packing plans. $DSEA_3$ also shows a significantly better result in 51-B and not worse than the others in 51-A. It shows that $DSEA_3$ has good performance in small bounded-strongly-correlated instances only, because in the other knapsack categories (uncorrelated and uncorrelated-similar-weights), it is outperformed by $DSEA_1$.

Meanwhile, $DSEA_2$, the proposed algorithm that uses Repack, barely obtains a better result, showing its inferior performance. This is because Repack is quite expensive. The results also show that Repack is less accurate than the integration of repair and optimizer approach ($DSEA_3$) in repairing the packing plan.

7.2 Impact of Tour Initialization

Table 4 presents the evaluation results of S4, S5, and C5 with the proposed tour initialization algorithm. Compared to Table 1 where the algorithms barely find feasible solutions, the proposed tour initialization algorithm is shown to help them find feasible solutions, except in 51-A and 225-A with $l = -1000$ where the time windows are too tight. They get feasible solutions consistently in most instances because the tour initialization algorithm finds the feasible tour at the beginning. In contrast, when the initialization algorithm cannot find the feasible tour, as in 51-A and 225-A with $l = -1000$, the final solutions are still infeasible.

Table 4 shows that all standard deviations are zero for S5-Initial, where the algorithm starts with Tour Initialization, then repeatedly only generating new solutions using CLK and PackIterative. This is because the CLK fails to find a feasible and better solution than the tour from the initialization. Similarly, C5-Initial gets the same results, except in 51-A with positive l. In these instances, C5-Initial outperforms S4-Initial and S5-Initial. It shows that C5-Initial with a complete pack of the proposed tour initialization, bitflip, insertion, and new tours by CLK can only perform well in small instances. The reason is that it costs too many FEs to carry out many operators, which results in not being able to explore better solutions in larger instances.

Meanwhile, in bigger instances that have more than 51 cities, S4-Initial obtains better results than S5-Initial and C5-Initial, although it also shares the same results as the others in 100-A with $l = 100$ and 150-A with $l = 1000$. Thus, the continuation of modifying the initial tour with insertion only (S4-Initial) yields better results than the continuation of generating new tours using CLK and not modifying the initial tour like S5-Initial.

Table 3. Comparison on Different Packing Plan Repair Algorithms

Instance	l	$DSEA_1$ (1)				$DSEA_2$ (2)				$DSEA_3$ (3)			
		mean OB	std OB	FR	stat	mean OB	std OB	FR	stat	mean OB	std OB	FR	stat
51-A	100	3834.78	28.99	100%	2(*),3(*)	3836.13	36.09	100%	1(*),3(*)	3847.39	57.81	100%	1(*),2(*)
	1000	3749.81	199.21	100%	2(*),3(−)	3715.53	208.95	100%	1(*),3(−)	3936.83	156.35	100%	1(+),2(+)
	−100	3468.07	0	100%	2(+),3(*)	3467.91	0.20	100%	1(−),3(−)	3524.47	87.85	100%	1(*),2(*)
	−1000	−581.81	78.99	0%	2(*),3(*)	−607.15	34.79	0%	1(*),3(*)	−645.55	111.49	0%	1(*),2(*)
100-A	100	4533.98	0	100%	2(*),3(*)	4533.98	0	100%	1(*),3(*)	4533.98	0	100%	1(*),2(*)
	1000	4533.98	0	100%	2(*),3(+)	4533.98	0	100%	1(*),3(+)	4531.47	9.45	100%	1(−),2(−)
	−100	4513.85	0	100%	2(*),3(*)	4513.85	0	100%	1(*),3(*)	4513.85	0	100%	1(*),2(*)
	−1000	4246.19	0	100%	2(*),3(*)	4245.44	2.86	100%	1(*),3(*)	4244.82	4.33	100%	1(*),2(*)
150-A	100	8772.66	12.16	100%	2(+),3(*)	8736.63	63.04	100%	1(−),3(*)	8755.52	32.82	100%	1(*),2(*)
	1000	8732.84	63.46	100%	2(+),3(+)	8681.79	82.66	100%	1(−),3(+)	8595.41	130.16	100%	1(−),2(−)
	−100	8646.24	12.16	100%	2(+),3(+)	8619.09	53.75	100%	1(−),3(*)	8619.19	49.44	100%	1(−),2(*)
	−1000	7487.09	58.05	100%	2(+),3(+)	7409.64	87.64	100%	1(−),3(*)	7411.04	69.37	100%	1(−),2(*)
225-A	100	14319.98	333.14	100%	2(+),3(+)	14049.71	145.76	100%	1(−),3(*)	14040.05	176.09	100%	1(−),2(*)
	1000	11916.29	989.30	100%	2(+),3(*)	9784.18	707.95	100%	1(−),3(−)	11886.63	1080.67	100%	1(*),3(+)
	−100	14822.83	11.48	100%	2(+),3(+)	14788.32	16.47	100%	1(−),3(*)	14788.32	16.79	100%	1(−),2(*)
	−1000	11393.85	438.19	100%	2(+),3(+)	10297.09	441.97	77%	1(−),3(*)	10297.43	559.47	87%	1(−),2(*)
575-A	100	29477.29	256.00	100%	2(+),3(+)	28771.41	92.60	100%	1(−),3(*)	28811.27	142.93	100%	1(−),2(*)
	1000	18712.86	2765.69	100%	2(+),3(+)	−8425.47	1464.20	100%	1(−),3(*)	−9557.94	1426.20	100%	1(−),2(*)
	−100	29488.05	83.12	100%	2(+),3(+)	29369.99	50.48	100%	1(−),3(*)	29352.19	35.33	100%	1(−),2(*)
	−1000	16297.26	1792.34	0%	2(+),3(+)	−9601.48	1409.96	0%	1(−),3(*)	−10408.44	1201.69	0%	1(−),2(*)
1000-A	100	137195	0	100%	2(*),3(−)	137195	0	100%	2(*),3(−)	143251.53	463.14	100%	1(+),2(+)
	1000	137195	0	100%	2(*),3(−)	137195	0	100%	2(*),3(−)	143273.77	467.19	100%	1(+),2(+)
	−100	137195	0	100%	2(*),3(−)	137195	0	100%	2(*),3(−)	143299.47	506.09	100%	1(+),2(+)
	−1000	137195	0	100%	2(*),3(−)	137195	0	100%	2(*),3(−)	143253.13	436.82	100%	1(+),2(+)
51-B	100	2877.25	6.88	100%	2(*),3(−)	3097.31	265.02	100%	1(*),3(−)	3614.04	128.92	100%	1(+),2(+)
	1000	3822.12	217.59	100%	2(*),3(−)	3771.86	239.99	100%	1(*),3(−)	3950.61	101.59	100%	1(+),2(+)
	−100	2955.17	3.59	100%	2(*),3(−)	2952.59	17.21	100%	1(*),3(−)	3198.42	78.35	100%	1(+),2(+)
	−1000	−45.98	176.82	100%	2(+),3(*)	−138.90	199.22	100%	1(−),3(−)	54.98	137.46	100%	1(*),2(+)
225-B	100	1934.61	624.41	100%	2(+),3(+)	1035.27	496.54	100%	1(−),3(*)	1264.81	506.77	100%	1(−),2(*)
	1000	1554.94	1166.95	100%	2(+),3(+)	−847.27	1726.95	100%	1(−),3(*)	−1348.48	2028.23	100%	1(−),2(*)
	−100	1563.13	614.52	100%	2(+),3(+)	994.48	529.71	100%	1(−),3(*)	1066.07	479.07	100%	1(−),2(*)
	−1000	228.87	451.60	100%	2(+),3(+)	−526.11	548.16	100%	1(−),3(*)	−775.67	547.38	100%	1(−),2(*)
575-B	100	302.09	1054.91	100%	2(+),3(+)	−7514.01	2240.66	100%	1(−),3(*)	−8422.34	2271.64	100%	1(−),2(*)
	1000	−663.51	906.37	100%	2(+),3(+)	−10275.33	1339.09	100%	1(−),3(*)	−10561.94	1456.25	100%	1(−),2(*)
	−100	1973.34	665.31	100%	2(+),3(+)	−2544.96	1015.65	100%	1(−),3(*)	−2869.41	1006.13	100%	1(−),2(*)
	−1000	−8863.53	2302.44	0%	2(+),3(+)	−32388.60	2188.13	0%	1(−),3(*)	−31948.07	2693.64	0%	1(−),2(*)
1000-B	100	134481	0	100%	2(*),3(−)	134792.97	826.06	100%	1(*),3(−)	142834.43	692.54	100%	1(+),2(+)
	1000	134481	0	100%	2(*),3(−)	134583.73	272.61	100%	1(*),3(−)	143007.67	692.54	100%	1(+),2(+)
	−100	134481	0	100%	2(*),3(−)	134836.07	804.90	100%	1(*),3(−)	143105.5	445.67	100%	1(+),2(+)
	−1000	134481	0	100%	2(*),3(−)	134597.57	483.35	100%	1(*),3(−)	142822.17	629.66	100%	1(+),2(+)
51-2	100	5176.47	0	100%	2(+),3(+)	5175.56	3.49	100%	1(−),3(*)	5175.56	3.49	100%	1(−),2(*)
51-u	100	2081.86	11.39	100%	2(+),3(+)	2019.91	96.81	100%	1(−),3(*)	2040.86	102.48	100%	1(−),2(*)
51-usw	100	1304.19	17.29	100%	2(+),3(+)	1281.79	27.36	100%	1(−),3(*)	1291.20	19.71	100%	1(−),2(*)
150-2	100	11934.29	42.99	100%	2(*),3(+)	11894.68	95.72	100%	1(*),3(*)	11887.35	75.02	100%	1(−),2(*)
150-u	100	7286.82	35.59	100%	2(+),3(+)	7268.77	18.39	100%	1(−),3(*)	7270.01	18.31	100%	1(−),2(*)
150-usw	100	3901.91	8.34	100%	2(*),3(+)	3885.13	24.21	100%	1(*),3(+)	3884.37	25.83	100%	1(−),2(−)
225-2	100	21025.80	132.98	100%	2(+),3(+)	20095.64	685.48	100%	1(−),3(*)	19912.64	582.11	100%	1(−),2(*)
225-u	100	9069.59	74.21	100%	2(+),3(+)	8207.10	430.28	100%	1(−),3(*)	8259.03	469.68	100%	1(−),2(*)
225-usw	100	7244.99	290.97	100%	2(+),3(+)	6910.00	278.98	100%	1(−),3(*)	6855.07	251.93	100%	1(−),2(*)

7.3 Comparison of DSEA with Other Algorithms

According to the results, DSEA outperforms S4-Init, S5-Init, and C5-Init. Table 4 shows that with the same tour initialization algorithm, $DSEA_1$ finds better or at least equal results compared to the other algorithms in almost all instances, except in 100-A with $l = -1000$. It shows that the combination of mutation, Topo and Rain helps DSEA to explore better than Insertion, CLK, and their combination with Bitflip. Furthermore, with tour initialization, all these algorithms do not appear to become increasingly challenged in optimizing solutions as the time windows shrink or the number of cities increases.

Table 4. Performance of S4, S5, and C5 with Tour Initialization on TTPTW Benchmark and Their Comparison to $DSEA_1$

Instance	l	S4-Initial			S5-Initial			C5-Initial			$DSEA_1$		
		mean OB	std OB	FR	mean OB	std OB	FR	mean OB	std OB	FR	mean OB	std OB	FR
51-A	100	1120.40	1047.96	100%	308.65	0	100%	1323.34	1069.58	100%	3834.78	28.99	100%
	1000	2638.16	366.31	100%	498.85	0	100%	2876.58	305.54	100%	3749.81	199.21	100%
	−100	1948.84	0	100%	1948.84	0	100%	1948.84	0	100%	3468.07	0	100%
	−1000	−2842.08	0	0%	−2842.08	0	0%	−2842.08	0	0%	−581.81	78.99	0%
100-A	100	4533.98	0	100%	4533.98	0	100%	4533.98	0	100%	4533.98	0	100%
	1000	2663.62	791.01	100%	−1456.52	0	100%	−1456.52	0	100%	4533.98	0	100%
	−100	4513.85	0	100%	4513.85	0	100%	4513.85	0	100%	4513.85	0	100%
	−1000	4266.41	0	100%	3437.13	0	100%	3437.13	0	100%	4246.19	0	100%
150-A	100	7629.30	0	100%	4749.73	0	100%	4749.73	0	100%	8772.66	12.16	100%
	1000	945.95	0	100%	945.95	0	100%	945.95	0	100%	8732.84	63.46	100%
	−100	8361.17	0	100%	3294.41	0	100%	3294.41	0	100%	8646.24	12.16	100%
	−1000	4498.31	44.84	100%	−3634.12	0	100%	−3634.12	0	100%	7487.09	58.05	100%
225-A	100	10262.16	46.57	100%	3322.41	0	100%	3322.41	0	100%	14319.98	333.14	100%
	1000	11406.45	50.71	100%	8305.97	0	100%	8305.97	0	100%	11916.29	989.30	100%
	−100	12499.20	0	100%	189.53	0	100%	189.53	0	100%	14822.83	11.48	100%
	−1000	4637.54	749.37	0%	−3722.13	0	0%	−3722.13	0	0%	11393.85	438.19	100%

Based on the feasible rate in all instances in Table 1 and 4, it is also clear that DSEA is better than S4, S5, C5, LKH-3, and VSR-LKH-3 that find difficulty obtaining feasible solutions. Thus, DSEA outperforms the other algorithms in this study.

8 Conclusion

As many real-world problems today involve more than one optimization component and have limited resources, we investigate the Traveling Thief Problem (TTP) which is a combination of the Traveling Salesman Problem (TSP) and the Knapsack Problem (KP) that interdependence to each other. Moreover, because time is often a critical and limited resource in many real-world applications, this study extends the TTP with time window constraints (TTPTW). We proposed the Dual Search Evolutionary Algorithm (DSEA) for TTPTW. In general, the DSEA algorithms outperform S4, S5, C5, S4-Initial, S5-Initial, C5-Initial, LKH-3, and VSR-LKH-3 to solve TTPTW and perform well in various settings of the time window. Moreover, the proposed tour initialization algorithm helps the algorithms find feasible tours. Based on the experimental result, the addition of the proposed tour initialization to S4, S5 and C5 resulted in a significant improvement compared to S4, S5 and C5 without initialization. DSEA algorithms without packing plan repair ($DSEA_1$) is the best variant among the DSEA variants. This newly defined problem and the benchmark may also be used in future work.

Acknowledgments. This work has been supported by the Australian Research Council through grant FT200100536 and by the Indonesia Endowment Fund for Education (LPDP) from the Ministry of Finance, Republic of Indonesia.

References

1. Alharbi, S.T.: The design and development of a modified artificial bee colony approach for the traveling thief problem. Int. J. Appl. Evol. Comput. **9**(3), 32–47 (2018)

2. Alharbi, S.T.: A hybrid genetic algorithm with tabu search for optimization of the traveling thief problem. Int. J. Adv. Comput. Sci. Appl. **11**, 276–287 (2018)
3. Alshaikh, R., Abdelfatah, A.: Optimization techniques in municipal solid waste management: a systematic review. Sustainability **16**(15) (2024)
4. Angmalisang, H.Y., Neumann, F.: The traveling thief problem with time windows: benchmarks and heuristics. CoRR **abs/2604.06724** (2026)
5. Bi, J., et al.: Learning to handle complex constraints for vehicle routing problems. In: NeurIPS (2024)
6. Bonyadi, M.R., Michalewicz, Z., Barone, L.: The travelling thief problem: the first step in the transition from theoretical problems to realistic problems. In: IEEE Congress on Evolutionary Computation, pp. 1037–1044. IEEE (2013)
7. Bonyadi, M.R., Michalewicz, Z., Przybylek, M.R., Wierzbicki, A.: Socially inspired algorithms for the travelling thief problem. In: GECCO, pp. 421–428. ACM (2014)
8. Chagas, J.B.C., Wagner, M.: A weighted-sum method for solving the bi-objective traveling thief problem. Comput. Oper. Res. **138**, 105560 (2022)
9. Deb, K.: An efficient constraint handling method for genetic algorithms. Comput. Methods Appl. Mech. Eng. **186**(2), 311–338 (2000)
10. Don, T.P., Neumann, A., Neumann, F.: The chance constrained travelling thief problem: problem formulations and algorithms. In: GECCO. ACM (2024)
11. Don, T.P., Neumann, A., Neumann, F.: Weighted-scenario optimisation for the chance constrained travelling thief problem. In: CEC, pp. 1–8. IEEE (2025)
12. Dumas, Y., Desrosiers, J., Gélinas, É., Solomon, M.M.: An optimal algorithm for the traveling salesman problem with time windows. Oper. Res. **43**(2), 367–371 (1995)
13. Faulkner, H., Polyakovskiy, S., Schultz, T., Wagner, M.: Approximate approaches to the traveling thief problem. In: GECCO, pp. 385–392. ACM (2015)
14. Helsgaun, K.: An extension of the lin-kernighan-helsgaun TSP solver for constrained traveling salesman and vehicle routing problems. Technical report. Technical Report, Roskilde University (2017)
15. Huang, S.H., Lin, P.C.: Vehicle routing-scheduling for municipal waste collection system under the “keep trash off the ground” policy. Omega **55**, 24–37 (2015)
16. Kim, W., Kim, H., Chang, G.L.: Design of real-time emergency response system for highway networks: application for high frequency of traffic emergency events during peak hours. Transp. Res. Rec. **2484**(1), 70–79 (2015)
17. Lin, S., Kernighan, B.W.: An effective heuristic algorithm for the traveling-salesman problem. Oper. Res. **21**(2), 498–516 (1973)
18. Liu, X., Liu, D., Yang, Q., Liu, X., Yu, W., Zhang, J.: Comparative analysis of five local search operators on visiting constrained multiple traveling salesmen problem. In: SSCI, pp. 1–8. IEEE (2021)
19. Maity, A., Das, S.: Efficient hybrid local search heuristics for solving the travelling thief problem. Appl. Soft Comput. **93**, 106284 (2020)
20. Martins Queiroz, F.H., et al.: Route optimization in home physiotherapy: a traveling salesman problem-based framework for health care logistics efficiency. Procedia Comput. Sci. **266**, 856–863 (2025)
21. Mei, Y., Li, X., Salim, F.D., Yao, X.: Heuristic evolution with genetic programming for traveling thief problem. In: CEC, pp. 2753–2760. IEEE (2015)
22. Mei, Y., Li, X., Yao, X.: Improving efficiency of heuristics for the large scale traveling thief problem. In: SEAL, pp. 631–643. LNCS. Springer, Cham (2014)
23. Nguyen, H., Le, N., Tran, K., Luong, N.H.: Simulated annealing with dynamic programming-based vertex insertion for efficiently solving the traveling thief problem. In: SoICT, pp. 165–172. ACM (2023)

24. Nikfarjam, A., Neumann, A., Bossek, J., Neumann, F.: Co-evolutionary diversity optimisation for the traveling thief problem. In: PPSN (1), pp. 237–249. LNCS. Springer, Cham (2022)
25. Nikfarjam, A., Neumann, A., Neumann, F.: Evolutionary diversity optimisation for the traveling thief problem. In: GECCO, pp. 749–756. ACM (2022)
26. Nikfarjam, A., Neumann, A., Neumann, F.: On the use of quality diversity algorithms for the traveling thief problem. In: GECCO, pp. 260–268. ACM (2022)
27. Polyakovskiy, S., Bonyadi, M.R., Wagner, M., Michalewicz, Z., Neumann, F.: A comprehensive benchmark set and heuristics for the traveling thief problem. In: GECCO, pp. 477–484. ACM (2014)
28. Pu, F., Li, Z., Wu, Y., Ma, C., Zhao, R.: Recent advances in disaster emergency response planning: integrating optimization, machine learning, and simulation. Saf. Emerg. Sci. **1**(1), 9590007 (2025)
29. Ravber, M., Liu, S., Mernik, M., Crepinsek, M.: Maximum number of generations as a stopping criterion considered harmful. Appl. Soft Comput. **128**, 109478 (2022)
30. Vieira, D.K.S., Soares, G.L., Vasconcelos, J.A., Mendes, M.H.S.: A genetic algorithm for multi-component optimization problems: The case of the travelling thief problem. In: EvoCOP, pp. 18–29. LNCS. Springer, Cham (2017)
31. Wagner, M., Lindauer, M., Misir, M., Nallaperuma, S., Hutter, F.: A case study of algorithm selection for the traveling thief problem. J. Heuristics **24**(3), 295–320 (2018)
32. Wang, X., Ji, C., Xu, H., Guo, K.: Research on dynamic optimization of takeout delivery routes considering food preparation time. Sustainability **17**(6) (2025)
33. Wuijts, R.H., Thierens, D.: Investigation of the traveling thief problem. In: GECCO, pp. 329–337. ACM (2019)
34. Yafrani, M.E., Ahiod, B.: Population-based vs. single-solution heuristics for the travelling thief problem. In: GECCO, pp. 317–324. ACM (2016)
35. Yafrani, M.E., Ahiod, B.: Efficiently solving the traveling thief problem using hill climbing and simulated annealing. Inf. Sci. **432**, 231–244 (2018)
36. Zhang, Z., Yang, L., Kang, P., Jia, X., Zhang, W.: Solving the traveling thief problem based on item selection weight and reverse-order allocation. IEEE Access **9**, 54056–54066 (2021)
37. Zheng, J., He, K., Zhou, J., Jin, Y., Li, C.: Combining reinforcement learning with lin-kernighan-helsgaun algorithm for the traveling salesman problem. In: AAAI, pp. 12445–12452. AAAI Press (2021)
38. Zheng, J., He, K., Zhou, J., Jin, Y., Li, C.: Reinforced lin-kernighan-helsgaun algorithms for the traveling salesman problems. Knowl. Based Syst. **260**, 110144 (2023)

Untangling the Tapestry: k-Bounded Problems and Local Optima Networks

Sarah L. Thomson[1](✉) and Michal Przewozniczek[2]

[1] Edinburgh Napier University, Edinburgh, UK
s.thomson4@napier.ac.uk
[2] Wroclaw University of Science and Technology, Wroclaw, Poland

Abstract. Local optima networks (LONs) have recently been used to analyse subfunction structure in problems which contain strong linkage. However, many questions remain: for example, it is not known how LON subfunction structure relates to optimiser performance on the underlying problems. In addition, the potential of using the associated features to distinguish between classes of problem is not well understood. Therefore we address these gaps. We construct LONs in two different ways, propose a new set of features (which are based upon the structure of subfunction argument sets), and investigate the associations with performance of three state-of-the-art optimisers. The results indicate that the subfunction-related features may depend on the LON construction algorithm and that some of them are useful for separating problem classes. Also, the proposed features seem to relate to the performance of some optimisers for some problem-operator combinations. The approach and the findings together not only provide new insight about the interplay between the studied problems and optimisers, but also lay the groundwork for possible online LON-based operator adaptation in the future.

Keywords: fitness landscape analysis · local optima networks · combinatorial optimisation · visualisation

1 Introduction

Local optima networks (LONs) [16] are a fitness landscape methodology which help to understand the relationship between optimisation problems and algorithms. LONs have brought insight across several domains [10,15,24,25].

Some problems, such as MAX3SAT or problems built from deceptive trap functions, have strong linkage structure. These problems have fitness functions which are composed of many sub-functions, taking as arguments a subset of the decision variables. If subfunctions are summarised, these problems are denoted as k-bounded [33], with k being the maximum number of arguments for a subfunction. It naturally follows that analysis of k-bounded fitness landscapes should specifically consider the interaction structure. However, most of the time LONs are analysed in a manner that does not consider the existence of linkage in the

G. Iacca et al. (Eds.): PPSN 2026, LNCS 16985, pp. 616–632, 2027.
https://doi.org/10.1007/978-3-032-36214-8_38

underlying problem. A few studies [4,15] have constructed LONs using linkage-aware optimisers. Even so, the obtained LONs were usually analysed in a way which did not directly account for the interaction structure. As a first step towards that aim, LONs were recently subject to a subfunction-focussed analysis for the first time [28]. However: that work did not link the observations to optimiser performance, did not consider whether LON subfunction features can separate problem classes, and only studied problems of toy size. Therefore, here we propose the following contributions:

- We propose a new set of LON features and show that they are related to the performance of some state-of-the-art linkage learning algorithms
- We verify and experimentally show that the proposed LON subfunction-related features can successfully separate problem classes
- Compared to the initial work [28], we consider a substantial increase in problem sizes

2 Preliminaries

2.1 Optimisation Problems

We consider the maximisation of $f(\boldsymbol{x})$, where $\boldsymbol{x} = [x_1, x_2, \ldots, x_n]$ is a binary solution vector of size n. Standard deceptive functions [6] are frequently employed to construct benchmark problems in black-box-oriented research [8,21,26]. We consider standard and bimodal deceptive functions with unitation [7] of order k defined as follows.

$$trap_k(u) = \begin{cases} k - u - 1 & , u < k \\ k & , u = k \end{cases} \tag{1}$$

where u is the *unitation*, i.e., the sum of binary values. Note that standard deceptive functions have two local optima (for $u = 0$) and (for $u = k$) from which one is global (for $u = k$). Such functions may be considered as hard to optimise because local search algorithms will usually converge to the suboptimal solution built only from zeroes. In contrast to the standard deceptive functions, bimodal functions are characterised by a large locally optimal solution set [7]

$$bimTrap_k(u) = \begin{cases} k/2 - |u - k/2| - 1 & , u \neq k \wedge u \neq 0 \\ k/2 & , u = k \vee u = 0 \end{cases} \tag{2}$$

Bimodal functions have two global optima built from all zeroes and ones, respectively. These problems have $\binom{k/2}{k}$ or $\binom{k/2+1}{k}$ locally optimal solutions for even and odd k, respectively. Deceptive functions are tools used to construct problems of various subfunction-based representations. Therefore, one can construct a problem with given features and verify if a given optimiser can successfully solve it. In concatenations, subfunctions (represented by deceptive functions) do not share arguments. Therefore, each subfunction can be optimised separately. Overlapping problems contain subfunctions that share some of their variables. We will denote shared variables as the *overlaps* [11,23]. These can be considered as significantly harder to solve because modifying the argument set of one subfunction may affect the other subfunctions. In cycling traps, each subfunction shares o of its variables with its two neighbours [37].

As well as deceptive functions, we have employed artificial MAX3SAT instances as benchmarks. In MAX3SAT, one has to satisfy 3-element logical clauses that share variables. The problem is NP-hard, and the employed instances can be considered as hard to solve [33]. We also consider Ising Spin Glass (ISG); here, each variable represents a spin and its value can be set to -1 or 1 [22]. Some spins may be connected with a quadratic interaction and their coupling constant requires setting them to the same or different values. We consider a $2D \pm J$ version of ISG. We maximise the following sum:

$$\sum_{i,j=0}^{n-1} x_i x_j J_{i,j} \tag{3}$$

where $J_{i,j}$ is the interaction strength between spins i and j.

2.2 Local Optima Networks

A solution is a local optimum if its fitness according to a fitness function f is better or equal to those of its neighbours [13]. We use a *sampled* neighbourhood in this work. So: the solution lo_i is a local optimum if its fitness is better than that of the neighbours in its *sampled* neighbourhood SN (we stipulate that it must be *strictly* better in order to support additional local optima filtering). Formally speaking: $\forall neigh \in SN(lo_i) : f(lo_i) >= f(neigh)$ (assuming maximisation, which is the case for all problems in the present work) where $SN(lo_i)$ is the neighbourhood of the local optimum and $neigh$ is a particular neighbour. The nodes in a LON can be referred to as LO. Two local optima lo_i and lo_j have a directed edge traced between them if lo_j can be reached by applying an operation (usually a random perturbation) to lo_i and then subsequently local search, and additionally $f(lo_j) \geq f(lo_i)$ (maximisation). This last criterion renders the edges *monotonic* in nature because they record only non-deteriorating, directed connections between local optima [17] i.e. fitness never worsens along an edge in this model. Edges have weights: the frequency of transition between the two local optima (i.e., the number of times a LON construction algorithm went from lo_i to lo_j). The set of edges is denoted by E. We can now define a LON $= (LO, E)$, which comprises nodes $lo_i \in LO$ i.e. the local optima, and edges $e_{ij} \in E$ between pairs of nodes lo_i and lo_j with weight w_{ij} iff $w_{ij} > 0$.

3 Related Work

3.1 Local Optima Networks

The majority of works on local optima networks have approached them in a black-box way i.e. subfunction structure was not directly considered [10,16,27,

31]. Nevertheless, some works consider the grey-box approach to construct LONs. In [4], a grey-box iterated local search with recombination was applied to optimise NK landscapes and construct LONs. These LONs were visualised but were not subject to analysis of their subfunction/linkage structure. In [15] grey-box operators were employed to construct LONs for MAX3SAT, but the presented metrics did not refer to the structure of variable interactions. In [3,24] LONs are extracted using grey-box operators for PUBOi and NKQ problems, respectively. Although subfunction structure is not considered in the LON analysis, the papers show a growing trend in the community towards utilising problem structure in landscape (and specifically LON) analysis. A recent work proposed to analyse LONs in a subfunction-aware manner [28] and demonstrated that doing so can bring additional insight. However, as discussed in Sect. 1, the study left many avenues open.

3.2 Variable Dependencies in Evolutionary Optimisation

In grey-box optimisation scenarios, variable dependencies are known *a-priori* [32]. They are used to improve the effectiveness of an optimiser, e.g., by constructing perturbation masks that tend to cover groups of subfunction arguments rather than randomly-selected genes [35]. The *additive form* can be employed to represent subfunction-based problem structure [33]:

$$f(\boldsymbol{x}) = \sum_{s=1}^{S} f_s(\boldsymbol{x}_{I_s}), \tag{4}$$

where I_s are subsets of $\{1, ..., n\}$ that may overlap, i.e., share variables, and S is the number of these subsets.

Problems which are k-bounded can be represented in the additive form in such a way that each subfunction takes no more than k arguments. Knowing the additive form, one can use so-called *partial evaluations*. In the general case, to evaluate a solution, the value of all subfunctions must be computed. However, if we know their values and modify only a small subset of variables, it may be enough to recompute only a small part of all subfunctions to get an exact evaluation—paying only a fraction of the overall evaluation cost [2,36].

In grey-box optimisation, non-linear dependencies are frequently considered. Variables x_g and x_h are dependent if there exists $\boldsymbol{x}$ for which [14]:

$$f(\boldsymbol{x}) + f(\boldsymbol{x}^{g,h}) \neq f(\boldsymbol{x}^g) + f(\boldsymbol{x}^h) \tag{5}$$

where $\boldsymbol{x}^g$, $\boldsymbol{x}^h$, and $\boldsymbol{x}^{g,h}$, are obtained from $\boldsymbol{x}$ by flipping x_g, x_h or both of them. One can construct the Walsh-based representation of the optimised function [9]

$$f(\boldsymbol{x}) = \sum_{i=0}^{2^n-1} w_i \varphi_i(\boldsymbol{x}) \tag{6}$$

where $w_i \in \mathbb{R}$ is the ith Walsh coefficient, $\varphi_i(\mathbf{x}) = (-1)^{\mathbf{i}^\mathrm{T}\mathbf{x}}$ generates a sign, and $\mathbf{i} \in \{0,1\}^n$ is the binary representation of index i. Any non-zero Walsh coefficient

w_i indicates that x_g and x_h are non-linearly dependent on each other [29]. Such knowledge is frequently represented in a form of the Variable Interaction Graph (VIG) [5]. VIGs are frequently represented by square matrices, in which entries equal one if two variables are dependent and zero, otherwise. VIGs are frequently utilised by grey-box operators.

Partition Crossover (PX). Uses a VIG to construct mixing masks [30]. For a pair of individuals, all edges referring to variables that are equal in both individuals are removed from the VIG. The differing variables are clustered using the remaining edges. PX masks are proven to preserve the subfunction argument sets even if they overlap [20]. This feature makes PX highly effective in solving overlapping problems. **VIG-based Perturbation (VIGbp)** is where Iterated Local Search (ILS) utilises the VIG to form perturbation masks of dependent variables. In such a scenario, for the randomly chosen variable x_r all variables dependent on x_r (or their subset) are clustered and then flipped or re-initialised [29]. After this operation, the solution is optimised with a greedy approach. Such a procedure might be highly effective for problems in which the overlaps are so severe that PX frequently yields full masks that are useless for mixing. The Granular Linkage Blocks Optimiser (GLBopt) [19] is a black-box-dedicated approach that constructs an empirical VIG (eVIG) during its run. It discovers dependencies using direct dependency checks, which assures discovering only the existing dependencies. Using eVIG, GLBopt constructs the maximal complete subgraphs (cliques). Later on, it collects the locally optimal variable value combinations for each clique and utilises this knowledge during the local search procedure. The cliques are named **Granular Linkage Blocks (GLBs)**.

4 Methods

4.1 LON Construction

In standard iterated local search (ILS), a single solution is iteratively improved using a chain of repeated cycles comprising perturbation followed by local search [12]. In ILS with PX for LON construction i.e. PX-LON, instead of perturbation chaining together the local searches, we use PX (see Sect. 3.2). This algorithm was proposed in [4] and is called deterministic recombination iterated local search (DRILS). The algorithm is grey-box: the VIG is provided a-priori.

In ILS with VIGbp for LON construction i.e. VIGBP-LON, we use VIGbp, the dependency-driven perturbation operator (see Sect. 3.2). The algorithm is based on the approach in a previous work [29]. We use a first-improvement hill climber with linkage learning (FIHCwLL) [21] to locally optimise solutions and learn dependencies simultaneously. The VIG is not provided a-priori but is learned during search. Notice this means that we have one construction method which is grey-box (PX-LON) and one which is not (VIGBP-LON, which actively learns linkage structure). This is because we want to conduct the investigation under substantially different LON construction settings. FIHCwLL optimises a solution by flipping its genes. If a gene flip improves fitness, then it is preserved, otherwise it is rejected. In each iteration, FIHCwLL considers all genes in a random order. Iterations are executed until at least one gene is modified during the preceding iteration. Fitness evaluations that are used for local optimisation are also used to perform variable dependency checks. Since we consider k-bounded problems that are additively decomposable, we use the non-linearity dependency check (Formula 5) that requires performing four fitness evaluations, i.e., $f(\boldsymbol{x})$, $f(\boldsymbol{x}^g)$, $f(\boldsymbol{x}^h)$, and $f(\boldsymbol{x}^{g,h})$. In FIHCwLL, three of them are the side-effect of the local search, and only one must be computed for linkage discovery purposes, which significantly reduces the cost of decomposition.

4.2 LON Metrics

We compute metrics relating to subfunction structure in LONs. To do this, we decompose the fitness function into a set of GLBs (recall Sect. 3.2). Walsh coefficients (recall Eq. 6) are assigned to GLBs according to coverage of their variables. This leads to a set of m additive subfunctions where $q \in \{1, \dots, m\}$ indexes the subfunctions and each of them is defined on a subset of variables S_q.

- `pairwise_mass_ratio` (*pmr*): the total Walsh coefficient amount coming from order-1 and order-2 Walsh terms (i.e. **low order Walsh mass**), divided by the total Walsh mass for the problem. This is not a LON metric; it captures the ratio of low-order interactions within a problem. It is formalised as $pmr = (w_0, ..., w_{2^n-1}) \sum_{w_s \in W_s} |w_s| / \sum_{i=1}^{2^n-1} |w_i|$ where W_s is a set of all w_i such that $\mathbf{i} \in \{0,1\}^n$ (the binary representation of index i) marks only one or two positions, i.e., $\mathbf{i}$ contains no more than 2 ones.
- `touched subf_count` (*tsc*): along a given LON edge, we consider which proportion of subfunctions have flipped variables when moving from the source to the destination. If $u_j \in \{0,1\}$ is the value of the variable at index j in solution u then for the source u and destination v of a LON edge $tsc(u,v) = \frac{1}{m} |\{q \in \{1, \dots, m\} : S_q \cap \{j : u_j \neq v_j\} \neq \emptyset\}|$ i.e. the proportion (over m total subfunctions) of subfunctions q (each of which are defined for a subset of variables S_q) that include at least one index j where $u_j \neq v_j$. We consider the mean *tsc* across all edges in the LON.
- `max_abs delta` (*mad*): for a given LON edge, we take the maximum absolute differential in subfunction value (across all subfunctions) from the source local optimum to the destination local optimum. For an edge from u to v, $mad(u,v) = \max_{q \in \{1,\dots,m\}} |s_q(v) - s_q(u)|$, where $s_q(u)$ is the value of subfunction q at solution u. We consider the mean *mad* across all edges in the LON.
- `mean_abs_delta touched` (*madt*): for a given LON edge, we consider only the subfunctions which had flipped variables. For those subfunctions only, we take the mean absolute subfunction value differential between the source and destination i.e. $madt(u,v) = \frac{1}{|T(u,v)|} \sum_{i \in T(u,v)} |s_i(v) - s_i(u)|$ where $T(u,v) = \{q \in \{1, \dots, m\} : S_q \cap \{j : u_j \neq v_j\} \neq \emptyset\}$. We consider the mean *madt* across all edges in the LON.
- `srcedge_energy` (*sre*): for a given LON edge, we compute the low-order Walsh contributions in the source u, i.e.

$sre(u) = \sum_{\substack{i\in\{0,1\}^n \\ |i|=1}} w_i\,\varphi_i(u) + \sum_{\substack{i\in\{0,1\}^n \\ |i|=2}} w_i\,\varphi_i(u)$. consider the mean *sre* across all edges in the LON.

- `abs_delta_energy` (*ade*): for a given LON edge, we compute the low-order Walsh contributions in the source and in the destination. We then take the absolute differential in low-order Walsh contributions i.e. $ade(u, v) = |sre(v) - sre(u)|$. We consider the mean *ade* across all edges in the LON. Both `srcedge_energy` and `abs_delta_energy` are only computed if the problem has a majority (70% or more) low-order Walsh mass, i.e. `pairwise_mass_ratio` is greater than or equal to 0.7.

5 Experimental Setup

5.1 Problem Instance Generation

We consider well-known benchmarks for linkage learning: deceptive functions, MAX3SAT, and ISG. The instances are substantially larger than those used in previous work [28]: while the previous sizes used n between 15 and 18, we here use n ranging from 200 to 1024. We have considered the following problem instances: MAX3SAT (200-bit, $cr = 4.27$), ISG (1024-bit), and problems of size 999–1002 bits built from standard deceptive functions (order-3 and order-5 subfunctions), bimodal deceptive functions (order-6 and order-10 subfunctions), and noised bimodal deceptive functions (order-10) [18]. For deceptive functions, we have considered their concatenations and cyclic traps of 100-bit length (or as close to it as possible), and overlap sizes varying from 1 to 9.

5.2 LON Construction

For each LON construction algorithm there are 30 independent runs per problem instance. The perturbation strength is set at $0.2 \cdot n$, where n is the problem dimension. ILS runs terminate when there has not been a strict improvement to the best-found solution fitness in 30 full cycles; this value was selected to help mitigate large computational expense. For the same reason, we set an upper limit of 30 new local optima discovered during individual runs. Only improving or equal local optima are accepted and logged.

5.3 Search Performance

We consider three state-of-the-art linkage-aware optimisers: 1) LT-GOMEA [1], 2) parameter-less population pyramid (P3) [8], and 3) GLB optimisation [19]. All considered optimisers are parameter-less and were joined in one C++ project. A time-based stop condition of 8 h of computation time was used. To assure fair comparisons, all optimisers shared parts of the source code where possible. Each experiment was executed in a clean environment and was single-threaded. The number of computation processes was kept constant and did not extend the number of physical processor cores. The experiments were conducted on the PowerEdge R6525 Dell Server AMD 2 Epyc 7H12 2.6 GHz 1 TB RAM. Thirty independent runs were performed for each algorithm-instance pair. Performance is measured as the mean number of evaluations required to reach the optimum, computed over successful runs only, where a run is considered successful if it reaches the optimum within an 8-hour computational budget[1]. This conditional measure focuses on convergence efficiency. Algorithm-instance pairs with zero successful runs are excluded from the analysis. After this filtering, an average of 24.22 instances per algorithm and problem-class combination remain. Across the included algorithm-instance pairs, the mean number of successful runs (out of 30) is 27.98.

5.4 Modelling

We conduct regression modelling. The predictors are LON features (Sect. 4.2); the response variable is the base-10 logarithm of the mean evaluations to reach the optimum across successful runs for each algorithm (Sect. 5.3). We employ Bayesian robust regression with Student-t likelihood (degrees of freedom = 4) to accommodate small sample sizes and potential outliers. Both predictors and response are standardised to zero mean and unit variance. Weakly informative priors are specified on this standardised scale: a normal distribution with mean 0 and standard deviation 2 for the intercept; normal distribution with mean 0 and standard deviation 1 for the slope; and half-normal distribution with scale 1 for the residual standard deviation. Posteriors are estimated via random-walk Metropolis-Hastings (40000 iterations, 10000 burn-in, thinning by 10, yielding 3000 draws).

5.5 Statistical Testing

When comparing the distribution of LON features between different problem types, we use Mann-Whitney U tests (which can handle unequal sample sizes) and we estimate effect sizes using cliff's δ. The p-values of the Mann-Whitney tests are adjusted using Benjamini-Hochberg correction (across all tests within a LON construction type) to account for the effect of multiple tests. Correlation heatmaps, which present the relationship between LON features and algorithm runtime, are produced using the Spearman's rank correlation coefficient and by applying the Benjamini-Hochberg procedure to adjust the p-values across all tests within the scope of a LON construction method.

6 Results

In Fig. 1, we present the VIGBP-LON and PX-LON for the concatenation of bim_6 functions which do not share variables i.e. do not overlap (Figs. 1a and

[1] Success rates (grouped by algorithm and problem class) are available in the supplemental material [see footnote 2].

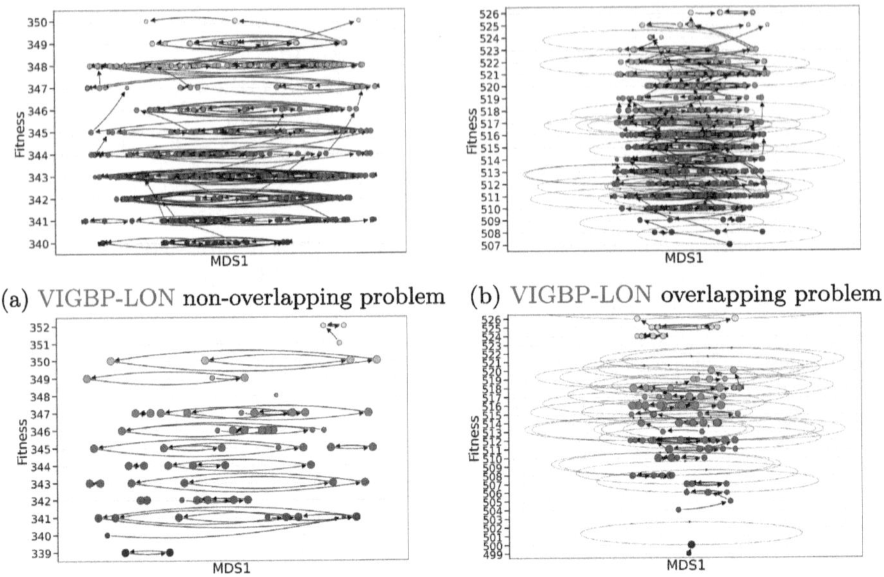

(a) VIGBP-LON non-overlapping problem (b) VIGBP-LON overlapping problem

(c) PX-LON non-overlapping problem (d) PX-LON overlapping problem

Fig. 1. LONs for two deceptive trap functions: (a) and (c) are for the concatenation of bim_6 functions which do not share variables i.e. do not overlap, while (b) and (d) are for the overlapping problem built from bim_6 blocks and with an overlap, o, of 2. The concatenation and the overlapping problem are both of the (almost) same size ($n = 1002$ and $n = 1000$, respectively). Position on the x-axis is obtained through multi-dimensional scaling of the binary solutions using Hamming distance.

1c), and the VIGBP-LON and PX-LON for the overlapping problem built from bim_6 (Figs. 1b and 1d, respectively). Even though the LONs in Figs. 1a and 1b consider functions built from the same blocks and using the same LON construction procedure, they differ significantly. In the concatenated case, the LON is sparser between different fitness levels because each VIGbp step affects only one bim_6 block, since blocks have no dependencies that connect them. Thus, the block is re-initialised and locally optimised. An improving edge is obtained if this operation improves a given bim_6 block. For the overlapping problem constructed from bim_6, the situation differs because every variable is directly or indirectly (if x_1 depends on x_2 and x_2 depends on x_3, then x_1 indirectly depends on x_3) dependent on any other variable in the genotype. Therefore, in each VIGbp iteration, many bim_6 blocks will be affected, and finding the improving move (especially for low-quality solutions) will be much more probable than in the case of concatenation. Therefore, the VIGBP-LON in Fig. 1b is relatively dense. In the considered PX-LON building procedure, we employ a single locally optimal solution for the PX-based mixing. Thus, for overlapping problems, PX masks may be too large. Therefore, the operation may be unlikely to yield an improving move. In effect, the PX-LON, similarly to the VIGBP-LON, has a

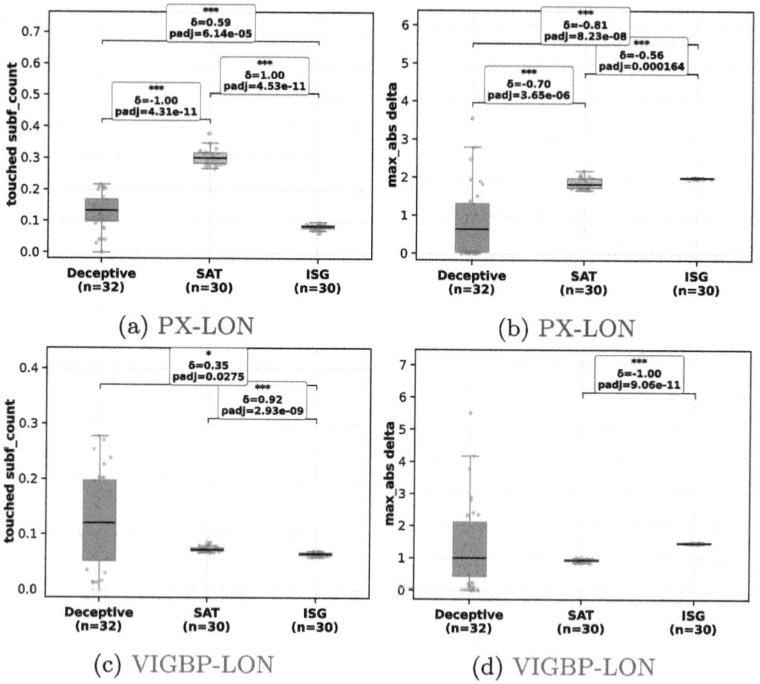

(a) PX-LON (b) PX-LON

(c) VIGBP-LON (d) VIGBP-LON

Fig. 2. Distribution of LON metrics. Each sub-plot is for a different metric. The individual boxes relate to problem classes: deceptive trap functions, MAX3SAT instances, and ISG instances (respectively).

large number of edges that refer to tunnelling moves [34] (this kind of move modifies a large subset of variables but leaves the fitness unchanged). However, the PX-LON reports a much lower number of improving moves.

The above comparison is not meant to present advantages or disadvantages of a given LON construction method. Its purpose is to show that each problem feature may have a measurable and visible influence on the LON structure. Thus, these features have the potential to be distinctive in choosing an optimiser suitable for solving the problem at hand.

6.1 LON Subfunction Metrics

Figure 2 displays distributions of LON subfunction metrics split by problem class. The significance result according to the (adjusted) p-value of Mann-Whitney tests are annotated as asterisks and the effect size (according to Cliff's δ) is written alongside. If there is no annotated box for a given pair of boxes (see e.g. Figure 2c) this is because the pair does not have a significant difference. We can see from Fig. 2 that most pairs of distributions (9 out of 12 pairs) have a significant difference: this indicates that the LON metrics can usually distinguish between the problem domains. For both of the presented metrics (`touched` subf_count and max_abs delta) the values relating to MAX3SAT instances are lower for VIGBP-LON than for PX-LON. Recall that VIGBP-LON is actively learning the linkage structure while PX-LON knows it a-priori. Lower values for these metrics represent less disruptive changes in subfunction structure along LON edges; it therefore seems that VIGBP-LON, not knowing the full linkage structure and less likely to know higher-order interactions, is making more conservative transitions. Note also that SAT PX-LONs have the highest values of the three domains for touched subf_count but this is not the case using the VIGBP-LON construction algorithm. This means that the ranking of distributions between problems can differ according to construction algorithm. Another observation is that ISG LONs have low variance among themselves in relation to these two metrics. Boxplots for all other LON metrics can be found in the supplemental material[2]. The overall trend from them is that there is often a significant difference between problem types (particularly in the case of PX-LON), and that often it is the case that the choice of LON construction algorithm can affect the LON feature distribution and/or the ranks for the domain distributions. Note that for the low-order Walsh contributions features, only three of 32 deceptive functions are included in the plots: this is because the other 29 did not meet the threshold specified in Sect. 4.2.

6.2 LON Subfunction Structure and Search

Figures 3a and 3b present correlations between PX-LON/VIGBP-LON metrics (on the rows) and the performance metric (on the columns; number of function evaluations to reach the optimum) for three algorithms and for different instance groups. The numeric values and colour of a square represent the Spearman rank correlation. There is also an indication of significance level according to (corrected) p-value: *** means $p < 0.001$; ** means $p < 0.01$; and * means $p < 0.05$. Note that there are NA values for the low-order interaction metrics in the case of deceptive functions: this is because the majority of the instances did not meet the minimum threshold for low-order Walsh mass as specified in Sect. 4.2. Across both LON construction methods the results show that some LON subfunction metrics are correlated to optimiser difficulty. For deceptive instances, some LON features exhibit correlations with algorithm runtime (in the case of P3[3] and GOMEA). This demonstrates that search difficulty in deceptive landscapes is partially related to how severely optimisation steps disrupt building blocks. There are no significant correlations between the deceptive problem LON features and GLB optimisation, however. This could be because the metrics largely quantify subfunction/linkage block disruption—and GLB, as an optimiser, is particularly careful not to disrupt the blocks. In contrast to the deceptive problems, ISG instances show less meaningful relationships between

[2] https://zenodo.org/records/19646468.

[3] Notice that the same value (0.87) is observed in both plots for the correlation between touched subf_count and P3-dec. We checked and the feature values do actually differ, but there is no rank difference.

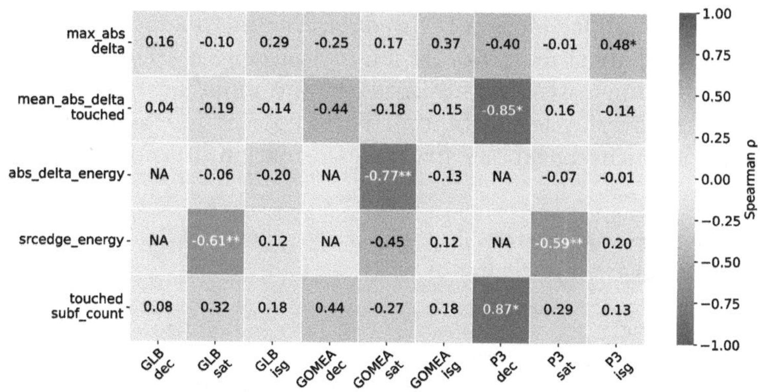

(a) PX-LON

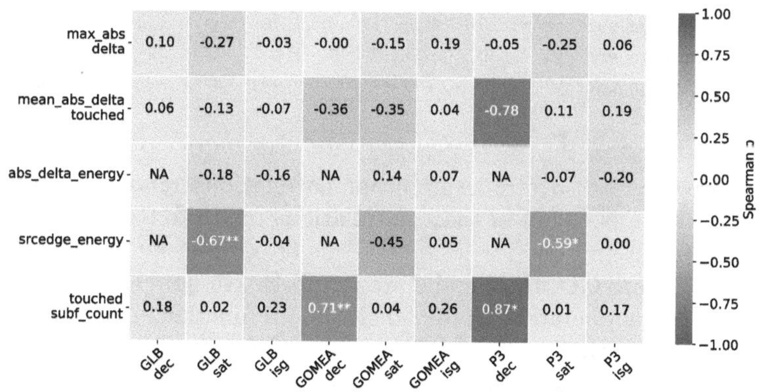

(b) VIGBP-LON

Fig. 3. Correlations between LON metrics and algorithm performance. Each column relates to the performance of the stated optimiser on the stated problem class: for example, **GLB dec** is the performance of the GLB optimisation on deceptive trap functions.

LON subfunction metrics and algorithm performance: there is only the PX-LON feature `max_abs delta`, which captures the maximum absolute change in any subfunction between the source and destination of a LON edge. This feature does not significantly correlate with search when computed on VIGBP LONs for ISG, however. The overall lack of other associations with search implies that ISG difficulty is dominated by collective variable rearrangements that may not be captured by edgewise LON subfunction metrics.

MAX3SAT instances exhibit different behaviour: some LON metrics which are focused on low-order interactions in the local optima seem to correlate to search performance (such as `abs_delta_energy`) computed on PX-LONs and correlating with GOMEA performance; this metric relates to the change in low-order fitness contributions along a LON edge). The relation between subfunction changes specifically affecting low-order interactions and search performance may be because the studied MAX3SAT instances are at or near the phase transition, where clause overlap is substantial. These specific LON metrics might be a proxy or lens capturing backbone structure/formation in some way.

Table 1. Robust Bayesian regression predicting $\log_{10}$ evaluations to the optimum on successful runs from single predictors

alg-problem	predictor	n	$\tilde{\beta}_{\text{std}}$	$\mathbb{P}(\beta < 0)$	$\Delta \log_{10}(\text{evals})$
P3-ISG	`max_abs delta`	30	0.111 [−0.027, 0.249]	0.051	0.252 [−0.061, 0.566]
GOMEA-SAT	`abs_delta_energy`	15	−0.608 [−1.015, −0.222]	0.997	−0.421 [−0.703, −0.154]
P3-SAT	`srcedge_energy`	29	−0.629 [−0.940, −0.325]	1.000	−0.043 [−0.065, −0.022]
GLB-SAT	`srcedge_energy`	29	−0.725 [−0.983, −0.449]	1.000	−0.037 [−0.050, −0.023]
GOMEA-DEC	`touched subf_count`	20	0.743 [0.418, 1.035]	0.000	1.258 [0.708, 1.752]
P3-SAT	`srcedge_energy`	29	−0.653 [−0.948, −0.357]	1.000	−0.042 [−0.061, −0.023]
GLB-SAT	`srcedge_energy`	29	−0.750 [−0.995, −0.494]	1.000	−0.035 [−0.047, −0.023]

Overall, we see that 6 of 10 significant correlations relate to PX-LONs; 4 relate to VIGBP-LONs. Half of these correlations relate to SAT problems; 4 for deceptive functions; and one for ISG. In terms of optimisers: 6 of these correlations are for P3; 2 are GOMEA; and 2 are for GLB. We note that this is the first evidence that the explanatory ability of LON subfunction metrics is problem-class dependent: metrics relating to the level of disruption along LON edges seem to be correlated with search in the case of deceptive and modular landscapes; for globally-constrained landscapes (MAX3SAT at the phase transition) only low-order interaction metrics seem relevant; the approach seems to break down almost entirely for ISG. Table 1 shows metrics related to Bayesian robust regression models using PX-LON and VIGBP-LON features. The choice of which models to build was informed by the findings from Fig. 3 (i.e. the most promising relationships which also had at least 15 observations across both predictor and response were considered). We report the standardised regression coefficient in the fourth column (i.e. when the predictor increases by one standard deviation this is the expected change in the response); the second-last column contains $\mathbb{P}(\beta < 0)$ i.e. the posterior probability of a negative slope; effects are also reported per +0.1 increase in the predictor (in the last column). The first four rows in the Table relate to PX-LON; the latter three rows to VIGBP-LON.

We can notice from Table 1 that for MAX3SAT the LON feature `srcedge_energy` appears to be related with a negative direction to algorithm runtime for both PX-LON and VIGBP-LON and also for P3 and GLB optimisers (notice rows 3, 4, 6, and 7). This association means that instances whose local optima are well aligned with low-order (linear and pairwise) Walsh structure are solved faster; that is, P3 and GOMEA seem to benefit when informative structure is already present at low interaction orders—which means that learned linkage sets reliably reflect improvements in the full MAX3SAT objective. On the other hand it appears that instances with lower `srcedge_energy` seem to require greater reliance on higher-order interactions; this may lead to increased search effort. The consistency of this relationship across two LON construction algorithms and also two optimisers indicates that `srcedge_energy` may capture a fundamental aspect of MAX3SAT landscape structure which affects search difficulty. The LON metric `abs_delta_energy` (in the case of MAX3SAT: see row 2 of Table 1) appears to have a negative-direction association with GOMEA runtime; this indicates faster convergence when transitions between local optima are energetically well-separated. Larger `abs_delta_energy` indicates stronger low-order energy differences between neighbouring local optima. The negative association with runtime suggests that GOMEA benefits from MAX3SAT landscapes in which transitions between local optima represent large structural changes; this might thereby reduce the number of evaluations required to reach high-quality solutions. For deceptive instances we have a model for `touched subf_count` (VIGBP-LON and GOMEA). Looking at Table 1 we can see that for deceptive-function LONs the aforementioned feature appears to have a positive-direction association with GOMEA runtime (see row 5), indicating increased difficulty when transitions between local optima require coordinated modification of many variables. Larger `touched subf_count` values indicate that neighbouring optima are separated by substantial, multi-variable jumps. This result suggests that deceptive landscapes hinder GOMEA when progress depends on coordinated and high-order changes; these may be poorly approximated by incremental recombination based on learned linkage sets—thereby leading to longer runtimes.

7 Conclusion

In this work, local optima networks (LONs) were constructed for high-dimensional linkage learning benchmark problems. We proposed a new set of features related to subfunction structure in the LONs and investigated the possibility of associations to the performance of state-of-the-art optimisers. The results indicate that the subfunction-related features may depend on the LON construction algorithm and that some of them are useful for separating problem classes. Also, the proposed features seem to be related to the convergence efficiency of some optimisers we studied for some problem-operator combinations. While this work considered only successful runs, future work will expand to the analysis of unsuccessful runs. The approach and the findings of this exploratory paper not only provide new insight about the interplay between the studied problems and optimisers, but also lay the groundwork for exploring the potential of LON-based online operator adaptation in response to local optima subfunction structure in the future. Code, data, and reproducibility instructions can be found at the Zenodo repository [footnote 2].

Acknowledgments. This work was sponsored by the Polish National Science Center (NCN) under Grant 2020/38/E/ST6/00370.

References

1. Bosman, P.A., Luong, N.H., Thierens, D.: Expanding from discrete cartesian to permutation gene-pool optimal mixing evolutionary algorithms. In: Proceedings of the Genetic and Evolutionary Computation Conference, pp. 637–644 (2016)
2. Bouter, A., Alderliesten, T., Bel, A., Witteveen, C., Bosman, P.A.N.: Large-scale parallelization of partial evaluations in evolutionary algorithms for real-world problems. In: Proceedings of the Genetic and Evolutionary Computation Conference. GECCO 2018, pp. 1199–1206. Association for Computing Machinery, New York (2018). https://doi.org/10.1145/3205455.3205610
3. Canonne, L., Derbel, B., Chicano, F., Ochoa, G.: To combine or not to combine graybox crossover and local search? In: Proceedings of the Genetic and Evolutionary Computation Conference, pp. 257–265 (2023)
4. Chicano, F., Whitley, D., Ochoa, G., Tinós, R.: Optimizing one million variable nk landscapes by hybridizing deterministic recombination and local search. In: Proceedings of the Genetic and Evolutionary Computation Conference, pp. 753–760 (2017)
5. Chicano, F., Whitley, D., Sutton, A.M.: Efficient identification of improving moves in a ball for pseudo-Boolean problems. In: Proceedings of the 2014 Annual Conference on Genetic and Evolutionary Computation, pp. 437–444 (2014)
6. Deb, K., Goldberg, D.E.: Sufficient conditions for deceptive and easy binary functions. Ann. Math. Artif. Intell. **10**(4), 385–408 (1993)
7. Deb, K., Horn, J., Goldberg, D.E.: Multimodal deceptive functions. Complex Syst. **7**(2) (1993)
8. Goldman, B.W., Punch, W.F.: Parameter-less population pyramid. In: Proceedings of the 2014 Annual Conference on Genetic and Evolutionary Computation. GECCO 2014, pp. 785–792. ACM (2014)
9. Heckendorn, R.B.: Embedded landscapes. Evol. Comput. **10**(4), 345–369 (2002)
10. Herrmann, S., Ochoa, G., Rothlauf, F.: Communities of local optima as funnels in fitness landscapes. In: Proceedings of the Genetic and Evolutionary Computation Conference, pp. 325–331 (2016)
11. Komarnicki, M.M., Przewozniczek, M.W., Tinós, R., Li, X.: Overlapping cooperative co-evolution for overlapping large-scale global optimization problems. In: Proceedings of the Genetic and Evolutionary Computation Conference. GECCO 2024, pp. 665–673. Association for Computing Machinery, New York (2024). https://doi.org/10.1145/3638529.3654171
12. Lourenço, H.R., Martin, O.C., Stützle, T.: Iterated local search. In: Handbook of Metaheuristics, pp. 320–353. Springer, Cham (2003)
13. Marmion, M.E., Dhaenens, C., Jourdan, L., Liefooghe, A., Verel, S.: Nils: a neutrality-based iterated local search and its application to flowshop scheduling. In: Proceedings of the 11th European Conference on Evolutionary Computation in Combinatorial Optimization, pp. 191–202 (2011)
14. Munetomo, M., Goldberg, D.: A genetic algorithm using linkage identification by nonlinearity check. In: IEEE SMC 1999 Conference Proceedings. 1999 IEEE International Conference on Systems, Man, and Cybernetics (Cat. No.99CH37028), vol. 1, pp. 595–600 (1999). https://doi.org/10.1109/ICSMC.1999.814159
15. Ochoa, G., Chicano, F.: Local optima network analysis for max-sat. In: Proceedings of the Genetic and Evolutionary Computation Conference Companion, pp. 1430–1437 (2019)

16. Ochoa, G., Tomassini, M., Vérel, S., Darabos, C.: A study of NK landscapes' basins and local optima networks. In: Proceedings of the 10th Annual Conference on Genetic and Evolutionary Computation, pp. 555–562 (2008)
17. Ochoa, G., Veerapen, N., Daolio, F., Tomassini, M.: Understanding phase transitions with local optima networks: number partitioning as a case study. In: Hu, B., López-Ibáñez, M. (eds.) EvoCOP 2017. LNCS, vol. 10197, pp. 233–248. Springer, Cham (2017). https://doi.org/10.1007/978-3-319-55453-2_16
18. Prusik, M., Frej, B., Przewozniczek, M.W.: Availability of perfect decomposition in statistical linkage learning for unitation-based function concatenations. In: Proceedings of the 18th ACM/SIGEVO Conference on Foundations of Genetic Algorithms. FOGA 2025, pp. 226–237. Association for Computing Machinery, New York (2025). https://doi.org/10.1145/3729878.3746631
19. Przewozniczek, M., et al.: Granular linkage blocks-the gates to the granular function form. Authorea Preprints (2025)
20. Przewozniczek, M.W., Tinós, R., Frej, B., Komarnicki, M.M.: On turning black - into dark gray-optimization with the direct empirical linkage discovery and partition crossover. In: Proceedings of the Genetic and Evolutionary Computation Conference. GECCO 2022, pp. 269–277. Association for Computing Machinery, New York (2022). https://doi.org/10.1145/3512290.3528734
21. Przewozniczek, M.W., Tinós, R., Komarnicki, M.M.: First improvement hill climber with linkage learning – on introducing dark gray-box optimization into statistical linkage learning genetic algorithms. In: Proceedings of the Genetic and Evolutionary Computation Conference. GECCO 2023, pp. 946–954. ACM (2023)
22. Saul, L., Kardar, M.: The 2D±J Ising spin glass: exact partition functions in polynomial time. Nucl. Phys. B **432**(3), 641–667 (1994)
23. Sun, Y., Li, X., Ernst, A., Omidvar, M.N.: Decomposition for large-scale optimization problems with overlapping components. In: IEEE Congress on Evolutionary Computation (CEC), pp. 326–333. IEEE (2019)
24. Tari, S., Ochoa, G., Basseur, M., Verel, S.: On the global structure of puboi fitness landscapes. In: Proceedings of the Companion Conference on Genetic and Evolutionary Computation, pp. 247–250 (2023)
25. Teixeira, M.C., Pappa, G.L.: Understanding automl search spaces with local optima networks. In: Proceedings of the Genetic and Evolutionary Computation Conference, pp. 449–457 (2022)
26. Thierens, D., Bosman, P.A.: Hierarchical problem solving with the linkage tree genetic algorithm. In: Proceedings of the 15th Annual Conference on Genetic and Evolutionary Computation. GECCO 2013, pp. 877–884. ACM (2013)
27. Thomson, S.L., Daolio, F., Ochoa, G.: Comparing communities of optima with funnels in combinatorial fitness landscapes. In: Proceedings of the Genetic and Evolutionary Computation Conference, pp. 377–384 (2017)
28. Thomson, S.L., Przewozniczek, M.W.: Subfunction structure matters: a new perspective on local optima networks. In: Proceedings of the Genetic and Evolutionary Computation Conference, pp. 85–93 (2025)
29. Tinós, R., Przewozniczek, M.W., Whitley, D.: Iterated local search with perturbation based on variables interaction for pseudo-Boolean optimization. In: Proceedings of the Genetic and Evolutionary Computation Conference. GECCO 2022, pp. 296–304. ACM (2022)
30. Tinós, R., Whitley, D., Chicano, F.: Partition crossover for pseudo-boolean optimization. In: Proceedings of the 2015 ACM Conference on Foundations of Genetic Algorithms XIII. FOGA 2015, pp. 137–149. Association for Computing Machinery, New York (2015). https://doi.org/10.1145/2725494.2725497

31. Verel, S., Daolio, F., Ochoa, G., Tomassini, M.: Local optima networks with escape edges. In: International Conference on Artificial Evolution (Evolution Artificielle), pp. 49–60. Springer, Cham (2011)
32. Whitley, D.: Next generation genetic algorithms: a user's guide and tutorial. In: Handbook of Metaheuristics, pp. 245–274. Springer, Cham (2018)
33. Whitley, D., Aguirre, H., Sutton, A.: Understanding transforms of pseudo-boolean functions. In: Proceedings of the 2020 Genetic and Evolutionary Computation Conference. GECCO 2020, pp. 760–768. Association for Computing Machinery, New York (2020)
34. Whitley, D., Hains, D., Howe, A.: Tunneling between optima: partition crossover for the traveling salesman problem. In: Proceedings of the 11th Annual Conference on Genetic and Evolutionary Computation. GECCO 2009, pp. 915–922. Association for Computing Machinery, New York (2009). https://doi.org/10.1145/1569901.1570026
35. Whitley, D., Ochoa, G., Chicano, F.: How partition crossover exposes parallel lattices and the fractal structure of k-bounded functions. In: Proceedings of the Genetic and Evolutionary Computation Conference. GECCO 2025, pp. 836–844. Association for Computing Machinery, New York (2025). https://doi.org/10.1145/3712256.3726340
36. Whitley, L.D., Chicano, F., Goldman, B.W.: Gray box optimization for MK landscapes (NK landscapes and max-ksat). Evol. Comput. **24**(3), 491–519 (2016)
37. Yu, T.L., Sastry, K., Goldberg, D.E.: Linkage learning, overlapping building blocks, and systematic strategy for scalable recombination. In: Proceedings of the 7th Annual Conference on Genetic and Evolutionary Computation. GECCO 2005, pp. 1217–1224. ACM (2005)

Author Index

G. Iacca et al. (Eds.): PPSN 2026, LNCS 16985, pp. 633–634, 2027.
https://doi.org/10.1007/978-3-032-36214-8

Kolibri 360 GmbH
Ferdinand-Jühlke-Straße 7
99095 Erfurt, Deutschland
produktsicherheit@kolibri360.de